Climate Change: In Context

Climate Change: In Context

Brenda Wilmoth Lerner & K. Lee Lerner, Editors

VOLUME 1

ABRUPT CLIMATE CHANGE TO GULF STREAM

GALE
CENGAGE Learning™

Detroit • New York • San Francisco • New Haven, Conn • Waterville, Maine • London

Climate Change: In Context

Brenda Wilmoth Lerner and K. Lee Lerner, Editors

Project Editor: Kathleen J. Edgar

Editorial: Madeline S. Harris, Debra Kirby, Kristine Krapp, Elizabeth Manar, Kimberley McGrath, Lemma Shomali

Production Technology: Paul Lewon

Rights and Acquisitions: Margaret Abendroth, Vernon English, Leitha Etheridge-Sims, Jackie Jones, Tim Sisler

Imaging and Multimedia: Lezlie Light

Art Director: Jennifer Wahi

Designer: Jeff Bane, CMB Design

Product Management: Janet Witalec

Composition and Electronic Capture: Evi Abou-El-Seoud

Manufacturing: Wendy Blurton, Dorothy Maki

LIBRARY OF CONGRESS CATALOGING-IN-PUBLICATION DATA

Climate change: in context / Brenda Wilmoth Lerner & K. Lee Lerner, editors.
p. cm.
Includes bibliographical references and index.
ISBN 978-1-4144-3614-2 (set) -- ISBN 978-1-4144-3615-9 (vol. 1) -- ISBN 978-1-4144-3616-6 (vol. 2) -- ISBN 978-1-4144-3708-8 (ebook)
1. Climatic changes--Encyclopedias. I. Lerner, Brenda Wilmoth. II. Lerner, K. Lee.

QC981.8.C5C5114123 2008
551.6--dc22 2007051762

Gale
27500 Drake Rd.
Farmington Hills, MI 48331-3535

ISBN-13: 978-1-4144-3614-2 (set)
ISBN-13: 978-1-4144-3615-9 (vol. 1)
ISBN-13: 978-1-4144-3616-6 (vol. 2)

ISBN-10: 1-4144-3614-9 (set)
ISBN-10: 1-4144-3615-7 (vol. 1)
ISBN-10: 1-4144-3616-5 (vol. 2)

This title is also available as an e-book.
ISBN-13: 978-1-4144-3708-8 (set) ISBN-10: 1-4144-3708-0 (set)
Contact your Gale sales representative for ordering information.

Printed in China by China Translation & Printing Services Limited
3 4 5 6 7 12 11 10 09

Contents

VOLUME 1

VOLUME 2

Advisors and Contributors

While compiling this volume, the editors relied upon the expertise and contributions of the following scientists, scholars, and researchers, who served as advisors and/or contributors for *Climate Change: In Context*:

Susan Aldridge, Ph.D.
Independent scholar and writer
London, United Kingdom

William Arthur Atkins, M.S.
Independent scholar and writer
Normal, Illinois

Julie Berwald, Ph.D.
Geologist (Ocean Sciences) and writer
Austin, Texas

Philip Chaney, Ph.D., P.S.
Associate Professor of Geography
Auburn University
Auburn, Alabama

Michele Chapman, M.S.
Independent scholar
Norfolk, Virginia

James Anthony Charles Corbett
Journalist
London, United Kingdom

Paul Davies, Ph.D.
Director, Science Research Institute
Adjunct Professor Université Paris
Paris, France

Sandra Dunavan, M.S.
Journalist
Saline, Michigan

Larry Gilman, Ph.D.
Independent scholar and journalist
Sharon, Vermont

Amit Gupta, M.S.
Journalist
Ahmedabad, India

Tony Hawas, M.A.
Writer and journalist
Brisbane, Australia

Brian D. Hoyle, Ph.D.
Microbiologist
Nova Scotia, Canada

Alexandr Ioffe, Ph.D.
Senior Scientist
Geological Institute of the Russian Academy of Sciences
Moscow, Russia

Kathleen R. Johnson, Ph.D.
Assistant Professor, Earth System Science
University of California
Irvine, California

David T. King Jr., Ph.D.
Professor, Department of Geology
Auburn University
Auburn, Alabama

Adrienne Wilmoth Lerner, J.D.
Independent scholar
Jacksonville, Florida

Miriam C. Nagel, M.S.
Independent scholar
Avon, Connecticut

Anna Marie Roos, Ph.D.
Research Associate, Wellcome Unit for the History of Medicine
University of Oxford
Oxford, United Kingdom

Acknowledgments

The editors are grateful to the truly global group of scholars, researchers, and writers who contributed to *Climate Change: In Context.* The editors wish to offer special thanks to author and advisor Dr. Larry Gilman, whose passion and commitment to the subject was surpassed only by his skill to summon understandable prose without loss of scientific accuracy.

The editors also wish to thank copyeditors Christine Jeryan, Kate Kretchmann, and Alicia Cafferty Lerner, whose keen eyes and sound judgments greatly enhanced the quality and readability of the text.

The editors gratefully acknowledge and extend thanks to Janet Witalec and Debra Kirby at the Gale Group for their faith in the project and for their sound content advice and guidance. Without the able guidance and efforts of talented teams in IT, rights and acquisition management, and imaging at the Gale Group, this book would not have been possible. The editors are especially indebted to Kimberley McGrath, Elizabeth Manar, Madeline Harris, Kristine Krapp, Lemma Shomali, and others at Gale for their invaluable help. The editors also wish to acknowledge the contributions of Marcia Schiff at the Associated Press for her help in securing archival images.

Deep and sincere gratitude is due Project Manager Kathleen Edgar who showed patience, experience, and steady skill in managing both content and tight deadlines. Ms. Edgar's knowledge and passion for the subject, along with keen editorial insights, greatly enhanced the quality, accuracy, and readability of this book.

Introduction

Editors' introductions to books usually attempt to offer words of motivation designed to inspire readers toward their studies. For *Climate Change: In Context*, however, the editors wish to stand aside a bit and ask readers, especially students just beginning their serious studies of science, to carefully read the special introductions by Dr. Wallace S. Broecker and Thomas Hayden that immediately follow. Together, these introductions serve as exemplary primary sources (personal narratives from experts in the field of climate change) and as both elegant motivation to readers to carefully consider the issues and impacts of climate change, and eloquent calls to actively engage in the challenge of finding solutions.

In the wake of the stunning 2007 Intergovernmental Panel on Climate Change (IPCC) reports, *Climate Change: In Context* is one of the first reference books designed to attempt to explain the complexities of those reports to younger students. *Climate Change: In Context* seeks to relate—in language accessible to younger students—the now well-established fact of climate change driven by human (anthropogenic) activity. Toward this goal, the editors incorporate, highlight, and explain the latest IPCC material and quote from the latest IPCC reports and National Academy of Sciences reports to provide evidence for climate change.

Written by a global array of experts in physics, geology, environmental science, sociology, and law, *Climate Change: In Context* further attempts to provide a comprehensive introduction to the scientific consensus on climate change within the context of new evidence, changing predictions, current events, and political decision-making.

The editors are indebted to these distinguished scientists for their generous contributions of time and compelling material.

The quest for technological advance dates to our most ancient ancestors' struggle for daily survival and instinctive need to fashion tools from which they could gain physical advantage beyond the strength of the relatively frail human body. Along with physical needs, humans also harbored an innate curiosity into the workings of the Cosmos, and this longing to understand our place in the natural world remains a key component upon which science is built. But science sometimes speaks truths we might not wish to hear, and at this time in human history science is speaking clearly, with a chorus of voices, that with regard to the human activities that drive climate change, it is now time to fuse our science and technology with our noblest qualities of caring, commitment, and sacrifice so that our children enjoy the pleasures of the good Earth.

Aux grands maux les grand remèdes (to great evils great cures).

K. Lee Lerner & Brenda Wilmoth Lerner, editors
PARIS, FRANCE, DECEMBER 2007

Primarily based in London and Paris, the Lerner & Lerner portfolio includes more than two dozen books and films that focus on science and science-related issues.

"Warming of the climate system is unequivocal, as is now evident from observations of increases in global average air and ocean temperatures, widespread melting of snow and ice, and rising global average sea level. . . ." —Statement of the Intergovernmental Panel on Climate Change (IPCC) as formally approved at the 10th Session of Working Group I of the IPCC in Paris, France, during February 2007.

A Special Introduction by Dr. Wallace S. Broecker

At the age of 20, I came to Columbia University's Lamont Geological Observatory to do a summer internship. Nearly 57 years later, I'm still here and I must say it's been an exciting and extremely rewarding run. I started doing lowly chores in the radiocarbon dating laboratory, but was lucky enough to catch the eye of the professor in charge. At the end of the summer, he convinced me to transfer to Columbia College for my senior year and then to continue on for my PhD degree. Further, when I finished my graduate work, he persuaded the faculty of the Geology Department to appoint me as Assistant Professor of Geochemistry. I now hold a "Chair" professorship in what has been renamed the Department of Earth and Environmental Sciences and conduct my research at an institute now called the Lamont-Doherty Earth Observatory. The name changes involve the substitution of "Earth" for "Geology" in one case and "Earth" for "Geological" in the other. The difference reflects an important shift in attitude. Instead of focusing on the history of our planet for its own sake, we now focus on the processes that influence our planet's environment. History remains important because it helps to put into context what is happening today. And, of course, what is happening today has an ever larger component of interference related to humankind's activities.

My personal awareness of the so-called "anthropogenic" impacts of industrialization arose through the studies of Earth's carbon cycle, especially as it involved the ocean. During the 1970s, I took part in a large program aimed at mapping the chemical and isotopic properties of the world ocean. It was through this that I came to realize that these measurements could be used to calculate the rate at which the CO_2 produced by the burning of fossil fuels would be sucked up by the ocean and hence about the magnitude of the warming CO_2 could eventually produce. What puzzled me at the time was that the year-by-year global temperature compilations showed that the slow warming that had occurred prior to World War II had petered out and 30 or so years had passed without any further warming. But physics demands that the extra greenhouse gases being added to the atmosphere must warm Earth. So what was going on? I pondered this and based on a newly published temperature record covering the last few thousands years based on isotopic measurement carried out on ice from a long core drilled in Greenland's cap, I decided that by chance the natural cycles seen in this record were canceling out the expected CO_2 warming. Further, if this were the explanation, then, as the natural cycle was due to turn from cooling to warming, I predicted that the situation would change and we would experience a dramatic temperature increase. The journal *Science* published my paper in 1975 and lo and behold a year later the warming commenced. It has continued ever since.

Inspired by the success of my prediction, I joined a small group of scientists concerned about the future impacts of fossil fuel burning. Somehow our joint efforts made their way to the top levels of the U.S. government. President Jimmy Carter instituted policies encouraging alternate energy sources. I was invited to brief a group of influential senators on the impacts of coal burning. The Department of Energy instituted a CO_2 program. Then, with the election of President Ronald Reagan, the bottom fell out. All the things started by the Carter administration were put on the back burner.

However, in 1980 a discovery was made that eventually came to play a big role in turning the tide from apathy back to serious concern. Measurements made on air trapped in Antarctic ice revealed that Earth's temperature had co-varied with atmospheric CO_2 content. During the peak of the last glacial period, the atmosphere's CO_2 content was 30% lower than it was prior to the Industrial Revolution. In the late 1980s, then-Senator Al Gore realized that this observation constituted a smoking gun. If low CO_2 was the driver of glaciation, then high CO_2 would surely bring about a serious warming.

As vice president, Gore, for political reasons, was silenced on this subject. But released from these constraints by his loss to George W. Bush, Gore turned his attention back to the pending problems that fossil fuel CO_2 posed. His movie *An Inconvenient Truth* woke up not only the American public but reinforced concern already in force elsewhere on the planet. For this effort, he received the 2007 Nobel Peace Prize.

The record kept in Greenland ice led me to an additional scientific discovery for which I subsequently received a number of prizes. During the time period corresponding to the last glacial period, roughly once each millennium, a large and abrupt temperature change occurred. I proposed that each such jump corresponded to either a shut down or a start up of a vast system of ocean currents that I dubbed the "Great Ocean Conveyor." Little did I realize at the time that this proposal would lead to an entirely new paradigm, namely, that Earth's climate system has a set of alternate states of operation and that it can in a decade or less jump from one of these to another. It also implied that Earth's climate system responded strongly to seemingly gentle nudges. As the CO_2 generated by a continuing business-as-usual use of fossil fuels would during the coming century constitute a stiff poke rather than a gentle nudge, my concern was intensified.

Because of this, in the last few years, I have become ever more involved in efforts to stem the rise in CO_2. This involvement has led me to the conclusion that no combination of conservation and alternate energy sources will be enough. Rather, if we are to avoid the awesome consequences of melting glaciers and sea level rise, wholesale changes in plant cover and animal habitat, further aridification of our drylands ... we must develop the means to recapture CO_2 from our atmosphere and either bury it or mineralize it. Although this is certainly doable, the task is a massive one and the time to do it is very short. Unless we greatly pick up the pace, we are destined to make changes in our planet that will certainly plague our descendants.

Our human activities have become so pervasive that in a sense we have wrested control of the habitability of our planet. With this power comes a huge responsibility. We must rapidly alter our behavior from that of despoilers to that of responsible stewards!

At 76 my days are numbered as are those of many who have pioneered the environmental movement. We've pointed out the problems, but most of the solutions await action by the young readers of this book. I guarantee you that nothing is more satisfying than a life spent doing science for the benefit of humanity. The planet needs you. Please join us!

Dr. Wallace S. Broecker
Newberry Professor of Geology
Lamont-Doherty Earth Observatory
Columbia University

[*Editor's note:* Dr. Broecker is a member of both the National Academy of Sciences and the British Royal Society. In 1996, he received the National Science Medal. A member of the Columbia faculty since 1959, Dr. Broecker serves as the Newberry Professor of Earth & Environmental Sciences and as member of the Earth Institute Academic Committee. His ongoing research interests include paleoclimatology, ocean chemistry, isotope dating, and other important facets of environmental science. He has authored or co-authored close to 500 journal articles, textbooks, and books, including *How to Build a Habitable Planet* (1987) and *Fixing Climate* (2008) with Robert Kunzig.

Among a lengthy list of respected honors and awards, Dr. Broecker is the recipient of the Alexander Agassiz Medal from the National Academy of Sciences, the Japanese Blue Planet Award, and the Swedish Crafoord Award. He has been honored by the Geological Society of America and Geological Society of London.]

A Special Introduction by Thomas Hayden

When I first started writing magazine articles about global warming in the mid-1990s, it was already clear that human activity could cause serious changes in Earth's atmosphere, and almost as certain that those changes had already begun. And yet it was a challenge to convince editors—and many readers—that global climate change was a real and immediate danger.

As the primary sources included in this book show, global climate change has had a particularly high profile in the print media over the last several years. But climate change has been an important story in journalism for more than two decades. Newspapers, magazines, television broadcasts and more recently, Web sites have played a crucial role in calling attention to the issue of our warming climate, and its implications. Journalists have also been central players in the spread of confusion surrounding the subject. There are two reasons for this problem.

The first has been called the issue of "false balance." Modern journalism and many of its rules and conventions grew out of covering politics, from local city council meetings to the intrigue and backroom dealings of national politics and international diplomacy. As a result, many journalists work with an assumption that the truth they seek is often hidden behind layers of secrecy—or indeed that their sources are actively trying to distort or "spin" the facts to fit a particular political or ideological position. One way to deal with this reality is to insist on balanced reporting—the idea that there are at least two equally valid sides to every story, and that it is a reporter's duty to find and present them as such.

Balance makes sense when you're covering a political race, and skepticism about sources is an essential skill for any reporter, as important for covering science as it is for reporting on politics. But what happens when the great majority of scientific evidence—and of reputable scientists—is all pointing in one direction, and only a small, if vocal, group of dissenters is saying "no way?"

The fact is that science works differently from politics, and the search for balance can end up distorting the story, rather than clarifying it. Nowhere has false balance been more obvious, or more persistent, than in media coverage of global climate change.

When the possibility of global warming first rose to national attention in the late 1970s, many scientists were skeptical—an entirely appropriate response to a fairly radical idea based on limited data. Researchers and reporters both did what they should; the journalists reported both the idea and the skepticism, and the scientists buckled down to study global climate change in every imaginable way.

Within a decade, the evidence was lining up in support of global warming as a real, significant and imminent danger—with humans as the cause. But journalists were still seeking "balance," giving equal time to a shrinking minority of skeptical scientists. By the time I left science for journalism in 1997 (I had studied ocean ecology, including its effects on the global climate), it was clear to me as a scientist that global climate change would be one of the most important stories of our time. For many journalists however, especially those with no scientific training, "balance" was still the ultimate goal.

That's where the second factor comes in—denial. And if false balance is a matter of good intentions gone wrong, the same cannot be said here. Despite the growing evidence that global climate change was a real danger, a small handful of scientists and others continued to produce reports—rather than original scientific studies—calling the whole idea of global warming into question. Most of the original skeptics had been convinced by the data, and by the mid-1990s, the few who remained became deniers—no longer poking and prodding an emerging scientific theory in search of the best answers, but stretching the truth and ignoring facts to obscure the increasingly certain reality of global warming.

To say that many of these deniers were funded by oil and coal companies may sound like a personal attack. But it is also true, and significant. It's not that the skeptics were paid off by big business—for the deniers, however, it may be another matter. But the business and political interests behind the most polluting industries had a knack for finding and promoting vocal opponents of global warming. And thanks to false balance, that was all it took to keep the "controversy" alive in the public mind long after it had been resolved in the scientific community.

Of course, even the strongest spin runs out of momentum eventually. By the late 1990s, the available pool of "skeptical" climate scientists had all but disappeared. And by the middle of the present decade, even newspaper and magazine editors could be convinced, for the most part, that giving equal time to global warming deniers had become akin to finding a doctor to make the case that tobacco is harmless every time they published a story about smoking.

So what now? With the broad story of global warming now written quite clearly, the most important question about global climate change may be this: Of the hundreds of potential solutions, from regulations to economic policies to new technologies, which will be most effective in slowing global warming and limiting its effects? There will be good ideas, bad ideas, and plenty of spin.

It is always crucial in science, journalism, and good citizenship to be well informed, to question assumptions, and to challenge the received wisdom. But it is equally important to understand the methods, data and analysis that inform a given conclusion or position—and to know that when good data contradicts a belief, no matter how fondly held, it is the belief that must change. The time for denial is long over, but as we enter the era of action on global climate change, critical thinking will be more important than ever—whether you're making the news, reporting it, or responding to it. This book, with its sources, context and analysis, makes a uniquely valuable starting point for doing just that.

Thomas Hayden

[*Editor's note:* Thomas Hayden writes about climate, science, and the environment from his home in San Francisco. A former staff writer at both *Newsweek* and *U.S. News & World Report*, he now writes for more than a dozen publications, including *National Geographic*, *Nature*, and *USA Today*.]

About the *In Context* Series

Written by a global array of experts yet aimed primarily at high school students and an interested general readership, the *In Context* series serves as an authoritative reference guide to essential concepts of science, the impacts of recent changes in scientific consensus, and the impacts of science on social, political, and legal issues.

Cross-curricular in nature, *In Context* books align with, and support, national science standards and high school science curriculums across subjects in science and the humanities, and facilitate science understanding important to higher achievement in the No Child Left Behind (NCLB) science testing. Inclusion of original essays written by leading experts and primary source documents serve the requirements of an increasing number of high school and international baccalaureate programs, and are designed to provide additional insights on leading social issues, as well as spur critical thinking about the profound cultural connections of science.

In Context books also give special coverage to the impact of science on daily life, commerce, travel, and the future of industrialized and impoverished nations.

Each book in the series features entries with extensively developed words-to-know sections designed to facilitate understanding and increase both reading retention and the ability of students to understand reading in context without being overwhelmed by scientific terminology.

Entries are further designed to include standardized subheads that are specifically designed to present information related to the main focus of the book. Entries also include a listing of further resources (books, periodicals, Web sites) and references to related entries.

In addition to maps, charts, tables and graphs, each *In Context* title has approximately 300 topic-related images that visually enrich the content. Each *In Context* title will also contain topic-specific timelines (a chronology of major events), a topic-specific glossary, a bibliography, and an index especially prepared to coordinate with the volume topic.

About This Book

The goal of *Climate Change: In Context* is to help high-school and early college-age students understand the essential facts and deeper cultural connections of topics and issues related to the scientific study of climate change and its impacts on humanity.

The relationship of science to complex ethical and social considerations is evident, for example, when considering the political and economic impacts of climate change related to human activity. The data clearly and strongly convince the vast majority of scientists that unnatural global climate change is happening and that it is almost certain that it is due to anthropogenic activity (caused by humans). This book reflects that scientific consensus and expands the discussion from the question of whether climate change is real to the question of what should be done about it.

In an attempt to enrich the reader's understanding of the mutually impacting relationship between science and culture, as space allows we have included primary sources that enhance the content of *In Context* entries. In keeping with the philosophy that much of the benefit from using primary sources derives from the reader's own process of inquiry, the contextual material introducing each primary source provides an unobtrusive introduction and springboard to critical thought.

General Structure

Climate Change: In Context is a collection of entries on diverse topics selected to provide insight into increasingly important and urgent topics associated with the study of climate change.

The articles in the book are meant to be understandable by anyone with a curiosity about topics related to the science of climate change, and the first edition of *Climate Change: In Context* has been designed with ready reference in mind:

- Entries are arranged alphabetically, rather than by chronology or scientific subfield.
- The **chronology** (timeline) includes many of the most significant events in the history of climate change science and advances of science. Where appropriate, related scientific advances are included to offer additional context.
- An extensive glossary section provides readers with a ready reference for content-related terminology. In addition to defining terms within entries, specific Words-to-Know sidebars are placed within each entry.
- A bibliography section (citations of books, periodicals, and Web sites) offers additional resources to those resources cited within each entry.
- A **comprehensive general index** guides the reader to topics and persons mentioned in the book.

Entry Structure

In Context entries are designed so that readers may navigate entries with ease. Toward that goal, entries are divided into easy-to-access sections:

- **Introduction**: A opening section designed to clearly identify the topic.
- **Words-to-know** sidebar: Essential terms that enhance readability and critical understanding of entry content.
- Established but flexible **rubrics** customize content presentation and identify each section, enabling the reader to navigate entries with ease. Inside *Climate Change: In Context* entries readers will find a general scheme of organization. All entries contain a brief introduction, words-to-know, and then a section describing the essential history and scientific foundations of the topic. The section titled "Impacts and Issues" then interrelates key scientific, political, or social considerations related to the topic.
- Many sidebars added by the editors enhance expert contributions by focusing on key areas, providing material for divergent studies, or providing evidence from key scientific reports (e.g., the Intergovernmental Panel on Climate Change (IPCC) reports, and writings of the National Academy of Sciences).
- If an entry contains a related primary source, it is appended to end of the author's text. Authors are not responsible for the selection or insertion of primary sources.
- **Bibliography:** Citations of books, periodicals, Web sites, and audio and visual material used in preparation of the entry or that provide a stepping stone to further study.
- **"See also" references** clearly identify other content-related entries.

Climate Change: In Context special style notes

Please note the following with regard to topics and entries included in *Climate Change: In Context*:

- Primary source selection and the composition of sidebars are not attributed to authors of signed entries to which the sidebars may be associated. Sources for sidebars containing external content (e.g., a quote from an Intergovernmental Panel on Climate Change (IPCC) report related to the entry) are clearly indicated.
- To the greatest extent practical for younger students, the editors have adopted the terminology of the Intergovernmental Panel on Climate Change (IPCC) and have provided an expanded glossary that includes IPCC terminology both used in the book and as a reference to help further the reader's own evaluation of IPCC reports.
- Equations are, of course, often the most accurate and preferred language of science, and are essential to scientists studying climate change. To better serve the intended audience of *Climate Change: In Context*, however, the editors attempted to minimize the inclusion of equations in favor of describing the elegance of thought or essential results such equations yield.
- A detailed understanding of physics and chemistry is neither assumed nor required for *Climate Change: In Context*. Accordingly, students and other readers should not be intimidated or deterred by the sometimes complex names of chemical molecules or biological classification. Where necessary, sufficient information regarding chemical structure or species classification is provided. If desired, more information can easily be obtained from any basic chemistry or biology reference.

Bibliography citation formats (How to cite articles and sources)

In Context titles adopt the following citation format:

Books

Gore, Al. *An Inconvenient Truth: The Planetary Emergency of Global Warming and What We Can Do About It.* Emmaus, PA: Rodale Press, 2006.

Metz, B., et al, eds. *Climate Change 2007: Mitigation of Climate Change: Contribution of Working Group III to the Fourth Assessment Report of the Intergovernmental Panel on Climate Change.* New York: Cambridge University Press, 2007.

Parry, M. L., et al, eds. *Climate Change 2007: Impacts, Adaptation and Vulnerability: Contribution of Working Group II to the Fourth Assessment Report of the Intergovernmental Panel on Climate Change.* New York: Cambridge University Press, 2007.

Solomon, S., et al, eds. *Climate Change 2007: The Physical Science Basis: Contribution of Working Group I to the Fourth Assessment Report of the Intergovernmental Panel on Climate Change.* New York: Cambridge University Press, 2007.

Weart, Spencer. *The Discovery of Global Warming.* Cambridge, MA: Harvard University Press, 2004.

Periodicals

Collins, William, et al. "The Physical Science Behind Climate Change." *Scientific American* (August 2007).

Oreskes, Naomi. "The Scientific Consensus on Climate Change." *Science* 306 (2004): 1686.

Thomas, Chris D. "Extinction Risk from Climate Change." *Nature* 427 (2004): 145–148.

Web Sites

"Climate Change." *U.S. Environmental Protection Agency*, November 19, 2007. <http://epa.gov/climatechange/index.html> (accessed December 9, 2007).

Intergovernmental Panel on Climate Change. <http://www.ipcc.ch> (accessed December 9, 2007).

United Nations Framework Convention on Climate Change. <http://unfccc.int/2860.php> (accessed December 9, 2006).

Alternative citation formats

There are, however, alternative citation formats that may be useful to readers and examples of how to cite articles in alternative formats are shown below.

APA Style

Books: Reisner, Marc. (1986). *Cadillac Desert.* New York: Viking Penguin. Excerpted in K. Lee Lerner and Brenda Wilmoth Lerner, eds. (2006) *Environmental Issues: Essential Primary Sources*, Farmington Hills, MI: Thomson Gale.

Periodicals: Leopold, Aldo. (October 1925). "Wilderness as a Form of Land Use." *The Journal of Land and Public Utility Economics,* 1: 398–404. Excerpted in K. Lee Lerner and Brenda Wilmoth Lerner, eds. (2006) *Environmental Issues: Essential Primary Sources*, Farmington Hills, MI: Thomson Gale.

Web Sites: U.S. Environmental Protection Agency. "How to Conserve Water and Use It Effectively." Retrieved January 17, 2006 from http://www.epa.gov/ow/you/chap3.html. Excerpted in K. Lee Lerner and Brenda Wilmoth Lerner, eds. (2006) *Environmental Issues: Essential Primary Sources*, Farmington Hills, MI: Thomson Gale.

Chicago Style

Books: Reisner, Marc. *Cadillac Desert.* New York: Viking Penguin, 1986. Excerpted in K. Lee Lerner and Brenda Wilmoth Lerner, eds. *Environmental Issues: Essential Primary Sources.* Farmington Hills, MI: Thomson Gale, 2006.

Periodicals: Leopold, Aldo. "Wilderness as a Form of Land Use." *The Journal of Land and Public Utility Economics,* 1 (October 1925): 398–404. Excerpted in K. Lee Lerner and Brenda Wilmoth Lerner, eds. *Environmental Issues: Essential Primary Sources.* Farmington Hills, MI: Thomson Gale, 2006.

Web sites: *U.S. Environmental Protection Agency.* "How to Conserve Water and Use It Effectively." <http://www.epa.gov/ow/you/chap3.html> (accessed January 17, 2006). Excerpted in K. Lee Lerner and Brenda Wilmoth Lerner, eds. *Environmental Issues: Essential Primary Sources.* Farmington Hills, MI: Thomson Gale, 2006.

MLA Style

Books: Reisner, Marc. *Cadillac Desert,* New York: Viking Penguin, 1986. Excerpted in K. Lee Lerner and Brenda Wilmoth Lerner, eds. *Environmental Issues: Essential Primary Sources,* Farmington Hills, MI: Thomson Gale, 2006.

Periodicals: Leopold, Aldo. "Wilderness as a Form of Land Use." *The Journal of Land and Public Utility Economics,* 1 (October 1925): 398–404. Excerpted in K. Lee Lerner and Brenda Wilmoth Lerner, eds. *Environmental Issues: Essential Primary Sources,* Farmington Hills, MI: Thomson Gale, 2006.

Web sites: "How to Conserve Water and Use It Effectively." U.S. Environmental Protection Agency. 17 January 2006. <http://www.epa.gov/ow/you/chap3.html>. Excerpted in K. Lee Lerner and Brenda Wilmoth Lerner, eds. *Environmental Issues: Essential Primary Sources,* Farmington Hills, MI: Thomson Gale, 2006.

Turabian Style (Natural and Social Sciences)

Books: Reisner, Marc. *Cadillac Desert,* (New York: Viking Penguin, 1986). Excerpted in K. Lee Lerner and Brenda Wilmoth Lerner, eds. *Environmental Issues: Essential Primary Sources,* (Farmington Hills, MI: Thomson Gale, 2006).

Periodicals: Leopold, Aldo. "Wilderness as a Form of Land Use." *The Journal of Land and Public Utility Economics,* 1 (October 1925): 398–404. Excerpted in K. Lee Lerner and Brenda Wilmoth Lerner, eds. *Environmental Issues: Essential Primary Sources,* (Farmington Hills, MI: Thomson Gale, 2006).

Web sites: U.S. Environmental Protection Agency. "How to Conserve Water and Use It Effectively." Available from http://www.epa.gov/ow/you/chap3.html; accessed January 17, 2006. Excerpted in K. Lee Lerner and Brenda Wilmoth Lerner, eds. *Environmental Issues: Essential Primary Sources,* (Farmington Hills, MI: Thomson Gale, 2006).

Using Primary Sources

The definition of what constitutes a primary source is often the subject of scholarly debate and interpretation. Although primary sources come from a wide spectrum of resources, they are united by the fact that they individually provide insight into the historical *milieu* (context and environment) during which they were produced. Primary sources include materials such as newspaper articles, press dispatches, autobiographies, essays, letters, diaries, speeches, song lyrics, posters, works of art—and in the twenty-first century, web logs—that offer direct, first-hand insight or witness to events of their day.

Categories of primary sources include:

- Documents containing firsthand accounts of historic events by witnesses and participants. This category includes diary or journal entries, letters, email, newspaper articles, interviews, memoirs, and testimony in legal proceedings.
- Documents or works representing the official views of both government leaders and leaders of other organizations. These include primary sources such as policy statements, speeches, interviews, press releases, government reports, and legislation.
- Works of art, including (but certainly not limited to) photographs, poems, and songs, including advertisements and reviews of those works that help establish an understanding of the cultural milieu (the cultural environment with regard to attitudes and perceptions of events).
- Secondary sources. In some cases, secondary sources or tertiary sources may be treated as primary sources. For example, if an entry written many years after an event, or to summarize an event, includes quotes, recollections, or retrospectives (accounts of the past) written by participants in the earlier event, the source can be considered a primary source.

Analysis of primary sources

The primary material collected in this volume is not intended to provide a comprehensive or balanced overview of a topic or event. Rather, the primary sources are intended to generate interest and lay a foundation for further inquiry and study.

In order to properly analyze a primary source, readers should remain skeptical and develop probing questions about the source. Using historical documents requires that readers analyze them carefully and extract specific information. However, readers must also read "beyond the text" to garner larger clues about the social impact of the primary source.

In addition to providing information about their topics, primary sources may also supply a wealth of insight into their creator's viewpoint. For example, when reading a news article about a likely impact of climate change, consider whether the reporter's words also indicate something about his or her origin, bias (an irrational disposition in favor of someone or something), prejudices (an irrational disposition against someone or something), or intended audience.

Students should remember that primary sources often contain information later proven to be false, or contain viewpoints and terms unacceptable to future generations. It is important to view the primary source within the historical and social context existing at its creation. If for example, a newspaper article is written within hours or days of an event, later developments may reveal some assertions in the original article as false or misleading.

Test new conclusions and ideas

Whatever opinion or working hypothesis the reader forms, it is critical that he or she then test that hypothesis against other facts and sources related to the incident. For example, it might be wrong to conclude that factual mistakes are deliberate unless evidence can be produced of a pattern and practice of such mistakes with an intent to promote a false idea.

The difference between sound reasoning and preposterous conspiracy theories (or the birth of urban legends) lies in the willingness to test new ideas against other sources, rather than rest on one piece of evidence such as a single primary source that may contain errors. Sound reasoning requires that arguments and assertions guard against argument fallacies that utilize the following:

- false dilemmas (only two choices are given when in fact there are three or more options);
- arguments from ignorance (*argumentum ad ignorantiam*; because something is not known to be true, it is assumed to be false);
- possibilist fallacies (a favorite among conspiracy theorists who attempt to demonstrate that a factual statement is true or false by establishing the possibility of its truth or falsity. An argument where "it could be" is usually followed by an unearned "therefore, it is.");
- slippery slope arguments or fallacies (a series of increasingly dramatic consequences is drawn from an initial fact or idea);
- begging the question (the truth of the conclusion is assumed by the premises);
- straw man arguments (the arguer mischaracterizes an argument or theory and then attacks the merits of his/her own false representations);
- appeals to pity or force (the argument attempts to persuade people to agree by sympathy or force);
- prejudicial language (values or moral goodness, good and bad, are attached to certain arguments or facts);
- personal attacks (*ad hominem*; an attack on a person's character or circumstances);
- anecdotal or testimonial evidence (stories that are unsupported by impartial observation or data that is not reproducible);
- *post hoc* (after the fact) fallacies (because one thing follows another, it is held to cause the other);
- the fallacy of the appeal to authority (the argument rests upon the credentials of a person, not the evidence).

Despite the fact that some primary sources can contain false information or lead readers to false conclusions based on the "facts" presented, they remain an invaluable resource regarding past events. Primary sources allow readers and researchers to come as close as possible to understanding the perceptions and context of events and thus to more fully appreciate how and why misconceptions occur.

Glossary

A

ABORIGINES: Native peoples or indigenous peoples: the cultural or racial group of the oldest inhabitants of a region.

ABRUPT CLIMATE CHANGE: Sometimes called rapid climate change, abrupt events, or surprises. "Abrupt" is not a precise term, but generally refers to changes occurring faster than the typical time scale of the cause (forcing) responsible for the change. For example, a sudden shift in climate (over a decade or two) due to slowly increasing carbon dioxide levels (over centuries) would be abrupt.

ABSOLUTE ZERO: The temperature at which significant molecular activity reaches its minimum. Molecular activity never stops because molecules vibrate even at absolute zero—but the motion is the smallest possible at absolute zero. Absolute zero is defined as 0 Kelvin (there are two degree units in the Kelvin scale) or approximately -459°F or -273°C.

ACID: Substance that when dissolved in water is capable of reacting with a base to form salts and release hydrogen ions.

ACID RAIN: A form of precipitation that is significantly more acidic than neutral water, often produced as the result of industrial processes.

ACTUALISM: The doctrine that past events in Earth's geological and biological history were the outcome of processes like those that we see in operation today. Actualism asserts uniformity of process throughout geological history, and is one of the elements of uniformitarianism.

ADDITIONALITY: The property in a carbon offsetting project, especially as defined by the Kyoto Protocol, that the project would not have been undertaken in any case, regardless of offset goals. For example, a project to burn off landfill methane lacks additionality if the methane would have to be burned off anyway for safety reasons.

AEROSOLS: Particles of matter, solid or liquid, larger than a molecule but small enough to remain suspended in the atmosphere. Natural sources include salt particles from sea spray and clay particles as a result of the weathering of rocks, both of which are carried upward by the wind. Aerosols can also originate as a result of human activities and in this case are often considered pollutants.

AFFORESTATION: Conversion of unforested land to forested land through planting, seeding, or other human interventions. Unforested land must have been unforested for at least 50 years for such intervention to qualify as afforestation; otherwise, it is termed reforestation.

AGENDA 21: Program initiated by most members of the United Nations in 1992 that seeks to foster sustainable development in poorer countries. Implemented by the U.N. Commission on Sustainable Development. The number 21 was chosen to refer to the twenty-first century.

AGROFORESTRY: The practice of sustainably combining forestry with agriculture by combining trees with shrubs, crops, or livestock. Two forms of agroforestry are alley cropping (strips of crops alternating with strips of woodland) and silvopasture (pasturing grazing livestock under widely spaced trees). Benefits include greater soil retention, biodiversity encouragement, higher monetary returns per acre, and more diverse product marketing.

AIR PRESSURE: The weight of the atmosphere over a particular point, also called barometric pressure. Average air exerts approximately 14.7 pounds (6.8 kg) of force on every square inch at sea level.

ALBEDO: A numerical expression describing the ability of an object or planet to reflect light.

ALGAE: Single–celled or multicellular plants or plant-like organisms that contain chlorophyll, thus making their own food by photosynthesis. Algae grow mainly in water.

ALGAL BLOOM: Sudden reproductive explosion of algae (single-celled aquatic green plants) in a large, natural body of water such as a lake or sea. Blooms near coasts are sometimes called red tides.

ALLOTROPE: One of two or more forms of an element.

ALTIPLANO: High plateau region in the Andes of Argentina, Bolivia, and Peru in South America, the second-largest high plateau on Earth after Tibet. Sediments in Lake Titicaca, into which the Altiplano drains, record tens of thousands of years of regional climate change.

AMINE: One of a family of compounds, the amines, containing carbon and nitrogen. When combined with a carboxyl group (CO_2H), an amine forms an amino acid. All living things build proteins by chaining together amino acids.

ANAEROBIC: Lacking free molecular oxygen (O_2). Anaerobic environments lack O_2; anaerobic bacteria digest organic matter such as dead plants in anaerobic environments such as deep water and the digestive systems of cattle. Anaerobic digestion releases methane, a greenhouse gas.

ANAEROBIC BACTERIA: Single-celled creatures that thrive in anaerobic environments, that is, environments lacking free molecular oxygen (O_2). They digest organic matter and release methane, a greenhouse gas.

ANNEX I AND NON-ANNEX I COUNTRIES: Groups of countries defined by the United Nations Framework Convention on Climate Change (UNFCCC) of 1992. Annex I countries are industrialized countries who agreed, under the treaty, to reduce their greenhouse emissions. All non-Annex I countries were developing (poorer, less-industrialized) countries. Annex II countries were wealthy Annex I countries who agreed to help pay for greenhouse reductions by developing (poorer) countries.

ANOXIC: Lacking free oxygen (O_2). Similar in meaning to anaerobic, but implies that oxygen has been depleted in a given organism or environment, rather than being naturally at a low level.

ANTARCTIC CIRCUMPOLAR CURRENT: An ocean current that circles Antarctica from west to east (clockwise, looking down on the continent), enabling mixing of the waters of the world's oceans. Also termed the West Wind Drift.

ANTHROPOGENIC: Made by people or resulting from human activities. Usually used in the context of emissions that are produced as a result of human activities.

AOSIS PROTOCOL: The Alliance of Small Island States (AOSIS) is a federation of 36 of the world's small island states formed in 1990 to influence international policy on climate change. The AOSIS Protocol was a proposal submitted to the first Conference of the Parties to the United Nations Framework Convention on Climate Change (UNFCCC) in 1995 that proposed that the developed nations reduce their greenhouse-gas emissions by 20% by 2005.

AQUIFER: Underground layer of sand, gravel, or spongy rock that collects water.

AROMATIC: In organic chemistry, a compound whose molecular structure includes some variation of the benzene ring.

ATLANTIC CONVEYOR BELT: The north-south circulation that dominates movement of water in the Atlantic Ocean and is a key part of the global meriodonal thermohaline circulation and of the climate system. Moves warm waters toward the Arctic, where they cool, sink, and move southward along the ocean floor.

ATLANTIC MERIDIONAL OVERTURNING CIRCULATION: The north-south circulation that dominates the movement of water in the Atlantic Ocean and is a key part of the global meriodonal thermohaline circulation and of the climate system. Moves warm waters toward the Arctic, where they cool, sink, and move southward along the ocean floor.

ATMOSPHERE: The air surrounding Earth, described as a series of shells or layers of different characteristics. The atmosphere—composed mainly of nitrogen and oxygen with traces of carbon dioxide, water vapor, and other gases—acts as a buffer between Earth and the sun. The layers—troposphere, stratosphere, mesosphere, thermosphere, and the exosphere—vary around the globe and in response to seasonal changes.

ATMOSPHERIC CHEMISTRY: Study of the chemistry of planetary atmospheres. The interactions of pollutants, greenhouse gases, Earth's natural atmosphere, solar radiation, and other factors are all studied under the aegis of atmospheric chemistry.

ATMOSPHERIC CO_2: Carbon dioxide (CO_2) that is found in the atmosphere. As of mid 2007, the concentration of atmospheric CO_2 was about 383 parts per million (ppm) (i.e., 383 molecules of CO_2 for every 1 million molecules of air, about 0.0383% of the atmosphere), an increase of 36% from its preindustrial value of 280 ppm.

ATOLL: A coral island consisting of a ring of coral surrounding a central lagoon. Atolls are common in the Indian and Pacific Oceans.

B

BANDED IRON FORMATION: Layered (banded) sedimentary rocks containing alternating layers of stone and iron. Most banded iron formations formed several billion years ago as a result of metabolic activity by bacteria in Earth's early, low-oxygen oceans. Some scientists have hypothesized that younger banded-iron formations may have formed during a snowball Earth period or periods about 600 million years ago.

BASE: One of the four chemicals (nitrogenous bases) that are a part of the deoxyribonucleic (DNA) molcule. The sequence of the bases determines the genetic code (the information contained within DNA). The first letters of each base provide the alphabet for the DNA code (A, C, G, and T [short for adenine, cytosine, guanine, and thymine]).

BASELINE GENERATION: Generation of electricity for baseline demand, that is, demand that is steady around the clock. Demand for electricity rises above baseline during the daytime and during heat waves, when power demand for air conditioning peaks.

BEACH: A gently sloping band of rock or shell particles that is bounded by land on one side and a river, lake, or ocean on the other.

BEACH FACE: Portion of a beach that slopes downward to the water.

BEHAVIORAL ADAPTABILITY: Capacity of an organism or society to survive altered environmental conditions by changing one or more behaviors. Changed behaviors might include timing of food-seeking, mating, egg-laying, and resting; use of alternative food resources, altered economic activities or building techniques; and many more.

BERM: A platform of wave-deposited sediment that is flat or slopes slightly landward.

BIAS: In a person, a tendency to omit or exaggerate facts to favor a predetermined opinion; in numerical data, an offset or constant error that changes the value of all measurements equally.

BIODIESEL: A fuel made from a combination of plant and animal fat. It can be safely mixed with petro diesel.

BIODIVERSITY: Literally, "life diversity": the number of different kinds of living things. The wide range of organisms—plants and animals—that exist within any given geographical region.

BIOENERGY: Energy for technological use derived from materials produced by living things. Wood, methane from anaerobic bacteria, and liquid fuels manufactured from crops are all forms of bioenergy.

BIOETHANOL: Ethanol produced by fermentation using yeast or bacteria. Most ethanol is bioethanol, but methods to produce it using purely chemical processes also exist.

BIOFUEL: A fuel derived directly by human effort from living things, such as plants or bacteria. A biofuel can be burned or oxidized in a fuel cell to release useful energy.

BIOGAS: Methane produced by rotting excrement or other biological sources. It can be burned as a fuel.

BIOGEOCHEMICAL CYCLE: The chemical interactions that take place among the atmosphere, biosphere, hydrosphere, and geosphere.

BIOGEOCHEMISTRY: The study of how substances and energy are exchanged between living things and the nonliving environment.

BIOMASS: The sum total of living and once-living matter contained within a given geographic area. Plant and animal materials that are used as fuel sources.

BIOME: Well-defined terrestrial environment (e.g., desert, tundra, or tropical forest). The complex of living organisms found in an ecological region.

BIOSPHERE: The sum total of all life-forms on Earth and the interaction among those life-forms.

BIPEDAL WALKING: Walking on two legs. The evolution of bipedal walking in early human ancestors freed up the hands for gripping and carrying, which may have triggered evolution of the enlarged human brain.

BIPOLAR SEESAW: The tendency of climate to warm at the North Pole while it cools at the South Pole and vice versa. Paleoclimatic data show that this seesaw effect has occurred many times in the geological past and is related to changes in ocean circulation, but scientists do not yet completely understand the mechanism of the seesaw.

BISTABILITY: The property of a system that it can exist in two stable states: a dinner plate, for example, is stable lying on either side, but not balanced on edge. A tipping point often exists in bistable systems, beyond which transition to the other stable state becomes automatic. A plate on edge is at a tipping point, where a slight push will quickly cause the system (i.e., the plate) to move all the way to one of its stable states. Some feedbacks in the climate system may be bistable.

BLACK CARBON: A type of aerosol (small, airborne particle) consisting mostly of carbon: includes soot, charcoal, and some other dark organic particles.

BOREAL FOREST: Type of forest covering much of northern Europe, Asia, and North America, composed mostly of coniferous evergreen trees. Although low in biodiversity, boreal forest covers more of Earth's land area than any other biome.

BOREHOLES: Hole or shaft drilled into solid ground or a glacier for the purpose of extracting information rather than some resource (such as petroleum). In climate research, temperature measurements are often made along the height of a borehole: this information can be mathematically processed (deconvolved) to give information about past changes of surface temperature up to several centuries past.

BREVETOXIN: Any of a class of neurotoxins produced by the algae that cause red tide (coastal algae blooms). Brevetoxins and other toxins from algal blooms can be concentrated by shellfish and poison humans who eat the shellfish.

BRINE: A solution of sodium chloride and water that may or may not contain other salts.

BROMOFLUOROCARBONS: A class of compounds, also called halons, that contain bromine, chlorine, and carbon or fluorine. Used as fire retardants in fire-extinguishing systems, these are potent greenhouse gases with high ozone depletion potential.

BUOYS: Tethered or free-floating devices that bear navigational aids, instruments, and in some cases radio equipment for automatically collecting and reporting data on oceanic and atmospheric conditions. Buoys may float on or beneath the surface, depending on their purpose.

BYRD-HAGEL RESOLUTION: U.S. Senate Resolution 98, proposed by Senators Robert Byrd and Chuck Hagel while the Kyoto Protocol was being negotiated. This resolution rendered it politically impossible that the United States would ratify Kyoto, stating the Senate's intention to oppose any climate treaty that injured the economy of the United States or did not mandate emissions limits for developing (non-Annex I) nations (as Kyoto did not). Passed 95 to 0 on July 25, 1997.

C

CALCINATION: An old term used to describe the process of heating metals and other materials in air.

CALIBRATION: Act of comparing an instrument's measuring accuracy to a known standard.

CALLENDAR EFFECT: Global warming due to atmospheric carbon dioxide released by burning fossil fuels. Named after Britsh engineer Guy Stewart Callendar (1898–1964), who proposed in the 1930s (erroneously) that weather records already showed such warming.

CALVING: Process of iceberg formation when huge chunks of ice break free from glaciers, ice shelves, or ice sheets due to stress, pressure, or the forces of waves and tides.

CAMBRIAN EXPLOSION: Relatively sudden evolution of a wide variety of multicellular forms of life at the beginning of the Cambrian period, about 530 million years ago, after billions of years during which Earth was inhabited almost entirely by single-celled organisms. A few forms of multicellular life did appear shortly before the Cambrian Explosion. The explosion may have been triggered by the ending of a major snowball Earth period or by the achievement of sufficiently high oxygen levels in the atmosphere.

CAP-AND-TRADE: The practice, in pollution-control or climate-mitigation schemes, of mandating an upper limit or cap for the total amount of some substance to be emitted (e.g., CO_2) and then assigning allowances or credits to polluters that correspond to fixed shares of the total amount. These allowances or credits can then be bought and sold by polluters, in theory allowing emission cuts to be bought where they are most economically rational.

CARBON: Chemical element with atomic number 6. The nucleus of a carbon atom contains 6 protons and from 6 to 8 neutrons. Carbon is present, by definition, in all organic substances; it is essential to life and, in the form of the gaseous compounds CO_2 (carbon dioxide) and CH_4 (methane), the major driver of climate change.

CARBON BANKING: Form of accounting that tracks the carbon emissions, reductions, and offsets of a client in a way that is analogous to the treatment of money in ordinary banking. A helpful adjunct to emissions trading schemes.

CARBON BLACK: A commercial useful form of nearly pure carbon consisting of tiny particles. Used as a black pigment in inks and plastics and in tire manufacture. Over 8 million metric tons are produced each year, mostly from natural gas or heavy oil feedstocks.

CARBON CALCULATOR: Software device, often accessed through a Web site, that allows an individual or business to calculate their carbon footprint, that is, how much greenhouse warming is generated to support the present mode of existence of that individual or business.

CARBON CREDIT: A unit of permission or value, similar to a monetary unit (e.g., dollar, euro, yen) that entitles its owner to emit one metric ton of carbon dioxide into the atmosphere.

CARBON CYCLE: All parts (reservoirs) and fluxes of carbon. The cycle is usually thought of as four main reservoirs of carbon interconnected by pathways of exchange. The reservoirs are the atmosphere, terrestrial biosphere (usually includes freshwater systems), oceans, and sediments (includes fossil fuels). The annual movements of carbon, the carbon exchanges between reservoirs, occur because of various chemical, physical, geological, and biological processes.

CARBON DIOXIDE: Odorless, colorless, non-poisonous gas, chemical formula CO_2. Carbon dioxide is released by natural processes and by burning fossil fuels. A minor but very important component of the atmosphere, carbon dioxide traps infrared radiation. Atmospheric CO_2 has increased dramatically since the early 1800s. Burning fossil fuels is the leading cause of increased CO_2 levels with deforestation the second major cause.

CARBON FOOTPRINT: The amount of carbon dioxide (or of any other greenhouse gas, counted in terms of the greenhouse-equivalent amount of CO_2) emitted to supply the energy and materials consumed by a person, product, or event. A concert, manufactured object, family, organization, or individual person may all have a carbon footprint. The more carbon is released, the larger the footprint.

CARBON NEUTRALITY: An arrangement by a person or group whereby as much carbon dioxide is being removed from or kept out of the atmosphere as is being released by the activities of that person or group. Carbon neutrality is achieved when offsets equal the person or group's carbon footprint.

CARBON OFFSETS: Reductions in emissions of CO_2 (or other greenhouse gases) or enhanced removals of such gases from the atmosphere that are arranged by polluters in order to compensate for their releases of greenhouse gases. Carbon offsets may be purchased by individuals or groups.

CARBON PRICING: Assignment of a market price to carbon credits on some emissions trading market, such as the Chicago Climate Exchange or European Union Emissions Trading Scheme.

CARBON SEQUESTRATION: Storage or fixation of carbon in such a way that it is isolated from the atmosphere and cannot contribute to climate change. Sequestration may occur naturally (e.g., forest growth) or artificially (e.g., injection of CO_2 into underground reservoirs).

CARBON SINKS: Carbon reservoirs such as forests or oceans that take in and store more carbon (carbon sequestration) than they release. Carbon sinks can serve to partially offset greenhouse-gas emissions.

CARBON TAX: Mandatory fee charged for the emission of a given quantity of CO_2 or some other greenhouse gas. Under a carbon taxation scheme, polluters who emit greenhouse gases must pay costs that are directly proportional to their emissions. The purpose of a carbon tax is to reduce greenhouse emissions. Carbon taxation is the main alternative to emissions trading.

CARBON TETRACHLORIDE: A compound of carbon that has all four bonding positions each filled with a chlorine atom.

CARBON TRADING: Buying and selling of carbon credits, abstract instruments (like money) that each represent the right to emit 1 ton of carbon dioxide or an equivalent amount of other greenhouse gases. Carbon trading presently takes place under the European Union Emission Trading Scheme and the Chicago Climate Exchange.

CARBON-BASED FUEL: Any substance composed mostly of carbon that is burned or otherwise chemically reacted to release energy. Most biofuels and all fossil fuels are carbon-based, although natural gas also contains a significant fraction of its energy in the form of hydrogen.

CARCINOGENS: Chemicals that induce, or increase the likelihood of, cancer. Carcinogens may be natural or artificial, and may have a carcinogenic effect in low or high concentrations, depending on the chemistry of the particular substance.

CATALYTIC CRACKING: Chemical process in which large organic molecules are broken into smaller molecules with the help of a catalyst, a substance that speeds up a chemical reaction without being changed by it. Used in oil refining; proposed (but not yet commercially viable) for the production of biofuels such as ethanol and biodiesel.

CATASTROPHISM: School of thought in geology which holds that the rates of processes shaping Earth have varied greatly in the past, occasionally acting with violent suddenness (catastrophes).

CAY: A low-lying reef of sand or coral.

CELLULOSIC ETHANOL: Ethanol (a liquid fuel, also found in alcoholic beverages) produced by chemical or biological digestion of cellulose, the main structural material of plant tissues. Commercial-scale production of cellulosic ethanol (not yet viable as of 2007) would allow fuel production from plant materials that presently cannot be used for that purpose because they consist almost entirely of cellulose, such as corn cobs and wood chips.

CELLULOSIC FERMENTATION: Digestion of high-cellulose plant materials (e.g., wood chips, grasses) by bacteria that have been bred or genetically engineered for that purpose. The useful product is ethanol, which can be burned as a fuel. Cellulosic fermentation is one way of producing cellulosic ethanol.

CENTRIFUGAL FORCE: The inertial reaction that causes a body to move away from a center about which it revolves.

CHANDLER WOBBLE: Small-scale cyclic wobble described by Earth's axis, typically on a 433-day schedule: caused by changes in water pressure at the ocean floor driven by Earth's climate cycle. Named after American astronomer Seth Carlo Chandler (1846–1913), who discovered the phenomenon in 1891, although its cause was not known until 2000.

CHINOOK WINDS: Föhn winds that occur in western North America where plains and prairies meet mountain ranges. Föhn winds are formed when air is forced over mountain ranges, losing its moisture and then warming as it flows downslope and is placed under higher atmospheric pressure.

CHLOROFLUOROCARBONS: Members of the larger group of compounds termed halocarbons. All halocarbons contain carbon and halons (chlorine, fluorine, or bromine). When released into the atmosphere, CFCs and other halocarbons deplete the ozone layer and have high global warming potential.

CIRRUS CLOUD: Thin clouds of tiny ice crystals that form at 20,000 ft (6 km) or higher. Cirrus clouds cover 20–25% of the globe, including up to 70% of tropical regions. Because they both reflect sunlight from Earth and reflect infrared (heat) radiation back at the ground, they can influence climate.

CLEAN AIR ACT OF 1970: Extension of the 1963 U.S. Clean Air Act that tasked the newly established Environmental Protection Agency with developing and enforcing regulations to reduce air pollution.

CLEAN DEVELOPMENT MECHANISM: One of the three mechanisms set up by the Kyoto Protocol to, in theory, allow reductions in greenhouse-gas emissions to be implemented where they are most economical. Under the Clean Development Mechanism, polluters in wealthy countries can obtain carbon credits (greenhouse pollution rights) by funding reductions in greenhouse emissions in developing countries.

CLIMATE: The average weather (usually taken over a 30-year time period) for a particular region and time period. Climate is not the same as weather, but rather, it is the average pattern of weather for a particular region. Weather describes the short-term state of the atmosphere. Climatic elements include precipitation, temperature, humidity, sunshine, wind velocity, phenomena such as fog, frost, and hail storms, and other measures of the weather.

CLIMATE CHANGE: Sometimes used to refer to all forms of climatic inconsistency, but because Earth's climate is never static, the term is more properly used to imply a significant change from one climatic condition to another. In some cases, climate change has been used synonymously with the term, global warming; scientists, however, tend to use the term in the wider sense to also include natural changes in climate.

CLIMATE MODELING: Mathematical representation of climate processes. Computer programs are written that describe the structure of Earth's land, ocean, atmospheric, and biological systems and the laws of nature that govern the behavior of those systems. Detail and accuracy are limited by scientific understanding of the climate system and computer power. Climate modeling is essential to understanding paleoclimate, present-day climate, and future climate.

CLIMATOLOGICAL WARFARE: Speculative use of deliberately engineered climate or weather changes as weapons of war: studied intensively in the United

States from 1945 to the 1970s, after which such research was abandoned.

CLIMATOLOGIST: Scientist who specializes in the study of climate.

CLOUD CONDENSATION NUCLEI: Tiny solid airborne particles around which the water droplets or ice crystals that form clouds condense. By providing additional cloud condensation nuclei, aerosol pollution may influence cloud formation and thus climate.

CLOUD SEEDING: The introduction of particles of (usually) dry ice or silver iodide with the hope of increasing precipitation from the cloud.

CO_2 FERTILIZATION EFFECT: Acceleration of plant growth by heightened atmospheric CO_2 concentrations. The effect has been suggested as a benefit to agriculture from artificially increased atmospheric CO_2 and as a source of negative climate-change feedback through the accelerated growth of forests.

COAL-TO-LIQUIDS: The production of liquid fuels from coal. A technology favored for development by U.S. energy policy but having many environmental drawbacks, including high greenhouse emissions.

COMBUSTION: When substances combine with oxygen and release energy.

COMMODIFY: To make something into a commodity, that is, something which is bought and sold. Some have argued that carbon markets are unethical because they commodify the well-being of Earth itself, on which all life depends.

COMPACT FLUORESCENT LIGHT BULB: Fluorescent light bulbs with tightly spiraled tubes that fit inside small, bulb-shaped glass envelopes. Can replace incandescent (hot-filament) light bulbs in most applications and use about a fourth as much energy to produce a given amount of light. They are most costly to install but can pay for themselves with lower energy use and longer lifetime.

COMPOSTING: The process by which organic waste, such as yard waste, food waste, and paper, is broken down by microorganisms and turned into a useful product for improving soil.

CONDENSATION NUCLEI: Tiny solid airborne particles around which the water droplets or ice crystals that form clouds condense. By providing additional cloud condensation nuclei, aerosol pollution may influence cloud formation and thus climate.

CONFERENCE OF THE PARTIES (COP): Signatories to the U.N. Framework Convention on Climate Change, a treaty drafted in 1992. The treaty entered into force in 1994, and COPs have been held ever since. It was at COPs that the Kyoto Protocol was drafted.

CONFERENCE OF THE PARTIES (COP) TO THE CLIMATE CHANGE CONVENTION: Annual meeting of representatives from nations that are signatories of (parties to) the United Nations Framework Convention on Climate Change, a treaty drafted in 1992. The treaty entered into force in 1994, and COPs have been held ever since. It was at COPs that the Kyoto Protocol was drafted.

CONTINENTAL DRIFT: A theory that explains the relative positions and shapes of the continents, and other geologic phenomena, by lateral movement of the continents. This was the precursor to plate tectonic theory.

CONTINENTAL PLATE: Solid segment of Earth's rocky crust that moves about independently. The motion of plates is one of the basic geological processes shaping Earth's surface and is the cause of continental drift, which over millions of years can radically reshape regional and global climate.

CONTINENTAL SHELF: A gently sloping, submerged ledge of a continent.

CONTRAIL: High-altitude cloud formed by the passage of an aircraft. Most contrails are formed by the condensation of water vapor in jet exhaust around small particles in the ambient air, the exhaust, or both. Contrails alter the cloud content of the atmosphere, but their contribution to global climate change is uncertain.

CONVECTION: The rising of warm air from an object, such as the surface of Earth.

CONVECTION CURRENT: Circular movement of a fluid in response to alternating heating and cooling.

CORAL ATOLL: Low tropical island, often roughly ring-shaped, formed by coral reefs growing on top of a subsiding island. The rocky base of the atoll may be hundreds of feet below present-day sea level. Atolls, like other low-lying islands, are threatened with submergence by rapid sea-level rise caused by anthropogenic climate change.

CORAL BLEACHING: Decoloration or whitening of coral from the loss, temporary or permanent, of symbiotic algae (zooxanthellae) living in the coral. The algae give corals their living color and, though photosynthesis, supply most of their food needs. High sea surface temperatures can cause coral bleaching.

CORAL POLYP: Living organism that, as part of a colony, builds the rocky calcium carbonate ($CaCO_3$)

skeleton that forms the physical structure of a coral reef.

CORE SAMPLE: Cylindrical, solid sample of a layered deposit, cut out of the deposit at right angles to its bedding planes. Core samples of lake-bottom and sea-bottom sediments, or of ice layers of the Greenland or Antarctic ice caps, can supply information about past climate variations and changes in atmospheric composition. For example, Antarctic ice cores supply climate data going back 800,000 years.

CORIOLIS EFFECT: A pseudoforce describing the deflection of winds due to the rotation of Earth, which produce a clockwise or counterclockwise rotation of storm systems in the Southern and Northern Hemispheres, respectively.

CORONAL MASS EJECTIONS: Events in which masses of hot gas (plasma) are ejected from the sun's outer atmosphere, its corona. After several days, if a coronal mass ejection is headed the right way, it may intercept Earth, causing auroras and in some cases damaging electronic equipment and power transmission grids.

COSMICS LEAVING OUTDOOR DROPLETS (CLOUD): Experiment at the CERN particle physics laboratory in Geneva, Switzerland, designed to investigate whether cosmic rays (high-energy particles from outer space) can influence Earthly weather and climate by triggering the formation of cloud particles. Began collecting data in 2006.

CRETACEOUS PERIOD: Geological period from 145 million years ago to 65 million years ago. During the Cretaceous, the supercontinent that geologists call Gondwana broke up, radically altering regional climates. The Cretaceous ended in the Cretaceous-Tertiary (K-T) extinction event, probably partly caused at least partly by an asteroid impact, which included the extinction of dinosaurs and about 85% of all other species.

CRUST: The hard, outer shell of Earth that floats upon the softer, denser mantle.

CRYOSPHERE: One of the interrelated components of Earth's system, the cryosphere is frozen water in the form of snow, permanently frozen ground (permafrost), floating ice, and glaciers. Fluctuations in the volume of the cryosphere cause changes in ocean sea-level, which directly impact the atmosphere and biosphere.

CYCLONE: An area of low pressure where winds blow counterclockwise in the Northern Hemisphere and clockwise in the Southern Hemisphere.

D

DANSGAARD-OESCHGER EVENTS: Global climate warming events that occur every 1,470 years or at some even multiple of this period (for example, 2×1470 = 2,940 years or 3×1470 = 4,410 years). The cause is uncertain; some scientists hypothesize shifts in ocean circulation triggered by regular changes in Earth's orbit.

DECOMPOSITION: The breakdown of matter by bacteria and fungi. It changes the chemical makeup and physical appearance of materials.

DEFORESTATION: Those practices or processes that result in the change of forested lands to non-forest uses. This is often cited as one of the major causes of the enhanced greenhouse effect for two reasons: 1) the burning or decomposition of the wood releases carbon dioxide; and 2) trees that once removed carbon dioxide from the atmosphere in the process of photosynthesis are no longer present and contributing to carbon storage.

DEGREE: A unit of angular measure represented by the symbol °. The circumference of a circle contains 360 degrees. When applied to the roughly spherical shape of Earth for geographic and cartographic purposes, degrees are each divided into 60 minutes.

DELTA: Triangular–shaped area where a river flows into an ocean or lake, depositing sand, mud, and other sediment it has carried along its flow.

DENIALISM: Insistent refusal to accept a scientific or historical fact in the face of overwhelming physical evidence and near-universal expert agreement.

DENIALISTS: People who insistently refuse to accept a scientific or historical fact in the face of overwhelming physical evidence and near-universal expert agreement.

DEOXYRIBONUCLEIC ACID (DNA): Large, complex molecules found in the nuclei of cells that carries genetic information for an organism's development.

DEPLETABLE ENERGY SOURCE: Energy source that diminishes as it is used: same as a non-renewable energy source (to "deplete" means to use up). All fossil fuels are depletable energy sources.

DEPOSITION: Process by which water changes phase directly from vapor into a solid without first becoming a liquid.

DESERT: A land area so dry that little or no plant or animal life can survive.

DESERTIFICATION: Transformation of arid or semi-arid productive land into desert.

DETERMINISTIC: Able to occur in only one way: determined by the laws of nature. In contrast to stochastic, random, or chaotic processes, which are inherently difficult to forecast even when the physical laws governing them are well understood.

DEVELOPING NATION: Country that is relatively poor, with a low level of industrialization and relatively high rates of illiteracy and poverty. Sometimes termed less-developed country. The use of "developing" as a euphemism for "poor" contains the built-in assumption that "development," in the sense of transition to a non-agricultural economy oriented toward perpetual growth, is innately good.

DEW POINT: The temperature to which air must be cooled for saturation to occur, exclusive of air pressure or moisture content change. At that temperature, dew begins to form, and water vapor condenses into liquid.

DIESEL FUEL: A liquid engine fuel used in diesel engines, which ignite fuel vapor by increasing pressure in a cylinder rather than by an electric spark. Diesel engines tend to be more efficient than gasoline engines. Most diesel fuel is refined from petroleum, but diesel may also be made from vegetable oil, in which case it is termed biodiesel.

DIFFERENTIAL EQUATION: In mathematics, any equation in which one or more derivatives (rates of change) of a variable appear along with the variable itself. Differential equations are needed to describe many physical processes in engineering, physics, and other sciences, and are essential to climate modeling.

DIGITAL MAPPING: Computerized production of maps, especially in geographic data systems.

DINOFLAGELLATE: Small organisms with both plant-like and animal-like characteristics, usually classified as algae (plants). They take their name from their twirling motion and their whip-like flagella.

DIPOLE: A system that has two equal but opposite characteristics separated by a distance.

DIRECT ENERGY USE: Consumption of energy for direct function, such as burning fuel to move a vehicle, heat food, or heat buildings or consuming electricity to run machines. Usually contrasted to indirect energy use, which is the consumption of energy via the consumption of products (food, clothes, computers, toys, etc.) that require energy for their manufacture, transport, and disposal. Actual energy use is the sum of direct and indirect use.

DISCOUNTING: The practice, in economics, of assigning lower value to the well-being of future generations than to that of the present one. A controversial aspect of efforts to calculate the social cost of carbon.

DISCRETIZE: To represent a phenonemon that occurs smoothly over some range, such as wind speed or temperature, as a set of numbers separated by fixed gaps. "Discrete" means separate, not connected. For instance, if temperature is measured only to the closest tenth of a degree, then temperature is discretized to jumps of 0.1 degree. All numerical representations of continuous natural phenomenon discretize them to some extent.

DISINTEGRATION: Spontaneous nuclear transformation characterized by the emission of energy and/or mass from the nucleus.

DISTILLATION: The process of separating liquids from solids or from other liquids with different boiling points by a method of evaporation and condensation, so that each component in a mixture can be collected separately in its pure form.

DIURNAL: Performed in twenty-four hours, such as the diurnal rotation of Earth. Diurnal also refers to animals and plants that are active during the day.

DOMOIC ACID: A neurotoxin produced by the algae that cause red tide (coastal algae blooms). Domoic acid and other toxins from algal blooms, such as brevetoxin, can be concentrated by shellfish and poisonous to humans who eat the shellfish.

DOPPLER RADAR: The weather radar system that uses the Doppler shift of radio waves to detect air motion that can result in tornadoes and precipitation, as previously developed weather radar systems do. It can also measure the speed and direction of rain and ice, as well as detect the formation of tornadoes sooner than older radars.

DROPSTONE: Mineral particles carried out to sea on icebergs and then dropped to the ocean floor when the icebergs melt. Dropstone layers in sea-floor sediments record episodes of strong glacial iceberg activity, for example, Heinrich events. Volcanic dropstones, thrown out to sea by volcanic eruptions, are also found.

DROUGHT: A prolonged and abnormal shortage of rain.

DUST BOWL: Megadrought that struck the U.S. and Canadian prairies from 1933 to 1939. Parched croplands lost soil to high winds, which blew the dust for long distances. Over half a million Americans became environmental refugees during the Dust Bowl period, many migrating to California.

E

EARTH DAY: An annual global commemoration (every April 22) of concerns about the environment, first celebrated in 1970.

EARTH LIBERATION FRONT (ELF): An underground group in North America that describes itself as "an international underground organization that uses direct action in the form of economic sabotage to stop the destruction of the natural environment." The group claims to have destroyed $100 million worth of property. The U.S. Federal Bureau of Investigation (FBI) declared it the number one domestic terror threat as of March 2001.

EARTH SUMMIT: Alternative name for the United Nations Conference on Environment and Development, Rio de Janeiro, Brazil, June 3–14, 1992, the meeting at which the United Nations Framework Convention on Climate Change (UNFCCC) was developed.

EARTHSHINE: Sunlight reflected from Earth and illuminating some other body, such as a spacecraft or the moon. The side of the moon that always faces Earth is illuminated by earthshine; by measuring the brightness of portions of the moon that are lit only by earthshine, the reflectivity (albedo) of Earth can be measured.

ECCENTRICITY: In the Keplerian orbit model, the satellite orbit is an ellipse, with eccentricity defining the 'shape' of the ellipse. When e=0, the ellipse is a circle. When e is very near 1, the ellipse is very long and skinny.

ECLIPSE: The partial or total apparent darkening of the sun when the moon comes between the sun and Earth (solar eclipse), or the darkening of the moon when the full moon is in Earth's shadow (lunar eclipse).

ECO-LEXICON: The group or class of words (lexicon) devised for, or especially connected with, speech about environmental concerns. Often has a derisive nuance, as in "Why have 'Civic' or 'Insight' not entered the eco-lexicon in the way that 'Prius' has?"

ECOLOGICAL COMMUNITY: System of species that live together in a given ecosystem and interact with each other. For example, all plants, animals, insects, and microorganisms living in or interacting with a lake form a single ecological community.

ECOLOGICAL FOOTPRINT: A term that refers to the effect of human activities on the environment, as determined by comparison of the consumption of natural resources and the natural ability to replace the resources. If consumption exceeds regeneration capacity, the footprint is negative.

ECOLOGICAL PYRAMID: Representation of the ascending levels of biomass productivity in an ecosystem, where each level eats the level below it. Green plants are the basis of a typical ecological pyramid, with top predators—predators on whom no other species preys—at the top.

ECOLOGY: The branch of science dealing with the interrelationship of organisms and their environments.

ECONOMIC LIBERALIZATION: Reduced government regulation of corporate behavior and state ownership of resources (e.g., oil, mines).

ECONOMIC MODEL: Mathematical representation of economic processes. Computerized economic models may be used to simulate the response of an economy to various changes or perturbations (e.g., discovery or exhaustion of a primary energy source).

ECONOMISTS' STATEMENT ON CLIMATE CHANGE: A statement released in 1997, endorsed by over 2,500 economists including eight Nobel Prize winners: the economists stated that "global climate change carries with it significant environmental, economic, social, and geopolitical risks, and that preventive steps are justified." They also emphasized that preventive steps would not necessarily be harmful to economies and might in fact "improve productivity in the long run."

ECOSYNTHESIS: May refer either to imagined techniques for implanting ecosystems in terraformed alien planetary environments (not practical in the foreseeable future), or to the introduction of non-native species into disrupted ecosystems with the intent of speeding recovery of those ecosystems.

ECO-TERRORISM: Criminal violence against persons or property carried out by an environmentally oriented group for symbolic purposes. As of 2007 no eco-terrorist act had entailed harm to persons, only to property (for example, sport utility vehicles), so the use of the word "terrorism" in this context is, critics of the term say, inflammatory and inaccurate.

ECO-TOURISM: Travel to experience particular ecosystems or view exotic wildlife. Eco-tourism was the tourism industry's fastest-growing sector in the early 2000s, increasing at 20% per year and accounting for $154 billion dollars of business as of 2000.

EL NIÑO: Regularly recurring warming of the surface waters of the eastern Pacific that has affects on global climate; part of the El Niño/Southern Oscillation

Cycle (ENSO). In some contexts, "El Niño" refers to the entire ENSO cycle.

EL NIÑO–SOUTHERN OSCILLATION: Global climate cycle that arises from interaction of ocean and atmospheric circulations. Every 2 to 7 years, westward-blowing winds over the Pacific subside, allowing warm water to migrate across the Pacific from west to east. This suppresses normal upwelling of cold, nutrient-rich waters in the eastern Pacific, shrinking fish populations and changing weather patterns around the world.

ELECTRICAL BOND: Nontechnical term for the bonds that hold chemical compounds together. Such bonds (ionic or covalent) are all electrical in the sense that they depend on the attraction between positive electrical charges in atomic nuclei and negative electrical charges carried by electrons in orbitals surrounding nuclei.

ELECTROLYSIS: The process by which an electrical current is used to cause a chemical change, usually the breakdown of some substance.

ELECTROMAGNETIC ENERGY: Energy conveyed by electromagnetic waves, which are paired electric and magnetic fields propagating together through space. X rays, visible light, and radio waves are all electromagnetic waves.

ELECTROMAGNETIC SPECTRUM: The entire range of radiant energies or wave frequencies from the longest to the shortest wavelengths—the categorization of solar radiation. Satellite sensors collect this energy, but what the detectors capture is only a small portion of the entire electromagnetic spectrum. The spectrum usually is divided into seven sections: radio, microwave, infrared, visible, ultraviolet, x-ray, and gamma-ray radiation.

EMERGENT DISEASE: Disease that is appearing for the first time or that is quickly increasing in incidence or geographic range.

EMISSIONS CAP: Government-mandated upper limit on total amount to be emitted of some pollutant (e.g., carbon dioxide) by all polluters in a country, region, or class.

EMISSIONS TRADING MARKET: System for trading in carbon credits, money-like units that entitle their owners to emit 1 ton of carbon dioxide or an equivalent amount of some other greenhouse gas. As of 2007, the most important emissions trading market was the European Union Emission Trading Scheme, with mandatory participation for large greenhouse polluters throughout the European Union (trading since 2005).

EMISSIVITY: The ratio of the radiation emitted by a surface to that emitted by a black body at the same temperature. The ability of a body to emit radiation.

ENDANGERED SPECIES ACT: U.S. federal law passed in 1973 that protects endangered animals and their habitats. It has been cited in climate disputes because an official government finding that polar bears are endangered would mandate, under the act, action to protect their habitat by mitigating climate change.

ENERGY POLICY ACT OF 2005: U.S. federal law passed in 2005 that offers tens of billions of dollars of subsidies and loan guarantees for energy technologies it categorizes as "clean," including renewables, some forms of coal-burning, and nuclear power. Most commentators believed that the majority of the loan guarantees authorized by the act would go to the nuclear industry.

ENHANCED OIL RECOVERY: Use of special technologies to increase the amount of oil that can be recovered from a given oil field. Enhanced oil recovery techniques include heating oil to make it more liquid or pressurizing it by injecting gas or liquid into the reservoir. At least 40% of oil remains in the ground even with enhanced recovery.

ENTERIC FERMENTATION: Digestion of organic matter by bacteria in an animal's digestive system. Enteric fermentation produces methane, a greenhouse gas.

ENTROPY: Measure of the disorder of a system.

EOCENE EPOCH: Geological period from 55.8 million years ago to 33.9 million years ago. Global climate was much warmer than today during most of the Eocene, with tropical conditions extending up into today's temperate latitude. The start of the Eocene was marked by the Paleocene-Eocene Thermal Maximum, a sudden rise in global temperature lasting only about 200,000 years that caused the extinction of many species and cleared the way for the evolution of modern mammals.

EPIDEMIC: From the Greek meaning prevalent among the people; most commonly used to describe an outbreak of an illness or disease in which the number of individual cases significantly exceeds the usual or expected number of cases in any given population.

EPOCH: Unit of geological time. From longest to shortest, the geological system of time units is eon, era, period, epoch, and stage. Epochs are generally about 500 million years in length.

EQUATORIAL: Concerning or related to Earth's equator, the line dividing the Northern and Southern Hemispheres.

EROSION: Processes (mechanical and chemical) responsible for the wearing away, loosening, and dissolving of materials of Earth's crust.

ESTUARY: Lower end of a river where ocean tides meet the river's current.

ETHANOL: Compound of carbon, hydrogen, and oxygen (CH_3CH_2OH) that is a clear liquid at room temperature; also known as drinking alcohol or ethyl alcohol. Ethanol can be produced by biological or chemical processes from sugars and other feedstocks and can be burned as a fuel in many internal-combustion engines, either mixed with gasoline or in pure form. Several governments, most notably Brazil and the United States, encourage the production of ethanol based fuels.

ETHICS: Beliefs about right and wrong. Ethics cannot be studied scientifically, since moral qualities cannot be observed or measured. However, ethics are inseparable from questions about how to deal with climate change, since human and natural well-being, suffering, and death are involved in action (or inaction) regarding climate.

ETHNOBOTANIST: Scientist who specializes in the study of relationships between human cultures and plants, whether wild or domesticated. Culture is intimately related to forests, herbal medicines, sacred or recreational mind-altering substances derived from various plants, crops, and the like.

EUSTATIC SEA LEVEL: Change in global average sea level caused by increased volume of the ocean (caused both by thermal expansion of warming water and by the addition of water from melting glaciers). Often contrasted to relative sea level rise, which is a local increase of sea-level relative to the shore.

EUTROPHICATION: The process whereby a body of water becomes rich in dissolved nutrients through natural or human-made processes. This often results in a deficiency of dissolved oxygen, producing an environment that favors plant over animal life.

EVAPOTRANSPIRATION: Transfer of water to the atmosphere from an area of land, combining water transfer from foliage (transpiration) and evaporation from non-living surfaces such as soils and bodies of water.

EVOLUTION: Biological change. The process by which all life forms have developed. In evolution, slight, inheritable variations occur in each new generation of plant, animal, or microorganism. A few of these random variations (mutations) happen to be helpful; individuals bearing these variations leave more offspring.

EXTERNALITIES: Costs of an economic activity that are not borne, for the most part, by participants in that activity. For example, the harms caused by pollution, including greenhouse pollution, are externalities for polluters.

EXTINCTION: The total disappearance of a species or the disappearance of a species from a given area.

EXTIRPATED: The condition in which a species is eliminated from a specific geographic area of its habitat.

EXTRUDE: To push out or force out a plastic substance such as lava or mud.

EYEWALL: Ring of severe storms surrounding the eye of a cyclonic storm (e.g., hurricane).

F

FACTION: A dissenting group within a larger group such as a political party. A faction is usually outnumbered by other members of the larger group.

FEEDBACK: Information that tells a system the results of its actions and thereby alters future actions.

FERMENTATION: Chemical reaction in which enzymes break down complex organic compounds (for example, carbohydrates and sugars) into simpler ones (for example, ethyl alcohol).

FISSION: Nuclear fission is a process in which the nucleus of an atom splits, usually into two daughter nuclei, with the transformation of tremendous levels of nuclear energy into heat and light.

FLUOROCARBON: Compounds containing both carbon and fluorine. Such compounds are extremely rare in nature but have many technological uses in refrigeration, microprocessor manufacture, and fire extinguishing. Many fluorocarbons damage the ozone layer and are powerful greenhouse gases as well.

FLUX: The measure of the flow of some quantity per unit area per unit time.

FOOD INSECURITY: Unreliable access by a person or group to adequate food. Persons who do not have adequate food and who are without prospects to find adequate nutritional sources.

FOOD SECURITY: Reliable access by a person or group to adequate food. Persons who do not have adequate food, or have adequate food at the moment but do not know how they will continue to get food, have low food security.

FORAMINIFERA: Single-celled marine organisms that produce small shells called tests in order to inhabit and float free in ocean surface waters. There are thousands of species of foraminifera. When the organism dies, its test sinks to the sea bottom as sediment, forming thick deposits over geological time. Because the numbers and types of foraminifera are climate-sensitive, analysis of these sediments gives data on ancient climate changes.

FOSSIL FUELS: Fuels formed by biological processes and transformed into solid or fluid minerals over geological time. Fossil fuels include coal, petroleum, and natural gas. Fossil fuels are non-renewable on the timescale of human civilization, because their natural replenishment would take many millions of years.

FOSSIL RECORD: The time-ordered mass of fossils (mineralized impressions of living creatures) that is found in the sedimentary rocks of Earth. The fossil record is one of the primary sources of knowledge about evolution and is also used to date rock layers (biostratigraphy).

FOSSIL WATER: Groundwater that has been in an underground aquifer for thousands or millions of years.

FOURTH ESTATE: The media. The term dates to eighteenth-century France, where political power was supposedly assigned to three "Estates" or groups, namely the clergy, nobles, and commoners. Some commentators suggested that the press was so powerful (through its ability to sway opinion) that it might be considered a fourth Estate.

FREE MARKET: Economic system in which price-setting and other behaviors are not constrained by special laws or regulations. Perfectly free markets are a philosophical ideal, not found in the real world.

FREQUENCY: The rate at which vibrations take place (number of times per second the motion is repeated), given in cycles per second or in hertz (Hz). Also, the number of waves that pass a given point in a given period of time.

FRIENDS OF THE EARTH: International network of environmental groups founded in 1969, having members in 70 countries as of 2007. Noted in the early 2000s for The Big Ask, its campaign for a new climate-change law in the United Kingdom that would cut U.K. greenhouse emissions by 3% per year.

FUEL CELL: Device that combines a fuel (e.g., hydrogen) and oxygen at a low or modest temperature to produce electricity directly. In a fuel cell, fuel is not burned (combined rapidly with oxygen) to produce heat, but combined ionically through a membrane. Fuel cells are more efficient users of fuel than thermal systems, but are also more expensive.

FUJITA SCALE: A scale of six categories that rates tornado wind speed based upon the observed destruction of the storm.

FUMAROLE: Opening in the ground that emits volcanic gases and steam. A gas commonly emitted is carbon dioxide.

G

GAIA HYPOTHESIS: The hypothesis that Earth's atmosphere, biosphere, and its living organisms behave as a single system striving to maintain a stability that is conducive to the existence of life.

GASOLINE: The fraction of petroleum that boils between 32°F (0°C) and 392°F (200°C).

GEOCODING: In geographic information systems, the assignment of geological data to particular features on maps or to other data records, such as photographs.

GEOLOGIC TIME: The period of time extending from the formation of Earth to the present.

GEOLOGICAL RECORD: Evidence of Earth's history left in rocks and sediments over thousands to billions of years. Events that can be inferred from the geological record include climate changes, biological evolution, continental drift, and asteroid impacts.

GEOMAGNETIC: Related to Earth's magnetic field.

GEOPHYSICAL SCIENCE: Scientific field that studies Earth as a physical system comprised of energy flows, energetic processes (e.g., radiation), and material flows such as currents and winds. Does not study biological phenomenon as such, but only their effects on the large-scale physical properties of Earth and its climate and other systems.

GEOTHERMAL ENERGY: Energy obtained from Earth's internal heat, which is maintained by the breakdown of radioactive elements. Geothermal means, literally, Earth-heat. Geothermal energy may be used either directly as heat (e.g., to heat buildings or industrial processes) or to generate electricity.

GEYSER: Hot spring that periodically sprays steam and hot water into the air. A geyser requires a pathway from the water table in contact with a geothermal heat source.

GLACIAL: Related to glaciers, a multi-year surplus accumulation of snowfall in excess of snowmelt on land and resulting in a mass of ice at least 0.04 mi^2 (0.1 km^2) in area that shows some evidence of movement in response to gravity. A glacier may terminate on land or in water. Glacier ice is the largest reservoir of freshwater on Earth and is second only to the oceans as the largest reservoir of total water. Glaciers are found on every continent except Australia.

GLACIAL BERGS: Icebergs that have formed by breaking off the edge of a glacier that is flowing to the sea.

GLACIAL CYCLE: Episode in Earth climate history in which temperatures decline and glaciers grow and spread, sometimes covering large parts of the northern and southern hemispheres. The most recent glacial cycle ended about 10,000 years ago.

GLACIAL LAKE: Lake fed by meltwater from a glacier or glaciers.

GLACIAL RETREAT: Melting and shrinkage of glaciers at the end of a glacial cycle.

GLACIAL TILL: Rock and soil scoured from Earth and transported by a glacier, then deposited along the glacier's sides or at its end.

GLACIATION: The formation, movement, and recession of glaciers or ice sheets.

GLACIER: A multi-year surplus accumulation of snowfall in excess of snowmelt on land and resulting in a mass of ice at least 0.04 mi^2 (0.1 km^2) in area that shows some evidence of movement in response to gravity. A glacier may terminate on land or in water. Glacier ice is the largest reservoir of freshwater on Earth and is second only to the oceans as the largest reservoir of total water. Glaciers are found on every continent except Australia.

GLOBAL BRIGHTENING: Any increase in the amount of sunlight reaching Earth's surface; in particular, any decrease in or reversal of global dimming, the blockage of sunlight by aerosols (fine particles) in Earth's atmosphere.

GLOBAL DIMMING: Decrease in amount of sunlight reaching Earth's surface caused by light blockage by clouds and aerosols. Global dimming increased from 1960 to 1990, reducing sunlight reaching Earth's surface by 4%, but this trend reversed after 1990 in most locations.

GLOBALIZATION: The integration of national and local systems into a global economy through increased trade, manufacturing, communications, and migration.

GRAVIMETRIC: Having to do with the measurement of gravity. Gravimetric data from satellites can reveal changes in large masses of ice on Earth's surface: in 2006, such data confirmed that Greenland's ice cap was melting faster than expected.

GRAVITATIONAL FORCE: The mutual force of attraction that exists between all particles or bodies with mass. A force dependent upon mass and the distance between objects. The English physicist and mathematician Sir Isaac Newton (1642–1727) set out the classical theory of gravity in his *Philosophiae Naturalis Principia Mathematica (Mathematical Principles of Natural Philosophy).*

GREAT BARRIER REEF: World's largest coral reef system, located along the northeastern coast of Australia. The reef system, which is approximately 1,600 miles (2,600 km) long, contains 3,000 individual reefs and 900 islands and is threatened by sea-level rise and ocean warming caused by anthropogenic climate change.

GREAT CONVEYOR BELT: The overturning circulation of the world's seas, driven by temperature and salinity differences between the poles and tropics; also called the thermohaline circulation or meridional overturning circulation. Because the great conveyor belt transports thermal energy from the tropics toward the poles, it is a central component of Earth's climate machine.

GREEN ENERGY: Energy obtained by any means that causes relatively little harm to the environment. There is no universal agreement on what constitutes a green energy source: for example, whether nuclear power is green is hotly debated, as is the question of whether windmills are by nature green, or only if sited in certain locations.

GREEN PARTY: Any of a number of political parties in the United States, Europe, and elsewhere, whose policies are centered on environmentalism, participatory democracy, and social justice. About 70 countries have Green Parties: in Europe, Australia, and New Zealand, a number of Greens have been elected to Parliament. The German Green Party is particularly powerful.

GREENHOUSE-GAS EMISSIONS: Releases of greenhouses gases into the atmosphere. To simplify discussion of how much warming is being caused by emissions of various greenhouse gases—each of which causes a different amount of warming, ton for ton—it is customary to translate emissions of gases other than carbon dioxide into the number of tons of CO_2 that would produce the same amount of warming, i.e., units of "tons CO_2 equivalent."

GREENHOUSE GAS MITIGATION TECHNOLOGIES: Technological methods for reducing (mitigating) releases of greenhouse gases. Such technologies include schemes to sequester carbon dioxide by injecting it into underground reservoirs, burnoff of methane from landfills, substitute low- or no-carbon fuels for carbon-based fuels, and more. As of the early 2000s, such technologies had made little impact on global greenhouse emissions.

GREENHOUSE GASES: Gases that cause Earth to retain more thermal energy by absorbing infrared light emitted by Earth's surface. The most important greenhouse gases are water vapor, carbon dioxide, methane, nitrous oxide, and various artificial chemicals such as chlorofluorocarbons. All but the latter are naturally occurring, but human activity over the last several centuries has significantly increased the amounts of carbon dioxide, methane, and nitrous oxide in Earth's atmosphere.

GREENHOUSE INTENSITY: Amount of greenhouse gas released by a national economy per unit of economic activity (i.e., of gross domestic product). Emissions of gases other than carbon dioxide are translated into equivalent emissions.

GREENLAND ICE CAP: Layer of ice covering about 80% of the island of Greenland; second-largest mass of ice on Earth after Antarctica, containing about 10% of Earth's ice, enough to raise sea levels 21 ft (6.5 m) if it were all to melt. Accelerated melting of the Greenland ice cap was confirmed by gravimetric measurements in 2006.

GREENLAND-SCOTLAND RIDGE: Underwater ridge connecting Greenland to Scotland in the North Atlantic that separates the Nordic seas, where North Atlantic deep water formation occurs. Below a depth of 2755 feet (840 meters), the ridge forms a continuous barrier between the two basins; in some areas it rises to shallower depths or islands. Formation of the North Atlantic deep water is an essential part of the great conveyor belt or thermohaline circulation of the oceans.

GREENPEACE: Nonprofit environmental group formed in 1971, originally to protest nuclear testing and whaling. The group remains active, now addressing a wide range of issues, including climate change.

GREENWASHING: Environmental whitewashing, or the practice of presenting information that makes the purveyor appear to be environmentally sensitive or to seem to be taking action toward sustainability in the absence of any substantive change.

GROSS DOMESTIC PRODUCT (GDP): A measure of total economic activity, whether of a nation, group of nations, or the world: slightly different from Gross National Product (GNP, used by economists until the early 1990s). Defined as the total monetary value of all goods and services produced over a given period of time (usually one year).

GROUNDING LINE: Underwater boundary or line along which a glacier that is flowing into the sea floats free of the ground. Grounding lines are typically many miles from the nominal shoreline. Retreat of grounding lines toward land accompanies speeded glacial flow. Retreat of grounding lines has occurred recently for some glaciers in Greenland and Antarctica.

GROUP OF 77 (G77): Group of developing countries, originally founded in 1964 with 77 members but having 130 members as of 2007. The group was formed as a counterpoint to the Group of 7 (G7) nations, a group of the world's wealthiest nations formed to advance their economic interests.

GULF STREAM: A warm, swift ocean current that flows along the coast of the Eastern United States and makes Ireland, Great Britain, and the Scandinavian countries warmer than they would be otherwise.

GYRE: A zone of spirally circulating oceanic water that tends to retain floating materials, as in the Sargasso Sea of the Atlantic Ocean.

H

HABER PROCESS: The chemical process by which nitrogen and hydrogen are combined with each other at high temperature and pressure over a catalyst to produce ammonia.

HABITABILITY: The degree to which a given environment can be lived in by human beings. Highly habitable environments can support higher population densities, that is, more people per square mile or kilometer.

HABITAT: The area or region where a particular type of plant or animal lives and grows.

HALF-LIFE: The time it takes for half of the original atoms of a radioactive element to be transformed into the daughter product.

HALINITY: Salt content of seawater: synonym for salinity or saltiness. Not all seawater is equally salty (haline). Halinity can be increased by evaporation and decreased by the addition of freshwater from rivers or melting glaciers. More haline water is denser and tends to sink.

HALOCARBON: Compound that contains carbon and one or more of the elements known as halons (chlorine, fluorine, or bromine). Halocarbons do not exist in nature; all are manufactured. When released into the atmosphere, many halocarbons deplete the ozone layer and have high global warming potential.

HALOGEN: Any one of the halogen family of elements with similar chemical properties, including fluorine, chlorine, bromine, and iodine.

HALONS: A class of compounds, also called bromofluorocarbons, that contain bromine, chlorine, and carbon or fluorine. Used as fire retardants in fire-extinguishing systems. Potent greenhouse gases with high ozone-depletion potential.

HIGH PRESSURE SYSTEM: In meteorology, an area or mass of air at higher pressure than surrounding air, also called a high pressure area. In a high pressure system, air is usually descending toward the surface. High pressure areas are associated with clear skies and light winds.

HOCKEY STICK CONTROVERSY: Controversy from 1998 to approximately 2006 over the validity of the hockey-stick graph, a chart of global average air temperatures indicating that recent climate warming is unprecedented for at least the last 1,100 years or so. Critics argued the graph was flawed and misleading, but in 2006, after a careful review of the evidence, the National Academy of Sciences affirmed that the graph is essentially accurate.

HOLOCENE EXTINCTION EVENT: The Holocene is the geological period from 10,000 years ago to the present; the Holocene extinction is the worldwide mass extinction of animal and plant species being caused by human activity. Global warming may accelerate the ongoing Holocene extinction event, possibly driving a fourth of all terrestrial plant and animal species to extinction.

HOLOCENE PERIOD: Geological period from 10,000 years ago to the present. Technically an epoch, not a period; the Pleistocene was the first epoch of the Quaternary Period (1.64 million years ago to present), and the Holocene was the second.

HOMEOSTASIS: State of being in balance; the tendency of an organism to maintain constant internal conditions despite large changes in the external environment.

HOTHOUSE EARTH: Hypothetical very hot global climate period occurring as rebound from a possible snowball Earth condition many millions of years ago. The term is sometimes used, loosely, to refer to the hotter Earth now developing because of anthropogenic climate change.

HUMIDITY: The amount of water vapor (moisture) contained in the air.

HURRICANE: Large, rotating system of thunderstorms whose highest windspeed exceeds 74 mph (119 km/h). Globally, such storms are termed tropical cyclones: the word "hurricane" is often reserved for tropical cyclones in the Atlantic.

HURRICANE MODEL: Computerized mathematical model of the physical structure of a hurricane.

HYBRID VEHICLES: Vehicles that use electric motors for their sole or primary motive power but produce most or all of their own electricity using internal combustion engines.

HYDROCARBON: A chemical containing only carbon and hydrogen. Hydrocarbons are of prime economic importance because they encompass the constituents of the major fossil fuels, petroleum and natural gas, as well as plastics, waxes, and oils. In urban pollution, these components—along with NO_x and sunlight—contribute to the formation of tropospheric ozone.

HYDROELECTRICITY: Electricity generated by causing water to flow downhill through turbines, fan-like devices that turn when fluid flows through them. The rotary mechanical motion of each turbine is used to turn an electrical generator.

HYDROGEN FUEL CELLS: Devices that combine hydrogen and oxygen, at a low or modest temperature, to produce electricity directly. In a fuel cell, hydrogen is not burned (combined rapidly with oxygen) to produce heat. Fuel cells are more efficient users of hydrogen than thermal systems such as hydrogen-burning internal combustion engines, but are also more expensive.

HYDROLOGIC CYCLE: The process of evaporation, vertical and horizontal transport of vapor, condensation, precipitation, and the flow of water from continents to oceans. It is a major factor in determining climate through its influence on surface vegetation, the clouds, snow and ice, and soil moisture. The hydrologic cycle is responsible for 25 to 30% of the mid-latitudes' heat transport from the equatorial to polar regions.

HYDROLOGY: The science that deals with global water (both liquid and solid), its properties, circulation, and distribution.

HYDROPOWER: Electricity generated by causing water to flow downhill through turbines, fan-like devices that turn when fluid flows through them. The rotary mechanical motion of each turbine is used to turn an electrical generator.

HYDROSPHERE: The totality of water encompassing Earth, comprising all the bodies of water, ice, and water vapor in the atmosphere.

HYPERTHERMIA: Hyperthermia is a condition in which internal body temperatures rise to dangerous and even lethal levels. These temperatures can reach anywhere between 104 and 115° F. Hyperthermia affects many people during hot summer seasons. It is also a condition that occurs during some surgeries. In these rare cases the patient has an unpredictable reaction to surgical anesthesia, causing a rapid rise in body temperature.

HYPOTHESIS: An idea phrased in the form of a statement that can be tested by observation and/or experiment.

I

ICE AGE: Period of glacial advance.

ICE BOREHOLE THERMOMETRY: Measurement of temperature at different depths of a vertical hole drilled in a large mass of ice, such as the Antarctic ice cap. By mathematical processing (deconvolution) of this spatially organized record of temperatures, a time record of temperature over the last several centuries can be inferred.

ICE CAP: Ice mass located over one of the poles of a planet not otherwise covered with ice. In our solar system, only Mars and Earth have polar ice caps. Earth's north polar ice cap has two parts, a skin of floating ice over the actual pole and the Greenland ice cap, which does not overlay the pole. Earth's south polar ice cap is the Antarctic ice sheet.

ICE CORE: A cylindrical section of ice removed from a glacier or an ice sheet in order to study climate patterns of the past. By performing chemical analyses on the air trapped in the ice, scientists can estimate the percentage of carbon dioxide and other trace gases in the atmosphere at that time.

ICE FLOES: Plates of floating sea ice, that is, ice formed by freezing of the top layer of the ocean; distinct from icebergs, which are thicker and are produced by the breakup of glaciers flowing into the sea. Any floating plate of ice wider than 6 mi (10 km) is termed an ice field, not an ice floe.

ICE SHEET: Glacial ice that covers at least 19,500 square mi (50,000 square km) of land and that flows in all directions, covering and obscuring the landscape below it.

ICE SHELF: Section of an ice sheet that extends into the sea a considerable distance and that may be partially afloat.

ICEBERG: A large piece of floating ice that has broken off a glacier, ice sheet, or ice shelf.

INCIDENT LIGHT: Light arriving at the surface of an object (for example, Earth).

INDIGENOUS PEOPLES: Human populations that migrated to their traditional area of residence some time in the relatively distant past, e.g., before the period of global colonization that begin in the late 1400s.

INDIRECT ENERGY USE: Energy consumption required to support the provision of a product or service that does not explicitly appear during the consumption of the product or service: for example, the energy used to manufacture a television is indirect energy use by the TV owner, while the electricity consumed by running the TV is direct energy use.

INDUSTRIAL REVOLUTION: The period, beginning about the middle of the 18th century, during which humans began to use steam engines as a major source of power.

INDUSTRIALIZATION: Shift of a large portion of a region or country's economy to mechanized manufacturing and away from agriculture. Industrialization has been an increasingly global process since the beginning of the Industrial Revolution, usually dated to about 1750.

INFRARED: Wavelengths slightly longer than visible light, often used in astronomy to study distant objects.

INFRARED RADIATION: Electromagnetic radiation of a wavelength shorter than radio waves but longer than visible light that takes the form of heat.

INSOLATION: Solar radiation incident upon a unit horizontal surface on or above Earth's surface.

INSTRUMENTAL DATA: Measurements of physical phenomena such as temperature, windspeed, and the like that are made directly using instruments so that numbers with well-defined units (e.g., degrees Celsius) can be assigned to the measurements.

INTEGRATED ASSESSMENT MODEL: A computerized, mathematical model of global climate change that combines results and models from the social, economic, biological, and physical sciences, attempting to capture the interactions of human and environmental factors. The goal is to understand current interactions and predict likely results of various policy choices.

INTERGLACIAL PERIOD: Sometimes called simply an interglacial: geological time period between glacial periods, which are periods when ice masses grow in the polar regions and at high elevations. The world is warmer during interglacials. The world is presently experiencing an interglacial that began about 11,000 years ago.

INTERGOVERNMENTAL PANEL ON CLIMATE CHANGE (IPCC): Panel of scientists established by the World Meteorological Organization (WMO) and the United Nations Environment Programme (UNEP) in 1988 to assess the science, technology, and socioeconomic information needed to understand the risk of human-induced climate change.

INTERNAL COMBUSTION ENGINE: An engine in which the chemical reaction that supplies energy to the engine takes place within the walls of the engine (usually a cylinder) itself.

INTERNATIONAL ENERGY AGENCY: International group established in 1974 by the Organisation of Economic Co-operation and Development (OECD), a group of 30 well-to-do Western countries formed after World War II to coordinate economic concerns. The IEA promotes nuclear energy and releases technical studies of the world energy situation and of particular energy issues.

INTERNATIONAL GEOPHYSICAL YEAR: Internationally coordinated, first-of-a-kind effort to enhance scientific understanding of Earth processes, including solar activity, the oceans, the atmosphere, polar regions, and more: lasted from July 1, 1957, to December 31, 1958.

INTERNATIONAL UNION OF GEOLOGICAL SCIENCES (IUGS): Non-governmental group of geologists founded in 1961 and headquartered in Trondheim, Norway. The group fosters international cooperation on geological research of an transnational or global nature.

INTERTROPICAL: Literally, between the tropics: usually refers to a narrow belt along the equator where convergence of air masses of the Northern and Southern Hemispheres produces a low-pressure atmospheric condition.

INVERSION: A type of chromosomal defect in which a broken segment of a chromosome attaches to the same chromosome, but in reverse position.

IONIZE: To add or remove electrons from an electrical neutral atom (i.e., one surrounded by as many electrons as it has protons in its nucleus, therefore electrically neutral). An ionized atom is termed an ion and has a positive or negative electrical charge. Molecules (clusters of atoms bound stably together) may also become ionized.

IONOSPHERE: A subregion within the thermosphere, extending from about 50 mi (80 km) to more than 250 mi (400 km) above Earth and containing elevated concentrations of charged atoms and molecules (ions).

IRON FERTILIZATION: Controversial speculative method for removing carbon from the atmosphere, in which adding powdered iron to ocean surface waters would cause single-celled aquatic plants (phytoplankton) to increase greatly in numbers, breaking down CO_2 to obtain carbon for their tissues. The dead organisms would then, ideally, sink to deep waters or the ocean floor, sequestering their carbon content from the atmosphere and so mitigating climate change.

J

JET STREAM: Currents of high-speed air in the atmosphere. Jet streams form along the boundaries of global air masses where there is a significant difference in atmospheric temperature. The jet streams may be several hundred miles across and 1 to 2 mi (1.6 to 3.2 km) deep at an altitude of 8 to 12 mi (12.9 to 19.3 km). They generally move west to east, and are strongest in the winter with core wind speeds as high as 250 mph (402 km/h). Changes in the jet stream indicate changes in the motion of the atmosphere and weather.

JOINT IMPLEMENTATION: Type of greenhouse mitigation project defined by the Kyoto Protocol (1997). In a joint implementation project, one developed country can finance a project in another developed country to reduce greenhouse-gas emissions. The goal, as with the other "flexible mechanisms" specified by Kyoto, is to reduce total greenhouse emissions.

JURASSIC PERIOD: Unit of geological time from 200 million years ago to 145 million years ago, famous in popular culture for its large dinosaurs. Global average temperature and atmospheric carbon dioxide concentrations were both much higher during the Jurassic than today.

K

KEELING CURVE: Plot of data showing the steady rise of atmospheric carbon dioxide from 1958 to the present, overlaid with annual sawtooth variations due to the growth of Northern Hemisphere plants in summer. Carbon dioxide began rising due to human activities in the 1800s, but direct, continuous measurements of atmospheric carbon dioxide were first made by U.S. oceanographer Charles David Keeling (1928–2005) starting in 1958.

KETTLE DRUM: Usually termed a kettle: depression or sinkhole left in a mass of rocky debris by a retreating glacier. A large, isolated chunk of ice buried in debris melts, causing a collapse of the overlying debris.

KILOWATT-HOUR: Amount of energy corresponding to a power flow of 1 kilowatt (1,000 watts) sustained for 1 hour. Often used as a unit of electricity consumption. A 100-watt light bulb left on for 10 hours consumes 1 kilowatt-hour (kWh). The average U.S. household purchases about 10,000 kWh of electricity per year.

KINETIC ENERGY: The energy due to the motion of an object.

KYOTO PROTOCOL: Extension in 1997 of the 1992 United Nations Framework Convention on Climate Change (UNFCCC), an international treaty signed by almost all countries with the goal of mitigating climate change. The United States, as of early 2008, was the only industrialized country to have not ratified the Kyoto Protocol, which is due to be replaced by an improved and updated agreement starting in 2012.

L

LA NIÑA: A period of stronger-than-normal trade winds and unusually low sea-surface temperatures in the central and eastern tropical Pacific Ocean; the opposite of El Niño.

LAKE-EFFECT SNOW: Snow that falls downwind of a large lake. Cold air moving over relatively warm lake water is warmed and moistened; after leaving the lake, the air cools again and some of its moisture precipitates out as snow.

LANDFALL: When the center of a storm system formed over the ocean (e.g., hurricane) reaches land, the storm is said to have made landfall. Some hurricanes never make landfall but exhaust themselves over the ocean. Those that do make landfall quickly lose power and dissipate, since they draw their energy from warm ocean surface waters.

LANDFILLS: Locations where garbage is dumped in pits and covered with soil. Anaerobic digestion by bacteria of organic matter in a landfill produces significant quantities of methane (a potent greenhouse gas), which must be vented lest it accumulate and possibly explode. In many countries (e.g., the United States), landfills are the single largest source of methane emissions.

LATITUDE: The angular distance north or south of Earth's equator measured in degrees.

LAW OF GRAVITY: Mathematical statement of the force of attraction between all material objects. The law of gravity states that the force pulling two objects together is proportional to the product of their masses divided by the square of the distance between them. Gravity keeps Earth's atmosphere from drifting away and is a crucial force in all circulations of wind and water.

LEAST DEVELOPED COUNTRIES: The world's poorest countries. The United Nations classifies a country as a Least Developed Country if per-capita income is less than $750 for three years running and if the country has low health and literacy rates. As of 2007, there were 48 Least Developed Countries, of which 33 were in Africa.

LEEWARD: Downwind of an object: opposite of windward.

LIGNOCELLULOSE: Any of several substances making up woody cell walls in plants, consisting of cellulose mixed with lignin (both organic polymers). Digestion of lignocellulose to produce ethanol is a sought-after technology for producing biofuel.

LIMESTONE: A carbonate sedimentary rock composed of more than 50% of the mineral calcium carbonate ($CaCO_3$).

LITHOSPHERE: The rigid, uppermost section of Earth's mantle, especially the outer crust.

LIVE EARTH: A popular music concert that occurred at 11 locations simultaneously around the world on July 7, 2007. The event was organized by former U.S. vice president Al Gore and others to publicize the problem of global climate change and to raise funds for projects to combat it.

LOW PRESSURE SYSTEM: In meteorology, a mass or area of air that is at a lower pressure than the air around it. In a low pressure system, air is often ascending from the surface. Low pressure systems are associated with cloudy skies and stronger winds.

LUNAR DAY: The length of time between one sunrise on the surface of Earth's moon and the next: the length of a day on the moon. A lunar day is approximately 28 days long; that is, the moon rotates once on its axis in the time it takes it to circle Earth, which means that one face of the moon is always kept facing Earth.

M

MACROECONOMIC STABILIZATION: Stabilization of an economy at the level of a whole nation (i.e., at the macroeconomic level). In many countries, governments manipulate certain financial variables to seek macroeconomic stabilization: for example, in the United States, the Federal Reserve Bank raises and lowers interest rates.

MACROECONOMICS: Economics (transfers of money and wealth) at the level of the nation as a whole: often distinguished from microeconomics, the economics of individual persons, households, or businesses.

MAGMA: Molten rock deep within Earth that consists of liquids, gases, and particles of rocks and crystals. Magma underlies areas of volcanic activity and at Earth's surface is called lava.

MALARIA: Group of parasitic diseases common in tropical and subtropical areas. According to 2004 statistics from the U.S. Centers for Disease Control and Prevention (CDC), approximately 300 million cases occur annually worldwide, and an estimated 700,000 to 2.7 million persons die of malaria each year. About 90% of these deaths occur in sub-Saharan Africa, with the majority being children.

MANGROVE FOREST: Coastal ecosystem type based on mangrove trees standing in shallow ocean water: also termed mangrove swamp. Mangrove forests support shrimp fisheries and are threatened by rising sea levels due to climate change.

MANTLE: Thick, dense layer of rock that underlies Earth's crust and overlies the core.

MARSHALL PLAN: Plan fostered by U.S. Secretary of State George Marshall (1880–1959) in the years following World War II, intended to rebuild the economies of Western Europe through loans and grants.

MATHEMATICAL MODEL: Used to develop descriptions of systems using numerical data and to make predictions of future states or events based upon current data and known changes in data over time.

MEAN: A measure of central tendency (average) found by adding all the terms in a set and dividing by the number of terms.

MECHANICAL ENERGY: Energy possessed by matter in motion. Winds and water currents, for example, possess mechanical energy. Mechanical energy can be transformed into other forms of energy by certain processes: for example, a hydropower generator or windmilll changes mechanical energy into electricity.

MEDIEVAL WARM EPOCH: Interval from 1000 to 1300 during which some parts of the Northern Hemisphere (e.g., Europe) were warmer than during the Little Ice Age, which followed. Global climate, contrary to claims made by some climate skeptics, was not as warm during the Medieval Warm Period as it has been since about 1950.

MEGA-DELTAS: Unusually large river deltas. Deltas are flat, low-lying, roughly triangular areas of sediment deposited where large rivers meet the ocean. Mega-deltas are found at the mouths of the Mississippi, Nile, Ganges, Yangtze, and many other large rivers.

MEGADROUGHT: Any unusually widespread, severe, and long-lasting drought. The term is generally reserved for droughts covering entire subcontinental regions and lasting for many years.

MELTWATER: Melted ice in a glacier's bottom layer, caused by heat that develops as a result of friction with Earth's surface.

MEMBRANE FILTRATION: The use of plastic sheets or tubes that have holes so small only very small molecules, like water molecules, will pass through them.

MERIDIONAL: Relating to the meridians, imaginary north-south lines that define longitudes. Meridional circulations of air or water are those that are predominantly north-south in character with one direction of flow located above the other.

MESOCYCLONE: A horizontal cylinder of rotating air produced by wind shear, which is basically friction between winds flowing at different rates of speed. A mesocyclone can tilt and become a vertical column of rotating air known as a funnel cloud and extend from the bottom of a thunderstorm. If the funnel cloud then reaches the ground it becomes known as a tornado.

METEOROLOGY: The science that deals with Earth's atmosphere and its phenomena and with weather and weather forecasting.

METHANE: A compound of one hydrogen atom combined with four hydrogen atoms, formula CH_4. It is the simplest hydrocarbon compound. Methane is a burnable gas that is found as a fossil fuel (in natural gas) and is given off by rotting excrement.

METHANE CLATHRATE: Form of water ice containing methane in its crystal structure; also called methane ice or methane hydrate. Methane clathrates form in sediments in many areas of the deep ocean floor. Estimates of total methane clathrate deposits have varied widely; such deposits are now thought to contain some hundreds of billion of tons of carbon, more than all natural gas resources put together. Methane clathrates, however, exist as scattered lumps mixed into sediment on the ocean floor .

METHANOGEN: Bacterium that produces methane as a waste product.

METHYL BROMIDE: A compound of carbon that is made of a methyl group, which is made of three hydrogen atoms bonded to one carbon atom; and in a fourth bonding position on the carbon atom there is a bromine atom.

MICA: Shiny mineral with a flaky or layered structure, composed primarily of silicon and oxygen; a component of granites and other stones, also found as a separate mineral.

MICROWAVE SOUNDING UNIT: Device carried by some Earth-observing satellites that measure microwave-band electromagnetic radiation emitted by Earth's atmosphere in order to measure temperature, humidity, precipitation rates, and snow and ice coverage.

MILANKOVITCH CYCLES: Regularly repeating variations in Earth's climate caused by shifts in its orbit around the sun and its orientation (i.e., tilt) with respect to the sun. Named after Serbian scientist Milutin Milankovitch (1879–1958), though he was not the first to propose such cycles.

MILANKOVITCH FORCING: Increases or decreases in the amount of thermal energy being absorbed by Earth's climate system governed by Milankovitch cycles, that is, regularly repeating variations in Earth's climate caused by shifts in its orbit around the sun and its orientation (i.e., tilt) with respect to the sun.

MONSOON: An annual shift in the direction of the prevailing wind that brings on a rainy season and affects large parts of Asia and Africa.

MORAINE: Mass of boulders, stones, and other rock debris carried along and deposited by a glacier.

MORBIDITY: The existence of disease or illness.

MOULIN: Vertical shaft or crevice in a glacier into which meltwater from the glacier's surface flows. Moulins allow liquid water to penetrate to the bottom of a glacier, lubricating its contact with the ground and accelerating its flow.

N

NAPHTHALENE: White aromatic hydrocarbon having chemical formula $C_{10}H_8$ and a chemical structure of two benzene rings linked to form a figure 8. Household mothballs are made of naphthalene, which is also used in some industrial chemical processes.

NATIONAL ENVIRONMENTAL POLICY ACT (NEPA): U.S. federal law signed by President Richard Nixon on January 1, 1970. The act requires federal agencies to produce Environmental Impact Statements describing the impact on the environment of projects they propose to carry out.

NATURAL SELECTION: Primary mechanism shaping biological evolution. Individual organisms that have inherited variations of form that help them to survive and thrive are naturally selected to leave more offspring: later generations will therefore be dominated by individuals inheriting the same variation. Accumulation of such changes over millions of years can produce new species and complex new structures such as the vertebrate eye.

NEOPROTEROZOIC PERIOD: Unit of geological time from 1 billion to 540 million years ago. The middle portion of the Neoproterozoic, from 850 to 630 million years ago, is named the Cryogenian period (from the Greek *kruos*, for frost) because during it the most severe ice age in Earth's history occurred, possibly freezing the entire planet during a snowball Earth episode.

NEPHOLOGISTS: Scientists who specialize in the study of clouds.

NEUROTOXIN: Poison that interferes with nerve function, usually by affecting the flow of ions through the cell membrane.

NITROGEN OXIDES: Compounds of nitrogen and oxygen such as those that collectively form from burning fossil fuels in vehicles.

NOBLE GAS: Elements of group 18 of the periodic table, including helium, neon, argon, krypton, xenon, and radon. These are called noble because they do not easily form chemical bonds with other elements (i.e., keep aloof, as nobles might be supposed to do).

NONLINEAR SYSTEM: Physical system in which changes are not always additive or smooth. Climate is a nonlinear system composed of numerous nonlinear subsystems: abrupt climate change occurs at tipping points where the nonlinearity of the system causes drastic change in response to a small additional change in conditions.

NORTH ATLANTIC OSCILLATION: Alternating annual variations of atmospheric (barometric) pressure near Iceland and the Azores (an island group in the eastern Atlantic): corresponds to fluctuations in the westerly winds over the Atlantic and influences other aspects of climate in the Northern Hemisphere. The Icelandic Low shifts westward while the Azores High shifts eastward, and vice versa: storms track eastward between these two rotating systems.

NORTH POLE: Point at which Earth's axis passes through its surface in the Arctic. The north pole is near, but not identical to, the northern magnetic pole, the point at which the field lines of Earth's magnetic field pass vertically through its surface.

O

OCEAN HEAT TRANSPORT: Movement by ocean currents of warm water from the tropics toward the poles, effectively transporting heat energy toward the poles where it is more quickly radiated into space. A crucial component of Earth's climate system. Shifts in ocean heat transport may have drastic effects on climate.

OCEANIC DEAD ZONES: Areas of the ocean where the deeper waters are devoid of life. Often occur where nutrients from agricultural fertilizers have been carried to the sea by rivers. Algae bloom (reproduce rapidly) in the nutrient-rich surface waters: when they die, they sink, and are digested by bacteria in deeper waters. These bacteria consume oxygen, resulting in low-oxygen deep waters in which all oxygen-dependent aquatic life, such as fish, dies.

OIL FIELD: Surface region above underground reservoirs of petroleum where wells have been drilled to extract the petroleum.

OPERATIONAL HYDROLOGY: Study of water movements (precipitation, surface flow, groundwater flow) in order to manage water resources for human use.

ORBITAL FORCING: Increases or decreases in the amount of thermal energy being absorbed by Earth's climate system governed by Milankovitch cycles, that is, regularly repeating variations in Earth's climate caused by shifts in its orbit around the sun and its orientation (i.e., tilt) with respect to the sun.

ORGANIC MATTER: Remains, residues, or waste products of any living organism.

OROGRAPHY: Branch of geology that deals with the arrangement and character of land altitude variations (hills and mountains); also, the average land altitude over a given region. In computer models of climate, orography in the second sense defines the lower bound of the atmosphere over land.

OSCILLATION: A repeated back–and–forth movement.

OUTCROPPING: Any rock formation that is accessible from the surface without digging: a protruding mass of rock, usually connected to a larger, buried mass of similar rock.

OXIDATION: A reaction involving a loss of electrons in which the oxidation number increases (becomes more positive).

OZONE: An almost colorless, gaseous form of oxygen with an odor similar to weak chlorine. A relatively unstable compound of three atoms of oxygen, ozone constitutes, on average, less than one part per million (ppm) of the gases in the atmosphere. (Peak ozone concentration in the stratosphere can get as high as 10 ppm.) Yet ozone in the stratosphere absorbs nearly all of the biologically damaging solar ultraviolet radiation before it reaches Earth's surface.

OZONE HOLE: A term invented to describe a region of very low ozone concentration above the Antarctic that appears and disappears with each austral (Southern Hemisphere) summer.

OZONE LAYER: The layer of ozone that begins approximately 9.3 mi (15 km) above Earth and thins to an almost negligible amount at about 31 mi (50 km) and shields Earth from harmful ultraviolet radiation from the sun. The highest natural concentration of ozone (approximately 10 parts per million by volume) occurs in the stratosphere at approximately 15.5 mi (25 km) above Earth. The stratospheric ozone concentration changes throughout the year as stratospheric circulation changes with the seasons.

P

PALEOCLIMATE RECORD: The history of Earth's climate prior to the beginning of instrumental weather records in the late 1800s. "Paleo" means old or ancient, from the Greek *palaiois* for old.

PALEOCLIMATOLOGY: The study of past climates, throughout geological history, and the causes of the variations among.

PALEONTOLOGY: The study of life in past geologic time.

PAN EVAPORATION RATE: Rate at which water evaporates from an open pan exposed to the sky. Pan evaporation rate depends on wind, sunlight, humidity, and temperature. Measurements of pan evaporation rates have declined globally over the last century or so: global dimming has been proposed as a possible contributor to this effect.

PANGEA: A supercontinent that was assembled from Gondwana, Euramerica, Siberia, and the Cathaysian and Cimmerian terranes. The assembly of Pangea lasted from the Late Carboniferous to the Middle Triassic, while the break up of this supercontinent began in the Jurassic and has continued to the present. The term was first used by Alfred Wegener to refer to the supercontinent of the Mesozoic. It can also be spelled "Pangaea" and comes from the Greek, meaning "all lands."

PARABOLIC TROUGH: A reflective trough or bent piece of smooth metal that is parabolic in cross-section. Light entering the trough is all concentrated along a single line, the focus. A pipe placed along the focus can be heated to high temperature if the trough is exposed to bright, direct sun. Steam generated from such pipes is used to turn electrical generating systems in some centralized solar-thermal generation systems.

PARAFFIN: Any member of a class of molecules having the formula C_nH_{2n+2}. Methane, with $n = 1$, is the simplest paraffin, CH_4.

PARATERRAFORMING: A speculative method of making planets other than Earth habitable by humans, in which enclosures resembling giant greenhouses are extended over increasingly large areas of land. Not feasible in the foreseeable future.

PARTICULATE MATTER: Matter consisting of small particles. Particulate matter that is airborne forms aerosol pollution; particulate matter may also mix with water or lie on the surface of snow, ice, or ground. On snow or ice, small quantities of dark particulate matter (e.g., soot from fossil-fuel burning) can greatly accelerate melting.

PATHOGEN: A disease-causing agent, such as a bacteria, virus, fungus, etc.

PAY TO POLLUTE: Principle that industries wishing to emit greenhouse gases or other pollutants should have to pay taxes or fees in proportion to how much they emit.

PEA SOUPER: Nineteenth-century term for a thick smog episode in London. London continued to suffer extreme smog episodes through the 1950s.

PEATLAND: An area where peat is forming or has formed. Peat is a dense, moist substance formed of compacted ground-cover plants, mainly moss and grasses. Over geologic time, peat can turn into coal.

PEER REVIEW: The standard process in science for reducing the chances that faulty or fraudulent claims will be published in scientific journals. Before publication of an article, scientists with expertise in the article's subject area review the manuscript, usually anonymously, and make criticisms that may lead to revision or rejection of the article.

PER CAPITA: Latin phrase meaning "for each person." The per capita greenhouse emissions of a country, for example, are the country's total emissions divided by the number of people in the country.

PERFLUOROCARBONS: A group of greenhouse gases made up of carbon, fluorine, and/or sulfur, which are effective at releasing frozen carbon dioxide as a gas into the atmosphere.

PERMAFROST: Perennially frozen ground that occurs wherever the temperature remains below 32°F (0°C) for several years.

PERMIAN PERIOD: Geological period from 299 to 250 million years ago. During the Permian, all of Earth's continental plates were united in the supercontinent Pangea.

PERUVIAN ANCHOVETA: A fish species related to anchovies, found in the southeastern Pacific, especially off the coasts of Chile and Peru. Populations vary with climate changes such as those produced by El Niño.

PETROCHEMICALS: Chemicals derived from petroleum, such as gasoline, propane, and most plastics.

PETROLEUM: A complex liquid mixture that is mostly composed of hydrocarbons, compounds of carbon and hydrogen, that is separated into different products with different boiling ranges by a process called cracking.

PH: Measures the acidity of a solution. It is the negative log of the concentration of the hydrogen ions in a substance.

PHOTIC ZONE: Region of the ocean through which light penetrates and the place where photosynthetic marine organisms live.

PHOTOCHEMICAL SMOG: A type of smog that forms in large cities when chemical reactions take place in the presence of sunlight; its principal component is ozone. Ozone and other oxidants are not emitted into the air directly but form from reactions involving nitrogen oxides and hydrocarbons. Because of its smog-making ability, ozone in the lower atmosphere (troposphere) is often referred to as "bad"ozone.

PHOTON: Smallest individual unit of electromagnetic radiation (light energy). These light particles are emitted by an atom as excess energy when that atom returns from an excited state (high energy) to its normal state.

PHOTOSYNTHESIS: The process by which green plants use light to synthesize organic compounds from carbon dioxide and water. In the process, oxygen and water are released. Increased levels of carbon dioxide can increase net photosynthesis in some plants. Plants create a very important reservoir for carbon dioxide.

PHOTOSYNTHETIC ORGANISMS: Plants and bacteria (cyanobacteria) that practice photosynthesis, that is, the storage of energy from light in chemical bonds. All animals depend on this energy, either directly (plant-eaters) or indirectly (predators). The only exceptions are certain deep-sea organisms that derive their energy from hot springs ultimately powered by Earth's internal radioactivity.

PHOTOVOLTAIC CELL: A device made of silicon that converts sunlight into electricity.

PHOTOVOLTAIC SOLAR POWER: Electricity produced by photovoltaic cells, which are semiconductor devices that produce electricity when exposed to light. Depending on the cell design, between 6% and (in laboratory experiments) 42% of the energy falling on a photovoltaic cell is turned into electricity.

PHYTOPLANKTON: Microscopic marine organisms (mostly algae and diatoms) that are responsible for most of the photosynthetic activity in the oceans.

PIEDMONT: A region of foothills along the base of a mountain range.

PLANKTON: Floating animal and plant life.

PLASMA: Matter in the form of electrically charged atomic particles that form when a gas becomes so hot that electrons break away from the atoms. Also, the colorless, liquid portion of the blood in which blood cells and other substances are suspended.

PLATE TECTONICS: Geological theory holding that Earth's surface is composed of rigid plates or sections that move about the surface in response to internal pressure, creating the major geographical features such as mountains.

PLEISTOCENE EPOCH: The geologic period characterized by ice ages in the Northern Hemisphere, from 1.8 million to 10,000 years ago.

POLAR CELLS: Air circulation patterns near the poles: relatively warm, moist air approaches the pole at a high altitude, cools, sinks at the pole, and flows southward at a lower altitude. Because of Earth's rotation, air approaching or receding from the poles flows eastward, producing the polar easterlies.

POLAR ICE CAP: Ice mass located over one of the poles of a planet not otherwise covered with ice. In our solar system, only Mars and Earth have polar ice caps. Earth's north polar ice cap has two parts, a skin of floating ice over the actual pole and the Greenland ice cap, which does not overlay the pole. Earth's south polar ice cap is the Antarctic ice sheet.

POSTDICTION: Prediction of past climate events by a computer model using information that predates the events; also called hindcasting. For example, postdiction of the climate of the late 1990s would be performed by feeding data on climate prior to 1990 into a global circulation model. Postdiction tests the realism of computerized climate models.

POTABLE: Potable—refers to water that is safe to drink

POWER: Energy being transferred from one system to another at a certain rate: in physics, the time rate of doing work. A common power unit is the watt (a joule per second). For example, a 100-watt light bulb dissipates 100 joules of energy every second, i.e., uses 100 watts of power. Earth receives power from the sun at a rate of approximately 1.75×10^{17} watts.

PRECESSION: The comparatively slow torquing of the orbital planes of all satellites with respect to Earth's axis, due to the bulge of Earth at the equator, which distorts Earth's gravitational field. Precession is manifest by the slow rotation of the line of nodes of the orbit (westward for inclinations less than 90 degrees and eastward for inclinations greater than 90 degrees).

PRECIPITATION: Moisture that falls from clouds. Although clouds appear to float in the sky, they are always falling, their water droplets slowly being pulled down by gravity. Because the water droplets are so small and light, it can take 21 days to fall 1,000 ft (305 m) and wind currents can easily interrupt their descent. Liquid water falls as rain or drizzle. All raindrops form around particles of salt or dust. (Some of this dust comes from tiny meteorites and even the tails of comets.)

PREVALENCE: The actual number of cases of disease (or injury) that exist in a population.

PRIMARY POLLUTANT: Any pollutant released directly from a source to the atmosphere.

PRIVATIZATION: Transfer of state-owned resources to private owners, whether by sale or outright gift.

PROPELLANT: Gas or liquid ejected by a rocket or other vehicle to make the vehicle move in the opposite direction.

PROXY DATA: Information (data) about past climate obtained indirectly from various long-lasting physical traces left by climate or weather, such as oxygen isotopes in ancient ice layers, tree-rings, coral reef layers, and more. "Proxy" means a thing that represents something else: in this case, directly measurable quantities stand for or represent ancient climate conditions.

R

RADAR: An acronym for *RA*dio *D*etection *A*nd *R*anging—the use of electromagnetic waves at sub-optical frequencies (i.e., less than about 10^{12} Hz) to sense objects at a distance.

RADIATIVE BALANCE: The balance between incoming solar radiation and outgoing infrared radiation.

RADIATIVE FORCING: A change in the balance between incoming solar radiation and outgoing infrared radiation. Without any radiative forcing, solar radiation coming to Earth would continue to be approximately equal to the infrared radiation emitted from Earth. The addition of greenhouse gases traps an increased fraction of the infrared radiation, reradiating it back toward the surface and creating a warming influence (i.e., positive radiative forcing).

RADIO WAVES: Electromagnetic waves that oscillate or vibrate between 3 and 300 billion times per second. Radio waves are physically identical to light waves, except that they do not vibrate as rapidly.

RADIOCARBON DATING: Method for estimating the age of a carbon-containing substance by measuring amounts of an unstable isotope of carbon.

RADIOMETRIC AGE: The age of an object as determined by the levels of certain radioactive substances in its substance. Radioactive substances break down into other substances (some radioactive and some not) over time in a regular way: measurements of their relative quantities can allow an estimate of an object's age (e.g., rock, piece of wood). Many techniques for determining radiometric age are used in the field of radiometric dating.

RAFTING: Also called ice rafting: the transport of mineral sediments by icebergs. Sediments rafted out to sea by an iceberg that has calved off a glacier sink when the iceberg melts; these fragments, called dropstone, become part of sea-bottom sediments and form a record of past iceberg activity.

RANGE SHIFT: Movement or shrinkage of the territory occupied by a given species of plant or animal due to climate change. When climate warms, species cease to occupy areas that were at the warm extreme of their ability to adapt and to colonize areas that were at the cool extreme. The result is a shifting range.

RATIONING: Mandatory distribution of fixed amounts of food, fuel, or other goods in order to conserve a scarce resource.

RECEIVING AREA: Geographical region to which refugees are fleeing, i.e., that receives the refugees.

RE-EMERGING DISEASE: Many diseases once thought to be controlled are reappearing to infect humans again. These are known as re-emerging diseases because they have not been common for a long period of time and are starting to appear again among large population groups.

REFLECTANCE SPECTROSCOPY: The study of the spectral (wavelength or frequency) properties of light that has been reflected or scattered from a solid, liquid, or gas. It has many uses in Earth sciences; for example, reflectance spectroscopy of near-infrared light is a rapid, inexpensive way of characterizing organic substances and has been used to infer past climate changes by examining the content of lake sediment layers.

REINSURANCE: Insurance purchased by insurance companies in case unexpectedly large claims, such as might be made after a large hurricane or other natural disaster, exceed the funds available to the insurance company. If this happens, the reinsurance company must pay out to the insurance company, just as the insurance company must pay out to the people who have made valid claims for loss of property, health-care costs, or the like.

RELATIVE SEA LEVEL: Sea level compared to land level in a given locality. Relative sea level may change because land is locally sinking or rising, not because the sea itself is sinking or rising (eustatic sea level change).

REMEDIATION: A remedy. In the case of the environment, remediation seeks to restore an area to its unpolluted condition, or at least to remove the contaminants from the soil and/or water.

RENEWABLE ENERGY: Energy obtained from sources that are renewed at once, or fairly rapidly, by natural or managed processes that can be expected to continue indefinitely. Wind, sun, wood, crops, and waves can all be sources of renewable energy.

RENEWABLE RESOURCE: Any resource that is renewed or replaced fairly rapidly (on human historical timescales) by natural or managed processes. No mined substance (e.g., copper, coal) is renewable. Some resources, such as crops or forests, are truly renewable only if exploited in a way that does not deplete soils faster than natural processes can replace them.

RESERVOIR: A natural or artificial receptacle that stores a particular substance for a period of time

RESIDENCE TIME: For a greenhouse gas, the average amount of time a given amount of the gas stays in the atmosphere before being absorbed or chemically altered. The residence time of a greenhouse gas is relevant to policy decisions about emitting the gas: emissions of a gas with a long residence time create global warming over decades, centuries, or millennia, until natural processes remove the emitted quantities of gas.

RESPIRATION: The process by which animals use up stored foods (by combustion with oxygen) to produce energy.

RICHTER SCALE: A scale used to compare earthquakes based on the energy released by the earthquake.

RIVER DELTA: Flat area of fine-grained sediments that forms where a river meets a larger, stiller body of water such as the ocean. Rivers carry particles in their turbulent waters that settle out (sink) when the water mixes with quieter water and slows down; these particles build the delta. Deltas are named after the Greek letter delta, which looks like a triangle. Very large deltas are termed megadeltas and are often thickly settled by human beings.

ROCK APRON: Area of crushed rock laid on the ground surface in order to prevent erosion by moving surface water or waves.

RUMINANT ANIMALS: Animals such as cows, deer, sheep, and buffalo, and many others, also called simply ruminants, that digest their food partly by bringing it from the stomach to the mouth (where it is called a "cud") and chewing it there before re-swallowing. The digestive systems of ruminant animals contain bacteria that produce methane: methane from domesticated ruminant animals contributes significantly to global warming.

RUNOFF: Water that falls as precipitation and then runs over the surface of the land rather than sinking into the ground.

S

SAHEL: The transition zone in Africa between the Sahara Desert to the north and tropical forests to the south. This dry land belt stretches across Africa and is under stress from land use and climate variability.

SALINITY: The degree of salt in water. The rise in sea level due to global warming would result in increased salinity of rivers, bays, and aquifers. This would affect drinking water, agriculture, and wildlife.

SALINIZATION: Increase in salt content. The term is often applied to increased salt content of soils due to irrigation: salts in irrigation water tend to concentrate in surface soils as the water quickly evaporates rather than sinking down into the ground.

SALT MOBILIZATION: Movement of salt under, through, or on the surface of the ground, usually by water.

SALTATION: Any sudden or jumping change. Particles of snow or sand lofted by wind undergo saltation, building dunes. Climate change may encourage saltation by drying out soils.

SANTA ANA WINDS: Warm, dry, westerly (westward-blowing) Föhn winds that occur in southern California in the winter. Föhn winds occur when air is forced over mountain ranges, losing its moisture and then warming as it flows downslope and is placed under higher atmospheric pressure.

SATELLITE IMAGING: Creating images of Earth from data collected by artificial satellites. Today, most satellite imaging consists of digital photographs collected by satellites and radioed to Earth stations. Other techniques include the collection of gravity measurements that are processed on Earth to yield imagery, such as the melting of the Greenland ice cap.

SAVANNA HYPOTHESIS: Educated scientific guess (hypothesis) that human upright posture evolved in response to mountain-building in eastern Africa 2 to 1.5 million years ago, which changed regional climate and replaced rainforest with a diverse ecosystem dominated by savanna (grassy plain with widely spaced trees).

SAXITOXIN: Neurotoxin found in a variety of dinoflagellates. If ingested, it may cause respiratory failure and cardiac arrest.

SEA ICE: Ice that forms from the freezing of ocean water. As the salt water freezes, it ejects salt, so sea ice is fresh, not salty. Sea ice forms in relatively thin layers, usually no more than 3 to 7 ft (1 to 2 m)

thick, but it can cover thousands of square miles of ocean in the polar regions.

SEA LEVEL: The datum against which land elevation and sea depth are measured. Mean sea level is the average of high and low tides.

SEA SEDIMENT CORE: Cylindrical, solid sample of a layered deposit of sediment on the ocean floor, cut out of the deposit at right angles to its bedding planes. Sediment cores have been obtained from the bottoms of the oceans and many large lakes, such as Titicaca in South America and Tanganyika in Africa. Information about past climate changes can be derived from sediment cores, as the types and abundances of dead marine organisms (e.g., algae and foraminifera) in each sediment layer reflect

SEDIMENT: Solid unconsolidated rock and mineral fragments that come from the weathering of rocks and are transported by water, air, or ice and form layers on Earth's surface. Sediments can also result from chemical precipitation or secretion by organisms.

SEDIMENT CORE: Cylindrical, solid sample of a layered deposit of sediment, cut out of the deposit at right angles to its bedding planes. Sediment cores have been obtained from the bottoms of the oceans and many large lakes, such as Titicaca in South America and Tanganyika in Africa. Information about past climate changes can be derived from sediment cores, as the types and abundances of dead marine organisms (e.g., algae and foraminifera) in each sediment layer reflect climate conditions at the time the layer was formed.

SEDIMENT LOADING: Presence of moving mineral particles in rivers or streams. Faster-moving water can carry more sediment. Erosion increases sediment loading, limited by stream capacity for increased sediment loading.

SEICHE: A standing wave in a body of water (pronounced SAYSH). In water, a standing wave is a stationary raised mass of water on the surface of an otherwise flat water mass that is sustained by some source of vibration agitating the body of water.

SEISMIC: Related to earthquakes.

SEMICONDUCTOR INDUSTRY: Global high-technology industrial sector that produces semiconductor-based electronic components, such as the microprocessor chips that are at the heart of most computers, personal music players, and other consumer electronics devices. The industry is a major emitter of the greenhouse gases perfluorocarbons, trifluoromethane, nitrogen trifluoride, and sulfur hexafluoride.

SHORELINE: The band or belt of land surrounding a large body of surface water, such as a lake or ocean.

SINK: The process of providing storage for a substance. For example, plants—through photosynthesis—transform carbon dioxide in the air into organic matter, which either stays in the plants or is stored in the soils. The plants are a sink for carbon dioxide.

SINK HOLE: Depression in the surface of the ground caused by removal of material beneath the surface. Sinkholes are often caused by depletion of groundwater or by the dissolving of minerals by groundwater. Hundreds of sinkholes from groundwater depletion riddle Mexico City; some have forced evacuations of neighborhoods.

SKEPTIC: Person who doubts a claim on the grounds that its truth has not been adequately proved. Skeptics, as opposed to denialists, may have reasonable grounds for their reluctance to believe: indeed, skepticism is essential to the scientific process of discovering new knowledge, in which claims are carefully tested before being accepted as correct.

SKEPTICISM: Doubt about the truth of a claim. Skepticism, as opposed to denialism, may have a reasonable basis: in fact, skepticism is essential to the scientific process of discovering new knowledge, in which claims are carefully tested before being accepted as correct.

SOCIAL COST: The cost of emitting a given quantity of a pollutant, translated into monetary terms. Usually used in reference to the social cost of carbon, the total future economic cost of emitting a ton of carbon (or an equivalent amount of some other greenhouse gas) at a given time.

SOLAR CELL: A device constructed from specially prepared silicon that converts radiant energy (light) into electrical energy.

SOLAR ENERGY: Any form of electromagnetic radiation that is emitted by the sun.

SOLAR INSOLATION: The measure of electromagnetic radiation from the sun that reaches Earth. When measured over a specific surface area, solar insolation is often expressed watts per square meter (W/m^2) or kilowatt-hours per square meter per day.

SOLAR RADIATION: Energy received from the sun is solar radiation. The energy comes in many forms, such as visible light (that which we can see with our eyes). Other forms of radiation include radio waves, heat (infrared), ultraviolet waves, and x-rays. These forms are categorized within the electromagnetic spectrum.

SONAR: Sound Navigation and Ranging (SONAR) is a remote sensing system with important military, scientific, and commercial applications. Active SONAR transmits acoustic (i.e., sound) waves. Passive SONAR is a listening mode to detect noise generated from targets. SONAR allows the determination of important properties and attributes of the target (i.e., shape, size, speed, distance, etc.).

SOUTH POLE: The geographically southernmost place on Earth.

SOUTHERN ANNULAR MODE: Pattern of oscillation (regular back-and-forth change) in atmospheric pressure and windspeeds that occurs in the extratropical Southern Hemisphere; the southern counterpart of the North Atlantic Oscillation. Also termed Antarctic Oscillation or High Latitude Mode. Scientists have attributed the increasing magnitude of the southern annular mode to stratospheric ozone depletion and greenhouse gas increases.

SOUTHERN OSCILLATION: A large-scale atmospheric and hydrospheric fluctuation centered in the equatorial Pacific Ocean. It exhibits a nearly annual pressure anomaly, alternatively high over the Indian Ocean and over the South Pacific. Its period is slightly variable, averaging 2.33 years. The variation in pressure is accompanied by variations in wind strengths, ocean currents, sea-surface temperatures, and precipitation in the surrounding areas.

SPATIAL DATA: Data (individual measurements with numerical values) that are associated with points in space, such as measurement stations dotted over the face of a continent. Spatial data give information about location and are complementary to temporal data, which record changes over time.

SPECIATION: The process by which new species arise. Although there are a variety of definitions, all involve an isolation event that separates an interbreeding population until the daughter lineages evolve in separate trajectories.

SPECIES-AREA RELATIONSHIP: In biology, the relationship between the size of an area and the number of species of plants and animals that can live in that area: the smaller the area, the fewer the species.

SPECTRAL: Relating to a spectrum, which is an ordered range of possible vibrational frequencies for a type of wave. The spectrum of visible light, for example, orders colors from red (slowest vibrations visible) to violet (fastest vibrations visible) and is itself a small segment of the much larger electromagnetic spectrum.

STEPPES: Treeless biome dominated by short grasses (as opposed to prairies, which are dominated by long grasses).

STERN REVIEW: *The Stern Review on the Economics of Climate Change*, a controversial report commissioned by the government of the United Kingdom and written by British economist Nicholas Stern (1946–). The *Stern Review* states that climate change, if not mitigated, will eventually severely damage world economic growth, causing disruptions comparable to those of the world wars and Great Depression.

STORM SURGE: Local, temporary rise in sea level (above what would be expected due to tidal variation alone) as the result of winds and low pressures associated with a large storm system. Storm surges can cause coastal flooding, if severe.

STRATIFICATION AND TURNOVER: Two processes in lakes that have to do with the mixing of waters at different depths. Stratification is layering, which occurs when upper layers are warmer and float stably on the deeper water. Stratification opposes mixing or turnover, which is the exchange of water (and thus of dissolved chemicals and suspended particles) between deep and surface waters.

STRATIGRAPHY: The branch of geology that deals with the layers of rocks or the objects embedded within those layers.

STRATOCUMULUS CLOUDS: Type of cloud that has rounded shapes and forms at low altitudes, i.e., below about 8,000 ft (2,400 m).

STRATOSPHERE: The region of Earth's atmosphere ranging between about 9 and 30 mi (15 and 50 km) above Earth's surface.

STRATOSPHERIC OZONE LAYER: Layer of Earth's atmosphere from about 9 to 22 mi (15 to 35 km) above the surface in which the compound ozone (O_3) is relatively abundant. Ozone absorbs ultraviolet light from the sun, shielding the surface from this biologically harmful form of light and becoming heated in the process. Heating by ultraviolet light warms the stratosphere.

STRATUS CLOUD: Extensive, layer-like cloud form ("stratum" means layer in Latin). Stratus clouds occur worldwide but play a particularly important role in Arctic climate, where they are prevalent and affect vertical exchanges of heat, moisture, and momentum; this, in turn, affects global climate.

SUBLIMATION: Transformation of a solid to the gaseous state without passing through the liquid state.

SUBSISTENCE FARMING: Agriculture carried on for the sake of the food it produces, rather than to produce crops to sell for cash.

SULFATE AEROSOLS: Tiny airborne particles (aerosols) containing sulfur that form from gases emitted to the atmosphere by oceanic phytoplankton, the burning of fossils fuels (especially coal), and volcanoes. Sulfate aerosols have a cooling effect on climate, though not enough to reverse the warming effect of greenhouse-gas emissions. About eighty percent of sulfate aerosols come from human activities such as coal-burning and copper smelting.

SUNSPOTS: Comparatively dark, cool patches that appear on the sun's surface in synchrony with increased solar activity every 11 years. By interacting with stratospheric ozone, sunspot activity affects Earth's climate, mostly at high altitudes but subtly at the surface (perhaps a few tenths of a degree of warming in the Northern Hemisphere).

SUN-SYNCHRONOUS ORBIT: Satellite orbit structured to keep the satellite directly above a fixed solar time as that time (area on the ground) moves around the rotating Earth. A sun-synchronous satellite will always look down on places that see the sun at a given height in the sky: shadows cast on the landscape just below the satellite will always be the same length for objects of identical height. Also called heliosynchronous orbit.

SUSPENSION: A temporary mixture of a solid in a gas or liquid from which the solid will eventually settle out.

SUSTAINABILITY: The quality, in any human activity (farming, energy generation, or even the maintenance of a society as a whole), of being sustainable for an indefinite period without exhausting necessary resources or otherwise self-destructing.

SUSTAINABLE: Capable of being sustained or continued for an indefinite period without exhausting necessary resources or otherwise self-destructing: often applied to human activities such as farming, energy generation, or even the maintenance of a society as a whole.

SUSTAINABLE DEVELOPMENT: Development (i.e., increased or intensified economic activity; sometimes used as a synonym for industrialization) that meets the cultural and physical needs of the present generation of persons without harming the ability of future generations to meet their needs.

SYMBIOSIS: A pattern in which two or more organisms live in close connection with each other, often to the benefit of both or all organisms.

SYMBIOTIC: A relationship or pattern of exchange that is mutually beneficial to two or more creatures. Algae are symbiotic with lichens and corals; intestinal bacteria are symbiotic with mammals.

T

TABULAR BERG: Flat-topped (i.e., like a table, hence "tabular") iceberg, usually formed by breaking off from the edge of a floating ice shelf.

TAR SANDS: Naturally occurring mixtures of thick crude oil (bitumen) and small mineral particles (e.g., clay or sand). Also called oil sands or bituminous sands. Large deposits exist in the Canadian province of Alberta, with smaller deposits in Venezuela and the United States. Petroleum can be extracted from tar sands, but only by strip-mining the landscapes underlain by the deposits.

TECTONIC: Relating to tectonics, the scientific study of the forces that shape planetary crusts (mountain ranges, continents, sea-beds, etc.).

TECTONIC FORCES: Forces that shape Earth's crust over geological time. The largest tectonic force is supplied by the slow convection of Earth's interior, driven by radiation of heat to space through the crust, much like the roiling of liquid in a boiling pot.

TECTONIC PLATE: Rigid unit of Earth's crust that moves about over geological time, merging with and separating from other tectonic plates as the continents rearrange but retaining its identity through these encounters. There are seven major tectonic plates on Earth and a number of smaller ones.

TEMPERATE ZONE: Band of moderate, seasonal climate located between the tropics and the polar regions. There is one temperate zone in the Northern Hemisphere and one in the Southern Hemisphere.

THAW LAKES: Ponds that form on thermokarsts, that is, landscapes where permafrost (soil frozen year-round, often to hundreds of feet in depth) is melting. Also called thermokarst lakes. Thaw lakes are forming in large numbers in the Arctic due to climate warming.

THERMOHALINE CIRCULATION: Large-scale circulation of the world ocean that exchanges warm, low-density surface waters with cooler, higher-density deep waters. Driven by differences in temperature and saltiness (halinity) as well as, to a lesser degree, winds and tides. Also termed meridional overturning circulation.

THERMONUCLEAR REACTION: A nuclear reaction that takes place only at very high temperatures, usually on the order of a few million degrees.

THERMOREGULATORY: Concerned with maintaining a constant internal temperature, as by many animals (or, rarely, plants). Thermoregulatory ability limits the climate adaptability of an organism: penguins, for example, cannot live in warm climates because they cannot discard enough heat to keep their internal temperature down. Decline in Galápagos penguins has been linked to anthropogenic climate warming.

THERMOSPHERE: The outermost shell of the atmosphere, between the mesosphere and outer space; where temperatures increase steadily with altitude.

THERMOSTERIC EXPANSION: Expansion due to warming, also called thermal expansion. From 1955 to 2003, thermosteric expansion of the top 2,300 ft (700 m) of the ocean, caused by global warming, was responsible for about 0.013 inch/yr (0.33 mm/year) of global sea level rise. About half of this expansion was due to warming of the Atlantic and a third due to warming in the Pacific. Most of the sea-level rise that has occurred since the beginning of the twentieth century is due to thermosteric expansion.

THREATENED: When a species is capable of becoming endangered in the near future.

TIDAL MIXING: Mixing of different layers of ocean water due to the rising and falling of the tides. Tidal mixing occurs along shorelines but also in some parts of the deep ocean, where it is crucial to the mixing of deep waters with warmer surface waters in the tropics; this allows water to flow to the surface as part of the global thermohaline circulation of the ocean, one of the defining systems of Earth's climate mechanism.

TIDAL POWER: Electrical power generated by harnessing the energy of rising and falling tides: a form of hydropower. Several countries have built or seek to build large tidal power facilities, but tidal power is not yet a major source of electricity.

TIDE GAUGE: Device, usually stationed along a coast, that measures sea level continuously. Measurements from tide gauges are the main source of sea-level data prior to the beginning of satellite measurements in the 1970s.

TIDES: Daily or twice-daily rise and fall of local sea level. Tides are caused by gravitational and centrifugal forces as the moon and Earth swing around their common center of gravity like a pair of dancers holding hands. The moon pulls water toward itself, creating a moon-facing tide, while centrifugal force dominates on the far side of Earth, causing a second tide directed away from the moon.

TIDEWATER GLACIER: Glacier that flows into the ocean, as opposed to melting on dry land. Tidewater glaciers in Greenland and the West Antarctica Peninsula have significantly speeded up their flow to the sea in recent years, contributing to accelerated sea-level rise.

TIPPING POINT: In climatology, a state in a changing system where change ceases to be gradual and reversible and becomes rapid and irreversible. Also termed a climate surprise.

TRACE AMOUNT: A relatively very small quantity of some substance in a mixture of substances. Carbon dioxide and water vapor, although having a large impact on Earth's climate through the greenhouse effect, are present in trace amounts in the atmosphere. CO_2 is about .037% of the atmosphere, water vapor about .25%.

TRACE GASES: Gases present in Earth's atmosphere in trace (relatively very small) amounts. All greenhouse gases happen to be trace gases, though some are more abundant than others; the most abundant greenhouse gases are CO_2 (0.037% of the atmosphere) and water vapor (0.25% of the atmosphere, on average).

TRADE WINDS: Surface air from the horse latitudes (subtropical regions) that moves back toward the equator and is deflected by the Coriolis Force, causing the winds to blow from the Northeast in the Northern Hemisphere and from the Southeast in the Southern Hemisphere. These steady winds are called trade winds because they provided trade ships with an ocean route to the New World.

TRANSFORMATION: The processes involved in the transfer of a substance from one reservoir to another.

TRANSPIRATION: The process in plants by which water is taken up by the roots and released as water vapor by the leaves. The term can also be applied to the quantity of water thus dissipated.

TREE RINGS: Marks left in the trunks of woody plants by the annual growth of a new coat or sheath of material. Tree rings provide a straightforward way of dating organic material stored in a tree trunk. Tree-ring thickness provides proxy data about climate conditions: most trees put on thicker rings in warm, wet conditions than in cool, dry conditions.

TREELINE: The highest-altitude (or highest-latitude) line along which trees can grow. As climate warms, treelines advance to higher altitudes and latitudes.

TRIASSIC PERIOD: Geological period from 251 million years ago to 199 million years ago. Global climate was particularly warm during the Triassic; the poles were ice-free and all Earth's continental plates were clumped into the supercontinent Pangea.

TROPHIC: Relating to feeding. Communities of living things can be imagined as forming a pyramid defined by trophic levels, with each level feeding on the one below it.

TROPIC OF CANCER: In geography, a tropic is one of two lines of latitude (lines of equal distance from the equator on Earth's surface) at 23° 26' north and 23° 26' south. The northern tropic is the Tropic of Cancer and the southern Tropic is the tropic of Capricorn. The belt of Earth's surface between these lines is the tropic.

TROPICAL CYCLONES: Large rotating storm systems characterized by a clear, low-pressure center surrounded by spiral arms of thunderstorms. Such storms form in the tropics because they are powered by the thermal energy of warm surface ocean waters. In the Atlantic, tropical cyclones are termed hurricanes.

TROPICAL DEPRESSION: Rotating system of thunderstorms with low atmospheric pressure in the center and maximum windspeed less than 39 mph (63 km/h). At higher speeds, such systems are characterized as tropical storms or cyclones (hurricanes).

TROPICAL STORM: Tropical storms generally form in the eastern portion of tropical oceans and track westward. Hurricanes, typhoons, and willy-willies all start out as weak low pressure areas that form over warm tropical waters (e.g., surface water temperature of at least 80°F).

TROPOPAUSE: The boundary between the troposphere and the stratosphere (about 5 mi [8 km] in polar regions and about 9.3 mi [15 km] in tropical regions), usually characterized by an abrupt change of lapse rate. The regions above the troposphere have increased atmospheric stability than those below. The tropopause marks the vertical limit of most clouds and storms.

TROPOSPHERE: The lowest layer of Earth's atmosphere, ranging to an altitude of about 9 mi (15 km) above Earth's surface.

TSUNAMI: Ocean wave caused by a large displacement of mass under the surface of the water, such as an earthquake or volcanic eruption.

TUNDRA: A type of ecosystem dominated by lichens, mosses, grasses, and woody plants. It is found at high latitudes (arctic tundra) and high altitudes (alpine tundra). Arctic tundra is underlain by permafrost and usually very wet.

TURBINE: An engine that moves in a circular motion when force, such as moving water, is applied to its series of baffles (thin plates or screens) radiating from a central shaft. Turbines convert the energy of a moving fluid into the energy of mechanical rotation.

U

UBIQUITOUS: Present more or less everywhere in a certain region. One might say, for example, that "in North American ecosystems, mice are ubiqutous."

ULTRAVIOLET RADIATION: The energy range just beyond the violet end of the visible spectrum. Although ultraviolet radiation constitutes only about 5% of the total energy emitted from the sun, it is the major energy source for the stratosphere and mesosphere, playing a dominant role in both energy balance and chemical composition.

UNIFORMITARIANISM: Doctrine of geology promoted by English geologist Charles Lyell (1797–1895), asserting three assumptions: (1) actualism (uniform processes acting throughout Earth's history), (2) gradualism (slow, uniform rate of change throughout Earth's history), and (3) uniformity of state (Earth's conditions have always varied around a single, steady state). Uniformitarianism is often contrasted to castrophism. Modern geology makes use of elements of both views.

UPWELLING: The vertical motion of water in the ocean by which subsurface water of lower temperature and greater density moves toward the surface of the ocean. Upwelling occurs most commonly among the western coastlines of continents, but may occur anywhere in the ocean. Upwelling results when winds blowing nearly parallel to a continental coastline transport the light surface water away from the coast. Subsurface water of greater density and lower temperature replaces the surface water.

URBAN HEAT ISLAND: Area of warm weather in and immediately around a built-up area. Pavement and buildings absorb solar energy while being little cooled by evaporation compared to vegetation-covered ground. Skeptics of global climate change at one time argued that the expansion of urban heat islands near and around weather stations has caused an illusion of global warming by biasing temperature measurements.

V

VAN ALLEN RADIATION BELTS: Two regions of space near Earth where fast-moving charged particles (electrons and protons) are confined by Earth's magnetic field. The belts are shaped somewhat like huge hollow doughnuts, one nested inside the other with Earth in the center hole.

VECTOR: Any agent, living or otherwise, that carries and transmits parasites and diseases. Also, an organism or chemical used to transport a gene into a new host cell.

VIENNA CONVENTION: In climate affairs, treaty signed in 1985 in Vienna, Austria, with the purpose of reducing emissions of chemicals that damage the ozone layer. The Montreal Protocol is an extension of the Vienna Convention that describes binding national obligations for the reduction of ozone-depleting emissions.

VORTEX: A rotating column of a fluid such as air or water.

VULNERABILITY: The degree to which an ecosystem or human community is susceptible to, or cannot adapt to or cope with, the negative effects of climate change. The type, intensity, and speed of climate change, the adaptive capacity of the system, and the sensitivity of the system to increased climate variability or climate change all determine vulnerability.

W

WATER CYCLE: The process by which water is transpired and evaporated from the land and water, condensed in the clouds, and precipitated out onto Earth once again to replenish the water in the bodies of water on Earth.

WATER INSECURITY: Lack of access to clean and dependable water supplies. Climate change alters the precipitation patterns on which water security largely depends; forecasted climate changes are likely to greatly increase the number of people suffering water insecurity worldwide, especially in parts of Africa, Asia, and South America.

WATER LENS: Body of fresh groundwater retained by a small ocean island. Rising sea levels shrink an island's water lens and may eliminate it.

WATER TABLE: Underground level or depth below which the ground is saturated with liquid water. Where the water table intersects the surface, water is found (e.g., lakes, springs, streams).

WATER VAPOR: The most abundant greenhouse gas, it is the water present in the atmosphere in gaseous form. Water vapor is an important part of the natural greenhouse effect. Although humans are not significantly increasing its concentration, it contributes to the enhanced greenhouse effect because the warming influence of greenhouse gases leads to a positive water vapor feedback.

WATERBORNE DISEASE: Infectious disease or parasite that is transmitted by unclean water.

WATERSHED: Ridge of high land that separates the catchment area of one river system from that of another.

WATT: Unit of power or rate of expenditure of energy. One watt equals 1 joule of energy per second. A 100-watt light bulb dissipates 100 joules of energy every second, i.e., uses 100 watts of power. Earth receives power from the sun at a rate of approximately 1.75×10^{17} watts.

WAVELENGTH: Distance between the peaks or troughs of a cyclic wave. The character and effects of electromagnetic radiation are determined by its wavelength: very short-wavelength rays (e.g., X rays) are biologically harmful, somewhat longer-wavelength rays are classified as ultraviolet light, rays of intermediate wavelength are visible light, and longer wavelengths are infrared radiation and radio waves.

WEATHERING: The natural processes by which the actions of atmospheric and other environmental agents, such as wind, rain, and temperature changes, result in the physical disintegration and chemical decomposition of rocks and earth materials in place, with little or no transport of the loosened or altered material.

WETLANDS: Areas that are wet or covered with water for at least part of the year.

WHITE PAPER: An authoritative, publicly available paper or report giving an overview of some important topic. Traditionally, only documents issued by governments were termed white papers, but corporate documents crafted to urge a certain point of view are increasingly labeled "white papers."

WIND CELLS: More commonly called convective or convection cells; vertical structures of moving air formed by warm (less-dense) air welling up in the center and cooler (more-dense) air sinking around the perimeter. Thunderstorms are shaped by convective cells.

WIND FARMS: Clusters of wind turbines generating electricity. Wind farms are the most efficient way to generate large amounts of electricity from wind because they can share a single, high-capacity line to transmit their power output to the long-distance electric-power network (grid).

WIND SHEAR: Variation in wind speed within a current of wind: that is, when wind shear is present, part of a wind current is moving at one speed while another, nearby part of the current, moving parallel to the first, is moving at a different speed.

WORLD BANK: International bank formed in 1944 to aid in reconstruction of Europe after World War II, now officially devoted to the eradication of world poverty through funding of development projects. Although ostensibly independent, the bank is always headed by a person appointed by the president of the United States.

WORLD HERITAGE CONVENTION: Treaty (the Convention Concerning the Protection of World Cultural and Natural Heritage) adopted by the United Nations Educational, Scientific and Cultural Organization (UNESCO) in 1972. Its goal is to encourage the protection of cultural and natural sites of "outstanding universal value." Such sites are designated as World Heritage Sites.

WORLD HERITAGE SITE: A site deemed by the World Heritage Committee of the United Nations Educational, Scientific and Cultural Organization (UNESCO) to be of "outstanding universal value." The site may be cultural, such as a building or neighborhood, or natural, such as a lake or mountain. Under the World Heritage Convention, the U.N. encourages (but cannot enforce) protection of World Heritage Sites.

WORLD WEATHER WATCH: Program established by the United Nations' World Meteorological Organization in 1963. Combines ground and satellite observing systems, telecommunications, and computerized data processing and forecasting to make meteorological information freely available to all countries. The global equivalent of the National Weather Service provided by the U.S. National Oceanographic and Atmospheric Administration (NOAA).

Y

YOUNGER DRYAS: A relatively recent episode of abrupt climate change. About 12,900 years ago, conditions in the Northern Hemisphere cooled in about a decade (extremely rapidly), in some locations by 27°F (15°C). The cold period persisted for about 1,300 years and then reversed, also suddenly. The causes of the Younger Dryas are not well understood, but the event does show that Earth's climate is capable of extremely rapid and dramatic shifts.

Z

ZOOPLANKTON: Animal plankton. Small herbivores that float or drift near the surface of aquatic systems and that feed on plant plankton (phytoplankton and nanoplankton).

ZOOXANTHELLAE: Algae that live in the tissues of coral polyps and, through photosynthesis, supply them with most of their food. The relationship is symbiotic, as the polyps supply the zooxanthellae with a hospitable environment. When water temperatures are too high, corals lose their zooxanthellae. If the loss is too pronounced for too long, the coral dies.

Intergovernmental Panel on Climate Change (IPCC) Abbreviations, Acronyms, and Definitions

The following abbreviations and acronyms, reprinted by permission and courtesy of the Intergovernmental Panel on Climate Change (IPCC), appear in IPCC reports and other scientific studies related to climate change. In addition to explaining abbreviations and acronyms that appear in this book, this list is provided as a ready reference to assist in further study of IPCC reports and other scientific documents related to climate change.

μMOL: Micromole

20C3M: 20th Century Climate in Coupled Models

A

A/R: Afforestation and Reforestation

AABW: Antarctic Bottom Water

AAIW: Antarctic Intermediate Water

AAO: Antarctic Oscillation

AATSR: Advanced Along Track Scanning Radiometer

ACC: Antarctic Circumpolar Current

ACCENT: Atmospheric Composition Change: A European Network

ACE: Accumulated Cyclone Energy or Aerosol Characterization Experiment

ACRIM: Active Cavity Radiometer Irradiance Monitor

ACRIMSAT: Active Cavity Radiometer Irradiance Monitor Satellite

ACTUAL NET GREENHOUSE GAS REMOVALS BY SINKS: The sum of the verifiable changes in carbon stocks in the carbon pools within the project boundary of an afforestation or reforestation project, minus the increase in GHG emissions as a result of the implementation of the project activity. The term stems from the Clean Development Mechanism (CDM) afforestation and reforestation modalities and procedures.

ACW: Antarctic Circumpolar Wave

ADAPTATION: Initiatives and measures to reduce the vulnerability of natural and human systems against actual or expected climate change effects.

ADAPTIVE CAPACITY: The whole of capabilities, resources and institutions of a country or region to implement effective adaptation measures.

ADB: Asian Development Bank

ADEC: Aeolian Dust Experiment on Climate

ADNET: Asian Dust Network

AEROCOM: Aerosol Model Intercomparison

AERONET: Aerosol Robotic Network

AGAG: Advanced Global Atmospheric Gases Experiment

AGCM: Atmospheric General Circulation Model

AGREEMENT: Reduction in emissions by sources or enhancement of removals by sinks that is additional to any that would occur in the absence of a Joint Implementation (JI) or a Clean Development Mechanism (CDM) project activity as defined in the Kyoto Protocol Articles on JI and CDM.

AGWP: Absolute Global Warming Potential

AIACC: Assessments of Impacts and Adaptations to Climate Change in Multiple Regions and Sectors

AIC: Aviation-Induced Cloudiness

AIJ: Activities Implemented Jointly

ALA: Alaska

ALAS: Autonomous Lagrangian Current Explorer

ALE: Atmospheric Lifetime Experiment

ALLIANCE OF SMALL ISLAND STATES (AOSIS): Formed at the Second World Climate Conference (1990). AOSIS comprises small-island and low-lying coastal developing countries that are particularly vulnerable to the adverse consequences of climate change, such as sea-level rise, coral bleaching, and the increased frequency and intensity of tropical storms. With more than 35 states from the Atlantic, Caribbean, Indian Ocean, Mediterranean, and Pacific, AOSIS share common objectives on environmental and sustainable development matters in the UNFCCC process.

ALM: Africa, Latin America, Middle East Region (SRES and post-SRES scenarios).

AMIP: Atmospheric Model Intercomparison Project

AMO: Atlantic Multi-Decadal Oscillation

AMSU: Advanced Microwave Sounding Unit

AMZ: Amazonia

ANCILLARY BENEFITS: Policies aimed at some target, e.g. climate change mitigation, may be paired with positive side effects, such as increased resource-use efficiency, reduced emissions of air pollutants associated with fossil fuel use, improved transportation, agriculture, land-use practices, employment, and fuel security. Ancillary impacts are also used when the effects may be negative. Policies directed at abating air pollution may consider greenhouse-gas mitigation an ancillary benefit.

ANNEX I COUNTRIES: The group of countries included in Annex I (as amended in 1998) to the UNFCCC, including all the OECD countries and economies in transition. Under Articles 4.2 (a) and 4.2 (b) of the Convention, Annex I countries committed themselves specifically to the aim of returning individually or jointly to their 1990 levels of greenhouse-gas emissions by the year 2000. By default, the other countries are referred to as Non-Annex I countries.

ANNEX II COUNTRIES: The group of countries included in Annex II to the UNFCCC, including all OECD countries. Under Article 4.2 (g) of the Convention, these countries are expected to provide financial resources to assist developing countries to comply with their obligations, such as preparing national reports. Annex II countries are also expected to promote the transfer of environmentally sound technologies to developing countries.

ANNEX B COUNTRIES: The countries included in Annex B to the Kyoto Protocol that have agreed to a target for their greenhouse-gas emissions, including all the Annex I countries (as amended in 1998) except for Turkey and Belarus.

ANT: Antarctic

ANTHROPOGENIC EMISSIONS: Emissions of greenhouse gases, greenhouse-gas precursors, and aerosols associated with human activities. These include the burning of fossil fuels, deforestation, land-use changes, livestock, fertilization, etc. that result in a net increase in emissions.

AO: Arctic Oscillation

AOGCM: Atmosphere-Ocean General Circulation Model

APE: Atmospheric Particulate Environment Change Studies

AR4: Fourth Assessment Report

ARC: Arctic

ARM: Atmospheric Radiation Measurement

ASOS: Automated Surface Observation Systems

ASSIGNED AMOUNT (AA): Under the Kyoto Protocol, the assigned amount is the quantity of greenhouse-gas emissions that an Annex B country has agreed to as its ceiling for its emissions in the first commitment period (2008 to 2012). The AA is the country's total greenhouse-gas emissions in 1990 multiplied by five (for the five-year commitment period) and by the percentage it agreed to as listed in Annex B of the Kyoto Protocol (e.g., 92% for the EU; 93% for the U.S.).

ASSIGNED AMOUNT UNIT (AAU): An AAU equals 1 tonne (metric ton) of CO_2-equivalent emissions calculated using the Global Warming Potential.

ASTEX: Atlantic Stratocumulus Transition Experiment

ATCM: Atmospheric Transport and Chemical Model

ATSR: Along Track Scanning Radiometer

AUTONOMOUS (EXOGENOUS) TECHNOLOGICAL CHANGE: Autonomous (exogenous) technological change is imposed from outside the model, usually in the form of a time trend affecting energy demand or world output growth. Endogenous technological change is the outcome of economic activity within the model, i.e., the choice of technologies is included within the model and affects energy demand and/or economic growth. Induced technological change implies endogenous technological change but adds further changes induced by policies and measures, such as carbon taxes triggering Ramp's efforts.

AVHRR: Advanced Very High Resolution Radiometer

B

BACKSTOP TECHNOLOGY: Models estimating mitigation often characterize an arbitrary carbon-free technology (often for power generation) that becomes available in the future in unlimited supply over the horizon of the model. This allows models to explore the consequences and importance of a generic solution technology without becoming enmeshed in picking the technology. This "backstop" technology might be a nuclear technology, fossil technology with capture and sequestration, solar, or something as yet unimagined. The backstop technology is typically assumed either not to currently exist, or to exist only at higher costs relative to conventional alternatives.

BANKING: According to the Kyoto Protocol [Article 3 (13)], parties included in Annex I to the UNFCCC may save excess AAUs from the first commitment period for compliance with their respective cap in subsequent commitment periods (post–2012).

BARRIER: Any obstacle to reaching a goal, adaptation or mitigation potential that can be overcome or attenuated by a policy, program, or measure. Barrier removal includes correcting market failures directly or reducing the transactions costs in the public and private sectors by e.g. improving institutional capacity, reducing risk and uncertainty, facilitating market transactions, and enforcing regulatory policies.

BASELINE: The reference for measurable quantities from which an alternative outcome can be measured, e.g. a non-intervention scenario is used as a reference in the analysis of intervention scenarios.

BATS: Bermuda Atlantic Time-Series Study

BAU: Business As Usual

BC: Black Carbon

BCC: Beijing Climate Center

BCCR: Bjerknes Centre for Climate Research

BECS: Bioenergy with CCS

BENCHMARK: A measurable variable used as a baseline or reference in evaluating the performance of an organization. Benchmarks may be drawn from internal experience, that of other organizations, or from legal requirement and are often used to gauge changes in performance over time.

BENEFIT TRANSFER: An application of monetary values from one particular analysis to another policy-decision setting, often in a geographic area other than the one in which the original study was performed.

BIOCHEMICAL OXYGEN DEMAND (BOD): The amount of dissolved oxygen consumed by micro-organisms (bacteria) in the bio-chemical oxidation of organic and inorganic matter in waste water.

BIOCOVERS: Layers placed on top of landfills that are biologically active in oxidizing methane into CO_2.

BIOFILTERS: Filters using biological material to filter or chemically process pollutants like oxidizing methane into CO_2.

BIOLOGICAL OPTIONS: Biological options for mitigation of climate change involve one or more of the three strategies: conservation—conserving an existing carbon pool, thereby preventing CO_2 emissions to the atmosphere; sequestration—increasing the size of existing carbon pools, thereby extracting CO_2 from the atmosphere; substitution—substituting biomass for fossil fuels or energy-intensive products, thereby reducing CO_2 emissions.

BIOME 6000: Global Palaeovegetation Mapping Project

BMRC: Bureau of Meteorology Research Centre

BOD: Biochemical Oxygen Demand / Biological Oxygen Demand

BOTTOM-UP MODELS: Models represent reality by aggregating characteristics of specific activities and processes, considering technological, engineering and cost details.

BRT: Bus Rapid Transport

BUBBLE: Policy instrument for pollution abatement named for its treatment of multiple emission points as if they were contained in an imaginary bubble. Article 4 of the Kyoto Protocol allows a group of countries to meet their target listed in Annex B jointly by aggregating their total emissions under one 'bubble' and sharing the burden (e.g. the EU).

C

C: Carbon

C2F6 OR C_2F_6: Perfluoroethane / Hexafluoroethane

C4MIP: Coupled Carbon Cycle Climate Model Intercomparison Project

CAA: Clean Air Act

$CaCO_3$: Calcium Carbonate

CAM: Central America

CAMS: Climate Anomaly Monitoring System (NOAA)

CAP: Mandated restraint as an upper limit on emissions. The Kyoto Protocol mandates emissions caps in a scheduled timeframe on the anthropogenic GHG emissions released by Annex B countries. By 2008 2012 the EU e.g. must reduce its CO_2-equivalent emissions of six greenhouse gases to a level 8% lower than the 1990 level.

CAPACITY BUILDING: In the context of climate change, capacity building is developing technical skills and institutional capabilities in developing countries and economies in transition to enable their participation in all aspects of adaptation to, mitigation of, and research on climate change, and in the implementation of the Kyoto Mechanisms, etc.

CAPE: Convective Available Potential Energy

CAR: Caribbean

CARBON CAPTURE AND STORAGE (CCS): A process consisting of separation of CO_2 from industrial and energy-related sources, transport to a storage location, and long-term isolation from the atmosphere.

CARBON DIOXIDE (CO_2): CO_2 is a naturally occurring gas, and a by-product of burning fossil fuels or biomass, of land-use changes, and of industrial processes. It is the principal anthropogenic greenhouse gas that affects Earth's radiative balance. It is the reference gas against which other greenhouse gases are measured and therefore it has a Global Warming Potential of 1.

CARBON DIOXIDE FERTILIZATION: The enhancement of the growth of plants because of increased atmospheric CO_2 concentration. Depending on their mechanism of photosynthesis, certain types of plants are more sensitive to changes in atmospheric CO_2 concentration than others.

CARBON INTENSITY: The amount of emissions of CO_2 per unit of GDP.

CARBON LEAKAGE: The part of emissions reductions in Annex B countries that may be offset by an increase of the emissions in the non-constrained countries above their baseline levels. This can occur through (1) relocation of energy-intensive production in non-constrained regions; (2) increased consumption of fossil fuels in these regions through decline in the international price of oil and gas triggered by lower demand for these energies; and (3) changes in incomes (thus in energy demand) because of better terms of trade. Leakage also refers to GHG-related effects of GHG-emission reduction or CO_2-sequestration project activities that occur outside the project boundaries and that are measurable and attributable to the activity. On most occasions, leakage is understood as counteracting the initial activity. Nevertheless, there may be situations where effects attributable to the activity outside the project area lead to GHG- emission reductions. These are commonly called spill-over. While (negative) leakage leads to a discount of emission reductions as verified, positive spill-over may not in all cases be accounted for.

CARBON POOL: Carbon pools are above-ground biomass, below-ground biomass, litter, dead wood, and soil organic carbon. CDM project participants may choose not to account one or more carbon pools if they provide transparent and verifiable information showing that the choice will not increase the expected net anthropogenic GHG removals by sinks.

CARBON PRICE: What has to be paid (to some public authority as a tax rate, or on some emission permit exchange) for the emission of 1 tonne of CO_2 into the atmosphere. In the models, the carbon price is the social cost of avoiding an additional unit of CO_2 equivalent emission. In some models it is represented by the shadow price of an additional unit of CO_2 emitted, in others by the rate of carbon tax, or the price of emission-permit allowances. It has also been used as a cut-off rate for marginal abatement costs in the assessment of economic mitigation potentials.

CAS: Central Asia

CBA: Cost Benefit Analysis

CCCm: Canadian Centre for Climate Modelling and Analysis

CCGT: Combined Cycle Gas Turbine

CCL4 OR CCL_4: Carbon Tetrachloride

CCM: Chemistry-Climate Model

CCN: Cloud Condensation Nuclei

CCSR: Centre for Climate System Research

CCS-READY: If rapid deployment of CCS is desired, new power plants could be designed and located to be "CCS-ready" by reserving space for the capture installation, designing the unit for optimal performance when capture is added and siting the plant to enable access to storage reservoirs.

CDIAC: Carbon Dioxide Information Analysis Center

CDM: Clean Development Mechanism

CDW: Circumpolar Deep Water

CERES: Clouds and Earth's Radiant Energy System

CERTIFIED EMISSION REDUCTION UNIT (CER): Equal to one metric ton of CO_2-equivalent emissions reduced or sequestered through a Clean Development Mechanism project, calculated using Global Warming Potentials. In order to reflect potential non-permanence of afforestation and reforestation project activities, the use of temporary certificates for Net Anthropogenic Greenhouse Gas Removal was decided by COP 9.

CF4 OR CF_4: Perfluoromethane / Tetrafluoromethane

CFCs: Chlorofluorocarbons

CFL: Compact Fluorescent Lamp

CGE: Computable General Equilibrium

CGI: East Canada, Greenland, and Iceland

CH2I2 OR CH_2I_2: Di-Iodomethane (Methylene Iodide)

CH2O OR CH_2O: Formaldehyde

CH3CCL3 OR CH_3CCL_3: Methyl Chloroform

CH3COOH OR CH_3COOH: Acetic Acid

CH4 OR CH_4: Methane

CHEMICAL OXYGEN DEMAND (COD): The quantity of oxygen required for the complete oxidation of organic chemical compounds in water; used as a measure of the level of organic pollutants in natural and waste waters.

CHP: Combined Heat and Power

CLAM: Chesapeake Lighthouse and Aircraft Measurements for Satellites

CLARIS: Europe-South America Network for Climate Change Assessment and Impact Studies

CLEAN DEVELOPMENT MECHANISM (CDM): Defined in Article 12 of the Kyoto Protocol, the CDM is intended to meet two objectives: (1) to assist parties not included in Annex I in achieving sustainable development and in contributing to the ultimate objective of the convention; and (2) to assist parties included in Annex I in achieving compliance with their quantified emission limitation and reduction commitments. Certified Emission Reduction Units from CDM projects undertaken in Non-Annex I countries that limit or reduce GHG emissions, when certified by operational entities designated by the Conference of the Parties/ Meeting of the Parties, can be accrued to the investor (government or industry) from parties in Annex B. A share of the proceeds from certified project activities is used to cover administrative expenses as well as to assist developing country parties that are particularly vulnerable to the adverse effects of climate change to meet the costs of adaptation.

CLIMATE CHANGE (CC): Climate change refers to a change in the state of the climate that can be identified (e.g. using statistical tests) by changes in the mean and/or the variability of its properties, and that persists for an extended period, typically decades or longer. Climate change may be due to natural internal processes or external forcings, or to persistent anthropogenic changes in the composition of the atmosphere or in land use. Note that UNFCCC, in its Article 1, defines "climate change" as "a change of climate which is attributed directly or indirectly to human activity that alters the composition of the global atmosphere and which is in addition to natural climate variability observed over comparable time periods." The UNFCCC thus makes a distinction between "climate change" attributable to human activities altering the atmospheric composition, and "climate variability" attributable to natural causes.

CLIMATE FEEDBACK: An interaction mechanism between processes in the climate system is a climate feedback when the result of an initial process triggers changes in secondary processes that in turn influence the initial one. A positive feedback intensifies the initial process; a negative feedback reduces the initial process. Example of a positive climate feedback: higher temperatures as initial process cause melting of the arctic ice leading to less reflection of solar radiation, what leads to higher temperatures. Example of a negative feedback: higher temperatures increase the amount of cloud cover (thickness or extent) that could reduce incoming solar radiation and so limit the increase in temperature.

CLIMATE SENSITIVITY: In IPCC reports: equilibrium climate sensitivity refers to the equilibrium change in annual mean global surface temperature following a doubling of the atmospheric CO2-equivalent concentration. The evaluation of the equilibrium climate sensitivity is expensive and often hampered by computational constraints. The effective climate sensitivity is a related measure that circumvents the computational problem by avoiding the requirement of equilibrium. It is evaluated from model output for evolving non-equilibrium conditions. It is a measure of the strengths of the feedbacks at a particular time and may vary with forcing history and climate state. The climate sensitivity parameter refers to the equilibrium change in the annual mean global surface temperature following a unit change in radiative forcing (K/W/m2). The transient climate response is the change in the global surface temperature, averaged over a 20-year period, centered at the time of CO2 doubling, i.e., at year 70 in a 1% per year compound CO2 increase experiment with a global coupled climate model. It is a measure of the strength and rapidity of the surface temperature response to green house gas forcing.

CLIMATE THRESHOLD: The point at which the atmospheric concentration of greenhouse gases triggers a significant climatic or environmental event, which is considered unalterable, such as widespread bleaching of corals or a collapse of oceanic circulation systems.

CMDL: Climate Monitoring and Diagnostics Laboratory (NOAA)

CMIP: Coupled Model Intercomparison Project

CNA: Central North America

CNR: Centre National de Recherches Météorologiques

CO: Carbon Monoxide

CO2 OR CO_2: Carbon Dioxide

CO_2-EQUIVALENT CONCENTRATION: The concentration of carbon dioxide that would cause the same amount of radiative forcing as a given mixture of carbon dioxide and other greenhouse gases.

CO_2-EQUIVALENT EMISSION: The amount of CO_2 emission that would cause the same radiative forcing as an emitted amount of a well mixed greenhouse gas, or a mixture of well mixed greenhouse gases, all multiplied with their respective Global Warming Potentials to take into account the differing times they remain in the atmosphere.

COADS: Comprehensive Ocean-Atmosphere Data Set

COARE: Coupled Ocean-Atmosphere Response Experiment

CO-BENEFITS: The benefits of policies implemented for various reasons at the same time, acknowledging that most policies designed to address greenhouse-gas mitigation have other, often at least equally important, rationales (e.g., related to objectives of development, sustainability, and equity). The term co-impact is also used in a more generic sense to cover both positive and negative side of the benefits.

COBE-SST: Centennial in-Situ Observation-Based Estimates of SSTs

COD: Chemical Oxygen Demand

CO-GENERATION: The use of waste heat from thermal electricity-generation plants. The heat is, for example, condensing heat from steam turbines or hot flue gases exhausted from gas turbines, for industrial use, buildings or district heating. Synonym for Combined Heat and Power (CHP) generation.

COMBINED-CYCLE GAS TURBINE (CCGT): Power plant that combines two processes for generating electricity. First, gas or light fuel oil feeds a gas turbine that inevitably exhausts hot flue gases (>800°C). Second, heat recovered from these gases, with additional firing, is the source for producing steam that drives a steam turbine. The turbines rotate separate alternators. It becomes an integrated CCGT when the fuel is syngas from a coal or biomass gasification reactor with exchange of energy flows between the gasification and CCGT plants.

COMPLIANCE: Compliance is whether and to what extent countries do adhere to the provisions of an accord. Compliance depends on implementing policies ordered, and on whether measures follow up the policies. Compliance is the degree to which the actors whose behavior is targeted by the agreement, local government units, corporations, organizations or individuals, conform to the implementing obligations.

CONCAWE: European Oil Company Organisation for Environment, Health, and Safety

CONFERENCE OF THE PARTIES (COP): The supreme body of the UNFCCC, comprising countries with right to vote that have ratified or acceded to the convention. The first session of the Conference of the Parties (COP-1) was held in Berlin (1995), followed by 2. Geneva (1996), 3. Kyoto (1997), 4. Buenos Aires (1998), 5. Bonn (1999), 6. The Hague/Bonn (2000, 2001), 7. Marrakech (2001), 8. Delhi (2002), 9. Milan (2003), 10 .Buenos Aires (2004), 11. Montreal (2005), and 12. Nairobi (2006).

CONTINGENT VALUATION METHOD (CVM): CVM is an approach to quantitatively assess values assigned by people in monetary (willingness to pay) and non monetary (willingness to contribute with time, resources etc.) terms. It is a direct method to estimate economic values for ecosystem and environmental services. A survey of people are asked their willingness to pay for access to, or their willingness to accept compensation for removal of, a specific environmental service, based on a hypothetical scenario and description of the environmental service.

COP: Conference of the Parties / Coefficient of Performance

COST: The consumption of resources such as labor time, capital, materials, fuels, and so on as the consequence of an action. In economics all resources are valued at their opportunity cost, being the value of the most valuable alternative use of the resources. Costs are defined in a variety of ways and under a variety of assumptions that affect their value. Cost types include: administrative costs of planning, management, monitoring, audits, accounting, reporting, clerical activities, etc. associated with a project or program; damage costs to ecosystems, economies and people due to negative effects from climate change; implementation costs of changing existing rules and regulation, capacity building efforts, information, training and education, etc. to put a policy into place; private costs are carried by individuals, companies or other private entities that undertake the action, where social costs include additionally the external costs on the environment and on society as a whole. Costs can be expressed as total, average (unit, specific) being the total divided by the number of units of the item for which the cost is being assessed, and marginal or incremental costs as the cost of the last additional unit. The negative of costs are benefits, and often both are considered together.

COST-BENEFIT ANALYSIS: Monetary measurement of all negative and positive impacts associated with a given action. Costs and benefits are compared in terms of their difference and/or ratio as an indicator of how a given investment or other policy effort pays off seen from the society's point of view.

COST-EFFECTIVENESS ANALYSIS: A special case of cost-benefit analysis in which all the costs of a portfolio of projects are assessed in relation to a fixed policy goal. The policy goal in this case represents the benefits of the projects and all the other impacts are measured as costs or as negative costs (co-benefits). The policy goal can be, for example, a specified goal of emissions reductions of greenhouse gases.

COWL: Cold Ocean-Warm Land

CPC: Climate Prediction Center (NOAA)

CREAS: Regional Climate Change Scenarios for South America

CREDITING PERIOD: The CDM crediting period is the time during which a project activity is able to generate GHG-emission reduction or CO_2 removal certificates. Under certain conditions, the crediting period can be renewed up to two times.

CRIEPI: Central Research Institute of Electric Power Industry

CRUTEM2V: Cru/Hadley Centre Gridded Land-Surface Air Temperature Version 2v

CRUTEM3: Cru/Hadley Centre Gridded Land-Surface Air Temperature Version 3

CSD: Commission for Sustainable Development

CSIRO: Commonwealth Scientific and Industrial Research Organization

CSP: Concentrating Solar Power

CTM: Chemical Transport Model

D

DAES: Domestic Alternative Energy Sources

DEMAND-SIDE MANAGEMENT (DSM): Policies and programs for influencing the demand for goods and/or services. In the energy sector, DSM aims at reducing the demand for electricity and energy sources. DSM helps to reduce greenhouse-gas emissions.

DEMATERIALIZATION: The process by which economic activity is decoupled from matter-energy throughput, through processes such as eco-efficient production or industrial ecology, allowing environmental impact to fall per unit of economic activity.

DEMETER: Development of a European Multimodel Ensemble System for Seasonal to Interannual Prediction

DEPOSIT-REFUND SYSTEM: A deposit or fee (tax) is paid when acquiring a commodity and a refund or rebate is received for implementation of a specified action (mostly delivering the commodity at a particular place).

DEVEGETATION: The loss of vegetation density within one land-cover class.

DEVELOPMENT PATH: An evolution based on an array of technological, economic, social, institutional, cultural, and biophysical characteristics that determine the interactions between human and natural systems, including production and consumption patterns in all countries, over time at a particular scale. Alternative development paths refer to different possible trajectories of development, the continuation of current trends being just one of the many paths.

DIC: Dissolved Inorganic Carbon

DJF: December, January, February

DLR: Deutsches Zentrum FÜr Luft- Und Raumfahrt

DMS: Dimethyl Sulphide

D-O: Dansgaard-Oeschger

DOC: Dissolved Organic Carbon

DORI: Determination D'orbite et Radiopositionnement IntÉgrÉs Par Satellite

DOUBLE DIVIDEND: The extent to which revenue-generating instruments, such as carbon taxes or auctioned (tradable) carbon emission permits can (1) limit or reduce GHG emissions and (2) offset at least part of the potential welfare losses of climate policies through recycling the revenue in the economy to reduce other taxes likely to cause distortions. In a world with involuntary unemployment, the climate change policy adopted may have an effect (a positive or negative "third dividend") on employment. Weak double dividend occurs as long as there is a revenue-recycling effect. That is, revenues are recycled through reductions in the marginal rates of distorting taxes. Strong double dividend requires that the (beneficial) revenue-recycling effect more than offsets the combination of the primary cost and in this case, the net cost of abatement is negative.

DSP: Dynamical Seasonal Prediction

DTR: Diurnal Temperature Range

DU: Dobson Unit

E

EAF: East Africa

EARLINET: European Aerosol Research Lidar Network

EAS: East Asia

EBMECMWF: Energy Balance Model European Centre for Medium Range Weather Forecasts

EBRD: European Bank for Reconstruction and Development

EC: European Commission

ECONOMIES IN TRANSITION (EITS): Countries with their economies changing from a planned economic system to a market economy.

ECONOMIES OF SCALE (SCALE ECONOMIES): The unit cost of an activity declines when the activity is extended (e.g., more units are produced).

ECOSYSTEM: A system of living organisms interacting with each other and their physical environment. The boundaries of what could be called an ecosystem are somewhat arbitrary, depending on the focus of interest or study. Thus, the extent of an ecosystem may range from very small spatial scales to the entire planet Earth ultimately.

ECS: Equilibrium Climate Sensitivity

EDGAR: Emission Database for Global Atmospheric Research

EEA: European Environmental Agency

EECCA: Countries of Eastern Europe, the Caucasus, and Central Asia

EIT: Economy in Transition

EMAS: ECO-Management and Audit Scheme

EMI: Earth System Model of Intermediate Complexity

EMISSION FACTOR: An emission factor is the rate of emission per unit of activity, output or input. For example, a particular fossil fuel power plant has a CO_2 emission factor of 0.765 kg/kWh generated.

EMISSION PERMIT: An emission permit is a non-transferable or tradable entitlement allocated by a government to a legal entity (company or other emitter) to emit a specified amount of a substance. A tradable permit is an economic policy instrument under which rights to discharge pollution—in this case an amount of greenhouse-gas emissions—can be exchanged through either a free or a controlled permit-market.

EMISSION QUOTA: The portion of total allowable emissions assigned to a country or group of countries within a framework of maximum total emissions.

EMISSION STANDARD: A level of emission that by law or by voluntary agreement may not be exceeded. Many standards use emission factors in their prescription and therefore do not impose absolute limits on the emissions.

EMISSION TRAJECTORIES: These are projections of future emission pathways, or observed emission patterns.

EMISSIONS DIRECT / INDIRECT: Direct emissions or "point of emission" are defined at the point in the energy chain where they are released and are attributed to that point in the energy chain, whether a sector, a technology or an activity. For example, emissions from coal-fired power plants are considered direct emissions from the energy supply sector. Indirect emissions or emissions "allocated to the end-use sector" refer to the energy use in end-use sectors and account for the emissions associated with the upstream production of the end-use energy. For instance, some emissions associated with electricity generation can be attributed to the buildings sector corresponding to the building sector's use of electricity.

EMISSIONS REDUCTION UNIT (ERU): Equal to one metric ton of CO_2-equivalent emissions reduced or sequestered arising from a Joint Implementation (defined in Article 6 of the Kyoto Protocol) project.

EMISSIONS TRADING: A market-based approach to achieving environmental objectives. It allows those reducing GHG emissions below their emission cap to use or trade the excess reductions to offset emissions at another source inside or outside the country. In general, trading can occur at the intra-company, domestic, and international levels. The Second Assessment Report by the IPCC adopted the convention of using permits for domestic trading systems and quotas for international trading systems. Emissions trading under Article 17 of the Kyoto Protocol is a tradable quota system based on the assigned amounts calculated from the emission reduction and limitation commitments listed in Annex B of the protocol.

ENA: Eastern North America

ENERGY: The amount of work or heat delivered. Energy is classified in a variety of types and becomes useful to human ends when it flows from one place to another or is converted from one type into another. Primary energy (also referred to as energy sources) is the energy embodied in natural resources (e.g., coal, crude oil, natural gas, uranium) that has not undergone any anthropogenic conversion. It is transformed into secondary energy by cleaning (natural gas), refining (oil in oil products), or by conversion into electricity or heat. When the secondary energy is delivered at the end-use facilities, it is called final energy (e.g., electricity at the wall outlet), where it becomes usable energy (e.g., light). Daily, the sun supplies large quantities of energy as rainfall, winds, radiation, etc. Some share is stored in biomass or rivers that can be harvested by people. Some share is directly usable such as daylight, ventilation, or ambient heat. Renewable energy is obtained from the continuing or repetitive currents of energy occurring in the natural environment and includes non-carbon technologies such as solar energy, hydropower, wind, tide and waves, and geothermal heat, as well as carbon-neutral technologies such as biomass. Embodied energy is the energy used to produce a material substance (such as processed metals or building materials), taking into account energy used at the manufacturing facility (zero order), energy used in producing the materials that are used in the manufacturing facility (first order), and so on.

ENERGY EFFICIENCY: The ratio of useful energy output of a system, conversion process or activity to its energy input.

ENERGY INTENSITY: The ratio of energy use to economic output. At the national level, energy intensity is the ratio of total domestic primary energy use or final energy use to Gross Domestic Product.

ENERGY SECURITY: The various security measures that a given nation, or the global community as a whole, must carry out to maintain an adequate energy supply.

ENERGY SERVICE COMPANY (ESCO): A company that offers energy services to end-users, guarantees the energy savings to be achieved tying them directly to its remuneration, as well as finances or assists in acquiring financing for the operation of the energy system, and retains an on-going role in monitoring the savings over the financing term.

ENHANCED GREENHOUSE EFFECT: Greenhouse gases effectively absorb infrared radiation, emitted by Earth's surface, by the atmosphere itself due to the same gases, and by clouds. Atmospheric radiation is emitted to all sides, including downward to Earth's surface. Thus, greenhouse gases trap heat within the surface-troposphere system. This is called the greenhouse effect. Thermal infrared radiation in the troposphere is strongly coupled to the temperature at the altitude at which it is emitted. In the troposphere, the temperature generally decreases with height. Effectively, infrared radiation emitted to space originates from an altitude with a temperature of, on average, −19°C, in balance with the net incoming solar radiation, whereas Earth's surface is kept at a much higher temperature of, on average, +14°C. An increase in the concentration of greenhouse gases leads to an increased infrared opacity of the atmosphere and therefore to an effective radiation into space from a higher altitude at a lower temperature. This causes a radiative forcing that leads to an enhancement of the greenhouse effect; the so-called enhanced greenhouse effect.

ENSO: El Niño-Southern Oscillation

ENVIRONMENTAL EFFECTIVENESS: The extent to which a measure, policy, or instrument produces a decided, decisive, or desired environmental effect.

ENVIRONMENTALLY SUSTAINABLE TECHNOLOGIES: Technologies that are less polluting, use resources in a more sustainable manner, recycle more of their wastes and products, and handle residual wastes in a more acceptable manner than the technologies that they substitute. They are also more compatible with nationally determined socio-economic, cultural, and environmental priorities.

EOF: Empirical Orthogonal Function

EOS: Earth Observing System

EPICA: European Programme for Ice Coring in Antarctica

EPRI: Electric Power Research Institute

ERBE: Earth Radiation Budget Experiment

ERBS: Earth Radiation Budget Satellite

ERS: European Remote Sensing Satellite

ESCO: Energy Service Company

ESRL: Earth System Research Library (NOAA)

ESTOC: European Station for Time-Series in the Ocean

ETS: Emission Trading Scheme

EU: European Union

EUROCS: European Cloud Systems

EVIDENCE: Information or signs indicating whether a belief or proposition is true or valid. In the IPCC Report, the degree of evidence reflects the amount of scientific/technical information on which the lead authors are basing their findings.

EXTERNALITY / EXTERNAL COST / EXTERNAL BENEFIT: Externalities arise from a human activity, when agents responsible for the activity do not take full account of the activity's impact on others' production and consumption possibilities, while there exists no compensation for such impact. When the impact is negative, so are external costs. When positive they are referred to as external benefits.

F

FACE: Free Air Co2 Enrichment

FAO: Food & Agriculture Organization

FAO FAR: Food and Agriculture Organization (UN) First Assessment Report

FC: Fuel Cell

FCCC: Framework Convention on Climate Change (UN)

FEED-IN TARIFF: The price per unit of electricity that a utility or power supplier has to pay for distributed or renewable electricity fed into the grid by non-utility generators. A public authority regulates the tariff.

FLARING: Open air burning of waste gases and volatile liquids, through a chimney, at oil wells or rigs, in refineries or chemical plants, and at landfills.

FOB: Free On Board

FOD: First-Order Decay

FORECAST: Projected outcome from established physical, technological, economic, social, behavioral, etc. patterns.

FOREST: Defined under the Kyoto Protocol as a minimum area of land of 0.05-1.0 ha with tree-crown cover (or equivalent stocking level) of more than 10-30% with trees with the potential to reach a minimum height of 2-5 m at maturity in situ. A forest may consist either of closed forest formations where trees of various storey and undergrowth cover a high proportion of the ground or of open forest. Young natural stands and all plantations that have yet to reach a crown density of 10-30% or tree height of 2-5 m are included under forest, as are areas normally forming part of the forest area that are temporarily un-stocked as a result of human intervention such as harvesting or natural causes but which are expected to revert to forest.

FRCGC: Frontier Research Center for Global Change

FRSGC: Frontier Research System for Global Change

FREE RIDER: One who benefits from a common good without contributing to its creation or preservation.

FUEL SWITCHING: In general, this is substituting fuel A for fuel B. In the climate-change discussion it is implicit that fuel A has lower carbon content than fuel B, e.g., natural gas for coal.

FULL-COST PRICING: Setting the final prices of goods and services to include both the private costs of inputs and the external costs created by their production and use.

G

GAGE: Global Atmospheric Gases Experiment

GARP: Global Atmospheric Research Program

GATE: Garp Atlantic Tropical Experiment

GATT: General Agreement on Trade and Tariffs

GAW: Global Atmosphere Watch

GCM: General Circulation Model or Global Circulation Model

GCOS: Global Climate Observing System

GCSS: Gewex Cloud System Study

GDP: Gross Domestic Product

GEF: Global Environment Facility

GEIA: Global Emissions Inventory Activity

GENERAL CIRCULATION (CLIMATE) MODEL (GCM): A numerical representation of the climate system based on the physical, chemical, and biological properties of its components, their interactions and feedback processes, and accounting for all or some of its known properties. The climate system can be represented by models of varying complexity, i.e. for any one component or combination of components a hierarchy of models can be identified, differing in such aspects as the number of spatial dimensions, the extent to which physical, chemical, or biological processes are explicitly represented, or the level at which the parameters are assessed empirically. Coupled atmosphere/ocean/sea-ice General Circulation Models provide a comprehensive representation of the climate system. There is an evolution towards more complex models with active chemistry and biology.

GENERAL EQUILIBRIUM ANALYSIS: General equilibrium analysis considers simultaneously all the markets and feedback effects among these markets in an economy leading to market clearance.

GEO-ENGINEERING: Technological efforts to stabilize the climate system by direct intervention in the energy balance of Earth for reducing global warming.

GEOS: Goddard Earth Observing System

GEWEX: Global Energy and Water Cycle Experiment

GFDL: Geophysical Fluid Dynamics Laboratory

GHCN: Global Historical Climatology Network

GHG: Greenhouse Gas, also written as Green House Gas in some reports. The preferred use is Greenhouse Gas.

GIA: Glacial Isostatic Adjustment

GIN Sea: Greenland-Iceland-Norwegian Sea

GISP2: Greenland Ice Sheet Project 2

GISS: Goddard Institute for Space Studies

GLACE: Global Land Atmosphere Coupling Experiment

GLAMAP: Glacial Ocean Mapping

GLAS: Geoscience Laser Altimeter System

GLOBAL ENVIRONMENTAL FACILITY (GEF): The Global Environment Facility (GEF), established in 1991, helps developing countries fund projects and programmes that protect the global environment. GEF grants support projects related to biodiversity, climate change, international waters, land degradation, the ozone layer, and persistent organic pollutants.

GLOBAL WARMING: Global warming refers to the gradual increase, observed or projected, in global surface temperature, as one of the consequences of radiative forcing caused by anthropogenic emissions.

GLOBAL WARMING POTENTIAL (GWP): An index, based upon radiative properties of well-mixed greenhouse gases, measuring the radiative forcing of a unit mass of a given well-mixed greenhouse gas in today's atmosphere integrated over a chosen time horizon, relative to that of CO_2. The GWP represents the combined effect of the differing lengths of time that these gases remain in the atmosphere and their relative effectiveness in absorbing outgoing infrared radiation. The Kyoto Protocol is based on GWPs from pulse emissions over a 100-year time frame.

GLODAP: Global Ocean Data Analysis Project

GLOSS: Global Sea Level Observing System

GMD: Global Monitoring Division (NOAA)

GMT: Global Mean Temperature GNP

GNP: Gross National Product

GOME: Global Ozone Monitoring Experiment

GOVERNANCE: The way government is understood has changed in response to social, economic, and technological changes over recent decades. There is a corresponding shift from government defined strictly by the nation-state to a more inclusive concept of governance, recognizing the contributions of various levels of government (global, international, regional, local) and the roles of the private sector, of non-governmental actors, and of civil society.

GPCC: Global Precipitation Climatology Centre

GPCP: Global Precipitation Climatology Project

GPS: Global Positioning System

GRACE: Gravity Recovery and Climate Experiment

GREEN ACCOUNTING: Attempts to integrate into macroeconomic studies a broader set of social welfare measures covering social, environmental, and development-oriented policy aspects. Green accounting includes both monetary valuations that attempt to calculate a 'green national product' with the economic damage by pollutants subtracted from the national product, and accounting systems that include quantitative non-monetary pollution, depletion, and other data.

GREENHOUSE EFFECT: Greenhouse gases effectively absorb infrared radiation, emitted by Earth's surface, by the atmosphere itself due to the same gases and by clouds. Atmospheric radiation is emitted to all sides, including downward to Earth's surface. Thus, greenhouse gases trap heat within the surface-troposphere system. This is called the greenhouse effect. Thermal infrared radiation in the troposphere is strongly coupled to the temperature at the altitude at which it is emitted. In the troposphere, the temperature generally decreases with height. Effectively, infrared radiation emitted to space originates from an altitude with a temperature of, on average, −19°C, in balance with the net incoming solar radiation, whereas Earth's surface is kept at a much higher temperature of, on average, +14°C. An increase in the concentration of greenhouse gases leads to an increased infrared opacity of the atmosphere and therefore to an effective radiation into space from a higher altitude at a lower temperature. This causes a radiative forcing that leads to an enhancement of the greenhouse effect; the so-called enhanced greenhouse effect.

GREENHOUSE GASES (GHGS): Greenhouse gases are those gaseous constituents of the atmosphere, both natural and anthropogenic, that absorb and emit radiation at specific wavelengths within the spectrum of infrared radiation emitted by Earth's surface, the atmosphere and clouds. This property causes the greenhouse effect. Water vapour (H_2O), carbon dioxide (CO_2), nitrous oxide (N_2O), methane (CH_4) and ozone (O_3) are the primary greenhouse gases in Earth's atmosphere. Moreover, there are a number of entirely human-made greenhouse gases in the atmosphere, such as the halocarbons and other chlorine- and bromine-containing substances, dealt with under the Montreal Protocol. Besides carbon dioxide, nitrous oxide, and methane, the Kyoto Protocol deals with the greenhouse gases sulphur hexafluoride, hydrofluorocarbons, and perfluorocarbons.

GRIP: Greenland Ice Core Project

GROSS NATIONAL PRODUCT (GNP): GNP is a measure of national income. It measures value added from domestic and foreign sources claimed by residents. GNP comprises Gross Domestic Product plus net receipts of primary income from non-resident income.

GROSS WORLD PRODUCT: An aggregation of each individual country's Gross Domestic Products to obtain the sum for the world.

GROUP OF 77 AND CHINA (G77/CHINA): Originally 77, now more than 130, developing countries that act as a major negotiating bloc in the UNFCCC process. Also referred to as Non-Annex I countries in the context of the UNFCCC.

GSA: Great Salinity Anomaly

Gt: Gigatonne (109 Tonnes)

GWE: Global Weather Experiment

GWP: Global Warming Potential

H

H / H2 OR H_2: Hydrogen (element / gas)

H2: Molecular Hydrogen

H2O OR H_2O: Water

HadAT: Hadley Centre Atmospheric Temperature Data Set

HadISST: Hadley Centre Sea Ice And Sea Surface Temperature Data Set

HadMA: Hadley Centre Marine Air Temperature Data Set

HadR: Hadley Centre Radiosonde Temperature Data Set

HADRT: Hadley Centre Radiosonde Temperature Data Set

HadSLP2: Hadley Centre MSLP Data Set Version 2

HadSST2: Hadley Centre SST Data Set Version 2

HALOE: Halogen Occultation Experiment

HCFC: Hydrochlorofluorocarbon

HCO_3^-: Bicarbonate

HDI: Human Development Index

HFC: Hydrofluorocarbon

HIRS: High Resolution Infrared Radiation Sounder

HLM: High Latitude Mode

HNO2 OR HNO_2: Nitrous Acid

HNO3 OR HNO_3: Nitric Acid

HO2 OR HO_2: Hydroperoxyl Radical

HOT: Hawaii Ocean Time-Series

HOT AIR: Under the terms of the 1997 Kyoto Protocol, national emission targets in Annex B are expressed relative to emissions in the year 1990. For countries in the former Soviet Union and Eastern Europe this target has proven to be higher than their current and projected emissions for reasons unrelated to climate-change mitigation activities. Russia and Ukraine, in particular, are expected to have a substantial volume of excess emission allowances over the period 2008–2012 relative to their forecast emissions. These allowances are sometimes referred to as hot air because, while they can be traded under the Kyoto Protocol's flexibility mechanisms, they did not result from mitigation activities.

hPa: Hectopascal

HSS: High Strength Steels

HVAC: Heating, Ventilation and Air Conditioning

HYBRID VEHICLE: Any vehicle that employs two sources of propulsion, especially a vehicle that combines an internal combustion engine with an electric motor.

HYDE: History Database of the Environment

HYDROFLUOROCARBONS (HFCs): One of the six gases or groups of gases to be curbed under the Kyoto Protocol. They are produced commercially as a substitute for chlorofluorocarbons. HFCs are largely used in refrigeration and semiconductor manufacturing. Their Global Warming Potentials range from 1,300 to 11,700.

I

IA: Integrated Assessment

IABP: International Arctic Buoy Programme

IAEA: International Atomic Energy Agency

IAES: Imported Alternative Energy Sources

ICESat: Ice, Cloud and Land Elevation Satellite

ICOAD: International Comprehensive Ocean-Atmosphere Data Set

ICST: Imperial College of Science, Technology and Medicine

IDP: Integrated Design Process (for buildings)

IET: International Emission Trading

IGBP: International Geosphere-Biosphere Programme

IGBP-DIS: Igbp Data and Information System

IGO: Intergovernmental Organization

IGRA: Integrated Global Radiosonde Archive

IIASA: International Institute for Applied System Analysis

IMO: Generally stands for the International Meteorological Organization, but also stands for International Maritime Organization in some reports and usage.

IMPLEMENTATION: Implementation describes the actions taken to meet commitments under a treaty and encompasses legal and effective phases. Legal implementation refers to legislation, regulations, judicial decrees, including other actions such as efforts to administer progress, which governments take to translate international accords into domestic law and policy. Effective implementation needs policies and programmes that induce changes in the behaviour and decisions of target groups. Target groups then take effective measures of mitigation and adaptation.

INCOME ELASTICITY (OF DEMAND): This is the ratio of the percentage change in quantity of demand for a good or service to a one percentage change in income. For most goods and services, demand goes up when income grows, making income elasticity positive. When the elasticity is less than one, goods and services are called necessities.

IND: Indian Ocean

INDOEX: Indian Ocean Experiment

INDUSTRIAL ECOLOGY: The relationship of a particular industry within its environment. It often refers to the conscious planning of industrial processes to minimize their negative externalities (e.g., by heat and materials cascading).

INERTIA: In the context of climate-change mitigation, inertia relates to the difficulty of change resulting from pre-existing conditions within society such as physical man-made capital, natural capital, and social non-physical capital, including institutions, regulations and norms. Existing structures lock in societies, making change more difficult.

InSAR: Interferometric Synthetic Aperture Radar

INTEGRATED ASSESSMENT: A method of analysis that combines results and models from the physical, biological, economic, and social sciences and the interactions between these components in a consistent framework to evaluate the status and the consequences of environmental change and the policy responses to it.

INTEGRATED DESIGN PROCESS (IDP) OF BUILDINGS: Optimizing the orientation and shape of buildings and providing high-performance envelopes for minimizing heating and cooling loads. Passive techniques for heat transfer control, ventilation, and daylight access reduce energy loads further. Properly sized and controlled, efficient mechanical systems address the left-over loads. IDP requires an iterative design process involving all the major stakeholders from building users to equipment suppliers, and can achieve 30-75% savings in energy use in new buildings at little or no additional investment cost.

INTELLIGENT CONTROLS: In this report, the notion of 'intelligent control' refers to the application of information technology in buildings to control heating, ventilation, air-conditioning, and electricity use effectively. It requires effective monitoring of parameters such as temperature, convection, moisture, etc., with appropriate control measurements (smart metering).

INTERACTION EFFECT: The consequence of the interaction of climate-change policy instruments with existing domestic tax systems, including both cost-increasing tax interaction and cost-reducing revenue-recycling effect. The former reflects the impact that greenhouse gas policies can have on labor and capital markets through their effects on real wages and the real return to capital. Restricting allowable GHG emissions raises the carbon price and so the costs of production and the prices of output, thus reducing the real return to labor and capital. With policies that raise revenue for the government, such as carbon taxes and auctioned permits, the revenues can be recycled to reduce existing distortional taxes.

INTERGOVERNMENTAL ORGANIZATION (IGO): Organizations constituted of governments. Examples include the World Bank, the Organization of Economic Co-operation and Development (OECD), the International Civil Aviation Organization (ICAO), the Intergovernmental Panel on Climate Change (IPCC), and other UN and regional organizations. The Climate Convention allows accreditation of these IGOs to attend negotiating sessions.

INTERNATIONAL ENERGY AGENCY (IEA): Established in 1974, the agency is linked with the OECD. It enables OECD member countries to take joint measures to meet oil supply emergencies, to share energy information, to coordinate their energy policies, and to cooperate in developing rational energy use programs.

IO: Iodine Monoxide

IOCI: Indian Ocean Climate Initiative

IOD: Indian Ocean Dipole

IOZM: Indian Ocean Zonal Mode

IPAB: International Programme for Antarctic Buoys

IPCC: Intergovernmental Panel on Climate Change

IPO: Inter-Decadal Pacific Oscillation

IPSL: Institut Pierre Simon Laplace

IS92: IPCC Scenarios 1992

ISCC: International Satellite Cloud Climatology Project

IT: Information Technology

ITC: Induced Technological Change

ITCZ: Inter-Tropical Convergence Zone

ITER: International Thermonuclear Experimental Reactor

IUCN: International Union for the Conservation of Nature and Natural Resources

J

J: Joule = Newton x meter (International Standard unit of energy)

JAMSTEC: Japan Marine Science and Technology Center

JI: Joint Implementation

JJA: June, July, August

JMA: Japan Meteorological Agency

JOINT IMPLEMENTATION (JI): A market-based implementation mechanism defined in Article 6 of the Kyoto Protocol, allowing Annex I countries or companies from these countries to implement projects jointly that limit or reduce emissions or enhance sinks and to share the Emissions Reduction Units. JI activity is also permitted in Article 4.2(a) of the UNFCCC. See also Activities Implemented Jointly and Kyoto Mechanisms.

JRC: Joint Research Centre (EU) L

K

ka: Kilo-annum (1,000 years ago)

KMA: Korea Meteorological Administration

KNMI: Royal Netherlands Meteorological Institute

KYOTO MECHANISMS (ALSO CALLED FLEXIBILITY MECHANISMS): Economic mechanisms based on market principles that parties to the Kyoto Protocol can use in an attempt to lessen the potential economic impacts of greenhouse gas emission-reduction requirements. They include Joint Implementation (Article 6), Clean Development Mechanism (Article 12), and Emissions trading (Article 17).

kyr: Kilo-year (1,000 years). Preferred term is ka, for kilo-annum.

L

LANDFILL: A landfill is a solid waste disposal site where waste is deposited below, at, or above ground level. Limited to engineered sites with cover materials, controlled placement of waste and management of liquids and gases. It excludes uncontrolled waste disposal.

LAND-USE: The total of arrangements, activities, and inputs undertaken in a certain land-cover type (a set of human actions). The social and economic purposes for which land is managed (e.g., grazing, timber extraction, and conservation). Land-use change occurs when, e.g., forest is converted to agricultural land or to urban areas.

LASG: National Key Laboratory of Numerical Modeling for Atmospheric Sciences and Geophysical Fluid Dynamics

LBA: Large-Scale Biosphere-Atmosphere Experiment in Amazonia

LBC LBL: Lateral Boundary Condition Line-By-Line

LCA: Life Cycle Assessment LHV

LEAPFROGGING: The ability of developing countries to bypass intermediate technologies and jump straight to advanced clean technologies. Leapfrogging can enable developing countries to move to a low-emissions development trajectory.

LEARNING BY DOING: As researchers and firms gain familiarity with a new technological process, or acquire experience through expanded production, they can discover ways to improve processes and reduce cost. Learning by doing is a type of experience-based technological change.

LEVELIZED COST PRICE: The unique price of the outputs of a project that makes the present value of the revenues (benefits) equal to the present value of the costs over the lifetime of the project.

LGM: Last Glacial Maximum

LIG: Last Interglacial

LKS: Lanzante-Klein-Seidel

LLGHG: Long-Lived Greenhouse Gas

LLJ: Low-Level Jet

LLNL: Lawrence Livermore National Laboratory

LMD: Laboratoire de Météorologie Dynamique

LOA LOSU: Laboratoire D'optique Atmospherique Level of Scientific Understanding

LOCK-IN EFFECT: Technologies that cover large market shares continue to be used due to factors such as sunk investment costs, related infrastructure development, use of complementary technologies, and associated social and institutional habits and structures.

LOW-CARBON TECHNOLOGY: A technology that over its life cycle causes less CO_2-eq. emissions than other technological options do.

LSCE: Laboratoire des Sciences du Climat et de L'environnement

LSM: Land Surface Model

LSW: Labrador Sea Water

LW: Longwave

LWP: Liquid Water Path

M

Ma: Megaannum (1,000,000 years ago)

MAC: Marginal Abatement Cost

MACROECONOMIC COSTS: These costs are usually measured as changes in Gross Domestic Product or changes in the growth of Gross Domestic Product, or as loss of welfare or consumption.

MAM: March, April, May

MARGINAL COST PRICING: The pricing of goods and services such that the price equals the additional cost arising when production is expanded by one unit. Economic theory shows that this way of pricing maximizes social welfare in a first-best economy.

MARGO mb: Multiproxy Approach for the Reconstruction of the Glacial Ocean Surface

MARKET BARRIERS: In the context of climate change mitigation, market barriers are conditions that prevent or impede the diffusion of cost-effective technologies or practices that would mitigate GHG emissions.

MARKET DISTORTIONS AND IMPERFECTIONS: In practice, markets will always exhibit distortions and imperfections such as lack of information, distorted price signals, lack of competition, and/or institutional failures related to regulation, inadequate delineation of property rights, distortion-inducing fiscal systems, and limited financial markets.

MARKET EQUILIBRIUM: The point at which the demand for goods and services equals the supply; often described in terms of price levels, determined in a competitive market, clearing the market.

MARKET EXCHANGE RATE (MER): This is the rate at which foreign currencies are exchanged. Most economies post such rates daily and they vary little across all the exchanges. For some developing economies official rates and black-market rates may differ significantly and the MER is difficult to pin down.

MARKET-BASED REGULATION: Regulatory approaches using price mechanisms (e.g., taxes and auctioned tradable permits), among other instruments, to reduce GHG emissions.

MATERIAL EFFICIENCY OPTIONS: Options to reduce GHG emissions by decreasing the volume of materials needed for a certain product or service.

mb: Millibar

MBT: Mechanical Biological Treatment

MDG: Millennium Development Goals

MDI: Michelson Doppler Imager

MEA: Multilateral Environmental Agreements

MEASURES: Measures are technologies, processes, and practices that reduce GHG emissions or effects below anticipated future levels. Examples of measures are renewable energy technologies, waste minimization processes, and public transport commuting practices, etc.

MED: Mediterrranean Basin

MEETING OF THE PARTIES (TO THE KYOTO PROTOCOL) (MOP): The Conference of the Parties (COP) of the UNFCCC serves as the Meeting of the Parties (MOP), the supreme body of the Kyoto Protocol, since the latter entered into force on 16 February 2005. Only parties to the Kyoto Protocol may participate in deliberations and make decisions.

METEOSAT: European Geostationary Meteorological Satellite

METHANE (CH4): Methane is one of the six greenhouse gases to be mitigated under the Kyoto Protocol. It is the major component of natural gas and associated with all hydrocarbon fuels, animal husbandry, and agriculture. Coal-bed methane is the gas found in coal seams.

METHANE RECOVERY: Methane emissions, e.g., from oil or gas wells, coal beds, peat bogs, gas transmission pipelines, landfills, or anaerobic digesters, are captured and used as a fuel or for some other economic purpose (e.g., chemical feedstock).

MFR: Maximum Feasible Reduction

MHT: Meridional Heat Transport

MILLENNIUM DEVELOPMENT GOALS (MDG): A set of time-bound and measurable goals for combating poverty, hunger, disease, illiteracy, discrimination against women, and environmental degradation, agreed at the UN Millennium Summit in 2000.

MINOS: Mediterranean Intensive Oxidants Study

MIP: Model Intercomparison Project

MIRAGE: Megacity Impacts on Regional and Global Environments

MISO: Monsoon Intra-Seasonal Oscillation

MISR: Multi-Angle Imaging Spectro-Radiometer

MITIGATION: Technological changes and substitutions that reduce resource inputs and emissions per unit of output. Although several social, economic and technological policies would produce an emission reduction with respect to climate change, mitigation means implementing policies to reduce GHG emissions and enhance sinks.

MITIGATIVE CAPACITY: This is a country's ability to reduce anthropogenic GHG emissions or to enhance natural sinks, where ability refers to skills, competencies, fitness, and proficiencies that a country has attained, and depends on technology, institutions, wealth, equity, infrastructure and information. Mitigative capacity is rooted in a country's sustainable development path.

MJO: Madden-Julian Oscillation

MLS: Microwave Limb Sounder

MMD: Multi-Model Data Set (At Pcmdi)

MOC: Meridional Overturning Circulation

MODIS: Moderate Resolution Imaging Spectrometer

mol: Mole

MONEX: Monsoon Experiment

MONTREAL PROTOCOL: The Montreal Protocol on Substances that Deplete the Ozone Layer was adopted in Montreal in 1987, and subsequently adjusted and amended in London (1990), Copenhagen (1992), Vienna (1995), Montreal (1997), and Beijing (1999). It controls the consumption and production of chlorine- and bromine-containing chemicals that destroy stratospheric ozone, such as chlorofluorocarbons, methyl chloroform, carbon tetrachloride, and many others.

MOPITT: Measurements of Pollution in the Tropospheret

MOZAIC: Measurement of Ozone by Airbus In-Service Aircraft

MPI: Max Planck Institute

MPIC: Max Planck Institute for Chemistry

MPLNET: Micro-Pulse Lidar Network

MRI: Meteorological Research Institute

MSLP: Mean Sea Level Pressure

MSU: Microwave Sounding Unit

MULTI-ATTRIBUTE ANALYSIS: Integrates different decision parameters and values in a quantitative analysis without assigning monetary values to all parameters. Multi-attribute analysis can combine quantitative and qualitative information.

MULTI-GAS: Along with CO_2, the other greenhouse gases (methane, nitrous oxide and fluorinated gases) are taken into account in achieving reduction of emissions (multi-gas reduction) or stabilization of concentrations (multi-gas stabilization).

Myr: Million Years

N

N / N2 OR N_2: Nitrogen element / gas

N_2O_5: Dinitrogen Pentoxide

NADW: North Atlantic Deep Water

NAFTA: North American Free Trade Agreement

NAH: North Atlantic subtropical High

NAM: Northern Annular Mode

NAMS: North American Monsoon System

NAO: North Atlantic Oscillation

NARCCAP: North American Regional Climate Change Assessment Program

NAS: Northern Asia

NASA: National Aeronautics and Space Administration

NATIONAL ACTION PLANS: Plans submitted to the COP by parties outlining the steps that they have adopted to limit their anthropogenic GHG emissions. Countries must submit these plans as a condition of participating in the UNFCCC and, subsequently, must communicate their progress to the COP regularly. The National Action Plans form part of the National Communications, which include the national inventory of GHG sources and sinks.

NAU: North Australia

NCAR: National Center for Atmospheric Research

NCDC: National Climatic Data Center

NCE: National Centers for Environmental Prediction

NEAQS NEP: New England Air Quality Study Net Ecosystem Production

NEDC: New European Driving Cycle NGO

NESDIS: National Environmental Satellite, Data and Information Service

NET ANTHROPOGENIC GREENHOUSE GAS REMOVALS BY SINKS: For CDM afforestation and reforestation projects, "net anthropogenic GHG removals by sinks" equals the actual net GHG removals by sinks minus the baseline net GHG removals by sinks minus leakage.

NEU: Northern Europe

NF_3: Nitrogen trifluoride

NGO: Non-Governmental Organization

NGRIP: North Greenland Ice Core Project

NH: Northern Hemisphere

NH3 OR NH_3: Ammonia

NH4+ OR NH_{4+}: Ammonium ion

NIES: National Institute for Environmental Studies

NITROUS OXIDE (N_2O): One of the six types of greenhouse gases to be curbed under the Kyoto Protocol.

NIWA: National Institute of Water and Atmospheric Research

NMAT: Nighttime Marine Air Temperature

NMHC/NMVOC: Non-Methane Hydrocarbon/Methane Volatile Organic Compound

NMVOC: Non-Methane Volatile Organic Compounds

NO: Nitric Oxide

NO_X: The sum of NO and NO_2 expressed in NO_2 mass equivalent

NO2 OR NO_2: Nitrogen dioxide

NO2NO3 OR NO_2NO_3: Nitrogen Dioxide Nitrate Radical

NOANOX: Nitrogen Oxides (the sum of NO and NO_2)

NON-ANNEX I COUNTRIES/PARTIES: The countries that have ratified or acceded to the UNFCCC but are not included in Annex I.

NON-ANNEX B COUNTRIES/PARTIES: The countries not included in Annex B of the Kyoto Protocol.

NO-REGRET POLICY (OPTIONS / POTENTIAL): Such policy would generate net social benefits whether or not there is climate change associated with anthropogenic emissions of greenhouse gases. No-regret options for GHG emissions reduction refer to options whose benefits (such as reduced energy costs and reduced emissions of local/regional pollutants) equal or exceed their costs to society, excluding the benefits of avoided climate change.

NORMATIVE ANALYSIS: Economic analysis in which judgments about the desirability of various policies are made. The conclusions rest on value judgments as well as on facts and theories.

NPA: North Pacific Ocean

NPI: North Pacific Index

NPIW: North Pacific Intermediate Water

NPP: Net Primary Productivity

O

O / O2 OR O_2: Oxygen (element / gas) O_3

O2: Molecular Oxygen

O3 OR O_3: Ozone

O&M: Operation and Maintenance

OASIS: Ocean Atmosphere Sea Ice Soil

OCTS: Ocean Colour and Temperature Scanner

ODA: Official Development Assistance

ODS: Ozone Depleting Substances

OECD: Organization for Economic Co-operation and Development

OGCM: Ocean General Circulation Model

OH: Hydroxyl Radical

OIL SANDS AND OIL SHALE: Unconsolidated porous sands, sandstone rock and shales containing bituminous material that can be mined and converted to a liquid fuel.

OIO: Iodine Dioxide

OLR: Outgoing Longwave Radiation

OPEC: Organization of Petroleum Exporting Countries

OPPORTUNITIES: Ozone, the tri-atomic form of oxygen, is a gaseous atmospheric constituent. In the troposphere, ozone is created both naturally and by photochemical reactions involving gases resulting from human activities. Troposphere ozone acts as a greenhouse gas. In the stratosphere, ozone is created by the interaction between solar ultraviolet radiation and molecular oxygen (O2). Stratospheric ozone plays a dominant role in the stratospheric radiative balance. Its concentration is highest in the ozone layer.

P

P: Phosphorus

PARETO CRITERION: A criterion testing whether an individual's welfare can be increased without making others in the society worse off. A Pareto improvement occurs when an individual's welfare is improved without making the welfare of the rest of society worse off. A Pareto optimum is reached when no one's welfare can be increased without making the welfare of the rest of society worse off, given a particular distribution of income. Different income distributions lead to different Pareto optima.

PASSIVE SOLAR DESIGN: Structural design and construction techniques that enable a building to utilize solar energy for heating, cooling, and lighting by non-mechanical means.

PCMDI: Program For Climate Model Diagnosis And Intercomparison

pCO_2 PDF: Partial Pressure of CO_2 Probability Density Function

PDI: Power Dissipation Index

PDO: Pacific Decadal Oscillation

PDSI: Palmer Drought Severity Index

PET: Potential Evapotranspiration

PETM: Palaeocene-Eocene Thermal Maximum

PFC: Perfluorocarbon

Pg: Petagram (1015 Grams)

PMIP: Paleoclimate Modelling Intercomparison Project

PMO: Physikalisch-Meteorologisches Observatorium Davos

PNA: Pacific-North American Pattern

PNNLPNV POA: Pacific Northwest National Laboratory Potential Natural Vegetationprimary Organic Aerosol

POLICIES: In UNFCCC parlance, policies are taken and/or mandated by a government - often in conjunction with business and industry within its own country, or with other countries - to accelerate mitigation and adaptation measures. Examples of policies are carbon or other energy taxes, fuel efficiency standards for automobiles, etc. Common and coordinated or harmonized policies refer to those adopted jointly by parties.

POM: Particulate Organic Matter

PORTFOLIO ANALYSIS: Deals with a portfolio of assets or policies that are characterized by different risks and pay-offs. The objective function is built up around the variability of returns and their risks, leading up to the decision rule to choose the portfolio with highest expected return.

POST-CONSUMER WASTE: Waste from consumption activities, e.g. packaging materials, paper, glass, rests from fruits and vegetables, etc.

POTENTIAL: In the context of climate change, potential is the amount of mitigation or adaptation that could be - but is not yet - realized over time.

ppb: Parts Per Billion

ppm: Parts Per Million

PR: Precipitation Radar

PREC/: Precipitation Reconstruction Over Land (Prec/L)

PRECAUTIONARY PRINCIPLE: A provision under Article 3 of the UNFCCC, stipulating that the parties should take precautionary measures to anticipate, prevent or minimize the causes of climate change and mitigate its adverse effects. Where there are threats of serious or irreversible damage, lack of full scientific certainty should not be used as a reason to postpone such measures, taking into account that policies and measures to deal with climate change should be cost-effective in order to ensure global benefits at the lowest possible cost.

PRECURSORS: Atmospheric compounds which themselves are not greenhouse gases or aerosols, but which have an effect on greenhouse gas or aerosol concentrations by taking part in physical or chemical processes regulating their production or destruction rates.

PRE-INDUSTRIAL: The era before the industrial revolution of the late 18th and 19th centuries, after which the use of fossil fuel for mechanization started to increase.

PRESENT VALUE: The value of a money amount differs when the amount is available at different moments in time (years). To make amounts at differing times comparable and additive, a date is fixed as the 'present.' Amounts available at different dates in the future are discounted back to a present value, and summed to get the present value of a series of future cash flows. Net present value is the difference between the present value of the revenues (benefits) with the present value of the costs.

PRICE ELASTICITY OF DEMAND: The ratio of the percentage change in the quantity of demand for a good or service to one percentage change in the price of that good or service. When the absolute value of the elasticity is between 0 and 1, demand is called inelastic; when it is greater than one, demand is called elastic.

PRIMARY MARKET TRADING: In commodities and financial exchanges, buyers and sellers who trade directly with each other constitute the 'primary market,' while buying and selling through exchange facilities represent the 'secondary market.'

PRODUCTION FRONTIER: The maximum outputs attainable with the optimal uses of available inputs (natural resources, labor, capital, information).

PROVOS: Prediction Of Climate Variations On Seasonal To Interannual Time Scales

PRP: Partial Radiative Perturbation

PSA: Pacific-South American Pattern

PSC: Polar Stratospheric Cloud

PSMSL: Permanent Service For Mean Sea Level

PSU psu: Pennsylvania State University Practical Salinity Unit

PURCHASING POWER PARITY (PPP): The purchasing power of a currency is expressed using a basket of goods and services that can be bought with a given amount in the home country. International comparison of, e.g., Gross Domestic Products of countries can be based on the purchasing power of currencies rather than on current exchange rates. PPP estimates tend to lower per capita GDPs in industrialized countries and raise per capita GDPs in developing countries. (PPP is also an acronym for polluter-pays-principle).

PV: Photovoltaïc

Q

QBO: Quasi-Biennial Oscillation

QELRCs: Quantified Emission Limitation or Reduction Commitments

R

RATPAC: Radiosonde Atmospheric Temperature Products For Assessing Climate

RCM: Regional Climate Model

RD&D: Research, Development and Demonstration

REA REML: Reliability Ensemble Average Restricted Maximum Likelihood

REBOUND EFFECT: After implementation of efficient technologies and practices, part of the savings is taken back for more intensive or other consumption, e.g., improvements in car-engine efficiency lower the cost per kilometer driven, encouraging more car trips or the purchase of a more powerful vehicle.

REFORESTATION: Direct human-induced conversion of non-forested land to forested land through planting, seeding and/or the human-induced promotion of natural seed sources, on land that was previously forested but converted to non-forested land. For the first commitment period of the Kyoto Protocol, reforestation activities will be limited to reforestation occurring on those lands that did not contain forest on 31 December 1989.

RF: Radiative Forcing

RSL: Relative Sea Level

RSS: Remote Sensing Systems

RTMIP: Radiative-Transfer Model Intercomparison Project

S

SA: Southern Annular Mode or Stratospheric Aerosol Measurement

SACZ: South Atlantic Convergence Zone

SAF: South Africa

SAFARI: Southern African Regional Science Initiative

SAFE LANDING APPROACH: See tolerable windows approach.

SAGE: Stratospheric Aerosol And Gas Experiment Or Centre For Sustainability And The Global Environment

SAH: Sahara

SAMS: South American Monsoon System

SAMW: Subantarctic Mode Water

SAR: Second Assessment Report Or Synthetic Aperture Radar with regard to IPCC reports this also refers to the IPCC Second Assessment Report

SARB: Surface And Atmosphere Radiation Budget

SARR: Space Absolute Radiometric Reference

SAS: South Asia

SAT SCA: Surface Air Temperature Snow-Covered Area

SAU: South Australia

SBSTA: Subsidiary Body for Scientific and Technological Advice

SCC: Social Cost of Carbon

SCENARIO: A plausible description of how the future may develop based on a coherent and internally consistent set of assumptions about key driving forces (e.g., rate of technological change, prices) and relationships. Note that scenarios are neither predictions nor forecasts, but are useful to provide a view of the implications of developments and actions.

SCIAMACHY: Scanning Imaging Absorption Spectrometer for Atmospheric Chartography

SCM: Simple Climate Model

SD: Sustainable Development

SEA: Southeast Asia

SeaWiFs: Sea-Viewing Wide Field-of-View Sensor

SECONDARY MARKET TRADING: In commodities and financial exchanges, buyers and sellers who trade directly with each other constitute the 'primary market,' while buying and selling through exchange facilities represent the 'secondary market.'

SEM: Southern Europe And Mediterranean

SEQUESTRATION: Carbon storage in terrestrial or marine reservoirs. Biological sequestration includes direct removal of CO_2 from the atmosphere through land-use change, afforestation, reforestation, carbon storage in landfills and practices that enhance soil carbon in agriculture.

SF6 OR SF_6: Sulphur hexafluoride

SH: Southern Hemisphere

SHADOW PRICING: Setting prices of goods and services that are not, or incompletely, priced by market forces or by administrative regulation, at the height of their social marginal value. This technique is used in cost-benefit analysis.

SINKS: Any process, activity or mechanism that removes a greenhouse gas or aerosol, or a precursor of a greenhouse gas or aerosol from the atmosphere.

SIO: Scripps Institution of Oceanography

SIS: Small Island States

SLE: Sea Level Equivalent

SLP: Sea Level Pressure

SMB: Surface Mass Balance

SMEs: Small and Medium Enterprises

SMM: Solar Maximum Mission

SMMR: Scanning Multichannel Microwave Radiometer

SO: Southern Oscillation

SO2 OR SO_2: Sulphur dioxide

SO4 OR SO_4: Sulphate

SOA: Secondary Organic Aerosol

SOCIAL COST OF CARBON (SCC): The discounted monetized sum (e.g. expressed as a price of carbon in $/t$CO_2$) of the annual net losses from impacts triggered by an additional ton of carbon emitted today. According to usage in economic theory, the social cost of carbon establishes an economically optimal price of carbon at which the associated marginal costs of mitigation would equal the marginal benefits of mitigation.

SOCIAL UNIT COSTS OF MITIGATION: Carbon prices in US$/t$CO_2$ and US$/tC-eq (as affected by mitigation policies and using social discount rates) required to achieve a particular level of mitigation (economic potential) in the form of a reduction below a baseline for GHG emissions. The reduction is usually associated with a policy target, such as a cap in an emissions trading scheme or a given level of stabilization of GHG concentrations in the atmosphere.

SOHO: Solar Heliospheric Observatory

SOI: Southern Oscillation Index

SOM: Soil Organic Matter

SON: September, October, November

SORCE: Solar Radiation And Climate Experiment

SOURCE: Source mostly refers to any process, activity or mechanism that releases a greenhouse gas, aerosol or a precursor of a greenhouse gas or aerosol into the atmosphere. Source can also refer to, e.g., an energy source.

SOx: Sulphur oxides expressed in SO_2 mass equivalent

SPA: South Pacific Ocean

SPARC: Stratospheric Processes And Their Role In Climate

SPCZ: South Pacific Convergence Zone

SPECIFIC ENERGY USE: The energy used in the production of a unit material, product or service.

SPILL-OVER EFFECT: The effects of domestic or sector mitigation measures on other countries or sectors. Spill-over effects can be positive or negative and include effects on trade, carbon leakage, transfer of innovations, and diffusion of environmentally sound technology and other issues.

SPM: With regard to IPCC reports this refers to the Summary For Policymakers (also written as Summary for Policy Makers)

SRALT: Satellite Radar Altimetry

SRCCS: Special Report on Carbon Capture and Storage (IPCC)

SRES: Special Report on Emission Scenarios

SRLULUCF: Special Report on Land-Use, Land-Use Change and Forestry (IPCC)

SRTT: Special Report on methodological and technological issues in Technology Transfer (IPCC)

SSA: Southern South America

SSM/I: Special Sensor Microwave/Imager

SST: Sea Surface Temperature

STABILIZATION: Keeping constant the atmospheric concentrations of one or more GHG (e.g., CO_2) or of a CO_2-equivalent basket of GHG. Stabilization analyses or scenarios address the stabilization of the concentration of GHG in the atmosphere.

STANDARDS: Set of rules or codes mandating or defining product performance (e.g., grades, dimensions, characteristics, test methods, and rules for use). Product, technology or performance standards establish minimum requirements for affected products or technologies. Standards impose reductions in GHG emissions associated with the manufacture or use of the products and/or application of the technology.

STARDEX: Statistical and Regional Dynamical Downscaling of Extremes for European Regions

STE: Stratosphere-Troposphere Exchange

STMW: Subtropical Mode Water

STORYLINE: A narrative description of a scenario (or a family of scenarios) that highlights the scenario's main characteristics, relationships between key driving forces, and the dynamics of the scenarios.

STRUCTURAL CHANGE: Changes, for example, in the relative share of Gross Domestic Product produced by the industrial, agricultural, or services sectors of an economy; or more generally, systems transformations whereby some components are either replaced or potentially substituted by other ones.

SUBSIDY: Direct payment from the government or a tax reduction to a private party for implementing a practice the government wishes to encourage. The reduction of GHG emissions is stimulated by lowering existing subsidies that have the effect of raising emissions (such as subsidies to fossil fuel use) or by providing subsidies for practices that reduce emissions or enhance sinks (e.g. for insulation of buildings or for planting trees).

SULPHUR HEXAFLUORIDE (SF6): One of the six greenhouse gases to be curbed under the Kyoto Protocol. It is largely used in heavy industry to insulate high-voltage equipment and to assist in the manufacturing of cable-cooling systems and semi-conductors. Its Global Warming Potential is 23,900.00.

SUNY: State University of New York

SUPPLEMENTARITY: The Kyoto Protocol states that emissions trading and Joint Implementation activities are to be supplemental to domestic policies (e.g. energy taxes, fuel efficiency standards) taken by developed countries to reduce their GHG emissions. Under some proposed definitions of supplementarity (e.g., a concrete ceiling on level of use), developed countries could be restricted in their use of the Kyoto Mechanisms to achieve their reduction targets. This is a subject for further negotiation and clarification by the parties.

SUSTAINABLE DEVELOPMENT (SD): The concept of sustainable development was introduced in the World Conservation Strategy (IUCN 1980) and had its roots in the concept of a sustainable society and in the management of renewable resources. Adopted by the WCED in 1987 and by the Rio Conference in 1992 as a process of change in which the exploitation of resources, the direction of investments, the orientation of technological development and institutional change are all in harmony and enhance both current and future potential to meet human needs and aspirations. SD integrates the political, social, economic and environmental dimensions.

SWH: Significant Wave Height

T

T/P: Topex/Poseidon

T12: Hirs Channel 12

T2: Msu Channel 2

T2LT: Msu Lower-Troposphere Channel

T3: Msu Channel 3

T4: Msu Channel 4

TAR: Third Assessment Report (IPCC)

TARFOX: Tropospheric Aerosol Radiative Forcing Experiment

TARGETS AND TIMETABLES: A target is the reduction of a specific percentage of GHG emissions from a baseline date (e.g., below 1990 levels) to be achieved by a set date or timetable (e.g., 2008–2012). Under the Kyoto Protocol the EU agreed to reduce its GHG emissions by 8% below 1990 levels by the 2008–2012 commitment period. Targets and timetables are an emissions cap on the total amount of GHG emissions that can be emitted by a country or region in a given time period.

TAX: A carbon tax is a levy on the carbon content of fossil fuels. Because virtually all of the carbon in fossil fuels is ultimately emitted as CO_2, a carbon tax is equivalent to an emission tax on each unit of CO_2-equivalent emissions. An energy tax - a levy on the energy content of fuels - reduces demand for energy and so reduces CO_2 emissions from fossil fuel use. An eco-tax is designed to influence human behavior (specifically economic behavior) to follow an ecologically benign path. An international carbon/emission/energy tax is a tax imposed on specified sources in participating countries by an international authority. The revenue is distributed or used as specified by this authority or by participating countries.

A harmonized tax commits participating countries to impose a tax at a common rate on the same sources, because imposing different rates across countries would not be cost-effective. A tax credit is a reduction of tax in order to stimulate purchasing of or investment in a certain product, like GHG emission reducing technologies. A carbon charge is the same as a carbon tax.

TBO: Tropospheric Biennial Oscillation

TCR: Transient Climate Response

TEAP: Technology and Economic Assessment Panel

TECHNOLOGICAL CHANGE: Mostly considered as technological improvement, i.e., more or better goods and services can be provided from a given amount of resources (production factors). Economic models distinguish autonomous (exogenous), endogenous and induced technological change.

TECHNOLOGY: The practical application of knowledge to achieve particular tasks that employs both technical artifacts (hardware, equipment) and (social) information ('software,' know-how for production and use of artifacts).

TECHNOLOGY TRANSFER: The exchange of knowledge, hardware and associated software, money and goods among stakeholders, which leads to the spreading of technology for adaptation or mitigation. The term encompasses both diffusion of technologies and technological cooperation across and within countries.

TGBM: Tide Gauge Bench Mark

TGICA: Task Group on Data and Scenario Support for Impact and Climate Analysis (IPCC)

THC: Thermohaline Circulation

THIR: Temperature Humidity Infrared Radiometer

TIB: Tibetan Plateau

TIM: Total Solar Irradiance Monitor

TIROS: Television Infrared Observation Satellite

TMI: Trmm Microwave Imager

TNE: Tropical Northeast Atlantic

TOA: Top of the Atmosphere

TOGA: Tropical Ocean Global Atmosphere

TOLERABLE WINDOWS APPROACH (TWA): This approach seeks to identify the set of all climate-protection strategies that are simultaneously compatible with 1) prescribed long-term climate-protection goals, and 2) normative restrictions on the emissions mitigation burden. The constraints may include limits on the magnitude and rate of global mean temperature change, on the weakening of the thermohaline circulation, on ecosystem losses and on economic welfare losses resulting from selected climate damages, adaptation costs and mitigation efforts. For a given set of constraints, and given a solution exists, the TWA delineates an emission corridor of complying emission paths.

TOM: Top of the Model

TOMS: Total Ozone Mapping Spectrometer

TOP-DOWN MODELS: Models applying macroeconomic theory, econometric and optimization techniques to aggregate economic variables. Using historical data on consumption, prices, incomes, and factor costs, top-down models assess final demand for goods and services, and supply from main sectors, such as the energy sector, transportation, agriculture, and industry. Some top-down models incorporate technology data, narrowing the gap to bottom-up models.

TOPEX: Topography Experiment

TOVS: Tiros Operational Vertical Sounder

TPES: Total Primary Energy Supply

TRACE GAS: A minor constituent of the atmosphere, next to nitrogen and oxygen that together make up 99% of all volume. The most important trace gases contributing to the greenhouse effect are carbon dioxide, ozone, methane, nitrous oxide, perfluorocarbons, chlorofluorocarbons, hydrofluorocarbons, sulphur hexafluoride and water vapor.

TransCom: Atmospheric Tracer Transport Model Intercomparison Project

TRMM: Tropical Rainfall Measuring Mission

TSI: Total Solar Irradiance

U

UAH: University of Alabama in Huntsville

UARS: Upper Atmosphere Research Satellite

UCDW: Upper Circumpolar Deep Water

UCI: University of California at Irvine

UEA: University of East Anglia

UHI: Urban Heat Island

UIO: University of Oslo

UK: United Kingdom

UKMO: United Kingdom Meteorological Office

ULAQ: University of L'aquila

UMD: University of Maryland

UMI: University of Michigan

UN: United Nations

UNCED: UN Conference on Environment and Development

UNCERTAINTY: An expression of the degree to which a value is unknown (e.g. the future state of the climate system). Uncertainty can result from lack of information or from disagreement about what is known or even knowable. It may have many types of sources, from quantifiable errors in the data to ambiguously defined concepts or terminology, or uncertain projections of human behavior. Uncertainty can therefore be represented by quantitative measures (e.g., a range of values calculated by various models) or by qualitative statements (e.g., reflecting the judgment of a team of experts).

UNDP: UN Development Programme

UNEP: United Nations Environment Programme

UNFCCC: United Nations Framework Convention on Climate Change

UNITED NATIONS FRAMEWORK CONVENTION ON CLIMATE CHANGE (UNFCCC): United Nations Framework Convention on Climate Change (UNFCCC). The Convention was adopted on 9 May 1992 in New York and signed at the 1992 Earth Summit in Rio de Janeiro by more than 150 countries and the European Economic Community. Its ultimate objective is the 'stabilization of greenhouse gas concentrations in the atmosphere at a level that would prevent dangerous anthropogenic interference with the climate system.' It contains commitments for all parties. Under the Convention parties included in Annex I aimed to return greenhouse gas emission not controlled by the Montreal Protocol to 1990 levels by the year 2000. The convention came into force in March 1994.

USA: United States of America

USHCN: U.S. Historical Climatology Network

UTC: Coordinated Universal Time

UTRH: Upper-Tropospheric Relative Humidity

UV: Ultraviolet

UVIC: University of Victoria

V

VALUE ADDED: The net output of a sector or activity after adding up all outputs and subtracting intermediate inputs.

VALUES: Worth, desirability or utility based on individual preferences. Most social science disciplines use several definitions of value. Related to nature and environment, there is a distinction between intrinsic and instrumental values, the latter assigned by humans. Within instrumental values, there is an unsettled catalogue of different values, such as (direct and indirect) use, option, conservation, serendipity, bequest, existence, etc. Mainstream economics define the total value of any resource as the sum of the values of the different individuals involved in the use of the resource. The economic values, which are the foundation of the estimation of costs, are measured in terms of the willingness to pay by individuals to receive the resource or by the willingness of individuals to accept payment to part with the resource.

VAT: Value Added Tax

VIRGO: Variability of Irradiance and Gravity Oscillations

VIRS: Visible Infrared Scanner

VOCs: Volatile Organic Compounds

VOLUNTARY ACTION: Informal programs, self-commitments and declarations, where the parties (individual companies or groups of companies) entering into the action set their own targets and often do their own monitoring and reporting.

VOLUNTARY AGREEMENT: An agreement between a government authority and one or more private parties to achieve environmental objectives or to improve environmental performance beyond compliance to regulated obligations. Not all voluntary agreements are truly voluntary; some include rewards and/or penalties associated with joining or achieving commitments.

VOS: Voluntary Observing Ships

VRGCM: Variable-Resolution General Circulation Model

W

W: Watt = Joule/second (International Standard unit of power)

WAF: West Africa

WAIS: West Antarctic Ice Sheet

WCRP: World Climate Research Programme

WDCGG: World Data Centre for Greenhouse Gases

WEC: World Energy Council

WGI: IPCC Working Group I

WGII: IPCC Working Group II

WGIII: IPCC Working Group III

WGMS: World Glacier Monitoring Service

WHO: World Health Organization

WMDW: Western Mediterranean Deep Water

WMO: World Meteorological Organization

WNA: Western North America

WOCE: World Ocean Circulation Experiment

WTO: World Trade Organization

WWF: World Wide Fund for nature

WWR: World Weather Records

Z

ZIA: 0°C Isotherm Altitude

Chronology

A chronology of events related to climate change studies in context with other scientific, environmental, and news milestones.

1827 French mathematician Joseph Fourier (1768–1830) suggests that atmospheric gases could raise the temperature of Earth's surface, creating a greenhouse-like effect.

1859 Irish physicist John Tyndall (1820–1893) discovers that some atmospheric gases block infrared radiation, and that changing the concentration of these gases could bring about climate change. Tyndall later describes how water vapor is the most efficient gas in maintaining the surface temperature of the planet, as it both absorbs heat from the atmosphere and inhibits heat from radiating back into space, in effect, describing the greenhouse effect.

1864 George Perkins Marsh (1801–1882) publishes *Man and Nature.*

1864 Yosemite in California becomes the first state park in the United States.

1869 Russian chemist Dmitri Ivanovich Mendeleyev (1834–1907) and German chemist Julius Lothar Meyer (1830–1895) independently put forth the Periodic Table of Elements, which arranges the elements in order of atomic weights. However, Meyer does not publish until 1870, nor does he predict the existence of undiscovered elements as does Mendeleyev.

1869 Ernst Heinrich Haeckel (1834–1919) coins the term ecology to describe "the body of knowledge concerning the economy of nature."

1871 Charles Darwin (1809–1882) published *The Descent of Man, and Selection in Relation to Sex.* This work introduces the concept of sexual selection and expands his theory of evolution to include humans.

1872 Yellowstone in Wyoming becomes the first national park in the United States.

1873 James Clerk Maxwell (1831–1879), Scottish mathematician and physicist, publishes *Treatise on Electricity and Magnetism* in which he identifies light as an electromagnetic phenomenon. Maxwell determines this when he finds his mathematical calculations for the transmission speed of both electromagnetic and electrostatic waves are the same as the known speed of light. This landmark work brings together the three main fields of physics—electricity, magnetism, and light.

1873 Dutch physicist Johannes Diderik van der Waals (1837–1923) offers an equation for the gas laws that contains terms relating to the volumes of the molecules themselves and the attractive forces between them. It becomes known as the Van der Waals equation.

1875 American Forestry Association founded to encourage wise forest management.

1879 U.S. Geological Survey established.

1880 German chemist Carl Oswald Viktor Engler (1842–1925) begins his studies on petroleum. He is the first to state that it is organic in origin.

1883 Frank Wigglesworth Clarke (1847–1931), American chemist and geophysicist, is appointed chief chemist to the U.S. Geological Survey. In this position, he begins an extensive program of rock analysis and is one of the founders of geochemistry.

1883 Danish chemist Johan Gustav Kjeldahl (1849–1900) devises a method for the analysis of the nitrogen content of organic material. His method uses concentrated sulfuric acid and is simple and fast.

1886 French metallurgist Paul Louis Toussaint Héroult (1863–1914) and American chemist Charles Martin Hall (1863–1914) independently invent an electrochemical process for extracting aluminum from its ore. This process makes aluminum cheaper and forms the basis of the huge aluminum industry. Hall makes the discovery in February of this year, and Héroult achieves his in April.

1887 German-born American chemist Herman Frasch (1851–1914) patents a method for removing sulfur compounds from oil. Once the foul sulfur smell is removed through the use of metallic compounds, petroleum becomes a marketable product.

1890 Yosemite becomes the third U.S. national park.

1892 John Muir (1838–1914) founds the Sierra Club to preserve the Sierra Nevada mountain chain.

1894 British physicist John William Strutt Rayleigh (1842–1919) and British chemist William Ramsay (1852–1916) succeed in isolating a new gas in the atmosphere that is denser than nitrogen and combines with no other element. They name it "argon," which is Greek for inert. It is the first of a series of rare gases with unusual properties whose existence had not been predicted.

1896 Swedish chemist Svante Arrhenius (1859–1927) calculates that carbon dioxide changes in the atmosphere from industry could result in the warming of Earth's surface temperature.

1897 American geologist Thomas C. Chamberlin (1843–1928) produces a model showing global carbon exchange and concludes (independently of Svante Arrhenius) that changes in carbon dioxide in the atmosphere can lead to changes in the temperature of Earth's surface.

1900 French physicist Paul Ulrich Villard (1860–1934) discovers what are later called gamma rays. While studying the recently discovered radiation from uranium, he finds that in addition to the alpha rays and beta rays, there are other rays, unaffected by magnets, that are similar to x rays, but shorter and more penetrating.

1902 Oliver Heaviside (1850–1925), British physicist and electrical engineer, and Arthur E. Kennelly (1861–1939), British-American electrical engineer, independently and almost simultaneously make the first prediction of the existence of the ionosphere, an electrically conductive layer in the upper atmosphere that reflects radio waves. They theorize correctly that wireless telegraphy works over long distances because a conducting layer of atmosphere exists that allows radio waves to follow Earth's curvature instead of traveling off into space.

1903 British chemists William Ramsay and Frederick Soddy (1877–1956) discover that helium is continually produced by naturally radioactive substances.

1904 British physicist Ernest Rutherford (1871–1937) postulates the age of Earth by radioactive dating.

1904 British chemist William Ramsay receives the Nobel Prize in Chemistry for the discovery of the inert gaseous elements in air and for determining their place in the periodic system.

1905 German-born American physicist Albert Einstein (1879–1955) develops a quantum theory of light, which explains the photoelectric effect. He suggests that light has a dual, wave-particle quality.

1907 French physicist Pierre Weiss (1865–1940) offers his theory explaining the phenomenon of ferromagnetism. He states that iron and other ferromagnetic materials form small *domains* of a certain polarity pointing in various directions. When some external magnetic field forces them to be aligned, they become a single, strong magnetic force.

1908 German geophysicist Alfred Wegener (1880–1930) proposes the theory of continental drift.

1908 British physicist Ernest Rutherford and German physicist Hans Wilhelm Geiger (1882–1945) develop an electrical alpha-particle counter. Over the next few years, Geiger continues to improve this device, which becomes known as the Geiger counter.

1908 British physicist Ernest Rutherford is awarded the Nobel Prize in Chemistry for his investigations into disintegration of the elements and the chemistry of radioactive substances.

1908 Tunguska event occurs when a comet or asteroid enters the atmosphere, causing major damage to a forested region in Siberia.

1911 British geologist Arthur Holmes (1890–1965) publishes the first geological time scale with dates based on radioactive measurements.

1911 British physicist Ernest Rutherford discovers that atoms are made up of a positive nucleus surrounded by electrons. This modern concept of the atom replaces the notion of featureless, indivisible spheres that dominated atomistic thinking for 23 centuries—since the time of Greek mathematician Democritus (c. 460–c. 370).

1911 American physicist Victor Hess (1883–1964) identifies high altitude radiation from space.

1912 German chemist Friedrich Bergius (1884–1949) discovers how to treat coal and oil with hydrogen to produce gasoline.

1913 French physicist Charles Fabry (1867–1945) first demonstrates the presence of ozone in the upper atmosphere. It is found later that ozone functions as a screen, preventing much of the sun's ultraviolet radiation from reaching Earth's surface. Seventy-five years after the discovery of ozone, in 1985, a hole in the ozone layer over Antarctica is discovered via satellite.

1913 Congress approves construction of the Hetch-Hetchy Valley dam in Yosemite National Park to provide water to San Francisco; however, the dam also floods areas of the park.

1913 Danish physicist Niels Bohr (1885–1962) proposes the first dynamic model of the atom. It is seen as a very dense nucleus surrounded by electrons rotating in orbitals (defined energy levels).

1914 British physicist Ernest Rutherford discovers a positively charged particle he calls a proton.

1915 German chemist Richard Martin Willstätter (1872–1942) is awarded the Nobel Prize in Chemistry for his research on plant pigments, especially chlorophyll.

1927 American geneticist Hermann Joseph Muller (1890–1967) induced artificial mutations in fruit flies by exposing them to X-rays. His work proved that mutations result from some type of physical-chemical change. Muller wrote extensively about the danger of excessive X-rays and the burden of deleterious mutations in human populations.

1928 Russian-American physicist George Gamow (1904–1968) develops the quantum theory of radioactivity, which is the first theory to successfully explain the behavior of radioactive elements, some of which decay in seconds and others after thousands of years.

1933 Dust Bowl conditions due to extended drought in United States exacerbate depression era economic and environmental woes.

1933 Tennessee Valley Authority created to assess impact of hydropower on the environment.

1934 American chemist Arnold O. Beckman invents the pH meter, which uses electricity to accurately measure a solution's acidity or alkalinity.

1936 National Wildlife Federation established.

1938 British engineer Guy Stewart Callendar (1898–1964) revives scientific interest in the role of carbon dioxide in climate change with the publication of his paper "The Artificial Production of Carbon Dioxide and Its Influence on Temperature." The paper states that global warming is already underway due to changes in the atmosphere resulting from burning fossil fuels. Climate changes brought on by fossil fuel combustion become known as the Callendar effect.

1939 American chemist Linus Pauling (1901–1994) publishes *The Nature of the Chemical Bond*, a classic work that becomes one of the most influential chemical texts of the twentieth century.

1943 First operational nuclear reactor is activated at the Oak Ridge National Laboratory in Oak Ridge, Tennessee.

1947 American chemist Willard Libby (1908–1980) introduces Carbon-14 dating.

1949 American conservationist Aldo Leopold (1186–1948) publishes *A Sand County Almanac*, in which he sets guidelines for the conservation movement and introduces the concept of a land ethic.

1950 The World Meteorological Organization (WMO) is established.

1952 Oregon becomes the first state to adopt a significant program to control air pollution.

1953 American biologists James D. Watson (1928–) and Francis H. C. Crick (1916–2004) publish two landmark papers in the journal *Nature*: "Molecular structure of nucleic acids: a structure for deoxyribose nucleic acid" and "Genetical implications of the structure of deoxyribonucleic acid." Watson and Crick propose a double helical model for DNA and call attention to the genetic implications of their model. Their model is based, in part, on the X-ray crystallographic work of Rosalind Elsie Franklin (1920–1958) and the biochemical work of Erwin Chargaff (1905–2002), and explains how the genetic material is transmitted.

1953 American chemist Stanley L. Miller (1930–2007) produces amino acids from inorganic compounds, similar to those in primitive atmosphere, with electrical sparks that simulate lightning.

1954 American chemist Linus Pauling receives the Nobel Prize in Chemistry for his research into the nature of the chemical bond and its applications to the elucidation of the structure of complex substances.

1956 American geologists William Maurice Ewing (1906–1974) and William L. Donn (1918–1987) explain rapid climate change, basing their hypothesis on a feedback mechanism between ice ages and changes in ocean circulation patterns. They predicted a new ice age within a few centuries. Although the hypothesis was soon proved wrong, it did stimulate scientists to picture the world in the grip of a dramatic climate change.

1957 Soviet Union launches Earth's first artificial satellite, Sputnik, into earth orbit.

1957 The International Geophysical Year (1957–1958), sponsored by the International Council of Scientific Unions, brings about increased funding for climate studies.

1957 American oceanographer Roger Revelle (1909–1991) warns that mankind is waging a "large-scale geophysical experiment" by continuingly releasing greenhouse gases into the environment. His colleague, American geochemist Charles David Keeling (1928–2005), begins the first continuous monitoring of atmospheric carbon dioxide levels at the Mauna Loa Observatory in Hawaii. Keeling observes a regular rise in the level of carbon dioxide.

1958 National Aeronautics and Space Administration (NASA) established.

1958 A greenhouse effect on the planet Venus is observed after telescopic studies show an elevated temperature of the atmosphere of Venus.

1961 American mathematician Edward Lorenz (1917–) advances chaos theory and offers possible implications on atmospheric dynamics and weather.

1962 *Silent Spring* is published by American biologist Rachel Carson (1907–1964) to document the effects of pesticides on the environment.

1963 First Clean Air Act passed in the United States.

1963 British geophysicists Frederick J. Vine (1939–) and Drummond Hoyle Matthews (1931–1997) offer important proof of plate tectonics by discovering that oceanic crust rock layers show equidistant bands of magnetic orientation on the sea floor.

1963 Nuclear Test Ban Treaty signed by the United States and the Soviet Union to stop atmospheric testing of nuclear weapons.

1965 American mathematician Edward Lorenz points out the chaotic nature of the climate system—possible sudden shifts may be due to sensitive dependency on initial conditions. This theory becomes known as the "Butterfly Effect."

1967 Environmental Defense Fund established.

1970 Environmental Protection Agency (EPA) created.

1970 First Earth Day celebrated on April 22.

1970 National Environmental Policy Act passes, requiring environmental impact statements for projects funded or regulated by the federal government.

1971 Greenpeace founded.

1972 Coastal Zone Management Act and Marine Protection, Research, and Sanctuaries Act passed.

1972 *Limits to Growth* published by the Club of Rome, calling for population control.

1972 United Nations Conference on the Human Environment is held in Stockholm, Sweden, to address environmental issues on a global level.

1972 The United Nations Environment Programme (UNEP) is created as a body to reflect the environmental conscience of the United Nations.

1973 British economist E.F. Schumacher (1911–1977) publishes *Small Is Beautiful: Economics as if People Mattered*, which advocates simplicity, self-reliance, and living in harmony with nature.

1976 Resource Conservation and Recovery Act passed, giving EPA authority to regulate municipal solid and hazardous waste.

1980 Comprehensive Environmental Response, Compensation, and Liability Act (Superfund) enacted to clean up abandoned toxic waste sites.

1980 *Global 2000 Report* published, documenting trends in population growth, natural resource depletion, and the environment.

1980 Mount St. Helens in Washington state explodes with a force comparable to 500 Hiroshima-sized bombs.

1980 American biologist Thomas Eugene Lovejoy (1941–) proposes the idea of the debt-for-nature swap, which helps developing countries alleviate national debt by implementing policies to protect the environment.

1984 Ozone hole over Antarctica discovered.

1985 Rainforest Action Network founded.

1985 International conference on the greenhouse effect at Villach, Austria. The panel concludes with a warning that greenhouse gases could lead to an increase in global mean temperature that is "greater than any in man's history."

1986 Chernobyl Nuclear Power Station undergoes nuclear core melt-down, spreading radioactive material over vast parts of the Soviet Union and northern Europe.

1987 Montreal Protocol on Substances that Deplete the Ozone Layer is signed by 24 nations, declaring their promise to decrease production of chlorofluorocarbons (CFCs).

1987 *Our Common Future* (The Brundtland Report) is published.

1987 World population reaches five billion.

1987 First in an almost two-decade-long string of frequent "warmest year" records. In fact, by 1990, seven of the ten warmest records ever recorded are attributed to the 1980s. This string of warmth begins to generate media interest.

1988 Global Relief program inaugurated with the motto "Plant a tree, cool the globe" to address the problem of global warming.

1988 Congressional hearings offer scientific testimony that ongoing U.S. drought is linked to global warming.

1988 Intergovernmental Panel on Climate Change (IPCC) established.

1989 Oil tanker *Exxon Valdez* runs aground in Prince William Sound, Alaska, spilling 11 million gallons of oil.

1989 American environmental activist Bill McKibben (1960–) publishes *The End of Nature.* It was the first book for a popular audience that proclaimed human beings were changing the global climate and called for reduced emissions of greenhouse gases.

1989 Global Climate Coalition (funded by energy interests) claims that scientific evidence of climate change is inconclusive to warrant changes that may adversely impact national economies—climate science is too uncertain to justify action.

1990 Clean Air Act amended to control emissions of sulfur dioxide and nitrogen oxides.

1990 Alliance of Small Island States (AOSIS), a lobbying and negotiating group that aims to raise awareness about the consequences of climate change, is formed.

1990 Oil Pollution Act is signed, setting liability and penalty system for oil spills as well as a trust fund for clean-up efforts.

1990 First Intergovernmental Panel on Climate Change (IPCC) report is issued. The IPCC asserts that the climate is warming and that the trend is expected to continue. Many of the warmest year records set in the 1980s are broken in the 1990s.

1990 The Global Change Research Act is passed in the United States.

1991 Mount Pinatubo in the Philippines erupts, shooting sulfur dioxide 25 mi (40 km) into the atmosphere.

1991 Persian Gulf War begins. During the war, Iraqi president Saddam Hussein (1937–2006) burns Iraqi oil fields, spewing dark smoke over the Gulf region.

1992 Mexico City, Mexico, suffers general shutdown as a result of incapacitating air pollution.

1992 United Nations Earth Summit is held in Rio de Janeiro, Brazil. The group issues the key document United Nations Framework Convention on Climate Change (UNFCCC), which calls for reductions in greenhouse gas emissions and is signed by 154 countries.

1993 Evidence that climate change can be both long-term and short-term (within decades) is found in Greenland ice core samples.

1994 Fearing inundation by potentially rising seas levels, the Alliance of Small Island States calls for deep cuts in greenhouse gases by industrialized countries by the year 2005.

1995 Nigerian activist Ken Saro-Wiwa (1941–1995) is executed in his homeland for protesting against oil industry practices in his country.

1995 Dutch meteorologist Paul Crutzen (1933–), American chemist Mario Molina (1943–), and American atmospheric chemist F. Sherwood Rowland (1927–) receive the Nobel Prize in Chemistry for their work in atmospheric chemistry, particularly concerning the formation and decomposition of ozone.

1995 Second IPCC report asserts that there is scientific evidence that global warming is being driven by human activity.

1995 Climate Change Convention in Berlin, Germany, produces *Berlin Mandate* that sets a 1997 goal for an international treaty to limit greenhouse gas emissions.

1996 Second meeting of the Climate Change Convention. U.S. delegates concede that IPCC data is valid and that binding cuts may be needed to avoid the most serious consequences of global warming due to human activity. After a four-year pause, global emissions of carbon dioxide resume their steep climb; scientists warn that by the year 2000, most industrialized countries will not meet the UNFCCC agreement to stabilize emissions at 1990 levels.

1997 Forest fires worldwide burn a total of five million hectares of forest.

1997 American environmental activist Julia "Butterfly" Hill (1974–) climbs a 180 ft (55 m) redwood tree in California to protest the logging of the surrounding forest as well as to protect the tree. Hill removed herself from the tree in 1999 after she negotiated a deal to save the tree and an additional three acres of the forest.

1997 Kyoto Protocol mandates a reduction of reported 1990 emissions levels by 6–8% by 2008.

1997 Monserrat volcano erupts.

1997 First consumer hybrid electric automobile is marketed in Japan.

1998 Intense El Niño is linked to then warmest year on record and several weather-related disasters.

1999 World population reaches six billion.

1999 World Trade Organization (WTO) conference in Seattle, Washington, is marked by heavy protests, highlighting WTO's weak environmental policies.

2000 IPCC scientists issue prediction that if "worst-case" scenarios happen, the world could see a 6°C increase in global mean temperature over the next century, a temperature increase that will produce dramatic and probably unprecedented ecological disasters.

2001 In February 2001, the complete draft sequence of the human genome was published. The public sequence data was published in the British journal *Nature* and the sequence obtained by Celera was published in the American journal *Science*.

2001 At the Bonn Conference, 178 countries agree to revisions of the Kyoto Protocol.

2001 The United States fails to ratify the Kyoto Protocol.

2001 On September 11, the World Trade Center towers in New York collapse after being struck by two commercial airplanes commandeered by Islamist fundamentalists belonging to Al Qaeda, a terrorist organization. A third airplane crashes into the Pentagon just outside Washington, D.C., causing loss of life and major damage. A fourth plane is forced down in a field in Pennsylvania. U.S. civilian aviation is grounded for three days.

2001 IPCC scientists issue a third report, asserting that unprecedented global warming is "very likely."

2001 Despite U.S. president George W. Bush's renouncement of the Kyoto Protocol, many signatories agree to abide by the provisions and to continue to seek ratification of enough countries to bring the accord into international law. With Australia's subsequent ratification in late 2007, by 2008 the United States is the only major industrialized nation not to have ratified the Kyoto Protocol.

2002 EPA adopts California emissions standards for off-road recreation vehicles to be implemented by 2004.

2002 EPA announces its Strategic Plan for Homeland Security to support the National Strategy for Homeland Security enacted after the September 11, 2001 terrorist attacks in the United States.

2002 U.S. president George W. Bush introduces the Clear Sky Initiative.

2002 Satellites capture images of icebergs more than ten times the size of Manhattan Island breaking off the Antarctic ice shelf.

2002 Severe Acute Respiratory Syndrome (SARS) virus is found in patients in China and other Asian countries. The newly discovered corona virus is not identified until early 2003. The spread of the virus reaches epidemic proportions in Asia and expands to the rest of the world.

2002 United Nations Earth Summit is held in Johannesburg, South Africa.

2002 Representatives from 185 nations meet in Delhi, India, and adopt the Delhi Declaration.

2003 Energy bill introduced in U.S. Congress includes ethanol use mandates.

2003 EPA rejects petition to regulate emissions from vehicles; EPA claims lack of authority under the Clean Air Act.

2003 Collapse of major ice sheets in Antarctica and Greenland raise new alarms about the impact of global warming.

2003 Europe records the hottest summer temperatures; some experts argue that the temperatures are the highest in more than 500 years.

2004 Russia ratifies Kyoto Protocol, putting it into effect worldwide even without ratification by the United States.

2004 American biologist Jared Diamond (1937–) publishes *Collapse: How Societies Choose to Fail or Succeed.* In the work, Diamond discusses environmental issues and their relationship to the fall of various civilizations.

2004 Kenyan environmentalist and human rights activist Wangari Maathai (1940–) wins the Nobel Peace Prize.

2004 On December 26, the most powerful earthquake in more than 40 years occurs underwater off the Indonesian island of Sumatra. The tsunami produced a disaster of unprecedented proportion in the modern era.

2005 American nun Dorothy Stang (1931–2005) is murdered in Brazil by contract killers after spending decades fighting efforts by loggers and ranchers to clear large areas of the Amazon rainforest. Less than a week later, Brazil's government awards a disputed patch of Amazon rainforest to a sustainable development project championed by Stang.

2005 Kyoto Protocol goes into effect on February 16, 2005.

2005 The United Nations reports that the hole in the ozone layer above Antarctica has grown to near record size, suggesting 20 years of attempted pollution controls have had little effect.

2005 The European Project for Ice Coring in Antarctica reports that carbon dioxide in the current atmosphere is greater than at any time during the last 650,000 years.

2005 California requests a waiver from the EPA so that it can impose stricter emissions standards. In 2007, after no response is received from the EPA, California announces it will file suit against the agency.

2005 H5N1 virus, responsible for avian flu, moves from Asia to Europe. The World Health Organization attempts to coordinate multinational disaster and containment plans. Some nations begin to stockpile antiviral drugs.

2005 Hurricane Katrina slams into the U.S. Gulf Coast, causing levees to break and massive flooding in New Orleans, Louisiana. Damage is extensive across the coasts of Alabama, Louisiana, and Mississippi. The Federal Emergency Management Agency (FEMA) is widely criticized for a lack of coordination in relief efforts. Three other major hurricanes make landfall in the United States within a two-year period, stressing relief and medical supply efforts. Long term heath studies of populations in devastated areas begin. Some climate models link the recent intensity in hurricane seasons to increased sea temperatures and global warming.

2005 A coalition of state governments, environmental groups, and the territory of American Samoa bring suit against the EPA in the U.S. federal court. The plaintiffs charge that the EPA is refusing to regulate carbon dioxide and other greenhouse gases known to be changing Earth's climate. The case, *Massachusetts et al. v. Environmental Protection Agency et al.* goes before the U.S. Supreme Court in 2007, and the court rules against the EPA.

2006 Norway announces plans to build a "doomsday vault" in a mountain close to the North Pole. The vault will house a two-million-crop seed bank in the event of catastrophic climate change, nuclear war, or rising sea levels.

2006 NASA research shows that the melting of the Antarctic ice sheet has raised global sea levels by 0.05 in (1.2 mm) since 2002.

2006 British economist Nicholas Stern (1946–) issues his 700-page *The Stern Review on the Economics of Climate Change,* which was commissioned by the government of the United Kingdom.

2006 NASA launches two satellites designed to provide the first 3-D views of Earth's clouds and help predict how cloud cover contributes to global warming.

2006 China's official news agency reports that glaciers in the Qinghai-Tibet plateau, also known as the "roof of the world," are melting at a rate of 7% annually due to global warming.

2006 U.S. scientists attribute a reported fourfold increase in the number of fires in the western United States to climate change.

2006 Researchers report that carbon dioxide from industrial emissions is raising the acidity of the world's oceans, threatening plankton and other organisms that form the base of the entire marine food chain.

2006 NASA satellite data shows Greenland's ice sheet is melting at a rate that exceeds scientists' estimates.

2006 Researchers report that Earth's temperature has been warming by 0.3°F (0.17°C) per year for the last three decades, and is at a current 12,000-year high.

2006 In a press conference about the Doomsday Clock, a symbol of the risk of atomic cataclysm, British physicist Stephen Hawking (1942–) describes climate change as a greater threat to Earth than terrorism.

2007 Scientists warn that glaciers could all but disappear from the European Alps by 2050—most will be gone by 2037.

2007 Fourth IPCC report is issued (the first segment in February, and the last in November). The panel of scientists from 113 countries issues a consensus, stating that global warming is caused by humans and that warmer temperatures and rises in sea levels will continue for centuries, no matter how much humans control their pollution.

2007 British airline tycoon Richard Branson (1950–) announces a $25 million prize for the first person to devise a way of removing greenhouse gases from the atmosphere.

2007 The United Nations Educational, Scientific, and Cultural Organization issues a report on World Heritage Sites that are in danger due to climate change.

2007 In August 2007, a month before annual Arctic sea-ice melting normally peaks, a new record is established for shrinkage of the north polar ice cap. Experts claim that the melting rate in June and July is unprecedented.

2007 In August, hundreds of demonstrators gather outside of Heathrow Airport in the United Kingdom to participate in a week-long Camp for Climate Action.

2007 World Meteorological Organization announces that during the first half of 2007, Earth showed significant increases in both high temperatures and frequency of extreme weather events (including heavy rainfalls, cyclones, and wind storms). The average global land temperatures for January and April were the warmest ever recorded for those two months.

2007 China overtakes the United States as the world's largest emitter of greenhouse gases.

2007 The World Wildlife Fund conservation group states that climate change, pollution, over extraction of water, and encroaching development are killing some of the world's major rivers including China's Yangtze, India's Ganges, and Africa's Nile.

2007 An international conference on desertification in Algeria closed with a call (dubbed the Algiers Appeal) to all African countries to ratify the Kyoto Protocol in an effort to help slow the rapid expansion of deserts on the continent.

2007 The United Nations reports that 16 of the most polluted cities in the world are located in China. Also, air pollution contributes to the premature death of more than 400,000 Chinese people each year.

2007 In August and September, the north polar sea-ice cap shrinks to the smallest size ever recorded. As of September 16, U.S. government scientists announce the ice was 1.59 million square mi (4.14 million square km) in size, about a fifth smaller than the previous record, which was set in September 2005. The Northwest Passage (the northern water route from the Atlantic to the Pacific) becomes ice-free for the first time in recorded history. Although surprised and concerned, climate scientists predict increased melting of the ice due to global climate change. Thus far, Arctic temperatures have warmed twice as fast as the rest of the world.

2007 The journal *Science* reports that the crucial "carbon sink" (Antarctica's Southern Ocean), which holds 15% of the world's excess carbon dioxide, is nearing saturation and soon may be unable to absorb more.

2007 Former U.S. Vice President Al Gore and the IPCC share the 2007 Nobel Peace prize for raising public awareness about global warming and climate change and for establishing the foundations to begin to solve the problem. Gore's film about climate change and global warming, *An Inconvenient Truth*, also wins major film awards.

2007 Live Earth concerts are held on all seven continents to raise awareness of global climate change and the need to act on it.

2007 World leaders at a G8 summit agree to "seriously consider" proposals to cut the emissions of greenhouse gases by 50% by 2050.

2007 In October, scientists announce that CO_2 levels have increased since 2000 faster than even the most pessimistic predicted forecasts of the late 1990s. Growth in atmospheric CO_2 was only 1.1% per year for 1990–1999, but accelerated sharply to more than 3% per year for 2000–2004. Climate scientists attribute most of the increased emissions to increases in human population and industrial activity.

2007 Researchers from the University of North Carolina report that the number of coral reefs in the Indo-Pacific, an area stretching from Sumatra to French Polynesia, dropped 20% since 1985 due largely to climate change and coastal development.

2007 In December, newly elected Australian Prime Minister Kevin Rudd (1957–) signs the paperwork to ratify the Kyoto Protocol.

2007 In December, the Bali Conference is hosted by the United Nations. Despite tensions running high, participants create the Bali roadmap, which sets forth agreement on a negotiating process for a post-Kyoto agreement to be completed by 2009.

2007 In December, the EPA denies California's request for a waiver to impose stricter emissions standards. In January 2008, California along with 15 other states and several environmental groups begins a lawsuit to get the waiver.

2008 Computer models of global climate change predict that East Antarctica will gain ice as snowfall increases, slowing the rise in sea levels. At the same time, West Antarctica will lose ice as the melting accelerates, raising sea levels. Both effects have been observed.

2008 In January, an article in *Nature Geoscience* reports that Antarctica is losing ice at an accelerating pace. Although ice loss is near zero in East Antarctica, in the West Antarctic Peninsula, it has increased from 1996 to 2006 by 140% (about 60 billion metric tons per year). In the rest of West Antarctica, ice loss increased by 59% (about 132 billion tons per year). Most of the increased loss happens along channels in which glaciers flow to the sea. According to the new figures, Antarctica is now losing ice almost as fast as Greenland.

2008 In January, the IPCC announces that its next report will study the possibility of accelerated melting of both the Greenland and Antarctic ice shelves, which could cause several meters of sea-level rise by the end of this century.

Abrupt Climate Change

Introduction

Large-scale natural disasters are sometimes capable of causing a rapid climate change. One example of abrupt climate change occurred in August 1883, when a series of huge eruptions of the Krakatoa volcano in Indonesia sent ash 50 mi (80 km) into the atmosphere. The penetration of sunlight through the ash-laden atmosphere to Earth's surface was restricted so much that the average global temperature during the next year was more than 2°F (1.1°C) below normal. Krakatoa is a good example of two aspects that define an abrupt climate change: the speed of the change, and a change in global temperature of at least a degree Fahrenheit or Celsius.

Global warming—the warming of the atmosphere that has been accelerating since the mid-twentieth century, and which may be a consequence of human activities—has generated worry among many people that Earth's climate is poised to undergo a drastic shift.

Since the mid-twentieth century, atmospheric warming has been gradually accelerating. The examination of past markers of climate—such as plant fossils, tree growth rings, and material in the sediment that is deposited over time on the bottom of lakes—has revealed abrupt climate change in the past. As such, the scientific consensus is that a drastic climate shift in the future is not only conceivable but could be likely. Humankind's role in this future, however, is unclear, as evidence for the role of global warming in tipping the scale toward an abrupt climate change is still controversial.

Historical Background and Scientific Foundations

Data that have been gathered from ice cores, growth rings of trees, sediment at the bottom of lakes, and other locations have revealed large changes in climatic factors such as precipitation and temperature. Some of the climate shifts occurred over hundreds of years, but others occurred over only one or several decades, and so represent abrupt climate changes. Furthermore, some of the climate shifts affected the entire Northern or Southern Hemisphere.

An abrupt climate change is considered to be a change that takes place within a period of time that is fast relative to the cause of the change. For example, the climate change caused by the eruptions of Krakatoa changed the global climate within a year. This was abrupt given that the change was only possible once the volcanic ash had circulated throughout the atmosphere.

The cause of other climate changes can be slower. For example, consider the southward migration of glaciers during the various ice ages that took tens of thousands of years. Relative to such a length of time, a climate change that took place over a century or more was abrupt.

The best-known abrupt climate change took place approximately 11,000 years ago, as the Northern Hemisphere was emerging from a glacial period to a warmer state. Over about one hundred years, temperatures returned to near glacial levels, where they remained for the next thousand years. The climate of the Northern Hemisphere then quickly warmed. The century-long period of cooling is called the Younger Dryas. Measurements have indicated that in some regions, such as present-day Greenland, the cooling to glacial temperatures occurred within 10 years rather than a century—a very abrupt climate change.

The reasons behind abrupt climate change are still not well understood. The abrupt climate changes associated with ice ages likely involved fluctuations in atmospheric carbon dioxide (CO_2). During the time when the land was covered by glacial ice, the level of CO_2 in the atmosphere decreased, since less vegetation was present on land and the gas was mostly stored by the ocean water. This lessened the trapping of sunlight by the atmosphere. With the return of vegetation, the atmospheric level of CO_2 rose and the atmosphere warmed.

In the twentieth century and particularly since the 1960s, the development of more sophisticated means of measurement, and the ability to obtain atmospheric data

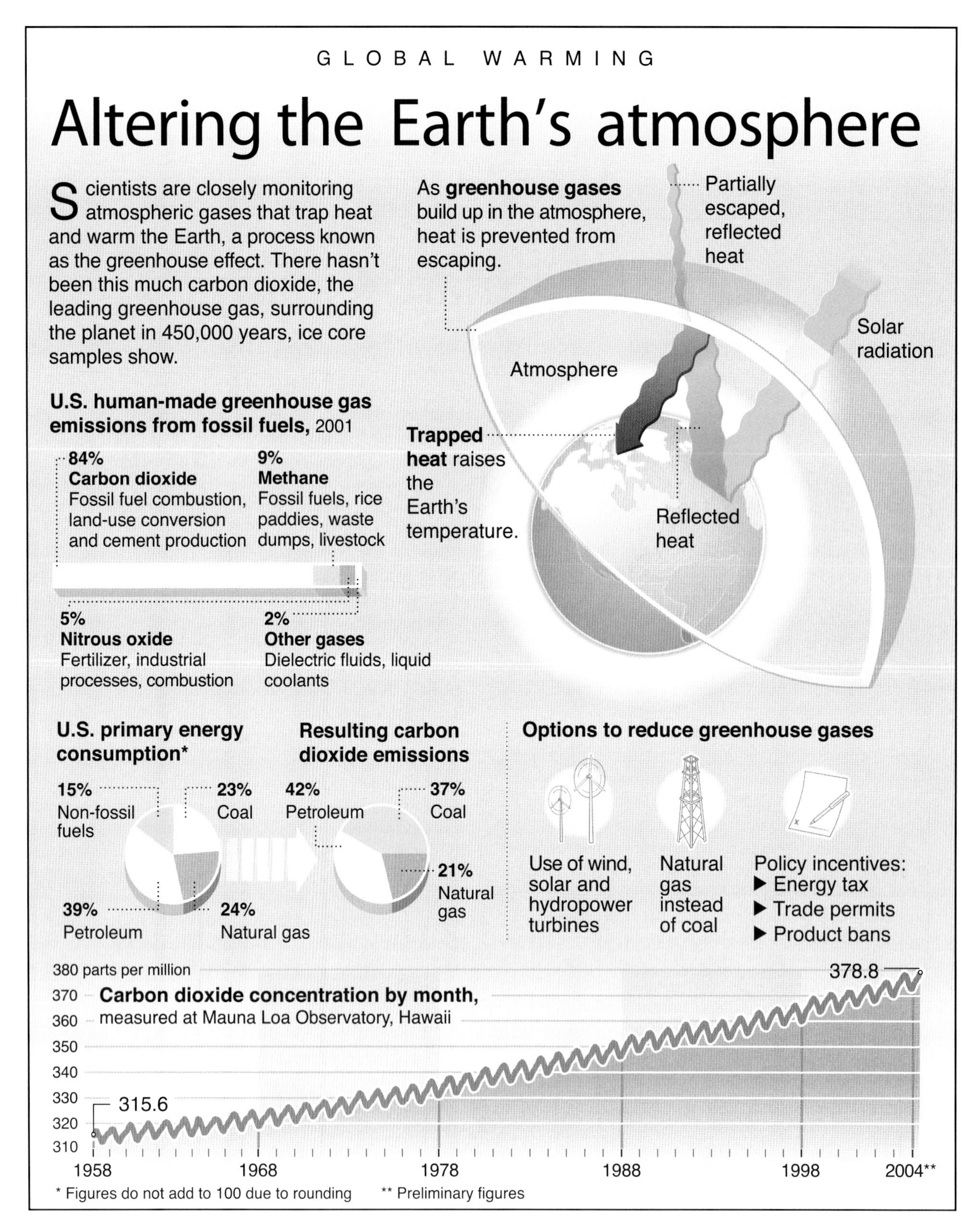

This graphic from the Associated Press discusses how global warming may be altering Earth's atmosphere. [Sources: "Facts at Your Fingertips," Reader's Digest; Scripps Institution of Oceanography; National Oceanic and Atmospheric Administration; Energy Information Administration; and Intergovernmental Panel on Climate Change.] *AP Images.*

WORDS TO KNOW

ATMOSPHERE: The air surrounding Earth, described as a series of shells or layers of different characteristics. The atmosphere—composed mainly of nitrogen and oxygen with traces of carbon dioxide, water vapor, and other gases—acts as a buffer between Earth and the sun. The layers—troposphere, stratosphere, mesosphere, thermosphere, and the exosphere—vary around the globe and in response to seasonal changes.

CLIMATE MODEL: A quantitative way of representing the interactions of the atmosphere, oceans, land surface, and ice. Models can range from relatively simple to quite comprehensive.

ECOSYSTEMS: According to the Intergovernmental Panel on Climate Change (and as published in IPCC reports): A system of living organisms interacting with each other and their physical environment. The boundaries of what could be called an ecosystem are somewhat arbitrary, depending on the focus of interest or study. Thus, the extent of an ecosystem may range from very small spatial scales to the entire planet Earth ultimately.

GREENHOUSE GASES: Gases that cause Earth to retain more thermal energy by absorbing infrared light emitted by Earth's surface. The most important greenhouse gases are water vapor, carbon dioxide, methane, nitrous oxide, and various artificial chemicals such as chlorofluorocarbons. All but the latter are naturally occurring, but human activity over the last several centuries has significantly increased the amounts of carbon dioxide, methane, and nitrous oxide in Earth's atmosphere, causing global warming and global climate change.

ICE AGE: Period of glacial advance.

ICE CORE: A cylindrical section of ice removed from a glacier or an ice sheet in order to study climate patterns of the past. By performing chemical analyses on the air trapped in the ice, scientists can estimate the percentage of carbon dioxide and other trace gases in the atmosphere at that time.

TREE RINGS: Marks left in the trunks of woody plants by the annual growth of a new coat or sheath of material. Tree rings provide a straightforward way of dating organic material stored in a tree trunk. Tree-ring thickness provides proxy data about climate conditions: most trees put on thicker rings in warm, wet conditions than in cool, dry conditions.

THERMOHALINE CIRCULATION: Large-scale circulation of the world ocean that exchanges warm, low-density surface waters with cooler, higher-density deep waters. Driven by differences in temperature and saltiness (halinity) as well as, to a lesser degree, winds and tides. Also termed meridional overturning circulation.

YOUNGER DRYAS: A relatively recent episode of abrupt climate change. About 12,900 years ago, conditions in the Northern Hemisphere cooled in about a decade (extremely rapidly), in some locations by 27°F (15°C). The cold period persisted for about 1,300 years and then reversed, also suddenly. The causes of the Younger Dryas are not well understood, but the event does show that Earth's climate is capable of extremely rapid and dramatic shifts.

by means of weather balloons and, beginning in the 1960s, the use of orbiting satellites, has provided much more information on the change in the atmosphere over time.

It is now clear that the global climate depends on the composition of the atmosphere and the circulation patterns in the atmosphere, and that these atmospheric aspects depend on the global circulation pattern of the ocean. In the ocean, the global cycling of water (known as the ocean conveyor and also as thermohaline circulation, among other terms) distributes the heat energy in the water to all regions of Earth and vertically through the ocean depths. If the ocean conveyor were disrupted, as has been proposed as a consequence of altered atmospheric circulation, the climate would change enormously. Modeling studies have indicated that the change would occur over years, not days.

Impacts and Issues

Modeling climate change is in its infancy. Climate involves many interactions between environmental forces, and some are not completely understood. So, building a computer model that mimics climate and thus can be used to investigate the likelihood of an abrupt climate change is not possible as of early 2008.

However, an understanding of the past is helpful in the present. The relationship between changes of atmospheric CO_2 and periods of cooling and heating remains relevant. CO_2 is a greenhouse gas—one of the compounds that increases the tendency of the atmosphere to retain heat. Measurements taken over the past 150 years clearly show that the CO_2 level in the atmosphere is increasing. This increase has not been constant over time, but rather has begun to accelerate. Whether this could help tip the atmosphere into an abrupt climate change is not clear and is very debatable. However, the majority of scientists now share the view that human activities increase the likelihood of the occurrence of climate change.

According to a 2002 report issued by the National Academy of Sciences, an abrupt climate change is not only possible but likely in the future. The report warned that such a rapid change of climate would greatly affect both ecosystems and societies.

An abrupt climate change would cause an upheaval in the lives of many people. For example, increased drought in agricultural regions could decrease food and water supplies.

Aside from the hardship to everyday survival, climate change could increase the animosity between "have" and "have-not" countries. The U.S. military has recognized the possible security implications of such a destabilized world.

Paradoxically, some scientists envision that the present-day warming of the atmosphere could tip Earth into another Ice Age, since the distribution of heat via the ocean conveyor would break down. This could cause cooling of the Northern Hemisphere.

The portion of the ocean conveyor that circulates through the North Atlantic region has been identified as particularly important, since the melting of the polar ice cap could send enough freshwater into this region to diminish or perhaps even halt ocean currents such as the Gulf Stream. A report in a 2002 issue of *Nature* documented the decreasing salinity of the North Atlantic since the 1960s, even to depths of 13,000 ft (about 4,000 m). Whether this has begun to affect the conveyor is unclear as of early 2008.

A plan spearheaded by the Woods Hole Oceanographic Institution and other agencies to deploy thousands of remote sensors throughout the global ocean could help to better understand the state of the ocean and the influence of global warming.

SEE ALSO *Atmospheric Circulation; Gaia; Great Conveyor Belt.*

BIBLIOGRAPHY

Books

Cox, James D. *Climate Crash: Abrupt Climate Change and What It Means for Our Future.* New York: Joseph Henry Press, 2005.

Diamond, Jared. *Collapse: How Societies Choose to Fail or Succeed.* New York: Penguin, 2006.

Web Sites

"Abrupt Climate Change: Should We Be Worried? A Report for a Panel on Abrupt Climate Change at the World Economic Forum." *Woods Hole Oceanographic Institution*, January 27, 2003. <http://www.whoi.edu/page.do?pid=12455&tid=282&cid=9986> (accessed November 13, 2007).

"Global Warming: Abrupt Climate Change." *Union of Concerned Scientists*, August 10, 2005. <http://www.ucsusa.org/global_warming/science/abrupt-climate-change-faq.html> (accessed November 13, 2007).

"What's After the Day After Tomorrow?" *Woods Hole Oceanographic Institution*, July 5, 2007. <http://www.whoi.edu/page.do?pid=12455&tid=282&cid=9948> (accessed November 13, 2007).

Acid Rain

Introduction

Rain is normally slightly acidic due to the presence of some carbon dioxide in the atmosphere. That normal acidity is increased significantly with the addition of sulfur and nitrogen oxides from burning fossil fuels. The amounts of these compounds entering the atmosphere as emissions from the burning of fossil fuels far exceed the amount released by natural sources into the atmosphere. Winds carry the damaging substances to regions far from their source, where the acid rain damages the ecosystem and corrodes buildings.

Historical Background and Scientific Foundations

Sulfur and nitrogen oxides released by the burning of fossil fuels react with water in the atmosphere to form strong acids. The acidity of solutions is measured in the

Acid rain takes a toll on many historic buildings and statues. The Colosseum in Rome, Italy, for example, is being damaged by acid rain. *Passport Stock/Royalty Free.*

The "Saints and Sinners" sculpture on the Oakland University campus in Michigan shows signs of streaking due to acid rain. Made of bronze, the artwork is being corroded. *© 2003 Kelly A. Quin.*

pH scale. The pH scale goes from 0 to 14; a pH of 7 is neutral. Any solution with a pH above 7 is described as basic, while a solution with a pH below 7 is acidic. The lower the pH number, the more acidic the solution. With the normal solubility of carbon dioxide in atmospheric water, a mildly acidic, unpolluted rainfall has a pH of 5.6. In the regions that are seriously impacted by acid rain from the sulfur and nitrogen oxides in the atmosphere, the precipitation may have a pH of 4.1 to 5.1.

In the early 1970s, a serious depletion of aquatic life was observed in water bodies in Scandinavia, Scotland, northern England, the northeastern United States, and Quebec, Canada. The cause of the problem was found to be high levels of acidity in these ecosystems. Scientists soon traced the source of this increased acidity to acid rain. The damage was related to air pollution from industrial development long distances down wind from the freshwater ecosystems affected. For example, industries located in Central Europe were found to be the cause of acid rain damage in Scandinavia.

Studies of lakes near major industrial areas showed that they were not as seriously impacted as those far away in the direction of the prevailing weather systems. By the 1980s, the effects of acid rain were also being observed in the forests of Germany. The impact of acid rain continued to be seen in the northern regions where aquatic ecosystems were already in trouble. By this time, salmon were extinct in some regions. Salmon are very sensitive to changes in pH and will die if the pH falls to 5.5 or below.

The way in which acid rain damages forests is complex and multi-faceted. It starts with soils that are too acidic. Important nutrients in the soil disappear as soil acidity increases. This increase in soil acidity is followed by the release of a soluble form of aluminum, which is very harmful to vegetation, into the soil. Especially at higher elevations, forests also can be damaged when acidic precipitation falls directly on tree needles or leaves.

■ Impacts and Issues

The ecological damage that results from acid rain was recognized as an international problem that required international cooperation to develop solutions. Since 1986, the International Cooperative Programme on Assessment and Monitoring of Air Pollution Effects on Forests has studied the impact of the release of sulfur and nitrogen oxides into the atmosphere on forests. Thirty-nine European countries have participated in these studies in close cooperation with the European Commission.

The European program has been extended to include the effects of climate change and carbon sequestering.

WORDS TO KNOW

ACID: Substance that when dissolved in water is capable of reacting with a base to form salts and release hydrogen ions.

BASE: One of the four chemicals (nitrogenous bases) that are a part of the deoxyribonucleic (DNA) molcule. The sequence of the bases determines the genetic code (the information contained within DNA). The first letters of each base provide the alphabet for the DNA code (A, C, G, and T [short for adenine, cytosine, guanine, and thymine]). Adenine and guanine are also classified as purine bases and are found in both DNA and RNA. Cytosine and thymine are classified as pyrimidine bases and while cytosine also occurs both in DNA and RNA (Ribonucleic acid), in RNA molecules another base (uracil) substitutes for the thymine found in DNA molecules. The bases bind together in very restricted and specific ways to form base pairs between DNA strands. In DNA, adenine always bonds with thymine (A-T bond) on the opposite strand and cytosine always bonds with guanine to form a (C-G) base pair between stands of the DNA helix. In electronics a base is also the middle slice of a transistor.

CARBON SEQUESTERING: Storage or fixation of carbon in such a way that it is isolated from the atmosphere and cannot contribute to climate change. Sequestration may occur naturally (e.g., forest growth) or artificially (e.g., injection of CO_2 into underground reservoirs).

FOSSIL FUELS: Fuels formed by biological processes and transformed into solid or fluid minerals over geological time. Fossil fuels include coal, petroleum, and natural gas. Fossil fuels are non-renewable on the timescale of human civilization, because their natural replenishment would take many millions of years.

GREENHOUSE GASES: Gases that cause Earth to retain more thermal energy by absorbing infrared light emitted by Earth's surface. The most important greenhouse gases are water vapor, carbon dioxide, methane, nitrous oxide, and various artificial chemicals such as chlorofluorocarbons. All but the latter are naturally occurring, but human activity over the last several centuries has significantly increased the amounts of carbon dioxide, methane, and nitrous oxide in Earth's atmosphere, causing global warming and global climate change.

PH: Measures the acidity of a solution. It is the negative log of the concentration of the hydrogen ions in a substance.

IN CONTEXT: ACID RAIN AND THE PH SCALE

The scale that is used to measure the acidity of a substance is called the pH scale. The pH scale runs from 0 to 14. If a material has a pH of 7, it is neutral, meaning that it is neither acidic nor alkaline (basic). Substances with pH values less than 7 are acidic and substances with pH values greater than 7 are alkaline. Distilled water is neutral, with a pH of 7.

The most important compounds that produce acids in the atmosphere are sulfur dioxide (SO_2) and nitrogen oxides, like nitrogen oxide (NO) and nitrogen dioxide (NO_2). These compounds combine with water in the atmosphere to form sulfuric acid (H_2SO_4), and nitric acid (HNO_3).

When describing acid rain, scientists actually refer to a more precise term, acid deposition. Scientists distinguish between two types of acid deposition: wet and dry. Dry deposition includes acidic gases and solid particles containing sulfuric and nitric acid that settle out of the air and land on the ground or other surfaces. Dry deposition usually occurs very close to the point where the air pollution occurs. Wet deposition occurs when precipitation—such as rain, sleet, fog, or snow—becomes acidic and falls to the ground. Wet deposition can occur hundreds of miles from the place where the air pollution originates.

There has been some increase in nitrogen oxide, but also a great increase in the amount of carbon dioxide that is entering the atmosphere from the use of gasoline in vehicles for transportation and from industrial development. What started out as a program to address the acid rain problem has expanded to include consideration of the adverse environmental effects of all polluting gases.

Some carbon dioxide is normal and necessary in the atmosphere, but excess carbon dioxide is a greenhouse gas that is contributing to global warming. Regulations have been passed to reduce the emissions of sulfur and nitrogen oxides and also to capture, or sequester, some of the extra carbon dioxide that is considered to be a prime offender in the issues of climate change beyond acid rain.

SEE ALSO *Climate Change; Climate Engineering; Environmental Policy; Environmental Pollution; Greenhouse Gases.*

BIBLIOGRAPHY

Periodicals

Sliggers, Johan, and Willem Kakebeeke, eds. "Clearing the Air: 25 Years of the Convention on Long-range Transboundary Air Pollution." New York: United Nations, 2005.

Web Sites

"Convention on Long-range Transboundary Air Pollution." *United Nations Economic Commission for Europe.* <http://www.unece.org/env/lrtap/> (accessed August 16, 2007).

"What Is Acid Rain?" *U.S. Environmental Protection Agency*, 2007 <http://www.epa.gov/acidrain/what/index.html> (accessed August 16, 2007).

Adaptation

Introduction

An adaptation to climate change is any adjustment in a natural or human system that decreases the harm caused by climate change or takes advantage of some opportunity it offers. Both natural and human systems may adapt to actually occurring climate change; humans are unique in being able to forecast changes and attempt to adapt to them in a planned way.

There are three types of adaptation, namely anticipatory, autonomous, and planned. Anticipatory adaptation takes place before predicted impacts are observed; autonomous or spontaneous adaptation takes place in direct response to ecological or economic impacts of climate change; and planned adaptation is based on deliberate policy decisions. How much adaptation will occur depends on several factors, including how much climate changes, how severe the impacts of that change are, where they are most severe, how adaptable impacted ecosystems and human communities are, and whether effective policies are followed for mitigating and adapting to climate impacts.

Some impacts of climate change are already being felt, and some adaptation by both natural and human systems has already begun. However, adaptation is not always possible. For ecosystems, failure to adapt means the collapse of existing ecologies and their replacement by others, along with species extinctions. For human communities, failure to adapt means partial or severe breakdown of existing ways of life, with consequences ranging from monetary losses to famine and mass death. In general, slower climate change is easier to adapt to than sudden change, and slight change is easier to adapt to than extreme change.

Historical Background and Scientific Foundations

Natural and Historic Adaptation

Over geologic time, Earth's climate has changed regionally and globally many times. Ice ages have spanned millions of years, dotted by glacial and interglacial cycles; the sun has gradually warmed throughout life's history; carbon dioxide (CO_2) concentrations were far higher than today's during most of Earth's history; and regional shifts in rainfall patterns and temperature have been common. Adaptation by natural systems—or failure to adapt, on a few occasions leading to mass extinctions involving the majority of living species—has thus always been a feature of life on Earth.

Since the evolution of human beings, regional and local climate changes have continued to occur, though not with the intensity of the most drastic changes of the deep past. Some climate changes, such as the El Niño–Southern Oscillation climate cycle, occur regularly and last for years, and adaptation to them is routine in affected parts of the world. Affected ecosystems and peoples have, historically, sought to adapt to climate changes, often successfully.

Most plants and animals are evolved to thrive in a certain range of climate conditions: too little or too much precipitation or warmth are not tolerable to them. Therefore no species is found everywhere on Earth, but only where climate conditions are favorable and where the species has managed to cross geographical barriers to colonize a given area. For example, the European downy birch, a tree species, requires cold temperatures. It thrives in northern Asia and Europe and is the only tree native to Greenland and Iceland; it is not found in tropical areas, even in high mountainous zones where it might be able to survive if its seeds could reach those locations.

Natural adaptation for plants consists mostly of range shifting. As global warming proceeds, a plant such as the European downy birch will gradually cease to be found in the southernmost parts of its historical range and will seed itself northward to areas that were formerly too cold. Since plants and animals tend to live in adapted communities, with animals depending on particular plants and vice versa, entire biomes (types of habitat) will shift location as climate changes, plants and animals

WORDS TO KNOW

ALTIPLANO: High plateau region in the Andes of Argentina, Bolivia, and Peru in South America, the second-largest high plateau on Earth after Tibet. Sediments in Lake Titicaca, into which the Altiplano drains, record tens of thousands of years of regional climate change.

BIOME: Well-defined terrestrial environment (e.g., desert, tundra, or tropical forest). The complex of living organisms found in an ecological region.

DEFORESTATION: Those practices or processes that result in the change of forested lands to non-forest uses. This is often cited as one of the major causes of the enhanced greenhouse effect for two reasons: 1) the burning or decomposition of the wood releases carbon dioxide; and 2) trees that once removed carbon dioxide from the atmosphere in the process of photosynthesis are no longer present and contributing to carbon storage.

EL NIÑO–SOUTHERN OSCILLATION: Global climate cycle that arises from interaction of ocean and atmospheric circulations. Every 2 to 7 years, westward-blowing winds over the Pacific subside, allowing warm water to migrate across the Pacific from west to east. This suppresses normal upwelling of cold, nutrient-rich waters in the eastern Pacific, shrinking fish populations and changing weather patterns around the world.

GEOLOGIC TIME: The period of time extending from the formation of Earth to the present.

GLACIAL CYCLE: Episode in Earth climate history in which temperatures decline and glaciers grow and spread, sometimes covering large parts of the northern and southern hemispheres. The most recent glacial cycle ended about 10,000 years ago.

ICE AGE: Period of glacial advance.

MANGROVE FOREST: Coastal ecosystem type based on mangrove trees standing in shallow ocean water: also termed mangrove swamp. Mangrove forests support shrimp fisheries and are threatened by rising sea levels due to climate change.

PALEOCLIMATE: The climate of a given period of time in the geologic past.

RANGE SHIFT: Movement or shrinkage of the territory occupied by a given species of plant or animal due to climate change. When climate warms, species cease to occupy areas that were at the warm extreme of their ability to adapt and to colonize areas that were at the cool extreme. The result is a shifting range.

RIVER DELTA: Flat area of fine-grained sediments that forms where a river meets a larger, stiller body of water such as the ocean. Rivers carry particles in their turbulent waters that settle out (sink) when the water mixes with quieter water and slows down; these particles build the delta. Deltas are named after the Greek letter delta, which looks like a triangle. Very large deltas are termed megadeltas and are often thickly settled by human beings. Rising sea levels threaten settlements on megadeltas.

together. Such shifts may happen quickly, on the scale of a human lifetime: range shifts have already been measured in many parts of the world.

For mountainous species, as warming causes cold climate zones to move to higher altitudes, habitats move upward and shrink. For example, the uppermost altitude for pine mistletoe in Switzerland was 656 ft (200 m) higher in 2004 than it was in 1910. Range shifts can also occur more slowly. In North America, almost all of present-day Canada was covered by ice about 21,000 years ago, and spruce trees were found across what is now the northern and central United States, down into Texas and the Southwest. Over the next 15,000 years, spruce forests migrated northward as the ice retreated. By 7,000 years ago, spruce was common only in northern areas that had formerly been covered by ice. This pattern persists today, and modern climate change will drive spruce forests even farther north.

Over longer time periods, adaptation may be genetic as well as geographic. However, evolving new climate tolerances is a far slower, chancier process than colonizing new regions. Failure to either migrate or evolve leads to extinction.

Human history shows varying levels of adaptation to climate swings. For example, the world's first civilized empire, the Akkadia, was established between 4300 and 4200 BC in Mesopotamia, that is, the plain between the Tigris and Euphrates rivers, much of which is located in what is Iraq today. After only about 100 years of prosperity, the empire collapsed. Soil cores from the region—cylinders of undisturbed soil extracted vertically from the ground—contain a layer of wind-blown silt that records a 300-year drought that began shortly before the Akkadian collapse. Archaeological evidence shows that the Akkadians tried to adapt to this change by building larger grain storage facilities and irrigation systems, but these adaptations were insufficient.

From paleoclimatic evidence, the Mesopotamian drought has been linked to a cooling event in the North Atlantic in which surface water temperatures declined by 1.8–3.6°F (1–2°C). Modern temperature records show that the year-to-year water supply of the Mesopotamian plain can be cut in half by unusual cooling of the Atlantic. Century-scale regional drought has also been implicated in the collapse of the classic Maya civilization around AD 800, along with overpopulation, deforestation, erosion, and war, and of the Tiwanaku culture in the Bolivian-Peruvian altiplano about AD 1000.

The collapse of the urban and military structures of these societies can be viewed as a form of adaptation: in response to increased scarcity brought on by climate change, they reduced their social complexity, abandoned cities, and reorganized their systems of production and supply. In all cases, human populations persisted, though decreased in number. However, such forced, autonomous adaptation is painful and often deadly for individuals caught up in it.

The global warming being experienced today is unique in Earth's history in that it is caused by human beings. Today's concern with anticipatory, planned adaptation to human-caused (anthropogenic) climate change is also a historical first. Not until the late twentieth century did humans understand the causes of climate change and possess the ability to make reasonably reliable predictions of its future course. The scientific study of adaptation to sea-level rise and the other effects of climate change began in the late 1980s and early 1990s, when extensive studies of adaptation strategies began to appear.

Modern adaptation, whether autonomous (reacting to existing changes) or anticipatory, will be based on the impacts of climate change. Hundreds of these potential impacts have been studied and many are being observed already, including rising sea levels, shifts in rainfall, increased heat waves, migrations of biomes (regional ecosystems), bleaching of corals, and many more. Impacts are often outlined by geographic region or by sector.

Freshwater Resources

Projected increases in air temperature, variability of rainfall, and sea level will impact freshwater systems in many parts of the world during the coming centuries. Over 16% of the world's population lives in basins fed by glacial melt or snowmelt flows, which will decline as warming shrinks glaciers, decreases precipitation in some areas, and shortens snow seasons. Rising sea level will extend salty water inland by penetrating groundwater aquifers, which will reduce freshwater availability for communities in coastal zones. Increased precipitation in some areas will increase flooding risk; decreased precipitation in others will increase drought risk.

Adaptations to impacts on freshwater resources can be either supply-side or demand-side. Supply-side adaptations seek to ensure supplies, while demand-side adaptations seek to reduce usage, allowing communities to prosper even when supplies decrease. Some supply-side adaptations include desalination of seawater, increasing storage capacity by building reservoirs, and expanding rainwater storage systems. Some demand-side adaptations include water recycling to increase water-use efficiency, reducing irrigation demands by changing crop mixes, planting calendars, irrigation methods, and providing incentives for water conservation such as metering and pricing.

Ecosystems

Increased atmospheric CO_2, along with the warming that it causes, will affect most ecosystems in the coming century and for long after. By the end of the twenty-first century, atmospheric CO_2 exceeds levels seen for at least 650,000 years. Even apart from warming, CO_2 has direct ecosystem effects, speeding the growth of terrestrial plants and making the oceans significantly more acidic. Some 20–30% of plant and animal species that had been specifically examined as of 2005 in an unbiased (random) sample were likely to be at increased extinction risk as the world warms by 3.6–5.4°F (2–3°C) above pre-industrial (pre-1750) levels.

Adaptations to impacts on ecosystems can be both natural and human. Natural adaptations will consist mostly of biome migration. Human responses will involve changes in management of natural resources and wild areas. There are a number of ways that human managers can increase the resilience of ecosystems, though these will likely become less effective or outright useless at higher levels of climate change. First, monitoring changes in climate and ecosystems is essential to allow for effective adjustments in management. Reducing harm to natural systems by other human activities—including pollution, habitat destruction, habitat fragmentation (where development and habitat destruction break up habitats into isolated pieces), and the introduction of invasive alien species—will almost always enhance the ability of ecosystems to adapt to climate change.

Expansion of national forests and parks and other kinds of reserve systems will reduce ecosystem vulnerability to climate change, especially if reserves are designed to consider long-term migrations of biomes and human settlements. For particular species, managing for connected populations (rather than isolated pockets of population), genetic diversity, and larger populations will increase survivability in response not only to climate change but to other challenges.

Agriculture and Forestry

In some mid-latitude and northerly regions, moderate global warming—that is, warming of 1.8–5.4°F (1–3°C)—is likely to benefit agriculture by increasing yields from pastures and crops. However, even mild warming will decrease yields in low-latitude regions, including the tropics, where most of the world's poor live. Higher levels of warming, which are quite possible, will have negative impacts on agriculture in all regions.

Although United Nations scenarios for possible world development show the number of malnourished persons in the world declining from about 820 million today to 100–380 million by 2080, this range is greater than it would be if global climate change were not occurring. Climate change also serves to shift the regional distribution of hunger, especially making hunger worse in sub-Saharan Africa. The productivity of commercial forestry is forecast to rise slightly to modestly in the short and medium term, partly due to the fertilizing affect of CO_2 on young trees.

There are many possible adaptive responses to climate change's impacts on agriculture. These include

IN CONTEXT: ABRUPT CLIMATE CHANGE HAPPENS

According to the National Academy of Sciences: "Evidence shows that the climate has sometimes changed abruptly in the past—within a decade—and could do so again. Abrupt changes, such as the Dust Bowl drought of the 1930s that displaced hundreds of thousands of people in the American Great Plains, take place so rapidly that humans and ecosystems have difficulty adapting to them."

SOURCE: *Staudt, Amanda, Nancy Huddleston, and Sandi Rudenstein.* Understanding and Responding to Climate Change. *National Academy of Sciences, 2006.*

altering the varieties or species planted to those better adapted to increased warmth or decreased rainfall, more efficient irrigation techniques, altering the timing and location of crop planting, and diversifying income generation (for example, by raising livestock). In a study of agricultural adaptation to hotter, drier summers in Modena, Italy, it was predicted that with unchanged farming practices sorghum crops would be reduced by 48–58%. With adjusted sorghum varieties and planting times, the impact could potentially be reduced to zero.

Coastal Systems

Rising sea levels and increased storm violence will impact coastal areas around the world. Over a billion people live in coastal areas today, with coastal population growing to over 5 billion by 2080 according to some global-development scenarios. The 300 million people living on large river deltas—close to, at, or even below present-day sea level—will be particularly at risk for impacts, especially flooding. Sea level is predicted to rise by 2 ft (0.6 m) or more by 2100, and to continue rising thereafter.

Over the next several centuries to a millennium, complete melting of the Greenland and West Antarctic Peninsula ice sheets could raise sea levels by 40 ft (12 m), radically altering coastlines. Hurricanes and tropical cyclones will probably increase in intensity, and the effects of these intensified storms will be amplified by higher sea levels. Without improved coastal protection, coastal flooding could increase by a factor of 10 by the 2080s, affecting over 100 million people, mostly in developing countries.

Adaptations to the coastal impacts of climate change take three basic forms: protection, accommodation, or retreat. Protection is practically synonymous with the building of dikes to keep back the sea, as in Holland and New Orleans, Louisiana. Accommodation may include flood-resistant building construction (e.g., buildings on pilings) or floating agricultural systems (now being tested in Holland). Retreat essentially means moving settlements back to higher ground. Astute combination of various forms of these adaptation options could reduce the impacts of sea-level rise by 10 to 100 times in many areas; at the other extreme, as for small islands in the Pacific, no adaptation may be feasible, and "retreat" may have to signify abandonment.

Human Health

Climate change is projected to increase malnutrition; injure child growth and development; increase the number of people suffering death and disease from extreme weather events such as storms, droughts, heat waves, and floods; shift the ranges of some infectious disease vectors; increase the amount of diarrheal disease (which presently causes 5–8 million deaths per year, mostly children); increase cardio-respiratory disease due to ground-level ozone; and decrease the number of deaths from cold in northern regions. Hurricane Katrina (2005) showed that even highly developed countries such as the United States may not be well-prepared for the health and other consequences of extreme weather events.

Adaptations to the impacts of climate change on health will mostly take the form of revised policies and procedures of national and international health organizations (in the United States, for example, the Centers for Disease Control and Prevention [CDC]). Climate-based early warning systems for heat waves and malaria outbreaks have already been implemented in some countries. Seasonal forecasts of events such as drought can allow the timely launching of public education campaigns on the prevention of diarrheal and other infectious diseases.

■ Impacts and Issues

The types and amounts of adaptations required—or possible—will depend on the amount of climate change that occurs. The more effective mitigation efforts, the less adaptation and failed adaptation will occur. Also, the pattern of impacts and therefore of demands on adaptation will be patchy. Tundra, boreal (northern-type) forests, and mountain and Mediterranean ecosystems will be highly vulnerable, as will mangroves and salt marshes along coasts. Low-lying coasts will be more vulnerable than steep coasts, and water resources will be more endangered in mid-latitude and dry tropical regions due to lessened rainfall and increased evaporation. Agriculture will be at risk in tropical regions. Africa, the Arctic, small islands, and large Asian river deltas are the areas most at risk from climate change.

As a rule, adaptation costs money. Therefore, wealthier nations would be better able to adapt even if they were facing equally grave impacts from climate change—but they are not. Impacts are likely to be much more severe in the developing countries that are less equipped to adapt to them. For example, by 2100, developing countries stand to lose about 10 times more land area to rising sea levels than developed countries, with four million people likely displaced versus a few hundred thousand, at most, in the developed countries.

Uncertainties about what climate change will occur makes planned adaptation more difficult, though there are many measures (called no-regrets measures) that carry benefits regardless of how much climate change occurs. For example, increased wildlands conservation will reduce species extinctions under any future climate scenario. Evaluation of which planned adaptations will be worth their cost depends sensitively on estimates of the social cost of carbon, that is, the total amount of future economic harm that will be caused by each ton of carbon (or equivalent amount of another greenhouse gas) emitted to the atmosphere. Calculation of this value is notoriously elusive and values-laden. The willingness or ability of governments to respond to threats that act on time-scales much greater than the election cycles or a ruler's personal lifespan also threaten timely and effective mitigation of and adaptation to climate change.

■ Primary Source Connection

Local ecosystems and the safety of human populations are put at risk as humans modify the environment around them. Local and national governments can minimize these effects by implementing adaptation responses. An adaptation response is a plan that addresses foreseeable environmental issues. Examples of adaptation responses include forest management, flood control, and habitat preservation and restoration. This article examines some of the adaptation responses taken by several European countries.

The European Environment Agency (EEA) is an agency of the European Union devoted to monitoring the environment in Europe and promoting sustainable development. The EEA has 32 member countries and six cooperating countries.

VULNERABILITY AND ADAPTATION TO CLIMATE CHANGE IN EUROPE

As a region of industrialised nations, Europe has a strong commitment to mitigating climate change by reducing greenhouse gas emissions. However, numerous scientific studies, as summarised in Section 3 of this report, and the considerable losses resulting from extreme weather events over recent years... demonstrated the vulnerability of Europe's natural environment and its society to projected climate change impacts. There is growing recognition that Europe should adapt to such impacts in order to maintain sustainable functioning of ecosystems and wellbeing of its population. Many EEA member countries have started to adjust their overall national climate policy framework to include climate change adaptation as an equally important component as mitigation. A wide range of adaptation responses have been initiated at varying governmental levels and in different sectors....

This section reviews adaptive responses in member countries in relation to natural ecosystems, water resources, coastal and river floods, natural hazards, human health and the business sector.

4.4.1 Maintaining the health of Europe's ecosystems

To address the wide ranging adverse impacts of climate change to Europe's terrestrial ecosystems, a variety of measures have been initiated or are being planned. This is often carried out in the context of nature conservation and sustainable resource management rather than deliberately directed at adapting to climate change.

Liechtenstein, is highly dependent on the stability of its ecosystems as it is a mountainous country. This has strongly motivated the introduction of an active national climate policy and the participation of the country in international processes, such as the Alpine Convention. Strategies to address climate change as a new risk are largely of a regulatory nature and are all designed primarily to address the issues related to sustainability, i.e., by introducing sector-oriented legal documents, such as the Nature and Landscape Protection Act (1996); the Forest Act (1991); the Preservation and Protection of Agricultural Lands Act (1992); the Ordinance on the integrated rehabilitation of the Alpine and mountain regions (1968) and the revised Tourism Act (2000).

In **France**, forest managers have been working on measures to improve the resilience of forests since the storms of December 1999 and the drought conditions associated with the 2003 heat wave. It is considered most important to develop a larger biodiversity among forest stands with more diverse varieties and through more genetic diversification. It is also necessary to install or regenerate species which are better adapted to present and future local conditions, so that they are more resilient to biotic and abiotic environmental conditions. Tools are being elaborated in order to improve the choice of species. More dynamic forest management practices with wider spacing and strong and early thinning can reduce vulnerability to wind storms and at the same time improve the water budget and therefore resistance to drought conditions. Better forest management practices and the presence of understory vegetation increases biodiversity and soil protection improves recycling of mineral elements and reduces mineral leaching. Development of heating plants or district heating using wood has been proposed in order to mitigate climate change. This would also help to utilise forest residues after storms and decrease the regeneration cost of damaged forests.

As a means to implement the Habitats Directive 92/43/EEC, the **German** Federal Nature Conservation Act (of 25 March 2002) states that the Federal Laender shall establish a network of interlinked biotopes covering at least 10% of the total area of each Federal Land. The required core areas connecting areas and connecting elements shall be legally secured via the designation of appropriate areas, detailed planning in accordance with the provisions of planning law, long-term arrangements (contractual nature conservation) or other appropriate measures. This is intended to safeguard an interlinked network of biotopes in a sustainable fashion. Such a network of interlinked biotopes is particularly important and represents a dynamic response of ecosystems to global change to protect biodiversity.

In their national communications, many countries reported their concerns over issues such as goods and services of terrestrial ecosystems (e.g., timber production, biodiversity etc.) that may be threatened under a changing climate. But very few countries go beyond the list of general adaptation options for forest management and biodiversity protection. Emerging from this is the inadequacy of knowledge on potential impacts in terrestrial ecosystems and practical guidance for adaptation.

In addition, some multi lateral initiatives have been taken in order to establish a stronger (i.e., 'climatically robust') network of ecological areas within Europe. An example of such an initiative is the Pan-European Ecological Network PEEN.

4.4.2 Managing Europe's water resources

Rising temperature and changing rainfall patterns are expected to change the availability of water resources in Europe.

The integrated management plans for water resources in **Spain** constitute one component for adaptation to climate change. The National Water Plan, Law 10/2001 of 5 July (Analysis of Water Systems) accounts for potential climate change induced reduction in water availability and analyses the effect of these reductions on management and planning.

Drought mitigation has been recognised as a national priority for Hungary. The improvement of drought forecast, for example, through development of reliable drought indices is recommended as an important adaptation measure. A national drought mitigation strategy is to be developed.

In **Greece**, an integrated water management plan is considered imperative to address the present day problem of low water use efficiency, which results from the combination of irrigation and cultivation practices. The preparation for the full implementation of this plan has started. Cross governmental departments have drafted a legislative framework for its implementation.

In the **Netherlands**, climate change and adaptation measures are explicitly integrated into the water policy agenda. Emphasis is placed on 'no-regret' strategies. Although flood risks seem to dominate the adaptation agenda in water policy, the increased risk of dry spells and water shortage are also recognised. The spatial implications of the Cabinet's position on water management and the associated adaptation measures have been incorporated in the Dutch Spatial Policy. Inclusion in the policy for Rural Areas offers an opportunity to combine the implementation of measures in rural areas for increased safety and flood prevention with measures for such objectives as improving water quality, combating dropping water-tables, reconstructing rural areas and improving the ecological infrastructure.

Organizations involved in providing water services also started to explore the implications of climate change in terms of vulnerability and options for adaptation. The **Norwegian** Water Resources and Energy Directorate is an example. In the **United Kingdom**, organizations in the water resources sector are also taking actions to prepare for a changing climate.

4.4.3 Protecting people and infrastructure from coastal and river floods

Coastal and low-lying areas constitute a substantial part of Europe. With changing rainfall pattern (including extremes) and global warming induced sea level rise, a number of countries will be facing increased risk of coastal and river flooding. Benefiting from the long tradition of dealing with extreme weather events, flood defence is among the areas with best developed adaptive measures. Policies, guidance documents, regulations, and even concrete technical adaptation actions have been developed at the EU, national, and sub-national levels. Some of these measures are not deliberately designed for adaptation to long-term climate change impacts, though. Instead, they are developed for addressing short-term extremes.

At **EU** level, a flood prevention and management action plan is being developed. EU Environment Ministers asked the European Commission to table a formal proposal for the plan, which will be based on solidarity and will make provision for an early warning system, integrated flood basin and flooding management plans and the development of flood risk maps. The plan includes a possible future Floods Directive.

At national level, many northern and western European countries have national flood management policies and guidelines. Integrating new information on climate change and its potential impacts, such policies and guidelines are being reviewed and adjusted periodically.

In the **United Kingdom**, the trauma of human misery and property loss caused by the 1953 coastal flooding alerted the Government to the potential dangers. A radical rethinking led to major new flood defence infrastructure being built and eventually the commissioning of the Thames Barrier in 1987. The first major IPCC Assessment (1991) led to changed approaches to coastal planning throughout the United Kingdom. This has incorporated an allowance for climate change and sea level rise built into all new coastal flooding infrastructure. Planning Policy Guidance 25, published by the Office of the Deputy Prime Minister (ODPM, 2001), takes a precautionary approach to managing development and flood risk. It aims to direct new development away from areas at highest risk of flooding and takes account of climate change....

In the **Netherlands**, criteria and boundary conditions for the safety features of all dykes and other protection infrastructure are periodically updated to incorporate the available information on climate change and other environmental changes. The Dutch coastal policy plan (3rd Coastal Policy, 2000) strongly emphasises the new challenges caused by climate change, especially sea level rise and an increase in the number of storms. Various national and sub-national policy plans like the Dutch Spatial Policy (Ministry of Housing, 2004) and the Water Policy Plan 21st Century (Nota Waterbeleid 21ste eeuw) (Ministry of Transport, 2000) recognize the need for adaptation in water management and coastal zone management. An example of an adaptation action for rivers is in Hengelo, where the peak flow of the Woolderbinnenbeek can be reduced by 60%, to prevent the downstream agricultural land and town centres from getting flooded.

EUROPEAN ENVIRONMENT AGENCY (EEA). "VULNERABILITY AND ADAPTATION TO CLIMATE CHANGE IN EUROPE" (2005) <HTTP://REPORTS.EEA.EUROPA.EU/TECHNICAL_REPORT_2005_1207_144937/EN/EEA_TECHNICAL_REPORT_7_2005.PDF> (ACCESSED NOVEMBER 21, 2007).

SEE ALSO *Coastal Populations; Coastlines, Changing; Economics of Climate Change; Energy Efficiency; Extinction; IPCC Climate Change 2007 Report: Impacts, Adaptation and Vulnerability; Lifestyle Changes; Refugees and Displacement; Social Cost of Carbon (SCC).*

BIBLIOGRAPHY

Books

Parry, M. L., et al, eds. *Climate Change 2007: Impacts, Adaptation and Vulnerability: Contribution of Working Group II to the Fourth Assessment Report of the Intergovernmental Panel on Climate Change.* New York: Cambridge University Press, 2007.

Periodicals

Davis, Margaret B., et al. "Range Shifts and Adaptive Responses to Quaternary Climate Change." *Science* 292 (2001): 673–679.

deMenocal, Peter B., et al. "Cultural Responses to Climate Change During the Late Holocene." *Science* 292 (2001): 667–674.

Giles, Jim. "How to Survive a Warming World." *Nature* 446 (2007): 716–717.

Kabat, Pavel, et al. "Climate Proofing the Netherlands." *Nature* 438 (2005): 283–284.

Parmesan, Camille, and Gary Yohe. "A Globally Coherent Fingerprint of Climate Change Impacts Across Natural Systems." *Nature* 421 (2003): 37–42.

Revkin, Andrew C. "Aid to Help Asia and Africa with Effects of Warming." *The New York Times* (August 9, 2007).

Web Sites

"China Now No. 1 in CO_2 Emissions; USA in Second Position." *Netherlands Environmental Assessment Agency*, June 22, 2007. <http://www.mnp.nl/en/dossiers/Climatechange/moreinfo/Chinanowno1inCO2emissionsUSAinsecondposition.html> (accessed November 10, 2007).

"Strategies for Adaptation to Sea Level Rise." *Intergovernmental Panel on Climate Change, Response Strategies Working Group, Coastal Systems Subgroup*, 1990. <http://yosemite.epa.gov/oar/globalwarming.nsf/UniqueKeyLookup/RAMR5EHLSJ/$File/adaption.pdf> (accessed November 10, 2007).

Larry Gilman

Aerosols

Introduction

Aerosols are naturally present microscopic airborne suspensions of liquid droplets or solid particles in a gas, usually air, which have tendencies to remain dispersed (floating) in the gas rather than to settle down. When the airborne suspensions are solid, such as dust and sea-salt, they are also commonly called particulate matter. A commonly seen natural event that spews out aerosols is an active volcano. Other examples of aerosols seen in Earth's atmosphere are mist, smog, and fog.

In addition, aerosols include all types of artificially made containers that hold a suspension of liquid or solid particles within a gaseous propellant under pressure. Such a container—sometimes called an aerosol dispenser, but more commonly called an aerosol can—seals in the suspension and, through a valve, dispenses it in a foam, liquid, or spray stream. Common products packaged in aerosol cans are cosmetics, detergents, foods, insecticides, and paints.

Historical Background and Scientific Foundations

The term aerosol is derived from the Greek term *aero*, which means pertaining to air, and the term *sol*, which is defined as any scattering, or dispersion, of microscopic or sub-microscopic particles in a liquid.

Naturally produced aerosol particles are considered by scientists to have a maximum size of one micrometer (or one millionth [10^{-6}] of a meter). They generally range from 4×10^{-8} to 4×10^{-5} in (10^{-7} to 10^{-4} cm) in diameter. However, violently moving suspensions can contain aerosol particles that are over one-hundred times larger than

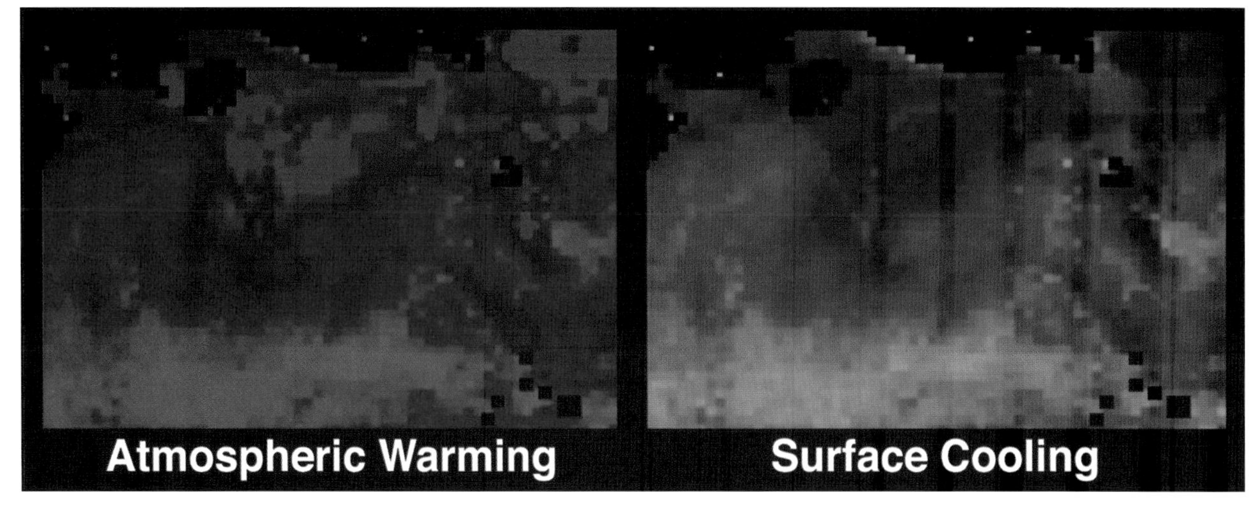

In this image based on NASA satellite data, the impact of black carbon pollution caused by humans is shown. The left image, taken from space, shows the absorption of black carbon aerosols in the atmosphere, with red indicating the highest levels of absorption and blue showing the lowest. The right image illustrates that aerosol particles reduce the amount of sunlight reaching Earth's surface, with the dark pixels showing where particles exert their cooling influence the most. *AP Images.*

WORDS TO KNOW

BIOMASS: The sum total of living and once-living matter contained within a given geographic area. Plant and animal materials that are used as fuel sources.

CONVECTION: The rising of warm air from an object, such as the surface of Earth.

DEFORESTATION: Those practices or processes that result in the change of forested lands to non-forest uses. This is often cited as one of the major causes of the enhanced greenhouse effect for two reasons: 1) the burning or decomposition of the wood releases carbon dioxide; and 2) trees that once removed carbon dioxide from the atmosphere in the process of photosynthesis are no longer present and contributing to carbon storage.

FOSSIL FUELS: Fuels formed by biological processes and transformed into solid or fluid minerals over geological time. Fossil fuels include coal, petroleum, and natural gas. Fossil fuels are non-renewable on the timescale of human civilization, because their natural replenishment would take many millions of years.

GLOBAL DIMMING: Decrease in amount of sunlight reaching Earth's surface caused by light blockage by clouds and aerosols. Global dimming increased from 1960 to 1990, reducing sunlight reaching Earth's surface by 4%, but this trend reversed after 1990 in most locations.

GREENHOUSE GASES: Gases that cause Earth to retain more thermal energy by absorbing infrared light emitted by Earth's surface. The most important greenhouse gases are water vapor, carbon dioxide, methane, nitrous oxide, and various artificial chemicals such as chlorofluorocarbons. All but the latter are naturally occurring, but human activity over the last several centuries has significantly increased the amounts of carbon dioxide, methane, and nitrous oxide in Earth's atmosphere, causing global warming and global climate change.

PARTICULATE MATTER: Matter consisting of small particles. Particulate matter that is airborne forms aerosol pollution; particulate matter may also mix with water or lie on the surface of snow, ice, or ground. On snow or ice, small quantities of dark particulate matter (e.g., soot from fossil-fuel burning) can greatly accelerate melting.

PRECIPITATION: Moisture that falls from clouds. Although clouds appear to float in the sky, they are always falling, their water droplets slowly being pulled down by gravity. Because the water droplets are so small and light, it can take 21 days to fall 1,000 ft (305 m) and wind currents can easily interrupt their descent. Liquid water falls as rain or drizzle. All raindrops form around particles of salt or dust. (Some of this dust comes from tiny meteorites and even the tails of comets.) Water or ice droplets stick to these particles, then the drops attract more water and continue getting bigger until they are large enough to fall out of the cloud. Drizzle drops are smaller than raindrops. In many clouds, raindrops actually begin as tiny ice crystals that form when part or all of a cloud is below freezing. As the ice crystals fall inside the cloud, they may collide with water droplets that freeze onto them. The ice crystals continue to grow larger, until large enough to fall from the cloud. They pass through warm air, melt, and fall as raindrops.

STRATOSPHERE: The region of Earth's atmosphere ranging between about 9 and 30 mi (15 and 50 km) above Earth's surface.

SUSPENSION: A temporary mixture of a solid in a gas or liquid from which the solid will eventually settle out.

TROPOSPHERE: The lowest layer of Earth's atmosphere, ranging to an altitude of about 9 mi (15 km) above Earth's surface.

normal, as is often the case with fog and cloud droplets and dust particles.

The climate of Earth has been affected by artificially produced aerosols. The combustion of fossil fuels is one detrimental way that sulfate aerosols are introduced into Earth's atmosphere. Other sources of artificially produced aerosols include agriculture, deforestation, industry, mining, and transportation.

Artificially produced aerosols have been made from such particulate matter as asbestos, diesel fuel, and natural silicon dioxide (silica). However, these substances have been found to be harmful to humans, causing a variety of diseases such as anthracosis, a lung disease of coal miners commonly called black lung. It has been found that humans who work around such dangerous aerosols can be largely protected from health problems when they wear respirators.

■ Impacts and Issues

Aerosols are naturally found in Earth's troposphere and the stratosphere. The particular layer in which each aerosol type is found depends on its size, chemical composition, origin, and other such physical factors. These factors can determine whether aerosol particles absorb or scatter solar radiation and, thus, how much radiation reaches Earth's surface from the sun. Research is ongoing to learn more about the wide types of aerosols in the atmosphere because of their affect on global weather patterns and the climate in general.

The study of aerosols in the atmosphere is important due to the number of artificially produced aerosols that have been, and continue to be, produced by humans. For instance, when humans burn fossil fuels, the aerosols produced by such activities absorb heat and, thus, contribute to global warming. Such aerosols are called atmosphere nuclei—that is, microscopic particles in the atmosphere that attract water droplets.

These aerosols tend to add more moisture to the atmosphere, which makes it more likely that precipitation will occur. Precipitation may occur outside of where the aerosols originated because it is easily transported by convection and wind currents. Consequently, problems around the world can occur due to localized aerosol use. Many of these atmosphere nuclei come from industrial environments where combustion forms particles such as sulfur dioxide and sulfur trioxide.

Some scientific research has found that many artificially produced aerosols have a cooling affect on the atmosphere. However, scientists contend, based on preliminary research, that such cooling (what is sometimes called global dimming) has a minimal effect on Earth's climate at best and does not counterbalance global warming, which is produced by greenhouse gases such as carbon dioxide, methane, and nitrous oxide.

Because of the increased presence of artificially produced aerosols in Earth's atmosphere, many organizations are actively seeking more and better information about aerosols. Almost continuously since 1978, the U.S. National Aeronautics and Space Administration (NASA) has provided aerosol maps of the world from observations made by a series of satellites called the Total Ozone Mapping Spectrometer (TOMS). The TOMS map provides a global map of aerosol particles found on any particular day on and above Earth. NASA also provides another map to show the cloud cover at the exact time the aerosol particle map is made.

The data from TOMS are being used by scientists around the world to better understand the behavior of aerosol particles within Earth's atmosphere. TOMS has been important to scientists because it was the first instrument to provide data of aerosol particles, especially as they move between land and water regions. Consequently, TOMS has provided valuable information about natural events such as forest fires and biomass burnings that affect the production of aerosol particles.

The TOMS satellites were replaced in January 1, 2006 by the Ozone Monitoring Instrument (OMI), which continues to provide even more advanced and detailed aerosol maps of the world as it orbits Earth onboard the *Auro* satellite, or Earth Observing Satellite (EOS) CH-1.

SEE ALSO *Carbon Dioxide (CO_2); Global Warming; Greenhouse Effect; Greenhouse Gases; Methane; Nitrous Oxide; Sulfate Aerosol; Volcanism.*

BIBLIOGRAPHY

Books

Friedlander, Sheldon K. *Smoke, Dust, and Haze: Fundamentals of Aerosol Dynamics.* New York: Oxford University Press, 2000.

Hinds, William C. *Aerosol Technology: Properties, Behavior, and Measurement of Airborne Particles.* Hoboken, NJ: Wiley-Interscience, 1999.

Web Sites

"Aerosols." *National Institute for Occupational Safety and Health (NIOSH), U.S. Centers for Disease Control and Prevention (CDC).* <http://www.cdc.gov/niosh/topics/aerosols/default.html> (accessed December 1, 2007).

"Aura: Atmospheric Chemistry." *Goddard Space Flight Center, U.S. National Aeronautics and Space Administration (NASA).* <http://aura.gsfc.nasa.gov/index.html> (accessed December 3, 2007).

"Data Product: AEROSOL INDEX." *National Aeronautics and Space Administration (NASA),* October 26, 2007. <http://toms.gsfc.nasa.gov/aerosols/aerosols_v8.html> (accessed December 3, 2007).

"Total Ozone Mapping Spectrometer." *National Aeronautics and Space Administration (NASA),* October 26, 2007. <http://toms.gsfc.nasa.gov/> (accessed December 3, 2007).

"Visible Earth: Aerosols." *National Aeronautics and Space Administration (NASA),* November 20, 2006. <http://visibleearth.nasa.gov/view_set.php?categoryID=109> (accessed December 3, 2007).

"Welcome to AAAR." *American Association for Aerosol Research (AAAR).* <http://www.aaar.org/> (accessed December 1, 2007).

Africa: Climate Change Impacts

■ Introduction

Africa, the world's second-largest continent, is inhabited by almost a billion people. Because of the interaction of climate change with other stresses in Africa—such as widespread poverty, population growth, acquired immunodeficiency syndrome (AIDS), and overgrazing and other ecosystem damage—it is considered to be the continent most vulnerable to the effects of climate change. Projected impacts of climate change on Africa include biodiversity loss (extinction of plant and animal species), diminished agriculture with increased hunger, increased disease, forced migration of populations (especially out of the Sahel), and more. About 70% of Africa's population lives by farming, often subsistence farming, with the poorest members of society tending to be most dependent on agriculture.

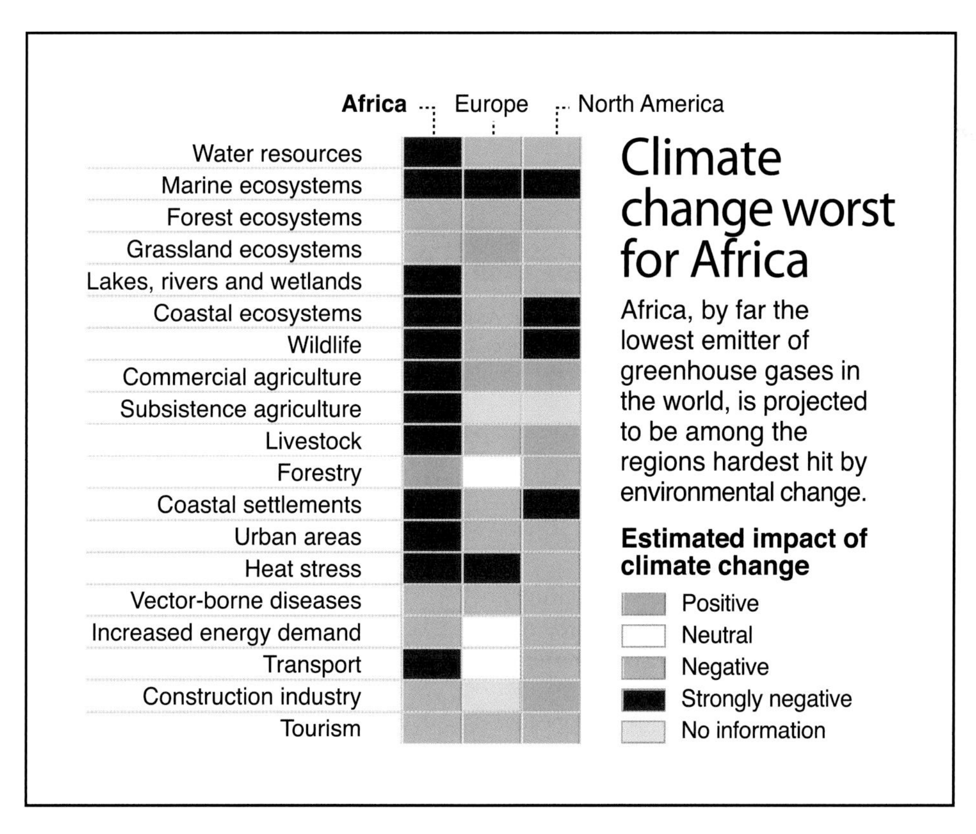

In this graphic from the Associated Press, the impact of climate change in Africa is compared to Europe and North America. [Source: Intergovernmental Panel on Climate Change.] *AP Images.*

Many scientists forecast that climate change will hit Africa especially hard. In Kenya, Maasai were among the climate change protesters in Nairobi in November 2006. According to the activists, the industrialized nations of the world are not doing enough to cut greenhouse gases. In particular, they criticized the United States and President George W. Bush. *AP Images.*

Historical Background and Scientific Foundations

For approximately 500 years, from the fifteenth century to the mid-twentieth century, Africa was the source of millions of slaves for European and Arab raiders, resulting in many millions of deaths. In the nineteenth and twentieth centuries, Africa was dominated by European colonial powers, who divided most of its territory and exploited the region's resources without regard for its inhabitants. One result of this colonial history is a legacy of weak social networks, chaotic political systems, and poverty. These problems have ongoing effects on ecosystems and impact the region's ability to adapt to climate change.

In 2007, the Intergovernmental Panel on Climate Change (IPCC), the world's largest and most authoritative body of climate and weather scientists, released the report of its Working Group II, *Climate Change 2007: Impacts, Adaptation and Vulnerability*. The report noted the following points, among others, about the impacts of climate change on Africa:

- Africa is one of the continents most vulnerable to global warming. This situation is made worse by Africa's low capacity to adapt to climate change. Poor access to loans, markets, technology,

WORDS TO KNOW

BIODIVERSITY: Literally, "life diversity": the number of different kinds of living things. The wide range of organisms—plants and animals—that exist within any given geographical region.

GREENHOUSE GASES: Gases that cause Earth to retain more thermal energy by absorbing infrared light emitted by Earth's surface. The most important greenhouse gases are water vapor, carbon dioxide, methane, nitrous oxide, and various artificial chemicals such as chlorofluorocarbons. All but the latter are naturally occurring, but human activity over the last several centuries has significantly increased the amounts of carbon dioxide, methane, and nitrous oxide in Earth's atmosphere, causing global warming and global climate change.

INTERGOVERNMENTAL PANEL ON CLIMATE CHANGE (IPCC): Panel of scientists established by the World Meteorological Organization (WMO) and the United Nations Environment Programme (UNEP) in 1988 to assess the science, technology, and socioeconomic information needed to understand the risk of human-induced climate change.

MALARIA: Group of parasitic diseases common in tropical and subtropical areas. According to 2004 statistics from the U.S. Centers for Disease Control and Prevention (CDC), approximately 300 million cases occur annually worldwide, and an estimated 700,000 to 2.7 million persons die of malaria each year. About 90% of these deaths occur in sub-Saharan Africa, with the majority being children. In 2002, in the United States (even though malaria has been eradicated from the country since the early 1950s), 337 cases of malaria, including eight deaths, were reported to the CDC.

SAHEL: The transition zone in Africa between the Sahara Desert to the north and tropical forests to the south. This dry land belt stretches across Africa and is under stress from land use and climate variability.

SUBSISTENCE FARMING: Agriculture carried on for the sake of the food it produces, rather than to produce crops to sell for cash.

IN CONTEXT: DISEASE PERILS

Epidemics can decimate the families and economics of a region. Epidemics of disease disrupt national economies, as large numbers of people become unable to work or care for themselves. For example, according to World Health Organization (WHO) estimates, in 2004 the number of healthy years of life lost due to premature death and disability caused by trypanosomiasis was 1.5 million. Trypanosomiasis (also known as African sleeping sickness because of the semi-conscious stupor and excessive sleep that can occur in someone who is infected) is an infection passed to humans through the bite of the tsetse fly. Thus, it is a vector-borne disease. If left untreated, trypanosomiasis is ultimately fatal. Since many African regions are still agricultural, the rural-based disease affects those who are most important to the economy. Of the 48,000 deaths that occurred in 2004, 31,000 were males, who are often the working family members.

Trypanosomiasis is a major health concern in approximately 20 countries in Africa. The WHO estimates that over 66 million people are at risk of developing the disease. However, fewer than 4 million people are being monitored and only about 40,000 people are treated every year. The proportion of people being monitored or treated is smaller than other tropical diseases, even though trypanosomiasis can increase to epidemic proportions and the death rate for those who are not treated is 100%.

The resurgence of trypanosomiasis during the 1970s was due to the interruptions in the monitoring of disease outbreaks, the displacement of people due to regional conflicts, and environmental changes. These problems are ongoing. In particular, the documented warming of the atmosphere will make Africa even more hospitable to the spread of the territory of the tsetse fly, which could increase the geographical distribution of trypanosomiasis.

machinery, and the like make Africa more vulnerable to climate change. A variety of disasters and conflicts already plague the region, including wars, overgrazing and other forms of ecosystem degradation, and AIDS. (About 25 million persons were infected with the human immunodeficiency virus [HIV] in Africa as of 2007, and about 2 million Africans die of AIDS each year.)

- African farmers have been adapting to current climate change, but these adaptations may be overwhelmed by future climate change. Food production may be "severely compromised" (in the IPCC's words) in many African countries. About one-third of all income in Africa is from agriculture, with one-third of household income coming from crops and livestock. Changes in agriculture due to loss of rainfall may be severe in some areas, especially in semi-arid areas, which are already marginal for agriculture due to limited rainfall. Semi-arid areas are projected to see a shortening of the growing season of more than 20% by 2050, according to some scenarios that address the continued emission of greenhouse gases. Large areas of land where agriculture is now marginal may be forced out of production.

- The water shortages faced by some countries will be made worse by climate change, while some countries that do not now have water shortages will be at risk for them. This is particularly true of northern Africa. About one-fifth of Africa's population—about 200 million people—presently experiences severe water shortages. By the 2050s, the population at risk for increased water stress—inadequate access to good-quality water for household use and agriculture—could rise to 350–600 million.
- Changes in African ecosystems are already seen, especially in southern Africa. About 25–40% of mammal species in national parks in sub-Saharan Africa (south of the Sahara desert) may become endangered in the twenty-first century.
- Human health will be threatened by global warming. For example, increased rainfall in eastern Africa will increase the land area where malaria is endemic. Presently, malaria kills about 900,000 people annually in Africa, 70% of them children. Pregnant women and children are particularly likely to be affected by climate change because of their greater vulnerability to heat, infectious diseases, and inadequate food.

Impacts and Issues

Most studies of climate impact focus on wealthier parts of the world, such as Europe and North America. For example, according to the IPCC, very little detailed information is available on the vulnerability of the energy sector (the part of the economy that produces heat and electricity) in Africa. Experts suggest that reductions of greenhouse-gas emissions in industrialized and developing countries, as well as assistance in mitigating the harmful effects of climate change, are essential to reducing climate-change impacts in Africa.

However, they note that it is difficult to persuade nations to modify their behavior for the benefit of others. It is likely that Africa will, due to its greater vulnerability, disproportionately experience the impacts of climate change, although its inhabitants have contributed only a tiny fraction of the greenhouse gases that are causing most of that change.

SEE ALSO *Arctic People: Climate Change Impacts; Asia: Climate Change Impacts; Europe: Climate Change Impacts; North America: Climate Change Impacts; Small Islands: Climate Change Impacts; South America: Climate Change Impacts.*

BIBLIOGRAPHY

Books

Parry, M. L., et al, eds. *Climate Change 2007: Impacts, Adaptation and Vulnerability: Contribution of Working Group II to the Fourth Assessment Report of the Intergovernmental Panel on Climate Change.* New York: Cambridge University Press, 2007.

Periodicals

Thomas, David S. G., et al. "Remobilization of Southern African Desert Dune Systems by Twenty-First Century Global Warming." *Nature* 435 (June 30, 2005): 1218–1221.

Web Sites

Desanker, Paul. "Impact of Climate Change in Africa." *WWF-UK*, August 2002. <http://www.wwf.org.uk/filelibrary/pdf/africa_climate_text.pdf> (accessed September 16, 2007).

"York Scientists Warn of Dramatic Impact of Climate Change on Africa" (press release). *University of York* (United Kingdom), June 9, 2005. <http://www.york.ac.uk/admin/presspr/pressreleases/climateafrica.htm> (accessed September 16, 2007).

Larry Gilman

Agriculture: Contribution to Climate Change

■ Introduction

Agriculture is the growing of plants (crops) and animals (livestock) for food and other purposes. In 2007, lands used for crop growing and animal grazing took up 40–50% of Earth's land surface, a 10% increase since 1961. Agriculture contributes to global climate change by releasing carbon dioxide (CO_2), methane (CH_4), and nitrous oxide (N_2O), the three gases presently causing the most greenhouse warming. Agriculture emitted 5.1–6.1 billion tons of CO_2 equivalent in 2005, that is, 10–12% of all human-caused (anthropogenic) releases of greenhouse gases in that year.

Its share of CH_4 and N_2O emissions was much greater: agriculture contributes about 47% of global CH_4 and 58% of N_2O. These figures do not count greenhouse emissions from electricity and fuel used in agriculture for machinery, buildings, processing, and transport. Also, emissions vary widely among countries, with more industrialized countries deriving much less of their greenhouse emissions from agriculture: for example, in the United States in 2006, direct emissions from agriculture accounted for only 6% of total emissions.

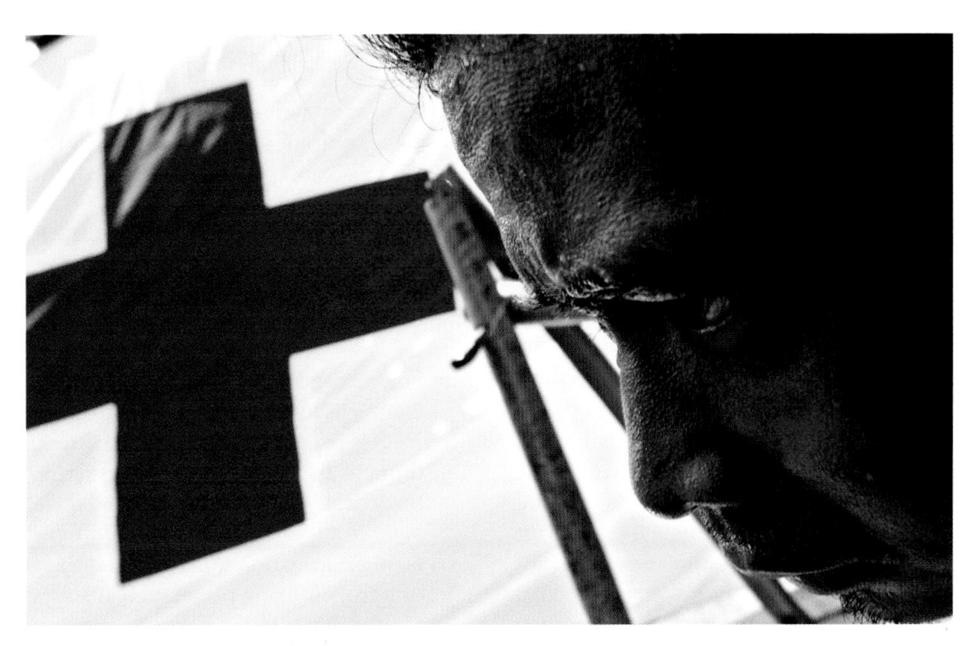

To raise awareness about global climate change, an activist in Indonesia takes part in a street theater production. With his face painted blue, he represents water, which is abundant in some parts of the world but not in others. In Indonesia, changing seasonal patterns have impacted how farmers plant and harvest rice. *AP Images.*

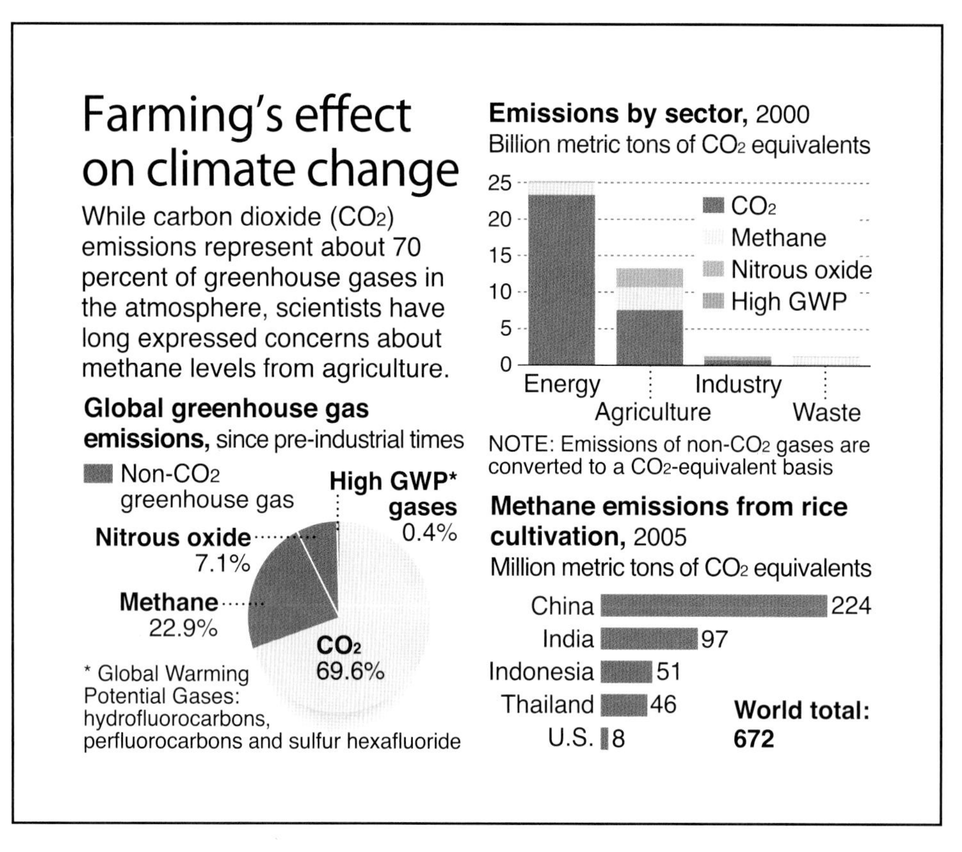

This graphic from the Associated Press charts farming's effect on climate change, particularly the cultivation of rice. [Source: Environmental Protection Agency.] *AP Images.*

Historical Background and Scientific Foundations

For most of the human race's existence, it did not practice agriculture. For many hundreds of thousands of years, humans and their near-human ancestors practiced various forms of hunting and gathering, finding edible plants and animals in the environment rather than raising them. The first known tools, made about 2.5 million years ago, were food processors, chipped stones devised to help butcher antelopes and horses in the part of Africa that is now Ethiopia. Cropping and livestock-raising began in the Middle East about 9500 BC and gradually spread to most peoples of the world, though there are still a few communities that live primarily through hunting and gathering.

Agriculture both releases and absorbs greenhouse gases. Plants absorb CO_2 from the atmosphere, extracting its carbon to build their tissues. Dead roots and other plant parts remaining in the soil after harvest increase the soil's carbon content, though depending on environmental conditions this carbon may be re-released to the atmosphere by bacterial decay. Not counting CO_2 releases from fuel and electricity usage associated with plowing, spraying, harvesting, transport, processing, and storage, agriculture's CO_2 release and uptake are about balanced. Almost all of agriculture's direct impact on climate comes from its releases of CH_4 and N_2O.

CH_4 and N_2O have a greater greenhouse impact, ton for ton, than does carbon dioxide. Methane is over 20 times as effective, by weight, at causing climate change than is CO_2, and nitrous oxide is about 296 times as effective. However, because much greater quantities of CO_2 are being released into the atmosphere, CO_2 accounts for most of the greenhouse warming now occurring.

Both CH_4 and N_2O are emitted, for the most part, by biological processes. These are as follows (where U.S. figures are for 2006 and global figures are for 2005):

- Soil management. When soils are disturbed, oxygen combines more readily with nitrogen in the soil and N_2O is released. Nitrogen is an essential plant nutrient that is added to soils as fertilizer in much of the world. Practices that increase formation of N_2O from soils include fertilization with nitrogen or manure, irrigation, draining, and the growing of nitrogen-fixing plants such as alfalfa. In the United States, 60% of direct agricultural emissions—in terms of greenhouse-warming efficacy, not by raw tonnage—consist of N_2O from soils. Globally, N_2O from soils comprises 38% of total non-CO_2 direct emissions from agriculture.

WORDS TO KNOW

ANAEROBIC: Lacking free molecular oxygen (O_2). Anaerobic environments lack O_2; anaerobic bacteria digest organic matter such as dead plants in anaerobic environments such as deep water and the digestive systems of cattle. Anaerobic digestion releases methane, a greenhouse gas.

BIOMASS: The sum total of living and once-living matter contained within a given geographic area. Plant and animal materials that are used as fuel sources.

DEFORESTATION: Those practices or processes that result in the change of forested lands to non-forest uses. This is often cited as one of the major causes of the enhanced greenhouse effect for two reasons: 1) the burning or decomposition of the wood releases carbon dioxide; and 2) trees that once removed carbon dioxide from the atmosphere in the process of photosynthesis are no longer present and contributing to carbon storage.

ENTERIC FERMENTATION: Digestion of organic matter by bacteria in an animal's digestive system. Enteric fermentation produces methane, a greenhouse gas.

FOSSIL FUELS: Fuels formed by biological processes and transformed into solid or fluid minerals over geological time. Fossil fuels include coal, petroleum, and natural gas. Fossil fuels are non-renewable on the timescale of human civilization, because their natural replenishment would take many millions of years.

KYOTO PROTOCOL: Extension in 1997 of the 1992 United Nations Framework Convention on Climate Change (UNFCCC), an international treaty signed by almost all countries with the goal of mitigating climate change. The United States, as of early 2008, was the only industrialized country to have not ratified the Kyoto Protocol, which is due to be replaced by an improved and updated agreement starting in 2012.

PER CAPITA: Latin phrase meaning "for each person." The per capita greenhouse emissions of a country, for example, are the country's total emissions divided by the number of people in the country.

RUMINANT ANIMALS: Animals such as cows, deer, sheep, and buffalo, and many others, also called simply ruminants, that digest their food partly by bringing it from the stomach to the mouth (where it is called a "cud") and chewing it there before re-swallowing. The digestive systems of ruminant animals contain bacteria that produce methane: methane from domesticated ruminant animals contributes significantly to global warming.

- Enteric fermentation. Bacteria that live in oxygen-free environments digest organic materials such as dead plants using a chemical process that produces methane. Oxygen-free (anaerobic) environments are found at the bottoms of bodies of water and in the intestines of ruminant animals such as sheep, goats, and cows. Methane generated by bacteria in ruminant animals is emitted as flatulence. Continued anaerobic digestion of animal manure after it has left the animal produces both CH_4 and N_2O (mostly methane). In the United States, 25% of direct agricultural emissions consist of methane from enteric fermentation. Globally, CH_4 from enteric fermentation constitutes 32% of non-CO_2 agricultural emissions.

- Manure management. Manure from livestock is often collected in ponds, lagoons, pits, or tanks. Anaerobic digestion of the waste in these systems releases methane and nitrous oxide. In the United States, 13% of direct agricultural emissions consists of CH_4 and N_2O emissions from enteric fermentation (9% CH_4, 4% N_2O). Globally, CH_4 from manure management constitutes 7% of non-CO_2 agricultural emissions.

- Rice farming. Anaerobic digestion of plant matter at the bottom of rice paddies (rice fields), which are flooded with water for much or all of the year, produces significant methane. In the United States, only 2% of direct agricultural emissions consist of methane from rice paddies, but globally, rice is a much more important food than it is in the United States and most other developed countries. Thus, CH_4 from rice paddies constitutes 11% of non-CO_2 agricultural emissions. Ninety-seven percent of world rice emissions occur in developing countries.

- Biomass residue burning. Burning crop residues such as cornstalks releases CO_2, CH_4, and N_2O. In the United States, less than 1% of direct agricultural emissions consist of greenhouse gases from residue burning; globally, however, it constitutes 12% of non-CO_2 agriculture emissions. Ninety-two percent of world biomass burning emissions occurs in developing countries.

Emissions of CH_4 and N_2O from agriculture are increasing with human population growth, which is mostly occurring in developing countries and entails corresponding growth in agriculture, and with rising per capita demand for meat in some developing countries. From 1990 to 2005, world agricultural emissions of CH_4 and N_2O increased by almost 17%. Direct agricultural emissions in developing countries (i.e., those not on the Annex I list attached to the Kyoto Protocol climate treaty) increased by 32% during this time period, while emissions from developed (Annex I) countries fell by 12% overall. By 2005, developing countries were accounting for about 75% of all agricultural direct emissions. Indirect agricultural emissions were higher in developed countries, where agriculture is more mechanized and therefore energy-intensive, and where food is far more intensively processed, packaged, marketed, and refrigerated. CO_2 emissions from on-farm fossil fuel use alone were equal to 12.7% (14 million metric tons CO_2 equivalent, MMTCE) of direct greenhouse emissions in the United States in 2006 (120 MMTCE).

Deforestation in the tropics is one of the main non-fossil-fuel contributors to global climate change, producing almost a third of global CO_2 emissions. Expansion of agricultural land area is one of the main drivers of such deforestation. In Brazil, where a third of the world's rainforest is found, the rainforest was being cleared in the early 2000s at a rate of about 4,000 mi^2 (10,000 km^2) per year, almost entirely for cattle ranching, soy farming, and small-scale subsistence farming.

Such deforestation is often referred to as slash-and-burn agriculture, because land is cleared by the simple expedient of cutting down all the trees, piling them up, and burning them. This immediately releases their stored carbon into the air—750 tons of CO_2 per acre (0.4 hectare) for an old-growth Indonesian rain forest. As of 2005, slash-and-burn deforestation to clear land for agriculture made Indonesia the world's third largest greenhouse-gas emitter and Brazil the world's fourth largest.

In some cases, the burning of cleared trees is only the beginning of the greenhouse emissions from slash-and-burn agriculture. In Indonesia in the 1990s, a government plan to expand rice production led to the clearance and drainage of almost 4,000 mi^2 (about 1 million hectares) of peat-swamp forests. The plan was a disaster: rice was never successfully produced, but much of the peat—a dense mat of partly decomposed plants rich in carbon, essentially young coal—dried out and caught fire, contributing to catastrophic fires in the late 1990s.

In 2007, the Australian and Indonesian governments announced a joint plan to preserve 270 mi^2 (70,000 hectares) of remaining Indonesian peat forests, re-flood 770 mi^2 (200,000 hectares) of dried peatland, and replant 100 million trees on the rehabilitated land. These measures were predicted to prevent 700 million tons of greenhouse emissions over 30 years. Annually, this would be more than the entire annual output of Australia. However, the joint project would still restore only a fraction of the land ravaged by the Mega Rice project of the 1990s.

■ Impacts and Issues

Is Meat the Number One Cause of Global Warming?

In 2007, vegan and animal-rights organizations ran ads in the U.S. media stating that emissions from meat-raising contribute more to global warming than cars do. For example, an ad by the group People for the Ethical Treatment of Animals (PETA) read: "Too Chicken to Go Vegetarian? Meat Is the Number One Cause of Global Warming."

According to the United Nations' Intergovernmental Panel on Climate Change's (IPCC's) 2007 Assessment Report, road transport (cars and trucks) contribute 17.8% of greenhouse emissions, with 10.2% coming from automobiles. In comparison, the IPCC report stated that agriculture contributed 10–12% of emissions. Some other U.N. sources attributed a much higher percentage of greenhouse emissions to agriculture. For example, a 2006 report from the Food and Agriculture Organization of the United Nations stated that the livestock sector of agriculture alone is responsible for 18% of greenhouse-gas emissions, "a higher share than transport." Although 18% is higher than 17.8%, the 0.2% difference between the two numbers is not statistically meaningful. It is therefore not clear whether, globally, agriculture or the livestock subsector of agriculture contribute more to climate change than does transport.

Moreover, research does not indicate that meat-eating by people in developed countries such as the United States contributes more to global warming than does driving. As noted earlier, U.S. direct agricultural emissions comprise only 6% of U.S. greenhouse emissions. Gases are also emitted to create food imported by the United States, but only 15% of U.S. food consumption was imported as of 2007. Therefore, assuming proportional emissions for imported food, all U.S. food consumption is responsible for less than 7% of U.S. emissions. Non-diesel cars and trucks, on the other hand, accounted for almost 20% of U.S. greenhouse emissions. The average U.S. citizen thus contributes about three times more to greenhouse warming by driving than by eating, and the meat portion of the American diet is proportionally even smaller.

Finally, the energy sector, not agriculture, is the largest contributor to global warming worldwide. Burning fossil fuels for heat and to generate electricity causes more than half of all global warming.

Nevertheless, it is true that, ounce for ounce, meat production contributes significantly more to global climate change than does the production of vegetable foods, and that agriculture is a major contributor to global warming. Reducing per-capita meat consumption in the developed world, medical and climate experts have argued, would have both health and climate benefits.

Mitigation of Agricultural Emissions

Altered agricultural practices can reduce agriculture's contribution to global warming. More efficient delivery of nitrogen to crops would reduce N_2O emissions and other ecological harms; livestock management for more efficient digestion of feeds would save money and reduce CH_4 emissions; crop residues and manures can be used as fuel sources (e.g., methane from decaying manure can be burned to generate electricity, rather than released to the atmosphere, and the fermentation residue can still be used as fertilizer). Many other measures are described by the IPCC. Cost-effective greenhouse mitigation measures are often friendly to sustainable-development goals, enhancing profitability and reducing soil losses.

■ Primary Source Connection

Methane (CH_4) and nitrous oxide (N_2O) are major greenhouse gases. Agriculture is a common, and often overlooked, contributor of methane and nitrous oxide. Both methane and nitrous oxide are produced naturally by livestock and soil management. Some human-controlled agricultural management techniques, however, increase methane and nitrous oxide production. This article examines methane and nitrous oxide production by various United States agricultural processes.

The Environmental Protection Agency (EPA) is an agency of the United States government that is devoted to protecting human health and working toward a cleaner environment. The EPA was founded in 1970 and currently employs over 17,000 people.

INVENTORY OF U.S. GREENHOUSE GAS EMISSIONS AND SINKS: 1990–2005

Agricultural activities contribute directly to emissions of greenhouse gases through a variety of processes, including the following source categories: enteric fermentation in domestic livestock, livestock manure management, rice cultivation, agricultural soil management, and field burning of agricultural residues. CH_4 and N_2O were the primary greenhouse gases emitted by agricultural activities. CH_4 emissions from enteric fermentation and manure management represented about 21 percent and 8 percent of total CH_4 emissions from anthropogenic activities, respectively, in 2005. Agricultural soil management activities such as fertilizer application and other cropping practices were the largest source of U.S. N_2O emissions in 2005, accounting for 78 percent. In 2005, emission sources accounted for in the Agricultural chapters were responsible for 7.4 percent of total U.S. greenhouse gas emissions....

Agricultural Soil Management

N_2O is produced naturally in soils through microbial nitrification and denitrification processes. A number of anthropogenic activities add to the amount of nitrogen available to be emitted as N_2O by microbial processes. These activities may add nitrogen to soils either directly or indirectly. Direct additions occur through the application of synthetic and organic fertilizers; production of nitrogen-fixing crops and forages; the application of livestock manure, crop residues, and sewage sludge; cultivation of high-organic-content soils; and direct excretion by animals onto soil. Indirect additions result from volatilization and subsequent atmospheric deposition, and from leaching and surface run-off of some of the nitrogen applied to or deposited on soils as fertilizer, livestock manure, and sewage sludge. In 2005, agricultural soil management accounted for 78 percent of U.S. N_2O emissions. From 1990 to 2005, emissions from this source decreased by 1.8 Tg CO_2 Eq. (0.5 percent); year-to-year fluctuations are largely a reflection of annual variations in weather, synthetic fertilizer consumption, and crop production.

Enteric Fermentation

During animal digestion, CH_4 is produced through the process of enteric fermentation, in which microbes residing in animal digestive systems break down food. Ruminants, which include cattle, buffalo, sheep, and goats, have the highest CH_4 emissions among all animal types because they have a rumen, or large fore-stomach, in which CH_4-producing fermentation occurs. Non-ruminant domestic animals, such as pigs and horses, have much lower CH_4 emissions. In 2005, enteric fermentation was the source of about 21 percent of U.S. CH_4 emissions, and about 70 percent of the CH_4 emissions from agriculture. From 1990 to 2005, emissions from this source decreased by 3.6 Tg CO_2 Eq. (3 percent). Generally, emissions have been decreasing since 1995, mainly due to decreasing populations of both beef and dairy cattle and improved feed quality for feedlot cattle.

Manure Management

Both CH_4 and N_2O result from manure management. The decomposition of organic animal waste in an anaerobic environment produces CH_4. The most important factor affecting the amount of CH_4 produced is how the manure is managed, because certain types of storage and treatment systems promote an oxygen-free environment. In particular, liquid systems tend to encourage anaerobic conditions and produce significant quantities of CH_4, whereas solid waste management approaches produce little or no CH_4. Higher temperatures and moist climatic conditions also promote CH_4 production.

CH_4 emissions from manure management were 41.3 Tg CO_2 Eq., or about 8 percent of U.S. CH_4 emissions in 2005 and 26 percent of the CH_4 emissions from agriculture. From 1990 to 2005, emissions from this source increased by 10.4 Tg CO_2 Eq. (34 percent). The bulk of this increase was from swine and dairy cow manure, and is attributed to the shift of the swine and dairy industries towards larger facilities. Larger swine and dairy farms tend to use liquid management systems.

N_2O is also produced as part of microbial nitrification and denitrification processes in managed and unmanaged manure. Emissions from unmanaged manure are accounted for within the agricultural soil management source category. Total N_2O emissions from managed manure systems in 2005 accounted for 9.5 Tg CO_2 Eq., or 2 percent of U.S. N_2O emissions. From 1990 to 2005, emissions from this source category increased by 0.9 Tg CO_2 Eq. (10 percent), primarily due to increases in swine and poultry populations over the same period.

Rice Cultivation

Most of the world's rice, and all of the rice in the United States, is grown on flooded fields. When fields are flooded, anaerobic conditions develop and the organic matter in the soil decomposes, releasing CH_4 to the atmosphere, primarily through the rice plants. In 2005, rice cultivation was the source of 1 percent of U.S. CH_4 emissions, and about 4 percent of U.S. CH_4 emissions from agriculture. Emission estimates from this source have decreased about 3 percent since 1990.

Field Burning of Agricultural Residues

Burning crop residues releases N_2O and CH_4. Because field burning is not a common debris clearing method in the United States, it was responsible for only 0.2 percent of U.S. CH_4 (0.9 Tg CO_2 Eq.) and 0.1 percent of U.S. N_2O (0.5 Tg CO_2 Eq.) emissions in 2005. Since 1990, emissions from this source have increased by approximately 28 percent.

EPA. "INVENTORY OF U.S. GREENHOUSE GAS EMISSIONS AND SINKS: 1990–2005." <HTTP://WWW.EPA.GOV/CLIMATECHANGE/EMISSIONS/DOWNLOADS06/07CR.PDF> (ACCESSED NOVEMBER 29, 2007).

SEE ALSO *Agriculture: Vulnerability to Climate Change; Biofuel Impacts; Cow Power; Sustainability.*

BIBLIOGRAPHY

Books

Parry, M. L., et al, eds. *Climate Change 2007: Impacts, Adaptation and Vulnerability: Contribution of Working Group II to the Fourth Assessment Report of the Intergovernmental Panel on Climate Change.* New York: Cambridge University Press, 2007.

Periodicals

Asner, Gregory P. "Grazing Systems, Ecosystem Responses, and Global Change." *Annual Review of Environment and Resources* 29 (2004): 261–299.

Deutsch, Claudia H. "Trying to Connect the Dinner Plate to Climate Change." *The New York Times* (August 29, 2007).

Izaurralde, R. César, et al. "Carbon Cost of Applying Nitrogen Fertilizer." *Science* 288 (2000): 809.

Web Sites

Johnson, Renée. "Climate Change: The Role of the U.S. Agriculture Sector." *Congressional Research Service*, March 6, 2007. <http://fpc.state.gov/documents/organization/81931.pdf> (accessed November 5, 2007).

Steinfeld, Henning, et al. "Livestock's Long Shadow: Environmental Issues and Options." *Food and Agriculture Organization of the United Nations*, 2006. <http://www.virtualcentre.org/en/library/key_pub/longshad/A0701E00.pdf> (accessed November 5, 2007).

Larry Gilman

Agriculture: Vulnerability to Climate Change

Introduction

Agriculture both contributes to and is vulnerable to climate change. It contributes to climate change because it is a source of the greenhouse gases carbon dioxide (CO_2), methane (CH_4), and nitrous oxide (N_2O). It is vulnerable to climate change because all agriculture depends on acceptable temperature ranges and patterns of rainfall for raising crops and livestock.

Climate change may create opportunities or benefits for some forms of agriculture in some parts of the world. For example, wine-grape growing (viticulture) has been enhanced throughout most of the wine regions of Europe and North America by climate changes since 1970. Also, by 2100, atmospheric CO_2 levels are likely to be about twice what they are in the first decade of the 2000s, a concentration often used by commercial greenhouse growers to speed plant growth. The benefits for open-air agriculture of this CO_2 fertilization effect, as it is called, have been questioned by recent research. Possible benefits to some agriculture must be weighed with shifts in precipitation patterns, more frequent droughts and floods, more frequent extreme weather, and rising temperatures in assessing the vulnerability of world agriculture to climate change.

A scientist checks rice samples at the International Rice Research Institute in Laguna, Philippines. Concerned about the impact of global warming on rice production, researchers are working to breed new varieties of rice—one of the world's major food staples. *Luis Liwanag/AFP/Getty Images.*

In Zambia, Africa, women are shown selling sweet potatoes along the side of a road in June 2007. Many farmers in the area have switched from growing maize to sweet potatoes because their rainy season has shortened, intensified, and become more erratic. *AP Images.*

■ Historical Background and Scientific Foundations

Agriculture is the most fundamentally life-sustaining of all human activities: sharp cutbacks in all other forms of well-being would be survivable for most people if only enough food could still be grown and distributed. Yet hundreds of millions of people—as of 2008, about 820 million—are chronically undernourished and at severe risk if their access to food diminishes any further. It is not surprising that the possible impacts of climate change on agriculture and food security has been of concern ever since the scientific study of global warming began in earnest in the 1960s.

In the 1970s and 1980s, studies suggested that increasing CO_2 levels would stimulate crop yields and increase the drought tolerance of crop plants. If true, this offered hope that the primary cause of global warming—high atmospheric CO_2—might offset some of its own ill-effects, at least in regard to the human food supply. It also became clear that the effects of global climate change on agriculture are bound to be a mixed bag regardless of CO_2 fertilization. Climate change will bring increased rainfall and longer growing seasons to some areas, helping agriculture in those areas even as it injures it in others. The question is, which parts of the world or which types of agriculture may benefit or be harmed by climate change, and whether the global balance will be positive.

Beginning in the early 1990s, scientific consultative groups commissioned by the United Nations, including the Intergovernmental Panel on Climate Change (IPCC), began to study the causes, likely impacts, and possible mitigations of climate change in a systematic way, assembling and sifting world scientific opinion to arrive at an informed consensus. These studies have been made available in a series of four Assessment Reports that outline in detail the state of scientific knowledge on climate change, including the vulnerabilities and possible opportunities of agriculture.

The increasing productivity of most agriculture worldwide—due to improvements in fertilization, breeding, pest and disease control, and other technical aspects—have long made it difficult to discern whether climate changes (or CO_2 levels) are yet affecting agriculture. Since 2001, however, evidence has appeared of climate changes that directly affect agriculture. These include changes in the length of the growing season (time between last spring frost and first autumn frost) and the number of growing-degree days per season.

Such changes have been observed in North America and in temperate Eurasia and have been, as far as they go, beneficial to many crops. However, increased droughts and heat are also damaging crops in other areas. In India, for example, the exceptionally severe 2007 monsoon season flooded nearly 8,000 square miles (20,700 square kilometers) of agricultural land, destroyed more than 130,000 homes, and killed at least 1,428 people.

Climate change does not affect agriculture by itself. Rather, it is only one of several fundamental problems besetting the current world agricultural system, under which inhabitants of developed countries are increasingly obese, while 820 million people elsewhere are chronically malnourished, including more than 150 million children under the age of five. These fundamental problems or vulnerabilities can be outlined as follows:

- *Social vulnerability.* Social vulnerability includes the effects on food supply of population growth, poverty, hunger itself (which reduces productivity and educational performance), gender inequality, and lack of services and resources, including technology. Social vulnerability affects much of Europe, Asia, and South America, but is most severe in sub-Saharan Africa (Africa south of the Sahara desert). Most of today's 820 million undernourished people live in 84 developing countries holding 4 billion people as of 2002, a number projected to grow to 7 billion by 2050.
- *Economic vulnerability.* Poorer countries find it difficult to compete in markets for oil, fertilizer, and food with richer countries. In 2002, 85% of the world's income went to the richest 20% of its population while only 1% of income went to the poorest 20%. Since 1990, under economic globalization, farmers in developing countries have experienced a decline in real income (buying power).
- *Environmental vulnerability.* Environmental threats to agriculture include salinization (increasing salt content of soil due to irrigation, which can render land useless for agriculture); loss of soils to erosion (an intractable problem with almost all forms of till agriculture, i.e., farming in which plows are used to turn the soil); water shortages; urban sprawl; and decreased biodiversity due partly to the replacement of locally adapted crop varieties with commercially produced green-revolution hybrids that have increased productivity while increasing dependence and insecurity. Many aspects of climate change will add to environmental vulnerability, especially in the tropical and subtropical parts of the world where most of the world's malnourished persons already live and where agriculture is already suffering most from social, economic, and environmental vulnerability.

A fundamental challenge to agriculture in much of the world could be changed precipitation patterns. Two-thirds of the world's people live in regions that receive, all told, only a fourth of the world's rain; worldwide,

WORDS TO KNOW

CO_2 FERTILIZATION EFFECT: Acceleration of plant growth by heightened atmospheric CO_2 concentrations. The effect has been suggested as a benefit to agriculture from artificially increased atmospheric CO_2 and as a source of negative climate-change feedback through the accelerated growth of forests. Studies have found that crops benefit only slightly from the effect and that its effect on forests is mostly limited to younger trees.

EROSION: Processes (mechanical and chemical) responsible for the wearing away, loosening, and dissolving of materials of Earth's crust.

FOOD SECURITY: Reliable access by a person or group to adequate food. Persons who do not have adequate food, or have adequate food at the moment but do not know how they will continue to get food, have low food security.

GREENHOUSE GASES: Gases that cause Earth to retain more thermal energy by absorbing infrared light emitted by Earth's surface. The most important greenhouse gases are water vapor, carbon dioxide, methane, nitrous oxide, and various artificial chemicals such as chlorofluorocarbons. All but the latter are naturally occurring, but human activity over the last several centuries has significantly increased the amounts of carbon dioxide, methane, and nitrous oxide in Earth's atmosphere, causing global warming and global climate change.

MONSOON: An annual shift in the direction of the prevailing wind that brings on a rainy season and affects large parts of Asia and Africa.

SALINIZATION: Increase in salt content. The term is often applied to increased salt content of soils due to irrigation: salts in irrigation water tend to concentrate in surface soils as the water quickly evaporates rather than sinking down into the ground.

IN CONTEXT: PREPARING AFRICA

A majority of climate change researchers assert that Africa's poorest regions are most vulnerable to increased threats of drought, flooding, severe weather, and disease associated with climate change. The World Bank estimates that crop yields in Africa could drop by 25% in the coming decade because of climate change. Such a significant crop failure could result in growing poverty and widespread famine in several regions of Africa. In October 2007, the World Bank called on developed nations and global businesses to invest in helping Africa's farms adapt to climate change. Acknowledging that much of Africa's agricultural infrastructure is fragile, the World Bank stressed the need for investment in farming technology to increase crop yield and prevent food scarcity. However, the World Bank also noted that investment in Africa's farms must coincide with increased access to less-expensive alternative energy sources in under-developed regions.

70% of freshwater supplies already go to agriculture (for irrigation), and this figure can be as high as 90% in drier countries. As of 2002, about 30 countries were facing chronic water shortages, a number expected to exceed 50 by 2050. Water shortages will impact both health and agriculture directly.

In temperate northerly regions, moderate warming—1.8–5.4°F (1–3°C)—will probably benefit agriculture by increasing yields. However, even higher levels of warming, which are possible, will have negative impacts on agriculture even in temperate and northerly regions. Experts warn that even mild warming will cause decreased crop yields in southerly regions. In 2007, Jacques Diouf, the director general of the United Nations Food and Agricultural Organization, warned that even slight global warming could reduce crop yields in southern regions, particularly due to increased droughts and floods. "Rain-fed agriculture in marginal areas in semiarid and subhumid regions is mostly at risk," Diouf was quoted in the *New York Times* as saying; "India ... could lose 125 million tons of its rain-fed cereal [grains] production, equivalent to 18% of its total production."

Despite all the vulnerabilities mentioned earlier, United Nations scenarios for possible world development predict that the number of malnourished people may more than halve by 2080, from 820 million to 100–380 million. However, even if these optimistic forecasts are correct, the IPCC notes that many fewer people would be malnourished were it not for the effects of climate change. Also, such forecasts do not take into account the possibility of abrupt climate change from unexpected feedbacks, such as many scientists warn have occurred numerous times in Earth's past.

■ Impacts and Issues

As mentioned earlier, it was long hoped that increased CO_2 in the atmosphere would offset some of the harms caused by global warming by accelerating plant growth and making plants more drought-resistant. Greenhouse (indoor) studies do indicate greatly improved yields with higher CO_2. Starting in the 1980s, a system of FACE (free air carbon dioxide enrichment) facilities has been set up around the world. At FACE installations, plants grow in open-air, real-world conditions surrounded by a ring of vertical pipes that can emit carbon dioxide. A computer uses readings of wind speed and direction to control which pipes emit how much carbon dioxide, assuring that the plants inside the ring experience fairly steady elevated CO_2.

Mixed Results

The results of FACE experiments have dimmed hopes for a neat offsetting by heightened CO_2 of global warming's harms to agriculture. Plants at FACE installations do produce larger yields. However, they also make less protein, and make proteins of different kinds (wheat, for example, makes about 20% less gluten proteins under FACE conditions), and contain fewer minerals such as calcium and zinc. Crops may therefore become less nourishing as CO_2 levels rise, and the grassy plants that feed grazing livestock may be less nourishing for them as well. There may also be effects on non-agricultural ecosystems. Experts are more or less evenly divided over the question of whether protein levels could be kept up in some crops by adding more nitrogen fertilizer to soils.

However, yield only increases about half as much under open-air conditions as in greenhouses, and for some plants, it does not increase at all. The FACE results are significant because computer models of the impact of climate change on global food supply, such as those on which the IPCC's predictions have been based, have assumed a 20–30% increase in yields with heightened CO_2. Observed yield increases are about half this large. Therefore, standard predictions of global agricultural output may be based on over-optimistic assumptions about CO_2 fertilization.

■ Primary Source Connection

Global climate change could have a major impact on agriculture by affecting crop production. Areas that were once suitable for growing a particular crop may become unsuitable as temperatures increase and droughts or floods become more common. This article examines the possible impact of climate change on agriculture in the United States in the twenty-first century.

The National Assessment Synthesis Team is an advisory committee of the United States Global Change Research Program (USGCRP). USGCRP is responsible for advising Congress and the president on issues related to global climate change.

CLIMATE CHANGE AND AGRICULTURE IN THE UNITED STATES

It is likely that climate changes and atmospheric CO_2 levels, as defined by the scenarios examined in this Assessment, will not imperil crop production in the US during the 21st century. The Assessment found that, at the national level, productivity of many major crops increased. Crops showing generally positive results include cotton, corn for grain and silage, soybeans, sorghum, barley, sugar beets, and citrus fruits. Pastures also showed increased productivity. For other crops including wheat, rice, oats, hay, sugar cane, potatoes, and tomatoes, yields are projected to increase under some conditions and decline under others.

Not all agricultural regions of the United States were affected to the same degree or in the same direction by the climates simulated in the scenarios. In general the findings were that climate change favored northern areas. The Midwest (especially the northern half), West, and Pacific Northwest exhibited large gains in yields for most crops in the 2030 and 2090 timeframes for both of the two major climate scenarios used in this Assessment, Hadley and Canadian. Crop production changes in other regions varied, some positive and some negative, depending on the climate scenario and time period. Yields reductions were quite large for some sites, particularly in the South and Plains States, for climate scenarios with declines in precipitation and substantial warming in these regions.

Crop models such as those used in this Assessment have been used at local, regional, and global scales to systematically assess impacts on yields and adaptation strategies in agricultural systems, as climate and/or other factors change. The simulation results depend on the general assumptions that soil nutrients are not limiting, and that pests, insects, diseases, and weeds, pose no threat to crop growth and yield. One important consequence of these assumptions is that positive crop responses to elevated CO_2, which account for one-third to one-half of the yield increases simulated in the Assessment studies, should be regarded as upper limits to actual responses in the field. One additional limitation that applies to this study is the models' inability to predict the negative effects of excess water conditions on crop yields. Given the "wet" nature of the scenarios employed, the positive responses projected in this study for rainfed crops, under both the Hadley and Canadian scenarios, may be overestimated.

Under climate change simulated in the two climate scenarios, consumers benefited from lower prices while producers' profits declined. For the Canadian scenario, these opposite effects were nearly balanced, resulting in a small net effect on the national economy. The estimated \$4-5 billion (in year 2000 dollars unless indicated) reduction in producers' profits represents a 13–17% loss of income, while the savings of \$3-6 billion to consumers represent less than a 1% reduction in the consumers food and fiber expenditures. Under the Hadley scenario, producers' profits are reduced by up to \$3 billion (10%) while consumers save \$9–12 billion (in the range of 1%). The major difference between the model outputs is that under the Hadley scenario, productivity increases were substantially greater than under the Canadian, resulting in lower food prices to the consumers' benefit and the producers' detriment.

At the national level, the models used in this Assessment found that irrigated agriculture's need for water declined approximately 5–10% for 2030 and 30-40% for 2090 in the context of the two primary climate scenarios, without adaptation due to increased precipitation and shortened crop-growing periods.

A case study of agriculture in the drainage basin of the Chesapeake Bay was undertaken to analyze the effects of climate change on surface-water quality. In simulations for this Assessment, under the two climate scenarios for 2030, loading of excess nitrogen into the Bay due to corn production increased by 17–31% compared with the current situation.

Pests are currently a major problem in US agriculture. The Assessment investigated the relationship between pesticide use and climate for crops that require relatively large amounts of pesticides. Pesticide use is projected to increase for most crops studied and in most states under the climate scenarios considered. Increased need for pesticide application varied by crop—increases for corn were generally in the range of 10–20%; for potatoes, 5–15%; and for soybeans and cotton, 2–5%. The results for wheat varied widely by state and climate scenario showing changes ranging from approximately −15 to +15%. The increase in pesticide use results in slightly poorer overall economic performance, but this effect is quite small because pesticide expenditures are in many cases a relatively small share of production costs.

The Assessment did not consider increased crop losses due to pests, implicitly assuming that all additional losses were eliminated through increased pest control measures. This could possibly result in underestimates of losses due to pests associated with climate change. In addition, this Assessment did not consider the environmental consequences of increased pesticide use.

Ultimately, the consequences of climate change for US agriculture hinge on changes in climate variability and extreme events. Changes in the frequency and intensity of droughts, flooding, and storm damage are likely to have significant consequences. Such events cause erosion, waterlogging, and leaching of animal wastes, pesticides, fertilizers, and other chemicals into surface and groundwater.

One major source of weather variability is the El Niño/Southern Oscillation (ENSO). ENSO effects vary widely across the country. Better prediction of these events would allow farmers to plan ahead, altering their choices of which crops to plant and when to plant them. The value of improved forecasts of ENSO events has been estimated at approximately $500 million per year. As climate warms, ENSO is likely to be affected. Some models project that El Niño events and their impacts on US weather are likely to be more intense. There is also a chance that La Niña events and their impacts will be stronger. The potential impacts of a change in frequency and strength of ENSO conditions on agriculture were modeled. An increase in these ENSO conditions was found to cost US farmers on average about $320 million per year if forecasts of these events were available and farmers used them to plan for the growing season. The increase in cost was estimated to be greater if accurate forecasts were not available or not used.

"CLIMATE CHANGE AND AGRICULTURE IN THE UNITED STATES." *CLIMATE CHANGE IMPACTS ON THE UNITED STATES: THE POTENTIAL CONSEQUENCES OF CLIMATE VARIABILITY AND CHANGE.* NATIONAL ASSESSMENT SYNTHESIS TEAM, EDS. CAMBRIDGE: CAMBRIDGE UNIVERSITY PRESS, 2001.

SEE ALSO *Agriculture: Contribution to Climate Change; Fisheries; Methane.*

BIBLIOGRAPHY

Books

Parry, M. L., et al, eds. *Climate Change 2007: Impacts, Adaptation and Vulnerability: Contribution of Working Group II to the Fourth Assessment Report of the Intergovernmental Panel on Climate Change.* New York: Cambridge University Press, 2007.

Periodicals

Asner, Gregory P. "Grazing Systems, Ecosystem Responses, and Global Change." *Annual Review of Environment and Resources* 29 (2004): 261–299.

Butt, Tanveer A., et al. "Policies for Reducing Agricultural Sector Vulnerability to Climate Change in Mali." *Climate Policy* 5 (2006): 583–598.

Deutsch, Claudia H. "Trying to Connect the Dinner Plate to Climate Change." *The New York Times*, August 29, 2007.

Izaurralde, R. César, et al. "Carbon Cost of Applying Nitrogen Fertilizer." *Science* 288, no. 5467 (May 5, 2000): 811–812.

Schimel, David. "Climate Change and Crop Yields: Beyond Cassandra." *Science* 312, no. 5782 (June 30, 2006): 188–189.

Sengupta, Somini. "Warming Threatens Farms in India, U.N. Official Says." *The New York Times*, August 8, 2007.

Stafford, Ned. "The Other Greenhouse Effect." *Nature* 448, no. 7153 (August 2, 2007): 526–528.

Web Sites

"Climate Change and Agricultural Vulnerability." *International Institute for Applied Systems Analysis*, 2002. <http://www.iiasa.ac.at/Research/LUC/JB-Report.pdf> (accessed November 5, 2007).

"Climate Change: The Role of the U.S. Agriculture Sector." *Congressional Research Service*, March 6, 2007. <http://fpc.state.gov/documents/organization/81931.pdf> (accessed November 5, 2007).

Larry Gilman

Albedo

Introduction

The albedo of an object or material is its tendency to reflect light. A bright, highly reflective surface, such as fresh snow, has a high albedo; a dark surface, such as asphalt, absorbs light and thus has a low albedo. Albedo is generally expressed as the fraction of incident light—the amount of light coming from the sun or some other source—that a surface reflects, with values ranging from zero (no reflection) to one (total reflection). Decreases in Earth's albedo brought on by melting snow, vegetation shifts, and other factors can exacerbate global warming, especially at high latitudes.

Historical Background and Scientific Foundations

Earth's albedo can be measured locally on the ground and used in computer models to predict large-scale effects, or measured on a broader scale using satellites and observations of earthshine, that is, light reflected

In the arctic tundra of Alaska, the difference in albedo is shown between a snow-covered surface and a vegetated surface during a snowmelt. *AP Images.*

WORDS TO KNOW

EARTHSHINE: Sunlight reflected from Earth and illuminating some other body, such as a spacecraft or the moon. The side of the moon that always faces Earth is illuminated by earthshine; by measuring the brightness of portions of the moon that are lit only by earthshine, the reflectivity (albedo) of Earth can be measured.

FOSSIL FUELS: Fuels formed by biological processes and transformed into solid or fluid minerals over geological time. Fossil fuels include coal, petroleum, and natural gas. Fossil fuels are non-renewable on the timescale of human civilization, because their natural replenishment would take many millions of years.

INCIDENT LIGHT: Light arriving at the surface of an object (for example, Earth).

RUNOFF: Water that falls as precipitation and then runs over the surface of the land rather than sinking into the ground.

SEA ICE: Ice that forms from the freezing of ocean water. As the salt water freezes, it ejects salt, so sea ice is fresh, not salty. Sea ice forms in relatively thin layers, usually no more than 3–7 ft (1–2 m) thick, but it can cover thousands of square miles of ocean in the polar regions.

TREELINE: The highest-altitude (or highest-latitude) line along which trees can grow. As climate warms, treelines advance to higher altitudes and latitudes.

TUNDRA: A type of ecosystem dominated by lichens, mosses, grasses, and woody plants. It is found at high latitudes (arctic tundra) and high altitudes (alpine tundra). Arctic tundra is underlain by permafrost and usually very wet.

from Earth onto the dark side of the moon. As little as a 1% decrease in Earth's overall albedo may be enough to spur significant climatic warming.

However, studies of Earth's global albedo have varied widely in their findings over the past several decades, complicating scientists' efforts to build the effects of albedo into a broader understanding of global warming. Part of this uncertainty likely stems from changes in the type and amount of cloud cover. The albedo-climate picture is generally clearer at smaller scales, such as on snow-covered tundra or sea ice, where albedo measurements can be more localized.

■ Impacts and Issues

Albedo may be involved in a number of different feedback mechanisms that amplify global warming. The best-known is perhaps the snow/ice albedo feedback, by which increasing temperatures shrink snow and ice cover at high latitudes and altitudes, exposing darker stretches of water or ground that more readily absorb solar radiation. This lowered surface albedo promotes warming that melts more snow and ice, promoting still more warming, and so on. Shrinking snow cover has played a significant role in the increasingly early onset of spring at high latitudes over the past several decades.

Changes in vegetation communities brought on by global warming, such as the advance of treeline to higher altitudes and latitudes and the loss of tundra to forests and shrublands, may also be darkening formerly bright, snowy expanses at high latitudes, increasing regional warming.

Human land-use practices can also change albedo. Particulate matter like dust kicked up from heavily grazed areas, deforested areas, farmland, growing deserts, new housing developments, and other impacted environs, as well as soot from the burning of fossil fuels, can collect on snow and ice and cause it to melt more quickly. Scientists from the National Snow and Ice Data Center in Boulder, Colorado, found that a single event spreading dust over the surface could cause snow to melt from a Rocky Mountain basin 18 days earlier than it would if the snow was dust-free. Changes to snowmelt regimes in mountain ranges around the world may lead to destructively powerful spring runoff as well as water shortages.

See Also *Arctic Melting: Greenland Ice Cap; Arctic Melting: Polar Ice Cap; Global Warming; Solar Illumination; Soot.*

BIBLIOGRAPHY

Books

Lemke, P., et al. "Observations: Changes in Snow, Ice and Frozen Ground." In *Climate Change 2007: The Physical Science Basis. Contribution of Working Group I to the Fourth Assessment Report of the Intergovernmental Panel on Climate Change*, edited by S. Solomon, et al. New York: Cambridge University Press, 2007.

Periodicals

Chapin , F. S. III, et al. "Arctic and Boreal Ecosystems of Western North America as Components of the Climate System." *Global Change Biology* 6 (2000): 211-223.

Curry, Judith A., and Julie L. Schramm. "Sea Ice-Albedo Climate Feedback Mechanism." *Journal of Climate* 8 (1995): 240-247.

Karl, Thomas R., and Kevin E. Trenberth. "Modern Global Climate Change." *Science* 302 (2003): 1719-1723.

Nijhuis, Michelle. "Dust and Snow: High in the Snowy San Juan Mountains, Tiny Particles Have Big Implications." *High Country News* (May 29, 2006).

Soja, Amber J., et al. "Climate-induced Boreal Forest Change: Predictions Versus Current Observations." *Global and Planetary Change* 56 (2007): 274-296.

Web Sites

"Albedo." *Arctic Coastal Ice Processes*, October 26, 2006. <http://www.arcticice.org/albedo.htm> (accessed October 31, 2007).

Britt, Robert Roy. "Baffled Scientists Say Less Sunlight Reaching Earth." *Live Science*, January 24, 2006. <http://www.livescience.com/environment/060124_ earth_albedo.html> (accessed October 31, 2007).

"Scientists Watch Dark Side of the Moon to Monitor Earth's Climate." *American Geophysical Union News*, April 17, 2001. <http://www.agu.org/sci_soc/prrl/prrl0113.html> (accessed October 31, 2007).

Sarah Gilman

Alliance of Small Island States (AOSIS)

Introduction

The Alliance of Small Island States (AOSIS) is a lobbying and negotiating group that aims to raise awareness about the consequences of climate change and to reduce carbon emissions so as to limit climate change. The small island states are recognized as being especially vulnerable to the negative effects of climate change. The extent of the concern and the potential harmful effects have resulted in the AOSIS being a powerful driving force in carbon emission reductions.

Historical Background and Scientific Foundations

The AOSIS formed in 1990. As of 2007, the alliance was made up of 39 states, of which 37 are members of the United Nations. The member states include islands in the Pacific Ocean, the Caribbean, the Atlantic Ocean, the Mediterranean, the Indian Ocean, and the South China Sea.

The goal of the AOSIS is to act as a lobbying group to raise awareness of issues impacting small island states and

Encroaching waves are seen toppling shoreline coconut trees on the Marshall Islands, a member of the AOSIS. Rising ocean waters, due to global warming, are eating away at beaches across the Pacific.
AP Images.

to act as a negotiating voice. The main focus of AOSIS is on reducing carbon emissions.

Impacts and Issues

Small island states are especially vulnerable to the negative effects of climate change. Sea-level rise is a major issue since many small island states have elevations of only 10-13 ft (3-4 m) above sea level.

Small island states are especially sensitive to climate change, and less likely than other larger nations to be able to adapt to change. Some of the specific threats noted by the Intergovernmental Panel on Climate Change (IPCC) include an increase in tropical cyclones; a reduction in tourism, which hurts the economy; reduced water resources; human health issues; and a decline in coastal industries.

The significance of climate change to small island states has made the AOSIS an important force in the climate change debate. Although the member states represent mainly small countries with limited political influence, their alliance into one larger group has given them greater power. Their influence is also increased due to the severity of the potential risks they face. They act as a voice for the worst-case scenarios of climate change, which can promote action on the part of other nations and push targets for reducing the impact of climate change to higher levels.

A major contribution by the AOSIS was what is now known as the AOSIS Protocol. This protocol sets the target of lowering the carbon emission levels by 20% to 1990 levels. This was originally presented at the 1995 First Conference of Parties in Berlin and was later adopted as part of the Kyoto Protocol.

See Also *Coastal Populations; Kyoto Protocol; Small Islands: Climate Change Impacts; Sea Level Rise.*

BIBLIOGRAPHY

Books

Gillespie, A., and W. C. G. Burns. *Climate Change in the South Pacific: Impacts and Responses in Australia, New Zealand, and Small Island States.* New York: Springer, 2000.

Web Sites

"The Alliance." *AOSIS: Alliance of Small Island States,* 2007. <http://www.sidsnet.org/aosis> (accessed October 25, 2007).

"Special Report on the Regional Impacts of Climate Change: An Assessment of Vulnerability." *IPCC: Intergovernmental Panel on Climate Change.* <http://www.grida.no/climate/ipcc/regional/index.htm> (accessed October 25, 2007).

WORDS TO KNOW

AOSIS PROTOCOL: The Alliance of Small Island States (AOSIS) is a federation of 39 of the world's small island states formed in 1990 to influence international policy on climate change. The AOSIS Protocol was a proposal submitted to the First Conference of the Parties to the United Nations Framework Convention on Climate Change (UNFCCC) in 1995 that proposed that the developed nations reduce their greenhouse-gas emissions by 20% by 2005. Although its specific demands were not fulfilled, it led to the adoption of the Berlin Mandate in 1995, which in turn set the stage for the negotiation of the Kyoto Protocol in 1997.

INTERGOVERNMENTAL PANEL ON CLIMATE CHANGE (IPCC): Panel of scientists established by the World Meteorological Organization (WMO) and the United Nations Environment Programme (UNEP) in 1988 to assess the science, technology, and socioeconomic information needed to understand the risk of human-induced climate change.

KYOTO PROTOCOL: Extension in 1997 of the 1992 United Nations Framework Convention on Climate Change (UNFCCC), an international treaty signed by almost all countries with the goal of mitigating climate change. The United States, as of early 2008, was the only industrialized country to have not signed the Kyoto Protocol, which is due to be replaced by an improved and updated agreement starting in 2012.

Annex I Parties

Introduction

The United Nations Framework Convention on Climate Change (UNFCCC) and the Kyoto Protocol split nations into Annex I and Non-Annex I parties. Annex I parties are industrialized nations and are legally bound to reduce greenhouse gas emissions once they have ratified the agreement. In contrast, Non-Annex I parties (developing nations) are only required to report emissions. The split is a significant issue, especially since Non-Annex I parties such as China and India are among the world's largest contributors to greenhouse gas emissions and have a high rate of increase in emissions. This is one of the reasons that the United States has not ratified the Kyoto Protocol, and it is also a point of debate for future agreements that will follow the Kyoto Protocol after it expires in 2012.

WORDS TO KNOW

GREENHOUSE GASES: Gases that cause Earth to retain more thermal energy by absorbing infrared light emitted by Earth's surface. The most important greenhouse gases are water vapor, carbon dioxide, methane, nitrous oxide, and various artificial chemicals such as chlorofluorocarbons. All but the latter are naturally occurring, but human activity over the last several centuries has significantly increased the amounts of carbon dioxide, methane, and nitrous oxide in Earth's atmosphere, causing global warming and global climate change.

INTERNATIONAL ENERGY AGENCY: International group established in 1974 by the Organisation of Economic Co-operation and Development (OECD), a group of 30 well-to-do Western countries formed after World War II to coordinate economic concerns. The IEA promotes nuclear energy and releases technical studies of the world energy situation and of particular energy issues.

KYOTO PROTOCOL: Extension in 1997 of the 1992 United Nations Framework Convention on Climate Change (UNFCCC), an international treaty signed by almost all countries with the goal of mitigating climate change. The United States, as of early 2008, was the only industrialized country to have not ratified the Kyoto Protocol, which is due to be replaced by an improved and updated agreement starting in 2012.

Historical Background and Scientific Foundations

The UNFCCC is an international treaty that entered into force in 1994. It seeks to find ways to address and reduce climate change. In 1997, the Kyoto Protocol was added to the UNFCCC and it went into force in February 2005. The Kyoto Protocol sets legally binding targets for greenhouse gas emissions.

The parties to the UNFCCC are divided into three groups. Annex I parties are industrialized nations and countries with economies in transition. Annex II parties are the Annex I countries but not the countries with economies in transition. Non-Annex I parties are mainly developing nations. Annex I parties are legally required to reduce their greenhouse gas emissions, while Non-Annex I parties are only required to report their emissions.

Impacts and Issues

As of 2007, most of the 41 Annex I parties have ratified the Kyoto Protocol, and so are legally obligated to reduce greenhouse emissions. The notable exception is the United States.

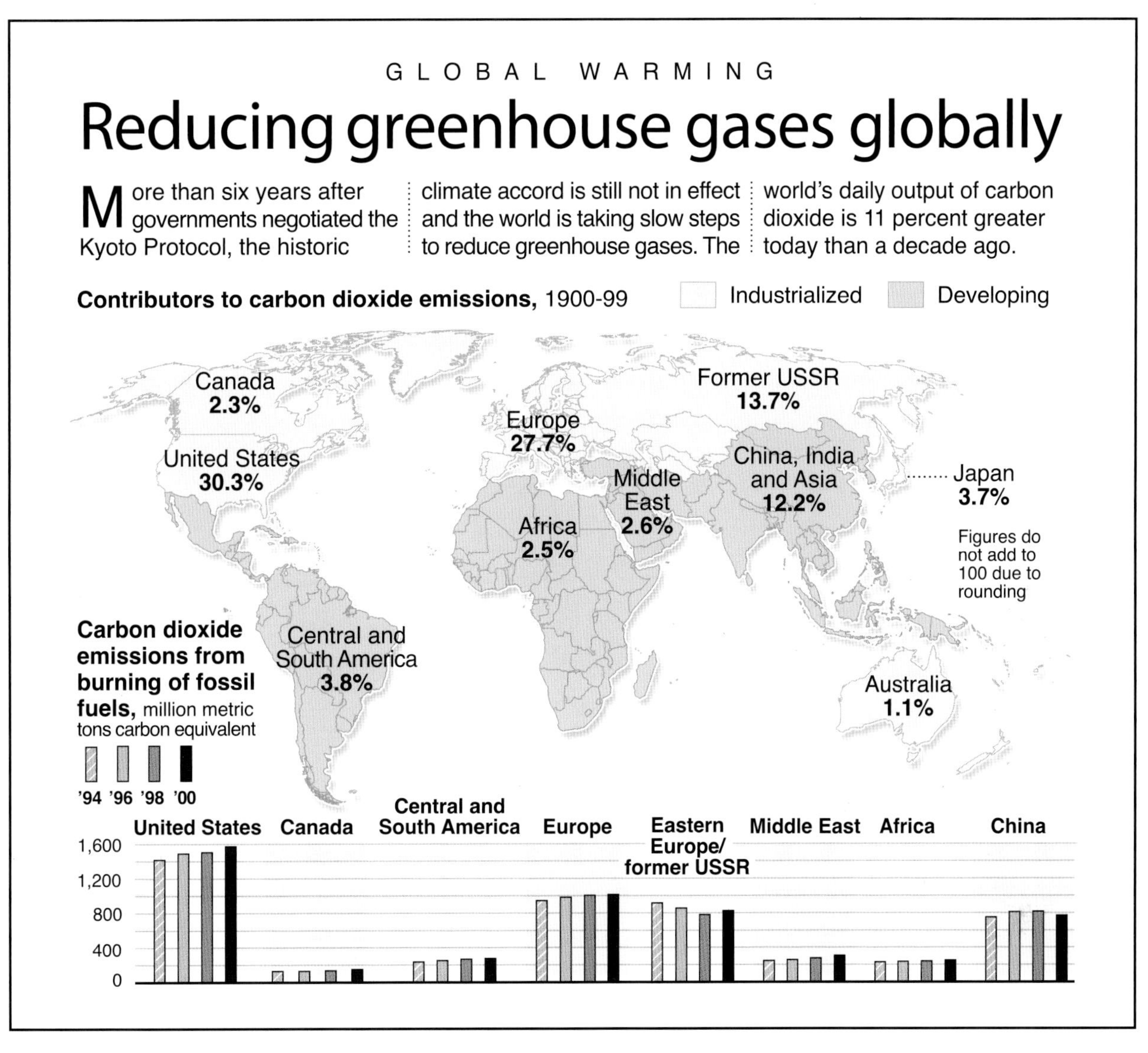

This graphic from the Associated Press, issued before the Kyoto Protocol went into effect in 2005, shows various nations' contributions to carbon dioxide emissions from 1900–1999. [Sources: World Resources Institute, Energy Information Administration, and the Intergovernmental Panel on Climate Change.] *AP Images.*

One reason for not ratifying the agreement is the negative economic effect that will likely result. A second reason is that China and India are not required to reduce greenhouse gas emissions although they contribute heavily to greenhouse gas emissions. China and India have both ratified the Kyoto Protocol, but as developing nations, they are Non-Annex I nations and so are not required to reduce emissions.

In 2006, the International Energy Agency predicted that China would overtake the United States to become the largest emitter of greenhouse gases by 2009. However, in June 2007, the Netherlands Environmental Assessment Agency reported that China had surpassed the United States to become the biggest producer of CO_2 emissions.

India, a fast developing nation, has also seen an increase in its greenhouse gas emissions. The inclusion of developing nations is emerging as one of the significant issues in negotiations for a new agreement to replace the Kyoto Protocol when it expires in 2012.

SEE ALSO *China: Climate and Energy Policies; China: Total Carbon Dioxide Emissions; Economics of Climate Change; Kyoto Protocol; United Nations Framework Convention on Climate Change (UNFCCC); United States: Climate Policy.*

BIBLIOGRAPHY

Books

Dessler, A. E., and E. A. Parson. *The Science and Politics of Global Climate Change: A Guide to the Debate.* New York: Cambridge University Press, 2006.

International Energy Agency. *World Energy Outlook 2006*. Paris: International Energy Agency, 2006.

Web Sites

"China Now No. 1 in CO_2 Emissions; USA in Second Position." *Netherlands Environmental Assessment Agency*, June 19, 2007. <http://www.mnp.nl/en/service/pressreleases/2007/20070619ChinanowNo1inCO2emissionsUSAinsecondposition.html> (accessed November 7, 2007).

"List of Annex I Parties to the Convention." *United Nations Framework Convention on Climate Change*, 2007. <http://unfccc.int/parties_and_observers/parties/annex_i/items/2774.php> (accessed October 25, 2007).

Antarctica: Melting

Introduction

Antarctica, the fifth-largest continent, is centered on the South Pole and is almost entirely covered with a thick sheet of ice. The Antarctic ice sheet is 1.5 mi (2.5 km) thick on average and contains about 7 million mi^3 (29 million km^3) of ice. Antarctica's ice holds 90% of the world's freshwater, enough to raise sea level worldwide by 200 ft (61 m), if it

This NASA image of the full Earth, as viewed from space during the Apollo 17 mission in 1972, shows part of Antarctica's south polar ice cap at that time. *AP Images.*

In 2002 scientists were shocked when the Larsen B ice shelf—an area in Antarctica the size of Rhode Island weighing some 500 billion tons—shattered and separated. It had existed since the last Ice Age, some 12,000 years ago. *AP Images.*

were all to melt. Although total melting is highly unlikely, some melting is occurring as global warming allows temperatures to rise above freezing across much of Antarctica during the summer months of the Southern Hemisphere. Scientists predicted as recently as 2001 that global warming might actually increase the amount of water stored on the Antarctic ice sheet by increasing snowfall, yet scientists announced in 2006 that the Antarctic ice sheet is, in fact, melting faster than it is being replenished by snowfall. Overall, Antarctic climate remains poorly understood as of 2008.

■ Historical Background and Scientific Foundations

Until the nineteenth century, Antarctica was not visited by human beings. A series of exploratory voyages starting in 1820 confirmed the presence of land near the South Pole, an idea that had long been a subject of speculation. In the early twentieth century, overland expeditions reached the South Pole, and in the 1930s airplanes began to be used to explore the continent. Year-round scientific stations have been maintained in Antarctica since the 1950s, and knowledge of the continent has been greatly increased by satellite observations since the 1970s.

Antarctica's inaccessibility, size, and thick ice covering make it difficult to study, and its role in global climate is only partly understood. In recent years there has been uncertainty about whether Antarctica was gaining overall ice mass or losing it, and about whether global warming would cause Antarctica to gain ice mass through increased snowfall faster than it lost ice mass by melting, or the reverse. Antarctica, unlike the other continents, has shown little change during the recent period of rapid global warming.

In 2005, data from NASA's QuikScat satellite, which can distinguish between frozen and melted water by bouncing radar pulses off Earth's surface, showed that an area the size of California had experienced surface snowmelt in January of that year. This was the largest Antarctic snowmelt seen in some three decades of satellite observation.

Although the 2005 melt event was temporary and did not directly contribute to overall loss of ice mass, scientists analyzing the QuikScat data asserted that it proved that parts of Antarctica are showing the first impacts of global warming.

In 2006, new evidence convinced many scientists that the Antarctic ice sheet is melting faster than it is growing. In 2005, scientists analyzed three years of data from satellites measuring the force of gravity over the Greenland and Antarctic ice sheets. Loss of ice mass due to melting, or gain due to snowfall, can be measured as decreases or increases in gravitational pull. Gravitational measurements showed that both the Greenland ice sheet and the Antarctic ice sheet are melting faster than had been thought. The Antarctic ice sheet is losing 36 mi^3 (152 km^3) of ice per year, despite increasing ice thickness over East Antarctica. If this trend is sustained over the long term, it will overturn the prediction, made by the Intergovernmental Panel on Climate Change (IPCC) in 2001, that Antarctica should gain mass during the twenty-first century.

WORDS TO KNOW

GRAVIMETRIC: Having to do with the measurement of gravity. Gravimetric data from satellites can reveal changes in large masses of ice on Earth's surface: in 2006, such data confirmed that Greenland's ice cap was melting faster than expected.

ICE SHEET: Glacial ice that covers at least 19,500 square mi (50,000 square km) of land and that flows in all directions, covering and obscuring the landscape below it.

SEA LEVEL: The datum against which land elevation and sea depth are measured. Mean sea level is the average of high and low tides.

■ Impacts and Issues

In 2006, taking the latest gravimetric and altitude-measurement data into account, the IPCC said that it was highly likely (more than 90% probable) that melting of the Greenland and Antarctic ice sheets had contributed to sea level rise from 1993 to 2003. Scientists also point out that uncertainties about the present speed and future course of melting remain, although these uncertainties are decreasing.

The main result to be feared from ice-sheet melting is sea-level rise, which could endanger coastal settlements around the world. Sea levels have been stable for most of the last 3,000 years, but began to rise at 0.04–0.08 in (1–2 mm) per year in the late 1800s. Since 1993, the average rate of rise has been faster, about 0.1 in (3 mm) per year, but as of 2007 it was still uncertain whether this acceleration would prove to be long-term or short-term.

Over the last 20,000 years, the West Antarctic ice sheet has lost two-thirds of its mass and raised sea levels by 33 ft (10 m). Complete melting of the sheet, where most of Antarctica's ice loss is presently happening, would take many years, but could raise sea level by another 20 ft (6 m).

SEE ALSO *Antarctica: Observed Climate Changes; Antarctica: Role in Global Climate.*

IN CONTEXT: MELTING AT EARTH'S POLES

Polar melting is a problem in both northern and southern polar regions. In August 2007, a month before annual Arctic sea-ice melting normally peaks, a record had already been set for shrinkage of the north polar ice cap. No summer since 1979, when satellites first allowed accurate tracking of the annual melting, recorded such a large loss of ice.

William Chapman, an expert in the region at the University of Illinois, Urbana-Champaign, said that the melting rate in June and July of 2007 had been incredible. Arctic sea-ice melting is predicted to have many impacts, including a much warmer Arctic region, increased Arctic coastal erosion, declines in polar bear populations, damage to traditional hunting practices of indigenous Arctic peoples, disruption of large-scale ocean circulations, and more.

BIBLIOGRAPHY

Periodicals

Cazenave, Anny. "How Fast Are the Ice Sheets Melting?" *Science* 314 (2006): 1250–1252.

Eilperin, Juliet. "Antarctic Ice Sheet Is Melting Rapidly." *Washington Post* (March 3, 2006): A01.

Shepherd, Andrew, and Duncan Wingham. "Recent Sea-Level Contributions of the Antarctic and Greenland Ice Sheets." *Science* 315 (2007): 1529–1532.

Velicogna, Isabella, and John Wahr. "Measurements of Time-Variable Gravity Show Mass Loss in Antarctica." *Science* 311 (2006): 1754–1756.

Web Sites

"Mission News: NASA Finds Vast Regions of West Antarctica Melted in Recent Past." *National Aeronautics and Space Administration (NASA)*, May 15, 2007. <http://www.nasa.gov/vision/earth/lookingatearth/arctic-20070515.html> (accessed August 10, 2007).

Larry Gilman

Antarctica: Observed Climate Changes

Introduction

Antarctica holds 90% of the world's freshwater in the form of a vast ice sheet thousands of feet thick. This ice covering also extends out to sea in the form of floating ice tongues or shelves connected to the land. The Antarctic climate is cold, with winter temperatures as low as -130°F (-90°C). The continent's climate has changed repeatedly in the distant past due to many factors, including cyclic changes in Earth's orbit. It is also showing effects from present-day human-caused climate change.

In contrast to the north polar region, which has suffered more rapid warming than the rest of the world, much of Antarctica has shown relatively little warming over the last half-century. This may be changing, however, as ice shelves many thousands of years old collapse, glaciers accelerate their flow to the sea, surface melting is observed on hitherto unknown scales, and satellite measurements show that the Antarctic ice sheet is losing mass faster than it gains mass from snowfall. Antarctica's climate remains difficult to characterize in detail because of the region's size and remoteness, and scientists caution that there are still many uncertainties about what changes are happening and are likely to happen.

In order to learn more about Antarctica and climate change, researchers study the region on land, from the air, and by ship. Because the area is remote and its weather severe, scientists only partly understand Antarctica's role in global climate today. *Image copyright Armin Rose, 2007. Used under license from Shutterstock.com.*

Emperor penguins jump up on an ice floe in Antarctica. The population of Antarctic emperor penguins has decreased some 50% in the last 50 years. Warming temperatures have greatly contributed to that decline. *Image copyright Jan Martin Will, 2007. Used under license from Shutterstock.com.*

Historical Background and Scientific Foundations

Observation of Antarctica's weather, ice bodies, and ecosystems began in earnest during the International Geophysical Year (IGY), a period of enhanced government-funded scientific observation of Earth that lasted from July 1957 to December 1958. In the years leading up to the IGY, scientific bases were established in Antarctica by the United States, the Soviet Union, and other governments. Continuous scientific monitoring of Antarctic climate from the ground, sea, and air has been supplemented since the 1970s by satellite observations.

Information about Antarctica's climate can be roughly divided into three categories: 1) the deep geological past; 2) the last 1 million years; and 3) the last half-century. Information about Antarctica's deep past is obtained from fossils and other geological evidence. Antarctica was once located near the equator. Fossils show that from about 250 to 65 million years ago, the continent was lushly forested, with a tropical climate, dinosaurs and amphibians, and tree ferns. Continental drift moved Antarctica toward the South Pole, after which it froze over and accumulated an ice cap.

Detailed records of Antarctic climate have been preserved for the last million years by its unique location. Despite the abundance of water on Antarctica's surface—the ice sheet is 1.5 mi (2.5 km) thick on average and contains about 7 million mi^3 (29 million km^3) of ice—it actually receives less annual precipitation per unit of area than any other continent and is considered, technically speaking, a desert. This apparent contradiction is explained by the fact that although little water falls on Antarctica, it tends to stay there. Each year's snowfall is relatively small (about 4 in [10 cm] is typical for a winter at the South Pole), but it does not melt. The snow packs down and freezes into a thin layer. Evaporation is slight because of the intense cold. Each new layer builds on the last. This accumulated ice is removed from the continent only by geologically rare melting episodes or by the slow creep of glaciers to the sea. Ice cores—2-mile-long cylinders of layered ice—have been drilled from Antarctic ice. These cores sample 800,000 continuous years of snowfall. Each layer contains air bubbles that were trapped when the layer was formed, so the core contains a continuous series of time-ordered samples of Earth's atmosphere.

Much information about both Antarctic climate and global climate is gathered from these cores. Atmospheric concentrations of dust and of the major greenhouse gases (carbon dioxide, methane, and nitrous oxide) are measured directly, and the overall temperature of Earth in each year is deduced from the abundance of certain isotopes of hydrogen and oxygen in the ice.

WORDS TO KNOW

ATLANTIC MERIDIONAL OVERTURNING CIRCULATION: The north-south circulation that dominates the movement of water in the Atlantic Ocean and is a key part of the global meridional thermohaline circulation and of the climate system. Moves warm waters toward the Arctic, where they cool, sink, and move southward along the ocean floor.

BIPOLAR SEESAW: The tendency of climate to warm at the North Pole while it cools at the South Pole and vice versa. Paleoclimatic data show that this seesaw effect has occurred many times in the geological past and is related to changes in ocean circulation, but scientists do not yet completely understand the mechanism of the seesaw.

GREAT CONVEYOR BELT: The overturning circulation of the world's seas, driven by temperature and salinity differences between the poles and tropics; also called the thermohaline circulation or meridional overturning circulation. Because the great conveyor belt transports thermal energy from the tropics toward the poles, it is a central component of Earth's climate machine.

GREENHOUSE GASES: Gases that cause Earth to retain more thermal energy by absorbing infrared light emitted by Earth's surface. The most important greenhouse gases are water vapor, carbon dioxide, methane, nitrous oxide, and various artificial chemicals such as chlorofluorocarbons. All but the latter are naturally occurring, but human activity over the last several centuries has significantly increased the amounts of carbon dioxide, methane, and nitrous oxide in Earth's atmosphere, causing global warming and global climate change.

MILANKOVITCH CYCLES: Regularly repeating variations in Earth's climate caused by shifts in its orbit around the sun and its orientation (i.e., tilt) with respect to the sun. Named after Serbian scientist Milutin Milankovitch (1879–1958), though he was not the first to propose such cycles.

Ice-core data show that over long periods of time, global and Antarctic climates warm and cool in response to changes in Earth's orbit. These climate cycles, called Milankovitch cycles after one of their discoverers, Russian physicist Milutin Milankovitch (1879–1958), last for about 100,000 years each. During the cold part of each cycle, ice expands from the Poles to cover much of the world; during the warm part of each cycle, the ice retreats. As of 2008, Earth is in a warm (interglacial) cycle that began between 18,000 and 10,000 years ago. This has caused massive melting of Antarctic ice. Over the last 20,000 years, the ice sheet covering the western part of Antarctica has shrunk by two-thirds, raising sea levels worldwide by 33 ft (10 m).

Comparing ice cores from the Greenland ice cap and Antarctica shows that on shorter time-scales—thousands of years rather than tens or hundreds of thousands of years—Antarctica's climate is interlocked with that of the Northern Hemisphere through a complex mechanism known as the thermal bipolar seesaw. That is, as the north polar region warms, Antarctica tends to cool, and vice versa. This effect is caused by changes to the Atlantic Meridional Overturning Circulation (also sometimes termed the Great Conveyor Belt or thermohaline circulation). Scientists are still trying to understand the implications of the bipolar seesaw for today's climate changes.

For the last 50 years, ice core data are not needed because direct observations have been made. As might have been guessed from the bipolar seesaw, although the north polar region recently has been warming rapidly, Antarctica as a whole has not. The climate picture over Antarctica is a patchwork: the high-altitude plateau at the center of the continent is dominated by a temperature inversion, in which cold air is found closer to the surface with warmer air occurring at higher altitudes. (Usually it is the other way around with valleys tending to be warmer than mountaintops.) The edges of the continent, especially the Antarctic Peninsula, are affected more by weather systems circulating around the continent in the polar vortex. As a result, the center of the continent has experienced slight cooling over the last 40 to 50 years, while the Antarctic Peninsula region has experienced significant warming.

■ Impacts and Issues

Warming of the Antarctic Peninsula is globally significant partly because it accelerates melting of the West Antarctic ice sheet, which raises sea levels. In 1995, a portion of the floating Larsen ice shelf called Larsen A, near the tip of the Antarctic Peninsula on the eastern side, broke up and melted in the sea. In 1998, scientists noticed early signs that the Larsen B ice shelf, much larger and older than Larsen A, might also be breaking up, undermined by warm ocean currents. These signs persisted, but scientists were shocked when, in 2002, the entire ice shelf—an area of ice 1,253 mi^2 (3,250 km^2, the size of the state of Rhode Island), 720 ft (220 m) thick, weighing on the order of 500 billion tons—disintegrated in a one-month period.

Although floating ice shelves do not raise sea levels when they melt, in 2003 scientists showed that after the 1995 breakup of Larsen A, glaciers on land that had been dammed up by the shelf accelerated their flow to the sea. A similar acceleration was found in 2004 for half a dozen glaciers in West Antarctica.

In 2006, scientists from the British Antarctic Survey (BAS) announced that they had established direct ties between anthropogenic global warming and the breakup of the Larsen B ice shelf. In 2007, BAS scientists announced that they had measured 300 West Antarctic glaciers and found that all were accelerating their rate of flow to the sea. Average glacier speed had increased by 12% from 1993 to 2003 as a result of more rapid melting where the glaciers meet the sea. Glacial ice does raise sea levels when it passes from the land to the sea.

The Intergovernmental Panel on Climate Change (IPCC) noted in early 2007 that it could not forecast a specific upper limit for sea-level rise due to Antarctic melting over the next several centuries because the ability of anthropogenic global warming to melt Antarctic ice is not well-enough understood. Complete melting of Antarctica's ice cover would raise sea levels by about 200 ft (61 m), but this much melting is unlikely. Complete melting of the West Antarctic ice sheet alone would raise sea levels by about 20 ft (6 m).

IN CONTEXT: ARID ANTARCTICA

Deserts are areas that receive little moisture. Although the most familiar image of a desert involves hot sand, the arctic north and Antarctica are also deserts, as they also receive very little moisture, usually in the form of snow. An arid climate is a desert-like dry climate. Arid climates receive less than 10 inches of rainfall in an entire year.

■ Primary Source Connection

Researchers often study Antarctica to observe changes in the global climate. Both the ozone layer over Antarctica and the continent's vast ice sheets respond to changes in atmospheric conditions. In this article scientists describe how global climate change may be contributing to the surprising growth of one of Antarctica's ice sheets.

Irene Brown is a science correspondent for United Press International.

RESEARCHERS CHART SURPRISING GROWTH OF ICE

PASADENA, Calif., Jan. 17 (UPI)—In a surprising discovery, researchers using satellite radar pictures have found the Western Antarctic Ice Sheet may no longer be shrinking and actually is growing due to blocked up rivers of ice.

Roughly 25 percent of the accumulation of snow is not being discharged in the ice streams, increasing the overall mass of the ice sheet by about 1.5 inches a year, say researchers in an article to be published in the Jan. 18 [2002] issue of the journal Science.

"We know from geological evidence that the ice sheet has been retreating, but we think the mass balance has switched recently," said Slawek Tulaczyk, an assistant professor of Earth Sciences at University of California-Santa Cruz. "We think we may have happened to capture a reversal of a geologic event."

Ian Joughin, with NASA's Jet Propulsion Laboratory in Pasadena, Calif., and Tulaczyk used Canada's Radarsat spacecraft to make maps charting ice flows in a 447,000-square mile tract in Western Antarctica. The velocity of the ice flows indicates how much ice is being discharged. Researchers use the data in conjunction with snowfall measures to determine trends in ice accumulation.

Until the satellite radar images were used to make measurements, researchers did not have an accurate portrayal of ice flow dynamics, Tulaczyk said.

"People have been going around digging holes and measuring (snow) accumulation rates over past decade. We know that pretty well. But the output has always been much more difficult to determine.—It's kind of hard to run around Antarctica and measure when everything is moving," said Tulaczyk.

"We now can make maps with extremely accurate ice displacement measurements," he added. "The satellite data really revolutionized the way we could look at ice sheets."

The scientists attribute the ice sheet growth to two events: the total blockage of a primary ice stream, which dissipated about 18 cubic miles of ice per year until about 150 years ago; and an ongoing slowdown in another ice stream, which blocks off the equivalent of about half the Missouri River. Tulaczyk said he expects that stream to come to a complete halt later this century.

The findings may be related to an overall global warming trend, said Reed Scherer with Northern Illinois University's Department of Geology and Environmental Sciences.

"You tend to think of warming in terms of melting, but in Antarctica warming can actually bring increased snowfall because the ocean is warmer and it puts more moisture in the air," Scherer said.

Tulaczyk says his research is not inconsistent with global warming trends.

"What has been the major driver of changes in ice mass—the production of water and how fast the ice moves—occurs at the base, not at the top of the ice sheet. If the climate is changing now, maybe in several thousand years it will affect the dynamics of water," he said.

Irene Brown

BROWN, IRENE. "RESEARCHERS CHART SURPRISING GROWTH OF ICE." UNITED PRESS INTERNATIONAL (UPI), NEWSWIRE. JANUARY 17, 2002.

SEE ALSO *Antarctica: Melting; Antarctica: Role in Global Climate.*

BIBLIOGRAPHY

Periodicals

Allison, Ian. "Antarctic Ice and the Global Climate System." *Australian Antarctic Magazine* (Autumn 2002): 3–4.

Cazenave, Anny. "How Fast Are the Ice Sheets Melting?" *Science* 314 (2006): 1250–1252.

Eilperin, Juliet. "Antarctic Ice Sheet Is Melting Rapidly." *Washington Post* (March 3, 2006): A01.

EPICA [European Project for Ice Coring in Antarctica] Community Members. "One-to-one Coupling of Glacial Climate Variability in Greenland and Antarctica." *Nature* 444 (2006): 195–198.

Jouzel, J. "Orbital and Millennial Antarctic Climate Variability Over the Past 800,000 Years." *Science* 317 (2007): 793–796.

Kerr, Richard A. "A Bit of Icy Antarctica Is Sliding Toward the Sea." *Science* 305 (2004): 1897.

Quayle, Wendy C., et al. "Extreme Responses to Climate Change in Antarctic Lakes." *Science* 295 (2002): 645.

Shepherd, Andrew, and Duncan Wingham. "Recent Sea-Level Contributions of the Antarctic and Greenland Ice Sheets." *Science* 315 (2007): 1529–1532.

Stocker, Thomas F. "North-South Connections." *Science* 297 (2002): 1814–1815.

Thompson, David W. J., and Susan Solomon. "Interpretation of Recent Southern Hemisphere Climate Change." *Science* 296 (2002): 895–899.

Velicogna, Isabella, and John Wahr. "Measurements of Time-Variable Gravity Show Mass Loss in Antarctica." *Science* 311 (2006): 1754–1756.

Web Sites

"Mission News: NASA Finds Vast Regions of West Antarctica Melted in Recent Past." *National Aeronautics and Space Administration (NASA),* May 15, 2007. <http://www.nasa.gov/vision/earth/lookingatearth/arctic-20070515.html> (accessed August 10, 2007).

Larry Gilman

Antarctica: Role in Global Climate

■ Introduction

The world's climate system is, in some ways, like a complex machine. Heat is moved from place to place by ocean currents and by winds; winds, ocean currents, sea ice, land ice, snow cover, vegetation, and other factors affect climate and are affected by climate. The chemical composition of the atmosphere, which is being changed by humans, is also a factor.

Antarctica is a unique part of the Earth's climate machine. It stores most of the world's freshwater, generates large amounts of sea ice, and is surrounded by pole-circling (circumpolar) currents of air and water. Antarctic conditions influence the pattern of ocean circulation called the thermohaline circulation or Great Conveyor Belt, which transports heat from the tropics toward the poles and increases the ability of the oceans to absorb carbon dioxide from the atmosphere.

Direct observations of Antarctica are difficult to make because of its remoteness and harsh weather. As a result, records of Antarctic climate date back less than 50 years. Satellite observation began in the 1970s and has greatly increased scientists' ability to understand what is happening in Antarctica and its role in global climate. Although uncertainties remain in the understanding of Antarctica's place in the climate system, scientists are beginning to unravel how human-caused climate changes are changing Antarctic weather, and how those changes might affect global climate.

■ Historical Background and Scientific Foundations

Mapmakers speculated about the existence of a south-polar continent for centuries, but human beings did not glimpse Antarctica until 1820. Although explorations and heroic dashes for the South Pole added to scientific knowledge over the following century, it was not until the International Geophysical Years (IGY) in 1957 to 1958 that modern scientific monitoring of the Antarctic continent began in earnest. In preparation for the IGY, governments established a number of year-round bases in Antarctica, some of which are still in operation. Today, Antarctica is surveyed constantly by ships, permanent bases, free-floating sea buoys, weather balloons, aircraft, and satellites.

In the last 10 to 20 years, understanding of Antarctic climate processes and their connection to the global climate system has advanced. The roles of sea ice, surface ice, melting, reflection of sunlight by ice and clouds, anthropogenic (human-caused) changes to the ozone layer, global warming, and circumpolar circulations of sea water and wind are all better understood now than a few decades ago. A few important aspects of Antarctic climate follow.

Sea Ice and Thermohaline Circulation

The area covered by sea ice in the Antarctic Ocean (also called the South Polar Ocean or Southern Ocean, namely all waters south of 60° south latitude) varies with the season. During the Antarctic summer, sea ice covers 1–1.54 million sq mi (3 or 4 million sq km); in the winter, it covers 6.6–7.73 million sq mi (17–20 million sq km).

Sea ice has several interactions with climate. First, snow and ice are highly reflective. They reflect solar energy to space like a mirror. Open seawater, however, is a comparatively efficient absorber of solar energy. Therefore, the long-term ice shelves and seasonal sea ice of Antarctica, which both consist of floating sheets of frozen sea-water, reduce the amount of energy Earth absorbs. When global warming melts sea ice, Earth becomes a better absorber of heat, which encourages further warming. This is a form of positive feedback: warming causing further warming.

Second, sea ice isolates the water from the air, reducing the transfer of energy between winds and ocean currents.

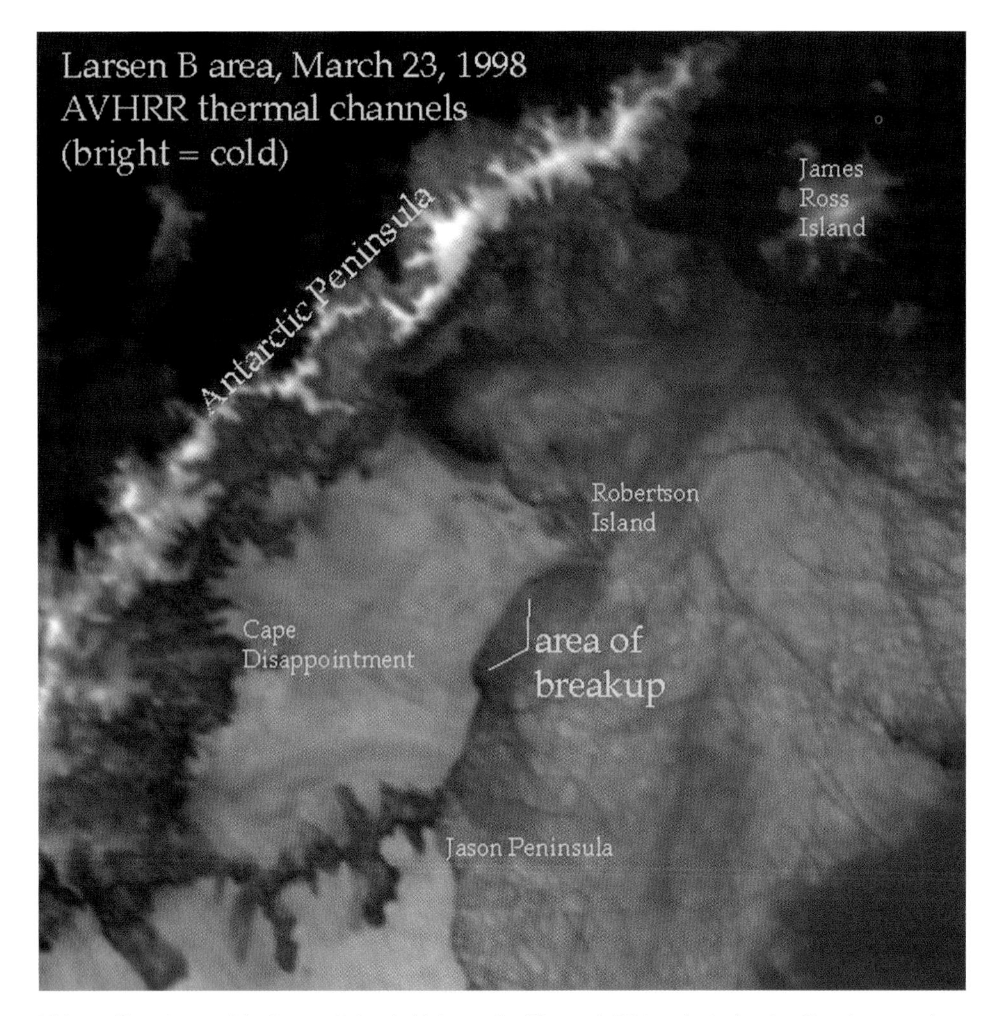

This satellite photo of the Larsen B ice shelf shows the 75 sq mi (194 sq km) chunk of ice sheet on the Antarctic Peninsula that snapped off in early 1998. Scientists cited global warming as the cause. *AP Images.*

Third, sea ice affects the global ocean circulations as it freezes and melts. When sea water freezes, its salt cannot be accommodated in the atomic structure of the forming water crystals and is pushed out. Small pockets of saltier (harder to freeze) water form in the ice. These tend to melt their way downward to unfrozen ocean water below. Freezing sea ice is therefore said to reject salt. Salt rejection makes the ocean near new sea ice saltier and, therefore, denser. (When salt dissolves in water it increases its weight without increasing its volume, so salty water is denser than freshwater.) This denser, saltier seawater sinks, pushing deep water aside and causing current to flow. However, when sea ice melts, it adds freshwater to the surrounding ocean, reducing its salinity (saltiness). Both of these processes affect the way in which deep and polar waters trade places with surface and tropical waters over time.

Since saltier water freezes at lower temperatures than less-salty water, the salty water near forming sea ice must be chilled to a lower temperature before it, too, freezes. This enables the wind to remove more heat from the sea near forming sea ice. Colder water, like saltier water, is denser than other water, so cooling also contributes to the sinking of water near the poles. As the area of ocean covered by Antarctic sea ice varies by thousands of square miles from winter to summer, the freezing and melting of sea ice are significant drivers of ocean circulation.

The overturning or circulation of the world's oceans driven by heat and saltiness is called the thermohaline circulation (*thermo* for heat, *haline* for salt). The thermohaline circulation transports heat from the equator to the polar regions, which has many effects on climate, including making Europe's climate milder. By bringing water from the bottom of the ocean to the top, it increases the ocean's ability to absorb carbon dioxide from the atmosphere.

Ice shelves in Western Antarctica, particularly along the shores of the Antarctic Peninsula, have been destabilized by anthropogenic global warming and are melting

WORDS TO KNOW

ANTARCTIC CIRCUMPOLAR CURRENT: An ocean current that circles Antarctica from west to east (clockwise, looking down on the continent), enabling mixing of the waters of the world's oceans. Also termed the West Wind Drift.

ANTHROPOGENIC: Made by people or resulting from human activities. Usually used in the context of emissions that are produced as a result of human activities.

BIPOLAR SEESAW: The tendency of climate to warm at the North Pole while it cools at the South Pole and vice versa. Paleoclimatic data show that this seesaw effect has occurred many times in the geological past and is related to changes in ocean circulation, but scientists do not yet completely understand the mechanism of the seesaw.

BUOYS: Tethered or free-floating devices that bear navigational aids, instruments, and in some cases radio equipment for automatically collecting and reporting data on oceanic and atmospheric conditions. Buoys may float on or beneath the surface, depending on their purpose.

ICE SHEET: Glacial ice that covers at least 19,500 square mi (50,000 square km) of land and that flows in all directions, covering and obscuring the landscape below it.

ICE SHELF: Section of an ice sheet that extends into the sea a considerable distance and that may be partially afloat.

OZONE LAYER: The layer of ozone that begins approximately 9.3 mi (15 km) above Earth and thins to an almost negligible amount at about 31 mi (50 km) and shields Earth from harmful ultraviolet radiation from the sun. The highest natural concentration of ozone (approximately 10 parts per million by volume) occurs in the stratosphere at approximately 15.5 mi (25 km) above Earth. The stratospheric ozone concentration changes throughout the year as stratospheric circulation changes with the seasons. Natural events such as volcanoes and solar flares can produce changes in ozone concentration, but man-made changes are of the greatest concern.

SOUTHERN ANNULAR MODE: Pattern of oscillation (regular back-and-forth change) in atmospheric pressure and windspeeds that occurs in the extratropical Southern Hemisphere; the southern counterpart of the North Atlantic Oscillation. Also termed Antarctic Oscillation or High Latitude Mode. Scientists have attributed the increasing magnitude of the southern annular mode to stratospheric ozone depletion and greenhouse gas increases.

THERMOHALINE CIRCULATION: Large-scale circulation of the world ocean that exchanges warm, low-density surface waters with cooler, higher-density deep waters. Driven by differences in temperature and saltiness (halinity) as well as, to a lesser degree, winds and tides. Also termed meridional overturning circulation.

at a faster rate. In a 35-day period in 2002, an area of floating ice called the Larsen B ice shelf, undermined by warm water currents, disintegrated completely into thousands of icebergs that floated away and melted. The shelf had been the size of the state of Rhode Island, 1,253 sq mi in extent (3,250 sq km) and 720 ft (220 m) thick. In 2006, scientists with the British Antarctic Survey announced that they had shown a definite climatic connection between this event and human-caused global warming. Warm westerly winds blowing across the peninsula, which climate models attribute to anthropogenic warming, were clearly responsible for the breakup of Larsen B.

The Antarctic Ice Sheet

Antarctica's ice sheet, which covers most of the continent, has an average thickness of 1.5 mi (2.5 km) and holds about 7 million cubic mi (29 million cubic km) of ice, about 90% of the world's freshwater. This is enough water to raise sea level worldwide by about 200 ft (61 m) if the entire ice sheet were to melt. Snowfall adds mass to the ice sheet while glaciers flowing to the sea and chunks of ice breaking off the edges of ice shelves remove it.

Ice that is floating in water does not deepen that water when it melts, so the melting Antarctic sea ice does not raise sea levels directly. However, melting of ice that rests on land does raise sea levels. Both of Earth's two great overland ice sheets, in Greenland and Antarctica, are now known to be melting. Further, ice shelves act as dams or barriers holding back (or at least slowing down) glaciers and ice rivers transporting ice from inland to the coast; therefore, when ice shelves shrink or disappear, melting of the Antarctic ice sheet can be accelerated.

Speedup of ice movement from this and other causes has been observed for several hundred glaciers in Western Antarctica, where most of Antarctica's recent warming has happened. This area has warmed by about 4.5°F (2.5°C) over the last 50 years, as much as anywhere on the planet. Eastern Antarctica is higher in altitude, and, therefore, colder than Western Antarctica. It also tends to be better isolated from the rest of the climate system by the ring-shaped circumpolar circulations of westerly winds and ocean currents. As a result, Eastern Antarctica has undergone slight cooling over the last several decades even as the rest of the world, including Western Antarctica, has seen significant warming.

Circumpolar Flows

Both air and water circle Antarctica from west to east like cars driving around a racetrack. These winds, like the general westward drift of weather all over the planet, are driven by the Earth's axial rotation and temperature

IN CONTEXT: ICE SHELVES

A Rhode Island-sized piece of one of Antarctica's floating ice shelves broke up into a fleet of thousands of icebergs over a few weeks in early 2002. Ice shelves are the floating edges of continental glaciers that form where a glacier flows out over the sea. The shelves that cover most of Antarctica's coastal inlets (narrow strips of water running into the land or between islands) and bays are the outlets of faster-moving currents called ice streams that drain ice from the interior of ice sheets.

differences between the poles and equator. Ocean currents flow in a similar pattern around Antarctica. Circumpolar winds also blow around the North Pole, but masses block the formation of a circum-Arctic ocean current. The westerly circling of water around Antarctica is called the Antarctic Circumpolar Current (ACC). The ACC plays an important role in the thermohaline circulation, acting as a blender for the waters of the world's oceans.

Short-term Climate Oscillations

Long-term climate cycles include ice ages lasting 100,000 years, caused by regularly recurring changes in Earth's orbit. Climate also shows much more rapid cycles of climate variability. The most important of these in the Antarctic region is the Southern Annular Mode or SAM, also called the Antarctic Oscillation. "Annular" means ring-shaped, and an oscillation occurs when a physical system switches repeatedly between two conditions. The SAM is an oscillation involving weather conditions over the South Pole and in the ring of eastward-moving air around Antarctica. In one state of the oscillation, pressure is low over the South Pole (central Antarctica) and high in the circumpolar wind ring; in this state, the westerly winds blow more strongly. At the other end of the oscillation, conditions are reversed. Pressure is high over the pole and low in the wind ring (also called a circumpolar vortex), which blows less strongly. The SAM shifts from one state to another over weeks, months, or years in an irregular way, but pressure always goes up over the center of the continent when it decreases in the circumpolar vortex and vice versa.

Changes in the SAM have been linked to climate changes in the Southern Hemisphere. Analysis of weather-balloon data measurements from 1969 to 1998 shows that the SAM has been spending more time at the positive or strong-wind end of its cycle. This, in turn, may be a result of human-caused changes in the atmosphere. Specifically, certain industrial chemicals (mostly chlorofluorocarbons) have caused a hole to appear in the ozone layer over Antarctica for part of every year since the 1980s. The hole reached a record size in 2000. Normally, the ozone layer absorbs solar energy and heats the stratosphere (upper atmosphere). Loss of ozone over central Antarctica, which occurs every year from September to December, therefore cools the stratosphere. September–December stratospheric temperatures dropped about 18°F (10°C) from 1985 to 2002.

In 2002, David W. J. Thompson and Susan Solomon, climate scientists working at the U.S. National Oceanic and Atmospheric Administration (NOAA), proposed a two-part theory concerning the SAM: 1) Changes in the SAM could, they said, account for about half of the observed warming of the Antarctic Peninsula and the southern tip of South America and about 90% of the slight cooling seen over central Antarctica. 2) Changes in the SAM could in turn be traced to ozone destruction caused by human activity. The second part of this theory was more controversial than the first.

How changes in the SAM might affect the rest of the global climate system and be affected in turn by it is still uncertain. However, scientists have increasingly been modifying their older view that the Antarctic climate is more or less irrelevant to global climate, isolated by the circumpolar vortex. A surprisingly strong connection between Antarctic and global climate has recently been discovered in the bipolar seesaw.

Bipolar seesaw

A connection between Antarctic climate and the climate of the north-polar region was first proposed in 1998. Comparison of climate information from ice cores from Antarctica and from Greenland, near the North Pole, showed that in case after case, over tens of thousands of years, while Greenland was undergoing a warming trend Antarctica was cooling, and while Antarctica was warming, Greenland was cooling. This back-and-forth pattern was dubbed the bipolar seesaw. ("Bipolar" means two poles, in this case the North and South Poles.)

The mechanism proposed for the bipolar seesaw is as follows. The thermohaline circulation of the oceans moves large amounts of heat northward in the Atlantic Ocean. This circulation is partly driven by the sinking of cooled waters near Antarctica. When the thermohaline circulation slows or stops, less heat travels northward through the Atlantic, while less cold water is removed from the vicinity of Antarctica. Thus, Greenland cools while Antarctica warms. When the thermohaline circulation starts up again, the pattern is reversed. Greenland warms as heat is transported northward and Antarctica is cooled.

In 2002, Thomas Stocker, a physicist at the University of Bern, Switzerland, suggested a challenge to the seesaw theory. Ice-core data showed that as Antarctica was warming steadily from about 19,000 years ago to

10,000 years ago, a sudden, relatively brief (1,800-year) Antarctic cooling event, named the Antarctic Cold Reversal, occurred. According to the seesaw hypothesis, Stocker said, this should have occurred at the same time as the sudden warming seen in Greenland 14,500 years ago. However, it actually occurred about 500 years too soon, while the climate was still warm in Greenland. This, Stocker said, was a serious challenge to the bipolar seesaw. Relatively rapid changes in Antarctic climate, he argued, on the scale of decades or centuries, were probably isolated from the rest of the world's climate by the circumpolar vortex and the Antarctic Circumpolar Current.

However, in 2006, a team of scientists from ten European countries, also using ice-core data, validated the bipolar seesaw for the period from 55,000 to 10,000 years ago. During this interval, even short, weak temperature changes in Antarctica are matched quickly by changes in Greenland and thus, presumably, in the thermohaline circulation. The apparent time-mismatch between Greenland and Antarctic temperatures during the Antarctic Cold Reversal that Stocker had noted was probably due to imprecise matching of the layers in Greenland and Antarctica ice cores. The 2006 seesaw conclusion was reached using a new ice core from Dronning Maud Land in Antarctica. This allowed for higher-resolution climate reconstruction than earlier ice cores because more snow falls each year in Dronning Maud Land, laying down thicker annual ice layers that are easier to count.

These results show that the Antarctic climate is in fact tightly linked with global climate. Given the circumpolar wind and ocean currents, which should have an isolating or decoupling effect between Antarctica and the rest of the world, this result is surprising. A group of Danish scientists from the Ice and Climate group at the Niels Bohr Institute, University of Copenhagen, that were involved in the 2006 confirmation of the bipolar seesaw released a statement that said it was "really astounding how systematically heat is moved between the north and south hemisphere with the Seesaw, causing really dramatic climate changes during the glacial period."

■ Impacts and Issues

Antarctic climate, like global climate change, is nothing new. Cylinders of ice up to 2 mi (3.2 km) long, drilled from thick deposits in Antarctica and Greenland, have supplied an 800,000-year archive of air samples that shows that Earth's climate has always varied, often quickly and dramatically. Changes in Earth's orbit around the sun and other factors have caused many periods of global cooling and warming.

However, the situation today is different. In the last few years, scientific evidence has indicated that the recent rapid warming of global climate is caused mostly by human activity. Changes in Antarctic climate, including the breakup of ice shelves, accelerated glacial movement, increased snowfall in some areas, and intensified climate oscillations have all been linked to anthropogenic climate changes, including ozone depletion and warming.

The consequences of large amounts of Antarctic melting could be catastrophic. Melting of the West Antarctic ice sheet alone (where most warming is seen to date, and where most future warming is expected to happen) would raise sea levels by 20 ft (6 m). About 100 million people live within about 3 ft (1 m) of today's sea level, and about 600 million live within about 20 ft (6 m) of sea level. Recent work has shown that although ice is accumulating in some parts of Antarctica, it is being lost in others, with loss outpacing gain. Overall, Antarctica is losing about 36 cubic mi (152 cubic km) of ice per year, causing about 0.016 in (0.4 mm) of sea level rise per year.

Despite recent progress in the understanding of Antarctica, its role in global climate change remains uncertain. For example, it is theoretically possible that the bipolar seesaw could be controlled by pushing on either its northern or southern end—that is, by causing climate changes in either the Arctic, the Antarctic, or both. Human-caused changes are indeed happening in both regions today, but it cannot yet be predicted whether these pushes might cause rapid or significant change in thermohaline circulation of the oceans.

■ Primary Source Connection

Seventy percent of the world's freshwater is contained in the Antarctic cryosphere. The cryosphere is the frozen part of the Earth's surface, which may be ice sheets, ice shelves, glaciers, snow cover, or any other frozen form. This article examines the response of global climate change on the Antarctic cryosphere and how that response contributes to further global climate change.

Ian Allison is a research scientist with the Australian government's Antarctic Division and a contributing writer for *Australian Antarctic Magazine.*

ANTARCTIC ICE AND THE GLOBAL CLIMATE SYSTEM

Planet Earth is a natural greenhouse. Some naturally occurring atmospheric trace gases, called greenhouse gases, permit incoming solar radiation to reach the Earth's surface but restrict the outward flow of infrared radiation. Carbon dioxide and water vapour absorb this outgoing infrared energy and re-radiate some of it back to ground level. This greenhouse effect is essential to most life on Earth. Without it the average temperature of the surface would be a frigid minus 18°C, rather than about 14°C as it is today.

But the concentration of greenhouse gases in the atmosphere, especially carbon dioxide, has been increased by human combustion of fossil fuels, exacerbated by deforestation. Since the Industrial Revolution began, carbon dioxide levels have risen from 280 parts per million by volume (ppmv) to 370 ppmv, and are reliably predicted to reach double pre-industrial levels in the second half of this century. Humans have also added other greenhouse gases such as methane, CFCs, and nitrous oxide to the atmosphere. The combined effect of these additional gases will be a rise in global temperatures, predicted by climate models to be 1°C to 4°C, by the end of the 21st century.

Global warming will not be uniform over the earth because of the complex interactions within and between oceans, atmosphere, land surface, clouds, biological systems and ice and snow. Some of the largest changes are predicted to occur at high latitudes. Exposed ocean or bare earth caused by the loss of ice and snow cover through melting will result in increased absorption of solar energy, which in turn will further reduce ice and snow cover leading to an amplified effect—a positive feedback. Against this, however, is an increase in heat fluxes from the ocean to the atmosphere—a negative feedback—caused by a decrease in sea ice.

A central objective of Australia's Antarctic program is to understand the role of Antarctica in the global climate system. This requires us to study Antarctic processes contributing to the climate system, determine the response of the Antarctic to climate change and seek evidence of past and present change in the region. Many important Antarctic climate-related processes involve ice. Ice and snow (the 'cryosphere') are important components of climate, with snow in particular limiting absorption of solar energy at the surface through its high reflectivity ('albedo'). Freezing of water and melting of ice involve latent heat exchange, and snow and ice on land or sea inhibit heat transfer. The water volume stored in ice sheets and glaciers is a major factor in considering sea level change. Ice and snow also provide evidence of past change from the ice core climate record and visual evidence of ongoing change due to melt.

All these factors make it important to understand the role of ice and snow in the climate system, a need recognised in the recent establishment of a new international research initiative, Climate and Cryosphere within the World Climate Research Programme. The Australian Antarctic glaciology program contributes to this program with cryosphere studies taking in Antarctic sea ice, the continental ice sheet including ice core climate records, subantarctic glaciers and abrupt change.

Sea ice

The extent of Southern Hemisphere sea ice (frozen sea) varies seasonally by a factor of five, from a minimum of 3–4 million km^2 in February to a maximum of 17–20 million km^2 in September. When the ice forms it ejects salt to the ocean, destabilising the water column and deepening the surface mixed layer. It can also influence formation of the global oceans' deep and bottom water and help drive overturning ocean circulation. Sea ice moved by wind and currents, as it melts, deposits freshwater onto the ocean surface to stabilise the water column. Sea ice has a dramatic effect on the physical characteristics of the ocean surface, modifying surface radiation balance due to its high albedo and influencing the exchange of momentum, heat, and matter between atmosphere and ocean. Through these effects, sea ice plays a key role in the global heat balance. A retreat of sea ice associated with climate warming could have global consequences through various feedback processes.

The Antarctic ice sheet

Antarctica's ice sheet covers 12.4 million km^2. It comprises 25.7 million km^3 of ice or 70 percent of the world's freshwater, which if melted would raise the sea level by nearly 65 m. Mass is continually added to the ice sheet from snowfall, and removed via melt and iceberg calving, particularly from ice shelves. Any change in the ice sheet's 'mass budget' caused by imbalance between these mass input and output terms affects sea level. However, with present Antarctic data a 20 percent imbalance, corresponding to about 10 cm of sea-level change per century, cannot be detected with confidence. The ice sheet is not a single dynamic entity, but comprises different drainage systems with both surface mass balance and dynamics responding differently to changing conditions. We need to be able to estimate the sensitivity of the mass budget to climate change before we can estimate Antarctica's future contribution to sea level change.

Most of the ice lost from the ice sheet comes from fast-flowing, wet-based outlet glaciers and ice streams, much of which passes through floating ice shelves. Up to 40 percent of the Antarctic coastline is composed of either large ice shelves in coastal embayments such as Filchner-Ronne, Ross and Amery or fringing shelves on the periphery of the ice sheet such as the West, Shackleton and Larsen shelves. Since ice shelves are floating on ocean waters at the freezing point, even a small change in ocean temperature (induced perhaps by changed ocean currents) can significantly affect the shelves' basal melt rate and cause them to thin much more quickly than rising air temperature. Ice shelves are already floating, which means their disintegration will by itself have no measurable impact on global sea level, but their depletion may lead to increased drainage of grounded ice 'buttressed' by the shelves which may cause sea-level rises.

Ice core records of past climate

Antarctica's ice sheet stores the Earth's longest and most representative record of atmospheric composition and temperature in times past. The ice sheet's layers of ice and snow, accumulated over tens or even hundreds of thousands of years, form a natural archive of global environmental information, accessible by drilling into the ice to sample past surface deposits. Analyses of the ice and the material trapped in it allow records to be made of both natural and man-made environmental variations over the time period during which the ice sheet has accumulated. Deep ice cores have yielded evidence of major interrelated climate and cryosphere fluctuations in glacial-interglacial cycles. Accurate information on local, regional and global climate change and potential changes in ice sheet surface elevation are available from ice cores.

Subantarctic glaciers

Like mountain glaciers in most parts of the world, subantarctic glaciers have been noticeably retreating over the past 50 years or more. Retreat of non-polar glaciers has contributed to sea level rise over the past century while also providing clear evidence of a changing climate. On Heard Island, for example, the Brown Glacier has decreased in area by 33 percent and in volume by 38 percent over the past 50 years.

Abrupt change

Palaeoclimate records from ice and ocean sediment cores show evidence of abrupt and widespread past climate changes—particularly, it seems, during periods of transition from one climate regime to another over glacial-interglacial cycles. While the causes and mechanisms of such rapid changes are by no means clear, a variety of roles have been suggested for ice sheets, glaciers and sea ice. These include effects of rapid glacial discharge and decomposition with a rise to melting point of basal ice temperature, massive iceberg discharge into the ocean delivering freshwater capable of modifying the overturning circulation of the global ocean, and changing sea ice formation causing change in brine release to the ocean. Greenhouse warming and other human alterations of the climate system may increase the possibility of large and abrupt regional or global climatic events.

A startling illustration of how abrupt some processes are is the recent rapid collapse of the Larsen B Ice Shelf on the eastern side of the Antarctic Peninsula in February and March 2002, when 3250 km^2 of ice 200 m thick disintegrated over a few weeks.... The break-up of this shelf into thousands of small icebergs is totally different from the normal episodic calving of giant icebergs from the front of ice shelves. Temperatures around the Antarctic Peninsula have risen by 2.5°C over the past 50 years. It is hypothesised that water from large surface melt ponds that formed on the ice shelf as a result of this warming forced open cracks and crevasses to completely fracture the shelf. The ice of the shelf was already floating, so the collapse has no measurable effect on sea level, and direct impacts are believed to be mostly local. However, a similar collapse of some other shelves could bring a significant increase in glacial discharge from the ice sheet.

Ian Allison

ALLISON, IAN. "ANTARCTIC ICE AND THE GLOBAL CLIMATE SYSTEM." *AUSTRALIAN ANTARCTIC MAGAZINE* (AUTUMN 2002): 3–4.

SEE ALSO *Antarctica: Melting; Antarctica: Observed Climate Changes; Anthropogenic Change; Arctic Melting: Greenland Ice Cap; Arctic Melting: Polar Ice Cap; Arctic People: Climate Change Impacts; Carbon Dioxide Concentrations; Climate Change; Environmental Policy; Environmental Pollution; Greenhouse Effect; Greenhouse Gases; Greenland: Global Implications of Accelerated Melting; Intergovernmental Panel on Climate Change (IPCC); Polar Bears; Polar Ice; Polar Wander.*

BIBLIOGRAPHY

Periodicals

Allison, Ian. "Antarctic Ice and the Global Climate System." *Australian Antarctic Magazine* (Autumn 2002): 3–4.

Cazenave, Anny. "How Fast Are the Ice Sheets Melting?" *Science* 314 (November 24, 2006): 1250–1252.

Eilperin, Juliet. "Antarctic Ice Sheet Is Melting Rapidly." *Washington Post* (March 3, 2006).

Kerr, Richard A. "A Bit of Icy Antarctica Is Sliding Toward the Sea." *Science* 305 (September 24, 2004): 1897.

Quayle, Wendy C., et al. "Extreme Responses to Climate Change in Antarctic Lakes." *Science* 295 (January 25, 2002): 645.

Stocker, Thomas F. "North-South Connections." *Science* 297 (September 13, 2002): 1814–1815.

Thompson, David W. J., and Susan Solomon. "Interpretation of Recent Southern Hemisphere Climate Change." *Science* 296 (May 3, 2002): 895–899.

Turner, John, et al. "Antarctic Climate Change During the Last 50 Years." *International Journal of Climatology* 25 (March 15, 2005): 279–294.

Web Sites

"Interactive Polar Ice Cap Melter" *Everybody's Weather.* <http://www.everybodysweather.com/

Static_Media/Polar_Ice_Cap_Melter/index.htm> (accessed November 2, 2007).

"Mission News: NASA Finds Vast Regions of West Antarctica Melted in Recent Past." *U.S. National Aeronautics and Space Administration (NASA)*, May 15, 2007. <http://www.nasa.gov/vision/earth/lookingatearth/arctic-20070515.html> (accessed August 10, 2007).

Larry Gilman

Anthropogenic Change

■ Introduction

Understanding anthropogenic change is crucial to understanding our world and its historical and modern transformations. Anthropogenic changes are alterations that result from human action or presence. They may be deliberate, such as when land is cleared for agriculture, modifying landscapes and introducing new species.

Anthropogenic changes may also be an unrecognized or poorly understood side-effect of human activity, as with the decreased biodiversity that accompanies increased urbanization or with much of the pollution resulting from industrialization and the technological advances of the twentieth century.

Increased production of carbon dioxide and other greenhouse gases and the resulting alteration of global climate is a good example of anthropogenic change that has been slowly revealed over the past several decades. Much of the difficulty in understanding and measuring anthropogenic climate change is caused by the complexity of Earth systems involved and by the challenge of differentiating natural variation from anthropogenic change.

■ Historical Background and Scientific Foundations

Although people have been altering their environment since prehistoric times, these changes were largely perceived

In 2007, regions in the United Kingdom were hit by severe floods. Claiming that anthropogenic climate change was to blame for the flooding, protesters in Oxford, U.K., held a demonstration demanding change. *AP Images.*

WORDS TO KNOW

BIODIVERSITY: Literally, "life diversity": the number of different kinds of living things. The wide range of organisms—plants and animals—that exist within any given geographical region.

DEFORESTATION: Those practices or processes that result in the change of forested lands to non-forest uses. This is often cited as one of the major causes of the enhanced greenhouse effect for two reasons: 1) the burning or decomposition of the wood releases carbon dioxide; and 2) trees that once removed carbon dioxide from the atmosphere in the process of photosynthesis are no longer present and contributing to carbon storage.

GREENHOUSE GASES: Gases that cause Earth to retain more thermal energy by absorbing infrared light emitted by Earth's surface. The most important greenhouse gases are water vapor, carbon dioxide, methane, nitrous oxide, and various artificial chemicals such as chlorofluorocarbons. All but the latter are naturally occurring, but human activity over the last several centuries has significantly increased the amounts of carbon dioxide, methane, and nitrous oxide in Earth's atmosphere, causing global warming and global climate change.

EROSION: Processes (mechanical and chemical) responsible for the wearing away, loosening, and dissolving of materials of Earth's crust.

RADIATIVE FORCING: A change in the balance between incoming solar radiation and outgoing infrared radiation. Without any radiative forcing, solar radiation coming to Earth would continue to be approximately equal to the infrared radiation emitted from Earth. The addition of greenhouse gases traps an increased fraction of the infrared radiation, reradiating it back toward the surface and creating a warming influence (i.e., positive radiative forcing because incoming solar radiation will exceed outgoing infrared radiation).

IN CONTEXT: HUMAN IMPACTS (ANTHROPOGENIC IMPACTS) ON GREENHOUSES GASES

"The dominant factor in the radiative forcing of climate in the industrial era is the increasing concentration of various greenhouse gases in the atmosphere. Several of the major greenhouse gases occur naturally but increases in their atmospheric concentrations over the last 250 years are due largely to human activities. Other greenhouse gases are entirely the result of human activities."

SOURCE: *Solomon, S., et al. "Technical Summary." In:* Climate Change 2007: The Physical Science Basis: Contribution of Working Group I to the Fourth Assessment Report of the Intergovernmental Panel on Climate Change. *New York: Cambridge University Press, 2007.*

as beneficial. Environmental degradation such as flooding, deforestation, pollution, and erosion were seen as temporary or small-scale problems that could be reversed by altering local practices or by moving to new areas, despite the concern of some nineteenth century scholars like George P. Marsh, who published *Man and Nature*—one of the first critical looks at anthropogenic change—in 1864.

By the 1950s, many scientists were beginning to understand that at least some anthropogenic changes were global, rather than local or regional in nature, and there was concern that some changes were irreversible. In the following decades, the term anthropogenic became more commonly used outside its original context in ecology (replacing the phrase man-made), and was used to describe much more extensive human influence than the landscape alteration and pollution highlighted before the 1980s.

■ Impacts and Issues

Anthropogenic climate change has become one of the foremost environmental issues facing the world. Richard Kerr points out in an article appearing in the July 6, 2007 issue of *Science* that understanding how human influences on climate interact with changing natural systems is vital.

Models of the feedback between deforestation, agriculture, air pollution, and climatic systems have emerged as critical areas of research and will shape regional predictions and government policies that attempt to address this human-caused crisis. Further research on the anthropogenic causes of climate change, and how they can be altered or mitigated, will continue to be critical in the next century.

SEE ALSO *Abrupt Climate Change; Acid Rain; Agriculture: Contribution to Climate Change; Antarctica: Melting; Arctic Melting: Greenland Ice Cap; Arctic Melting: Polar Ice Cap; Atmospheric Pollution; Automobile Emissions; Aviation Emissions and Contrails; Baseline Emissions; Biodiversity; Biogeochemical Cycle; Biosphere; Carbon Dioxide Concentrations; Climate Change; Dams; Desert and Desertification; Drought; Dust Storms; Economics of Climate Change; Endangered Species; Enhanced Greenhouse Effect;*

Environmental Pollution; Extinction; Fisheries; Forests and Deforestation; Greenhouse Effect; Greenhouse Gases; Human Evolution; Temperature Record; Urban Heat Islands; Waste Disposal; Wetlands; Wildfires.

BIBLIOGRAPHY

Books

Hughes, J. Donald. *An Environmental History of the World: Humankind's Changing Role in the Community of Life.* New York: Routledge, 2001.

Kolbert, Elizabeth. *Field Notes from a Catastrophe: Man, Nature, and Climate Change.* New York: Bloomsbury, 2006.

Marsh, George P. *Man and Nature*, edited by David Lowenthal. Seattle: University of Washington Press, 2003.

McNeill, J. R. *Something New Under the Sun: An Environmental History of the Twentieth-Century World.* New York: W. W. Norton, 2000.

Thomas, William L., Jr., ed. *Man's Role in Changing the Face of the Earth.* Chicago: University of Chicago Press, 1956.

Turner, B. L., II, ed. *The Earth as Transformed by Human Action: Global and Regional Changes in the Biosphere over the Past 300 Years.* Cambridge: Cambridge University Press, 1990.

Weart, Spencer R. *The Discovery of Global Warming.* Cambridge, MA: Harvard University Press, 2003.

Periodicals

Kellogg, William W. "Mankind's Impact on Climate: The Evolution of an Awareness." *Climatic Change* 10, no. 2 (April 1987): 113–136.

Kerr, Richard A. "Humans and Nature Duel Over the Next Decade's Climate." *Science* 317 (July 6, 2007): 113–136.

Schiermeier, Quirin. "What We Don't Know About Climate Change." *Nature* 445 (February 8, 2007): 580–581.

Arctic Melting: Greenland Ice Cap

Introduction

Greenland is an island in the North Atlantic Ocean. With an area of about 836,000 sq mi (2,166,000 sq km), mostly above the Arctic Circle, Greenland is by far the world's largest island, over twice the size of New Guinea. Politically, it is a territory of Denmark and has a population of about 57,000 people. Over 80% of the island is covered by an ice sheet with an average thickness of about 1 mi (1.6 km), but the ice sheet is twice this thick in its central area. There are also a number of large glaciers and smaller ice caps around the periphery of the island. The total amount of ice stored in the cap is 964,000 cubic mi (4 million cubic km), 10% of all the ice on the planet (most of the rest is in the Antarctic ice sheet). If the Greenland ice cap were to melt completely, scientists estimate that the resulting meltwater would raise global sea levels by about 21 ft (6.5 m, with many experts estimating 7 m). Although most scientists presently assert that the complete melting of the Greenland ice cap is not likely to occur in the next century or so, in recent years greatly accelerated melting of this ice cap due to global warming has been measured.

Historical Background and Scientific Foundations

The Greenland ice sheet, like that of Antarctica, is formed of layers of annual snowfall that pack down into solid ice, each year building on the previous year. Eventually, after many millennia, the ice sheet becomes heavy enough to spread or flow under the force of its own weight, causing glaciers to flow downhill to the sea. As of 2007, the oldest ice that had been recovered by drilling into the Greenland ice sheet was about 123,000 years old, but the ice cap as a whole is much older. It is a leftover from the Pleistocene glaciation, which covered a large part of the Northern Hemisphere with thick glaciers from about 1,800,000 years ago to 11,500 years ago.

Scientific knowledge of Greenland's ice was limited until the 1970s, when Denmark, the United States, and Switzerland organized the Greenland Ice Sheet Program to obtain ice cores from the sheet. Ice cores are cylinders of ice obtained by drilling vertically downward with a hollow device especially designed to return undamaged pieces of ice to the surface. As of 2007, the longest such cores from Greenland recorded 123,000 years of snow layering; in Antarctica, the longest cores recorded some 880,000 years. Ice bubbles trapped in these cores give data on local snowfall rates, and ratios of atomic isotopes in the trapped water indicate global warmth over time. Evidence of global lead pollution by the Greek and Roman civilizations has even been found in Greenland ice cores.

Ice floating on the seas, such as the north polar ice pack, does not raise sea level when it melts. Each piece of melting ice can be visualized as simply filling in the hollow it makes in the water as it floats. However, the ice locked in the Greenland and Antarctic ice sheets, because it is sitting on land, does raise sea levels when it melts. The amount of freshwater running off of Greenland into the sea can also affect the thermohaline or conveyor-belt circulation of the oceans, which transports warm water in surface currents to the Arctic and Antarctic and returns it to the tropics in cool currents deeper in the ocean. This, in turn, affects climate patterns worldwide.

Measuring the amount of ice in Greenland is not an easy matter. The ice sheet is vast (almost as large as Mexico), accessible only by air over large areas, and of uneven thickness. In the 1990s, surface-height changes of the Greenland ice mass, which show gain or loss of ice, were made using laser altimeters (altitude-measuring devices) carried in airplanes. Repeated measurements were compared to each other to detect changes over time. As of 2000, these data indicated that the higher-altitude (above 6,560 ft/2,000 m), thicker part of the ice sheet—which totals 70% of the ice sheet—was in approximate balance, gaining mass from snow and losing it from melting at about equal rates. Seventy percent of

WORDS TO KNOW

CALVING: Process of iceberg formation when huge chunks of ice break free from glaciers, ice shelves, or ice sheets due to stress, pressure, or the forces of waves and tides.

GLACIER: A multi-year surplus accumulation of snowfall in excess of snowmelt on land and resulting in a mass of ice at least 0.04 mi^2 (0.1 km^2) in area that shows some evidence of movement in response to gravity. A glacier may terminate on land or in water. Glacier ice is the largest reservoir of freshwater on Earth and is second only to the oceans as the largest reservoir of total water. Glaciers are found on every continent except Australia.

GREENHOUSE GASES: Gases that cause Earth to retain more thermal energy by absorbing infrared light emitted by Earth's surface. The most important greenhouse gases are water vapor, carbon dioxide, methane, nitrous oxide, and various artificial chemicals such as chlorofluorocarbons. All but the latter are naturally occurring, but human activity over the last several centuries has significantly increased the amounts of carbon dioxide, methane, and nitrous oxide in Earth's atmosphere, causing global warming and global climate change.

GROUNDING LINE: Underwater boundary or line along which a glacier that is flowing into the sea floats free of the ground. Grounding lines are typically many miles from the nominal shoreline. Retreat of grounding lines toward land accompanies speeded glacial flow. Retreat of grounding lines has occurred recently for some glaciers in Greenland and Antarctica.

ICE CORE: A cylindrical section of ice removed from a glacier or an ice sheet in order to study climate patterns of the past. By performing chemical analyses on the air trapped in the ice, scientists can estimate the percentage of carbon dioxide and other trace gases in the atmosphere at that time.

ICEBERG: A large piece of floating ice that has broken off a glacier, ice sheet, or ice shelf.

PALEOCLIMATE: The climate of a given period of time in the geologic past.

PLEISTOCENE EPOCH: Geologic period characterized by ice ages in the Northern Hemisphere, from 1.8 million to 10,000 years ago.

THERMOHALINE CIRCULATION: Large-scale circulation of the world ocean that exchanges warm, low-density surface waters with cooler, higher-density deep waters. Driven by differences in temperature and saltiness (halinity) as well as, to a lesser degree, winds and tides. Also termed meridional overturning circulation.

the sheet area below 6,560 ft (2,000 m) was losing mass, but not very rapidly. These data from the 1990s indicated that the ice mass of Greenland was in approximate balance, with snowfall gains in the southwest balanced by melting in the southeast. The amount of water added to the oceans by Greenland ice loss was calculated to be contributing about 7% of observed sea-level rise, that is, about 0.005 in (0.13 mm) per year.

Gaining More Accurate Data

More accurate measurements of the changing situation in Greenland became possible starting in 2002, when the U.S. National Aeronautics and Space Administration (NASA) launched its Gravity Recovery and Climate Experiment (GRACE). GRACE, which remained in operation as of late 2007, is a pair of Earth-orbiting satellites that fly a common orbit about 136 mi (220 km) apart, one satellite ahead of the other. As the leading satellite dips into and out of regions of stronger gravity, such as exist in the vicinity of mountain ranges or thicker areas of ice sheets, the distance between the satellites changes. The satellites continuously measure this distance to extremely high accuracy. Computer analysis of the distance measurements recorded over many orbits provides a sensitive picture of gravity anomalies (areas of higher- or lower-than-average gravity) over Earth's surface. Not only does GRACE allow continuous surveying of Greenland's ice sheet with high sensitivity, but it covers the whole sheet at equal resolution, which had been impractical with airplanes.

GRACE's gravity measurements are sensitive enough to detect Greenland's seasonal changes in ice mass, with gains from snowfall in the winter and losses from melting in the summer. The results from GRACE startled scientists when enough data had been accumulated to show a trend. The new data revealed that the situation in Greenland has changed rapidly in just a few years. While the high-elevation interior has continued to gain mass from increased snowfall—about 2–2.36 in (5–6 cm) per year over the area above 6,560 ft (2,000 m) elevation—the coastal regions are losing far more than the interior gains. From 2002 to 2006, Greenland's ice sheet lost a total of between 211 million and 284 million tons (192 million and 258 million metric tons) of ice each year (about 50–68 mi^3 or 212–284 km^3).

This loss contributed to an annual sea-level rise of 0.02 in (0.5 mm) per year, much more than previous studies had found. Furthermore, it found that the mass loss was accelerating rapidly. The loss rate during 2004 to 2006 was 2.5 times the rate during 2002 to 2004. Other satellite data showed that the Greenland ice sheet had experienced summertime surface melting over a greater part of its surface in 2006 than at any time at least since continuous satellite surveillance began in 1979.

In recent years, scientists have observed rapid melting in Greenland due to global warming.

Mass loss from the ice sheet occurs not only from melting, but also from speedier flow of coastal glaciers to the sea, where they calve off as floating icebergs that eventually melt. Many of Greenland's outlet glaciers have accelerated their transport of ice to the sea. Ice thinning along Greenland's periphery is concentrated along the deep channels in the ground that the glaciers flow along. The three largest glaciers—Jakobshavn Isbrae, Kangerdlugssuaq, and Helheim—have all accelerated and thinned. Jakobshavn Isbrae on the west coast, the largest, has been thinning by 49 ft (15 m) per year since 1997, while the other two, on the east coast, have been thinning by 131 ft (40 m) and 82 ft (25 m) per year, respectively. Jakobshavn Isbrae accelerated by 95% from 1996 to 2005; ice discharged almost doubled from 1996 to 2005, from 5.7 mi^3 (24 km^3) of ice per year to 11 mi^3 (46 km^3) per year. Kangerdlugssuaq flowed at a stable speed from 1962 to 1996, but from 2000 to 2005 it accelerated by 210%, at which time it was flowing at about 8.7 mi (14 km) per year at the calving front, the area where the glacier breaks up into icebergs.

So far, little velocity change in glacier outflow has been seen in northern Greenland. The changes described earlier are all in the south. However, as global warming proceeds, similar changes are expected to extend farther and farther northward in Greenland.

Impact of Global Warming

Global warming is accelerating the discharge of ice from both Greenland and Antarctica by several means. First, there is direct melting. Second, as floating ice tongues or shelves break up due to warmer water, they uncork or unblock the glaciers behind them, allowing them to flow more quickly to the sea. Third, downward infiltration of surface meltwater in the summer months can lubricate the glacial channel, accelerating it seasonally. Fourth, there is a lubricant effect from warmer ocean waters getting under the glacier. Many large glaciers, such as Jakobshavn Isbrae and the other large glaciers of Greenland, run along deep channels in the earth that are actually below sea level near their outlets. (In eastern Antarctica, submarine flow beds extend all the way to the center of the ice sheet.) Eventually, the glacial ice flowing down the channel floats free of the ground along a front called the grounding line. Scientists have proposed that warmer sea waters are melting ice along the grounding lines of Greenland's glaciers, pushing the grounding lines upstream. This makes the glaciers' channels more slippery and accelerates the ice flow.

Surface water temperatures in the vicinity of Greenland, however, are not high enough to account for the shifting of glacial grounding lines. Scientists explain how warmer water can nevertheless get to Greenland's glaciers in the following way. Efforts to account for Earth's heat budget—how much comes in from the sun, and where it goes—have shown that atmospheric warming cannot account for all the heat Earth is absorbing because of its increased greenhouse gases. The extra heat has been absorbed by the oceans, mostly in the top 3,280 ft (1,000 m). However, in the North Atlantic

Ocean, the heat is carried to greater depths. Warm, salty water is denser than cold, freshwater from glacial melting, so at the outflow of the glacier the warmer, mid-depth ocean water slips down to the grounding line.

Glacial acceleration has been observed not only in Greenland but in the western peninsula of Antarctica, which is the only part of the world to have been warmed as much by global climate change as the Arctic (at about twice the rate of the rest of the planet).

So heavy are the ice flows of glaciers that when they slip suddenly, moving forward several feet or more in a quick jerk, a small earthquake is produced. The existence of these glacial earthquakes was only detected in 2003. By 2006, careful analysis of recordings of vibrations in Earth's crust had revealed that from 1993 to the late 1990s, there was a modest increase in the number of glacial earthquakes. This was followed by a rapid increase from 2002 onward, with nearly as many glacial earthquakes in 2005 as in all of 1993–1996. Seventy-two percent of the earthquakes came from Greenland's three largest glaciers: Jakobshavns Isbrae, Kangerdlugssuaq, and Helheim. Although it is not clear whether the accelerated motion of these glaciers is producing more glacial earthquakes or simply more glacial earthquakes that are easy to detect, scientists take this activity as another sign of changing conditions under the Greenland ice sheet.

The earthquake and ice-flow data do indicate that modest changes in temperature in the area of the ice sheet, such as have already been seen, may lead to large increases in the amount of glacial ice discharged to the sea. Computer models of ice sheets used until 2006, including those used by contributors to the Intergovernmental Panel on Climate Change (IPCC), the world's largest and most authoritative body of climate and weather scientists, have not taken this newly discovered sensitivity into account. As such, they may have underestimated how much future warming may accelerate Greenland's melting and thus its contribution to sea-level rise.

Finally, another form of melting has recently been observed in Greenland on an unprecedented scale—superficial melting, the direct melting of ice on the surface of the sheet as opposed to the shifting of ice directly into the sea by glaciers. A re-analysis of satellite and other data in 2007 showed that over the last 25 years, superficial melting in Greenland has accelerated twice as fast as was previously thought. From 1979 to 2005, the average year-round temperature rose 3.6°F (2°C), and the area of Greenland that saw superficial melting at least one day per year grew by 42%. The earlier studies had underestimated the melting because they were based on microwave radar waves beamed at Earth from satellites. However, clouds weaken such reflections, and the increased melting was releasing more water into the air, thus creating more clouds. In effect, the melting was partly hiding behind a screen of clouds. As with accelerated glacial movement, the increased superficial-melt area is around the edges of the ice cap, not at the center. Surface melt area set an all-time record in 2002, then broke that record in 2005. The annual difference between snowfall (which adds mass) and superficial melting (which removes it) accounted for about a third of Greenland's current mass loss; the rest is from glaciers, which dump icebergs into the sea. Like an ice cube dropped in a drink, an iceberg raises sea level as soon as it is put in, not when it finally melts.

■ Impacts and Issues

The impact of complete melting of Greenland's ice could be a 21–23 ft (6.5–7 m) rise in sea levels that would radically alter coastlines and force the migration of hundreds of millions of people. Although complete melting of the cap in the near future seems unlikely, there is scientific dispute over just how much of the cap may melt and when, under the conditions of global warming likely to be seen.

The IPCC's Third Assessment Report, in 2001, suggested a sea-level rise of 0.3–2.9 ft (0.09–0.88 m) by 2100 without large reductions in greenhouse-gas emissions. Continuing unabated emissions, the report said, could lead to a sea-level rise of over 16 ft (5 m) over the next thousand years. In subsequent years, other models showed that the Greenland ice sheet could melt completely over the next thousand years or so. In 2007, the IPCC issued an even more conservative estimate of sea-level rise, lowering the upper end of its estimate of sea-level rise by 2100 to 1.9 ft (0.59 m).

A number of scientists have criticized the IPCC's reports as being too conservative, that is, understating the likely or possible effects of climate change. For example, the rules for drafting the 2007 report excluded all scientific papers published after December 2005. This means that those studies which used satellite gravity measurements to reveal much faster melting of the Greenland ice cap from 2004–2006 were not taken into account in the new, lower sea-level rise estimates of 2007.

Even the higher 2001 estimate was criticized as too conservative. In a 2006 paper in *Science*, Jonathan T. Overpeck and colleagues examined the paleoclimate (prehistoric climate) record and found that warming of the Arctic and Antarctic regions might reach levels seen 130,000 to 127,000 years ago by 2100. At that time, sea levels were several meters higher than present-day levels, so much higher sea-level rises than a half-meter or so should not be ruled out or characterized as extremely unlikely.

The IPCC's computer models contain only a small contribution from glaciers' dynamic response, that is, the ability of glaciers to respond to warming by flowing faster, which turns out to be important in Greenland.

Overpeck and his co-workers therefore argue that the record of "past ice-sheet melting indicates that the rate of future melting and related sea-level rise could be faster than widely thought." A threshold might be crossed, they warned, before the end of the twenty-first century, a point beyond which ice-sheet melting would accelerate greatly beyond what is already seen. Human-released soot is darkening the ice, making it absorb more solar energy and melt faster, and climate models indicate that the Antarctic might melt much more than it did 129,000 years ago. Further, climate scientists agree that warming will continue for many years even if humans greatly reduce their greenhouse-gas emissions, since the gases are already in the atmosphere.

■ Primary Source Connection

Data gathered by NASA satellites show that the melting of Greenland's ice sheet more than doubled from 2002–2004 to 2004–2006, causing oceans to rise faster worldwide. This NASA press release describes the technology used to gather the data and presents a lead scientist's view of the melting's significance.

NASA'S GRACE FINDS GREENLAND MELTING FASTER, 'SEES' SUMATRA QUAKE

In the first direct, comprehensive mass survey of the entire Greenland ice sheet, scientists using data from the NASA/German Aerospace Center Gravity Recovery and Climate Experiment (Grace) have measured a significant decrease in the mass of the Greenland ice cap. Grace is a satellite mission that measures movement in Earth's mass.

In an update to findings published in the journal Geophysical Research Letters, a team led by Dr. Isabella Velicogna of the University of Colorado, Boulder, found that Greenland's ice sheet decreased by 162 (plus or minus 22) cubic kilometers a year between 2002 and 2005. This is higher than all previously published estimates, and it represents a change of about 0.4 millimeters (.016 inches) per year to global sea level rise.

"Greenland hosts the largest reservoir of freshwater in the northern hemisphere, and any substantial changes in the mass of its ice sheet will affect global sea level, ocean circulation and climate," said Velicogna. "These results demonstrate Grace's ability to measure monthly mass changes for an entire ice sheet—a breakthrough in our ability to monitor such changes."

Other recent Grace-related research includes measurements of seasonal changes in the Antarctic Circumpolar Current, Earth's strongest ocean current system and a very significant force in global climate change. The Grace science team borrowed techniques from meteorologists who use atmospheric pressure to estimate winds. The team used Grace to estimate seasonal differences in ocean bottom pressure in order to estimate the intensity of the deep currents that move dense, cold water away from the Antarctic. This is the first study of seasonal variability along the full length of the Antarctic Circumpolar Current, which links the Atlantic, Pacific and Indian Oceans.

Dr. Victor Zlotnicki, an oceanographer at NASA's Jet Propulsion Laboratory in Pasadena, Calif., called the technique a first step in global satellite monitoring of deep ocean circulation, which moves heat and salt between ocean basins. This exchange of heat and salt links sea ice, sea surface temperature and other polar ocean properties with weather and climate-related phenomena such as El Niños. Some scientific studies indicate that deep ocean circulation plays a significant role in global climate change.

The identical twin Grace satellites track minute changes in Earth's gravity field resulting from regional changes in Earth's mass. Masses of ice, air, water and solid Earth can be moved by weather patterns, seasonal change, climate change and even tectonic events, such as this past December's Sumatra earthquake. To track these changes, Grace measures micron-scale changes in the 220-kilometer (137-mile) separation between the two satellites, which fly in formation. To limit degradation of Grace's satellite antennas due to atomic oxygen exposure and thereby preserve mission life, a series of maneuvers was performed earlier this month to swap the satellites' relative positions in orbit.

"NASA'S GRACE FINDS GREENLAND MELTING FASTER, 'SEES' SUMATRA QUAKE" (NEWS RELEASE). *NATIONAL AERONAUTICS AND SPACE ADMINISTRATION (NASA)*, DECEMBER 20, 2005. <HTTP://WWW.JPL.NASA.GOV/NEWS/NEWS.CFM?RELEASE=2005-176> (ACCESSED NOVEMBER 29, 2007).

SEE ALSO *Antarctica: Melting; Arctic Melting: Polar Ice Cap; Great Conveyor Belt; Greenland: Global Implications of Accelerated Melting.*

BIBLIOGRAPHY

Books

Solomon, S., et al, eds. *Climate Change 2007: The Physical Science Basis: Contribution of Working Group I to the Fourth Assessment Report of the Intergovernmental Panel on Climate Change.* New York: Cambridge University Press, 2007.

Periodicals

Chen, J. L., et al. "Satellite Gravity Measurements Confirm Accelerated Melting of Greenland Ice Sheet." *Science* 313 (September 29, 2006): 1958–1960.

Luthcke, S. B., et al. "Recent Greenland Ice Mass Loss by Drainage System from Satellite Gravity Observations." *Science* 314 (November 24, 2006): 1286–1289.

Murray, Tavi. "Greenland's Ice on the Scales." *Nature* 443 (September 21, 2006): 277–278.

North Greenland Ice Core Project Members. "High-Resolution Record of Northern Hemisphere Climate Extending into the Last Interglacial Period." *Nature* 431 (September 9, 2004): 147–151.

Overpeck, Jonathan T., et al. "Paleoclimatic Evidence for Future Ice-Sheet Instability and Rapid Sea-Level Rise." *Science* 311 (March 24, 2006): 1747–1750.

Pollitz, Fred F. "A New Class of Earthquake Observations." *Science* 313 (August 4, 2006): 619–620.

Quadfasel, Detlief. "The Atlantic Heat Conveyor Slows." *Nature* 438 (December 1, 2005): 555–556.

Rignot, Eric, and Pannir Kanagaratnam. "Changes in the Velocity Structure of the Greenland Ice Sheet." *Science* 311 (February 17, 2006): 986–990.

Velicogna, Isabella, and John Wahr. "Acceleration of Greenland Ice Mass Loss in Spring 2004." *Nature* 443 (September 21, 2006): 329–331.

Web Sites

"The Greenland Ice Sheet Is Melting More Rapidly than Previously Thought." *Centre Nationale de la Recherche Scientifique* (National Center for Scientific Research), April 18, 2005. <http://www2.cnrs.fr/en/902.htm?&debut=16> (accessed September 20, 2007).

Steffen, Konrad, and Russell Huff. "Greenland Melt Extent, 2005." *Steffen Research Group,* September 28, 2005. <http://cires.colorado.edu/science/groups/steffen/greenland/melt2005/> (accessed September 15, 2007).

Larry Gilman

Arctic Melting: Polar Ice Cap

Introduction

The North and South Poles of Earth are both covered with ice because the light of the sun strikes them at a glancing angle or not at all, depending on the time of year. The South Pole is occupied by a large continent, Antarctica, which bears a thick ice sheet containing about 90% of all the ice on Earth. The North Pole is covered by ocean that is skinned over by frozen seawater. This patch of floating ice expands and contracts with the seasons. The Arctic has been more strongly affected by climate change than any other region except the western peninsula of Antarctica, and in recent years has shown rapid thinning of its ice and reduction of its ice area, especially in summer. Within a few decades the Arctic may be completely ice-free in summer.

Historical Background and Scientific Foundations

Although indigenous peoples such as the Inuit (Eskimos) have inhabited the region around the Arctic for thousands of years, they had no reason or ability to reach its more isolated areas, such as the vicinity of the North Pole or the interior of the Greenland ice cap. The Arctic interior was first explored by Europeans in the nineteenth century, when a number of expeditions tried to reach the North Pole itself. This was finally achieved in 1909 by American engineer and explorer Robert E. Peary (1856–1920). The North Pole has since been visited by many expeditions, including nuclear-powered submarines, aircraft, adventurers on foot, and icebreakers.

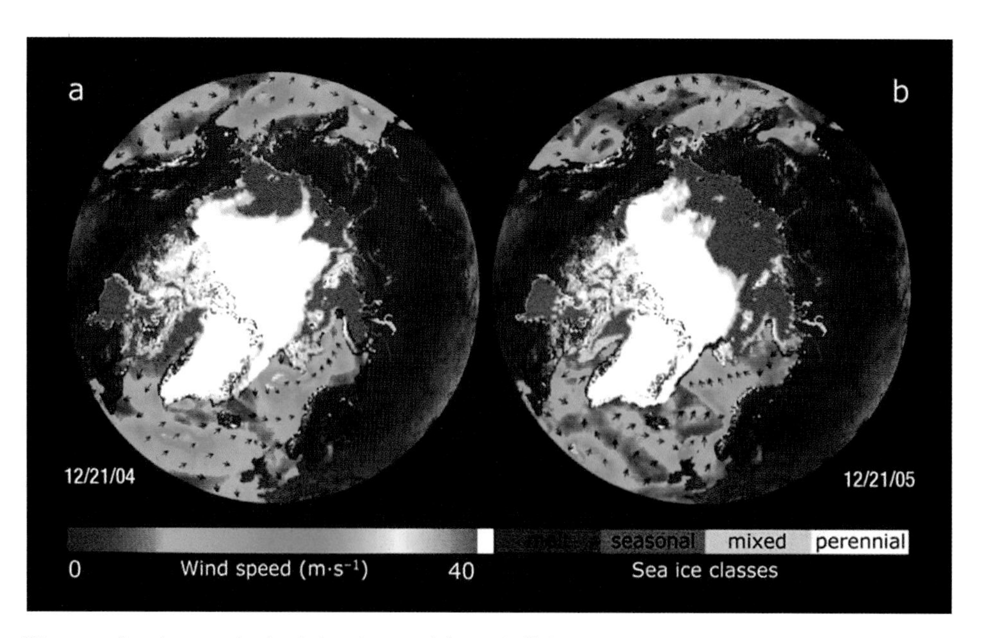

The area of sea ice over the Arctic has decreased dramatically in recent years. A comparison of the image on the right, taken a year later than the image on the left, reveals the extent of the change. *AP Images.*

For decades, scientific knowledge of the Arctic depended on aircraft and ground stations. However, simultaneous observing of the whole region is practical only from orbit. Satellite data was first obtained from the Landsat satellites in the late 1970s. Infrared radiation emitted by all objects reveals the object's temperature, so infrared observations from space have allowed tracking of temperatures all over the Arctic since 1979.

The Arctic sea-ice cap is an important part of the global climate system. It reflects solar energy back into space like a mirror, thus helping control circulation of water in the oceans via the North Atlantic branch of the Great Conveyor Belt or thermohaline overturning circulation by which all the oceans of the world remain in constant cyclic motion. When seawater freezes in the winter, it rejects salt into the water, increasing the ocean's local density. This denser, saltier water then sinks, helping to force the global ocean conveyor. Rainfall patterns in the Northern Hemisphere and the livelihoods of humans and animals living in the Arctic are also strongly influenced by the Arctic ice.

The Arctic has been particularly vulnerable to global warming. Warming is started by increased atmospheric concentrations of greenhouse gases, such as carbon dioxide and methane. Aerosol pollutants, tiny particles from incomplete burning of fossil fuels, forests, and other combustibles, also cause Arctic warming. Arctic haze has been observed by aircraft pilots since the 1950s, but it was not until the 1970s that scientists realized that the haze was air pollution, namely aerosols. Aerosol particles, which are dark, absorb energy from sunlight, warming the air. They also can settle on ice and snow darkening them and so making melting more likely. Over the last 50 years, temperatures in Alaska, western Canada, and eastern Russia have increased by 4–7°F (2–4°C), twice the average global warming. As of 2007, scientists were forecasting that during the twenty-first century temperatures in these areas would rise by another 7–13°F (4–7°C).

Snow and ice have a high albedo (brightness) and reflect about 90% of incoming solar energy to space, but open sea and ground have low albedo, absorbing much more energy than they reflect. Positive feedback occurs, accelerating warming, when melting of snow cover or sea ice exposes dark ground or ocean. This greatly increases the amount of solar energy absorbed and so speeds up the warming that started the melting initially.

The vulnerability of the Arctic to warming has been seen in increased atmospheric temperatures, melting of permafrost, and—most dramatically—the rapid shrinkage of the north polar ice cap.

The size of the polar sea-ice cap varies greatly, between a maximum in winter and a minimum in summer. Therefore, to see changes in sea-ice area from year to year, each month's ice area must be compared to the ice area of the same month in the previous year. Satellite data show that while Antarctic sea ice has been growing, Arctic ice has been declining in every month since at least 1979, when reliable data began to be collected. Although there had been ups and downs, the trend from 1979 to 2004 was a decrease in September (annual minimum) ice area by 7.7% per decade.

Moreover, the pace of shrinkage has been rapidly accelerating in the early 2000s. Polar cap area reached historic minimum lows in 2002, 2005, and 2007. The ice-area low of September 2007 was particularly shocking to scientists, about 25% less area than the previous record low in 2005. Arctic specialist William L. Chapman of the

WORDS TO KNOW

AEROSOL: Particles of liquid or solid dispersed as a suspension in gas.

ALBEDO: A numerical expression describing the ability of an object or planet to reflect light.

GREAT CONVEYOR BELT: The overturning circulation of the world's seas, driven by temperature and salinity differences between the poles and tropics; also called the thermohaline circulation or meridional overturning circulation. Because the great conveyor belt transports thermal energy from the tropics toward the poles, it is a central component of Earth's climate machine.

GREENHOUSE GASES: Gases that cause Earth to retain more thermal energy by absorbing infrared light emitted by Earth's surface. The most important greenhouse gases are water vapor, carbon dioxide, methane, nitrous oxide, and various artificial chemicals such as chlorofluorocarbons. All but the latter are naturally occurring, but human activity over the last several centuries has significantly increased the amounts of carbon dioxide, methane, and nitrous oxide in Earth's atmosphere, causing global warming and global climate change.

NORTH POLE: Point at which Earth's axis passes through its surface in the Arctic. The north pole is near, but not identical to, the northern magnetic pole, the point at which the field lines of Earth's magnetic field pass vertically through its surface.

PERMAFROST: Perennially frozen ground that occurs wherever the temperature remains below 32°F (0°C) for several years.

SOUTH POLE: The geographically southernmost place on Earth.

THERMOHALINE CIRCULATION: Large-scale circulation of the world ocean that exchanges warm, low-density surface waters with cooler, higher-density deep waters. Driven by differences in temperature and saltiness (halinity) as well as, to a lesser degree, winds and tides. Also termed meridional overturning circulation.

IN CONTEXT: LOSSES FROM ICE SHEETS

"New data since the TAR [IPCC Third Assessment Report, 2001] now show that losses from the ice sheets of Greenland and Antarctica have very likely contributed to sea level rise over 1993 to 2003. Flow speed has increased for some Greenland and Antarctic outlet glaciers, which drain ice from the interior of the ice sheets. The corresponding increased ice sheet mass loss has often followed thinning, reduction or loss of ice shelves or loss of floating glacier tongues. Such dynamical ice loss is sufficient to explain most of the Antarctic net mass loss and approximately half of the Greenland net mass loss. The remainder of the ice loss from Greenland has occurred because losses due to melting have exceeded accumulation due to snowfall."

Statement of the Intergovernmental Panel on Climate Change (IPCC) as formally approved at the 10th Session of Working Group I of the IPCC in Paris, France, during February 2007.

SOURCE: *Solomon, S., et al, eds.* Climate Change 2007: The Physical Science Basis: Contribution of Working Group I to the Fourth Assessment Report of the Intergovernmental Panel on Climate Change. *New York: Cambridge University Press, 2007.*

University of Illinois at Urbana-Champaign characterized the extreme shrinkage as "simply incredible." The Northwest Passage—the sea route around the northern perimeter of North America from the Pacific Ocean to the Atlantic Ocean—was free of ice and open to navigation by ships for the first time in recorded history.

The polar ice has also been getting thinner since the late 1970s. Changing winds have moved older, thicker ice out of the Arctic bit by bit, driving it into the North Atlantic (where it melts) and replacing it with younger, thinner ice. As of 2007, the Arctic ice pack had thinned to about half its average thickness of 50 years earlier.

■ Impacts and Issues

One of the most common scientific criticisms of the authoritative 2007 report on climate change from the Intergovernmental Panel on Climate Change (IPCC), *Climate Change 2007*, is that its predictions are too conservative, particularly because they disregard new knowledge about accelerated melting in the Arctic and the western peninsula of Antarctica. The IPCC barred the use of scientific papers published after December 2005 in its 2007 report, so recent data about accelerated Arctic melting were not used. Scientists at the National Center for Atmospheric Research (NCAR) in Boulder, Colorado, announced in April 2007 that Arctic sea-ice melting was happening more rapidly than predicted by all 18 of the computer climate models used by the IPCC in preparing its 2007 report. In 2006, NCAR scientists announced that if greenhouse gases continue to build up in the atmosphere at their current rate, the Arctic could be completely free of ice in summer as early as 2040.

Such drastic changes in sea-ice patterns may slow the ocean conveyor and accelerate melting of Greenland's ice cap by making the polar ocean a much more effective absorber of solar energy. Polar bear populations would almost certainly be decimated, since polar bears depend on sea ice for access to seals, one of their main sources of food. In 2007, the U.S. Geological Survey announced that the world population of polar bears could be reduced by two-thirds by 2050 even under conservative (less drastic) assumptions about global warming.

The livelihoods of indigenous peoples in the Arctic are also threatened by Arctic warming. Radical changes in Arctic ice may trigger changes in rainfall patterns at the temperate latitudes, including reduced rainfall in the American West and increased rainfall in parts of Europe.

■ Primary Source Connection

One of the most debated and least predictable effects of global climate change is the issue of rising sea levels. Scientists expect sea levels to rise over the next several centuries as the ice sheets of Antarctica and Greenland melt. However, predicting the extent of rise in sea levels has been difficult for scientists because of the complex relationship between the ice sheets and air temperature and sea temperature. This article details some of the difficulties that scientists face in trying to construct computer models that accurately estimate the rise in sea level.

Juliet Eilperin is a staff writer for the *Washington Post*.

CLUES TO RISING SEAS ARE HIDDEN IN POLAR ICE

Few consequences of global warming pose as severe a threat to human society as sea-level rise. But scientists have yet to figure out how to predict it.

And not knowing what to expect, policymakers and others are hamstrung in considering how to try to prevent it or prepare for it.

To calculate sea-level rise, the key thing researchers need to understand is the behavior of the major ice sheets that cover Greenland and Antarctica. The disintegration of one would dramatically raise the ocean. But while computer models now yield an increasingly sophisticated understanding of how a warming atmosphere would behave, such models have yet to fully encapsulate the complex processes that regulate ice sheet behavior.

"The question is: Can we predict sea level? And the answer is no," said David Holland, who directs New York University's Center for Atmosphere Ocean Science. Holland, an oceanographer, added that this may mean researchers will just have to watch the oceans to see what happens: "We may observe the change much more than we ever predict it."

In its executive summary report for policymakers in February, the Intergovernmental Panel on Climate Change, composed of hundreds of leading climate scientists, barely hazarded a guess on sea level, predicting that it would rise between 7.8 inches and two feet by the end of the century. However, the United Nations-sponsored panel—which operated under the assumption that, by 2100, the Greenland ice sheet would lose some mass but that the Antarctic ice sheet would gain some—did not venture a best estimate or an upper limit for possible sea-level rise.

The panel could agree to say only there is a 50–50 chance that a global temperature increase of between 1.8 and 7.2 degrees Fahrenheit would lead to a partial melting of the ice sheets over a period of several hundred to several thousand years.

Because so much is at stake—a three-foot increase in sea level could turn at least 60 million people into refugees, the World Bank estimates—ice sheet modelers are working furiously to try to unravel the mystery of how these sheets accumulate and lose mass.

Michael Oppenheimer, a Princeton University professor of geosciences and international affairs, does make a prediction: He figures that if the Greenland ice sheet disintegrates, sea level would rise about 23 feet. If the West Antarctic sheet melts, as well, it would add an additional 17 feet or so.

"If either of these ice sheets were to disintegrate, it would destroy coastal civilization as we know it," Oppenheimer said.

One of the biggest challenges facing researchers is that ice sheets are under "attack from the edges," in the words of Richard B. Alley, a Pennsylvania State University geosciences professor. Each sheet amounts to a pile of snow compressed over time into a two-mile thick, continent-spanning sheet of ice, which spreads out under its own weight, Alley said.

Near the coast, the pile develops quick-moving "ice streams," which flow between slower-moving sections of ice and float out onto the ocean in an "ice shelf." While recent satellite data have indicated that these ice streams are flowing faster and delivering more water to the oceans, many uncertainties remain.

David Vaughan, a glaciologist with the British Antarctic Survey in Cambridge, said the terrain beneath the ice streams helps determine how they move, but the contours of the land are largely unknown because it is buried so far under the ice. The streams may run aground on elevated bedrock, slow down as they move past rocky fjord walls or speed up as they move over mud.

"There's a continent of topography sitting under Antarctica," Vaughn said. "Everything there has an impact on how the ice sheet flows, and very little of that has been mapped."

Researchers are also trying to measure the layer of water that lies under the ice sheets, as that also helps regulate ice stream flows.

"They're essentially afloat on their own sub-glacial water, even if there's not much water there," said Garry Clarke, a glaciology professor at the University of British Columbia. "We don't know very much about how water flows underneath ice sheets."

Another uncertainty is how much the oceans surrounding the ice sheets are warming, something that is difficult to measure because the areas are remote. Vaughan and his colleagues suspect that warmer waters around Antarctica have contributed to melting the Western Antarctic ice sheet, but there is little good data because few ships venture there.

Researchers are now going to extraordinary lengths to collect the data they need. Holland at NYU recently returned from a trip to Greenland, where he was collecting information about the Ilulissat glacier, which has doubled its speed over the past decade as it flows toward the ocean and melts. To test the temperature and salinity of the water surrounding the glacier, Holland and other researchers had to hover in a helicopter and lower their instruments into an opening in the ice.

"It's kind of beautiful, and scary and fun," he said.

Even with better data, scientists find it difficult to enter the information into computer models. Most models do not attempt to calculate what could happen to ice sheets at their edges.

Adding to the challenge, Oppenheimer said, is that models "are only good at explaining things that happen at a large scale. Ice sheets are very complex beasts, and the water moves at a very small scale."

Ice streams move along narrow channels, and plugging such detail into a computer model takes a long time. But without that level of detail, the results are incomplete.

Researchers have made some progress in ice sheet science over the past decade by using satellites to measure the sheets' changing mass.

Last month, for example, a team of NASA and university scientists used readings from NASA's QuikScat satellite to measure snow accumulation and melt in Antarctica from July 1999 through July 2005. They discovered that broad areas of snow had melted in west Antarctica in January 2005 in response to warmer temperatures. The finding was surprising because Antarctica had shown relatively little warming in the recent past.

Konrad Steffen, director of the Cooperative Institute for Research in Environmental Sciences at the University of Colorado at Boulder, who led the study, said increases in snowmelt "definitely could have an impact on larger-scale melting of Antarctica's ice sheets if they were severe or sustained over time."

Because ice sheet modeling has not ranked as a high priority for government laboratories and has not been integrated into large-scale climate models, scientists from around the world are now collaborating to develop more sophisticated models to inform policymakers about potential sea-level rise. The researchers have convened two major meetings this year, one at the NOAA Geophysical Fluid Dynamics Laboratory at Princeton University and one at the University of Texas at Austin, in an effort to generate a new generation of ice sheet models.

Vaughan, who attended both conferences, said he is hopeful that he and others will solve the question of ice sheet modeling by the time he ends his career: "It will be 15 years before I retire, and I want it nailed by then."

But other researchers are less optimistic. Holland, who like Vaughan is in his mid-40s, doubts that scientists will master the problem before greenhouse gas emissions trigger significant melting of the ice sheets that he studies.

"We will get there eventually, but it won't be for a long time. It won't be in my lifetime," Holland said. "There's no plan; there's no program. There's no one responsible for sea-level rise."

Juliet Eilperin

EILPERIN, JULIET. "CLUES TO RISING SEAS ARE HIDDEN IN POLAR ICE," *WASHINGTON POST* (JULY 16 2007): A06.

SEE ALSO *Albedo; Antarctica: Melting; Arctic Melting: Greenland Ice Cap; Arctic People: Climate Change Impacts; Great Conveyor Belt; Greenland: Global Implications of Accelerated Melting; Polar Bears; Soot.*

BIBLIOGRAPHY

Books

Solomon, S., et al, eds. *Climate Change 2007: The Physical Science Basis: Contribution of Working Group I to the Fourth Assessment Report of the Intergovernmental Panel on Climate Change.* New York: Cambridge University Press, 2007.

Periodicals

Johannessen, Ola M., et al. "Satellite Evidence for an Arctic Sea Ice Cover in Transformation." *Science* 286 (December 3, 1999): 1937–1939.

Kennedy, Donald, and Brooks Hanson. "Ice and History." *Science* 311 (March 24, 2006): 1673.

Law, Kathy S., and Andreas Stohl. "Arctic Air Pollution: Origins and Impacts." *Science* 315 (March 16, 2007): 1537–1540.

Revkin, Andrew C. "Analysts See 'Simply Incredible' Shrinking of Floating Ice in the Arctic." *New York Times* (August 10, 2007).

Scheiermeier, Quirin. "The New Face of the Arctic." *Nature* 446 (March 8, 2007): 133–135.

Serreze, Mark C. "Perspectives on the Arctic's Shrinking Sea-Ice Cover." *Science* 315 (March 16, 2007): 1533–1536.

Web Sites

"Abrupt Ice Retreat Could Produce Ice-Free Arctic Summers by 2040." *National Center for Atmospheric Research,* December 11, 2006. <http://www.ucar.edu/news/releases/2006/arctic.shtml> (accessed September 21, 2007).

"Arctic Ice Retreating More Quickly Than Computer Models Project." *National Center for Atmospheric Research,* April 30, 2007. <http://www.ucar.edu/news/releases/2007/seaice.shtml> (accessed September 21, 2007).

"Arctic Sea Ice News, Fall 2007." *National Snow and Ice Data Center,* September 20, 2007. <http://nsidc.org/news/press/2007_seaiceminimum/20070810_index.html> (accessed September 21, 2007).

Larry Gilman

Arctic People: Climate Change Impacts

■ Introduction

Two regions on Earth are showing much more rapid climate change than the rest of the world—the Western Peninsula of Antarctica and the Arctic. The Arctic is often considered to include all land and sea from the North Pole southward to the Arctic Circle at 66° 33' north latitude. This region has warmed at about twice the rate of the rest of the planet, causing many disruptions to natural and human patterns. Unlike the South Pole, where the polar ice cap is extremely thick and rests mostly on solid ground, the ice in the Arctic has two major components: the land-based Greenland ice cap and the polar ice proper, a skin of ice floating on the ocean above and around the Pole. The floating polar ice has recently retreated at a rate that has surprised scientists. Warming and sea-ice shrinkage are having numerous impacts on the peoples of the Arctic region and have implications for the course of global warming around the world.

Near the Eureka Weather Station in the Arctic Canadian territory of Nunavut stands an Inuit *inukshuk*. Such stone figures, resembling humans, were used to guide the Arctic aboriginal people for centuries. Today, the Inuit are among the people most affected by global warming. *AP Images.*

The changes brought on by global warming have impacted how some people hunt and fish in the Arctic. Here, an Inupiak Eskimo is shown bringing a seal he just harpooned aboard his boat. The spring seal hunt was traditionally done on ice, but warmer weather has forced people to use boats, rather than dog sleds or snowmobiles. *Gilles Mingasson/Getty Images.*

■ Historical Background and Scientific Foundations

Archaeological research shows that human beings have lived in the Arctic for many thousands of years. These peoples have evolved traditional ways of life adapted to the extreme cold and scarcity of plants, usually basing their livelihoods on hunting and fishing. Although these ways of life have already been widely changed by technology—snowmobiles having largely replaced dog teams in Canada and Alaska, for example—many Arctic groups continue to depend on whales, seals, caribou, fish, and other Arctic animals for their livelihoods.

Arctic sea ice grows and shrinks on a seasonal cycle, reaching its greatest coverage around the end of March and its minimum in September. In 2007, scientists were astonished by the unprecedented melting of the Arctic sea ice. Arctic climate researcher William L. Chapman described it as "simply incredible." As of early August 2007, Arctic sea ice had already reached its minimum extent in recorded history. The 2007 melt was also unique in that the sea ice showed unprecedented shrinkage in all sectors, that is, all around its perimeter and even its central region, not just in the north Atlantic Ocean or Bering Sea.

Over the last 50 years or so, temperatures in much of the Arctic—that is, in Alaska, western Canada, and eastern Russia—have increased by 4–7°F (2–4°C), twice the global average warming. Over the next century, temperatures are predicted to rise by another 7–13°F (4–7°C).

The effects of climate change in the Arctic are not likely to be all bad, and the Arctic peoples are not likely to be wiped out by climate change. A reduction in cold-related deaths is probable, for example. In Finland, deaths from cold-related disease and injury each year outnumber deaths from heat and car crashes by about a factor of 10. Respiratory disease rates also are expected to decrease overall for Arctic peoples with warming climate. Early spring melting gives Inuit hunters a longer season in which to hunt beluga whales, and boat travel may be easier with less ice.

However, the bulk of the impacts are likely to be negative. Walruses, Arctic cod, and polar bears are all threatened by climate change. Melting of permafrost destabilizes structures and renders roads impassable by converting them to rivers of mud. Harvesting wildlife on unstable ice is now more dangerous and access to hunting areas is becoming more difficult. As ancestral knowledge becomes rapidly less relevant to present-day conditions, the transfer of skills to younger generations breaks down, weakening social networks.

■ Impacts and Issues

Impacts of global warming on Arctic peoples have already been profound. Chief Gary Harrison of the Arctic

Athabaskan Council, an international group of American and Canadian native peoples belonging to the Athabaskan family, said in 2005, "Arctic indigenous peoples are threatened with the extinction or catastrophic decline of entire bird, fish and wildlife populations, including species of caribou, seals, and fish critical to our food security.... This has the potential for catastrophic damage to millennia-old Arctic indigenous cultures."

Representatives of Arctic peoples emphasize that their troubles are of global significance. Sheila Watt-Cloutier of the Inuit Circumpolar Conference, a multinational nongovernmental organization representing the Inuit (Eskimo) family of indigenous peoples, said in 2007: "The Arctic is the early warning for the rest of the world. What happens to the planet happens first in the Arctic.... We must all take what action we can to slow the pace of climate change, while there is still time." Watt-Cloutier, who has worked to raise awareness of the impact of climate change on Arctic people, was nominated for the 2007 Nobel Peace Prize.

But Arctic peoples also emphasize that they intend to adapt to climate change if at all possible. "We won't quit existing," says Frank Pokiak, an Inuit and chair of the Inuvialuit Game Council in Tuktoyaktuk, Alaska.

Scientific forecasts of global climate change and observed sea-ice melting trends show that the Arctic Ocean may be completely free of ice in summer starting in just a few decades. Or, it could happen sooner, as projections of future change always involve uncertainties and the melting in the Arctic Ocean has recently been more rapid than predicted by computer models. This would be the first time the Arctic has been ice-free in about one million years. Since the Arctic is "the air conditioner of the world," as global warming scientist Susan Hassol put it in 2004, the seasonal melting of the Arctic Ocean—which absorbs far more solar energy when open to the sun, rather than covered with reflecting ice—could have profound effects on world climate, accelerating global warming and changing weather patterns. In 2007, the U.S. Geological Survey forecast that two-thirds of the Arctic's 20,000 or so polar bears (which depend on sea ice) will be gone by 2050 even under moderate global-warming scenarios.

WORDS TO KNOW

GREENLAND ICE CAP: Layer of ice covering about 80% of the island of Greenland; second-largest mass of ice on Earth after Antarctica, containing about 10% of Earth's ice, enough to raise sea levels 21 feet (6.5 meters) if it were all to melt. Accelerated melting of the Greenland ice cap was confirmed by gravimetric measurements in 2006.

INDIGENOUS PEOPLES: Human populations that migrated to their traditional area of residence some time in the relatively distant past, e.g., before the period of global colonization that begin in the late 1400s.

POLAR ICE CAP: Ice mass located over one of the poles of a planet not otherwise covered with ice. In our solar system, only Mars and Earth have polar ice caps. Earth's north polar ice cap has two parts, a skin of floating ice over the actual pole and the Greenland ice cap, which does not overlay the pole. Earth's south polar ice cap is the Antarctic ice sheet.

SEA ICE: Ice that forms from the freezing of ocean water. As the salt water freezes, it ejects salt, so sea ice is fresh, not salty. Sea ice forms in relatively thin layers, usually no more than 3–7 ft (1–2 m) thick, but it can cover thousands of square miles of ocean in the polar regions.

■ Primary Source Connection

The Arctic is warming twice as fast as the rest of the world. The rapid warming of the Arctic has created quickly changing conditions for native peoples who have been adapted to extreme cold for thousands of years. In this *New York Times* article, natives of the Arctic describe the changes that global warming has caused to their environment and way of life.

OLD WAYS OF LIFE ARE FADING AS THE ARCTIC THAWS

... In Russia, 20 percent of which lies above the Arctic Circle, melting of the permafrost threatens the foundations of homes, factories, pipelines. While the primary causes are debated, the effect is an engineering nightmare no one anticipated when the towns were built, in Stalin's time.

Coastal erosion is a problem in Alaska as well, forcing the United States to prepare to relocate several Inuit villages at a projected cost of $100 million or more for each one.

Across the Arctic, indigenous tribes with traditions shaped by centuries of living in extremes of cold and ice are noticing changes in weather and wildlife. They are trying to adapt, but it can be confounding.

Take the Inuit word for June, *qiqsuqqaqtuq*. It refers to snow conditions, a strong crust at night. Only those traits now appear in May. Shari Gearheard, a climate researcher from Harvard, recalled the appeal of an Inuit hunter, James Qillaq, for a new word at a recent meeting in Canada.

One sentence stayed in her mind: "June isn't really June any more."

IN CONTEXT: WHO OWNS THE ARCTIC?

Global climate change is not only affecting Arctic ice, it is also changing many neighboring nation's perspectives on Arctic resources. A warming Arctic has the potential to make valuable resources increasingly accessible. An estimated quarter of the world's undiscovered oil reserves lies in the Arctic. A longer ice-free season in arctic waters has renewed the possibility of a northwest shipping passage. In August 2007, a Russian submarine planted a Russian flag on the sea floor beneath the North Pole, symbolically claiming the region and its resources as its own. Canada vowed to begin building military bases in the Arctic. The United States sent Coast Guard ships to explore Arctic waters and map the sea floor. Denmark has also launched polar mapping expeditions.

The rush for potential Arctic resources has made some nations reevaluate their positions on international laws and treaties governing the Arctic region. For example, the United States initially rejected the international Law of the Sea Treaty in 1982. The treaty called for defined borders of national waters and set forth a panel to evaluate new claims of national waters, fishing and drilling rights, and sea floor ownership. President George W. Bush began calling for the United States to ratify the longstanding treaty to improve the government's ability to negotiate new and existing U.S. claims in the Arctic.

Changing Traditions

In Finnmark, Norway's northernmost province, the Arctic landscape unfolds in late winter as an endless snowy plateau, silent but for the cries of the reindeer and the occasional whine of a snowmobile.

A changing Arctic is felt there, too. "The reindeer are becoming unhappy," said Issat Heandarat Eira, a 31-year-old reindeer herder and one of 80,000 Samis, or Laplanders, who live in the northern reaches of Scandinavia and Russia.

Few countries rival Norway when it comes to protecting the environment and preserving indigenous customs. The state has lavished its oil wealth on the region, and Sami culture has enjoyed something of a renaissance. There is a Nordic Sami Institute, a Sami College, a state-sponsored film festival and a drive-in theater where moviegoers watch from snowmobiles.

And yet no amount of government support can convince Mr. Eira that his livelihood, intractably entwined with the reindeer, is not about to change. Like a Texas cattleman, he keeps the size of his herd secret. But he said warmer temperatures in fall and spring were melting the top layers of snow, which then refreeze as ice, making it harder for his reindeer to dig through to the lichen they eat. He worries, too, about the encroachment of highways and industrial activity on his once isolated grazing lands.

"The people who are making the decisions, they are living in the south and they are living in towns," said Mr. Eira, sitting inside his home made of reindeer hides. "They don't mark the change of weather. It is only people who live in nature and get resources from nature who mark it."

Other Arctic cultures that rely on nature report similar disruptions. For 5,000 years, the Inuit have lived on the fringe of the Arctic Ocean, using sea ice as a highway, building material and hunting platform. In recent decades, their old ways have been fading under forced relocations, the erosion of language and lore and the lure of modern conveniences, steady jobs and a cash economy.

Now the accelerating retreat of the sea ice is making it even harder to preserve their connections to "country food" and tradition. In Canada, Inuit hunters report that an increasing number of polar bears look emaciated because the shrinking ice cover has curtailed their ability to fatten up on seals. In Alaska, whale hunters working in unusually open seas have seen walruses try to climb onto their white boats, mistaking them for ice floes.

Hank Rogers, a 54-year-old Inuvialuit who helps patrol Canada's Far North, said the pelts of fox, marten and other game he trapped were thinning. As for the flesh of fish caught in coastal estuaries of the Yukon, "they're too mushy," he said. Slushy snow and weaker ice has made traveling by snowmobile impossible in places.

"The next generation coming up is not going to experience what we did," he said. "We can't pass the traditions on as our ancestors passed on to us."

Even seasoned hunters have been betrayed by the thaw, stepping in snow that should be covering ice but instead falling into water. And on Shingle Point, a sandy strip inhabited by Inuvialuit at the tip of the Yukon in Canada, Danny A. Gordon, 70, said it was troubling that fewer icebergs were reaching the bay. It has become windier, too, for reasons people here cannot explain.

"In the summer 40 years ago, we had lots of icebergs, and you could land your boat on them and climb on them even in summer," Mr. Gordon said. "Now in the winter they are tiny. The weather has changed. Everyone knows it. It's global warming"....

A Less Wild Future

One day last summer, the 1,200 residents of Pangnirtung, a windswept outpost on a fjord in Nunavut, Canada's Inuit-administered Arctic territory, were startled to see a 400-foot European cruise ship drop anchor unannounced and send several hundred tourists ashore in small boats.

While small ships have stopped in the Canadian Arctic, visits from large liners are increasing as interest grows in the opening Northwest Passage, said Maureen Bundgaard, chief executive of Nunavut Tourism, a trade association.

Ms. Bundgaard has been training villagers how to stage cultural shows, conduct day tours and sell crafts and traditional fare—without being overrun. "We're not prepared to deal with the huge ships, emotionally or in other ways," she said.

Inuit leaders say they are trying to balance tradition with the inevitable changes that are sweeping their lands. The Inuit Circumpolar Conference, which represents 155,000 Inuit scattered across Canada, Greenland, Russia and the United States, has enlisted lawyers and movie stars like Jake Gyllenhaal and Salma Hayek to draw attention to its imperiled traditions.

The group's leaders hope to submit a petition to the Inter-American Commission on Human Rights in December, claiming that the United States, by rejecting a treaty requiring other industrialized countries to cut emissions linked to warming, is willfully threatening the Inuit's right to exist.

The commission, an investigative arm of the Organization of American States, has no enforcement powers. But legal analysts say that a declaration that the United States has violated the Inuit's rights could create the foundation for a lawsuit either against the United States in international court or American companies in federal courts.

But some Inuit question the wisdom of the petition. They ask, how can they push countries to stem global warming when the Inuit's own prosperity in places like Nunavut is tied to revenues from oil and gas, which are sources of greenhouse gases when burned?

Sheila Watt-Cloutier, the elected chairwoman of the group, said the goal was not to stop development but to make sure that native cultures had a say in how development was carried out.

"It's how we do the business that's more important," she said. "There are more environmentally friendly ways in which we can do development and still live a certain way, with a way of life and business that can balance both." While it is the people of the Arctic who will feel the melt and the rush for development most directly, the world, too, will have to give up something—its treasured notion of the Far North as a place of wilderness, simplicity and unspoiled cultures.

In a report on Arctic development, the United Nations Environment Program estimated that 15 percent of the region's lands were affected in 2001 by mining, oil and gas exploration, ports or other industrial incursions. But that figure is likely to reach 80 percent in 2050, it said.

The Arctic, then, is probably making the same transition that swept the coastal plains of the North Slope of Alaska starting 38 years ago when the first oil was struck in Prudhoe Bay, said Charles Wohlforth, an Alaskan and author of "The Whale and the Supercomputer," describing Arctic climate change.

Since then, a lacework of pipelines and wells has steadily spread west and east from that central field, ending the sweeping sense of emptiness that defined the Arctic landscape through the ages.

"Even if you support oil development and think it makes sense, there's a point at which it becomes West Texas or the Gulf of Mexico and is not really the Arctic any more," Mr. Wohlforth said.

Steven Lee Myers, et al.

MYERS, STEVEN LEE, ANDREW C. REVKIN, SIMON ROMERO, CLIFFORD KRAUSS, AND CRAIG DUFF. "OLD WAYS OF LIFE ARE FADING AS THE ARCTIC THAWS." *THE NEW YORK TIMES*. OCTOBER 20, 2005: A(6)1.

SEE ALSO *Africa: Climate Change Impacts; Arctic Melting: Greenland Ice Cap; Arctic Melting: Polar Ice Cap; Asia: Climate Change Impacts; Australia: Climate Change Impacts; Europe: Climate Change Impacts; North America: Climate Change Impacts; Small Islands: Climate Change Impacts; South America: Climate Change Impacts.*

BIBLIOGRAPHY

Books

Parry, M. L., et al, eds. *Climate Change 2007: Impacts, Adaptation and Vulnerability: Contribution of Working Group II to the Fourth Assessment Report of the Intergovernmental Panel on Climate Change.* New York: Cambridge University Press, 2007.

Periodicals

Ford, James D., et al. "Vulnerability to Climate Change in the Arctic: A Case Study from Arctic Bay, Canada." *Global Environmental Change* 16 (2006): 145–160.

Scheiermeier, Quirin. "The New Face of the Arctic." *Nature* 446 (2007): 133–135.

Serreze, Mark C. "Perspectives on the Arctic's Shrinking Sea-Ice Cover." *Science* 315 (2007): 1533–1536.

Web Sites

"Arctic Indigenous Peoples Unveil Statement on Climate Change" (news release). *Arctic Athabaskan Council,* June 12, 2005. <http://www.arcticathabaskancouncil.com/press/20051206.php> (accessed September 16, 2007).

Handwerk, Brian. "Arctic Melting Fast: May Swamp U.S. Coasts by 2099." *National Geographic*

News, November 9, 2004. <http://news.nationalgeographic.com/news/2004/11/1109_041109_polar_ice_2.html> (accessed September 16, 2007).

"Study: Arctic Warming Threatens People, Wildlife: Eight-Nation Report Faults Fossil Fuels; U.S. in Wait-and-See Mode." *MSNBC*, Nov. 8, 2004. <http://www.msnbc.msn.com/id/6433717/> (accessed September 16, 2007).

Larry Gilman

Asia: Climate Change Impacts

Introduction

Asia—not including its western peninsula, Europe, traditionally considered a separate continent for cultural reasons—is the world's largest continent, with about 4 billion inhabitants. Like all other continents, it is presently experiencing observable effects of anthropogenic (human-caused) climate change and will experience more such effects as global warming continues in the coming decades. Asia is a vast and diverse landmass, stretching from Siberia in the north, to the Middle East in the west, to the Indian subcontinent in the south, and to China, Southeast Asia, Japan, the Philippines, and Indonesia in the east and southeast. It is difficult to make general statements about Asia regarding climate change or anything else. However, a number of regional impacts have been observed and predicted for the region.

Historical Background and Scientific Foundations

In recent decades, parts of Asia, including India, Korea, and China, have been undergoing unprecedented economic growth. China's economy, for example, is already

To urge people and corporations to take action against climate change, Greenpeace activists projected a message on a cooling tower at the National Thermal Power Corporation in Dadri, India, in March 2007. *AP Images.*

WORDS TO KNOW

ANTHROPOGENIC: Made by people or resulting from human activities. Usually used in the context of emissions that are produced as a result of human activities.

GREENHOUSE GASES: Gases that cause Earth to retain more thermal energy by absorbing infrared light emitted by Earth's surface. The most important greenhouse gases are water vapor, carbon dioxide, methane, nitrous oxide, and various artificial chemicals such as chlorofluorocarbons. All but the latter are naturally occurring, but human activity over the last several centuries has significantly increased the amounts of carbon dioxide, methane, and nitrous oxide in Earth's atmosphere, causing global warming and global climate change.

PERMAFROST: Perennially frozen ground that occurs wherever the temperature remains below 32°F (0°C) for several years.

RIVER DELTA: Flat area of fine-grained sediments that forms where a river meets a larger, stiller body of water such as the ocean. Rivers carry particles in their turbulent waters that settle out (sink) when the water mixes with quieter water and slows down; these particles build the delta. Deltas are named after the Greek letter delta, which looks like a triangle. Very large deltas are termed megadeltas and are often thickly settled by human beings. Rising sea levels threaten settlements on megadeltas.

SUSTAINABLE DEVELOPMENT: Development (i.e., increased or intensified economic activity; sometimes used as a synonym for industrialization) that meets the cultural and physical needs of the present generation of persons without damaging the ability of future generations to meet their own needs.

the third-largest in the world and is growing faster than the economy of any other large country. In the first quarter of 2007 alone, China's economy grew by 11%. At the same time, most of the poor people in the world live in south Asia. In 2007, more than 65% of all people living without sanitation in rural areas, of underweight children, and of people living on less than $1 a day reside in Asia. All these groups are at greater risk from water stress, increased food costs, and other changes likely to be the indirect result of climate change.

In addition, Asia's population is growing. From 1995 to 2050, its population is projected to grow by about 2 billion. Poverty, development, and population growth all place stress on the Asian environment and make both that environment and the human societies that depend on it more vulnerable to injury by climate change.

In 2007, the Intergovernmental Panel on Climate Change (IPCC) released the report of its Working Group II, *Climate Change 2007: Impacts, Adaptation and Vulnerability*. The report noted the following points, among others, about the impacts of climate change on Asia:

- Climate change is already affecting life in Asia. For example, rising temperatures and more frequent extreme weather events have caused crop yields to decline in many Asian countries. Permafrost and glaciers are both retreating in an unprecedented fashion.
- Future climate change in Asia will probably affect agriculture and reduce water availability. Compared to 1990 levels, parts of Asia will probably see a 2.5–10% decrease in crop yields by the 2020s and a 5–30% decrease by the 2050s. Accelerated melting of glaciers will cause an increase in downstream flooding, but then a decrease in overall flow as the glaciers disappear. The Himalayan glaciers are receding faster than any others on Earth, and many are very likely to be gone by 2035 or sooner if warming continues at its current rate. If this occurs, rivers on which over half a billion people depend for water will cease to flow for part of the year.
- Marine and coastal ecosystems will probably be affected by sea-level rise and warming. Human settlements will likely be affected, with a million people along the coasts of south and Southeast Asia at increased risk from flooding. In large river deltas, which are flat and densely populated, flooding of land by rising ocean waters will affect aquaculture and endanger infrastructure (roads, buildings, bridges, etc.). Between 24% and 30% of the coral reefs of Asia are likely to disappear in the next 10 to 30 years.
- Future climate change in Asia will probably cause human health to deteriorate. In south and Southeast Asia, increased rates of illness and death from diarrhea are likely to increase. (Diarrhea now causes about 4% of all deaths, globally, each year; the percentage is higher in Asia and Africa.) Also, the habitats for disease vectors (insects or other animals that spread diseases) will likely expand in north Asia.
- Various environmental problems will be made worse by climate change. As in Africa, Asia's environment is presently being stressed by a number of problems, including population growth, the vast expanse of cities, and the rapid growth of industry and industrial pollution. Climate change makes it more difficult to sustain these processes or to reduce the damage they cause. For example, rapid industrialization in China is causing that country to open about one new coal-fired electricity-generating plant each week, accelerating global warming and contributing to the deaths of many of the 650,000 people who die every year in China from diseases associated with air pollution.

■ Impacts and Issues

China and India are the largest countries in the world, with over 1 billion inhabitants each. Both are rapidly industrializing, leading them to rapidly increase the amount of greenhouse gases they emit and causing great hardship to poor rural populations. In 2007, China became the world's largest emitter of greenhouse gases, surpassing the United States (although per-person U.S. emissions remain far higher than those of any Asian country).

According to the IPCC, sustainable development policies would likely reduce pressure on natural resources in Asia and make it easier to adapt to global warming. Such measures would include: 1) government policies against diverting land from agriculture even when it becomes more profitable to devote that land to industry, as such diversion can undermine food production; 2) sustainable forestry practices; 3) increased efficiency of energy use for various purposes; 4) generation of energy by non-greenhouse-polluting methods such as wind mills, and more.

IN CONTEXT: CHINA ISSUES POLICY

In June 2007, China announced its first official policy on global climate change. China's policy aims to reduce energy use by one-fifth by 2010. The policy statement also advocated expanding alternative and renewable energy sources. As of 2007, China was adding an average of one coal-fired power plant per week. Typically, those plants had fewer pollution controls than their counterparts in more developed nations.

Although the Chinese government has pledged to contribute to greenhouse-gas reduction, it firmly asserts that as a developing nation, China is not obligated to reduce its emissions. China's policy states that the most-developed western nations, as the largest historical emitters of greenhouse gases over the past century, carry the primary responsibility for reducing current emissions and preventing climate change. In 2007, China became the world's top emitter of greenhouse gases by total volume, eclipsing the United States. However, the United States remains the top per-person emitter of greenhouse gases worldwide.

SEE ALSO *Africa: Climate Change Impacts; Arctic People: Climate Change Impacts; Australia: Climate Change Impacts; Europe: Climate Change Impacts; North America: Climate Change Impacts; Small Islands: Climate Change Impacts; South America: Climate Change Impacts.*

BIBLIOGRAPHY

Books

Parry, M. L., et al, eds. *Climate Change 2007: Impacts, Adaptation and Vulnerability: Contribution of Working Group II to the Fourth Assessment Report of the Intergovernmental Panel on Climate Change.* New York: Cambridge University Press, 2007.

Web Sites

Bagla, Pallava. "Melting Himalayan Glaciers May Doom Towns.' *National Geographic News*, May 7, 2002. <http://news.nationalgeographic.com/news/2002/05/0501_020502_himalaya.html> (accessed September 16, 2007).

Black, Richard. "Climate Change 'Ruining' Everest." *BBC News*, November 17, 2004. <http://news.bbc.co.uk/2/hi/south_asia/4018261.stm> (accessed September 16, 2007).

"Climate Change Impacts in Asia." *The Climate Institute*. <http://www.climate.org/topics/climate/impacts_as.shtml> (accessed September 16, 2007).

Larry Gilman

Atmospheric Chemistry

Introduction

Since the beginning of industrial development and its associated increase in the burning of fossil fuels, humans have been altering the composition and chemistry of the atmosphere. More recently, the introduction of human-made halocarbons has also had an impact on atmospheric composition. The atmosphere is a mixture of gases that includes about 78% nitrogen and 21% oxygen at Earth's surface. The rest of the atmosphere near Earth's surface is made up of carbon dioxide and trace amounts of a number of other gases. It is the increase in carbon dioxide and trace gases that has altered atmospheric chemistry.

Historical Background and Scientific Foundations

The carbon dioxide content of the atmosphere has grown significantly in the last 50 years as a byproduct of the burning of fossil fuels. There has been a 36% increase in atmospheric carbon dioxide since the mid-1800s with almost all of the increase due to human activities.

Methane, a trace gas, is naturally found in the atmosphere as the product of decaying organic matter. The concentration of methane in the atmosphere has risen to 148% above its pre-industrial age level. Anthropogenic (human-caused) sources of methane include

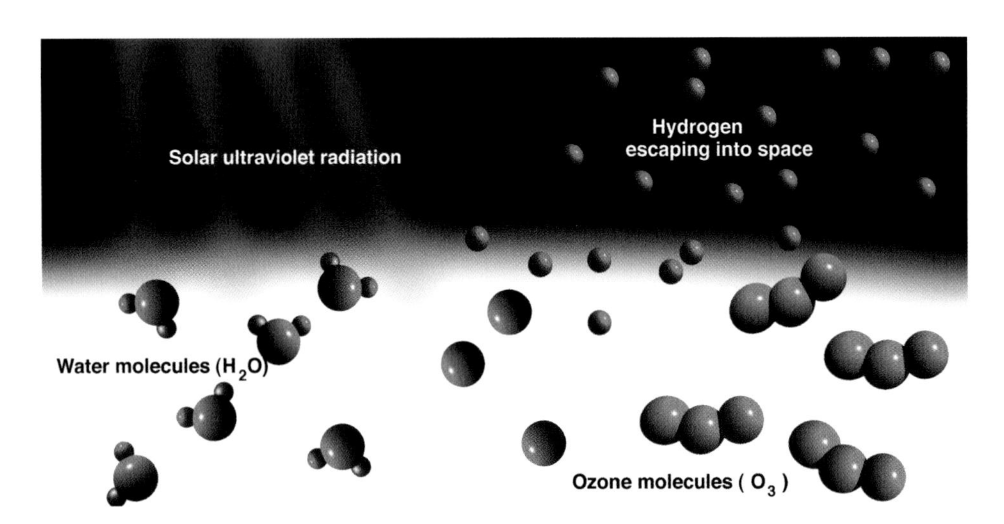

This illustration depicts the removal of water from the upper atmosphere. Energetic solar ultraviolet (UV) radiation breaks up water molecules (H_2O) at high altitude. The products are hydrogen and atomic oxygen (O). The high temperatures at high altitude, and the low mass of hydrogen, means that the hydrogen escapes to space and so the water cannot reform. Atomic oxygen can form oxygen (O_2) and eventually ozone (O_3). Water in the lower atmosphere (troposphere) is shielded from the UV radiation (mostly by the ozone) and is trapped under the temperature inversion that marks the tropopause. This results in a wet troposphere and a dry upper atmosphere. *Jon Lomberg/Photo Researchers, Inc.*

WORDS TO KNOW

ACID RAIN: A form of precipitation that is significantly more acidic than neutral water, often produced as the result of industrial processes.

FOSSIL FUELS: Fuels formed by biological processes and transformed into solid or fluid minerals over geological time. Fossil fuels include coal, petroleum, and natural gas. Fossil fuels are non-renewable on the timescale of human civilization, because their natural replenishment would take many millions of years.

HALOCARBON: Compound that contains carbon and one or more of the elements known as halons (chlorine, fluorine, or bromine). Halocarbons do not exist in nature; all are manufactured. When released into the atmosphere, many halocarbons deplete the ozone layer and have high global warming potential.

OZONE LAYER: The layer of ozone that begins approximately 9.3 mi (15 km) above Earth and thins to an almost negligible amount at about 31 mi (50 km) and shields Earth from harmful ultraviolet radiation from the sun. The highest natural concentration of ozone (approximately 10 parts per million by volume) occurs in the stratosphere at approximately 15.5 mi (25 km) above Earth. The stratospheric ozone concentration changes throughout the year as stratospheric circulation changes with the seasons. Natural events such as volcanoes and solar flares can produce changes in ozone concentration, but man-made changes are of the greatest concern.

PHOTOCHEMICAL SMOG: A type of smog that forms in large cities when chemical reactions take place in the presence of sunlight; its principal component is ozone. Ozone and other oxidants are not emitted into the air directly but form from reactions involving nitrogen oxides and hydrocarbons. Because of its smog-making ability, ozone in the lower atmosphere (troposphere) is often referred to as "bad"ozone.

TRACE GASES: Gases present in Earth's atmosphere in trace (relatively very small) amounts. All greenhouse gases happen to be trace gases, though some are more abundant than others; the most abundant greenhouse gases are CO_2 (0.037% of the atmosphere) and water vapor (0.25% of the atmosphere, on average).

ULTRAVIOLET RADIATION: The energy range just beyond the violet end of the visible spectrum. Although ultraviolet radiation constitutes only about 5 percent of the total energy emitted from the sun, it is the major energy source for the stratosphere and mesosphere, playing a dominant role in both energy balance and chemical composition.

landfills, the management of livestock, and natural gas and petroleum processing.

Nitrous oxide is another gas that is naturally found in trace amounts in the atmosphere. The concentration of nitrous oxide is largely derived from agricultural sources, with vehicle emissions also contributing nitrous oxide to the atmosphere. The concentration of nitrous oxide in the atmosphere has risen approximately 18% in the last 200 years.

Carbon dioxide, methane, and nitrous oxide are greenhouse gases. Their increased concentration in the atmosphere is contributing to global warming. Greenhouse gases cause global warming by absorbing reflected heat from Earth's surface thereby warming the atmosphere.

In addition to carbon dioxide, methane, and nitrous oxide, there are entirely human-made trace gases in the atmosphere that have added significantly to global warming. Halocarbons, particularly chlorofluorocarbons (CFCs) and hydrochlorofluorocarbons (HCFCs), are found in a variety of products and substances, including aerosol propellants, solvents, and refrigerants, and are released into the atmosphere from these sources. Other fluorinated gases—such as sulfur hexafluoride that has escaped from applications in electric power equipment and perfluorocarbons that have escaped from aluminum processing and the manufacture of semiconductors—are also found in trace amounts in the atmosphere.

Earth's atmosphere is actually comprised of four layers. The troposphere starts at ground level and extends up approximately 10 mi (16 km). The next layer is the stratosphere, which is roughly 10–30 mi (16–48 km) above Earth's surface. Above the stratosphere is the mesosphere (30–60 mi/48–97 km above Earth's surface); and beyond the mesosphere is the thermosphere. Although each layer has a distinct composition, the troposphere and stratosphere are the layers that are critical to climate change issues.

In the troposphere, the concentration of nitrogen is around 78% and oxygen is about 21%. This is the atmospheric layer where carbon dioxide, methane, nitrous oxide, sulfur hexafluoride, and the halocarbons collect. In the stratosphere, a naturally occurring layer of ozone is formed from the reaction of sunlight with oxygen. Ozone is a form of oxygen that has three atoms in each molecule instead of the usual two atoms, and it plays an important protective role by diminishing the amount of potentially damaging ultraviolet radiation reaching Earth. CFCs and HCFCs are drifting from the troposphere into the stratosphere and destroying some of this natural ozone layer. The chlorine in the halocarbons

IN CONTEXT: HUMAN ACTIVITY IMPACTS NATURAL BALANCE

"The current concentration of a greenhouse gas in the atmosphere is the net result of the history of its past emissions and removals from the atmosphere. The gases and aerosols considered [by the Intergovernmental Panel on Climate Change] are emitted to the atmosphere by human activities or are formed from precursor species emitted to the atmosphere. These emissions are offset by chemical and physical removal processes. With the important exception of carbon dioxide (CO_2), it is generally the case that these processes remove a specific fraction of the amount of a gas in the atmosphere each year and the inverse of this removal rate gives the mean lifetime for that gas. In some cases, the removal rate may vary with gas concentration or other atmospheric properties (e.g., temperature or background chemical conditions)."

"Long-lived greenhouse gases (LLGHGs), for example, CO_2, methane (CH_4) and nitrous oxide (N_2O), are chemically stable and persist in the atmosphere over time scales of a decade to centuries or longer, so that their emission has a long-term influence on climate."

"Because these gases are long lived, they become well mixed throughout the atmosphere much faster than they are removed and their global concentrations can be accurately estimated from data at a few locations. Carbon dioxide does not have a specific lifetime because it is continuously cycled between the atmosphere, oceans and land biosphere and its net removal from the atmosphere involves a range of processes with different time scales."

"Short-lived gases (e.g., sulphur dioxide and carbon monoxide) are chemically reactive and generally removed by natural oxidation processes in the atmosphere, by removal at the surface or by washout in precipitation; their concentrations are hence highly variable. Ozone is a significant greenhouse gas that is formed and destroyed by chemical reactions involving other species in the atmosphere. In the troposphere, the human influence on ozone occurs primarily through changes in precursor gases that lead to its formation, whereas in the stratosphere, the human influence has been primarily through changes in ozone removal rates caused by chlorofluorocarbons (CFCs) and other ozone-depleting substances."

SOURCE: *Solomon, S., et al. "Technical Summary." In* Climate Change 2007: The Physical Science Basis. Contribution of Working Group I to the Fourth Assessment Report of the Intergovernmental Panel on Climate Change. *New York: Cambridge University Press. 2007.*

reacts with ozone to convert molecules of ozone to oxygen, thereby depleting the ozone layer.

Also in the troposphere, carbon dioxide reacts with water vapor to make a very weak acid (carbonic acid), a reaction that depends on the temperature. Carbon dioxide gas is more soluble at colder temperatures, but saturation of the gas in water vapor is quickly reached with normal amounts of the gas in the atmosphere.

Increasing the concentration of the carbon dioxide in the atmosphere does not make the water vapor any more acidic. However, the introduction into the atmosphere of sulfur and nitrogen oxides, which react with water to form strong acids, does make the water vapor and the precipitation derived from it more acidic. This acid rain was the first recognized negative climate change that is caused by human activities.

Satellite observations have provided data on the chemical composition of the stratosphere and changes in the ozone layer. In the troposphere, no satellites are needed to observe the damaging chemical changes that have periodically occurred as a result of human activities. Photochemical smog is a serious air pollution problem that occurs when sunlight acts on nitrogen oxides, hydrocarbons, and other pollutants that are created from fossil fuel combustion. This type of smog is seen most often in highly populated areas with large numbers of vehicles during periods of high temperature and intense sunlight. In such situations, concentrations of the offending gases near ground level increase.

■ Impacts and Issues

Observing atmospheric chemistry requires international efforts. The World Meteorological Organization (WMO), an agency of the United Nations, has 188 participating member states and territories cooperating in a continuous study of all aspects of the atmosphere. The WMO working with the United Nations Environmental Programme established the Intergovernmental Panel on Climate Change (IPCC) in 1988 to assess information being gathered about atmospheric composition and chemistry. The IPCC is not a research organization. It issues reports on the data collected from the WMO and other sources.

In 1987, 145 countries signed the Montreal Protocol to ban the production of ozone-depleting CFCs by the year 2000. Although the use of CFCs was banned by the Montreal Protocol in 1987, the existing CFCs will remain in the atmosphere for hundreds of years. HCFCs are less damaging to the ozone layer, but their use is also being phased out because they are greenhouse gases.

In 1992, 189 countries joined the United Nations Framework Convention on Climate Change (UNFCCC) to set general goals and rules to address climate change. One result of the activities of UNFCCC was the 1997 Kyoto Protocol, which included an outline of controls for the industrial nations to contain six greenhouse gases—carbon dioxide, methane, nitrous oxide, hydrofluorocarbons (HFCs), perfluorocarbons (PFCs), and sulfur hexafluoride.

SEE ALSO *Acid Rain; Climate Change; Greenhouse Gases.*

BIBLIOGRAPHY

Web Sites

"Atmosphere Changes." *U.S. Environmental Protection Agency.* <http://www.epa.gov/climatechange/science/recentac.html> (accessed August 16, 2007).

"International Global Atmospheric Chemistry Science Plan and Implementation Strategy." *IGBP Secretariat.* <http://www.igac.noaa.gov/> (accessed August 16, 2007).

"The Nobel Prize in Chemistry 1995." *Nobelprize.org.* <http://nobelprize.org/nobel_prizes/chemistry/laureates/1995/index.html> (accessed August 16, 2007).

"Photochemical Smog—What It Means for Us." *U.S. Environmental Protection Agency*, March 2004. <http://www.epa.sa.gov.au/pdfs/info_photosmog.pdf> (accessed August 16, 2007).

"WMO in Brief." *World Meteorological Organization.* <http://www.wmo.ch/pages/about/index_en.html> (accessed August 16, 2007).

Atmospheric Circulation

Introduction

Atmospheric circulation includes the movement of air on a global scale. It is the manner in which that heat is distributed throughout the atmosphere, from equatorial regions that are warmer to polar regions that are colder. The circulation of air in the atmosphere varies somewhat from year to year, but overall, the basic mechanism of circulation remains the same. This helps produce a stable global climate.

Atmospheric circulation is linked to ocean temperature and winds. An example of this relationship are two naturally occurring variations in the temperature of the tropical Pacific Ocean known as El Niño and La Niña. The warming sea temperature in the case of El Niño and cooling temperature in the case of La Niña can persist for several years. These alterations affect wind patterns, which in turn alter weather. For example, the greater frequency and severity of tropical storms that occurred in the equatorial Atlantic Ocean in 2005 and 2006 (the best known example from the U.S. perspective was Hurricane Katrina) has been linked to the increased upper atmosphere easterly blowing winds that were stimulated by a La Niña. These winds reduce vertical wind shear—wind changes with altitude—which increases the likelihood of the formation of thunderstorms and tropical storms.

Whether global warming is altering atmospheric circulation is still debatable, as of 2007. However, measurements of circulation of air over the tropical Pacific have provided evidence of human-induced changes.

Combining mathematical models and weather data, these computer-generated simulations depict atmospheric and oceanic circulation. *Hank Morgan/Photo Researchers, Inc.*

Historical Background and Scientific Foundations

Circulation of air in the atmosphere has likely occurred ever since the formation of the modern-day atmosphere following the appearance of life on Earth. The basis for atmospheric circulation is the differential heating of Earth. Tropical regions are heated more so than the polar regions because the thickness of the atmosphere that the

WORDS TO KNOW

CORIOLIS EFFECT: A pseudoforce describing the deflection of winds due to the rotation of Earth, which produce a clockwise or counterclockwise rotation of storm systems in the Southern and Northern Hemispheres, respectively.

GREENHOUSE GASES: Gases that cause Earth to retain more thermal energy by absorbing infrared light emitted by Earth's surface. The most important greenhouse gases are water vapor, carbon dioxide, methane, nitrous oxide, and various artificial chemicals such as chlorofluorocarbons. All but the latter are naturally occurring, but human activity over the last several centuries has significantly increased the amounts of carbon dioxide, methane, and nitrous oxide in Earth's atmosphere, causing global warming and global climate change.

ICE AGE: Period of glacial advance.

INTERTROPICAL: Literally, between the tropics: usually refers to a narrow belt along the equator where convergence of air masses of the Northern and Southern Hemispheres produces a low-pressure atmospheric condition.

POLAR CELLS: Air circulation patterns near the poles: relatively warm, moist air approaches the pole at a high altitude, cools, sinks at the pole, and flows southward at a lower altitude. Because of Earth's rotation, air approaching or receding from the poles flows eastward, producing the polar easterlies.

TRADE WINDS: Surface air from the horse latitudes (subtropical regions) that moves back toward the equator and is deflected by the Coriolis Force, causing the winds to blow from the Northeast in the Northern Hemisphere and from the Southeast in the Southern Hemisphere. These steady winds are called trade winds because they provided trade ships with an ocean route to the New World.

WIND CELLS: More commonly called convective or convection cells; vertical structures of moving air formed by warm (less-dense) air welling up in the center and cooler (more-dense) air sinking around the perimeter. Thunderstorms are shaped by convective cells.

sun's rays penetrate through is greater at the poles. As a result, the tropical atmosphere is warmer than the atmosphere over the poles, which causes the movement of warm air northward or southward from the equator.

Winds are essential to atmospheric circulation. Aside from winds that occur temporarily, the general circulation of the atmosphere involves surface winds that blow regularly. There are three so-called belts of wind in each hemisphere. Polar easterlies are found from 60 to 90 degrees latitude. In the Northern Hemisphere, these blow from the northeast to the southwest, and from the southeast to the northwest in the Southern Hemisphere. The prevailing winds from 30 to 60 degrees latitude are called prevailing westerlies. In the Northern Hemisphere, the prevailing westerlies blow from the southwest to the northeast, and from the northwest to the southeast in the Southern Hemisphere. Finally, the region from the equator to 30 degrees latitude north and south is the area where tropical easterlies are found. In the Northern Hemisphere, they blow from the northeast to the southwest, and from the southeast to the northwest in the Southern Hemisphere.

Tropical easterlies are also called trade winds. The northern and southern trade winds converge near the equator in a zone; this intertropical convergence zone is an area of cloud and thunderstorms that encircles Earth. It is also an area where winds can be light and variable. Centuries ago, mariners could be calmed for days in the doldrums.

The directions of the winds in these belts are influenced by what are known as cells. The Northern and Southern Hemisphere polar winds are guided by northern and southern polar cells, which are created because of Earth's rotation. The hemispheric trade winds are guided by a cell called the Ferrel cell, while the tropical easterlies are guided by the Hadley cell.

The polar and Hadley cells are closed loops; warm, more southerly air rises and moves northward. As the air cools, it sinks and moves southward where it heats and rises again, completing the loop. The reason that the winds in the belts do not move north-south and south-north is because of the Coriolis effect—the influence that Earth's rotation exerts.

In addition to these three air circulation cells, other cells operate horizontally, from west-to-east to east-to-west. The horizontal circulation occurs because Earth's surface is composed of land and water. Water absorbs heat more slowly than does the land, and loses heat to the atmosphere more slowly than land. On a small scale, the result is clear along the ocean coast, where the winds blow onto shore during the day, as cooler air over the sea migrates toward land, and blows out to sea at night, when the air over the ground is cooler than the sea air. On a much larger scale, this back and forth flow of air occurs over a period of months or even years.

Atmospheric circulation is crucial for the global climate and the global pattern of precipitation. The movement of air from regions of low pressure, which tends to encourage precipitation, to regions of high pressure, which do not favor precipitation, helps to distribute moisture through the atmosphere.

Impacts and Issues

Climate changes due to altered atmospheric circulation are not new. Scientists have evidence that altered air circulation in the tropical Pacific Ocean, similar to that which occurs during El Niño, triggered a global climate change about one million years ago. Then, the changed circulation of air caused the polar ice sheets to grow in area, which lengthened the periods of glaciation (the ice ages).

This research has relevance in modern times, for it indicates that tropical regions are very influential to global climate. Thus, conditions that alter atmospheric circulation can change the global climate.

It is known that Earth's atmosphere is warming, due to the increased retention of heat. One reason for this has been suggested to be the gradual accumulation of greenhouse gases—gases produced by human activities. The link between human activity and atmospheric change used to be very contentious. In 2007, however, only a small minority of scientists still argued that atmospheric warming is free from human influence.

The question of whether human activities are influencing atmospheric circulation, however, remains contentious. As one example, a paper published in *Nature* in 2006 reported on data gathering from 1861 to the early years of the 21st century, which revealed that the difference in pressure between the higher pressure of the western Pacific to the lower pressure of the eastern Pacific has declined over the past 150 years. The data were used in several computer models of climate; some models factored in the influence to pressure change of only natural conditions, and others had the added influence of human activities. The model that incorporated human-influenced atmospheric change most closely matched the actual data.

Other scientists are skeptical of the concluded link between human activities and atmospheric change, because data collected on sea surface temperatures for a much longer time do not support the air pressure data. Whether or not human activities have influenced the changed environment over the Pacific Ocean, however, it is clear that change has occurred, and that such changes in atmospheric circulation do affect global climate.

SEE ALSO *Abrupt Climate Change; Atmospheric Chemistry; Atmospheric Pollution; Atmospheric Structure; El Niño and La Niña; Global Warming; Greenhouse Effect.*

BIBLIOGRAPHY

Books

Barry, Roger G. *Atmosphere, Weather and Climate.* Oxford, United Kingdom: Routledge, 2003.

Lutgens, Frederick K., Edward J. Tarbuck, and Dennis Tasa. *The Atmosphere: An Introduction to Meteorology.* New York: Prentice Hall, 2006.

Trefil, Calvo. *Earth's Atmosphere.* Geneva, IL: McDougal Littell, 2005.

Periodical

Vecchi, Gabriel A., Brian J. Soden, Andrew T. Wittenberg, et al. "Weakening of Tropical Pacific Atmospheric Circulation Due to Anthropogenic Forcing." *Nature* 441, no. 7089 (May 4, 2006): 73–76.

Atmospheric Pollution

Introduction

Earth's atmosphere is a layer of gases that surrounds the planet, reaching a thickness of about 300 mi (480 km). This same distance may be considerable to a human traveling by car, but relative to the infinite vastness of space, our atmosphere is vanishingly thin.

This thin blue line cocooning Earth, which is so apparent from orbiting spacecraft and which makes life on the planet possible, is susceptible to change and damage. Contamination of the atmosphere by noxious compounds has occurred ever since prehistoric peoples used fire. However, the tremendous growth in the use of machinery that began in the late eighteenth century spawned the growth of industry and increased air pollution.

A legacy of the twentieth century is the use of compounds such as chlorofluorocarbons (CFCs) in a variety of household and personal products. The escape of these compounds into the atmosphere has led to the depletion of atmospheric ozone (O_3), which has increased the amount of damaging ultraviolet radiation reaching Earth's surface. These and other human activities far outweigh natural forms of atmospheric pollution.

Historical Background and Scientific Foundations

Atmospheric pollution can occur naturally. For example, the Hawaiian volcano Kilauea spews out approximately 2,000 tons (1,800 metric tons) of sulfur dioxide (SO_2) every day during eruptions. Even more spectacularly, the series of massive eruptions of the Krakatoa Volcano on August 27, 1883 propelled ash 50 mi (80 km) into the atmosphere. The ash restricted the penetration of sunlight to such an extent that the average global temperature the next year was up to 2°F (1.2°C) below normal.

Nonetheless, these and other natural causes of atmospheric pollution are nowhere near as influential as human activities, which have occurred for hundreds of thousands of years, ever since prehistoric peoples learned to use fire for cooking and warmth. Examination of ice cores recovered from the Greenland ice sheet has revealed higher levels of lead, mercury, and nickel beginning

In this satellite image released by NASA, aerosol pollution (grey haze) is shown over India and Bangladesh. The suspension of liquid and solid particles lies over the southern edge of the Himalayan mountains (white, upper frame) in India and south over Bangladesh (center right) and the Bay of Bengal (blue, lower frame). *NASA/ Photo Researchers, Inc.*

Although some atmospheric pollution occurs naturally, the amount generated through human industrial activity has the greatest impact on the atmosphere, particularly in regard to global warming.
AP Images.

about 5,000 years ago, when mining and smelting of metal ores began in Europe.

Despite this long history of atmospheric abuse, the present problems with atmospheric pollution date from the end of the eighteenth century and early nineteenth century, the period that is known as the Industrial Revolution. Then, a number of technological advances including the introduction of the steam engine spurred the growth of manufacturing plants in or near major cities (a source of cheap and abundant labor for the factories). The concentration of industries and their use of coal as the power source instead of flowing water increased air pollution.

Although the air in the atmosphere is still made up predominantly of oxygen and nitrogen, atmospheric data collected since the Industrial Revolution have shown that the content of other, so-called trace gases has changed. Furthermore, new compounds have appeared. The best example are chlorofluorocarbons (CFCs), which are a product of twentieth century technology.

The atmospheric release of CFCs and hydrochlororfluorocarbons (which are used in air conditioners, refrigerators, and aerosol cans), halons (an ingredient of fire extinguishers), methyl chloroform, and methyl bromide have depleted the level of the atmospheric gas called ozone. Ozone consists of three oxygen molecules. The gas is found mainly in an upper layer of the atmosphere called the stratosphere.

Ozone is able to absorb the ultraviolet (UV) wavelengths of sunlight. UV light has sufficient energy to penetrate into the upper layers of skin, causing sunburn and, more ominously, to break apart the chains of genetic material inside cells. This genetic damage can lead to the development of some cancers. UV light can also damage vision. CFCs and the other compounds chemically destroy ozone, allowing more UV radiation to reach Earth's surface.

Scientists began to monitor atmospheric ozone levels in the 1970s. Over the next three decades, the declining levels of ozone were recognized. The decline is not evenly spread throughout the atmosphere. Rather, the depletion is more pronounced in some regions, in particular over the Antarctic as an "ozone hole."

Since the time of the Industrial Revolution, trace gases such as carbon dioxide (CO_2), nitrous oxide (N_2O), and sulfur dioxide (SO_2) have built up in concentration in the atmosphere. These gases have been dubbed greenhouse gases because, analogous to the way a greenhouse traps the sun's heat, they cause the heat from the sun to be retained by the atmosphere. The result has been the warming of the atmosphere that is popularly known as global warming.

■ Impacts and Issues

Human industrial activity is by far the greatest contributor to atmospheric pollution. According to the U.S. Environmental Protection Agency (EPA), approximately 6.5 billion lbs (3 billion kg) of toxic compounds (including 100 million lbs, or 45 million kg, of cancer-causing chemicals) are released to the atmosphere every year in the United States alone.

A recent illustration of the influence of human activity on atmospheric quality is Beijing, China. Satellite monitoring of China as part of the European Space Agency's Dragon Programme has revealed that the development of Beijing into an industrially important mega-city containing over 10 million people and nearly three million vehicles has been accompanied by an accumulation of the planet's highest levels of nitrogen dioxide. This gas is a respiratory irritant and, paradoxically to ozone depletion higher in the atmosphere, causes ozone build-up near the ground, which triggers smog. Similar findings have been found over developing areas of India.

WORDS TO KNOW

CHLOROFLUOROCARBONS: Members of the larger group of compounds termed halocarbons. All halocarbons contain carbon and halons (chlorine, fluorine, or bromine). When released into the atmosphere, CFCs and other halocarbons deplete the ozone layer and have high global warming potential.

GREENHOUSE GASES: Gases that cause Earth to retain more thermal energy by absorbing infrared light emitted by Earth's surface. The most important greenhouse gases are water vapor, carbon dioxide, methane, nitrous oxide, and various artificial chemicals such as chlorofluorocarbons. All but the latter are naturally occurring, but human activity over the last several centuries has significantly increased the amounts of carbon dioxide, methane, and nitrous oxide in Earth's atmosphere, causing global warming and global climate change.

ICE CORE: A cylindrical section of ice removed from a glacier or an ice sheet in order to study climate patterns of the past. By performing chemical analyses on the air trapped in the ice, scientists can estimate the percentage of carbon dioxide and other trace gases in the atmosphere at that time.

INDUSTRIAL REVOLUTION: The period, beginning about the middle of the eighteenth century, during which humans began to use steam engines as a major source of power.

OZONE: An almost colorless, gaseous form of oxygen with an odor similar to weak chlorine. A relatively unstable compound of three atoms of oxygen, ozone constitutes, on average, less than one part per million (ppm) of the gases in the atmosphere. (Peak ozone concentration in the stratosphere can get as high as 10 ppm.) Yet ozone in the stratosphere absorbs nearly all of the biologically damaging solar ultraviolet radiation before it reaches Earth's surface, where it can cause skin cancer, cataracts, and immune deficiencies, and can harm crops and aquatic ecosystems.

STRATOSPHERE: The region of Earth's atmosphere ranging between about 9 and 30 mi (15 and 50 km) above Earth's surface.

As much a concern as ozone depletion is, there is good news. The effect is reversible in the absence of the ozone-destroying compounds. More than 165 nations are signatories to the Montreal Protocol, an international agreement drafted in 1987 that commits them to phase out the use of ozone-depleting substances according to a timetable.

See Also *Carbon Dioxide (CO_2); Climate Change; Cow Power; Global Warming; Kyoto Protocol; Nitrous Oxide.*

BIBLIOGRAPHY

Books

DiMento, Joseph F. C., and Pamela M. Doughman. *Climate Change: What It Means for Us, Our Children, and Our Grandchildren.* Boston: MIT Press, 2007.

Gore, Al. *An Inconvenient Truth: The Planetary Emergency of Global Warming and What We Can Do About It.* New York: Rodale Books, 2006.

Seinfeld, John H., and Spyros N. Pandis. *Atmospheric Chemistry and Physics: From Air Pollution to Climate Change.* New York: Wiley Interscience, 2006.

Periodicals

Chow, J. C., J. G. Watson, J. J. Shah, et al. "Megacities and Atmospheric Pollution." *Journal of the Air and Waste Management Association* 54 (2004): 1226-1236.

Web Sites

"Breath of the Dragon: ERS-2 and Envisat Reveal Impact of Economic Growth on China's Air Quality." *European Space Agency*, September 1, 2005. <http://www.esa.int/esaEO/SEMEE6A5QCE_environment_0.html> (accessed November 5, 2007).

"The 500-Meter Wheeze: Will Air Pollution Affect the Athletes at the 2008 Olympics in Beijing?" *Slate.com*, October 25, 2007. <http://www.slate.com/id/2176636/nav/fix> (accessed November 5, 2007).

Atmospheric Structure

■ Introduction

The late astronomer and author Carl Sagan (1934–1996) famously described Earth when viewed from deep space as "a pale blue dot." His description, which was intended to highlight the fragility of the planet, points out the visual effect of Earth's atmosphere. Viewed from space, the optical properties of the atmosphere surrounding Earth haloes the planet in a thin film of blue.

Compared to Earth's diameter, which averages about 7,800 mi (12,550 km), the atmosphere, which peters out to the near-vacuum of space at an altitude of approximately 620 mi (1,000 km), is paper-thin. Furthermore, most of the planet's weather is accounted for by the regions of the atmosphere within 35 mi (56 km) of the surface.

The atmosphere of the primordial Earth was different in composition from the atmosphere that cocoons the planet now. The appearance and evolution of life on Earth influenced atmospheric structure. The susceptibility of the atmosphere to change is the root of global warming, which all but a small minority of climate researchers now concede is a consequence of human activities.

■ Historical Background and Scientific Foundations

Earth is about 4.5 billion years old. The newly formed planet had no atmosphere, but as the planet cooled, the release of various gases created an atmosphere that was likely very different from that of the present day. Although the composition of this primordial atmosphere is still debatable, the majority of scientists who study the early climate agree that the atmosphere was probably rich in carbon (C) and nitrogen (N), and lacked oxygen (O). As life began, the atmosphere changed, with carbon dioxide (CO_2) decreasing and oxygen appearing and accumulating.

The present-day atmosphere consists predominantly of nitrogen (an average of 78% of the total material) and oxygen (average of 21%). The remaining 1% of the atmosphere consists of the so-called trace gases—argon (Ar), helium (He), hydrogen (H), krypton (Kr), neon (Ne), methane (CH_4), ozone (O_3), and xenon (Xe)—as well as carbon dioxide and water vapor.

The atmosphere is not a single layer. Rather, it consists of regions that are separated from one another by narrow zones of transition. The atmosphere gives way to space at an altitude of approximately 620 mi (1,000 km).

The atmosphere is also not uniform in the density of the constituent gases. Instead, over 99% of the mass of the atmosphere is concentrated within 25 mi (40 km) of Earth's surface. Finally, the atmosphere is not uniform in temperature. As anyone who has climbed a mountain can attest, air temperature decreases with altitude, as the heat-absorbing gases become more dilute. Atmospheric temperature drops by about 11°F (6°C) for every 0.6 mi (1 km) of altitude in the atmospheric layer immediately above Earth's surface.

This layer is called the troposphere. The upper range of the troposphere varies depending on latitude. At higher latitudes, it is about 5 mi (8 km) high, while at the equator it is upwards of 11 mi (18 km) high. The troposphere contains almost all (99%) of the atmospheric water vapor. Again, there is geographic variation, with water vapor concentration being up to 3% of total atmospheric content above the equator, but less toward the poles.

Weather occurs exclusively in the troposphere. Indeed, the meaning of the word troposphere ("region of mixing") reflects the importance of air currents in this layer. Pollutants that enter the troposphere will be evenly dispersed within days; some of the chemicals will return to the surface in precipitation, as occurs in acid rain.

The troposphere is separated from the next layer of the atmosphere, the stratosphere, by a thin transition region called the tropopause. The stratosphere is approximately 25 mi (40 km) thick. It begins about 6.3 mi (10 km) above Earth's surface, 1.5 mi (2.4 km) above the peak of Mt. Everest. Commercial aircraft

WORDS TO KNOW

ACID RAIN: A form of precipitation that is significantly more acidic than neutral water, often produced as the result of industrial processes.

CHLOROFLUOROCARBONS: Members of the larger group of compounds termed halocarbons. All halocarbons contain carbon and halons (chlorine, fluorine, or bromine). When released into the atmosphere, CFCs and other halocarbons deplete the ozone layer and have high global warming potential.

FOSSIL FUELS: Fuels formed by biological processes and transformed into solid or fluid minerals over geological time. Fossil fuels include coal, petroleum, and natural gas. Fossil fuels are nonrenewable on the timescale of human civilization, because their natural replenishment would take many millions of years.

INVERSION: A type of chromosomal defect in which a broken segment of a chromosome attaches to the same chromosome, but in reverse position.

OZONE: An almost colorless, gaseous form of oxygen with an odor similar to weak chlorine. A relatively unstable compound of three atoms of oxygen, ozone constitutes, on average, less than one part per million (ppm) of the gases in the atmosphere. (Peak ozone concentration in the stratosphere can get as high as 10 ppm.) Yet ozone in the stratosphere absorbs nearly all of the biologically damaging solar ultraviolet radiation before it reaches Earth's surface, where it can cause skin cancer, cataracts, and immune deficiencies, and can harm crops and aquatic ecosystems.

TRACE GASES: Gases present in Earth's atmosphere in trace (relatively very small) amounts. All greenhouse gases happen to be trace gases, though some are more abundant than others; the most abundant greenhouse gases are CO_2 (0.037% of the atmosphere) and water vapor (0.25% of the atmosphere, on average).

WATER VAPOR: The most abundant greenhouse gas, it is the water present in the atmosphere in gaseous form. Water vapor is an important part of the natural greenhouse effect. Although humans are not significantly increasing its concentration, it contributes to the enhanced greenhouse effect because the warming influence of greenhouse gases leads to a positive water vapor feedback. In addition to its role as a natural greenhouse gas, water vapor plays an important role in regulating the temperature of the planet because clouds form when excess water vapor in the atmosphere condenses to form ice and water droplets and precipitation.

cruise at altitudes that are in the lower to middle portions of the stratosphere.

The temperature within the stratosphere also varies with height, but in a different pattern to that of the troposphere. The temperature does not vary up to an altitude of about 15 mi (24 km), after which it gradually increases until reaching the next atmospheric transition zone, which is called the stratopause. This temperature pattern, with warmer air overlying colder air, is known as an inversion. Glimpsing a towering thunderhead on a summer's day provides a visual example of the influence of the inversion; the thunderhead flattens off when the warm rising air in the cumulus cloud contacts the cooler air in the lower stratosphere, which halts the rising of the air.

The increasing temperature with altitude in the stratosphere acts to make this layer more stable than the underlying troposphere. Another contributor to this stability, and the reason for the stratospheric temperature inversion, is ozone. Ozone is a three-oxygen compound that absorbs incoming ultraviolet radiation from sunlight. This retention of heat is what maintains temperature with increasing altitude.

Beyond the stratopause lies the mesosphere. This atmospheric layer extends to approximately 50 mi (80 km) above the surface. There is little water vapor or ozone in this layer, hence, temperatures are low and keep decreasing with altitude. As well, the levels of oxygen and nitrogen are far less than in the troposphere and stratosphere; mesopheric air pressure (the number of atoms per given area) is 1,000 times less than air pressure at sea level.

A transition layer called the mesopause separates the mesosphere from the thermosphere. The thermosphere extends to about 75 mi (121 km) above Earth's surface. The thermosphere is home to the International Space Station and orbits of the space shuttle.

The final layer of the atmosphere is the exosphere. Beyond lies the near-vacuum of space.

Impacts and Issues

Because regions of the atmosphere determine weather patterns and the global climate, atmospheric changes can be profound. The documented increase in the atmospheric levels of carbon dioxide, chlorofluorocarbons (CFCs), methane, and nitrous oxide (N_2O), which are collectively known as greenhouse gases, is driving an increase in the temperature of the troposphere that has been termed global warming.

The final greenhouse compound is ozone. Degradation of ozone in the stratosphere has been accelerated from the naturally occurring rate due to the presence of human-made compounds including CFCs and hydrochlorofluorocarbons (which are used in air conditioners, refrigerators, and aerosol cans), halons (an ingredient of fire extinguishers), methyl chloroform ($C_2H_3Cl_3$), and methyl bromide (CH_3Br) is allowing more ultraviolet light to reach Earth's surface.

The energy of ultraviolet light is sufficient to permit the light to penetrate into the upper layers of the skin and even to slice apart the genetic material inside cells. Consequences include skin damage such as sunburn and, more ominously, the increased tendency of the genetically damaged cells to become cancerous.

Although in the past it was argued that global warming was a natural phenomenon, only a tiny minority of scientists continue to hold this view. The vast majority of scientists now accept that human activities are at the heart of global warming today.

For example, carbon dioxide released into the atmosphere by the burning of fossil fuels and the burning of felled lumber from deforested regions, as two examples, account for almost half of the atmospheric warming caused by human activity. In another example, the build-up of CFCs not only stimulates ozone breakdown, but increases the retention of heat, since CFCs are a powerful greenhouse gas. Indeed, one molecule of CFC has about 20,000 times the heat-trapping power as a molecule of carbon dioxide.

The pollution of the atmosphere near Earth's surface with noxious compounds can be unhealthy. An example from 2007 is Beijing, China. Air pollution in Beijing, which is mainly caused by the millions of vehicles operating daily in the mega-city, has become a great concern to officials of the International Olympic Committee responsible for ensuring that Beijing is ready to host the Summer Olympics in 2008. Events such as the marathon may need to be shifted to early morning, when air pollution is less. Alternatively, the government has proposed a ban on all vehicles in Beijing during the games.

See Also *Atmospheric Circulation; Greenhouse Effect; Jet Stream.*

BIBLIOGRAPHY

Books

Barry, Roger G. *Atmosphere, Weather and Climate.* Oxford, U.K.: Routledge, 2003.

Lutgens, Frederick K., Edward J. Tarbuck, and Dennis Tasa. *The Atmosphere: An Introduction to Meteorology.* New York: Prentice Hall, 2006.

Trefil, Calvo. *Earth's Atmosphere.* Geneva, IL: McDougal Littell, 2005.

Ward, Peter. *Out of Thin Air: Dinosaurs, Birds, and Earth's Ancient Atmosphere.* Washington, DC: Joseph Henry Press, 2006.

Australia: Climate Change Impacts

■ Introduction

Australia, the smallest continent, is the only one completely occupied by a single nation-state, the Commonwealth of Australia. In discussions of global climate, the island nation of New Zealand, about 1,250 mi (2,000 km) to the southeast, is sometimes grouped with Australia. Effects of human-caused climate change have been observed in Australia and New Zealand, including more heat, less cold, less rain, rising sea level, and damage to coral reefs, with larger changes in the future predicted by climate scientists.

Both natural and human systems are seeking to adapt to the ongoing changes. Human systems in this region, which is industrialized and technologically advanced, are relatively adaptable, but natural systems have limited capacity to adapt. Forecast climate changes are more than 90% likely to exceed the adaptive ability of many species, and may cause a number of these species to become extinct. The world's largest reef system, the Great Barrier Reef, 1,300 mi (2,100 km) long, is located along the northeastern coast of Australia and will likely be severely damaged by global warming.

While the leaders of the Asia-Pacific Economic Cooperation (APEC) were meeting in Sydney, Australia, in September 2007 to discuss issues including climate change, local surfers and other residents gathered at Bondi Beach to demand climate targets be set at the summit. Here, two young protesters play on the group's target banner. *AP Images.*

In the early 2000s, Australia began suffering the effects of a six-year drought. The result was a severe impact on the availability of drinking water and water available for agricultural use. *Image copyright Alecia Scott, 2007. Used under license from Shutterstock.com.*

Historical Background and Scientific Foundations

About 96 million years ago, Australia's continental plate separated from Antarctica and began to drift toward its present position. Surrounded by wide expanses of ocean, its land-dwelling animals, plants, and many of its birds have been isolated from other populations ever since. This has led to the evolution of thousands of unique Australian species, including such famous animals as koalas and kangaroos. Australia has also been inhabited by human beings, the peoples known as the Aborigines, for about 42,000 years. The continent was settled by Europeans starting in the late 1700s, leading to the extinction of at least 140 plant and animal species.

The climate of Australia is dry in the interior, with lush or semi-arid regions around much of the coast. Half the continent receives less than 11 in (300 mm) of rain per year. This makes Australian agriculture marginal in many places and particularly vulnerable to drought caused by climate change. Most of the human population lives near the southern and eastern coasts.

In 2007, the Intergovernmental Panel on Climate Change (IPCC) released the report of its Working Group II, *Climate Change 2007: Impacts, Adaptation and Vulnerability*. The report noted the following points, among others, about the impacts of climate change on Australia and New Zealand:

- Regional climate change has already begun (stated by the IPCC with "very high confidence"). There has been 0.29–1.26°F (0.4–0.7°C) warming in the region. Australian droughts are more intense. There is less rain in southern and eastern Australia and northeastern New Zealand. There are more heat waves and fewer frosts.
- Impacts from climate change are already occurring. These include stressed water supplies and agriculture, reduced seasonal snow cover, shrinking glaciers, and changes in ecosystems.
- In the twenty-first century, Australia's climate is virtually certain to be warmer. There will be more fires, heat waves, floods, landslides, droughts, and storm surges. Western New Zealand is likely to receive more rain.
- Significant loss of biodiversity (number of species) is likely in the Great Barrier Reef, the Queensland Wet Tropics, and other ecologically rich areas.

Impacts and Issues

In the early 2000s, Australia experienced a six-year drought that imperiled water supplies and Australian agriculture, prompting the country's prime minister to announce in early 2007 that irrigation water for farms might be cut off to preserve drinking-water supplies for cities. The official suggested that people begin to pray for rain.

Rivers dried up in the large Murray-Darling basin, where much Australian agriculture is located, causing the country to consider importing food for the first time in its history. Some Australian climatologists, while warning that individual weather events can never be linked definitively to global climate change, said that there was a high chance that the drought did reflect global warming. Wayne Meyer, professor of natural resources science at Adelaide University, said that "on the balance of evidence from southern Australia, rainfall patterns appear to have shifted.... There's no question about the evidence in terms of increased temperature. We have seen this persistent increase in temperature over the last 30 or 50 years. All the projections are that that will continue." Meyer also said that Australia's geography, including warm climate, deserts, and flatness, made it particularly vulnerable to climate change. "We are the ones that are going to be at the forefront because we're less buffered."

Australian public opinion in favor of the reality and dangerousness of global climate change runs higher than in any other country. About 69% of Australians believe that global warming is a critical problem, compared to 46% in the United States, according to a 2007 poll of 17 countries by the Chicago Council on Public Affairs and WorldPublicOpinion.org.

WORDS TO KNOW

CONTINENTAL PLATE: Solid segment of Earth's rocky crust that moves about independently. The motion of plates is one of the basic geological processes shaping Earth's surface and is the cause of continental drift, which over millions of years can radically reshape regional and global climate.

DROUGHT: A prolonged and abnormal shortage of rain.

GREAT BARRIER REEF: World's largest coral reef system, located along the northeastern coast of Australia. The reef system, which is approximately 1,600 mi (2,600 km) long, contains 3,000 individual reefs and 900 islands and is threatened by sea-level rise and ocean warming caused by anthropogenic climate change.

■ Primary Source Connection

This extract from a publication of the Australian government shows the Australian government's official view on some of the impacts of climate change on Australia's ecosystems. Although this report acknowledges these likely impacts, the Australian government was initially one of the most reluctant industrialized nations to take action to reduce greenhouse emissions. In December 2007, Kevin Rudd, the newly elected prime minister of Australia, signed the paperwork to ratify the Kyoto Protocol in his first official act as prime minister.

POTENTIAL IMPACTS OF CLIMATE CHANGE: AUSTRALIA

4.1 Water Supply and Hydrology...

4.1.3 Water allocation and policy

Until recently, water planning in Australia was driven by demand and controlled by engineers, not by economics or ecological considerations. This situation has changed with growing population and demand (rural and urban/industrial), including rapid growth in irrigation of high-value crops such as cotton and vineyards. There also is an increasing awareness of stress on riverine ecosystems resulting from reduced mean flows, lower peak flows, and increasing salinity and algal blooms. Higher temperatures and changed precipitation due to climate change would generally exacerbate these problems and sharpen competition among water users. In 1995, the Council of Australian Governments reviewed water resource policy in Australia and agreed to implement a strategic framework to achieve an efficient and sustainable water industry through processes to address water allocations, including provision of water for the environment and water-trading arrangements....

4.1.4 Inland and coastal salinisation

Natural salinity and high water tables have been present in Australia for centuries. However, because of changes in land management—notably land clearing and irrigation—salinity is now a major environmental issue in Australia. About 2.5 Mha are affected in Australia, with the potential for this to increase to 12.5 Mha in the next 50 years. Much of this area covers otherwise productive agricultural land. The area damaged by salinity to date represents about 4.5% of presently cultivated land, and known costs include A$130 million annually in lost agricultural production, A$100 million annually in damage to infrastructure (such as roads, fencing, and pipes), and at least A$40 million in lost environmental assets. The average salinity of the lower Murray River (from which Adelaide draws much of its water supply) is expected to exceed the 800 EC threshold for desirable drinking water about 50% of the time by 2020.

Although climate is a key factor affecting the rate of salinisation and the severity of impacts, a comprehensive assessment of the effects of climate change on this problem has not yet been carried out. Revegetation policies and associated carbon credit motivational policies designed to increase carbon sinks are likely to have a significant impact on recharge and runoff. Global warming and dryland

IN CONTEXT: AUSTRALIA ADOPTS KYOTO

The Kyoto Protocol of the United Nations Framework Convention on Climate Change (UNFCCC) is part of an international treaty on climate change. The Kyoto Protocol (or Kyoto Amendment) binds participating nations to reduce overall greenhouse-gas emissions or limit emission growth. At that time the protocol entered into force, the United States and Australia were the two largest, developed nations that did not ratify it. Australia made a commitment to lower greenhouse-gas emissions at the Kyoto conference, but originally declined to be bound by the treaty amendment.

Under the Kyoto Protocol, Australia was to cut its greenhouse-gas emission growth to 108% of its stated 1990 levels. Australian delegates successfully negotiated for Australia to be permitted an increase in emissions, arguing that the nation occupies a large, isolated land area. Furthermore, unlike the European Union countries, Australia did not have partner nations with which it could finance and share its reduction burden.

Australia remains one of the largest per capita emitters of greenhouse gases in the world. The UNFCCC reports that Australia's greenhouse-gas emissions in 2004 were 125% of their 1990 levels. However, Australian government and several independent environmental watch groups assert that changes in pollution laws and new emissions trading schemes will help lower Australia's greenhouse-gas emissions by 2012.

When Australia's new government took office in December 2007, one of the first acts of Prime Minister Kevin Rudd was to adopt the Kyoto Protocol. Australia's shift in policy and ratification comes nearly three years after the protocol entered into force, but Australia is committed to meeting the protocol's 2012 emissions reduction targets.

salinity policies need to be coordinated to maximise synergistic impacts. . . .

4.1.5 Water quality

Water quality would be affected by changes in biota, particularly microfauna and flora; water temperature; carbon dioxide concentration; transport processes that place water, sediment, and chemicals in streams and aquifers; and the timing and volume of water flow. More intense rainfall events would increase fast runoff, soil erosion, and sediment loadings, and further deforestation and urbanisation would tend to increase runoff amounts and flood wave speed. These effects would increase the risk of flash flooding, sediment load, and pollution (Basher et al., 1998). On the other hand, increases in plantation and farm forestry—in part for carbon sequestration and greenhouse mitigation purposes—would tend to reduce soil erosion and sediment loads . . ..

4.2 Ecosystems and Conservation

. . . .Until recent settlement, Australia was isolated for millions of years, and its ecosystems have evolved to cope with unique climate and biological circumstances. Australia has a very limited altitude range and is bounded to the south by ocean, which limits the potential for migration of species. Despite large year-to-year climatic variability, many Australian terrestrial species have quite limited ranges of long-term average climate, of about 1 to 2°C in temperature and 20% in rainfall. Thus, many Australian species have evolved to cope with large year-to-year variability, but not to long term change in the average climate. Australian ecosystems are therefore vulnerable to climatic change, as well as to other threats including invasion by exotic animals and plants.

Rapid land clearance and land-use change have been occurring as a result of human activity in Australia, subsequent to Aboriginal arrival tens of thousands of years ago, and especially since European settlers arrived a little over 200 years ago. This has led to loss of biodiversity in many ecosystems. One of the major impacts has been an increase in weedy species and animal pests; this is likely to continue, and to be exacerbated by climate change. Land-use change also has led to fragmentation of ecosystems and to salinisation through rising water tables. These trends can inhibit natural adaptation to climate change via the dispersal/migration response. Some systems and species may therefore become more vulnerable, and some might become extinct. Such problems have been identified by Pouliquen-Young and Newman (1999) in relation to fragmented habitat for endangered species in the south-west of Western Australia, which suggests that survival of threatened species may require human intervention and relocation. In the highly diverse tropical rainforests of northern Queensland, Hilbert et al. (2001a) predict highland rainforest environments will decrease by 50% with only a 1°C increase in temperature.

Many of Australia's wetlands, riverine environments, and coastal and marine systems are also sensitive to climate variations and changes. A key issue is the effect on Australia's coral reefs of greenhouse-related stresses in addition to non-climatic features, such as overexploitation and increasing pollution and turbidity of coastal waters from sediment loading, fertilisers, pesticides, and herbicides. An increase in temperature of 2°C will likely modify tropical near-shore communities from coral to algal dominated communities with major implications for reef biodiversity.

Hughes (2003) presents a comprehensive review of climate change impacts on Australian ecosystems projected to result from climate change. She finds that climate change may have a significant impact on most vegetation types that have been modelled to date, although the generally positive effect of increases in atmospheric carbon dioxide has often not been included in models....

4.2.2 Forests and woodlands

In Australia, some 50% of the forest cover in existence at the time of European settlement still remains, although about half of that has been logged....

The present temperature range for 25% of Australian Eucalyptus trees is less than 1°C in mean annual temperature. Similarly, 23% have ranges of mean annual rainfall of less than 20% variation. The actual climate tolerances of many species are wider than the climate envelope they currently occupy (due to effects of soil, competition and other factors) and may be affected by increasing carbon dioxide concentrations. Nevertheless, if present-day boundaries even approximately reflect actual thermal or rainfall tolerances, substantial changes in Australian native forests may be expected with climate change. Climate change-induced modifications to vegetation composition, structure and productivity will likely have flow on effects to other components of biodiversity through alterations in the quality and quantity of habitat available to vertebrate and invertebrate fauna. Howden and Gorman (1999) suggest that adaptive responses would include monitoring of key indicators, flexibility in reserve allocation, increased reserve areas and reduced fragmentation....

Climate change and elevated levels of atmospheric carbon dioxide may have important impacts on the ecosystems and biodiversity in the rainforests of northern Australia. For example, the extent of highland rainforest environments may decrease by 50% with a 1°C temperature increase. Lowland rainforest environments, however, are expected to increase. Modelling suggests that many of the region's endemic vertebrates, which mostly occur in the uplands, will experience habitat fragmentation and eventually lose all climatically suitable habitat. For example, the golden bowerbird loses 63% of its current habitat with 1°C of warming and only 2% of its habitat remains after 3°C of warming. Habitat for fauna using the drier parts of the landscape may be more affected by changes in rainfall. Modelling indicates that habitat for the endangered northern bettong would decline in the tropical north if climate warming is accompanied by greater precipitation. Conversely, available habitat for this marsupial may increase if rainfall decreases with warming. The effects of elevated carbon dioxide in conjunction with warming are largely unknown but Kanowski (2001) suggests that populations of native folivores (leaf-eaters) may decline in abundance due to changes in foliage chemistry and composition that accompany increased atmospheric carbon dioxide. Analyses of long term permanent forest plots suggest that tree mortality rates increase and that the stocks of biomass and carbon in rainforests decrease with increased temperature....

4.2.3 Rangelands

Rangelands are important for meat and wool production in Australia. In their natural state, rangelands are adapted to relatively large short term variations in climatic conditions (mainly rainfall and temperature). However, they are also under stress from human activity, mostly as a result of animal production, introduced animals such as rabbits, inappropriate management, and interactions between all of these factors....

According to the review by Hughes (2003), the interaction between elevated carbon dioxide and water supply will be especially critical for grasslands and rangelands where about 90% of the variance in primary production can be accounted for by annual precipitation. Sensitivity studies by Hall et al. (1998) have indicated that a doubling of carbon dioxide may reduce the potentially negative effects of a combined higher temperature/ reduced rainfall scenario on the carrying capacity of rangelands. Simulations by Howden et al. (1999b, d) for native pastures showed that the beneficial effects of doubling CO_2 are relatively stronger in dry years, but that nitrogen limitations may reduce the potential benefits. Positive effects of carbon dioxide are predicted to balance a 10% reduction in rainfall but greater rainfall decreases would result in reduced productivity. Some limited changes in the distributions of C3 and C4 grasses are also suggested, although this will be moderated by any temperature change. Any increase in pasture growth, especially after high rainfall events, is likely to increase burning opportunities that in turn will affect carbon stores and future greenhouse gas emissions....

4.2.4 Alpine systems

Basher et al. (1998) conclude that alpine systems in Australia are particularly susceptible to climate change. Despite the fact that they cover only a small area, they are important for many plant and animal species, many of which are listed as threatened. These systems are also under pressure from tourism activities. The Australian Alps have a relatively low altitude (maximum about 2,000 m), and much of the Alpine ecosystem area and ski fields are marginal. Most year-to-year variability is related to large fluctuations in precipitation, but interannual temperature variations are small compared to warming anticipated in the 21st century. Studies by Hewitt (1994), Whetton et al. (1996b) and Whetton (1998) all point to a high degree of sensitivity of seasonal snow cover duration and depth. For Australia, Whetton (1998) estimates, for the full range of CSIRO (1996a) scenarios, an 18–66% reduction in the total area of snow cover by 2030 and a 39–96% reduction by

2070. This would seriously affect the range of particular alpine ecosystems and species. Decreases in precipitation and increased fire danger would also adversely affect alpine ecosystems. . . .

4.2.5 Wetlands

The State of the Environment Report (1996) states, "Wetlands continue to be under threat, and large numbers are already destroyed." For example, Johnson et al. (1999) estimate wetland loss of about 70% in the Herbert River catchment of Northern Queensland between 1943 and 1996. In the Murray-Darling Basin, the quality of wetlands has been significantly reduced, particularly between the Hume Dam and Mildura. Hydrological condition in the river channel is poor for all areas, with the extent, timing and duration of floodplain inundation all significantly affected.

Wetland loss is caused by many processes, including water storage; hydroelectric and irrigation schemes; dams, weirs, and river management works; desnagging and channelisation; changes to flow, water level, and thermal regimes; removal of instream cover; increased siltation; toxic pollution and destruction of nursery and spawning or breeding areas; and use of wetlands for agriculture. Climate change will add to these factors through changes in inflow, increased water losses, and changes to soil and bank erosion rates due to increases in drought and heavy rainfall events.

Specific threats to wetlands from climate change and sea level rise have been studied as part of a national vulnerability assessment. The best example is provided for Kakadu National Park in northern Australia. World Heritage and Ramsar Convention-recognised freshwater wetlands in this park could become saline, given current projections of sea level rise and climate change. Projected impacts for Kakadu raise the possibility that many other Australian coastal wetlands could be similarly affected. Some of these wetlands may be unable to migrate upstream because of physical barriers in the landscape. . . .

4.2.7 Coastal and marine systems

Australia has some of the finest and most extensive coral reefs in the world. These reef systems stretch for two thousand kilometres along the north-east coast. Others exist off-shore of the Kimberley coast and further south at the Monte Bello and Dampier Islands. Coral reefs in the Australian region are subject to greenhouse related stresses including increasingly frequent bleaching episodes, changes in sea level, and probable decreases in calcification rates as a result of changes in ocean chemistry.

Mass bleaching has occurred on several occasions in Australia's Great Barrier Reef (GBR) and elsewhere since the 1970s. Particularly widespread bleaching, leading to the death of some corals, occurred globally in 1997–98 in association with a major El Niño event, and again in 2002. Bleaching was severe on the inner GBR but less severe on the outer reef in 1997–98. This episode was associated with generally record-high sea surface temperatures (SSTs) over most of the GBR region. This was a result of global warming trends resulting from the enhanced greenhouse effect and regional summer warming from the El Niño event, the combined effects of which caused SSTs to exceed bleaching thresholds. Three independent databases support the view that 1997–98 SST anomalies were the most extreme in the past 95 years and that average SSTs off the north-east coast of Australia have significantly increased from 1903 to 1994. Lowered seawater salinity as a result of flooding of major rivers between Ayr and Cooktown early in 1998 is also believed to have been a major factor in exacerbating the effects in the inshore GBR. Solar radiation, which is affected by changes in cloud cover and thus by El Niño, may also have been a factor.

"POTENTIAL IMPACTS OF CLIMATE CHANGE: AUSTRALIA." *CLIMATE CHANGE—AN AUSTRALIAN GUIDE TO THE SCIENCE AND POTENTIAL IMPACTS*. <HTTP://WWW.GREENHOUSE.GOV.AU/SCIENCE/GUIDE/PUBS/CHAPTER4.PDF> (ACCESSED NOVEMBER 27, 2007).

SEE ALSO *Africa: Climate Change Impacts; Arctic People: Climate Change Impacts; Asia: Climate Change Impacts; Europe: Climate Change Impacts; North America: Climate Change Impacts; Small Islands: Climate Change Impacts; South America: Climate Change Impacts.*

BIBLIOGRAPHY

Books

Parry, M. L., et al, eds. *Climate Change 2007: Impacts, Adaptation and Vulnerability: Contribution of Working Group II to the Fourth Assessment Report of the Intergovernmental Panel on Climate Change.* New York: Cambridge University Press, 2007.

Periodicals

Noticewala, Sonal. "At Australia's Bunny Fence, Variable Cloudiness Prompts Climate Study." *New York Times*, August 14, 2007.

Web Sites

"Climate Change—An Australian Guide to the Science and Potential Impacts." *Australian Greenhouse Office*, 2003. <http://www.greenhouse.gov.au/science/guide/index.html> (accessed September 16, 2007).

Coorey, Madeleine. "Australian Drought Linked to Global Warming." *TerraDaily*, April 20, 2007. <http://www.terradaily.com/reports/Australian_Drought_Linked_To_Global_Warming_999.html> (accessed September 16, 2007).

Larry Gilman

Automobile Emissions

■ Introduction

Automobiles and trucks are major contributors to global climate change—17.8% of the world's energy-related greenhouse gas emissions come from road vehicles, including 10.2% from automobiles. There are over half a billion automobiles in the world today, almost all running on gasoline. Burning 1 gallon (3.8 liters) of gasoline releases 19.4 lb (8.8 kg) of carbon dioxide (CO_2) into the atmosphere, along with a number of other pollutants. The average car emits its own weight in CO_2 every year.

Emissions from automobiles are rising worldwide, as car ownership increases in the developing world and average vehicle size increases in the developed world. Increasing efficiency through decreased size, mileage increases, hybrid vehicles, and diesel engines is one way of decreasing greenhouse emissions from automobiles. Also, automobile usage may be reduced through public and alternative transport, zoning cities and other settlements so as to minimize commuting and traffic jams, and other measures.

Biofuels, which do not make a direct contribution to greenhouse gas emissions, are another option for reducing automobile emissions. Biofuels are controversial, however, because the agricultural sector (which would provide the feedstocks with which to make the biofuels) is itself fossil-fuel-intensive and causes other environmental damage as well, such as soil loss and oceanic dead zones.

■ Historical Background and Scientific Foundations

The first experimental automobiles were built in the mid-nineteenth century, and the first practical models were built in the 1880s. They remained a rare luxury item for several more decades, but in the early twentieth century a rapid shift to personal automobile ownership and reliance on motor-powered trucks for short- and mid-range transport of goods occurred. The lowering of automobile prices—accomplished through assembly-line techniques for mass production first applied on a large scale by American industrialist Henry Ford (1863–1947)—was crucial to this transition. Ford's Model T automobile was first marketed in 1908. Eight years later Ford's company had sold half a million Model Ts and the price of the car was down to a mere $316 ($6,100 in 2006 dollars).

Although some early cars were powered by hydrogen, coal, or alcohol, gasoline soon came to completely dominate the automobile and truck world and continues to do so today with rare exceptions. Industrial society's dependence on road transport, and of road transport on petroleum, has resulted both in large-scale emissions of greenhouse gases and in strained international relations resulting occasionally in conflict in order to maintain access to petroleum resources. (Many of these petroleum resources are not located in the industrialized countries that consume the most oil.)

Analysts divide the world's economic activity into a number of sectors, such as buildings, industry, energy, forestry, agriculture, and transportation. Transportation, in turn, is broken down into rail, air, sea shipping, and ground transport, and ground transport is broken into five vehicle categories, namely light-duty vehicles (automobiles and small trucks and vans), two-wheelers (motorcycles, scooters, and the like), heavy freight trucks, medium freight trucks, and buses. In 2004, the transport sector as a whole accounted for 26% of world primary energy use and 23% of world energy-related greenhouse-gas emissions. (Other emissions came primarily from deforestation and agriculture.) All types of ground vehicles together accounted for 74% of transport emissions, that is, 17% of world energy-related greenhouse gas emissions. Automobiles accounted for 44.5% of transport emissions, that is, 10.2% of world energy-related greenhouse gas emissions.

Transport emissions consist mostly of CO_2, but methane (CH_4) and nitrous oxide (N_2O) are also

To raise awareness about CO_2 emissions from automobiles, Greenpeace members filled up a balloon to represent how much CO_2 is released when each full tank of gas is used in a medium-sized car. *AP Images.*

emitted. CH_4 accounts for 0.1–.3% of transport greenhouse gas emissions and N_2O for 2–2.8%. In 2004, fluorine-containing compounds (F-gases), leaked from vehicular air conditioning systems, produced an amount of global warming equal to 5–10% of the warming produced by CO_2 emitted by vehicles.

About 95% of transport energy comes from petroleum-based fuels, with most of the other 5% coming from biodiesel and ethanol. The three major petroleum-based transport fuels are diesel, which supplies 31% of total transport energy; gasoline, which supplies 47%; and aviation fuel, which supplies 9%.

The transport sector's contribution to global greenhouse gas emissions almost doubled from 1971 to 2000. Most of this increase was attributable to increased emissions from road transport. Two main factors led to this increase. One was the increase in vehicle numbers due to urbanization (increased numbers of people living in cities) and to decentralization (spreading-out) of cities. The sprawling patterns of urban transport that result from decentralization are not easily served by public transportation systems, which are best at transporting large numbers of people between a relatively small number of destinations. Two-wheelers and cars have, therefore, increased rapidly in many parts of the world. From 1950 to 1997, the world automobile fleet grew five times faster than population, from 50 million vehicles to 580 million. Some parts of the world, notably China, are poised for even more explosive growth. Chinese automobile sales grew from 2.4 million in 2001 to 7.2 million in 2006.

Nevertheless, 64% of transport emissions come not from developing countries, such as China, but from the world's 30 wealthiest states, all members of the Organisation for Economic Co-operation and Development (OECD). About 20% of the world's population resides in the OECD countries. The other 80% of the planet's people produced, as of 2007, only 36% of greenhouse emissions from transport. The non-OECD share of transport emissions was forecast by the United Nations Intergovernmental Panel on Climate Change (IPCC) to grow to 46% by 2030, if present trends continued.

The second main reason for the rise of transport emissions since 1971 has been growth in the size, weight, and power of automobiles, especially in the industrialized world. This growth has been largely driven by the surge in sport utility vehicle (SUV) sales since the 1970s. In 1975, SUVs accounted for only 2% of U.S. vehicle sales; by 2003, the share was 23%. SUVs are heavier than cars, carry no more passengers, and get much lower fuel economy. In 2007, the American SUV fleet average was 19.2 miles per gallon (8.2 km/L), compared to 26.9 miles per gallon (11.5 km/L) for family sedans.

The average U.S. new automobile of 2005 was 27% heavier than that of 1987; was able to accelerate 30% more quickly from 0–60 miles per hour (0–97 km/h); and got 5% poorer mileage. Note that mileage did not shrink in proportion to weight and performance, thanks to improvements in technological efficiency. However, weight and performance outpaced efficiency for a net loss. If the 2005 new-car fleet had remained at 1987

WORDS TO KNOW

BIOFUEL: A fuel derived directly by human effort from living things, such as plants or bacteria. A biofuel can be burned or oxidized in a fuel cell to release useful energy.

DEFORESTATION: Those practices or processes that result in the change of forested lands to non-forest uses. This is often cited as one of the major causes of the enhanced greenhouse effect for two reasons: 1) the burning or decomposition of the wood releases carbon dioxide; and 2) trees that once removed carbon dioxide from the atmosphere in the process of photosynthesis are no longer present and contributing to carbon storage.

FOSSIL FUELS: Fuels formed by biological processes and transformed into solid or fluid minerals over geological time. Fossil fuels include coal, petroleum, and natural gas. Fossil fuels are non-renewable on the timescale of human civilization, because their natural replenishment would take many millions of years.

GREENHOUSE GASES: Gases that cause Earth to retain more thermal energy by absorbing infrared light emitted by Earth's surface. The most important greenhouse gases are water vapor, carbon dioxide, methane, nitrous oxide, and various artificial chemicals such as chlorofluorocarbons. All but the latter are naturally occurring, but human activity over the last several centuries has significantly increased the amounts of carbon dioxide, methane, and nitrous oxide in Earth's atmosphere, causing global warming and global climate change.

HYBRID VEHICLES: Vehicles that use electric motors for their sole or primary motive power but produce most or all of their own electricity using internal combustion engines.

HYDROGEN FUEL CELLS: Devices that combine hydrogen and oxygen, at a low or modest temperature, to produce electricity directly. In a fuel cell, hydrogen is not burned (combined rapidly with oxygen) to produce heat. Fuel cells are more efficient users of hydrogen than thermal systems such as hydrogen-burning internal combustion engines, but are also more expensive.

OCEANIC DEAD ZONES: Areas of the ocean where the deeper waters are devoid of life. Often occur where nutrients from agricultural fertilizers have been carried to the sea by rivers. Algae bloom (reproduce rapidly) in the nutrient-rich surface waters: when they die, they sink, and are digested by bacteria in deeper waters. These bacteria consume oxygen, resulting in low-oxygen deep waters in which all oxygen-dependent aquatic life, such as fish, dies. One of the largest hypoxic or dead zones is in the Gulf of Mexico: it is about 7,000 square mi (18,000 square km) in size, as large as the U.S. state of New Jersey, and is caused by pollution from U.S. agriculture draining into the Mississippi River.

weight and performance levels, it would have had 24% better fuel economy due to efficiency improvements. In 2006, due to higher gas prices, sales of SUVs plummeted and several of the largest, most fuel-inefficient models were discontinued.

Impacts and Issues

According to the nonprofit Pew Center on Climate Change, the fuel economy of passenger road vehicles could be increased by one-third by 2030, without reducing size or performance, through adoption of already proven technologies. (Reducing size, performance, or both would result in even greater improvements.) These technologies include gasoline or diesel hybrid cars, advanced diesel engines, and electric vehicles powered by hydrogen fuel cells. However, such technological changes are likely to occur only gradually. The speed with which more efficient vehicles become common will depend on both market forces and government regulations.

In the United States, attempts by California and other states to further mandate better mileage for automobiles and trucks have met with resistance. The federal Clean Air Act (first passed 1963) gives California the unique ability to set automobile emissions standards that are stricter than federal standards, as long as the state receives a waiver (special permission) from the U.S. Environmental Protection Agency (EPA). In 2002, California passed a state law requiring automakers who wished to sell cars in the state to produce models emitting fewer greenhouse gases. Mandatory emissions reductions would start in 2009 and increase to 25% by 2030. Eleven other states (Connecticut, Maine, Maryland, Massachusetts, New Jersey, New York, Oregon, Pennsylvania, Rhode Island, Vermont, and Washington) adopted versions of the California law that would only take effect if California's law did.

In 2005, California requested that the EPA give it a waiver so that it could impose its stricter emissions standards. In the spring of 2007, automobile manufacturers brought suit against Vermont, arguing that its pending emissions law, modeled on California's, was illegal. In September 2007, a federal judge in Vermont ruled against the automakers and in favor of the state, making it likely that all the other states' emissions laws would be upheld. As of October 2007, a suit against the new standards brought by the Association of International Automobile Manufacturers (Honda, Nissan, Toyota, and 11 other foreign car makers) was still pending in a California federal court.

Automobiles are large contributors to greenhouse gas emissions, which cause global climate change. *Image copyright Wrangler, 2007. Used under license from Shutterstock.com.*

In April 2007, after more than a year of silence from the EPA in response to California's request for a waiver, Governor Arnold Schwarzenegger announced that if the EPA had not issued the waiver in six more months the state would file suit against the EPA. In October, the deadline expired and the California attorney general announced plans to sue. The EPA rejected the waiver in December, prompting California and other states to sue in January 2008. The outcome of such legal struggles may have a significant effect on automobile emissions around the world, since manufacturers that develop low-emissions technologies for the large U.S. market would be likely to market them globally.

■ Primary Source Connection

This source from the California Environmental Protection Agency (EPA) explains the process by which the California Air Resources Board (ARB) adopts emissions standards for automobiles that are stricter than the national standard. The proposed California emissions standard would greatly reduce greenhouse gas emissions, but the United States Environmental Protection Agency has yet to grant approval to California's proposed standards.

CLIMATE CHANGE EMISSIONS STANDARDS FOR VEHICLES

What are California's Motor Vehicle Greenhouse Gas Emissions Standards?

In September 2004, the California Air Resources Board approved regulations to reduce greenhouse gas emissions from new motor vehicles, based on a law by former Assemblywoman Fran Pavley, signed in 2002. The law directed the Board to adopt regulations that achieve the maximum feasible and cost effective reduction in greenhouse gas emissions from motor vehicles. The regulations establish emission standards for new passenger vehicles and light duty trucks beginning with the 2009 model year.

What are the greenhouse gas benefits expected from the motor vehicle greenhouse gas emissions standards?

In California alone, officials estimate that the standard will reduce climate change emissions by approximately 30 million metric tons, in 2020 and over 50 million metric tons in 2030. This equates to an overall 18% reduction in climate 'change' emissions from passenger cars in 2020 and a 27% reduction in 2030. In addition, staff estimates that the regulation will reduce "upstream" smog-forming emissions of hydrocarbons and oxides of nitrogen by approximately 6 tons per day in 2020 and 10 tons per day in 2030.

Reductions from the standards for vehicles in all 12 states that have adopted California's standards will reduce greenhouse gas emissions by 74 million metric tons in 2020. Adoption by the six additional states that are considering the policy would increase the total emissions reduction to 100 million metric tons in 2020.

How many other states have adopted California[']s Motor Vehicle Greenhouse Gas Emissions Standards?

Eleven states: Connecticut, Maine, Maryland, Massachusetts, New Jersey, New York, Oregon, Pennsylvania, Rhode Island, Vermont and Washington. Six additional states are actively considering adopting the standards: Arizona, Colorado, Illinois, New Hampshire, New Mexico and North Carolina.

Why does California need a waiver from the U.S. EPA to enforce its own vehicle standards?

California is the only state, under the Federal Clean Air Act, with the unique ability to set stricter-than-federal standards for vehicles, as long as it gets a waiver from the federal government. The Federal Clean Air Act preempts state and local governments from adopting or enforcing standards to control emissions from new motor vehicles or engines. However, once California receives a waiver of preemption from the federal government, then other states can adopt California's standards.

How long has California been waiting for the U.S. EPA to grant a waiver?

In December 2005, the California Air Resources Board requested a waiver from the U.S. EPA. In April 2006 and again in October 2006, Governor Arnold Schwarzenegger followed up the request with letters urging swift action. In April 2007, Governor Schwarzenegger met with U.S. EPA Administrator Stephen Johnson to personally request assistance granting California its waiver. In April 2007, the Supreme Court ruled that the U.S. EPA must take action regarding greenhouse gas emissions. In April 2007, the U.S. EPA finally announced two public hearings to consider California's request for a waiver, both in May, one scheduled in Washington D.C. and one in Sacramento.

How does the U.S. EPA waiver process work?

Once California has determined that its state standards are as protective of the public health and safety as applicable federal standards, and has applied for a waiver, the Clean Air Act directs U.S. EPA to provide the opportunity for a hearing and then grant the waiver.

What are the reasons that the U.S. EPA would deny a waiver request?

The U.S. EPA must grant a waiver unless the federal agency makes one of three findings:

1. That California's "protectiveness finding" is arbitrary and capricious;
2. That California does not need its state standards to meet compelling and extraordinary conditions; and
3. That the state standards are not consistent with section 202(a)-part of the Clean Air Act provisions on U.S. EPA's adoption of motor vehicle emission standards. Section 202(a) indicates that (i) the California Standards must provide manufacturers with adequate lead time, taking the cost of compliance into account, and (ii) manufacturers must be able to put a vehicle through one set of tests to determine compliance with both the state and federal standards.

How many waiver requests has U.S. EPA approved or denied?

Since 1968, EPA has granted about 50 new (or "full") waivers and about 40 determinations that amendments were within the scope of prior waivers. Five waivers have been denied, the last in 1975. Since 1975, there have been instances where ARB made modifications to regulations where a waiver was pending or, in the case of the Zero Emission Vehicle regulations, where a waiver was granted only until model year 2011, meaning ARB will have to do a new waiver application for model years 2012 and beyond.

Does the technology exist to meet the vehicle emissions standards?

Yes. Technology is in use today to meet the motor vehicle greenhouse gas emissions standards. Even in 2002 when the law was signed, technology was available that could be used by auto manufacturers to reduce their fleet average emissions.

Does the U.S. EPA have the authority to regulate greenhouse gases?

Yes. The U.S. Supreme Court recently ruled that the U.S. EPA had the authority to regulate greenhouse gases, and the obligation to review the mounting scientific data. The court decision countered the U.S. EPA's argument that it could not regulate greenhouse gas emissions because of the "substantial scientific uncertainty" about the harmful effects of greenhouse gas emissions.

In its ruling, the Supreme Court stated that, gases such as carbon dioxide "act like a ceiling of a greenhouse, trapping solar energy and retarding the escape of reflected heat." The Supreme Court indicated that the effects of greenhouse gases on climate and weather are covered under the Clean Air Act as such effects threaten human welfare. Following the ruling, Governor Schwarzenegger expected the U.S. EPA to move quickly now in granting the ARB's request for a waiver.

Does the President's Executive Order from 5/14/07 affect California[']s request for a waiver of preemption?

No. California's standing under the Clean Air Act does not relieve the U.S. EPA from taking swift action to grant a waiver. Although it appears the Executive Order is an effort by the Federal government to interfere with California's clear right under the Clean Air Act to regulate greenhouse gas pollutants from vehicles, the ARB remains optimistic that the U.S. EPA will grant the waiver.

CALIFORNIA EPA. "*CLIMATE CHANGE EMISSIONS STANDARDS FOR VEHICLES: ACTIONS TO REDUCE GREENHOUSE GASES FROM CARS AND TRUCKS.*" MAY 2007. <HTTP://WWW.ARB.CA.GOV/CC/FACTSHEETS/CCFAQ.PDF> (ACCESSED NOVEMBER 30, 2007).

SEE ALSO *Aviation Emissions and Contrails; Biofuel Impacts; Petroleum: Economic Uses and Dependency; United States: State and Local Greenhouse Policies.*

BIBLIOGRAPHY

Books

Metz, B., et al, eds. *Climate Change 2007: Mitigation of Climate Change: Contribution of Working Group III to the Fourth Assessment Report of the Intergovernmental Panel on Climate Change*. New York: Cambridge University Press, 2007.

Periodicals

Egelko, Bob. "California to Sue Bush Administration Over Law to Limit Emissions." *San Francisco Chronicle*, October 20, 2007.

Freeman, Sholnn. "States Adopt California's Greenhouse Gas Limits." *Washington Post*, January 3, 2006.

Web Sites

"California AG to sue EPA to Force Emissions Decision." *MSNBC*, October 23, 2007. <http://www.msnbc.msn.com/id/21428148/> (accessed November 5, 2007).

"Climate Change Emissions Standards for Vehicles." *California Environmental Protection Agency*. May 2007. <http://www.arb.ca.gov/cc/factsheets/ccfaq.pdf> (accessed November 5, 2007).

"Emission Facts: Greenhouse Gas Emissions from a Typical Passenger Vehicle." *U.S. Environmental Protection Agency (EPA)*, February 2005. <http://www.epa.gov/otaq/climate/420f05004.htm> (accessed November 5, 2007).

Greene, David L., and Andreas Schafer. "Reducing Greenhouse Gas Emissions from U.S. Transportation." *Pew Center on Global Climate Change*, May 2003. <http://www.pewclimate.org/docUploads/ustransp.pdf> (accessed November 5, 2007).

Meltz, Robert. "The Supreme Court's Climate Change Decision: *Massachusetts v. EPA*." *Congressional Research Service Report for Congress*, May 18, 2007. <http://www.ncseonline.org/NLE/CRSreports/07Jun/RS22665.pdf> (accessed October 15, 2007).

Larry Gilman

Aviation Emissions and Contrails

Introduction

Jet aircraft contribute to global climate change by producing carbon dioxide, ozone, and aerosol particles and by triggering the formation of thin, high-altitude clouds termed cirrus clouds. Cirrus clouds reflect sunlight back up into space, which tends to cool Earth, but also reflect infrared radiation back down toward the ground, which tends to warm Earth. The warming effect is greater than the cooling effect even during the day. At night, there is only warming, no cooling.

Although aviation is presently a relatively small contributor to global climate change, the number of commercial flights is growing rapidly in much of the world. Barring technical breakthroughs or changes in the timing and number of flights, aviation will contribute much more significantly to climate change over the next several decades. Unrestrained increases in air traffic have the potential to offset or completely overwhelm reductions in greenhouse-gas emissions in other industries.

Historical Background and Scientific Foundations

Civilian passenger jets began flying in 1948, replacing most propeller-driven passenger aircraft in the 1960s.

Contrails are shown streaming from a jumbo jet as it soars across the sky. *Image copyright Lowell Sannes, 2007. Used under license from Shutterstock.com.*

Environmental activists gathered near the site of a proposed new runway at London's Heathrow Airport in August 2007 to protest against the expansion. They cited concerns about the impact of increased aviation on the environment. *AP Images.*

Air travel grew at an average rate of 9% per year from 1960 to 2000; continued growth at 5% per year is forecast through at least 2015. As of 2007, the world aviation industry was carrying more than 2 billion passengers per year, almost all on jet aircraft.

Jet aircraft burn large amounts of fuel. For example, a Boeing 767 carrying 180 passengers emits about 1 ton of carbon dioxide (CO_2) per passenger during a flight from London to Washington, D.C. About 7% to 8% of jet engine exhaust is CO_2 and water vapor, both products of fuel combustion. About 0.5% of the exhaust is made of various compounds of nitrogen and oxygen (NO_x), compounds of sulfur and oxygen (SO_x), carbon monoxide (CO), sooty particles consisting mostly of unburned carbon, and trace pollutants such as metals. The remaining 91.5% to 92.5% of the exhaust consists of the major ingredients of normal air, oxygen, and nitrogen, and does not contribute to climate change.

Under the right atmospheric conditions, when hot jet exhaust mixes with the surrounding air, its water content condenses on the surfaces of small, airborne particles. These particles may be present in the atmosphere, the jet exhaust, or both. This causes large numbers of tiny ice crystals to form behind the engine, which are often visible as a narrow trail of white cloud. This cloud is termed a condensation trail or contrail for short. Contrails can also be formed by the passage of humid air over the wings, but engine exhaust is the dominant cause of contrails.

When a contrail forms in humid air, it can grow and persist for hours. Such a contrail is called a persistent contrail. Persistent contrails can be hundreds of miles long and extend more than a mile in width and 650–1,300 ft (200–400 m) in thickness.

Persistent contrails have contradictory affects on the temperature of Earth. On one hand, because they are white, they tend to reflect sunlight back into space, which cools Earth. On the other hand, they tend to reflect infrared radiation back down toward Earth, which keeps Earth warmer. Scientists calculate that the warming effect is stronger than the cooling effect even during the day (at night, only the warming effect occurs). Therefore, persistent contrails contribute to global warming.

The effect of clouds on global warming is one of the largest remaining sources of uncertainty in climate science. Furthermore, because it is impossible to distinguish fully developed persistent contrails from natural cirrus clouds—whose formation may be influenced by natural causes, climate change, aerosols from non-aviation sources, and other factors—it is hard to know exactly how much cloudiness is being caused by jets. In 2007, the United Nations' Intergovernmental Panel on Climate Change (IPCC) cut its 2001 estimate of how much global warming is caused by aviation in half. It also said that there was still much uncertainty over how much cirrus-cloud formation should be attributed to aviation.

When a contrail forms in dry air, it does not form a persistent contrail, but quickly evaporates into invisible water vapor. In this case, the jet engine's contribution to global climate change is governed only by the amount of CO_2, tiny particles (aerosols), and other pollutants it

WORDS TO KNOW

AEROSOL: Particles of liquid or solid dispersed as a suspension in gas.

CIRRUS CLOUD: Thin clouds of tiny ice crystals that form at 20,000 ft (6 km) or higher. Cirrus clouds cover 20–25% of the globe, including up to 70% of tropical regions. Because they both reflect sunlight from Earth and reflect infrared (heat) radiation back at the ground, they can influence climate.

CONTRAIL: High-altitude cloud formed by the passage of an aircraft. Most contrails are formed by the condensation of water vapor in jet exhaust around small particles in the ambient air, the exhaust, or both. Contrails alter the cloud content of the atmosphere, but their contribution to global climate change is uncertain.

GREENHOUSE GASES: Gases that cause Earth to retain more thermal energy by absorbing infrared light emitted by Earth's surface. The most important greenhouse gases are water vapor, carbon dioxide, methane, nitrous oxide, and various artificial chemicals such as chlorofluorocarbons. All but the latter are naturally occurring, but human activity over the last several centuries has significantly increased the amounts of carbon dioxide, methane, and nitrous oxide in Earth's atmosphere, causing global warming and global climate change.

NITROGEN OXIDES: Compounds of nitrogen and oxygen such as those that collectively form from burning fossil fuels in vehicles.

STRATOSPHERIC OZONE LAYER: Layer of Earth's atmosphere from about 9–22 mi (15–35 km) above the surface in which the compound ozone (O_3) is relatively abundant. Ozone absorbs ultraviolet light from the sun, shielding the surface from this biologically harmful form of light and becoming heated in the process. Heating by ultraviolet light warms the stratosphere. Depletion of ozone in the ozone layer is caused by chlorofluorocarbons and some other long-lived artificial chemicals. Effects of ozone depletion on Antarctic climate have been documented.

WATER VAPOR: The most abundant greenhouse gas, it is the water present in the atmosphere in gaseous form. Water vapor is an important part of the natural greenhouse effect. Although humans are not significantly increasing its concentration, it contributes to the enhanced greenhouse effect because the warming influence of greenhouse gases leads to a positive water vapor feedback. In addition to its role as a natural greenhouse gas, water vapor plays an important role in regulating the temperature of the planet because clouds form when excess water vapor in the atmosphere condenses to form ice and water droplets and precipitation.

emits. Aerosols from jet exhaust may linger at high altitudes, encouraging later cloud formation, but no good estimate of the size of this effect was available as of fall 2007. Also, the NO_x in jet exhaust encourages the formation of ozone, a greenhouse gas and toxic pollutant. (Ozone is only beneficial when it resides in the stratospheric ozone layer, which forms at about 9.3–21.7 mi [15–35 km]—higher than commercial aircraft go). These non-contrail effects on climate also occur when a contrail does form.

Because of its chemical composition, jet exhaust is two to four times more effective at causing global warming than if it consisted of carbon dioxide alone. Because it is released at high altitude—mostly between 5.6 and 8 mi (9 and 13 km)—it is two times more effective at causing global warming than if it were burned at ground level.

Impacts and Issues

As of 2007, scientists estimated that by 2050 aviation would contribute about 5% of the human-caused energy input to global warming.

Aviation's rapid growth threatens to overwhelm efforts to cut greenhouse-gas emissions in other areas. The British aviation industry, for example, is expected to double the number of passengers it carries to 475 million by 2030, with emissions growing at about 7% per year, and China's airline passenger count grew 28% in 2003 and 2004. British analysts said that for Britain to meet its official target of cutting greenhouse-gas emissions in 2050 to 60% below 1990 levels, all non-aviation sectors of the economy would have to emit zero greenhouse gases to make up for aviation growth.

There are a number of strategies for reducing aviation's impact on climate. Contrail formation, for example, could be reduced by operating aircraft differently. Reducing average cruise altitude by 6,000 ft (1,830 m) would reduce contrail formation by about 45%. However, because air is denser at lower altitudes and presents more resistance to aircraft, this would also increase fuel consumed by about 6%. In 2006, British scientists concluded that 60–80% of climate warming from contrails is caused by night flights and that about 50% of warming from contrails is caused by winter flights. Night flights are only 25% of all flights, winter flights only 22%. Rescheduling flights from night to day and discouraging winter flying could reduce aviation's contribution to global warming.

Aviation experts are also urging the more rapid introduction of more efficient jet engines that burn less fuel.

IN CONTEXT: CLIMATE CAMP PROTESTS EXPANSION

In August 2007, hundreds of demonstrators gathered outside of Heathrow Airport in London, England, as part of the week-long Camp for Climate Action. The climate camp participants sought to highlight commercial aviation's greenhouse-gas emissions and the overall impact of using airplanes for trade and tourism. In addition to living outside of the airport's perimeter fence for a week, camp participants employed a series of direct action protects. Demonstrators chained and posted closed signs on dozens of London area travel agencies. Other participants blocked entrances or occupied buildings of the transit authorities, carbon offset companies, private jet airports, and commercial aviation companies. Some of the camp's leaders initially expressed their intent to use direct action tactics to block runways and disrupt airport operations within Heathrow, sparking government concern over passenger and airline security.

However, the demonstration remained outside of the airport's perimeter fence and instead focused on urging people to limit airline leisure travel, use technology such as videoconferencing to replace business travel, and eat locally grown foods whenever possible. A major target of the camp was a proposal to add another runway as part of the ongoing Heathrow expansion project. Camp for Climate Action participants argued that expanding Heathrow's facilities would further increase the number of flights emitting pollutants into the upper atmosphere.

During the week of the climate camp, an estimated 1.5 million passengers passed through Heathrow Airport. Many critics of the Camp for Climate Change noted that although improvements should be made in the aviation industry to reduce emissions, tourism and trade continue to fuel many of the world's economies.

However, increased engine efficiency can have paradoxical results: more-efficient engines produce exhaust with higher relative humidity, which can increase contrail formation. They also tend to produce more NO_x because they burn fuel at higher temperatures and pressures. To significantly reduce aviation's contribution to global climate change, radically more efficient aircraft may be needed. One such design, the SAX-40, was unveiled by the Silent Aircraft Initiative of Cambridge University in the United Kingdom and the Massachusetts Institute of Technology in the United States in 2006. The 215-passenger plane, still only a concept, featured a blended wing-body design and efficient engines.

Even such radically novel aircraft might not reduce aviation's impact on climate change sufficiently. The European Commission, executive branch of the European Union, proposed in 2006 to include aviation in its Europe-wide carbon-trading scheme. As of 2007, the U.S. government had announced no plan to mandate decreased greenhouse-gas emissions or contrail formation by the aviation industry.

■ Primary Source Connection

As evidenced by excerpts from the U.S. Environmental Protection Agency (EPA) factsheet that follows, the EPA uses and relies on the data, reports, and assessments of the Intergovernmental Panel on Climate Change (IPCC). For example, in forming its public statements, the EPA relied on a 1999 IPCC report, "Aviation and the Global Atmosphere," to help form its key assertions that "Aviation's overall potential for influencing climate was recently assessed to be approximately 3.5 percent of the potential from all human activities."

The IPCC was established by the World Meteorological Organization (WMO) and the United Nations Environment Programme (UNEP) in 1988 to assess the science, technology, and socioeconomic information needed to understand the risk of human-induced climate change.

AIRCRAFT CONTRAILS FACTSHEET

What are contrails?

Contrails are line-shaped clouds or "condensation trails," composed of ice particles, that are visible behind jet aircraft engines, typically at cruise altitudes in the upper atmosphere. Contrails have been a normal effect of jet aviation since its earliest days. Depending on the temperature and the amount of moisture in the air at the aircraft altitude, contrails evaporate quickly (if the humidity is low) or persist and grow (if the humidity is high). Jet engine exhaust provides only a small portion of the water that forms ice in persistent contrails. Persistent contrails are mainly composed of water naturally present along the aircraft flight path.

How are aircraft emissions linked to contrail formation?

Aircraft engines emit water vapor, carbon dioxide (CO_2), small amounts of nitrogen oxides (NO_x), hydrocarbons, carbon monoxide, sulfur gases, and soot and metal particles formed by the high-temperature combustion of jet fuel during flight. Of these emittants, only water vapor is necessary for contrail formation. Sulfur gases are also of potential interest because they lead to the formation of small particles. Particles suitable for water droplet formation are necessary for contrail formation. Initial contrail particles, however, can either be already present in the atmosphere or formed in the exhaust gas. All other engine emissions are considered nonessential to contrail formation.

How do contrails form?

For a contrail to form, suitable conditions must occur immediately behind a jet engine in the expanding engine exhaust plume. A contrail will form if, as exhaust gases cool and mix with surrounding air, the humidity becomes high enough (or, equivalently, the air temperature becomes low enough) for liquid water condensation to occur. The level of humidity reached depends on the amount of water present in the surrounding air, the temperature of the surrounding air, and the amount of water and heat emitted in the exhaust. Atmospheric temperature and humidity at any given location undergo natural daily and seasonal variations and hence, are not always suitable for the formation of contrails.

If sufficient humidity occurs in the exhaust plume, water condenses on particles to form liquid droplets. As the exhaust air cools due to mixing with the cold local air, the newly formed droplets rapidly freeze and form ice particles that make up a contrail. Thus, the surrounding atmosphere's conditions determine to a large extent whether or not a contrail will form after an aircraft's passage. Because the basic processes are very well understood, contrail formation for a given aircraft flight can be accurately predicted if atmospheric temperature and humidity conditions are known.

After the initial formation of ice, a contrail evolves in one of two ways, again depending on the surrounding atmosphere's humidity. If the humidity is low (below the conditions for ice condensation to occur), the contrail will be short-lived. Newly formed ice particles will quickly evaporate as exhaust gases are completely mixed into the surrounding atmosphere. The resulting line-shaped contrail will extend only a short distance behind the aircraft.

If the humidity is high (greater than that needed for ice condensation to occur), the contrail will be persistent. Newly formed ice particles will continue to grow in size by taking water from the surrounding atmosphere. The resulting line-shaped contrail extends for large distances behind an aircraft. Persistent contrails can last for hours while growing to several kilometers in width and 200 to 400 meters in height. Contrails spread because of air turbulence created by the passage of aircraft, differences in wind speed along the flight track, and possibly through effects of solar heating....

Why are persistent contrails of interest to scientists?

Persistent contrails are of interest to scientists because they increase the cloudiness of the atmosphere. The increase happens in two ways. First, persistent contrails are line-shaped clouds that would not have formed in the atmosphere without the passage of an aircraft. Secondly, persistent contrails often evolve and spread into extensive cirrus cloud cover that is indistinguishable from naturally occurring cloudiness. At present, it is unknown how much of this more extensive cloudiness would have occurred without the passage of an aircraft. Not enough is known about how natural clouds form in the atmosphere to answer this question.

Changes in cloudiness are important because clouds help control the temperature of the Earth's atmosphere. Changes in cloudiness resulting from human activities are important because they might contribute to long-term changes in the Earth's climate. Many other human activities also have the potential of contributing to climate change. Our climate involves important parameters such as air temperature, weather patterns, and rainfall. Changes in climate may have important impacts on natural resources and human health. Contrails' possible climate effects are one component of aviation's expected overall climate effect. Another key component is carbon dioxide (CO_2) emissions from the combustion of jet fuel. Increases in CO_2 and other "greenhouse gases" are expected to warm the lower atmosphere and Earth's surface. Aviation's overall potential for influencing climate was recently assessed to be approximately 3.5 percent of the potential from all human activities.

Persistent line-shaped contrails are estimated to cover, on average, about 0.1 percent of the Earth's surface.... The estimate uses:

- meteorological analysis of atmospheric humidity to specify the global cover of air masses that are sufficiently humid (low enough atmospheric temperature) for persistent contrails to form
- data from 1992 reported aircraft operations to specify when and where aircraft fly

IN CONTEXT: A HISTORY OF ASSESSMENTS AND PREDICTIONS

The "Aircraft Contrails Factsheet," a September 2000 publication produced by the U.S. Environmental Protection Agency (EPA), stated the following with regard to whether persistent contrails are harmful to the public:

"Persistent contrails pose no direct threat to public health. All contrails are line-shaped clouds composed of ice particles. These ice particles evaporate when local atmospheric conditions become dry enough (low enough relative humidity). The ice particles in contrails do not reach the Earth's surface because they fall slowly and conditions in the lower atmosphere cause ice particles to evaporate.

"Contrail cloudiness might contribute to human-induced climate change. Climate change may have important impacts on public health and environmental protection."

- an estimated average for aircraft engine characteristics that affect contrail formation
- satellite images of certain regions of the Earth in which contrail cover can be accurately measured

The highest percentages of cover occur in regions with the highest volume of air traffic, namely over Europe and the United States. This estimate of contrail cloudiness cover does not include extensive cirrus cloudiness that often evolves from persistent line-shaped contrails. Some evidence suggests that this additional cirrus cloudiness might actually exceed that of line-shaped cloudiness.

How is contrail coverage expected to change in the future?

Contrail cover is expected to change in the future if changes occur in key factors that affect contrail formation and evolution. These key factors include aircraft engine technologies that affect emissions and conditions in the exhaust plume; amounts and locations of air traffic; and background atmospheric humidity conditions. Changes in engine fuel efficiency, for example, might change the amount of heat and water emitted in the exhaust plume, thereby affecting the frequency and geographical cover of contrails. Changes in air traffic might also affect persistent contrail formation. It is currently estimated that regions of the atmosphere with sufficient humidity to support the formation of persistent contrails cover about 16 percent of the Earth's surface. If air traffic in these regions increases in the future, persistent line-shaped contrail cover there will also increase. Overall, based on analysis of current meteorological data and on assumptions about future air traffic growth and technological advances, persistent contrail cover is expected to increase between now and the year 2050.

"AIRCRAFT CONTRAILS FACTSHEET." *U.S. ENVIRONMENTAL PROTECTION AGENCY (EPA)*, SEPTEMBER 2000.

SEE ALSO *Atmospheric Chemistry; Atmospheric Circulation; Atmospheric Pollution; Atmospheric Structure; Environmental Protection Agency (EPA); Intergovernmental Panel on Climate Change (IPCC); IPCC Climate Change 2007 Report.*

BIBLIOGRAPHY

Books

Solomon, S., et al, eds. *Climate Change 2007: The Physical Science Basis: Contribution of Working Group I to the Fourth Assessment Report of the Intergovernmental Panel on Climate Change.* New York: Cambridge University Press, 2007.

Periodicals

Giles, Jim. "Europe Set for Tough Debate on Curbing Aircraft Emissions." *Nature* 436 (2005): 764–765.

Lee, Joosung J., et al. "Historical and Future Trends in Aircraft Performance, Cost, and Emissions." *Annual Review of Energy and Environment* 26 (2001): 167–200.

Stuber, Nicola, et al. "The Importance of the Diurnal and Annual Cycle of Air Traffic for Contrail Radiative Forcing." *Nature* 441 (2006): 864–867.

Web Sites

"Aircraft Contrails Factsheet." *U.S. Environmental Protection Agency*, September 2000. <http://www.epa.gov/otaq/regs/nonroad/aviation/contrails.pdf> (accessed August 9, 2007).

Greener by Design: Dedicated to Sustainable Aviation. <http://www.greenerbydesign.org.uk/home/index.php> (accessed August 15, 2007).

Larry Gilman

Bali Conference

■ Introduction

The Bali Conference, hosted by the United Nations on the Indonesian island of Bali between December 3–14, 2007, brought together delegates from more than 180 nations—accompanied by thousands of observers from intergovernmental and non-governmental organizations—to start negotiations on a new climate change treaty to succeed the Kyoto Protocol, which is set to expire in 2012.

Prior to the conference, there was recognition from the U.N. and a majority of governments that the main goal of the conference was to initiate a timetable for negotiations on a new international climate change agreement, rather than to deliver a fully negotiated and agreed climate accord.

■ Historical Background and Scientific Foundations

The Kyoto Protocol, which came into effect in February 2005, marked the most significant intergovernmental treaty aimed at mitigating the effects of climate change. The agreement set out to lower overall emissions from a group of six greenhouse gases in industrialized nations by an average of 5.2% by 2012 (when compared to their 1990 levels).

Although hailed by a majority of politicians, scientists, and environmentalists as a landmark agreement, Kyoto was simultaneously criticized by many within that same constituency for being replete with flaws. It lacked the support of the United States, the world's largest polluter; there were no targets for developing nations, such as China and India; and the overall greenhouse-gas reductions simply scratched the surface of the climate change problem instead of delivering the radical cuts necessary to mitigate it. Furthermore, endless political wrangling had meant that the protocol—first agreed to in December 1997—had taken more than seven years to come into force, giving it a lifetime of less than the period of haggling that led to its passage into law. Indeed, beginning from the 1992 Earth Summit, at which the United Nations Framework Convention on Climate Change first emerged, the Kyoto process had taken nearly 13 years to pass from conception to reality.

Almost as soon as the Kyoto Protocol came into effect, therefore, the search for a viable successor treaty, to take effect in 2012 when Kyoto ended, began. Led by European Union (EU) member countries, ambitious targets for greenhouse-gas emissions were often mooted, as scientific evidence of the devastating extent of human-made climate change became overwhelming. Pressure for developing countries to be bound by emissions reductions in a post–2012 deal, also began to mount. And as the presidency of George W. Bush neared its conclusion, there were hopes that the United States would sign up to a post-Kyoto Treaty.

Staged in Bali in December 2007, the thirteenth meeting of the conference of parties of the United Nations Framework Convention on Climate Change (COP–13), was the occasion on which a timetable for negotiations for a Kyoto successor treaty was laid out. "Bali must advance a negotiating agenda to combat climate change on all fronts, including adaptation, mitigation, clean technologies, deforestation and resource mobilization," said U.N. Secretary General Ban Ki-moon in the build up to the conference. "Bali must be the political response to the recent scientific reports by the IPCC [Intergovernmental Panel on Climate Change]. All countries must do what they can to reach agreement by 2009, and to have it in force by the expiry of the current Kyoto protocol commitment period in 2012." Ban, a South Korean career diplomat and former foreign minister, succeeded Kofi Annan as U.N. secretary general on January 1, 2007. Ban identified climate change as one of the key issues facing his administration.

WORDS TO KNOW

CONFERENCE OF THE PARTIES (COP) TO THE CLIMATE CHANGE CONVENTION: Signatories to the U.N. Framework Convention on Climate Change, a treaty drafted in 1992. The treaty entered into force in 1994, and COPs have been held ever since. It was at COPs that the Kyoto Protocol was drafted.

KYOTO PROTOCOL: Extension in 1997 of the 1992 United Nations Framework Convention on Climate Change (UNFCCC), an international treaty signed by almost all countries with the goal of mitigating climate change. The United States, as of early 2008, was the only industrialized country to have not ratified the Kyoto Protocol, which is due to be replaced by an improved and updated agreement starting in 2012.

Impacts and Issues

The Bali Conference came at the end of a year in which climate change had, perhaps for the first time, been forced to the absolute forefront of the international political agenda. Earlier in 2007, former U.S. Vice President Al Gore and the Intergovernmental Panel on Climate Change shared the Nobel Peace Prize. Gore's film about global warming, *An Inconvenient Truth*, won an Oscar. Hundreds of millions of people watched or attended the Live Earth concerts. The IPCC's Fourth Assessment Report, published in June, made newspaper headlines and led news broadcasts the world over with its near apocalyptic predictions of the effects of global warming should greenhouse-gas emissions continue unimpeded.

This fed a fresh wave of resolve in government ministries across the world. In the build up to Bali, British Prime Minister Gordon Brown, German chancellor Angela Merkel, and French President Nicholas Sarkozy all expressed their desire to agree to a new international deal that would keep global warming under the widely accepted danger level of two degrees centigrade. Brown succeeded Tony Blair as British prime minister in June 2007 after a decade as chancellor of the exchequer and had expressed a longstanding determination for Britain to lead the global agenda on climate change.

On the eve of the conference, the new Australian government of Prime Minister Kevin Rudd signed up to the Kyoto Protocol after years of opposition. Even in the United States, where the Bush administration had achieved global notoriety for its refusal to agree to binding emissions caps, more than twenty U.S. states and hundreds of cities had enacted legislation to curb carbon emissions.

By the end of the first week of negotiations, such optimism had begun to be translated into demands for the significant reductions necessary to mitigate climate change, although the wrangling which had so defined and undermined the Kyoto negotiations was also much in evidence. An initial road map for a post-Kyoto agreement had been put forward and backed by the European Union, calling on dramatic greenhouse-gas reductions in developed nations. The first draft demanded countries respond to the "unequivocal scientific evidence that preventing the worst effects of climate change will require [developed nations] to reduce emissions in a range of 25-40% below 1990 levels by 2020 and that global emissions of greenhouse gases need to peak in the next 10 to 15 years." Such bold targets, argued the draft's supporters, were needed to reflect the urgency of the problem and in keeping with the radical action demanded by scientific bodies, such as the IPCC. It would also encourage industry to invest in green technology, which economists including Sir Nicholas Stern, say is crucial to nullifying the economic costs of reducing greenhouse gas emissions.

However, the United States dismissed the proposal as unrealistic and unhelpful, and made clear from an early stage that it would not agree to a firm target, presented either as an emissions reduction or as a maximum temperature rise. Canada and Japan also expressed an unwillingness to support such dramatic targets.

Another issue, not dealt with by the Kyoto Protocol, regarded carbon emissions from international aviation and shipping. Environmentalists and scientists had long expressed concerns that increases in aviation and shipping emissions could cancel out gains made elsewhere (although there was recognition by a majority of scientists that calculating to which nation international transport emissions belonged was replete with difficulties). After initial moves by the EU to get a commitment on international transport emissions reductions included in Bali's draft framework failed, EU negotiators expressed hopes that their efforts would initiate serious negotiations on the issue in the future.

Prior to the Bali Conference, the Bush administration had hinted at a new era of cooperation on climate change, even convening a meeting to discuss its economic consequences subsequent to the Bali gathering. However, the U.S. delegation present at the Bali Conference did not depart from previous U.S. policies expressed at climate change negotiations. The United States pressed for language in the conference's "road map" to make its terms voluntary and non-binding.

Speaking on the penultimate day of the Bali Conference, former U.S. Vice President Al Gore asserted that the United States was "principally responsible" for blocking progress toward an agreement on launching negotiations for a post-Kyoto deal. He further urged delegates to reach unanimous agreement before the conference's end, even if it meant putting aside goals for emissions cuts. He said: "You can do one of two things here. You can feel anger and frustration and direct it at the United States of America, or you can make a second choice. You can decide to move forward and do all of the difficult work that needs to be done."

In the end, as time ran out on the Bali Conference, a compromise was finally reached. However, the agreement's lack of ambition for dramatic emissions reduction targets prompted anger from several critics, including environmental groups and climate scientists. International criticism was again directed toward the United States. "The United States in particular is behaving like passengers in first class in a jumbo jet, thinking a catastrophe in economy class won't affect them," said Tony Juniper, a spokesman for the environmentalist coalition present in Bali. "If we go down, we go down together, and the United States needs to realize that very quickly." Other critics noted that heavy-emitting developing nations must also play a larger role in future agreements.

Tensions were high during the last days of the conference and the sessions ran into overtime. Frustrations were evident as a representative from Papau New Guinea, borrowing from early American Revolutionary patriot Thomas Paine's writings, lectured the American delegation to "Lead, follow, or get out of the way." The American delegation's subsequent policy u-turn and agreement to compromise resulted in the adoption of the Bali roadmap.

The Bali roadmap sets forth agreement on a negotiating process for a post-Kyoto agreement to be completed by 2009. In addition to agreements to further meetings, the Bali conference facilitated the launch of the Adaptation Fund and key decisions on technology transfer and resolutions on specifically reducing emissions from deforestation. UNFCCC Executive Secretary Yvo de Boer asserted, "This [the Bali agreement] is a real breakthrough, a real opportunity for the international community to successfully fight climate change.... Parties have recognized the urgency of action on climate change and have now provided the political response to what scientists have been telling us is needed."

SEE ALSO *Kyoto Protocol; Intergovernmental Panel on Climate Change (IPCC); United States: Climate Policy.*

BIBLIOGRAPHY

Web Sites

"United Nations Climate Change Conference in Bali." *United Nations*, 2007. <http://unfccc.int/meetings/cop_13/items/4049.php> (accessed December 10, 2007).

James Corbett

Baseline Emissions

Introduction

Baseline emissions refer to the production of greenhouse gases that have occurred in the past and which are being produced prior to the introduction of any strategies to reduce emissions. The baseline measurement is determined over a set period of time, typically one year. This historical measurement acts as a benchmark to evaluate the success of subsequent efforts to reduce emissions. Without the knowledge of baseline emissions, it is impossible to reliably judge the success of any remediation efforts.

Baseline emissions are important in some climate protocols. For example, countries that adopted the greenhouse-gas reduction guidelines of the Kyoto Protocol were obligated to provide information on their nation's 1990 greenhouse-gas emissions. By doing so, the progress of the participating countries in meeting the protocol's emission-reduction timelines could be judged.

Baseline emission information is also valuable when nations or industries seek to negotiate with other jurisdictions to trade emissions so that both parties can meet their overall emission-reduction targets.

Historical Background and Scientific Foundations

The concept of establishing baseline measurements has been a fundamental aspect of science for centuries. The

In April 2007, Canada's environment minister John Baird explained what the nation would need to do in order to comply with the emissions goals set in the Kyoto Protocol. He explained that prices (particularly gasoline) would soar and jobs would be lost as Canada entered a recession. *AP Images.*

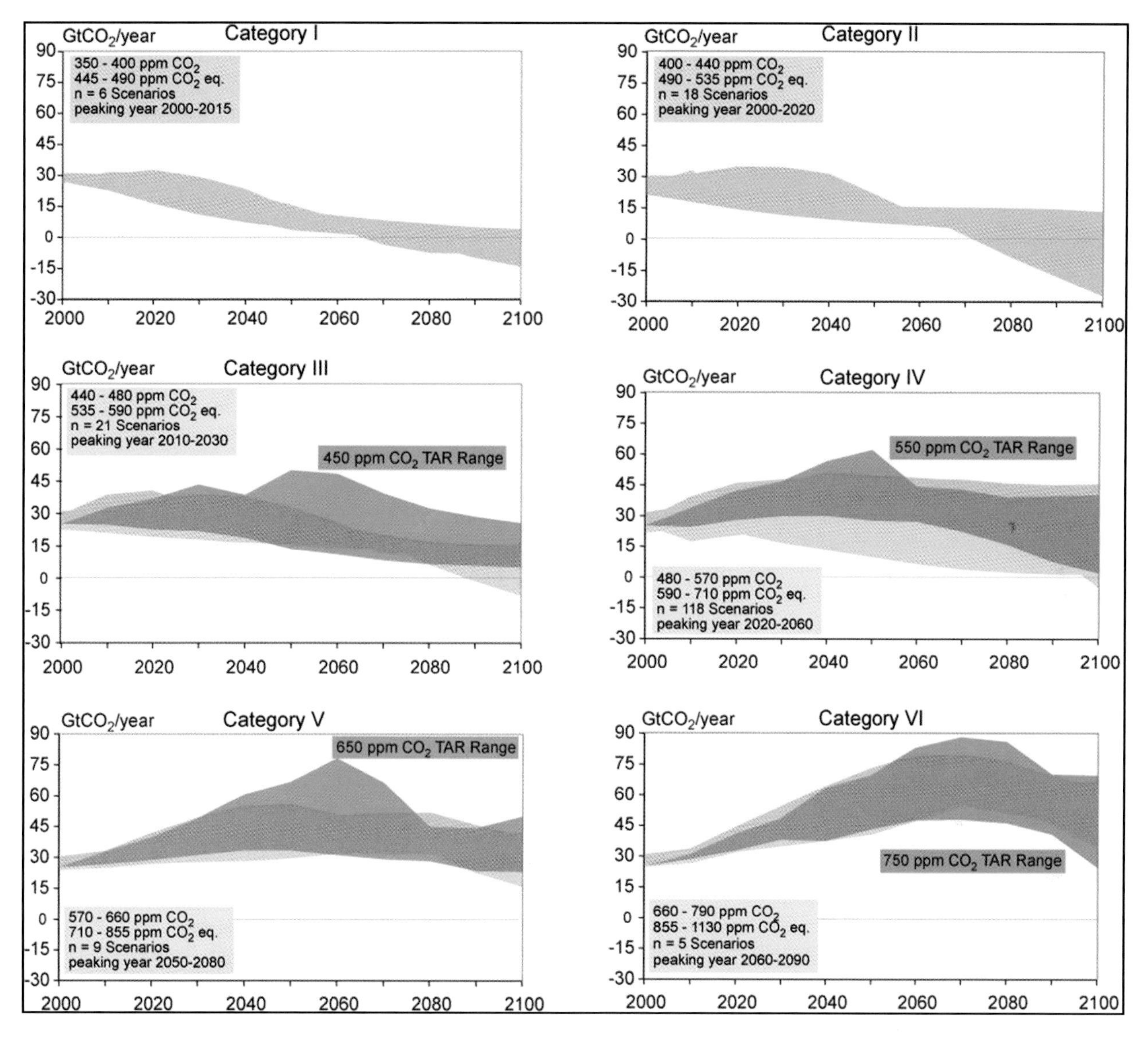

Various projections, such as this IPCC Summary for Policymakers Figure SPM7, can be drawn from baseline emissions. This figure shows emission pathways for alternative categories of stabilization levels, 2000–2100 (Categories I–VI as defined in the box in each panel). The pathways are for CO_2 emissions only. Light brown shaded areas give the CO_2 emissions for the post-Third Assessment Report (TAR) emissions scenarios. Green shaded and hatched areas depict the range of more than 80 TAR stabilization scenarios. Base year emissions may differ between models due to differences in sector and industry coverage. To reach the lower stabilization levels some scenarios deploy removal of CO_2 from the atmosphere (negative emissions) using technologies such as biomass energy production utilizing carbon capture and storage. *Climate Change 2007: Mitigation of Climate Change, Summary for Policymakers, Intergovernmental Panel on Climate Change.*

progress of an experiment or a strategy is impossible to evaluate unless there is a starting point. In the case of climate change, knowledge of the emissions of the compounds of concern at the start of the process is essential.

In climate change, baseline emissions are typically concerned with greenhouse gases—the gases, such as carbon dioxide, produced mainly due to human industrial and other activities that accumulate in the atmosphere and increase the retention of the heat energy of the sun. Greenhouse gases are the basis of global warming—the increasing temperature of the atmosphere that has been occurring for about 150 years, and which has been accelerating since the mid-twentieth century.

Emission reduction agreements such as the Kyoto Protocol mandate that the baseline emission data go back to 1990. This acts to put all participating nations at a common starting point, enabling similar reduction guidelines to be imposed on all nations, rather than negotiating numerous individual reduction timetables.

Other reduction programs have different requirements for baseline emissions. For example, in Canada, the Pilot Emission Reduction Trading (PERT) and

WORDS TO KNOW

GREENHOUSE GASES: Gases that cause Earth to retain more thermal energy by absorbing infrared light emitted by Earth's surface. The most important greenhouse gases are water vapor, carbon dioxide, methane, nitrous oxide, and various artificial chemicals such as chlorofluorocarbons. All but the latter are naturally occurring, but human activity over the last several centuries has significantly increased the amounts of carbon dioxide, methane, and nitrous oxide in Earth's atmosphere, causing global warming and global climate change.

KYOTO PROTOCOL: Extension in 1997 of the 1992 United Nations Framework Convention on Climate Change (UNFCCC), an international treaty signed by almost all countries with the goal of mitigating climate change. The United States, as of early 2008, was the only industrialized country to have not ratified the Kyoto Protocol, which is due to be replaced by an improved and updated agreement starting in 2012.

Greenhouse Gas Emission Reduction Trading (GERT) programs that have been established to develop the concept of emission trading require baseline data from the previous five years (for PERT) or since the beginning of 1997 (for GERT).

Whatever the date of the baseline emissions, the intent is the same—to provide a common starting point for all participants that allows evaluation of the subsequent emission-reduction programs.

■ Impacts and Issues

Baseline emission data are crucial for nations participating in the Kyoto Protocol, which mandates a 5.2% reduction in greenhouse-gas emissions by industrialized nations compared to the baseline 1990 levels by 2008–2012. This target is collective, so by trading emissions with other nations, a country may need to reduce emissions of one type of greenhouse gas less than the reductions required for another gas.

Although debate has arisen concerning whether these targets can be achieved, with nations such as Canada actively backing away from their Kyoto obligations, baseline data will continue to be useful as a benchmark for alternative emission-reduction plans that nations such as Canada ultimately adopt.

Baseline emission data are valuable on a local level. For example, the city of Berkeley, California, compiled a baseline greenhouse-gas emissions inventory during 2007. The results of the inventory are being used to guide municipal legislators and industries in forming reduction guidelines as part of the city's participation in the International Council for Local Environmental Initiatives (ICLEI) Cities for Climate Protection Campaign. As of 2007, more than 800 local governments worldwide are participants in the campaign.

SEE ALSO *Acid Rain; Emissions Trading; Offsetting.*

BIBLIOGRAPHY

Books

DiMento, Joseph F. C., and Pamela M. Doughman. *Climate Change: What It Means for Us, Our Children, and Our Grandchildren.* Boston: MIT Press, 2007.

Gore, Al. *An Inconvenient Truth: The Planetary Emergency of Global Warming and What We Can Do About It.* New York: Rodale Books, 2006.

Web Sites

"GLOBE-Net Special Feature: A Primer on Climate Change and Carbon Trading." *GLOBE-Net.* <http://www.globe-net.ca/special_features/primer_section5.cfm> (accessed November 25, 2007).

Beach and Shoreline

Introduction

The beach, also known as the shoreline or strand, is the line of intersection between a standing body of water (an ocean, gulf, lagoon, bay, estuary, or lake) and the surrounding land. The beach may consist of rock fragments (pebbles), sand, or mud (a mixture of sand and clay). The pebbles, sand, and mud may be derived from local bedrock, sediment, or soil, or the beach may be made of local bedrock, which is usually referred to as a rocky shoreline. In some instances, the sand and mud at the shoreline may be organically derived from biologic components in the nearby ocean (for example, broken up shells or coral fragments). There are various parts to a typical beach, including the berm, beach face, and shallow offshore bars. The location of a beach changes with the changing level of the adjacent ocean or lake, or can change with changing land level as well.

Coastal erosion, melting permafrost, and rising sea levels continue to take their toll on villages such as Shaktoolik, Alaska. Shown here in 2006, the village is facing the same problems that caused it to relocate to this area in the 1960s. *AP Images.*

Historical Background and Scientific Foundations

Beaches and Their Features

Beaches, especially oceanic beaches and those on large lakes, have a profile that usually consists of a berm (high ridge above the water level), which slopes toward the water. The berm gives way to the beach face, which continues the slope down to the water. Waves break on the beach face and the slope of the beach face changes with the intensity of waves associated with different seasons of the year.

Beyond the beach face is a wide area of sand known as the lower shoreface, which includes one or more parallel sets of offshore bars. The lower shoreface grades imperceptibly into the shallow shelf region of oceans or the more basinal parts of large lakes.

Beaches and Longshore Drift

Waves approaching the beach break upon the beach face at an angle that sends water up the beach face (called

WORDS TO KNOW

BEACH FACE: Portion of a beach that slopes downward to the water.

BERM: A platform of wave-deposited sediment that is flat or slopes slightly landward.

SEDIMENT: Solid unconsolidated rock and mineral fragments that come from the weathering of rocks and are transported by water, air, or ice and form layers on Earth's surface. Sediments can also result from chemical precipitation or secretion by organisms.

SEICHE: A standing wave in a body of water (pronounced SAYSH). In water, a standing wave is a stationary raised mass of water on the surface of an otherwise flat water mass that is sustained by some source of vibration agitating the body of water.

TSUNAMI: Ocean wave caused by a large displacement of mass under the surface of the water, such as an earthquake or volcanic eruption.

swash) and then runs back down (a process called backwash). Because waves approach the beach at an angle most of the time on most beaches, the swash is at an angle. However, backwash, which is controlled by gravity, flows directly back to the water. Continual swash and backwash over time moves sediment comprising the beach in a flow direction that is parallel to the beach. This is called longshore drift.

Longshore drift is essential to the continued existence of the beach because sediment is moved from points where it is first placed on the beach (for example at the mouth of a river or stream) to points farther down the beach. Anything that interrupts this flow can deprive the beach of sediment and cause its destruction.

Beaches and Other Processes

In addition to changes in relative water-land level and sediment supply, beaches are affected by other processes including large cyclonic storms (hurricanes and tropical storms), tides, and tsunamis. Cyclonic storms most often affect beaches on the eastern sides of continental landmasses and areas at moderate to low latitudes. The passage of such storms, which is usually accompanied by a surge of high water and strong waves, can redistribute beach sediment in such a way as to reconfigure the beach in a very short time.

This is particularly true with barrier-island beaches, a type of beach that exists on a narrow sand island, which is detached from the mainland by a lagoon or bay. Such low-lying sand islands are particularly susceptible to devastation by storm surges and waves. Tides, the continual ebb and flow of water due to celestial gravity, also affect beaches and their processes. Tsunamis, or seismic sea waves (called seiche in large lakes), can also cause sudden erosion of beaches and their reconfiguration.

Impacts and Issues

Beaches have existed on Earth since the first water accumulated on the surface a few billion years ago. There are many examples of ancient beaches, which are preserved in ancient sedimentary rock. The action of waves washing on the beach face produces distinctive fine layers in the beach sediment, which is in turn preserved in sedimentary rock. Continuous changes in the world's sea level over time have resulted in widespread rock formations that contain ancient beach deposits. Today, sea-level change is causing movement and relocation of modern beaches. To a lesser extent and in smaller areas, land-level changes affect beaches as well.

As Earth's climates change, so do Earth's beaches and shorelines. Beaches and shorelines change location as sea level rises or falls, which may be a response to either climatic warming or cooling. Global warming may intensify tropical storms and hurricanes (typhoons in the Pacific), which impacts beaches and shorelines as well.

See Also *Coastal Populations; Coastlines, Changing; Hurricanes; Tides.*

BIBLIOGRAPHY

Books

Bascom, W. *Waves and Beaches.* Garden City, NY: Anchor Press/Doubleday, 1980.

Davis, R. A., and D. M. Fitzgerald. *Beaches and Coasts.* Oxford, UK: Blackwell, 2004.

Web Sites

"Coastal Zones and Sea Level Rise." *U.S. Environmental Protection Agency,* October 19, 2006. <http://www.epa.gov/climatechange/effects/coastal/index.html> (accessed November 30, 2007.

David T. King Jr.

Berlin Mandate

Introduction

The Berlin Mandate was an agreement made in April 1995 between signatories to the United Nations Framework Convention on Climate Change (UNFCCC). The mandate acknowledged that the existing convention did not go far enough in its efforts to mitigate global warming and initiated a period of negotiation to agree on binding targets.

Historical Background and Scientific Foundations

The Framework Convention on Climate Change, signed at the Earth Summit held in Rio de Janeiro, Brazil, in 1992, set in place the internationally recognized principle that greenhouse-gas emissions should be set at a level that would not interfere with the planet's climate.

At the first general meeting of signatories held in Berlin, Germany, between March 28 and April 7, 1995, the Berlin Mandate was a compromise reached unanimously on the final day. It acknowledged that the existing convention was too limited in merely urging signatories to return greenhouse-gas emissions to 1990 levels by the year 2000. Specifically, the signatories agreed to begin a two-year negotiation process to establish legally binding targets and timetables for greenhouse-gas emissions after 2000.

Impacts and Issues

The Berlin Mandate, which in its final form was brokered by then-German Environment Minister Angela Merkel, ultimately reflected the position of the European Union. It also allowed the U.S. government to demonstrate its eco-credentials internationally by offering its support to the mandate, without the maelstrom of domestic criticism that signing binding targets would have invited.

However, the mandate disappointed an array of government delegations, particularly from the developing world, as well as environmental and other lobbying groups, for not initiating immediate action on reducing greenhouse emissions. In particular, it had been hoped that a binding protocol with targets and timetables would emerge from the meeting.

Instrumental in opposing specific targets for reductions in emissions were the United States, along with most Organization of the Petroleum Exporting Countries (OPEC) members. U.S. businesses, particularly those represented by the Global Climate Coalition, a lobby group opposing immediate action on greenhouse emissions and representing some of the country's biggest companies,

WORDS TO KNOW

GREENHOUSE GASES: Gases that cause Earth to retain more thermal energy by absorbing infrared light emitted by Earth's surface. The most important greenhouse gases are water vapor, carbon dioxide, methane, nitrous oxide, and various artificial chemicals such as chlorofluorocarbons. All but the latter are naturally occurring, but human activity over the last several centuries has significantly increased the amounts of carbon dioxide, methane, and nitrous oxide in Earth's atmosphere, causing global warming and global climate change.

KYOTO PROTOCOL: Extension in 1997 of the 1992 United Nations Framework Convention on Climate Change (UNFCCC), an international treaty signed by almost all countries with the goal of mitigating climate change. The United States, as of early 2008, was the only industrialized country to have not ratified the Kyoto Protocol, which is due to be replaced by an improved and updated agreement starting in 2012.

were particularly vigorous in urging its government to oppose any set targets.

The Berlin Mandate set in process the period of negotiation that culminated in the Kyoto Protocol two years later. But the criticism that Berlin did not initiate immediate action still rang true almost a decade later. For it was not until February 2005 that binding targets for reducing greenhouse-gas emissions finally came into place.

SEE ALSO *Energy Industry Activism; Kyoto Protocol; United Nations Framework Convention on Climate Change (UNFCCC).*

BIBLIOGRAPHY

Periodicals

Abbott, Alison. "Meeting Agrees on Need for New Targets for Greenhouse Gas Emissions." *Nature* 374 (1995): 584–585.

Web Sites

Purvis, Andrew. "Heroes of the Environment" *Time.* <http://www.time.com/time/specials/2007/article/0,28804,1663317_1663319_1669897,00.html> (accessed November 21, 2007).

James Corbett

Biodiversity

Introduction

Biodiversity is the number of distinct varieties or types within a group of living systems: distinct genes in a species, species in an ecosystem, or ecosystems in a biome. The term is often used to mean the total number of species living in a given ecosystem or on Earth as a whole. Climate change affects biodiversity primarily by shifting the boundaries of ecosystems, by altering the timing of seasonal events such as hatching and budding, and altering the temperature and chemical characteristics of lakes, rivers, and oceans. Atmospheric carbon dioxide (CO_2), which is a main cause of global warming, also has direct effects on ecosystems, acidifying the oceans and encouraging the growth of some plants more than others. Such stresses inevitably cause extinctions, that is, loss of biodiversity.

Only evolution can create new species, and this occurs only over geologic time (usually millions of years). On human or historical time scales, extinction decreases biodiversity irreversibly. Some extinctions from climate change have already been recorded, although to date most have been caused by other human activities such as pollution, over-hunting, and deforestation. The rate of extinctions caused by climate change is predicted to be greater later in the twenty-first century than today.

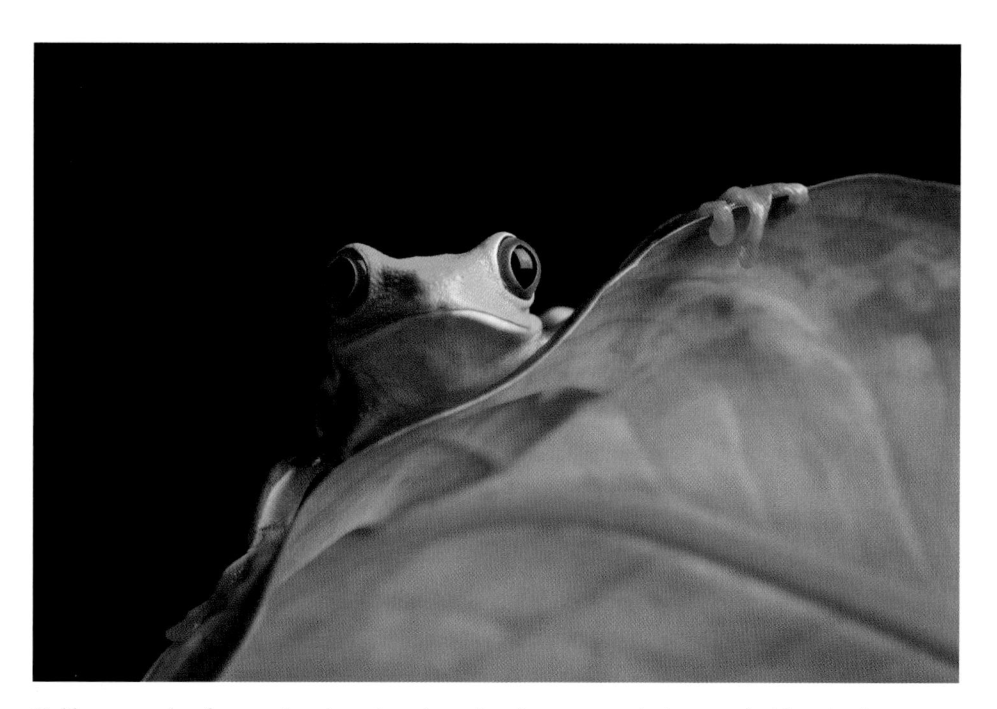

Unlike some other frog species, the red-eyed tree frog is not presently threatened with extinction. Scientists continue to carefully study the tropical-rainforest dweller, whose survival can be impacted by deforestation and lack of rain. *Image copyright Sascha Burkard, 2007. Used under license from Shutterstock.com.*

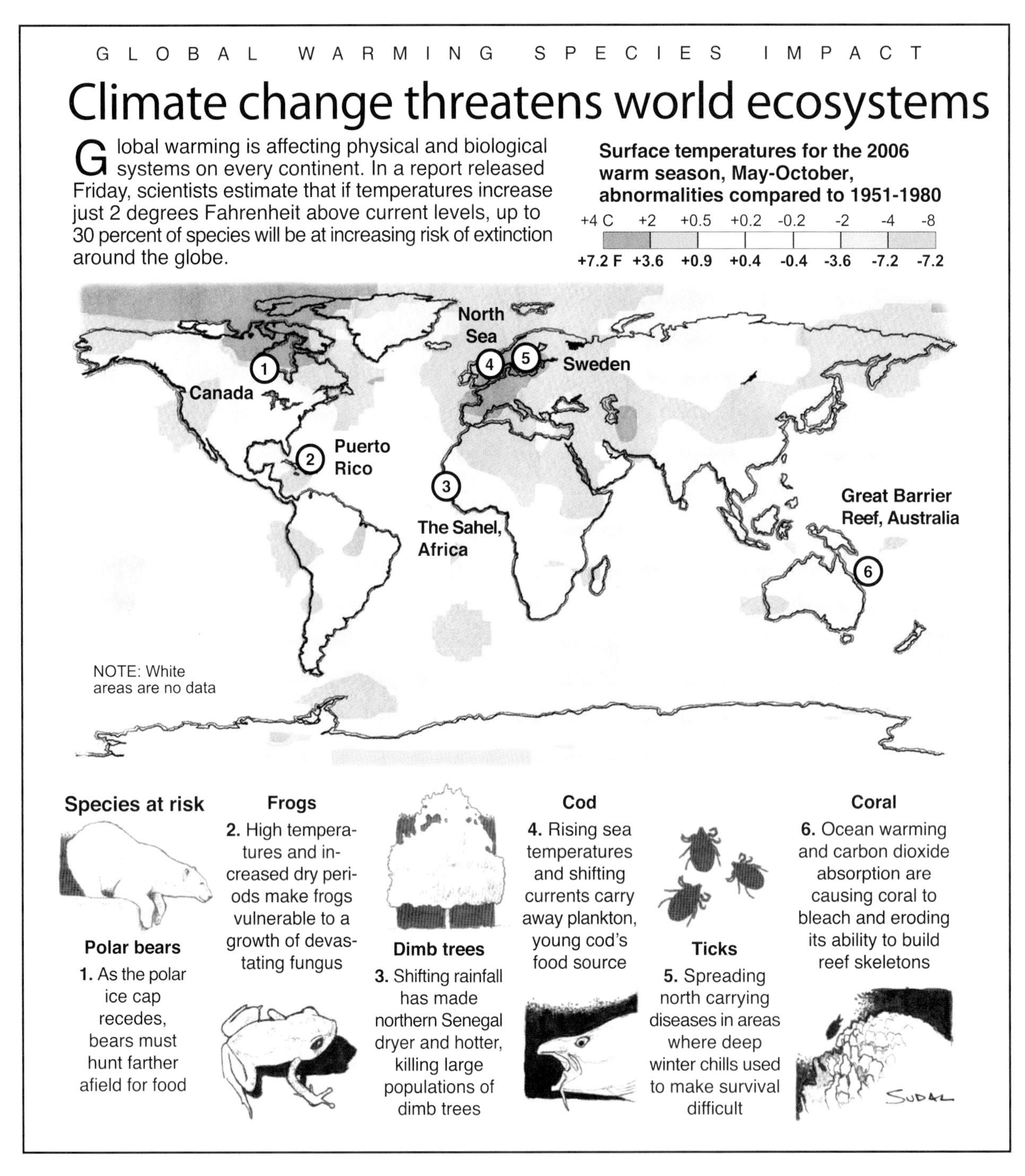

This Associated Press graphic describes species at risk due to climate change. [Sources: Intergovernmental Panel on Climate Change and National Aeronautics and Space Administration (NASA)]. *AP Images.*

Historical Background and Scientific Foundations

Context

The modern systematic classification of species was invented by Swedish naturalist Carl Linnaeus (1707–1788) in the eighteenth century. In following decades, European naturalists scoured Earth looking for and classifying new species of plants and animals, greatly expanding scientific knowledge of just how diverse life on Earth is. Such knowledge was crucial to the development of evolutionary biology by Charles Darwin (1809–1882) and others in the mid-nineteenth century. Today, biologists estimate that there are more than 280,000 species of plants and over 1,250,000 species of animals (including insects). Nineteenth- and twentieth-century paleontology has shown that extinction is a normal process: over 99% of all species that have ever lived are now extinct. The typical lifespan of a species is between 1 and 10 million years.

Awareness that species biodiversity is a key feature of ecosystems was not common until the 1970s, however, when the term "biological diversity" first came into frequent use. The word "biodiversity," a shortening of "biological diversity," first appeared in print in 1988. In the 1990s and beyond, most biologists have agreed that human beings are causing the first mass extinction in 65 million years. This modern pulse of extinctions is sometimes called the Holocene extinction event. (The Holocene is the geological term for the era from 11,500 years ago to the present.)

Humans are causing extinctions through land development, destruction of rainforests, over-hunting and over-fishing, pollution, and other activities. Some of the more famous extinctions of recent times include the dodo bird, the passenger pigeon, and (announced in 2007) the baiji, a white river dolphin of China. Hunting has caused about 23% of known animal extinctions since 1600; the introduction of invasive species, 39%; and habitat destruction, 36%. These numbers are approximate because many of the species that are being destroyed are uncatalogued insects and plants living in rainforests. Estimates of how many extinctions have already occurred due to human activities range from several tens of thousands to over a million.

Although impacts by large asteroids have caused or at least contributed to some ancient mass extinctions, climate change has been the most common cause of natural mass extinction. Ancient climate changes were brought about by shifts in Earth's orbit, continental drift, volcanism, and other processes. Today's episode of climate change is unique in being caused by a single species, humans. Moreover, today's mass extinction is unique in that human beings, by taking action to mitigate (reduce the severity of) climate change, can influence the overall severity of the event.

Mechanisms by which Climate Change Affects Biodiversity

As warming continues, other forms of human pressure on biodiversity will continue and will be, in most cases, amplified by the effects of climate change. Although effects may vary from region to region, the overall effect of global warming is to cause the cooler zones of the world—the regions around the poles (especially the North Pole) and on mountains—to shrink. Shrinkage of habitat puts species at risk because smaller habitats support smaller populations, and smaller populations are always at higher risk of extinction.

Climate change also has many other effects on ecosystems. Some are not immediately obvious. For example, Lake Tanganyika in Africa, the world's second-largest and second-deepest lake, harbors at least 350 species of fish, most unique to its own waters. Lake Tanganyika is not typical of ecosystems most vulnerable to climate change: being near the equator it is likely to see less drastic warming than, say, the Arctic, while its large thermal mass (4,526 cubic mi of water; 18,900 cubic km) will resist temperature shifts and thus might be expected to moderate climate-change impacts on the lake ecosystem.

However, regional climate warming by 1.08°F (0.6°C), along with lessened wind speeds, has had rapid effects on Tanganyika's ecosystem. Warming of surface waters more than deep waters has decreased mixing between the two: warm water is less dense than cool water, so the bigger the temperature difference between the layers, the more stably the warm upper water floats on top of the cool deeper water. Since the deeper waters are more nutrient-rich, reduced mixing has meant that fewer planktonic organisms (tiny, floating organisms, both plants and animals) can thrive in the upper water, where energy from the sun is abundant but nutrients are poor. As of 2003, plankton density in Lake Tanganyika had declined to less than one third what it was 25 years before; algae density had declined by 30% from values 80 years before. Since plankton are the basis of the marine food chain, fish stocks declined along with the plankton: fish stocks in the lake were 30% smaller than they were 80 years earlier.

The reduction of water mixing due to climate warming has made other changes in the lake's chemistry: for one, oxygen dissolved from the air no longer mixes as well in deeper waters. As a result, the habitat has shrunk for some of the lake's endemic species, such as the snail *Tiphobia horei*, which in 1890 lived at depths down to

WORDS TO KNOW

BIOME: Well-defined terrestrial environment (e.g., desert, tundra, or tropical forest). The complex of living organisms found in an ecological region.

CORAL BLEACHING: Decoloration or whitening of coral from the loss, temporary or permanent, of symbiotic algae (zooxanthellae) living in the coral. The algae give corals their living color and, through photosynthesis, supply most of their food needs. High sea surface temperatures can cause coral bleaching.

GEOLOGIC TIME: The period of time extending from the formation of Earth to the present.

HOLOCENE EXTINCTION EVENT: The Holocene is the geological period from 10,000 years ago to the present; the Holocene extinction is the worldwide mass extinction of animal and plant species being caused by human activity. Global warming may accelerate the ongoing Holocene extinction event, possibly driving a fourth of all terrestrial plant and animal species to extinction.

PALEONTOLOGY: The study of life in past geologic time.

PH: Measures the acidity of a solution. It is the negative log of the concentration of the hydrogen ions in a substance.

PLANKTON: Floating animal and plant life.

IN CONTEXT: PRESERVING BIODIVERSITY

"Reducing both loss of natural habitat and deforestation can have significant biodiversity, soil and water conservation benefits, and can be implemented in a socially and economically sustainable manner. Forestation and bioenergy plantations can lead to restoration of degraded land, manage water runoff, retain soil carbon and benefit rural economies, but could compete with land for food production and may be negative for biodiversity, if not properly designed."

SOURCE: *Metz, B., et al, eds.* Climate Change 2007: Mitigation of Climate Change: Contribution of Working Group III to the Fourth Assessment Report of the Intergovernmental Panel on Climate Change. *New York: Cambridge University Press, 2007.*

1,000 ft (300 m) but as of 2003 lived only down to 330 ft (100 m). The snail's habitat has thus shrunk by about two thirds, even though the lake itself has not shrunk and its bulk average temperature has changed only slightly.

Tanganyika surface plankton loss has been reflected in declining food-fish harvests from the lake (about 400,000 tons per year in 2003). Since Tanganyika supplies 25–40% of the protein needs of the four nations bounding the lake, such declines can have direct impacts on human populations as well as on biodiversity. Losses in biodiversity have not yet been measured directly in Tanganyika, but smaller populations will put some species at risk of extinction, especially as warming in the region continues to about 2.7°F (1.5°C), possibly higher, with even more drastic stabilization of the lake's waters and consequent effects on its ecosystem.

Slight changes in climate can lead to pressures on biodiversity by other mechanisms. For example, a 2006 study by J. Alan Pounds and colleagues found that global warming has almost certainly caused the recent extinction of about 67% of the 110 or so species of the Monteverde harlequin tree frog of the mountains of Costa Rica. The scientists saw the extinctions as validating the climate-linked epidemic hypothesis, according to which shifts in temperature, rainfall, and other climate variables make populations more vulnerable to disease and therefore to extinction. In the case of the Monteverde frogs, more frequent warm years shifted conditions toward the growth optimum of the *Bactrachochytrium* fungus, which infects the frogs. The researchers found that extinctions of the frogs consistently followed temperature peaks that were favorable to growth of the fungal disease.

Other effects are not strictly changes in climate, in the sense of temperature or precipitation, but chemical changes to air and water. Increased carbon dioxide (CO_2) in the atmosphere has two major effects that are likely to decrease biodiversity: 1) Heightened atmospheric CO_2 causes increased levels of dissolved CO_2 in the ocean. When CO_2 dissolves in water, it produces a weak acid, carbonic acid. Rising atmospheric CO_2 thus acidifies the oceans. 2) Green plants extract carbon from the air by breaking up CO_2, constructing their tissues using the carbon, and releasing the oxygen. CO_2 is plant food. Thus, increasing atmospheric CO_2 tends to cause more rapid growth in most plant species, an effect called CO_2 enrichment.

Acidification of the oceans by dissolved excess CO_2 will impact biodiversity by making survival more difficult for organisms that form shells of calcium carbonate. This includes bivalves such as clams, mollusks such as periwinkles and conches, microscopic plankton species, and corals. Corals, which are also subject to bleaching in excessively warm waters, form large, shallow communities in tropical waters that have been compared to rainforests because of their high level of biodiversity. A typical large reef may support on the order of a million species of plants and animals.

Over the last two centuries, the average pH of the oceans has fallen by 0.1, corresponding to a rise in acidity and a 30% reduction in the number of carbonate ions (CO_3^{2-}) available to shell-making organisms as building material. When carbonate ions fall below a certain level, corals have difficulty making their skeletons. This threshold may be reached if the atmospheric CO_2 level, today about 375 parts per million, rises to over 500 parts per million, as may occur by the end of the twenty-first century.

Increasing atmospheric CO_2 will also affect plant growth. Farmers today often add CO_2 to the air inside greenhouses, because under indoor conditions, extra CO_2 speeds plant growth and increases crop yields. Under outdoor conditions, however, the gain in yield is about half as much and the foods produced are significantly lower in protein and minerals. In the wild, rising CO_2 will favor some species over others, depending on rooting depth, woodiness, and photosynthetic chemistry; this will impact biodiversity by altering competitive balances. The CO_2 fertilization effect will be strongest in biomes where plant growth is limited by water availability, such as grasslands, savanna, and desert. The biodiversity impact of a 2.5-fold increase in CO_2 would likely be only about a third as great on a boreal (northern pine) forest as on savanna or grassland, and half as great as on desert.

■ Impacts and Issues

Global Patterns

Most plants require a specific range of temperature, moisture, and seasonal change to thrive; most animals require certain plants or other animals to thrive, and also have a limited range of tolerance for temperature and moisture. As climate warms, a typical ecosystem will tend to migrate away from areas where it was at the warm edge of its tolerance range and toward places where it was formerly at the cool edge. The most general effect of global climate change is thus to move ecological zones toward the poles and toward higher altitudes. For each 1.8°F (1°C) of warming, terrestrial (on-land) ecosystems typically shift poleward by 100 mi (160 km): for example, if climate warms by 5.4°F (3°C) by 2100, plant and animal communities in the Northern Hemisphere will migrate an average of 300 mi (480 km) northward—if they can—to stay in a suitable climate zone. This effect is observed, not only predicted.

In the Northern Hemisphere, terrestrial animal and plant ranges have been observed to shift northward, on average, by 3.8 mi (6.1 km) per decade over the last 50 years. In mountainous terrain, plant and animal ranges have shifted upward by 20 ft (6.1 m) over the same time period. Fragmentation of landscapes by human activity such as agriculture and city-building makes ecosystem migration more difficult today than during past climatic shifts, such as glacial periods. Species that fail to colonize new areas as the climate changes may go extinct.

Although climate change has so far been most intense in the Arctic and the West Antarctic Peninsula, where warming has been about twice the global average and dramatic effects such as retreating sea ice and melting tundra are readily visible, biodiversity is low in these regions compared to the tropics, where rainforests, coral reefs, and other particularly diverse communities are found. Thus, a smaller climate shift can have a greater impact on biodiversity in the tropics than a larger shift in boreal or temperate regions.

For marine ecosystems, changes in circulation patterns, ocean temperature, and ocean chemistry all influence biodiversity. For example, over the last 40 years or so, warm-water plankton species have shifted about 620 mi (1,000 km) in the North Atlantic due to warming.

Not all observed climate effects on biological systems are consistent with climate warming; a few are consistent with cooling. Also, some observed effects are consistent with natural climate shifts rather than those attributed to human-caused (anthropogenic) global warming. However, mathematical analysis shows that the very great majority of changes are consistent with warming trends, and that a combination of natural and anthropogenic climate changes describe observed changes in physical and biological systems better than either natural or anthropogenic changes alone. Anthropogenic changes have been added to or laid over those caused by natural processes, and are gradually becoming more dominant.

Future Impacts

The five main drivers of biodiversity change, ranked from most severe impact to least severe between now and 2100, are: 1) land-use changes (including deforestation); 2) climate change; 3) nitrogen deposition (from fertilizer use); 4) biotic exchange (the introduction of invasive species); and 5) direct effects of increasing atmospheric CO_2, apart from climate change.

The result of these combined, continuing, and growing pressures will be an irreversible loss of biodiversity in many parts of the world. However, many uncertainties remain. Ecologists do not understand the relationship between ecosystem structure and rapid climate change well enough to predict the exact effects of current climate changes on biomes. It is also unknown whether efforts to mitigate climate change will occur or succeed, and if so, to what extent.

Despite these uncertainties, scientists have estimated the likely impact that climate change will have on biodiversity. In 2004, Chris Thomas and colleagues published their study of an unbiased or representative sample of 1,103 animal and plant species. They found that climate change was likely to commit 15–37% of all species examined to extinction by 2050. "Committed to extinction" does not mean that a species would necessarily be extinct by that time, but that the population of each species would be so reduced that its species' extinction becomes highly likely. In many or most ecological regions, climate change will become the greatest threat to biodiversity by 2050. There are 5 to 15 million species of creatures on Earth (the large range arises from the difficulty of counting insect, bacterial, and fungal species). If only 15% of all species are committed to extinction by climate change—the lower end of the range given by Thomas and colleagues—then 750,000 to 2,250,000 million species will eventually become extinct as a result of global climate change.

■ Primary Source Connection

Human activities and the resulting global climate change have had, and will continue to have, a major impact on biodiversity across the globe. Biodiversity is the variety of all living organisms that exist in an ecosystem. This paper from the Intergovernmental Panel on Climate Change (IPCC) discusses the effect of human activity on biodiversity and possible adaptation and mitigation strategies.

The IPCC is a scientific panel that was founded by the United Nations in 1988 as part of the United Nations Environment Program and the U.N.'s World Meteorological Organization.

CLIMATE CHANGE AND BIODIVERSITY

At the global level, human activities have caused and will continue to cause a loss in biodiversity through, inter alia, land-use and land-cover change; soil and water pollution and degradation (including desertification), and air pollution; diversion of water to intensively managed ecosystems and urban systems; habitat fragmentation; selective exploitation of species; the introduction of non-native species; and stratospheric ozone depletion. The current rate of biodiversity loss is greater than the natural background rate of extinction. A critical question for this Technical Paper is how much might climate change (natural or human-induced) enhance or inhibit these losses in biodiversity?

Changes in climate exert additional pressure and have already begun to affect biodiversity. The atmospheric concentrations of greenhouse gases have increased since the pre-industrial era due to human activities, primarily the combustion of fossil fuels and land-use and land-cover change. These and natural forces have contributed to changes in the Earth's climate over the 20th century: Land and ocean surface temperatures have warmed, the spatial and temporal patterns of precipitation have changed, sea level has risen, and the frequency and intensity of El Niño events have increased. These changes, particularly the warmer regional temperatures, have affected the timing of reproduction in animals and plants and/or migration of animals, the length of the growing season, species distributions and population sizes, and the frequency of pest and disease outbreaks. Some coastal, high-latitude, and high-altitude ecosystems have also been affected by changes in regional climatic factors.

Climate change is projected to affect all aspects of biodiversity; however, the projected changes have to take into account the impacts from other past, present, and future human activities, including increasing atmospheric concentrations of carbon dioxide (CO_2). For the wide range of Intergovernmental Panel on Climate Change (IPCC) emissions scenarios, the Earth's mean surface temperature is projected to warm 1.4 to 5.8°C by the end of the 21st century, with land areas warming more than the oceans, and the high latitudes warming more than the tropics. The associated sea-level rise is projected to be 0.09 to 0.88 m. In general, precipitation is projected to increase in high-latitude and equatorial areas and decrease in the subtropics, with an increase in heavy precipitation events. Climate change is projected to affect individual organisms, populations, species distributions, and ecosystem composition and function both directly (e.g., through increases in temperature and changes in precipitation and in the case of marine and coastal ecosystems also changes in sea level and storm surges) and indirectly (e.g., through climate changing the intensity and frequency of disturbances such as wildfires). Processes such as habitat loss, modification and fragmentation, and the introduction and spread of non-native species will affect the impacts of climate change. A realistic projection of the future state of the Earth's ecosystems would need to take into account human land- and water-use patterns, which will greatly affect the ability of organisms to respond to climate change via migration.

The general effect of projected human-induced climate change is that the habitats of many species will move poleward or upward from their current locations. Species will be affected differently by climate change: They will migrate at different rates through fragmented landscapes, and ecosystems dominated by long-lived species (e.g., long-lived trees) will often be slow to show evidence of change. Thus, the composition of most current ecosystems is likely to change, as species that make up an ecosystem are unlikely to shift together. The most rapid changes are expected where they are accelerated by changes in natural and anthropogenic non-climatic disturbance patterns.

Changes in the frequency, intensity, extent, and locations of disturbances will affect whether, how, and at which rate the existing ecosystems will be replaced by new plant and animal assemblages. Disturbances can increase the rate of species loss and create opportunities for the establishment of new species.

Globally by the year 2080, about 20% of coastal wetlands could be lost due to sea-level rise. The impact of sea-level rise on coastal ecosystems (e.g., mangrove/coastal wetlands, seagrasses) will vary regionally and will depend on erosion processes from the sea and depositional processes from land. Some mangroves in low-island coastal regions where sedimentation loads are high and erosion processes are low may not be particularly vulnerable to sea-level rise.

The risk of extinction will increase for many species that are already vulnerable. Species with limited climatic ranges and/or restricted habitat requirements and/or small populations are typically the most vulnerable to extinction, such as endemic mountain species and biota restricted to islands (e.g., birds), peninsulas (e.g., Cape Floral Kingdom), or coastal areas (e.g., mangroves, coastal wetlands, and coral reefs). In contrast, species with extensive, non-patchy ranges, long-range dispersal mechanisms, and large populations are at less risk of extinction. While there is little evidence to suggest that climate change will slow species losses, there is evidence it may increase species losses. In some regions there may be an increase in local biodiversity—usually as a result of species introductions, the long-term consequences of which are hard to foresee.

Where significant ecosystem disruption occurs (e.g., loss of dominant species or a high proportion of species, or much of the species redundancy), there may be losses in net ecosystem productivity (NEP) at least during the transition period. However, in many cases, loss of biodiversity from diverse and extensive ecosystems due to climate change does not necessarily imply loss of productivity as there is a degree of redundancy in most ecosystems; the contribution to production by a species that is lost from an ecosystem may be replaced by another species. Globally, the impacts of climate change on biodiversity and the subsequent effects on productivity have not been estimated.

Changes in biodiversity at ecosystem and landscape scale, in response to climate change and other pressures (e.g., changes in forest fires and deforestation), would further affect global and regional climate through changes in the uptake and release of greenhouse gases and changes in albedo and evapotranspiration. Similarly, structural changes in biological communities in the upper ocean could alter the uptake of CO_2 by the ocean or the release of precursors for cloud condensation nuclei causing either positive or negative feedbacks on climate change.

Modeling the changes in biodiversity in response to climate change presents some significant challenges. The data and models needed to project the extent and nature of future ecosystem changes and changes in the geographical distribution of species are incomplete, meaning that these effects can only be partially quantified.

Impacts of climate change mitigation activities on biodiversity depend on the context, design, and implementation of these activities. Land-use, land-use change, and forestry activities (afforestation, reforestation, avoided deforestation, and improved forest, cropland, and grazing land management practices) and implementation of renewable energy sources (hydro-, wind-, and solar power and biofuels) may affect biodiversity depending upon site selection and management practices. For example, 1) afforestation and reforestation projects can have positive, neutral, or negative impacts depending on the level of biodiversity of the non-forest ecosystem being replaced, the scale one considers, and other design and implementation issues; 2) avoiding and reducing forest degradation in threatened/vulnerable forests that contain assemblages of species that are unusually diverse, globally rare, or unique to that region can provide substantial biodiversity benefits along with the avoidance of carbon emissions; 3) large-scale bioenergy plantations that generate high yields would have adverse impacts on biodiversity where they replace systems with higher biological diversity, whereas small-scale plantations on degraded land or abandoned agricultural sites would have environmental benefits; and 4) increased efficiency in the generation and/or use of fossil-fuel-based energy can reduce fossil-fuel use and thereby reduce the impacts on biodiversity resulting from resource extraction, transportation (e.g., through shipping and pipelines), and combustion of fossil fuels.

Climate change adaptation activities can promote conservation and sustainable use of biodiversity and reduce the impact of changes in climate and climatic extremes on biodiversity. These include the establishment of a mosaic of interconnected terrestrial, freshwater, and marine multiple-use reserves designed to take into account projected changes in climate, and integrated land and water management activities that reduce non-climate pressures on biodiversity and hence make the systems less vulnerable to changes in climate. Some of these adaptation activities can also make people less vulnerable to climatic extremes.

The effectiveness of adaptation and mitigation activities can be enhanced when they are integrated with broader strategies designed to make development paths more sustainable. There are potential environmental and social synergies and tradeoffs between climate adaptation and mitigation activities (projects and policies), and the objectives of multilateral environmental agreements (e.g., the conservation and sustainable use objective of the Convention on Biological Diversity) as well as other aspects of sustainable development. These synergies and tradeoffs can be evaluated for the full range of potential activities—inter alia, energy and land-use, land-use change, and forestry projects and policies through the application of project, sectoral, and regional level environmental and social impact assessments—and can be compared against a set of criteria and indicators using a range of decision making frameworks. For this, current assessment methodologies, criteria, and indicators for evaluating the impact of mitigation and adaptation activities on biodiversity and other aspects of sustainable development will have to be adapted and further developed.

IPCC. "CLIMATE CHANGE AND BIODIVERSITY." *2001 IPCC TECHNICAL PAPER V.* <HTTP://WWW.GRIDA.NO/CLIMATE/IPCC_TAR/BIODIV/PDF/BIO_ENG.PDF> (ACCESSED NOVEMBER 29, 2007).

SEE ALSO *Endangered Species; Extinction.*

BIBLIOGRAPHY

Books

Parry, M. L., et al, eds. *Climate Change 2007: Impacts, Adaptation and Vulnerability: Contribution of Working Group II to the Fourth Assessment Report of the Intergovernmental Panel on Climate Change.* New York: Cambridge University Press, 2007.

Periodicals

Araúo, Miguel B., and Carsten Rahbek. "How Does Climate Change Affect Biodiversity?" *Science* 313 (2006): 1,396–1,397.

Higgins, Paul A. T. "Biodiversity Loss Under Existing Land Use and Climate Change: An Illustration Using Northern South America." *Global Ecology and Biogeography* 16 (2007): 197–204.

Jenkins, Martin. "Prospects for Biodiversity." *Science* 302 (2003): 1,175–1,177.

Livingstone, Daniel A. "Global Climate Change Strikes a Tropical Lake." *Science* 301 (2003): 468–469.

Pounds, J. Alan, et al. "Widespread Amphibian Extinctions from Epidemic Disease Driven by Global Warming." *Nature* 439 (2006): 161–167.

Sala, Enric, and Nancy Knowlton. "Global Marine Biodiversity Trends." *Annual Review of Energy and the Environment* 231 (2006): 93–122.

Sala, Osvaldo E., et al. "Global Biodiversity Scenarios for the Year 2100." *Science* 287 (2000): 1,770–1,774.

Thomas, Chris D. "Extinction Risk from Climate Change." *Nature* 427 (2004): 145–148.

Thuiller, Wilfried. "Climate Change and the Ecologist." *Nature* 448 (2007): 550–552.

Willis, K. J., and H. J. B. Birks. "What Is Natural? The Need for a Long-Term Perspective in Biodiversity Conservation." *Science* 314 (2006): 1,261–1,265.

Zimmer, Carl. "Predicting Oblivions: Are Existing Models Up to the Task." *Science* 317 (2007): 892–893.

Web Sites

"Climate Change and Biodiversity." *Intergovernmental Panel on Climate Change,* IPCC Technical Paper V. <http://www.ipcc.ch/pub/tpbiodiv.pdf> (accessed November 8, 2007).

Larry Gilman

Biofuel Impacts

Introduction

Fossil fuels are used in more than 95% of the global transportation markets, but the demand for biofuels is growing at a rate ten times faster than the demand for oil. The use of biofuels, ethanol, and biodiesel for the transportation industries increases energy independence and reduces greenhouse-gas emissions. Biofuels are produced from renewable resources, largely from agricultural products.

The production of biofuels in the quantities that are needed to make a significant impact on reducing oil imports and greenhouse-gas emissions requires increased agricultural production. That change in agricultural management has the potential to impact food supplies and degrade soil and water resources.

Historical Background and Scientific Foundations

The use of biofuels in the transportation industry is not new. In 1900, German inventor Rudolf Diesel (1858–1913) exhibited his new engine at the World Exhibition in Paris, France. He ran the engine on peanut oil. In 1912 Diesel suggested that vegetable oils would be the fuel of the future for his diesel engine. In 1925, Henry Ford (1863–1947), a leader in the U.S. automobile industry, promoted corn-based ethanol as the fuel of the future for the automotive industry.

It took most of the 20th century for these predictions to be revived. During World War I (1914–1918) and

Former British Prime Minister Tony Blair visited with students in South Africa in June 2007 to hear about biofuel crops and research being conducted at an experimental farm. *AP Images.*

Some consumers have turned to biodiesel as a way to reduce reliance on fossil fuels, which are major contributors to greenhouse-gas emissions. *AP Images.*

World War II (1939–1945) there was some mixing of ethanol with gasoline in the United States to extend supplies but the ready availability of cheap petroleum postponed all serious efforts to produce biofuels for the transportation markets until the supply crises of the 1970s.

The Brazilian government first started promoting the use of sugarcane-based ethanol for the transportation market in the 1970s to boost the economy of a distressed sugarcane industry and to reduce petroleum importing. Sweden was also impacted severely by the 1970s oil import crises and provided government incentives to move toward biofuels. The rest of the world did very little to switch to biofuels until recent political instability and environmental concerns about greenhouse-gas emissions prompted a shift in governmental policies toward the promotion of biofuels.

Ethanol production more than doubled to nine billion gallons between 2000 and 2005. Biodiesel production did not start as early, but production of biodiesel quadrupled during the same period. The leading producer of ethanol today is Brazil with close to 5,000 million gallons produced in 2005 from sugarcane. The United States is second in ethanol production with close to 4,500 million gallons produced from corn in 2005. Biodiesel is mainly produced in Europe with Germany leading at 507 million gallons from rapeseed in 2005 and France was second with 135 million gallons from soybeans.

Ethanol is produced by a fermentation process, which is an energy intensive process. However, because growing plants remove carbon dioxide from the atmosphere, even the fact that fossil-fuel energy is used in the production of ethanol, and carbon dioxide is emitted when ethanol is used as fuel, there is a net reduction in atmospheric carbon dioxide when ethanol is used to replace gasoline. Depending on the specific processes used in the production of corn-based ethanol, greenhouse-gas emissions could be reduced by 15–40% compared to using an equivalent amount of petroleum-based gasoline.

Biodiesel is produced in chemical processes that are not as energy intensive as fermentation. The net gain of reducing carbon dioxide emissions from using pure biodiesel is over 43% and for hydrocarbon emissions is over 55%. Petroleum-based diesel fuels have had problems with particulate emissions that are reduced by 55% when biodiesel is used. Only nitrogen oxides increase in the emissions of burning biodiesel fuel. Biodiesel is not as useful in cold weather because it thickens more than petroleum-based diesel.

Both biodiesel and ethanol are used more as blends with petroleum-based fuels. Ethanol cannot be used as a blended fuel above 10% ethanol except in specially designed engines. FlexFuel gasoline engines are being promoted to use 85% ethanol–15% gasoline mixtures.

■ Impacts and Issues

In Europe it would take approximately 3 million acres planted specifically in crops for biofuel production to reduce the use of petroleum-based fuel by 1%. The issue

WORDS TO KNOW

BIODIESEL: A fuel made from a combination of plant and animal fat. It can be safely mixed with petro diesel.

BIOMASS: The sum total of living and once-living matter contained within a given geographic area. Plant and animal materials that are used as fuel sources.

DIESEL FUEL: A liquid engine fuel used in diesel engines, which ignite fuel vapor by increasing pressure in a cylinder rather than by an electric spark. Diesel engines tend to be more efficient than gasoline engines. Most diesel fuel is refined from petroleum, but diesel may also be made from vegetable oil, in which case it is termed biodiesel.

ETHANOL: Compound of carbon, hydrogen, and oxygen (CH_3CH_2OH) that is a clear liquid at room temperature; also known as drinking alcohol or ethyl alcohol. Ethanol can be produced by biological or chemical processes from sugars and other feedstocks and can be burned as a fuel in many internal-combustion engines, either mixed with gasoline or in pure form. Several governments, most notably Brazil and the United States, encourage the production of ethanol from corn, switchgrass, algae, or other crops to substitute for imported fossil fuels. Ethanol is criticized by some as being based on environmentally destructive agriculture, putting human populations into competition with automobiles for the produce of arable land, and providing little more energy (depending on the manufacturing process used) than is required to produce it.

FERMENTATION: Chemical reaction in which enzymes break down complex organic compounds (for example, carbohydrates and sugars) into simpler ones (for example, ethyl alcohol).

FOSSIL FUELS: Fuels formed by biological processes and transformed into solid or fluid minerals over geological time. Fossil fuels include coal, petroleum, and natural gas. Fossil fuels are non-renewable on the timescale of human civilization, because their natural replenishment would take many millions of years.

GASOLINE: The fraction of petroleum that boils between 32°F (0°C) and 392°F (200°C).

PETROLEUM: A complex liquid mixture that is mostly composed of hydrocarbons, compounds of carbon and hydrogen, that is separated into different products with different boiling ranges by a process called cracking.

of available land for biofuel crops that would provide enough biofuel to make a significant difference in the use of petroleum-based fuels is one that is not resolved.

The world population is growing at the rate of about 90 million people a year. Demands for both food and energy are increasing. There is competition for land to produce biofuel crops and land for food crops. There has been considerable progress in developing cost-effective technologies to produce ethanol from biomass such as the corn stalks and other woody waste materials in place of corn, but that technology is not expected to enter the biofuel market until at least 2012.

In addition to competition between food crops that are also sources of biofuels, the increased production of the crops could have a negative environmental effect as more fertilizer increases the nitrogen content of the run-off, compounding the negative environmental effects such as the dead zone that has been created in the Gulf of Mexico. Soil and groundwater degradation from increased crop production is a serious concern.

Palm oil, the second largest source of edible oil, is increasingly being used for biodiesel. The source of 90% of the world's palm oil is Malaysia and Indonesia. Palm oil would have to be imported to Europe for biodiesel production, which would defeat the goal of sustainable energy from local renewable resources.

SEE ALSO *Agriculture: Vulnerability to Climate Change; Carbon Sinks; Europe: Climate Policy.*

BIBLIOGRAPHY

Web Sites

"Biological Ways of Producing Ethanol." *Oak Ridge National Laboratory*, 2007. <http://www.ornl.gov/info/ornlreview/v33_3_00/ethanol.htm> (accessed August 16, 2007).

"International Conference on Biofuels 2007." *European Commission, External Relations*, 2007. <http://ec.europa.ey/external_relations/energy/biofuels/indes.htm> (accessed August 16, 2007).

"March 2007 Monthly Update: Global Biofuel Trends." *Earth Trends, World Resource Institute*, 2007. <http://www.earthtrends.wri.org/updates/node180> (accessed August 16, 2007).

Biogeochemical Cycle

■ Introduction

A biogeochemical cycle describes the transformations that occur in a substance that is fundamental to the environment as it cycles through Earth's lithosphere (upper mantle), biosphere (life-supporting areas), hydrosphere (water and water vapor), and atmosphere (layer of gases). The substances most often studied in biogeochemical cycling include carbon, nitrogen, phosphorous, hydrogen, oxygen, and sulfur.

■ Historical Background and Scientific Foundations

Biogeochemical cycles are characterized by reservoirs, also called sinks, and transformations. A substance can remain trapped or fixed in a reservoir for some period of time. Examples of reservoirs are rocks for phosphorous, oceans for hydrogen and oxygen, and forests for carbon. Eventually, something will happen to the reservoir that will cause a transformation of the substance. In some cases, these transformations are state changes, as when water in the ocean evaporates and becomes vapor in the atmosphere. In other cases, the transformation may involve breaking and forming chemical bonds, as when photosynthesis by plants converts carbon dioxide in the atmosphere into carbohydrates.

In the biogeochemical cycle for carbon, for example, the atmosphere serves as a reservoir of carbon dioxide. Atmospheric carbon dioxide is removed by photosynthesis of plants on land and algae in water, both of which represent significant carbon sinks. When plants and algae are buried, they can become fossilized and the carbon becomes transformed into fossil fuel. This reservoir is another significant carbon sink. When fossil fuels or forests are burned, carbon is transformed back into carbon dioxide and returns to the atmosphere.

The differences in biogeochemical cycles result from differences in the chemical properties of the different substances. These differences affect the types of reservoirs, the residence times in the different reservoirs, as well as the rates of transformation.

■ Impacts and Issues

As the global climate undergoes changes, both the reservoirs and transformations between reservoirs of all the biogeochemical cycles are affected. In particular, reliance on petroleum products is causing widespread changes to the carbon biogeochemical cycle. The enormous reservoir of carbon bound in fossilized plant material is being used at rapid rates through the burning of petroleum products. This combustion adds carbon in the form of carbon dioxide to the atmosphere. The size of the atmospheric carbon reservoir has increased significantly since the Industrial Revolution began in the late 1700s. In addition, many forests, which compose another large carbon reservoir, have been harvested without being replanted, transferring a portion of the carbon that was bound in trees to the atmosphere. The additional carbon in the atmosphere from the depletion of carbon in fossil fuels and forests is the major suspected cause of global climate change.

The carbon cycle is not the only one impacted by human practices since the Industrial Revolution. Oxygen and hydrogen biogeochemical cycles are affected by climate change because of their role in the hydrologic cycle. Global warming has caused melting of the ice sheets at the poles at a rapid rate. The cycling of nitrogen and sulfur are linked to acid rain, which results from burning fossil fuels. Both the phosphorus and nitrogen cycles are also significantly affected by the spread of fertilizer on agricultural fields, contributing to high nutrient runoff and eutrophication (over-enrichment with minerals and nutrients) of lakes and rivers.

WORDS TO KNOW

ACID RAIN: A form of precipitation that is significantly more acidic than neutral water, often produced as the result of industrial processes.

BIOSPHERE: The sum total of all life-forms on Earth and the interaction among those life-forms.

CARBON SINKS: Carbon reservoirs such as forests or oceans that take in and store more carbon (carbon sequestration) than they release. Carbon sinks can serve to partially offset greenhouse-gas emissions.

EUTROPHICATION: The process whereby a body of water becomes rich in dissolved nutrients through natural or human-made processes. This often results in a deficiency of dissolved oxygen, producing an environment that favors plant over animal life.

HYDROLOGIC CYCLE: The process of evaporation, vertical and horizontal transport of vapor, condensation, precipitation, and the flow of water from continents to oceans. It is a major factor in determining climate through its influence on surface vegetation, the clouds, snow and ice, and soil moisture. The hydrologic cycle is responsible for 25 to 30% of the mid-latitudes' heat transport from the equatorial to polar regions.

HYDROSPHERE: The totality of water encompassing Earth, comprising all the bodies of water, ice, and water vapor in the atmosphere.

INDUSTRIAL REVOLUTION: The period, beginning about the middle of the eighteenth century, during which humans began to use steam engines as a major source of power.

LITHOSPHERE: The rigid, uppermost section of Earth's mantle, especially the outer crust.

RESERVOIR: A natural or artificial receptacle that stores a particular substance for a period of time

TRANSFORMATION: The processes involved in the transfer of a substance from one reservoir to another.

SEE ALSO *Acid Rain; Atmospheric Circulation; Carbon Cycle; Carbon Sequestration Issues; Carbon Sinks; Forests and Deforestation; Greenhouse Gases; Hydrologic Cycle; Melting; Ocean Circulation and Currents; Rainfall; Sea Level Rise; Sequestration; Sink; Water Vapor.*

BIBLIOGRAPHY

Books

Raven, Peter H., and Linda R. Berg. *Environment.* Hoboken, NJ: John Wiley and Sons, 2006.

Web Sites

"Biogeochemical Cycles." *Environmental Literacy Council,* October 30, 2006. <http://www.enviroliteracy.org/subcategory.php/198.html> (accessed October 17, 2007).

"Global Biogeochemical Cycles & the Physical Climate System." *University Corporation for Atmospheric Research.* <http://www.ucar.edu/communications/gcip/m4bgchem/m4overview.html> (accessed October 17, 2007).

Juli M. Berwald

Biomass

Introduction

Biomass is the total mass of biological material, both living and recently dead, in a defined area. In an ecological context, biomass often refers to the amount of biological material in different parts of an ecological pyramid or in different ecological communities. In terms of energy supply, biomass refers to plant material that is grown as a source of energy.

Historical Background and Scientific Foundations

Nearly all of the biomass on Earth is produced by photosynthesis performed by plants and algae. Photosynthesis converts solar energy into the energy in the chemical bonds of molecules. Inherent in the idea of biomass is the notion that energy is stored in organic materials. Transformations to organic materials release energy that can then be converted to other forms of energy and matter.

In ecology, the concept of biomass is often used to describe ecological pyramids. These pyramids summarize the various types of organisms in a particular environment. The bottom of the pyramid usually consists of organisms that convert sunlight into stored chemical energy through photosynthesis. These organisms are producers. The next level of the pyramid consists of organisms that consume the producers. These organisms are the primary consumers. The next level consists of secondary consumers, those

In Stockton, California, a worker bulldozes biomass fuel to be used for electricity called biopower.
Lawrence Migdale/Photo Researchers, Inc.

WORDS TO KNOW

BIODIESEL: A fuel made from a combination of plant and animal fat. It can be safely mixed with petro diesel.

BIOENERGY: Energy for technological use derived from materials produced by living things. Wood, methane from anaerobic bacteria, and liquid fuels manufactured from crops are all forms of bioenergy.

BIOETHANOL: Ethanol produced by fermentation using yeast or bacteria. Most ethanol is bioethanol, but methods to produce it using purely chemical processes also exist.

BIOFUEL: A fuel derived directly by human effort from living things, such as plants or bacteria. A biofuel can be burned or oxidized in a fuel cell to release useful energy.

ECOLOGICAL COMMUNITY: System of species that live together in a given ecosystem and interact with each other. For example, all plants, animals, insects, and microorganisms living in or interacting with a lake form a single ecological community.

ECOLOGICAL PYRAMID: Representation of the ascending levels of biomass productivity in an ecosystem, where each level eats the level below it. Green plants are the basis of a typical ecological pyramid, with top predators—predators on whom no other species preys—at the top.

FOSSIL FUELS: Fuels formed by biological processes and transformed into solid or fluid minerals over geological time. Fossil fuels include coal, petroleum, and natural gas. Fossil fuels are non-renewable on the timescale of human civilization, because their natural replenishment would take many millions of years.

PHOTOSYNTHESIS: The process by which green plants use light to synthesize organic compounds from carbon dioxide and water. In the process, oxygen and water are released. Increased levels of carbon dioxide can increase net photosynthesis in some plants. Plants create a very important reservoir for carbon dioxide.

IN CONTEXT: BIOENERGY FEEDSTOCKS

"Biomass from agricultural residues and dedicated energy crops can be an important bioenergy feedstock, but its contribution to mitigation depends on demand for bioenergy from transport and energy supply, on water availability, and on requirements of land for food and fibre production. Widespread use of agricultural land for biomass production for energy may compete with other land uses and can have positive and negative environmental impacts and implications for food security."

SOURCE: *Metz, B., et al, eds.* Climate Change 2007: Mitigation of Climate Change: Contribution of Working Group III to the Fourth Assessment Report of the Intergovernmental Panel on Climate Change. *New York: Cambridge University Press, 2007.*

that eat the primary consumers. Because the transfer of energy from one level of an ecological pyramid to another is inefficient, the biomass at each level also decreases. In general, the biomass of a given level is only 10% of the preceding level.

Industrial engineers use the concept of biomass to describe the quantity of a plant crop that can be used as an energy source. Related terms are biofuels and bioenergy. Some of the plants most often used for biomass are corn, sugarcane, grasses, and hemp. Garbage, wood, landfill gases, and alcohol fuels may also be considered biomass.

■ Impacts and Issues

As fossil fuels become more expensive and concern about the increase in carbon dioxide and other pollutants produced by fossil fuels in the atmosphere intensifies, industrial researchers have become more invested in developing alternative energy sources. Biomass is one of these alternatives.

Many industrial scientists argue that burning of biomass is a carbon-neutral contributor to greenhouse gases in the environment. In its simplest form, this argument states that because the photosynthesis that produces the crops draws carbon dioxide out of the environment, burning the plants for energy releases the same carbon dioxide back into the atmosphere. Critics argue that the fertilizer and farm equipment required to produce the biomass add significant carbon dioxide to the atmosphere.

Research continues into developing biomass that both creates less pollution than fossil fuels and provides a more economical source of energy. Some ideas for accomplishing these goals include collecting methane gas produced by biomass decomposition in landfills, burning municipal waste in order to slow its accumulation in landfills, and developing technologies to burn wood in a cleaner fashion. In addition, using bioethanol and biodiesel may result in fewer pollutants than the combustion of fossil fuels.

See Also *Biofuel Impacts; Biosphere; Carbon Dioxide (CO_2); Carbon Sequestration Issues; Carbon Sinks; Energy Efficiency; Energy Industry Activism; Ethanol; Industry (Private Action and Initiatives); Methane; Natural Gas; Petroleum; Petroleum: Economic Uses and Dependency; Renewable Energy; Social Cost of Carbon (SCC).*

BIBLIOGRAPHY

Books

Raven, Peter H., and Linda R. Berg. *Environment.* Hoboken, NJ: John Wiley and Sons, 2006.

Web Sites

"BIOMASS—Renewable Energy from Plants and Animals." *Energy Information Administration,* October 2006. <http://www.eia.doe.gov/kids/energyfacts/sources/renewable/biomass.html> (accessed October 16, 2007).

"Introduction to Biogeochemical Cycles." *University of Colorado.* <http://www.colorado.edu/GeolSci/courses/GEOL1070/chap04/chapter4.html> (accessed October 16, 2007).

Juli Berwald

Biosphere

Introduction

The biosphere is that part of Earth comprising the living ecosystem, which includes all living matter of the world and the inorganic components that they rely upon. The biosphere is a part of the world on par with the lithosphere (Earth's rocky crust), the atmosphere (Earth's gaseous envelope), and the hydrosphere (Earth's water realm). As scientists have come to know more about the deeper reaches of Earth's crust and the sea floor, the depth of the biosphere has grown greater. The term biosphere was first used by Austrian geologist Eduard Suess (1831–1914), who said it was "the place on the Earth's surface where life dwells."

Historical Background and Scientific Foundations

The biosphere may be divided into biomes, or subparts with respect to latitude (climate) and elevation (depth in the ocean, depth in the crust, and altitude on mountains). Polar biomes are found at high latitude. Mid-latitude biomes and equatorial biomes are at lower latitudes respectively. In the ocean, biomes are found at shallow depths (for example, near shore or on continental shelves), at abyssal depths (for example on the sea floor), and in oceanic trenches, where water may be over 5 mi (8 km) in depth. In the crust, biomes exist at shallow depths (for

According to the World Wildlife Fund, the tufted puffin is especially vulnerable to changes in its ecosystem. A study in Canada revealed that warmer ocean surface temperatures have led to significantly lower growth rates of the puffins. *Image copyright Kerry L. Werry, 2007. Used under license from Shutterstock.com.*

example, soils and surface rocks) or up to 5.6 mi (9 km) or more in crustal rocks. In the atmosphere, some birds fly at heights up to 6.8 mi (11 km) or more. On land, microscopic biomes exist from sea level to the highest mountain areas. There are several main ecosystems within the biome of the land, including forests, deserts, grasslands, and wetlands.

No matter how the biosphere is subdivided, all living organisms are interrelated in some way. Within the biosphere and its biomes are regional and local ecosystems. Each of these ecosystems consists of living organisms and the non-living components upon which they depend. The living organisms depend upon one another as parts of the food web. The food web has a hierarchy: producers (organisms that make food), consumers (organisms that consume producers), and decomposers (organisms that recycle dead organisms). Local ecosystems are parts of the regional ecosystem and, in turn, are components in the biomes and global biosphere.

Biosphere and Earth History

The biosphere has existed and continued to expand, with brief lapses, over most of the 4.6-billion-year history of Earth. It is not known exactly when life arose on Earth, but it is a consensus view that life was probably in existence by about 3.8 billion years ago. The diversity of life has increased over time, both in the number of different groups of organisms and the complexity of some of the groups.

The biosphere has experienced some crises during Earth history in which large numbers of organic groups have been extinguished over a brief span of time. These events are called mass extinctions and represent episodes of reorganization of ecosystems and biomes of Earth. For example, the mass extinction that claimed the dinosaurs and many other species about 65 million years ago was such an event. There have been about four such other events in Earth history over the past 500 million years.

■ Impacts and Issues

Biosphere is a term associated with some biology experiments that have been conducted over the past few decades. Biosphere 2 (so named because the Earth is viewed as Biosphere 1) was a large laboratory experiment in Arizona. Bios-3 was an experimental facility operated by the government of the former Soviet Union in the late 1960s. Biosphere J is a more recent Japanese experiment. The biospheres were restricted environments where a local ecosystem was simulated.

WORDS TO KNOW

BIOME: Well-defined terrestrial environment (e.g., desert, tundra, or tropical forest). The complex of living organisms found in an ecological region.

CRUST: The hard, outer shell of Earth that floats upon the softer, denser mantle.

FOSSIL RECORD: The time-ordered mass of fossils (mineralized impressions of living creatures) that is found in the sedimentary rocks of Earth. The fossil record is one of the primary sources of knowledge about evolution and is also used to date rock layers (biostratigraphy).

HYDROSPHERE: The totality of water encompassing Earth, comprising all the bodies of water, ice, and water vapor in the atmosphere.

LITHOSPHERE: The rigid, uppermost section of Earth's mantle, especially the outer crust.

Earth's biosphere necessarily changes in response to climatic change. Climate change can spur migrations and relocations of many organisms. Likewise, climate change can spell doom for species that cannot adapt to such changes. The demise of species opens the way for new species to arise to fill those niches. As the fossil record shows, over the long span of geological time, many species have adapted, become extinct, or arisen as a result of climatic change.

SEE ALSO *Biodiversity; Extinction; Gaia.*

BIBLIOGRAPHY

Books

Lovelock, James. *Gaia: A New Look at Life on Earth.* Oxford, UK: Oxford University Press, 2000.

Parry, M. L., et al, eds. *Climate Change 2007: Impacts, Adaptation and Vulnerability: Contribution of Working Group II to the Fourth Assessment Report of the Intergovernmental Panel on Climate Change.* New York: Cambridge University Press, 2007.

Web Sites

Pidwirny, Michael. "Fundamentals of Physical Geography. Chapter 9: Introduction to the Biosphere." *PhysicalGeography.net*, 2007. <http://www.physicalgeography.net/fundamentals/chapter9.html> (accessed November 30, 2007).

David T. King Jr.

Blizzards

Introduction

Blizzards are extreme snow events that occur when high winds whip snow cover and falling snow into suspension in the air for prolonged periods, causing deep drifts, extreme wind chill, and severely reduced visibility. Blizzards typically occur from late fall to early spring at high latitudes and altitudes, and less frequently in temperate regions.

The effect of human-caused climate change on blizzards, which tend to be associated with cold, low pressure zones in storm systems called cyclones, has not been thoroughly studied and will likely vary from region to region. However, as with other types of snowstorms, blizzards are, in general, predicted to occur less frequently over the course of the century. This decrease in total number of storms will likely be accompanied by an increase in the intensity of blizzards that do occur, due to elevated levels of humidity as temperatures climb, resulting in more precipitation and more violent storms.

Historical Background and Scientific Foundations

The occurrence of blizzards has varied widely over the past 50 years in different locations. On the prairies of western Canada, for example, the number of blizzards decreased between 1953 and 1997. In the contiguous United States,

The Himalayas have experienced a consistent increase in temperatures while receiving significantly less snowfall in recent years. Less snowfall and fewer storms may jeopardize ecosystems that rely on snow for freshwater. *Image copyright Alexey Fateev, 2007. Used under license from Shutterstock.com.*

the number of blizzards has increased slightly since the 1950s, while the overall number of severe, damaging snowstorms has shown no significant trend in either direction. But the incidence of more extreme precipitation events like severe blizzards, as well as above normal temperatures and drought, rose considerably in different parts of the United States between 1980 and 1994. Meanwhile, subsistence reindeer herders in eastern Russia have also reported anecdotal increases in intense snowstorms and other extreme weather events.

As temperatures warm due to increased greenhouse-gas emissions over the coming century, storms may become more infrequent and more extreme. General circulation computer models show that extra-tropical cyclones, which tend to spawn severe snowstorms like blizzards in the wintertime at mid and high latitudes, will likely decrease. Rising temperatures will also mean that more precipitation falls as rain. These models also predict an increase in cyclone intensity, which may, in turn, result in an increase in the intensity of accompanying snowstorms. Climate models that factor in greater atmospheric concentrations of carbon dioxide and warming tend to predict the greatest increases in cyclone intensity.

WORDS TO KNOW

CLIMATE MODEL: A quantitative way of representing the interactions of the atmosphere, oceans, land surface, and ice. Models can range from relatively simple to quite comprehensive.

LATITUDE: The angular distance north or south of Earth's equator measured in degrees.

PRECIPITATION: Moisture that falls from clouds. Although clouds appear to float in the sky, they are always falling, the water droplets slowly being pulled down by gravity. Because the water droplets are so small and light, it can take 21 days to fall 1,000 ft (305 m) and wind currents can easily interrupt their descent. Liquid water falls as rain or drizzle. All raindrops form around particles of salt or dust. (Some of this dust comes from tiny meteorites and even the tails of comets.) Water or ice droplets stick to these particles, then the drops attract more water and continue getting bigger until they are large enough to fall out of the cloud. Drizzle drops are smaller than raindrops. In many clouds, raindrops actually begin as tiny ice crystals that form when part or all of a cloud is below freezing. As the ice crystals fall inside the cloud, they may collide with water droplets that freeze onto them. The ice crystals continue to grow larger, until large enough to fall from the cloud. They pass through warm air, melt, and fall as raindrops.

■ Impacts and Issues

Blizzards can pose significant impediments to travel by ground and air, hamper economic activity, knock out electricity, damage structures, and wipe out livestock, as well as cause human injuries and death through vehicle crashes and exposure. But big storms can also boost soil moisture and water supplies, and be an economic boon to winter recreation industries like ski resorts.

A decrease in the overall frequency of severe snowstorms may benefit certain industries, but an increase in the intensity of storms that do occur will increase the danger and damage resulting from individual storms. Meanwhile, less overall snowfall from fewer storms may jeopardize ecosystems that rely on snow for freshwater, such as those in big mountain ranges like the Himalayas or the Rocky Mountains.

SEE ALSO *Anthropogenic Change; Catastrophism; Extreme Weather; General Circulation Model (GCM); Lake Effect Snows; Melting; Rainfall; Tourism and Recreation.*

BIBLIOGRAPHY

Books

Weart, Spencer. *The Discovery of Global Warming.* Cambridge, MA: Harvard University Press, 2004.

Periodicals

Changnon, Stanley A., and David Changnon. "A Spatial and Temporal Analysis of Damaging Snowstorms in the United States." *Natural Hazards* 37 (March 2006): 373–389.

Easterling, D. R., et al. "Observed Variability and Trends in Extreme Climate Events: A Brief Review." *Bulletin of the American Meteorological Society* 81 (March 2000): 417–425.

Lambert, Steven J. "The Effect of Enhanced Greenhouse Warming on Winter Cyclone Frequencies and Strengths." *Journal of Climate* 8 (May 1995): 1447–1452.

Lambert, Steven J., and John C. Fife. "Changes in Winter Cyclone Frequencies and Strengths Simulated in Enhanced Greenhouse Warming Experiments: Results from the Models Participating in the IPCC Diagnostic Exercise." *Climate Dynamics* 26 (June 2006): 713–728.

Lawson, Bevan D. "Trends in Blizzards in Selected Locations on the Canadian Prairies." *Natural Hazards* 29 (June 2003): 123–138.

McCabe, Gregory J., et al. "Trends in Northern Hemisphere Surface Cyclone Frequency and Intensity." *Journal of Climate* 14 (June 2001): 2763–2768.

Schwartz, Robert M., and Thomas W. Schmidlin. "Climatology of Blizzards in the Conterminous United States, 1959–2000." *Journal of Climate* 15 (July 2002): 1765–1772.

Stevens, William K. "Blame Global Warming for the Blizzard." *The New York Times* (January 14, 1996).

Web Sites

"Witnessing Climate Change in Russia's Far East." *WWF*, May 10, 2006. <http://www.panda.org/news_facts/newsroom/features/index.cfm?uNewsID=67580> (accessed November 18, 2007).

Sarah Gilman

Bonn Conference (2001)

■ Introduction

In July 2001, at the Bonn Conference, 178 countries agreed to revisions to the Kyoto Protocol. Several participating nations claimed the changes made the protocol more workable, ending nearly four years of limbo following its initial drafting. The Bonn Conference set in motion the process leading to the Kyoto Protocol's implementation in February 2005.

■ Historical Background and Scientific Foundations

Prior to the Bonn Conference, the Kyoto Protocol established binding targets for reductions in six key greenhouse gases. Initially drafted in December 1997, the landmark agreement formalized the principles of the United Nations Framework Convention on Climate Change (UNFCCC), and was later signed by most

While the United Nations Framework Convention on Climate Change (UNFCCC) was meeting in Bonn, Germany, in 2001, protesters demonstrated for change. Some activists tried to block the street and were arrested by the German police. *AP Images.*

WORDS TO KNOW

CARBON SINKS: Carbon reservoirs such as forests or oceans that take in and store more carbon (carbon sequestration) than they release. Carbon sinks can serve to partially offset greenhouse-gas emissions.

FOSSIL FUELS: Fuels formed by biological processes and transformed into solid or fluid minerals over geological time. Fossil fuels include coal, petroleum, and natural gas. Fossil fuels are non-renewable on the timescale of human civilization, because their natural replenishment would take many millions of years.

GREENHOUSE GASES: Gases that cause Earth to retain more thermal energy by absorbing infrared light emitted by Earth's surface. The most important greenhouse gases are water vapor, carbon dioxide, methane, nitrous oxide, and various artificial chemicals such as chlorofluorocarbons. All but the latter are naturally occurring, but human activity over the last several centuries has significantly increased the amounts of carbon dioxide, methane, and nitrous oxide in Earth's atmosphere, causing global warming and global climate change.

JOINT IMPLEMENTATION: Type of greenhouse mitigation project defined by the Kyoto Protocol (1997). In a joint implementation project, one developed country can finance a project in another developed country to reduce greenhouse-gas emissions. The goal, as with the other "flexible mechanisms" specified by Kyoto, is to reduce total greenhouse emissions.

KYOTO PROTOCOL: Extension in 1997 of the 1992 United Nations Framework Convention on Climate Change (UNFCCC), an international treaty signed by almost all countries with the goal of mitigating climate change. The United States, as of early 2008, was the only industrialized country to have not ratified the Kyoto Protocol, which is due to be replaced by an improved and updated agreement starting in 2012.

LEAST DEVELOPED COUNTRIES: The world's poorest countries. The United Nations classifies a country as a Least Developed Country if per-capita income is less than $750 for three years running and if the country has low health and literacy rates. As of 2007, there were 48 Least Developed Countries, of which 33 were in Africa.

nations at the Earth Summit of 1992. Kyoto marked the culmination of several years of intense international negotiations.

Without the participation of the United States, which rejected the Kyoto Protocol mainly due to the lack of including developing countries in the process, along with a lack of specified means and costs for the United States to comply, many observers considered the Kyoto agreement essentially dead. The Bonn Conference was arranged to reach accord among the countries agreeing to the Kyoto Protocol, so that it could proceed with its schedule of implementation.

■ Impacts and Issues

At the Bonn Conference in Germany, two climate change funds, including one for "least developed countries," were established to assist in the switch to cleaner technologies. More money was earmarked for the training of researchers to monitor emissions, and also to encourage OPEC countries to diversify their fossil fuel centered economies.

Among the most contentious developments were new reduction mechanisms. The amendments permitted emissions trading, enabling countries to sell spare greenhouse-gas emission allotments to other signatory countries. Although limitations on trading were placed on most countries, significantly, neither Russia nor Ukraine were party to such restrictions. Following the collapse of the Soviet Union, both countries witnessed a collapse in carbon emissions compared to their 1990 benchmarks, giving each a vast—although likely temporary—surplus. Financial incentives offered by emissions trading promised to open the way for both countries to ratify the Kyoto Protocol.

The joint implementation program was also controversial at the Bonn Conference. Under this program, participating countries bound to reduce emissions were permitted to claim credit for investment in clean technology projects in developing countries. An international U.N. regulatory panel was established to assess qualifying projects as they were implemented.

Heavily forested countries such as Russia and Canada wanted to use their carbon sinks as an offset against reduction measures—essentially claiming credit for the absorption of carbon through trees in dense forests. The European Union (EU) nations opposed an amendment allowing consideration of carbon sinks. A compromise was reached by which countries were allowed to take credit for forestry management schemes, but with limitations.

Penalties were also put in place for those countries that failed to meet their Kyoto or amended Bonn emissions-reduction targets. For every ton of carbon a country emits over its limit, it will be obliged to reduce an additional 1.3 tons after 2012 while also suffering a ban from emissions trading.

Bonn also served as final recognition that the United States would never ratify the Kyoto Protocol. Other developed nations, especially those of the European Union, and leading international environmental groups criticized U.S. policy on climate change. The U.S. rejection of Kyoto was largely unpopular in international public opinion, especially in Europe and South America. America's chief delegate, Paula Dobriansky, was jeered by observers in the public gallery at the conference's closing ceremony when she said that "the Bush administration takes the issue of climate change very seriously." However, the United States was not the only developed nation and leading emitter of greenhouse gases to fail to ratify; Australia also rejected Kyoto at that time.

The Bonn Agreement reaffirmed many of the goals of the Kyoto Protocol. But while hailed by many environmental groups and participating nations as a breakthrough, the Bonn agreement was still subject to heavy criticism. In particular, under Kyoto industrialized countries were to reduce greenhouse gases by an average of 5.2% below their 1990 level by 2012; under the Bonn Agreement, this target was reduced to about 2%.

SEE ALSO *Berlin Mandate; Kyoto Protocol; United States: Climate Policy.*

BIBLIOGRAPHY

Periodicals

Giles, Jim. "Political Fix Saves Kyoto from Collapse." *Nature* 412 (2001): 365.

Web Sites

Lynas, Mark. "Kyoto Could Make Things Even Worse." *The Guardian*, July 27, 2001. <http://education.guardian.co.uk/higher/comment/story/0,,528315,00.html> (accessed November 21, 2007).

Official Website of the Sixth Conference of Parties: July 16–27, 2001. <http://unfccc.int/cop6_2/> (accessed November 21, 2007).

"Q & A: The U.S. and Climate Change." *BBC News*, February 14, 2002. <http://news.bbc.co.uk/2/hi/americas/1820523.stm> (accessed November 21, 2007).

James Corbett

Carbon Credits

Introduction

Carbon credits are financial abstractions, similar to money, that entitle their possessor to emit a certain amount of carbon dioxide (CO_2) or other greenhouse gases. They are created by governments and bought and sold on emissions trading markets. Participation by polluters in emissions trading markets may be voluntary, as with the Chicago Climate Exchange, or mandatory, as with the European Climate Exchange. The largest system of carbon credits and emissions trading was established by the Kyoto Protocol in 2005.

The term carbon credits is sometimes used interchangeably with the term carbon offsets, but the latter often refers to voluntary purchase by businesses or individuals of decreased CO_2 emissions in order to cancel out their own emissions (for example, paying a company to plant trees to cancel out the carbon released by one's air travel). Offsetting of this kind is not a tradable commodity but a one-time purchase made for reasons of conscience or public relations. It is discussed in the entry "Offsetting" in this set. In this article, only carbon credits that are a tradable commodity in a legally defined emissions trading system are discussed.

Historical Background and Scientific Foundations

In 1992, the United Nations Framework Convention on Climate Change (UNFCCC) committed all its signatories—-virtually every country in the world—to "formulate, implement, publish and regularly update national and, where appropriate, regional programmes containing measures to mitigate climate change by addressing anthropogenic emissions by sources and removals by sinks." In 1997, to fulfill this earlier commitment, most countries signed the Kyoto Protocol to the UNFCCC.

The Kyoto Protocol defined targets for reduced greenhouse emissions from its industrialized signatories, namely, all the industrialized countries except the United States and Australia. Specifically, industrialized countries signing the protocol agreed to reduce their greenhouse emissions to 5.2% below 1990 levels by 2012. Countries could meet this goal either by actually reducing their emissions or by acquiring carbon credits that represented reductions elsewhere.

Kyoto provided three mechanisms by which signatories could obtain carbon credits: joint implementation, the Clean Development Mechanism, and emissions trading. Reductions or removals in one country can count as credits in another country that has paid or traded for them. These credits can be applied toward the receiving country's emission targets or quotas. For example, if a country is obliged under the protocol to emit no more than 1 million units of carbon but is actually emitting 1.1 million units, it can still meet its treaty obligation by obtaining 100,000 carbon credits through one or more of the following three mechanisms:

- Joint implementation: Refers to implementation by a developed country of an emissions-reducing project or sink-enhancing project in another developed country. According to the United Nations, in practice, joint implementation means the building of emissions-reducing projects in countries formerly part of the Soviet Union, termed "transition economies," using funds from Western European or other industrialized countries. Countries that sponsor projects that reduce emissions or increase uptake by carbon sinks receive credits—the more carbon is kept out of the atmosphere, the more credits are earned.
- Clean Development Mechanism: Allows industrialized countries to earn carbon credits by funding projects in developing countries that either lower greenhouse emissions or enhance removals through sinks, particularly through increasing forests.

Some companies, like this nursery in Madagascar, plant forests of native tree species, which ingest fossil fuel emissions in the atmosphere. Because the forests are able to consume carbon, they are offered for sale to companies and individuals looking to purchase carbon credits. *AP Images.*

- International emissions trading: Allows Kyoto signatory countries to buy and sell carbon credits internationally. In principle, a country that can keep a ton of carbon out of the atmosphere for less than the market price of a carbon credit can then sell its extra credits to countries for whom reducing emissions is more expensive than buying credits. In this way, emission reductions will happen where they are cheapest and all parties will benefit. The United States is the only industrialized country to have signed but not ratified (agreed formally to abide by) the Kyoto Protocol.

The Kyoto Protocol is an international cap-and-trade scheme. That is, it defines a limit on total emissions by the world's industrialized economies, then permits trade and transfer of carbon credits so that savings can happen where they are most profitable. The basic unit or tradable credit in all three of these mechanisms is the Kyoto Allocation Allowance Unit (or "unit") and is equivalent to one metric ton (2,204 lb, 1,000 kg) of CO_2.

The word "equivalent" means that carbon credits do not necessarily stand only for emissions of CO_2. They may represent emissions of other gases, some containing carbon (e.g., methane, CH_4) and some not, such as nitrogen-oxygen compounds. Since CO_2 is the most important greenhouse gas, accounting for 63% of anthropogenic (human-caused) global warming, it is customary in climate science to translate emissions of non-CO_2 gases into the number of tons of CO_2 that would create an equal amount of global warming. For example, since methane is 21 times more powerful a greenhouse gas than CO_2, the right to emit 1/21 of a ton of methane would be marketed as a single carbon credit.

There are several different species of Kyoto carbon credits:

- Assigned amount unit (AAU). Carbon credit issued to a polluter by an Annex I country from its treaty-assigned budget of credits. An AAU is, in effect, permission to emit one ton of carbon dioxide.
- A removal unit (RMU). Carbon credit issued by an Annex I party for carbon reclaimed from the atmosphere, or not emitted to the atmosphere, by forestry and land-use practices. An operator who keeps one ton of greenhouse gas (carbon equivalent) out of the atmosphere through land use practices may, in effect, be issued the right to emit a ton of gas elsewhere.
- Emission reduction unit (ERU). Carbon credit generated by a joint implementation project.
- Certified emission reduction (CER). Carbon credit generated by an emissions-reduction project sponsored by an Annex I country in a non-Annex I country such as China.

The Kyoto Protocol is not the only legal setup for defining and trading carbon credits. The European Union and other groups of countries established smaller carbon markets before the protocol entered into force in February 2005. These systems were designed, however,

WORDS TO KNOW

ANNEX I AND NON-ANNEX I COUNTRIES: Groups of countries defined by the United Nations Framework Convention on Climate Change (UNFCCC) of 1992. Annex I countries are industrialized countries who agreed, under the treaty, to reduce their greenhouse emissions. All non-Annex I countries were developing (poorer, less-industrialized) countries. Annex II countries were wealthy Annex I countries who agreed to help pay for greenhouse reductions by developing (poorer) countries.

ANTHROPOGENIC: Made by people or resulting from human activities. Usually used in the context of emissions that are produced as a result of human activities.

CAP AND TRADE: The practice, in pollution-control or climate-mitigation schemes, of mandating an upper limit or cap for the total amount of some substance to be emitted (e.g., CO_2) and then assigning allowances or credits to polluters that correspond to fixed shares of the total amount. These allowances or credits can then be bought and sold by polluters, in theory allowing emission cuts to be bought where they are most economically rational.

CARBON SINK: Any process or collection of processes that is removing more carbon from the atmosphere than it is emitting. A forest, for example, is a carbon sink if more carbon is accumulating in its soil, wood, and other biomass than is being released by fire, forestry, and decay. The opposite of a carbon sink is a carbon source.

CARBON TAX: Mandatory fee charged for the emission of a given quantity of CO_2 or some other greenhouse gas. Under a carbon taxation scheme, polluters who emit greenhouse gases must pay costs that are directly proportional to their emissions. The purpose of a carbon tax is to reduce greenhouse emissions. Carbon taxation is the main alternative to emissions trading.

EMISSIONS TRADING MARKET: System for trading in carbon credits, money-like units that entitle their owners to emit 1 ton of carbon dioxide or an equivalent amount of some other greenhouse gas. As of 2007, the most important emissions trading market was the European Union Emission Trading Scheme, with mandatory participation for large greenhouse polluters throughout the European Union (trading since 2005); in the United States, a smaller, voluntary market called the Chicago Climate Exchange (trading since 2003).

GREENHOUSE GASES: Gases that cause Earth to retain more thermal energy by absorbing infrared light emitted by Earth's surface. The most important greenhouse gases are water vapor, carbon dioxide, methane, nitrous oxide, and various artificial chemicals such as chlorofluorocarbons. All but the latter are naturally occurring, but human activity over the last several centuries has significantly increased the amounts of carbon dioxide, methane, and nitrous oxide in Earth's atmosphere, causing global warming and global climate change.

KYOTO PROTOCOL: Extension in 1997 of the 1992 United Nations Framework Convention on Climate Change (UNFCCC), an international treaty signed by almost all countries with the goal of mitigating climate change. The United States, as of early 2008, was the only industrialized country to have not ratified the Kyoto Protocol, which is due to be replaced by an improved and updated agreement starting in 2012.

to create regional markets that would link up with the global market defined by the protocol once it became operational, and so define their tradable carbon credits using Kyoto's terms. As of 2007, carbon credits were being traded on five European exchanges. On the European carbon market, the price of a carbon credit was, in mid-2007, around 20 euros (US$28).

After the ratification of the Kyoto Protocol by Russia in October 2004, the Kyoto-defined trade in carbon credits was forecast by investors in the financial industry to become one of the world's largest commodity markets, trading somewhere between $60 billion and $250 billion annually by 2008. By mid-2007, the market was still trading only $30 billion, but investment analysts were predicting that it could grow to as much as $1 trillion by 2017.

Impacts and Issues

The Kyoto Protocol has been criticized by most conservatives and libertarians as an ineffectual system for reducing greenhouse-gas emissions. Cap-and-trade schemes defining carbon credits as an international commodity, these critics say, are a form of top-down interference in markets and should be replaced by voluntary schemes that encourage, but do not require, reductions in emissions.

Carbon credits, as well as pollution credits of other kinds, have been criticized by environmentalists and political progressives as granting a right to pollute and so being essentially unethical. The Durban Declaration of 2004, signed by over 150 popular organizations, mostly from undeveloped nations in the global South, states that carbon trading "turns the Earth's recycling capacity into property to be bought or sold in a global market. Through this process of creating a new commodity—carbon—the Earth's ability and capacity to support a climate conducive to life and human societies is now passing into the same corporate hands that are destroying the climate" (quoted in Vallette et al., 2004).

Such critics urge that commodification through carbon credits should be replaced by what they term the "polluter pays principle," whereby direct taxation or fines would be imposed on emitters in proportion to how much they emit. Carbon taxes are a form of polluter-pays system. Some environmentalists, as well as some fiscal conservatives, have accused the European carbon market and Kyoto global carbon market of being a scam that funnels profits to private investors without significantly reducing greenhouse emissions. These critics charge that emissions caps have been set so high that industries are not actually reducing their emissions significantly, if at all.

■ Primary Source Connection

In this press release, the Environment Secretary of the United Kingdom discusses a novel proposal under which each citizen of the U.K. would be issued personal carbon allowances. The goal of the plan is to reduce carbon emissions in the U.K. Individuals would have carbon credits deducted from their allowances for any activity that results in carbon emissions. People who exhaust their carbon allowance would be forced to buy credits from other individuals on an open market.

GOVERNMENT TO CONSIDER PERSONAL CARBON ALLOWANCES

The Government is contemplating issuing tradeable personal carbon allowances to the public to combat rising emissions from the domestic sector, according to Environment Secretary David Miliband.

Mr Miliband explained the merits of such a system at the Audit Commissions annual lecture on Wednesday evening.

He said he believes it is worthwhile giving people a limited amount of carbon allowances, of which they can sell any surplus for cash should they opt to reduce their emissions.

The scheme would be fairer than tax increases as it offers free entitlements and only penalises those that exceed their entitlement, he added.

Personal carbon allowances cover people's direct use of energy through their electricity, gas, petrol and air travel and make up to 44% of the economy's total emissions.

"Imagine a country where carbon becomes a new currency," he said. "We carry bank cards that store both pounds and carbon points.

"When we buy electricity, gas and fuel, we use our carbon points, as well as pounds. To help reduce carbon emissions, the Government would set limits on the amount of carbon that could be used.

"People on low incomes are likely to benefit as they will be able to sell their excess allowances. People on higher incomes tend to have higher carbon emissions due to higher car ownership and usage, air travel and tourism, and larger homes," he added.

"It is more empowering than many forms of regulation because instead of banning particular products, services or activities, or taxing them heavily, a personal carbon allowance enables citizens to make trade-offs.

"It is also empowering because many citizens want to be able to do their bit for the environment and tackle climate change, but there is no measurable way of guiding their decisions."

Personal carbon tradeable allowances is one of a number of ways the Government is looking at how individuals can be better informed and involved in tackling climate change.

Carbon loyalty cards, league tables, the use of carbon offsets at point of purchase for certain sectors, awareness raising through labelling, and carbon calculators are all being explored as potential long term measures.

GOVERNMENT OF THE UNITED KINGDOM OF GREAT BRITAIN AND NORTHERN IRELAND. "GOVERNMENT TO CONSIDER PERSONAL CARBON ALLOWANCES," PRESS RELEASE. JULY 19, 2006. <HTTP://WWW.DEFRA.GOV.UK/NEWS/LATEST/2006/CLIMATE-0719.HTM> (ACCESSED NOVEMBER 29, 2007).

SEE ALSO *Clean Development Mechanism; Emissions Trading; Kyoto Protocol; Offsetting; Social Cost of Carbon (SCC).*

BIBLIOGRAPHY

Periodicals

Chameides, William, and Michael Oppenheimer. "Carbon Trading Over Taxes." *Science* 315 (2007): 1670.

Kanter, James. "Carbon Trading: Where Greed Is Green." *International Herald Tribune*, June 20, 2007.

Web Sites

"Emission Impossible: Access to JI/CDM Credits in Phase II of the EU Emissions Trading Scheme." *World Wide Fund for Nature, UK*, June 2007. <http://assets.panda.org/downloads/emission_impossible__final_.pdf> (accessed November 2, 2007).

"Emissions Trading." *United Nations Framework Convention on Climate Change Secretariat*, June 22, 2007. <http://unfccc.int/kyoto_protocol/mechanisms/emissions_trading/items/2731.php> (accessed November 2, 2007).

Vallette, Jim, et al. "A Wrong Turn from Rio: The World Bank's Road to Climate Catastrophe." *Sustainable Energy and Economy Network*, December 2004. <http://www.seen.org/PDFs/Wrong_turn_Rio.pdf> (accessed November 2, 2007).

Larry Gilman

Carbon Cycle

■ Introduction

The carbon cycle describes the reservoirs and transformations that the atom carbon undergoes as it cycles through the atmosphere, lithosphere, biosphere, and hydrosphere (Earth's air, land, water, and other areas that support life). The carbon cycle plays a key role in climate change because carbon dioxide is perhaps the most important gas contributing to global warming in the atmosphere. In addition, carbon is one of the most basic molecules used by organisms to store energy in chemical bonds.

■ Historical Background and Scientific Foundations

One of the largest reservoirs of carbon is carbon dioxide in the atmosphere. Plants on land and algae in water-based ecosystems incorporate carbon dioxide from the atmosphere into carbohydrates during photosynthesis. Some algae sink to the lakebed or seafloor, and over millennia, algal cells are converted into natural gas and oil. Similarly, plant material can be buried and converted into coal. These fossilized plants, or fossil fuels, represent a large carbon reservoir. When fossil fuels and plant biomass are

Using a no-till drill, an Ohio farmer plants soybeans. By planting crops without plowing the soil, he hopes to slow global climate change by trapping carbon in the soil instead of releasing carbon dioxide into the air. *AP Images.*

burned, carbon dioxide is released from the chemical bonds of these materials and returned to the atmosphere.

Chemically mediated transformations of carbon within the larger carbon cycle described above also occur. Carbon dioxide from the atmosphere can be dissolved into water as carbonate and bicarbonate. Similarly, it can be released from its dissolved form back into the atmosphere.

Biologically mediated cycles are also present within the larger carbon cycle. Plants and algae respire carbon dioxide directly into the atmosphere. When an animal consumes a plant, it incorporates the carbon from the plant's carbohydrates into its own biomolecules. During animal respiration, carbon dioxide is released into the atmosphere.

Some marine organisms incorporate dissolved carbon into their shells. When these organisms die, they sink to the bottom of the ocean and leave their shells behind. Over long periods of time, these shells are transformed into limestone rock. When limestone is uplifted, it becomes land. As the limestone weathers, carbon is returned to the atmosphere and dissolved in water.

Impacts and Issues

Because of industrial dependence on fossil fuels, larger concentrations of carbon dioxide are input to the atmosphere than can be taken up by biologically and chemically mediated transformations. In addition, burning large sections of forests without replanting has released carbon dioxide into the atmosphere and decreased the size of a significant carbon reservoir.

Carbon dioxide has a property that allows it to trap heat. Most scientists agree that the increase in carbon dioxide in the atmosphere has caused an increase in the global temperature and will likely continue to do so for quite some time. Such an increase has significant impacts, including the melting of the polar ice sheets contributing to sea level rise, changes in the patterns and intensity of weather systems, the increased spread of insect-born disease, and challenges for agriculture.

SEE ALSO *Arctic Melting: Greenland Ice Cap; Atmospheric Chemistry; Atmospheric Circulation; Atmospheric Pollution; Atmospheric Structure; Biogeochemical Cycle; Biosphere; Carbon Credits; Carbon Dioxide (CO_2); Carbon Dioxide Concentrations; Carbon Sequestration Issues; Carbon Sinks; Coal; Extreme Weather; Forests and Deforestation; Glacier Retreat; Global Warming; Industry (Private Action and Initiatives); Infectious Disease and Climate Change; Melting; Natural Gas; Petroleum; Petroleum: Economic Uses and Dependency; Polar Ice; Sea Level Rise; Sequestration; Sink; Social Cost of Carbon (SCC).*

WORDS TO KNOW

BIOSPHERE: The sum total of all life-forms on Earth and the interaction among those life-forms.

CARBON: Chemical element with atomic number 6. The nucleus of a carbon atom contains 6 protons and from 6 to 8 neutrons. Carbon is present, by definition, in all organic substances; it is essential to life and, in the form of the gaseous compounds CO_2 (carbon dioxide) and CH_4 (methane), the major driver of climate change.

FOSSIL FUELS: Fuels formed by biological processes and transformed into solid or fluid minerals over geological time. Fossil fuels include coal, petroleum, and natural gas. Fossil fuels are non-renewable on the timescale of human civilization, because their natural replenishment would take many millions of years.

HYDROSPHERE: The totality of water encompassing Earth, comprising all the bodies of water, ice, and water vapor in the atmosphere.

LIMESTONE: A carbonate sedimentary rock composed of more than 50% of the mineral calcium carbonate ($CaCO_3$).

LITHOSPHERE: The rigid, uppermost section of Earth's mantle, especially the outer crust.

PHOTOSYNTHESIS: The process by which green plants use light to synthesize organic compounds from carbon dioxide and water. In the process, oxygen and water are released. Increased levels of carbon dioxide can increase net photosynthesis in some plants. Plants create a very important reservoir for carbon dioxide.

RESERVOIR: A natural or artificial receptacle that stores a particular substance for a period of time

TRANSFORMATION: The processes involved in the transfer of a substance from one reservoir to another.

BIBLIOGRAPHY

Books

Raven, Peter H., and Linda R. Berg. *Environment.* Hoboken, NJ: John Wiley and Sons, 2006.

Web Sites

"Carbon Cycle." *Environmental Literacy Council,* September 25, 2006. <http://www.enviroliteracy.org/article.php/478.html> (accessed October 17, 2007).

"The Carbon Cycle." *NASA's Earth Observatory.* <http://earthobservatory.nasa.gov/Library/CarbonCycle/> (accessed October 17, 2007).

"Understanding the Global Carbon Cycle." *Woods Hole Research Center,* 2007. <http://www.whrc.org/carbon/index.htm> (accessed October 17, 2007).

Juli Berwald

Carbon Dioxide (CO_2)

Introduction

Carbon dioxide consists of one carbon (C) atom doubly bonded to two oxygen (O) atoms to produce the molecular compound O=C=O, with the chemical symbol CO_2. It does not have color or odor, but carbon dioxide does have a slight sour taste. This acidic flavor is found in carbonated beverages because carbon dioxide gives such liquids their fizz. Under normal temperatures and pressures found on Earth, carbon dioxide is a gas but can also be found in liquid and solid states.

Scientists have learned that carbon dioxide is found in Earth's atmosphere in a proportion of about 300 to 400 parts per million (ppm). However, as the global climate has been studied, an increase in the concentration of carbon dioxide has been noticed by scientists at a rate of about 0.4% annually. This increase in concentration is important to the study of climate change because scientists contend that a portion of this increase is due to the combustion of fossil fuels and vegetation such as wood, petroleum, and coal.

Historical Background and Scientific Foundations

Flemish chemist Johannes van Helmont (1580-1644) observed that the final mass was less than the initial mass when charcoal burned in a closed container. Van Helmont

Pollution in China is extreme in some areas as the country undergoes industrialization. Here, a cyclist wears a face mask as she travels in Gansu Province in December 2006. The pollution caused visibility to fall to 984 ft (300 m) that morning. *AP Images.*

WORDS TO KNOW

BIOSPHERE: The sum total of all life-forms on Earth and the interaction among those life-forms.

CARBON CYCLE: All parts (reservoirs) and fluxes of carbon. The cycle is usually thought of as four main reservoirs of carbon interconnected by pathways of exchange. The reservoirs are the atmosphere, terrestrial biosphere (usually includes fresh-water systems), oceans, and sediments (includes fossil fuels). The annual movements of carbon, the carbon exchanges between reservoirs, occur because of various chemical, physical, geological, and biological processes. The ocean contains the largest pool of carbon near the surface of Earth, but most of that pool is not involved with rapid exchange with the atmosphere.

FOSSIL FUELS: Fuels formed by biological processes and transformed into solid or fluid minerals over geological time. Fossil fuels include coal, petroleum, and natural gas. Fossil fuels are non-renewable on the timescale of human civilization, because their natural replenishment would take many millions of years.

GREENHOUSE GAS: A gaseous component of the atmosphere contributing to the greenhouse effect. Greenhouse gases are transparent to certain wavelengths of the sun's radiant energy, allowing them to penetrate deep into the atmosphere or all the way into Earth's surface. Greenhouse gases and clouds prevent some infrared radiation from escaping, trapping the heat near Earth's surface where it warms the lower atmosphere. Alteration of this natural barrier of atmospheric gases can raise or lower the mean global temperature of Earth.

INDUSTRIAL REVOLUTION: The period, beginning about the middle of the eighteenth century, during which humans began to use steam engines as a major source of power.

PHOTOSYNTHESIS: The process by which green plants use light to synthesize organic compounds from carbon dioxide and water. In the process, oxygen and water are released. Increased levels of carbon dioxide can increase net photosynthesis in some plants. Plants create a very important reservoir for carbon dioxide.

RESPIRATION: The process by which animals use up stored foods (by combustion with oxygen) to produce energy.

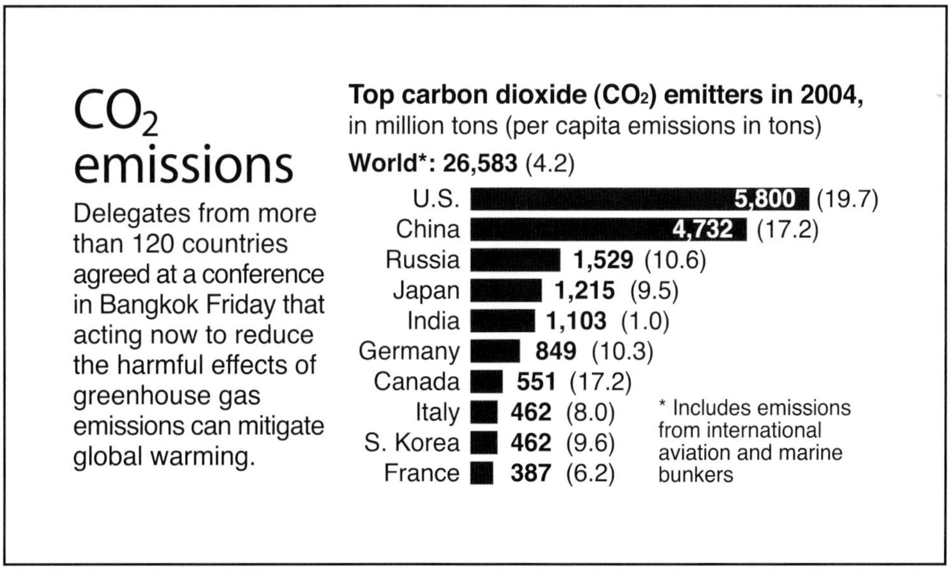

CO_2 emissions

Delegates from more than 120 countries agreed at a conference in Bangkok Friday that acting now to reduce the harmful effects of greenhouse gas emissions can mitigate global warming.

Top carbon dioxide (CO_2) emitters in 2004, in million tons (per capita emissions in tons)

World*: 26,583 (4.2)

Country	Million tons	Per capita (tons)
U.S.	5,800	(19.7)
China	4,732	(17.2)
Russia	1,529	(10.6)
Japan	1,215	(9.5)
India	1,103	(1.0)
Germany	849	(10.3)
Canada	551	(17.2)
Italy	462	(8.0)
S. Korea	462	(9.6)
France	387	(6.2)

* Includes emissions from international aviation and marine bunkers

This graphic from the Associated Press shows the top carbon dioxide emitters with their per capita emissions in 2004. [Source: International Energy Agency.] *AP Images.*

conjectured the difference was the removal of an invisible substance (eventually identified as carbon dioxide).

Later, Scottish chemist Joseph Black (1728–1799) produced an unknown substance—called "fixed air"—when he broke down chalk and limestone. Black is generally considered the first scientist to discover carbon dioxide.

In addition, Swedish chemist Svante Arrhenius (1859–1927) was the first scientist to publish a paper about carbon dioxide emissions from the burning of fossil fuels. Arrhenius published "On the influence of carbonic acid in the air upon the temperature of the ground" in 1896.

IN CONTEXT: CO_2 CAPTURE AND STORAGE

"CCS [CO_2 Capture and Storage] in underground geological formations is a new technology with the potential to make an important contribution to mitigation by 2030. Technical, economic and regulatory developments will affect the actual contribution."

SOURCE: *Metz, B., et al. "IPCC, 2007: Summary for Policymakers." In:* Climate Change 2007: Mitigation of Climate Change. Contribution of Working Group III to the Fourth Assessment Report of the Intergovernmental Panel on Climate Change *New York: Cambridge University Press. 2007.*

■ Impacts and Issues

Carbon dioxide plays an important role in the carbon cycle, which involves an exchange of carbon dioxide between the biosphere (living beings), geosphere (land masses), hydrosphere (water bodies), and atmosphere (air).

Under photosynthesis, green plants convert carbon dioxide and water into food such as oxygen and glucose. Conversely, plants and animals release carbon dioxide in a process called respiration.

However, humans affect the amount of carbon dioxide released into the atmosphere with their artificially produced activities such as the burning of fossil fuels, clearing of forests, and the use of fuel-powered vehicles. Because of such activity, the concentration of carbon dioxide in the atmosphere has been steadily increasing since about 1850, the start of the second Industrial Revolution. As reported by the United Nations' Intergovernmental Panel on Climate Change (IPCC), between 1970 and 2004 the increase in CO_2 emissions has grown by about 80%.

The environmental problem called global warming is a worry to humans because of the noticed increase of greenhouse gases such as carbon dioxide. Scientists are investigating this greenhouse effect to see if the melting of glaciers and icebergs, increased storm activity, and warmer-than-normal temperatures are due to an exaggerated greenhouse effect. By 2050, the IPCC predicts that carbon dioxide in the atmosphere could reach 450 to 550 ppm.

SEE ALSO *Carbon Cycle; Carbon Dioxide Concentrations; Climate Change; Greenhouse Effect; Greenhouse Gases; Intergovernmental Panel on Climate Change (IPCC).*

BIBLIOGRAPHY

Books

Gunter, Valerie Jan. *Volatile Places: A Sociology of Communities and Environmental Controversies.* Thousand Oaks, CA: Pine Forge Press, 2007.

Mackenzie, Fred T. *Carbon in the Geobiosphere: Earth's Outer Shell.* Dordrecht, Netherlands: Springer, 2006.

National Academy of Engineering, National Research Council of the National Academies. *The Carbon Dioxide Dilemma: Promising Technologies and Policies.* Washington, DC: National Academies Press, 2003.

Web Sites

"Contributions of Working Group III to the Fourth Assessment Report of the Intergovermental Panel on Climate Change." *Intergovermental Panel on Climate Change*, May 4, 2007. <http://www.ipcc.ch/SPM040507.pdf> (accessed November 5, 2007).

Carbon Dioxide Concentrations

Introduction

Carbon dioxide, symbolized by CO_2, is a compound in which each carbon atom binds with two oxygen atoms. CO_2 is released into Earth's atmosphere primarily by volcanoes, the burning of fuels containing carbon, and the decay of plant matter. It is removed from the atmosphere mostly by plants, which take carbon from CO_2 to build their tissues, and by the oceans, in which CO_2 dissolves.

CO_2 is found in the atmospheres of Earth, Mars, and Venus. On Earth, it is normally an invisible, odorless gas. Because it is opaque to some infrared radiation (the electromagnetic waves emitted by warm objects), carbon dioxide in the atmosphere slows the loss of heat energy from Earth into space.

CO_2 is the most important of the greenhouse gases that are causing Earth to warm, changing climate and weather patterns. Atmospheric CO_2 has increased greatly since humans began burning large amounts of coal and petroleum in the nineteenth century. As of mid-2007, CO_2 comprised about 383 parts per million (ppm) or 0.0383% of the atmosphere, an increase of more than 36% over its pre-industrial level of about 280 ppm.

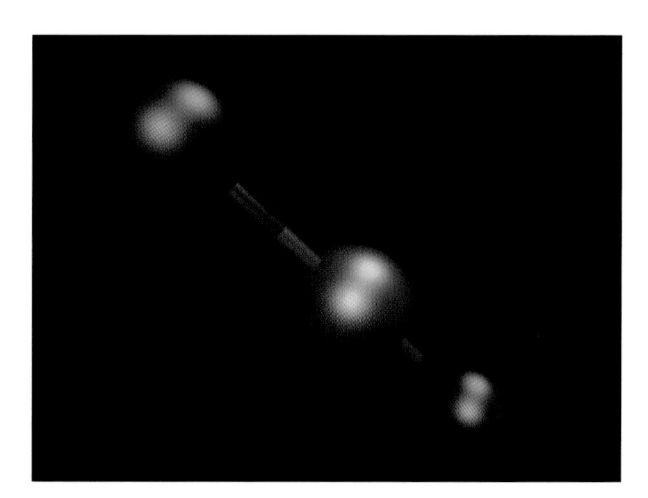

A carbon dioxide molecule, with red spheres representing oxygen atoms and a blue sphere representing the carbon atom. *Image copyright Mark Lorch, 2007. Used under license from Shutterstock.com.*

Historical Background and Scientific Foundations

One source of information about atmospheric carbon dioxide concentration is direct measurement of the air. Such data have been gathered steadily since 1958, when American geochemist Charles David Keeling (1928–2005) made the first atmospheric CO_2 measurements at Mauna Loa Observatory in Hawaii. Keeling found that atmospheric CO_2 tracks the growing season in the Northern Hemisphere, which holds most of the world's vegetation. In the spring and summer, as green plants grow, they remove CO_2 from the air; in the winter, fuel burning and plant decay continue to release CO_2, while plants absorb little.

The result is a series of peaks and valleys in atmospheric CO_2 concentration. From the top of each winter peak to the bottom of each summer valley, CO_2 concentration decreases by about 5 ppm. The terminology "*N* parts per million" means that out of every 1,000,000 gas molecules in a sample of air, *N* are CO_2 molecules. Most of the others are nitrogen (N_2) and oxygen (O_2).

On average, not counting these seasonal variations, the average CO_2 concentration is slowly increasing by about 2 ppm per year. A chart of these changes, plotting time horizontally and CO_2 concentration vertically, is called a Keeling curve and looks somewhat like an upward-curved saw-blade. In 1958, the CO_2 concentration, apart from seasonal ups and downs, was about 315 ppm; today it is over 380 ppm.

In Antarctica and Greenland, annual snowfall has been packing down into thin layers of ice for many years. Small air bubbles trapped in this ice are samples of ancient air. By counting ice layers like tree rings, scientists know

WORDS TO KNOW

FOSSIL FUELS: Fuels formed by biological processes and transformed into solid or fluid minerals over geological time. Fossil fuels include coal, petroleum, and natural gas. Fossil fuels are non-renewable on the timescale of human civilization, because their natural replenishment would take many millions of years.

GREENHOUSE GASES: Gases that cause Earth to retain more thermal energy by absorbing infrared light emitted by Earth's surface. The most important greenhouse gases are water vapor, carbon dioxide, methane, nitrous oxide, and various artificial chemicals such as chlorofluorocarbons. All but the latter are naturally occurring, but human activity over the last several centuries has significantly increased the amounts of carbon dioxide, methane, and nitrous oxide in Earth's atmosphere, causing global warming and global climate change.

ICE CORE: A cylindrical section of ice removed from a glacier or an ice sheet in order to study climate patterns of the past. By performing chemical analyses on the air trapped in the ice, scientists can estimate the percentage of carbon dioxide and other trace gases in the atmosphere at that time.

INFRARED: Wavelengths slightly longer than visible light, often used in astronomy to study distant objects.

INTERGOVERNMENTAL PANEL ON CLIMATE CHANGE (IPCC): Panel of scientists established by the World Meteorological Organization (WMO) and the United Nations Environment Programme (UNEP) in 1988 to assess the science, technology, and socioeconomic information needed to understand the risk of human-induced climate change.

KEELING CURVE: Plot of data showing the steady rise of atmospheric carbon dioxide from 1958 to the present, overlaid with annual sawtooth variations due to the growth of Northern Hemisphere plants in summer. Carbon dioxide began rising due to human activities in the 1800s, but direct, continuous measurements of atmospheric carbon dioxide were first made by U.S. oceanographer Charles David Keeling (1928–2005) starting in 1958.

how old these samples are. Thus, cylinders of ice (ice cores) drilled out of such deposits reveal the amount of CO_2 in the air over long periods of time. The Vostok ice core, drilled in East Antarctica from 1990 to 1994, supplied a continuous series of CO_2 samples from 420,000 years ago to the present. In 2004, scientists announced results drilled from an Antarctic location called Dome C that pushed the record back to 740,000 years ago. In 2006, the Dome C ice-core record was increased to 800,000 years—a cylinder of ice some 2 mi (3.2 km) long.

These and other data show that human activity over the last 200 years has raised atmospheric CO_2 to a level substantially higher than any seen in the last 800,000 years. CO_2 is also rapidly increasing at an unprecedented rate. The most rapid rate of increase observed in the last 800,000 years was 30 ppm over 1,000 years, while the most recent increase of 30 ppm has occurred in only the past 17 years.

Impacts and Issues

Most scientists agree that human activity is causing most of the observed increase in atmospheric CO_2 concentrations. They also agree that increased CO_2 is the primary cause of the recent warming of the world's climate, which they predict will continue.

International gatherings of scientists such as the Intergovernmental Panel on Climate Change (IPCC) advise that reducing the amount of CO_2 that humans add to the atmosphere from this time forward, will result in smaller changes in global climate. There is political disagreement over what steps should be taken to reduce human CO_2 emissions. Burning less fossil fuel would reduce emissions, but as modern industrial economies are still deeply dependent on relatively cheap and abundant coal and petroleum, this is a difficult step to take.

Primary Source Connection

The Intergovernmental Panel on Climate Change (IPCC) is not a research organization, but a scientific body established by the World Meteorological Organization (WMO) and the United Nations Environment Programme (UNEP). The IPCC was awarded the Nobel Peace Prize in 2007 for its efforts to educate global policymakers about human-made climate change. The following source from the IPCC's 2007 report highlights scientific data on the increase over time of atmospheric CO_2, CH_4, and other greenhouse gases (GHGs).

IPCC SCIENCE BASIS TECHNICAL SUMMARY

Current concentrations of atmospheric CO_2 and CH_4 far exceed pre-industrial values found in polar ice core records of atmospheric composition dating back 650,000 years. Multiple lines of evidence confirm that the post-industrial rise in these gases does not stem from natural mechanisms.

The total radiative forcing of the Earth's climate due to increases in the concentrations of the LLGHGs CO_2, CH_4 and N_2O, and very likely the rate of

increase in the total forcing due to these gases over the period since 1750, are unprecedented in more than 10,000 years. It is *very likely* that the sustained rate of increase in the combined radiative forcing from these greenhouse gases of about +1 W m-2 over the past four decades is at least six times faster than at any time during the two millennia before the Industrial Era, the period for which ice core data have the required temporal resolution. The radiative forcing due to these LLGHGs has the highest level of confidence of any forcing agent.

The concentration of atmospheric CO_2 has increased from a pre-industrial value of about 280 ppm to 379 ppm in 2005. Atmospheric CO_2 concentration increased by only 20 ppm over the 8000 years prior to industrialisation; multi-decadal to centennial-scale variations were less than 10 ppm and *likely* due mostly to natural processes. However, since 1750, the CO_2 concentration has risen by nearly 100 ppm. The annual CO_2 growth rate was larger during the last 10 years (1995–2005 average: 1.9 ppm yr^{-1}) than it has been since continuous direct atmospheric measurements began (1960–2005 average: 1.4 ppm yr^{-1}).

Increases in atmospheric CO_2 since pre-industrial times are responsible for a radiative forcing of $+1.66 \pm 0.17$ W m^{-2}; a contribution which dominates all other radiative forcing agents considered in this report. For the decade from 1995 to 2005, the growth rate of CO_2 in the atmosphere led to a 20% increase in its radiative forcing.

Emissions of CO_2 from fossil fuel use and from the effects of land use change on plant and soil carbon are the primary sources of increased atmospheric CO_2. Since 1750, it is estimated that about 2/3rds of anthropogenic CO_2 emissions have come from fossil fuel burning and about 1/3rd from land use change. About 45% of this CO_2 has remained in the atmosphere, while about 30% has been taken up by the oceans and the remainder has been taken up by the terrestrial biosphere. About half of a CO_2 pulse to the atmosphere is removed over a time scale of 30 years; a further 30% is removed within a few centuries; and the remaining 20% will typically stay in the atmosphere for many thousands of years.

In recent decades, emissions of CO_2 have continued to increase. Global annual fossil CO_2 emissions increased from an average of 6.4 ± 0.4 GtC yr^{-1} in the 1990s to 7.2 ± 0.3 GtC yr^{-1} in the period 2000 to 2005. Estimated CO_2 emissions associated with land use change, averaged over the 1990s, were 0.5 to 2.7 GtC yr^{-1}, with a central estimate of 1.6 Gt yr^{-1}.

Since the 1980s, natural processes of CO_2 uptake by the terrestrial biosphere and by the oceans have removed about 50% of anthropogenic emissions. These removal processes are influenced by the atmospheric CO_2 concentration and by changes in climate. Uptake by the oceans and the terrestrial biosphere have been similar in magnitude but the terrestrial biosphere uptake is more variable and was higher in the 1990s than in the 1980s by about 1 GtC yr^{-1}. Observations demonstrate that dissolved CO_2 concentrations in the surface ocean have been increasing nearly everywhere, roughly following the atmospheric CO_2 increase but with large regional and temporal variability.

Carbon uptake and storage in the terrestrial biosphere arise from the net difference between uptake due to vegetation growth, changes in reforestation and sequestration, and emissions due to heterotrophic respiration, harvest, deforestation, fire, damage by pollution and other disturbance factors affecting biomass and soils. Increases and decreases in fire frequency in different regions have affected net carbon uptake, and in boreal regions, emissions due to fires appear to have increased over recent decades. Estimates of net CO_2 surface fluxes from inverse studies using networks of atmospheric data demonstrate significant land uptake in the mid-latitudes of the Northern Hemisphere (NH) and near-zero land-atmosphere fluxes in the tropics, implying that tropical deforestation is approximately balanced by regrowth.

Short-term (interannual) variations observed in the atmospheric CO_2 growth rate are primarily controlled by changes in the flux of CO_2 between the atmosphere and the terrestrial biosphere, with a smaller but significant fraction due to variability in ocean fluxes. Variability in the terrestrial biosphere flux is driven by climatic fluctuations, which affect the uptake of CO_2 by plant growth and the return of CO_2 to the atmosphere by the decay of organic material through heterotrophic respiration and fires. El Niño-Southern Oscillation (ENSO) events are a major source of interannual variability in atmospheric CO_2 growth rate, due to their effects on fluxes through land and sea surface temperatures, precipitation and the incidence of fires.

Susan Solomon et al.

SOLOMON, SUSAN, DAHE QIN, AND MARTIN MANNING, ET AL. "IPCC SCIENCE BASIS TECHNICAL SUMMARY." *IPCC REPORT 4TH TECHNICAL SUMMARY.* UNITED NATIONS, 2007.

SEE ALSO *Carbon Cycle; Climate Change; Greenhouse Effect; Greenhouse Gases; Intergovernmental Panel on Climate Change (IPCC).*

BIBLIOGRAPHY

Books

Solomon, S., et al, eds. *Climate Change 2007: The Physical Science Basis: Contribution of Working Group I to the Fourth Assessment Report of the Intergovernmental Panel on Climate Change.* New York: Cambridge University Press, 2007.

Periodicals

European Project for Ice Coring in Antarctica. "Eight Glacial Cycles from an Antarctic Ice Core." *Nature* 429 (June 10, 2004): 623–628.

Maseh, Betsy. "The Hot Hand of History." *Nature* 427 (February 12, 2004): 582–583.

Web Sites

Amos, Jonathan. "Deep Ice Tells Long Climate Story." *BBC News*, September 4, 2006. <http://news.bbc.co.uk/2/hi/science/nature/5314592.stm> (accessed August 5, 2006).

"Climate Change Affecting Earth's Outermost Atmosphere." *University Corporation for Atmospheric Research*, December 11, 2006. <http://www.ucar.edu/news/releases/2006/thermosphere.shtml> (accessed August 5, 2007).

"Trends in Atmospheric Carbon Dioxide—Mauna Loa; Trends in Atmospheric Carbon Dioxide—Global." *U.S. National Oceanic and Atmospheric Administration (NOAA)*. <http://www.esrl. noaa.gov/gmd/ccgg/trends/> (accessed August 5, 2007).

Larry Gilman

Carbon Dioxide Equivalent (CDE)

Introduction

A carbon dioxide equivalent (CDE, or CO_2E) is the standard measure used to report greenhouse gas emissions. It provides a meaningful comparison of greenhouse gas emissions by providing a measure of the quantity of the gas emitted, while taking into account the relative effect of the gas on global warming compared to the effect of carbon dioxide.

Historical Background and Scientific Foundations

Greenhouse gases are gases that contribute to Earth's greenhouse effect by absorbing and emitting infrared radiation. Greenhouse gases directly affect Earth's atmosphere by absorbing and trapping infrared radiation, thereby contributing to global warming.

Greenhouse gases can also have an indirect effect on the atmosphere. This occurs when a greenhouse gas undergoes a chemical reaction to produce a second greenhouse gas, influences other greenhouse gases (for example, increasing their lifetime), or alters some other aspect of the atmosphere that affects radiation, such as promoting cloud formation.

Differences in the effect of each greenhouse gas means that the impact each gas has on the greenhouse effect cannot be measured only by reporting the quantity of each gas released. Instead, greenhouse gas emissions must be reported based on the impact the gas has on the greenhouse effect.

Traffic jams, a common sight in Los Angeles and other major cities, contribute significantly to greenhouse-gas emissions. *Image copyright egd, 2007. Used under license from Shutterstock.com.*

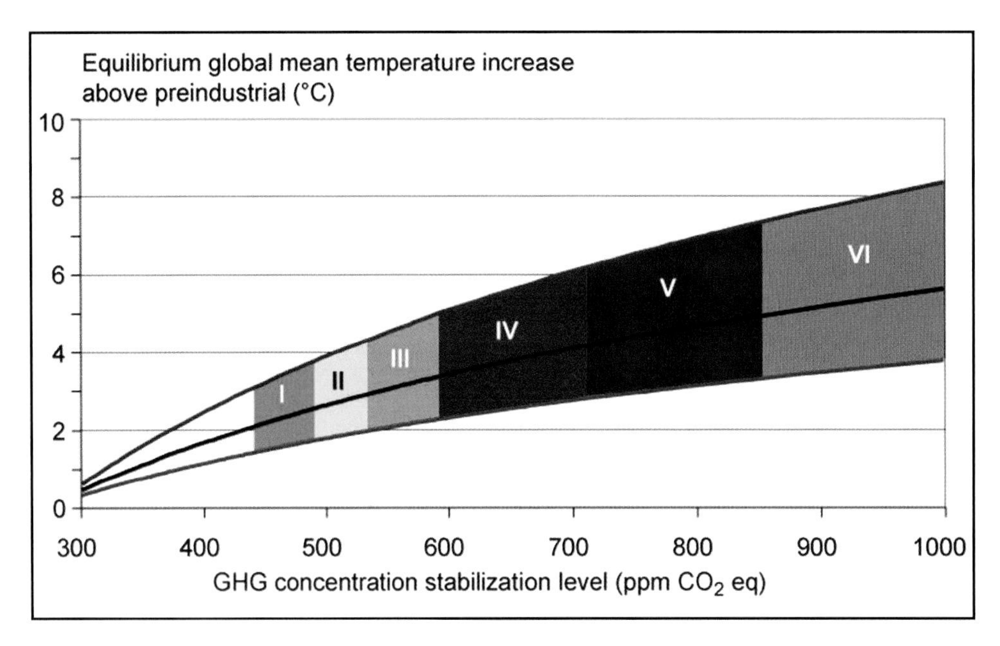

Stabilization scenario categories for global mean temperature increase above preindustrial temperatures in degrees Celsius (°C) in relation to stabilization of greenhouse gas concentrations in the atmosphere. To stabilize the concentration of GHGs in the atmosphere, emissions would need to peak and decline thereafter. The lower the stabilization level, the more quickly this peak and decline would need to occur. Mitigation efforts over the next two to three decades will have a large impact on opportunities to achieve lower stabilization levels. *Climate Change 2007: Mitigation of Climate Change, Summary for Policymakers, Intergovernmental Panel on Climate Change.*

WORDS TO KNOW

GREENHOUSE GASES: Gases that cause Earth to retain more thermal energy by absorbing infrared light emitted by Earth's surface. The most important greenhouse gases are water vapor, carbon dioxide, methane, nitrous oxide, and various artificial chemicals such as chlorofluorocarbons. All but the latter are naturally occurring, but human activity over the last several centuries has significantly increased the amounts of carbon dioxide, methane, and nitrous oxide in Earth's atmosphere, causing global warming and global climate change.

INFRARED RADIATION: Electromagnetic radiation of a wavelength shorter than radio waves but longer than visible light that takes the form of heat.

INTERGOVERNMENTAL PANEL ON CLIMATE CHANGE (IPCC): Panel of scientists established by the World Meteorological Organization (WMO) and the United Nations Environment Programme (UNEP) in 1988 to assess the science, technology, and socioeconomic information needed to understand the risk of human-induced climate change.

Impacts and Issues

Global warming potential (GWP) is the measure used to indicate the ability of a greenhouse gas to trap heat energy in the atmosphere. The GWP of a greenhouse gas is given as compared to carbon dioxide, which has a GWP of 1. The GWP can also be given over different time horizons. The Intergovernmental Panel on Climate Change (IPCC) has standardized the measurements for 20-year, 100-year, and 500-year GWPs for greenhouse gases.

The 100-year GWPs of methane, nitrous oxide, sulfur hexafluoride, and HFC-23 are 23, 296, 22,000, and 12,000 respectively. This indicates that each gram of methane has the same global warming effect as 23 grams of carbon dioxide, while each gram of HCF-23 emitted has the same global warming effect as 12000 grams of carbon dioxide.

Measures of greenhouse gas emissions are reported by multiplying the GWP of a greenhouse gas by the mass of the emissions of that gas. This measure is known as the carbon dioxide equivalent (CO_2E) and allows the amount of each greenhouse gas to be compared based on its effect on global warming.

SEE ALSO *Carbon Dioxide (CO_2); Global Warming; Greenhouse Effect; Greenhouse Gases.*

BIBLIOGRAPHY

Books

Houghton, J. T., et al. *Climate Change 2001: The Scientific Basis.* Cambridge: Cambridge University Press, 2001.

Web Sites

"Fourth Climate Action Report to the UN Framework Convention on Climate Change." *U.S. Department of State*, 2007 <http://www.state.gov/g/oes/rls/rpts/car/> (accessed November 6, 2007).

Tony Hawas

Carbon Footprint

Introduction

A carbon footprint is the total amount of greenhouse emissions that result directly and indirectly either from an individual's lifestyle, a company's operations, or the full life cycle of a product or service.

Historical Background and Scientific Foundations

Since the late 1990s, the term carbon footprint has become widely used in the public debate on personal and corporate responsibility in response to global climate change. Its

Great Britain's Prince Charles (right) and Dr. Nader Al Awadhi of the Kuwait Institute for Scientific Research discuss the effects of climate change and ways to reduce carbon emissions. Prince Charles invests in tree planting and sustainable energy projects in an effort to reduce his carbon footprint. *AP Images.*

WORDS TO KNOW

CARBON CALCULATOR: Software device, often accessed through a Web site, that allows an individual or business to calculate their carbon footprint, that is, how much greenhouse warming is generated to support the present mode of existence of that individual or business.

CARBON OFFSETS: Reductions in emissions of CO_2 (or other greenhouse gases) or enhanced removals of such gases from the atmosphere that are arranged by polluters in order to compensate for their releases of greenhouse gases. Carbon offsets may be purchased by individuals or groups.

CARBON SINKS: Carbon reservoirs such as forests or oceans that take in and store more carbon (carbon sequestration) than they release. Carbon sinks can serve to partially offset greenhouse-gas emissions.

COMMODIFY: To make something into a commodity, that is, something which is bought and sold. Some have argued that carbon markets are unethical because they commodify the well-being of Earth itself, on which all life depends.

ECO-LEXICON: The group or class of words (lexicon) devised for, or especially connected with, speech about environmental concerns. Often has a derisive nuance, as in "Why have 'Civic' or 'Insight' not entered the eco-lexicon in the way that 'Prius' has?"

FOSSIL FUELS: Fuels formed by biological processes and transformed into solid or fluid minerals over geological time. Fossil fuels include coal, petroleum, and natural gas. Fossil fuels are non-renewable on the timescale of human civilization, because their natural replenishment would take many millions of years.

GREENHOUSE GASES: Gases that cause Earth to retain more thermal energy by absorbing infrared light emitted by Earth's surface. The most important greenhouse gases are water vapor, carbon dioxide, methane, nitrous oxide, and various artificial chemicals such as chlorofluorocarbons. All but the latter are naturally occurring, but human activity over the last several centuries has significantly increased the amounts of carbon dioxide, methane, and nitrous oxide in Earth's atmosphere, causing global warming and global climate change.

ubiquity is largely because it appears to be an easy way of commodifying an individual or company's CO_2 emissions.

But despite its frequent popular use, there is no clear scientific definition of the term, and there remains confusion as to what it actually means and measures. Definitions range from direct CO_2 emissions to full life-cycle greenhouse-gas emissions. For example, an individual's carbon footprint for a New York to London return flight would be 1.7 metric tons.

Impacts and Issues

In "A Definition of a Carbon Footprint," Thomas Wiedmann and Jan Minx showed several discrepancies between different U.K. organizations' definitions of carbon footprint. For example, ETAP (an environmental body under the jurisdiction of the European Commission), defines it as: "a measure of the impact human activities have on the environment in terms of the amount of greenhouse gases produced, measured in tonnes [metric tons] of carbon dioxide." The U.K. Parliamentary Office for Science and Technology defines it as: "the total amount of CO_2 and other greenhouse gases, emitted over the full life cycle of a process or product. It is expressed as grams of CO_2 equivalent per kilowatt-hour of generation (gCO_2eq/kWh), which accounts for the different global warming effects of other greenhouse gases."

Wiedmann and Minx suggested the following conglomerate definition, to include activities of people, governments, and industries, as well as the impact of goods and services: "the carbon footprint is a measure of the exclusive total amount of carbon dioxide emissions that is directly and indirectly caused by an activity or is accumulated over the life stages of a product."

Despite the differing definitions, a carbon footprint is intended to measure the impact of people and products on the planet. Many important questions remain. Should a carbon footprint include other greenhouse gases? Should it be defined by an area unit (ha, m^2, km^2, etc.) as opposed to a mass unit (kg, t, etc.)? Should it include all sources of CO_2 emissions (e.g., those that come naturally from the soil) and not just those from fossil fuels? Should it take into account carbon sinks that help Earth reabsorb CO_2? Should there be a separate "ecological footprint" that uses a wider range of measures?

Part of the attraction of a carbon footprint is that it gives a clear marker as to an individual or organization's contribution to and responsibility for greenhouse-gas emissions. Best Foot Forward, a U.K.-based consultancy that specializes in measuring and communicating environmental impact, assesses that the average annual CO_2 footprint of a U.S. resident is 25.9 metric tons (28.5 tons). The average annual footprint of a U.K. resident, by comparison, is 11.6 metric tons; and an African, 0.9 metric tons. This is broken down five ways: food 4% (from farming, processing, and shipping); personal travel 34% (road, air, rail, and water); housing 18% (electricity and heating); services 6% (shops, gyms, etc.); and manufacture and construction 38% (factory goods, road building, etc.).

Another reason why the term has entered the eco-lexicon so quickly is because it gives conscientious individuals and companies a figure that they can move to neutralize. Once an entity is aware of its carbon footprint, it can take steps to reduce it.

A popular way of neutralizing a carbon footprint is through an offset or carbon-neutral program. One of a host of offsetting companies will use a carbon calculator to estimate the emissions related to whatever activity an individual or company wishes to neutralize. This is then translated into a fee that the offsetting organization will use to offset or "soak up" an equivalent amount from the atmosphere. Different companies operate different schemes to achieve this: some simply plant trees; others invest in green energy sources or cleaner and more energy efficient industrial and household technologies. Popular offset programs permit individuals and companies to purchase offsets for their homes, cars, offices, and travel.

However, offset schemes are not without criticism. Critics assert that offsetting merely masks the unsustainability of carbon-intensive activities and is often less beneficial or effective than reducing or stopping emissions in the first place. Another criticism of carbon offset programs is their expense—few individuals can afford to offset every—or even a significant portion—of their greenhouse gas emitting activities. Critics further note that some offset schemes may not produce the claimed carbon savings or may do so only after a number of years. For example, tree planting programs can take years to soak up the emissions one has paid to offset.

Despite these criticisms, carbon-neutral programs continue to increase in popularity. Participation in carbon-neutral programs has increased since their inception, and some offsets have become less expensive as offset providers have become profitable.

SEE ALSO *Economics of Climate Change; Mitigation Strategies; Sustainability.*

BIBLIOGRAPHY

Periodicals

Kleiner, Karl. "The Corporate Race to Cut Carbon." *Nature Reports: Climate Change* 3 (August 2007).

"Please Do Not Sponsor This Tree." *New Internationalist* 391 (June 2006).

Wiedmann, Thomas, and Jan Minx. "A Definition of a Carbon Footprint." *ISA Research* (June 2007).

Web Sites

"Carbon Down, Profits Up." *The Climate Group*, 2007. <http://www.resourcesaver.org/file/toolmanager/CustomO16C45F54513.pdf> (accessed December 3, 2007).

James Corbett

Carbon Sequestration Issues

Introduction

Carbon sequestration is the collection and storage of carbon dioxide (CO_2) to keep it out of the atmosphere. CO_2 can be sequestered in the oceans, underground reservoirs, carbon compounds, biomass, or soil. A variety of ways to sequester carbon have been proposed since the 1980s. Some, such as planting trees and changing agricultural practices to sequester carbon in soil, are for the most part uncontroversial but may not be able to sequester enough CO_2 to stabilize global climate.

Other methods, such as carbon capture and storage (CCS), have not yet proved to be effective or affordable. Some, such as iron fertilization of phytoplankton (single-celled green plants) in the seas, have not yet proved to be effective and might, according to their scientific critics, do more ecological harm than good even if they are capable of sequestering large quantities of carbon. The leading candidate technology for large-scale carbon sequestration in the near future is CCS.

Historical Background and Scientific Foundations

Human activities have increased the atmospheric concentration of CO_2 from 380 parts per million in 1750, before the widespread burning of fossil fuels, to 383 parts per million in 2007, a 37% increase. Other gases have increased as well, but CO_2 is the most important greenhouse gas, accounting for 63% of anthropogenic (human-caused) global warming. The other 37% of warming is due to other greenhouse gases and to changes in albedo (global brightness).

About 84% of CO_2 emissions were, as of 2007, from the burning of fossil fuels, with the other 16% coming mostly from deforestation. In the United States in 2006, CO_2 emissions were about 86% of greenhouse pollution, and globally about 29.8 billion tons (27 billion metric tons) of CO_2 were being emitted yearly. Many climate scientists agree that in order to stabilize global average temperature at no more than about 3.6°F (2°C) above its preindustrial value, CO_2 emissions must be greatly reduced by mid-century. To limit warming this much, a 2003 study in *Science* stated that by 2050, 75% to 100% of total power demand would have to be met by non-CO_2-releasing sources. Nuclear power, renewables, and energy efficiency have all been proposed as ways to meet energy demand, but many experts believe that energy generation from fossil-fuel point sources such as coal-fired power plants will continue to grow in the coming decades. As more such plants are constructed, their CO_2 emissions are bound to increase unless sequestration is used.

Italian energy specialist E. Marchetti suggested geologic or capture-and-store sequestration of CO_2 from-fossil-fuel burning in 1976. In 1977, he suggested the alternative possibility of pumping CO_2 into deep ocean waters, where it would dissolve and remain for centuries until making its way to the atmosphere. However, no action was taken on Marchetti's ideas, as there did not yet seem any urgent need for sequestration of carbon. Analysis of Marchetti's ideas in the mid 1980s seemed to show that sequestration would be prohibitively expensive.

In the 1990s, public and scientific concern over anthropogenic climate change intensified. In a little over ten years, from about 1995 to 2005, carbon sequestration went from being an idea of interest to only a few specialists to a widely discussed possibility for mitigating climate change, with hundreds of millions of dollars of government funding for demonstration projects from Europe, the United States, and elsewhere. By the late 1990s, the United States was funding studies of CCS in various American geological formations. By 2007, several industrial-scale CCS pilot projects were underway.

Technologies

Carbon sequestration occurs naturally. Plants extract carbon from the atmosphere to build their tissues, and this carbon may be sequestered as dead plant matter in soils (hundreds of billions of tons are locked up in this form in the permafrost soils of the Arctic region) or even

WORDS TO KNOW

ALBEDO: A numerical expression describing the ability of an object or planet to reflect light.

AMINE: One of a family of compounds, the amines, containing carbon and nitrogen. When combined with a carboxyl group (CO_2H), an amine forms an amino acid. All living things build proteins by chaining together amino acids.

BIOMASS: The sum total of living and once-living matter contained within a given geographic area. Plant and animal materials that are used as fuel sources.

DEFORESTATION: Those practices or processes that result in the change of forested lands to non-forest uses. This is often cited as one of the major causes of the enhanced greenhouse effect for two reasons: 1) the burning or decomposition of the wood releases carbon dioxide; and 2) trees that once removed carbon dioxide from the atmosphere in the process of photosynthesis are no longer present and contributing to carbon storage.

ENHANCED OIL RECOVERY: Use of special technologies to increase the amount of oil that can be recovered from a given oil field. Enhanced oil recovery techniques include heating oil to make it more liquid or pressurizing it by injecting gas or liquid into the reservoir. At least 40% of oil remains in the ground even with enhanced recovery.

FOSSIL FUELS: Fuels formed by biological processes and transformed into solid or fluid minerals over geological time. Fossil fuels include coal, petroleum, and natural gas. Fossil fuels are non-renewable on the timescale of human civilization, because their natural replenishment would take many millions of years.

GREENHOUSE GASES: Gases that cause Earth to retain more thermal energy by absorbing infrared light emitted by Earth's surface. The most important greenhouse gases are water vapor, carbon dioxide, methane, nitrous oxide, and various artificial chemicals such as chlorofluorocarbons. All but the latter are naturally occurring, but human activity over the last several centuries has significantly increased the amounts of carbon dioxide, methane, and nitrous oxide in Earth's atmosphere, causing global warming and global climate change.

IRON FERTILIZATION: Controversial speculative method for removing carbon from the atmosphere, in which adding powdered iron to ocean surface waters would cause single-celled aquatic plants (phytoplankton) to increase greatly in numbers, breaking down CO_2 to obtain carbon for their tissues. The dead organisms would then, ideally, sink to deep waters or the ocean floor, sequestering their carbon content from the atmosphere and so mitigating climate change.

PERMAFROST: Perennially frozen ground that occurs wherever the temperature remains below 32°F (0°C) for several years.

PHYTOPLANKTON: Microscopic marine organisms (mostly algae and diatoms) that are responsible for most of the photosynthetic activity in the oceans.

turned, over millions of years, into trillions of tons of oil, coal, and gas deposits. The carbon now being released from fossil fuels was sequestered by green plants hundreds of millions of years ago. The oceans also sequester carbon by absorbing it from the atmosphere: about half of the CO_2 emitted by human activities is absorbed in this way, preventing it from enhancing global warming but making the oceans significantly more acidic.

Today's debates over carbon sequestration focus on allowing natural sequestration to proceed (i.e., by cutting down fewer forests or planting new ones), enhancing natural sequestration processes (i.e., by feeding iron to the phytoplankton floating in ocean surface waters, which transport carbon to the deep ocean when they die and sink), and building CCS systems. Iron fertilization is the most controversial of these options and CCS is the most mature.

Carbon Capture and Storage (CCS)

All CCS technologies have three major parts. First is the capture of CO_2, either before or after fuel is burned. Second is transportation of the captured CO_2 to the site where it is to be sequestered. Third is the disposal or sequestration itself.

There are three ways of capturing CO_2 from fossil fuels, namely pre-combustion, post-combustion, and oxy-fuel combustion. In the first method, pre-combustion capture, chemical reactions are used to extract CO_2 from the fuel before it is burned. For example, reacting the fuel with steam (water, H_2O) produces two molecules of hydrogen (H_2) and one of CO_2 for each atom of carbon in the original fuel. The hydrogen may then be burned, producing water as the only byproduct. A steam reaction of this type is already the standard industrial technique for manufacturing hydrogen from natural gas.

The second method of capturing CO_2 from fuel, post-combustion capture, first burns the fuel, allowing CO_2 to form, then uses chemical reactions to capture the CO_2 from the flue gas (hot gas normally sent right up the chimney or flue). The usual method of post-combustion capture is amine scrubbing. In this technique, the flue gas is bubbled through a watery solution of one or more of the chemicals called amines. This amine solution absorbs the CO_2 from flue gas (which consists mostly of nitrogen, because air itself is 78% nitrogen). When heated in another part of the machinery, the amine solution gives up the CO_2 in pure form, which is then collected.

A technology for pre-combustion capture that many energy experts view as promising is the integrated gasification combined cycle (IGCC). In IGCC, fuel is mixed with oxygen and steam to produce a burnable gas consisting mostly of hydrogen and carbon monoxide (CO). This first mixture is then reacted with steam to make CO_2 and hydrogen. This second mixture can then be burned as-is directly, or the CO_2 can be separated, leaving the hydrogen. Typical IGCC heat-to-electric efficiency is about 40% today and could, in theory, be made much higher (around 60%), which would make the energy cost of CO_2 capture more affordable. IGCC plants—several of which were already in operation as of 2007, though without carbon capture—cost at least 20% more to build than conventional power plants, but also produce much less air pollution than comparable coal-fired plants.

Oxy-fuel combustion, the third major option for carbon capture in fuel burning, combusts fuel with pure oxygen, which produces flue gas that is mostly CO_2 and steam and which can then be separated. Oxy-fuel combustion is already used in a number of power plants today because it produces more efficient combustion, but is not yet used with carbon capture.

Carbon capture is the most expensive part of CCS, and depending on the process can consume a good deal of energy. For example, a typical large, centralized, electricity-generation plant converts about 30% of the heat released from its fuel into electricity; the other 70% is wasted to the atmosphere. The best conventional plants (e.g., IGCC plants) operate at about 40% conversion efficiency. CCS might decrease the efficiency of an efficient plant from 40% to about 30%.

After capture, transportation is the next major phase of CCS. CO_2 is nontoxic, so the main danger with transporting large quantities of it is that it can displace air near accident scenes and so cause suffocation. Also, CO_2 that is to be shipped through pipelines must contain almost no water vapor because water and CO_2 combine to form carbonic acid (H_2CO_3), which corrodes ordinary metal parts.

The third phase is sequestration or storage. Once transported by pipeline or tanker ship to an appropriate location, CO_2 can be pumped at high pressure into deep boreholes made by standard drilling equipment. There are four basic options for underground (geological) storage. The first is depleted oil and gas reservoirs. Having been emptied of their fossil fuels, these underground pockets can be refilled with CO_2. The second option is enhanced gas and oil recovery. In this method, CO_2 is injected into gas or oil wells to squeeze out new fuel. These operations typically pay for themselves through the market value of the recovered fuel.

The third option is storage in deep saline aquifers (bodies of porous rock permeated with salty water), either onshore or offshore. The fourth is methane (CH_4) recovery from coal seams that are too deep to mine. Coal naturally exudes methane, the main ingredient of natural gas. This option, Enhanced Coal Bed Methane recovery, consists of pumping CO_2 into deep coal seams, capturing the methane that is forced out through other boreholes, and burning that methane as a fuel. CCS could be used to capture the CO_2 from the burning methane and inject it back into the ground to force out still more methane, and so on. Deep coal deposits in the United States are estimated to have storage potential for about 37 billion tons (33.5 billion metric tons) of CO_2, or six years' worth of emissions. The capacity of deep saline aquifers is uncertain, but the high end of the range of figures quoted by the U.S. Department of Energy is 500 billion tons (454 billion metric tons).

There is little doubt that CCS with geological sequestration can work, at least for the geological short term. Since 1986, Chevron Oil's Rangely Field project in the United States has used CO_2 injection into underground oil reservoirs to help force petroleum to the surface. As of 2006, the project had sequestered over 26.5 million tons (24 million metric tons) of CO_2.

In 1991, Norway became the first country to impose a per-ton tax on carbon emissions. In 1996, in order to avoid paying this tax on CO_2 pumped from its Sleipner natural gas well in the North Sea, the Norwegian oil company Statoil began disposing of CO_2 by injecting it into an aquifer 3,280 ft (1,000 m) below the sea floor. The natural gas obtained from the Sleipner well is unusual in that it consists of 9% CO_2, which has to be separated from the gas to make it marketable. The Norwegian tax on CO_2 emissions is about US$50 a ton, but it costs Statoil only about US$30 a ton to inject the CO_2 into the deep aquifer, making the operation profitable. Statoil sequesters 1 million tons of CO_2 per year at Sleipner, enough to increase Norway's CO_2 emissions by 3% if it were all released into the atmosphere. Statoil plans to dispose, all told, about 20 million metric tons of CO_2 at the Sleipner well.

In 2007, this was still the largest project for sequestering CO_2 that was not part of an enhanced-oil-recovery (EOR) scheme. The Weyburn project in Canada, an EOR project, was also injecting about 1 million tons (907,000 metric tons) of CO_2 per year. At the In Salah natural gas well in Algeria, CO_2 separated from the gas was being injected back into the ground much as at the Sleipner field. The In Salah plan called for the eventual sequestration of 17 million tons (15.4 million metric tons) of CO_2.

The alternative to geological storage is ocean storage. There are two basic options. The first is dissolution-type storage, in which a pipeline is run out to sea at a depth of approximately 1 mi (1.5 km). Liquid or gaseous CO_2 is then forced through the pipeline, after which it bubbles up through the ocean in a rising plume. The gas

U.S. President George W. Bush (2nd from right) holds a briefing on carbon sequestration with (l-r) Secretary of State Colin Powell, Secretary of Energy Spencer Abraham, and Environment Protection Agency Administrator Christie Whitman in early 2003. The U.S. government has spent millions of dollars researching the feasibility of carbon sequestration. *Alex Wong/Getty Images.*

dissolves in the ocean before it reaches the surface layer. An alternative form of dissolution-type storage is to run a pipe down to a depth of at least 2 mi (3 km) and dissolve a sinking plume of liquid CO_2 in deeper water that will sequester the carbon for a longer period of time.

The second basic type of ocean storage is lake-type storage. In this method, liquid CO_2 is deposited on the deep ocean floor at about 2.5 mi (4 km) or greater depth in the form of a cold pool. Some experiments indicate that a skin of stable ice-like hydrates may form over such a pool, slowing its eventual dissolution in the ocean. Ocean disposal of CO_2 was in the research phase in the early 2000s.

Other

Proposed methods of carbon capture other than CCS include the manufacture of carbonate minerals; reformed agricultural practices to encourage carbon uptake by soils; the manufacture of carbon powder and its addition to soils as a fertilizer; planting forests and preventing deforestation; the addition of powdered iron compounds to ocean surface waters to encourage the growth of phytoplankton that will then die and sink, transporting their carbon to deep waters; and more. Most of these methods depend on sequestering carbon after it has been released, rather than capturing it before release as CCS does. As of 2007, pilot projects were underway for all these technology-dependent sequestration schemes.

■ Impacts and Issues

The goal of carbon sequestration is to reduce the amount of CO_2 entering (or staying in) the atmosphere. For any given method or combination of methods, how much carbon is captured depends on several factors, namely the fraction of CO_2 captured, the increase in CO_2 production needed to achieve a given real output (e.g., amount of electricity) due to lowered efficiency with CCS, leakage of CO_2 during transport, and the fraction of CO_2 that leaks out of storage over a given time period. Today's CCS technologies capture about 85–95% of CO_2 from gasified fuel or flue gas. Depending on the type of combustion, power plants would need to produce 10–40% more energy to capture this amount of CO_2 and compress it into liquid form for disposal, whether underground, at sea, or by some other method.

As for retention of CO_2 in storage, the disposal of CO_2 underground may, in many cases, sequester carbon from the atmosphere for many thousands of years. Disposal in ocean waters would be relatively temporary, because CO_2 added to the ocean must eventually migrate into the atmosphere until atmospheric and oceanic CO_2 are in equilibrium or balance. Scientists believe that CO_2 injected in the oceans would equilibrate with atmospheric CO_2 in 2,000 years or less.

All methods of sequestering CO_2 have expert advocates and expert critics. However, some methods are more widely criticized in the scientific community than others. Iron fertilization of the oceans is probably the most criticized method, followed by oceanic CCS.

Critics of iron fertilization charge that although tests have shown that adding finely ground iron particles to surface ocean waters can indeed cause plankton blooms, these tests have not shown that the method significantly increases the transport of carbon to deep waters, where it would be sequestered.

Concerns about CCS, the most mature of the carbon sequestration technologies, include doubts about whether the CO_2 will leak out through old boreholes or other geological defects, perhaps cracks caused by earthquakes. Sudden release of a large amount of CO_2 could be disastrous for local populations (as well as defeating the greenhouse-abatement goal of putting the CO_2 underground originally). A natural demonstration of this possibility occurred in 1986, when a large bubble of CO_2 escaped from a volcanic lake in Cameroon, Africa, killing 1,700 people by suffocation. Storing CO_2 in aquifers may be unstable because CO_2 combines with water to form carbonic acid, which can weaken rock over time. However, no CO_2 leakage has yet been observed from any of the pilot CCS sequestration projects now being conducted worldwide.

Ocean storage of CO_2 would, according to biologists, injure deep-sea life by making the deep ocean more acidic. Biological communities near injection sites would be devastated at once, while the effect on living things farther away would be more gradual. CO_2 from the atmosphere is already making the oceans significantly more acidic. By the end of the twenty-first century, acidification of the oceans may make it difficult for shelly marine creatures such as corals and clams to build their shells at all, even without additional acidity from oceanic CO_2 sequestration.

Despite doubts, carbon sequestration may prove to be an indispensable technology in mitigating climate change. As of 2007, the European Union was considering requiring geologic CCS systems for all new coal-burning power plants. In the United States, a government-industry partnership called FutureGen began considering proposals in 2006 for the construction of a $1 billion power plant that would produce hydrogen and 275 megawatts of electricity while using CCS to yield near-zero greenhouse emissions. The plant was supposed to start construction in 2009 and enter service in 2012, and would act as a demonstration model for the feasibility of CCS on a commercial scale. Other governments worldwide were also participating in various CCS demonstration projects.

■ Primary Source Connection

The Environmental Protection Agency (EPA) is the lead environmental agency of the United States government. Here, EPA research explores the feasibility and efficiency of various methods of carbon sequestration for reducing greenhouse gas emissions.

GEOLOGIC SEQUESTRATION

Geologic sequestration, a type of carbon dioxide (CO_2) capture and storage (CCS) process, is a promising technology for stabilizing atmospheric greenhouse gas concentrations. Instead of releasing CO_2 to the atmosphere, geologic sequestration involves separating and capturing CO_2 from an industrial or energy-related source, transporting it to a storage location, and injecting it deep underground for long-term isolation from the atmosphere.

Capture

The goal of CO_2 capture is to produce a concentrated stream of CO_2 that can be readily transported to a geologic sequestration site. Capture of CO_2 can be applied to large stationary sources such as power plants, cement or ammonia production or natural gas processing. Several technologies, in different stages of development, exist for CO_2 capture. Although these technologies are currently used in a limited number of facilities, research is still needed to improve the efficiency and cost.

Transport

After the CO_2 is captured from the source and compressed, it can be geologically sequestered on-site or transported to a separate injection site. CO_2 can be transported as a liquid in ships, road or rail tankers, but pipelines are the most efficient and cost-effective approach for transporting large volumes of CO_2. In the U.S., there is a network of CO_2 pipelines that supply CO_2 to oil and gas fields, where it is used to enhance oil recovery. The majority of the 40 Tg CO_2 (Tg = 10^9 kg = 10^6 metric tons = 1 million metric tons) transported in these pipelines today is produced from natural CO_2 reservoirs; however, the same pipelines can carry CO_2 captured from industrial facilities. In fact, a synfuels plant located in North Dakota (Dakota Gasification) has been transporting captured CO_2 via pipeline to a sequestration site hundreds of miles away in Canada since 2000.

Injection and Sequestration

Once a suitable geologic formation has been identified through detailed site characterization, CO_2 is injected into that formation at a high pressure and to depths generally greater than 2625 feet (800 meters). Below this depth, the pressurized CO_2 remains "supercritical" and behaves like a liquid. Supercritical CO_2 is denser and takes up less space. Once underground, the CO_2 occupies pore spaces in the surrounding rock, like water in a sponge. Saline water which already resides in the pore space will compress under pressure and/or move to allow room for the CO_2. Over time, the CO_2 also dissolves in water and chemical reactions between the dissolved CO_2 and rock can create solid carbonate minerals, more permanently trapping the CO_2.

Suitable geologic storage sites have a caprock, which is an overlying impermeable layer that prevents CO_2 from escaping back towards the surface. Target formations for sequestration include geologic formations, both on and off-shore, that can demonstrate their ability to retain CO_2 for very long periods of time. Well-suited formations include the following:

- Deep saline formations, rock units containing water with a high concentration of salts, are thought to have the largest storage capacity.
- Depleted oil and gas reservoirs are also targeted for CO_2 sequestration and have a history of retaining fluids and gases underground for geologic time-scales. There is also more data available on these formations which may help characterize and better predict the long-term fate of injected CO_2.
- Unminable coal beds, which are either too thin or too deep to be mined economically, offer less storage capacity but they have the benefit of enhancing the production of methane, a valuable fuel source. Less is known about the efficacy of using these formations as targets for sequestration, but research is underway to evaluate them.

Storage Capacity

With proper site selection and management, geologic sequestration could play a major role in reducing emissions of CO_2 (IPCC, 2005). Current assessments indicate that the storage capacity of these geologic formations is extremely large and widespread, with a significant proportion of storage opportunities in the U.S. In the U.S., an evaluation of CO_2 sources and potential storage sites suggests that 95% of the largest 500 point sources (i.e., power plants and other industrial facilities), accounting for 82% of annual CO_2 emissions, are within 50 miles of a candidate CO_2 reservoir. . . .

Risk Management

There is limited experience with commercial-scale geologic sequestration today. However, closely related and well-established industrial experience and scientific knowledge can serve as the basis for appropriate risk management strategies. Key components of a risk management strategy include appropriate site selection based on thorough geologic characterization, a monitoring program to detect problems during or after injection, appropriate remediation methods if necessary and a regulatory system to protect human health and the environment. . . .

Potential pathways exist for CO_2 to migrate from the target geologic formation to shallower zones or back to the atmosphere. These conduits for CO_2 leakage could be largely avoided through proper site characterization and selection. Pathways for CO_2 leakage include escape through the caprock (if it is compromised by high pressures or chemical degradation), an undetected or reactivated fault or an artificial penetration such as a poorly plugged abandoned well. In addition to careful site selection, a proper monitoring program can help ensure that CO_2 does not escape from the storage site. A monitoring system would detect movement of CO_2 into shallower formations and allow significant time to take corrective action in order to reduce potential impacts to human health and the environment.

Ground water could be affected both by CO_2 leaking directly into an aquifer and by saline ground water that enters an aquifer as a result of being displaced by injected CO_2. The risk of these impacts can be minimized through appropriate management strategies. Underground injection of CO_2 for the purpose of sequestration is regulated by the Underground Injection Control (UIC) Program under the Safe Drinking Water Act (SDWA). The UIC program ensures that injection activities are performed safely and do not endanger current or future sources of drinking water.

Existing and Planned Projects

Internationally, commercial-scale geologic sequestration (greater than 1 Tg CO_2 per year) is occurring or planned in various locations. Projects that are underway include the Weyburn CO_2 Flood Project (Canada), Sleipner (Norway), and In Salah (Algeria).

The Weyburn CO_2 Flood Project in Canada is the first international CO_2-enhanced oil recovery (EOR) project to be studied extensively. The CO_2 source is the Dakota Gasification plant near Great Plains, North Dakota. Unlike traditional EOR operations, the Weyburn operator will not use conventional end of projects techniques, which can release CO_2, but will maintain the site in order to test and monitor long-term sequestration.

Commercial-scale geologic CO_2 sequestration is also occurring at the Sleipner West field in the North Sea. Sleipner West is a natural gas/condensate field located about 500 miles off the coast of Norway. The CO_2 is compressed and injected via a single well into a 500 foot thick, saline formation located at a depth of about 2,000 feet below the seabed.

In 2004, a CO_2 capture and storage project was launched at the In Salah gas field, in the Algerian desert. Approximately 10% of the produced gas is made up of CO_2. Rather than venting the CO_2, a common practice on projects of this type, this project is compressing and injecting it 5900 feet deep into a lower level of the gas reservoir where the reservoir is filled with water. Around one million tons of CO_2 will be injected into the reservoir every year.

Additionally, commercial scale projects are planned throughout the world. More information can be found on international projects through the Carbon Sequestration Leadership Forum (CSLF), an international climate change initiative focused on the development of improved cost-effective technologies for the separation and capture of CO_2 for its transport and long-term safe storage.

In the U.S., the Department of Energy (DOE) is the lead federal agency on research and development of geologic sequestration technologies. The Department of Energy's Fossil Energy program is developing a portfolio of technologies that can capture and permanently store greenhouse gases. As part of this portfolio, DOE and an industry alliance recently launched FutureGen, an initiative to complete the world's first near-zero emissions, coal-based power plant with sequestration by 2012. DOE is also sponsoring a number of small-scale CO_2 pilot projects designed to learn more about how CO_2 behaves in the sub-surface and answer practical technical questions on how to design and operate geologic sequestration projects.

EPA. "GEOLOGIC SEQUESTRATION." <HTTP://WWW.EPA.GOV/CLIMATECHANGE/EMISSIONS/CO2_GEOSEQUEST.HTML> (ACCESSED NOVEMBER 29, 2007).

SEE ALSO *Carbon Cycle; Carbon Sinks; Iron Fertilization.*

BIBLIOGRAPHY

Periodicals

Berstein, Lenny, et al. "Carbon Dioxide Capture and Storage: A Status Report." *Climate Policy* 6 (2006): 241–246.

Caldeira, Ken, et al. "Climate Sensitivity Uncertainty and the Need for Energy Without CO_2 Emission." *Science* 299 (2003): 2052–2054.

Holloway, Sam. "Storage of Fossil Fuel–Derived Carbon Dioxide Beneath the Surface of the Earth." *Annual Review of Energy and the Environment* 26 (2001): 145–166.

Jackson, Robert B., et al. "Trading Water for Carbon with Biological Carbon Sequestration." *Science* 310 (2005): 1944–1947.

Kintisch, Eli. "Report Backs More Projects to Sequester CO_2 from Coal." *Science* 315 (2007): 1481.

Lackner, Klaus S. "Carbonate Chemistry for Sequestering Fossil Carbon." *Annual Review of Energy and the Environment* 27 (2002): 193–232.

Lackner, Klaus S. "A Guide to CO_2 Sequestration." *Science* 300 (2003): 1677–1678.

Lal, R. "Soil Carbon Sequestration Impacts on Global Climate Change and Food Security." *Science* 3024 (2004): 1623–1627.

Schiermeier, Quirin. "Putting the Carbon Back: The Hundred Billion Tonne Challenge." *Nature* 442 (2006): 620–623.

Schlesinger, William H. "Carbon Sequestration in Soils." *Science* 284 (1999): 2095.

Web Sites

Metz, Bert H., et al. "IPCC Special Report on Carbon Dioxide Capture and Storage." *Intergovernmental Panel on Climate Change*, 2005. <http://www.ipcc.ch/activity/srccs/SRCCS.pdf> (accessed November 6, 2007).

Sharp, Philip, et al. "The Future of Coal." *Massachusetts Institute of Technology*, 2007. <http://web.mit.edu/coal/The_Future_of_Coal.pdf> (accessed November 6, 2007).

Larry Gilman

Carbon Sinks

Introduction

A carbon sink is any system, natural or artificial, that takes carbon out of the atmosphere. As of the early 2000s, artificial carbon sequestration schemes were still only in the testing and study stage. This article is concerned only with natural carbon sinks.

Carbon is removed from the atmosphere by two basic mechanisms, namely: 1) photosynthesis by green plants, which obtain carbon with which to build their tissues by breaking up carbon dioxide (CO_2) molecules in the air releasing the O_2; and 2) absorption of CO_2 by the oceans. Natural carbon sinks remove slightly more than half of the carbon from the atmosphere that is placed there every year by the burning of fossil fuels. Scientists refer to two sinks, the land sink and the ocean sink. The land sink consists of green plants, while the ocean sink consists of direct absorption of CO_2 by ocean water and secondary absorption by organisms living in the sea followed by transport of the carbon in the dead organisms to deep waters or the ocean floor.

Anthropogenic (human-caused) climate change would be much greater today if it were not for natural carbon sinks. However, human activities are imperiling the efficacy of carbon sinks. The more CO_2 the oceans absorb, the more slowly they absorb CO_2. Also, ozone, a common air pollutant, decreases the ability of green plants to act as carbon sinks, while cutting down forests and replacing them with other land uses (such as farmland) removes acreage from the land sink.

Historical Background and Scientific Foundations

Several hundred millions of years ago, large amounts of carbon were removed from Earth's atmosphere by large swamp forests and by tiny organisms floating in the seas. Some of the dead swamp growth accumulated in thick blankets, was covered by other sediments, and was eventually transformed into coal by geological processes. Some of the dead ocean organisms drizzled to the bottom of the sea, were buried, and were eventually transformed into oil and natural gas. These billions of tons of ancient carbon, which were collected at a time when atmospheric CO_2 was far higher than today, are now being returned to Earth's atmosphere in the space of only a few centuries. Human beings are adding about 29.7 billion tons (27 billion metric tons) of CO_2 to the atmosphere every year, and the rate is increasing. The two major sources of CO_2 in the modern atmosphere are fossil-fuel burning and land-use changes, especially the cutting down of forests.

Starting in the late 1950s, American geochemist Charles David Keeling (1928–2005) was the first scientist to measure the steadily increasing concentration of CO_2 in Earth's atmosphere. Keeling found that CO_2 decreases during the growing season in the Northern Hemisphere, where most of the world's vegetation—its land sink—is located. In the spring and summer, growing plants remove CO_2 from the air, while in the winter, plant decay and fuel burning release more CO_2 than is absorbed. The result is a series of sawtooth spikes in atmospheric CO_2 concentration on top of a steadily rising line that reflects increasing average CO_2 concentration.

Thanks to natural carbon sinks, the rate at which CO_2 is increasing is about half of what it would be if all the CO_2 being added to the atmosphere by human beings was staying there. What is more, as the amount of CO_2 released by human activities has grown, tripling from about 1950 to the present, the amount taken up by sinks has grown proportionally. The whole global system of carbon sources and sinks, which is continuously releasing and absorbing carbon around the globe, is termed the carbon cycle. Carbon sinks are only one part of the carbon cycle.

The global carbon-sink picture is difficult to characterize. The atmosphere itself is the only reservoir of

WORDS TO KNOW

ANTHROPOGENIC: Made by people or resulting from human activities. Usually used in the context of emissions that are produced as a result of human activities.

BIOSPHERE: The sum total of all life-forms on Earth and the interaction among those life-forms.

DEFORESTATION: Those practices or processes that result in the change of forested lands to non-forest uses. This is often cited as one of the major causes of the enhanced greenhouse effect for two reasons: 1) the burning or decomposition of the wood releases carbon dioxide; and 2) trees that once removed carbon dioxide from the atmosphere in the process of photosynthesis are no longer present and contributing to carbon storage.

FOSSIL FUELS: Fuels formed by biological processes and transformed into solid or fluid minerals over geological time. Fossil fuels include coal, petroleum, and natural gas. Fossil fuels are non-renewable on the timescale of human civilization, because their natural replenishment would take many millions of years.

OZONE: An almost colorless, gaseous form of oxygen with an odor similar to weak chlorine. A relatively unstable compound of three atoms of oxygen, ozone constitutes, on average, less than one part per million (ppm) of the gases in the atmosphere. (Peak ozone concentration in the stratosphere can get as high as 10 ppm.) Yet ozone in the stratosphere absorbs nearly all of the biologically damaging solar ultraviolet radiation before it reaches Earth's surface, where it can cause skin cancer, cataracts, and immune deficiencies, and can harm crops and aquatic ecosystems.

PHOTOSYNTHESIS: The process by which green plants use light to synthesize organic compounds from carbon dioxide and water. In the process, oxygen and water are released. Increased levels of carbon dioxide can increase net photosynthesis in some plants. Plants create a very important reservoir for carbon dioxide.

carbon that is easy to study, because it is so well-mixed that an air sample taken anywhere on Earth's surface gives information about global conditions. In contrast, the upper and lower layers of the ocean mix slowly—water in the deepest parts of the North Pacific has been out of contact with the atmosphere for about 1,000 years—and the composition of the ocean is not uniform around the globe. As a result, many thousands of measurements must be taken at various depths and around the world to characterize the carbon content of the oceans. Determining how much carbon goes where is even more difficult in the case of the land sink, which changes constantly and varies over different climates and landscapes.

Ocean Sink

In the early 2000s, about a third of annual anthropogenic carbon emissions were being absorbed by the ocean sink. The ocean sink has two components: the biological pump and the solubility pump. Each component transfers or pumps CO_2 out of the atmosphere. The biological pump consists of tiny marine organisms, both plants and animals, which incorporate carbon into their tissues and shells and then die, sinking to deeper waters. There they either decompose, in which case their carbon is dissolved in deep waters, or settle to the bottom as sediment, where their carbon may remain isolated from the atmosphere for much longer.

The solubility pump is driven by the overturning global circulation of the oceans. Surface waters move toward the polar regions, cooling as they go. As they cool, they become capable of absorbing more CO_2 from the atmosphere. Near the poles, they sink and begin to journey back toward the tropics along the ocean floor. Eventually, after as many as 1,500 years, the water rises to the surface in the tropics and is heated. When heated, the water gives up CO_2 to the atmosphere again.

The Southern Ocean, the ring-shaped body of water that surrounds Antarctica south of 60° south latitude, accounts for about half of all absorption by the oceans, that is, about 15% of annual anthropogenic carbon releases.

Land Sink

The land sink is about the same size as the ocean sink, but there are many uncertainties about its size and nature. For decades, most scientists assumed that the land sink's increasing uptake of CO_2 was being driven by the fertilizing effect of increased CO_2 in the atmosphere (most plants grow faster when there is more CO_2). However, in the early 2000s, studies of forest growth in the United States showed that this fertilization effect was far too small to account for the large size of the land sink in North America. In the United States, at least, it now seems more likely that the regrowth of abandoned farmland and formerly logged lands probably accounts for the relatively large size of the land carbon sink. Increased tree growth in areas where forest fires have been suppressed also contributes.

More than half the total (land plus ocean) sink for anthropogenic carbon is in the Northern Hemisphere, and most of this northern-hemispheric sink is terrestrial (on land). Partly due to global warming, which has made for longer growing seasons, the amount of carbon being taken up by the terrestrial biosphere increased from about 220 million tons (200 million metric tons) per year in the 1980s, with large uncertainty, to about 1.5 billion tons (1.4 billion metric tons) per year in the 1990s.

■ Impacts and Issues

Despite the enhancement of the land carbon sink in the 1980s and 1990s due to longer growing seasons, scientists predict that the negative effects of climate change on the land biosphere will soon be dominant, and that global warming will slow CO_2 uptake by both the ocean and land carbon sinks. This will increase the fraction of anthropogenic CO_2 that remains in the atmosphere, making climate change more severe and rapid, other factors being equal.

The global scientific consensus as expressed in 2007 by the United Nations' Intergovernmental Panel on Climate Change (IPCC) is that it is more than 90% likely that terrestrial ecosystems will become net sources of CO_2 between 2050 and 2100. That is, the land carbon sink will shrink, and land-based carbon sources, such as deforestation, will grow until they are emitting more CO_2 than the land sink is absorbing. Deforestation, higher temperatures, and shifts in rainfall patterns will all contribute to the shrinkage of the land sink. Ozone (O_3) pollution is also reducing the efficacy of the land sink by slowing plant growth.

The ocean sink will slow its absorption of carbon as the amount of CO_2 dissolved in the water increases and lowers the water's ability to take up still more CO_2. In 2007, an international science team announced that the Southern Ocean's absorption of CO_2 has decreased by about 15% per decade since 1981. This decrease was caused by global climate change, but not (yet) by increased carbon dissolved in the ocean; rather, increased wind strength due to climate shifts was the cause, altering ocean mixing patterns and decreasing carbon uptake.

■ Primary Source Connection

Forests, prairies, marshlands, and other densely vegetated areas that compose "carbon sinks" help Earth reabsorb carbon dioxide. John Roach, a correspondent for *National Geographic News*, reports on new research on the North American carbon sink and why it may not help offset human-made emissions in the future.

STUDIES MEASURE CAPACITY OF 'CARBON SINKS.'

After years of wide disagreement, scientists are getting a better grip on how much carbon Earth's forests and other biological components suck out of the atmosphere, thus acting as "carbon sinks." New research in this area may be highly useful in efforts to devise international strategies to address global warming.

The emission of carbon dioxide from the combustion of fossil fuels is the leading cause of the buildup of greenhouse gases in the atmosphere, which many people believe is the main culprit behind an increase in Earth's temperatures.

For a long time, scientists have known that forests, crops, soils, and other organic matter soak up some of that carbon, thereby slowing down the rate of global warming. Yet their calculations of how much carbon is absorbed have differed, in some cases significantly.

A team of scientists led by Stephen Pacala, a professor of ecology and evolutionary biology at Princeton University in New Jersey, set out to resolve this discrepancy in calculations. Their research is reported in the June 22 issue of *Science*.

Different Measuring Techniques

While some carbon is absorbed by organic matter such as trees and shrubs, carbon is also regularly emitted into the atmosphere by activities on land such as the burning of fossil fuels.

Researchers' lack of agreement on how much carbon is "stored" has been rooted in the use of two different methods of measurement—one atmosphere based, the other land based.

The first method involves measuring concentrations of carbon dioxide in the air as the air moves across landmasses from Point A to Point B. The second method entails making an inventory of all the carbon in a given area of ground and calculating the difference between the levels of carbon recorded from year to year.

Although there is wide variation among different atmospheric models of carbon measurement, their results have consistently indicated that higher levels of carbon are absorbed than the land-based models show.

Pacala said his team's land-based analysis was more thorough than earlier studies. "We did the first exhaustive analysis of the land sink," he said.

Previous land-based models inventoried mainly the amount of carbon absorbed by trees, he explained. He and his colleagues included measures of carbon absorbed by landfills, soils, houses, and even silt at the bottom of reservoirs.

"We found out that the land sink was bigger than had been reported by other analyses, about twice as big, and the atmosphere [models] gave numbers that were consistent," he said.

The researchers used their results to help answer a major question that has been a subject of much contention:

How big is the entire "carbon sink" of the continental United States?

According to their findings, the scientists estimate that U.S. forests and other terrestrial components absorb from one-third to two-thirds of a billion tons of carbon each year.

At the same time, reliable figures indicate that the United States emits more than two to four times that amount of carbon each year, about 1.4 billion tons.

Taking into account the carbon sink effect, 800 million to 1.1 billion tons of carbon accumulates annually in the atmosphere, the researchers say. This refutes the idea that the U.S carbon sink is big enough to equal the amount of carbon that U.S. factories emit through the burning of fossil fuels, as some studies have concluded.

The results of the Princeton-led study are particularly interesting because the 23 scientists who participated in the research and agreed on the conclusions initially held strongly differing views about the size of the U.S. carbon sink.

Diminishing Effect

Pacala and his colleagues say the main reason the United States is drawing in a large volume of carbon is because many forests and areas of land that were logged or converted to agriculture in the last 100 years are now recovering with the growth of new vegetation.

These trees and shrubs absorb carbon dioxide from the air and channel it into the growth of massive tree trunks, branches, and foliage. This, in turn, gradually expands the overall size of the U.S. carbon sink.

Pacala emphasizes, however, that the U.S. absorption of carbon does not fully offset the emissions of carbon from fossil fuels and should not be seen as a license to release more carbon. A large part of the current sink effect, he said, is the land re-absorbing large quantities of carbon that were released during heavy farming and logging of the past.

"When we chopped down the forests, we released carbon trapped in the trees into the atmosphere. When we plowed up the prairies, we released carbon from the grasslands and soils into the atmosphere," said Pacala. "Now the ecosystem is taking some of that back." But, he added, the sink effect will steadily decrease and eventually disappear—as U.S. ecosystems complete their recovery from past land use.

"The carbon sinks are going to decrease at the same time as our fossil fuel emissions increase," he said. "Thus, the greenhouse problem is going to get worse faster than we expected."

Carbon Sink in China

In a separate study in *Science*, researchers reported on a similar carbon sink effect in China, which they attribute to the regrowth of logged forests and intensive planting of new forests.

Jingyun Fang, an ecology professor at Peking University in Beijing, and his colleagues noted that Chinese forests were heavily exploited from 1949 to the end of the 1970s. Since then, however, the government has undertaken wide-scale forest planting and reforestation, mainly to combat erosion, flooding, desertification, and loss of biodiversity.

An unintended consequence of this increase in vegetation was the growth of a carbon sink that is estimated to be on par with that of North American forests.

John Roach

ROACH, JOHN. "STUDIES MEASURE CAPACITY OF 'CARBON SINKS.'" *NATIONAL GEOGRAPHIC NEWS*, JUNE 21, 2001.

SEE ALSO *Carbon Cycle; Carbon Sequestration Issues; Forests and Deforestation; Sink.*

BIBLIOGRAPHY

Periodicals

Baker, David F. "Reassessing Carbon Sinks." *Science* 316 (2007): 1708–1709.

Field, Christopher B., and Inez Y. Fung. "The Not-So-Big U.S. Carbon Sink." *Science* 285 (1999): 544–545.

Hopkin, Michael. "Carbon Sinks Threatened by Ozone." *Science* 448 (2007): 396–397.

Kaiser, Jocelyn. "Soaking Up Carbon in Forests and Fields." *Science* 290 (2000): 922.

Martine, Philippe, et al. "Carbon Sinks in Temperate Forests." *Annual Review of Energy and the Environment* 26 (2001): 435–465.

Reay, Dave, et al. "Spring-time for Sinks." *Nature* 446 (2007): 727–728.

Wofsy, Steven. "Where Has All the Carbon Gone?" *Science* 292 (2001): 2261–2263.

Web Sites

Sarmiento, Jorge, and Nicolas Gruber. "Sinks for Anthropogenic Carbon." *Physics Today*, August, 2002. <http://www.aip.org/pt/vol-55/iss-8/p30.html> (accessed November 6, 2007).

Larry Gilman

Carbon Tax

Introduction

A carbon tax is a levy on sources that emit carbon dioxide (CO_2) into the atmosphere. The purpose is to place a financial disincentive on carbon-emitting activities and encourage investment in cleaner technologies and practices. Although no internationally levied carbon tax is in operation, several nations apply variants of this pollution tax.

Historical Background and Scientific Foundations

On January 1, 1991, Sweden became the first country to implement a carbon tax, when it introduced a levy of 0.25 SEK/kg (approximately 0.11c/kg) on the use of carbon-based fuels used in domestic travel. A rate of 0.125 SEK/kg was placed on the use of such fuels by Swedish industry. Other countries—such as Finland,

Smoke engulfs an area that was set on fire to clear a tract of forested land in Indonesia in 2005. The pollution from such fires places the nation among the world's top producers of greenhouse gases. Indonesia's environment minister said that developing nations such as his will demand money from rich countries in order to preserve such forests as part of any new plan that will replace the Kyoto Protocol. *AP Images.*

Norway, the Netherlands, and the United Kingdom—have since implemented their own variants of a carbon tax.

The economic principles underlying a carbon tax are simple. By placing an economic disincentive on carbon-emitting activity, primarily the burning of fossil fuels, it simultaneously discourages the discharge of CO_2 into Earth's atmosphere, while encouraging technological innovation to both reduce CO_2 and provide an alternative energy to carbon-based fuels.

Impacts and Issues

A lack of international consensus on the efficacy of carbon taxes remains the main reason they are not more ubiquitous. Some argue that unless applied more universally, individual governments likely remain reluctant to place the additional economic burden of carbon taxation on domestic industries and businesses, claiming that foreign competitors not subject to a carbon tax in their home nations would gain an economic advantage. Others assert that the selective carbon taxes should target only the heaviest polluters—often industries with relatively high margins of profit, such as fossil fuel companies.

Supporters of carbon taxes maintain that they are essential to reduce emissions in order to prevent atmospheric concentrations of CO_2 from reaching an irreversible "tipping point." They assert that internationally applied carbon taxes would help transform the planet's fossil fuels-based energy system to reliance on energy efficiency, renewable energy, and sustainable fuels. It also encourages "green" innovation in other areas.

A significant share of opposition to carbon taxes stems from business and industrial interests. Critics note the additional cost burden placed on businesses, also the lack of viability without an international framework. Unless all countries adopt a carbon tax, it places polluting nations in an economically advantageous position; doubly penalizing business rivals elsewhere that are compelled to pay a carbon tax.

The Swedish experience shows some of the difficulties of maintaining a carbon tax when other countries refrain from introducing such a levy. From the outset, Swedish businesses, already subject to relatively high taxation, complained that a carbon tax lessens their competitiveness in the marketplace. As a concession, the tax was halved for Swedish businesses, and halved again during the financial recession of the mid-1990s. A further concession fully exempted certain high energy using industries such as commercial horticulture, mining, manufacturing, and the paper industry.

The environmental benefits of Sweden's carbon tax are mixed. Annual CO_2 emissions have fallen only slightly since 1990. Because energy costs represent a relatively small percentage of a businesses' total costs, companies were slow to modify or upgrade existing plants as a result of the new taxes. On the other hand, carbon taxes have prompted technical innovation, for example in the home-heating industry. Sweden is now a world leader in the manufacture of biomass and geothermal energy products. This has not just helped reduce domestic carbon emissions, but has created a new export industry and economic boon.

WORDS TO KNOW

BIOMASS: The sum total of living and once-living matter contained within a given geographic area. Plant and animal materials that are used as fuel sources.

CARBON-BASED FUEL: Any substance composed mostly of carbon that is burned or otherwise chemically reacted to release energy. Most biofuels and all fossil fuels are carbon-based, although natural gas also contains a significant fraction of its energy in the form of hydrogen.

FOSSIL FUELS: Fuels formed by biological processes and transformed into solid or fluid minerals over geological time. Fossil fuels include coal, petroleum, and natural gas. Fossil fuels are non-renewable on the timescale of human civilization, because their natural replenishment would take many millions of years.

RENEWABLE ENERGY: Energy obtained from sources that are renewed at once, or fairly rapidly, by natural or managed processes that can be expected to continue indefinitely. Wind, sun, wood, crops, and waves can all be sources of renewable energy.

TIPPING POINT: In climatology, a state in a changing system where change ceases to be gradual and reversible and becomes rapid and irreversible. Also termed a climate surprise.

SEE ALSO *Europe: Climate Policy; Mitigation Strategies; Sustainability.*

BIBLIOGRAPHY

Periodicals

Mankiw, N. Gregory. "One Answer to Global Warming: A New Tax." *The New York Times* (September 16, 2007).

Web Sites

"Carbon Taxes: An Introduction." *Carbon Tax Center.* <http://www.carbontax.org/introduction/#what> (accessed November 21, 2007).

"Emission Possible." *The Age*, June 17, 2007. <http://www.theage.com.au/news/in-depth/emission-possible/2007/06/17/1182018934799.html?page=fullpage#contentSwap2;> (accessed November 21, 2007).

Frank, Robert H. "A Way to Cut Fuel Consumption That Everyone Likes, Except the Politicians." *The New York Times*, February 16, 2006. <http://query.nytimes.com/gst/fullpage.html?res=9D05E7DA133EF935A25751C0A9609C8B63> (accessed November 21, 2007).

James Corbett

Catastrophism

Introduction

Catastrophism is the general concept that the history of Earth has been profoundly affected by sudden violent events. In the Biblical creationist view, these sudden, violent events are typically viewed as supernatural in origin and are global events of great devastation. In the modern holistic view of Earth history, catastrophism is viewed as the concept that sudden, violent, but entirely explicable events have occurred in Earth's past and may have had an effect upon the rock and fossil record of Earth. In this view, a catastrophe may have been, for instance, a colossal volcanic eruption, a comet or asteroid impact on Earth, the burst of a large glacial lake's dam, or a very powerful earthquake.

Catastrophism stands in contrast to a long-standing but somewhat simplistic view in geology, that Earth processes were more nearly uniform and gradual over geologic time (uniformitarianism). Careful geological research in many subfields of geology over many decades has shown that geological history may be characterized by episodes of uniform and gradual conditions, which in turn are punctuated by episodes with relatively sudden, violent events. In other words, the notion of catastrophes in Earth history is a sound one, but the notion that Earth history is largely one of profound catastrophe is not correct.

Historical Background and Scientific Foundations

Catastrophism as a point of view in geology was founded in the nineteenth century during a time when it was popular to invoke Biblical, supernatural explanations for geological features that were confounding to practitioners of the infant science of geology. One of the first generalizations coming out of catastrophism was the notion of a Biblical flood-related origin for such things as the separation of the continents, mass mortality in the fossil record, glacial erratic boulders and other features, and some aspects of inter-regionally distributed rock

The eruption of Mount St. Helens on May 18, 1980, in Washington state caused rapid change throughout its environment. The explosion is a recent example of how Earth can be affected by sudden violent events. *U.S. Geological Survey.*

WORDS TO KNOW

EPOCH: Unit of geological time. From longest to shortest, the geological system of time units is eon, era, period, epoch, and stage. Epochs are generally about 500 million years in length.

GEOLOGIC TIME: The period of time extending from the formation of Earth to the present.

UNIFORMITARIANISM: Doctrine of geology promoted by English geologist Charles Lyell (1797–1895), asserting three assumptions: (1) actualism (uniform processes acting throughout Earth's history), (2) gradualism (slow, uniform rate of change throughout Earth's history), and (3) uniformity of state (Earth's conditions have always varied around a single, steady state). Uniformitarianism is often contrasted to castrophism. Modern geology makes use of elements of both views, acknowledging that change can be drastic or slow, and that while some processes operate steadily over many millions of years, others (such as asteroid impacts) may happen rarely and cause castatrophic, sudden changes when they do.

formations. Catastrophism was generally closely associated with a young-Earth viewpoint, specifically that Earth was not more than a few thousand years old. Therefore, in order for the many features we see to exist, they must have all developed in a short time span, hence catastrophes as the key to understanding Earth history.

With the advent of the view of deep time (or vast geological history), an opposing viewpoint called uniformitarianism emerged. In this contrary view, Earth history is gradual and changes are uniform with time. The view of uniformitarianism was intended more as a counterpoint to catastrophism, but was taken quite literally for many decades. Today, both gradual and catastrophic origins of features and rock formations are embraced by geologists according to the interpretations of those features and rock formations warranted by the facts at hand.

Catastrophism and the Fossil Record

Some nineteenth-century paleontologists (scientists who study fossils) who were also catastrophists thought that episodes of mass extinction of fossil groups showed evidence of supernatural catastrophes. Today, we understand that some fossil groups disappeared over relatively short intervals of geologic time, but in each instance we can see evidence of readily explainable causes. Further, careful study shows that even the most seemingly instantaneous extinction event probably occurred over hundreds or perhaps thousands of years, thus showing that these apparent catastrophes were not so sudden. Over the evolutionary history of many fossil groups, it can be shown that the development of new species and the death of other ones is relatively sudden (but not instantaneous). This is called punctuated gradualism in evolution, and is thought to be the result of rapid shifts in the fossil group's environment.

Impacts and Issues

Modern catastrophism is held by many creationists and others who hold to what they consider to be a fundamentalist view of Earth history. For example, in the modern catastrophist view held by creationists, Earth history can be divided into pre-flood and post-flood epochs. All of Earth history is divided according to one catastrophic event. Modern catastrophists also hold to the young-Earth view that all of Earth history occurred in a few thousand years. In this sense, most of Earth's history had to be sudden and violent, or Earth had to be formed as we know it without any significant evolutionary history.

From a catastrophist point of view, the rapidity of modern climate change on Earth would be viewed as supporting evidence of interpreted climate changes in Earth's past that may have been relatively rapid. That modern climate change is relatively rapid is a view that is not out of step with modern scientific understandings.

SEE ALSO *Abrupt Climate Change; Extinction; Geologic Time Scale.*

BIBLIOGRAPHY

Books

Palmer, T. *Catastrophism, Neocatastrophism and Evolution.* Nottingham, UK: Nottingham Trent University, 1994.

Rudwick, M. J. S. *The Meaning of Fossils.* Chicago: University of Chicago Press, 1972.

Web Sites

Baker, Victor. "Catastrophism and Uniformitarianism: Logical Roots and Current Relevance in Geology." *Geological Society of London*, 1998. <http://sp.lyellcollection.org/cgi/content/abstract/143/1/171> (accessed December 4, 2007).

David T. King Jr.

Cement Industry

Introduction

Cement is a rocky powder that can be mixed with water and molded to any desired shape, after which it hardens to a rock-like consistency. It is used worldwide both as a mortar to join bricks and blocks together and as an ingredient of concrete, which is mixture of crushed rock, sand, and cement that is used to construct buildings, bridges, roads, pipes, dams, and other structures. Cement manufacture releases large amounts of carbon dioxide (CO_2), the most important greenhouse gas. The cement manufacturing industry is the third-largest source of anthropogenic (human-caused) carbon dioxide emissions after forest destruction and the burning of fossil fuels for transportation and electricity generation. About 5% of global anthropogenic carbon dioxide emissions are from cement manufacture.

Historical Background and Scientific Foundations

Roman builders used an early form of cement to construct the dome of the Pantheon temple in Rome, which is still standing after almost 1,900 years. Cement was

Children in Cairo, Egypt, play in a debris-filled field near a cement factory. At the time this photo was taken in 1996, Cairo had the highest levels of lead and other pollutants in the world. As a result, the government sought to fit the factories with filters to provide less toxic pollution rather than halt development. *AP Images.*

WORDS TO KNOW

ANTHROPOGENIC: Made by people or resulting from human activities. Usually used in the context of emissions that are produced as a result of human activities.

CALCINATION: An old term used to describe the process of heating metals and other materials in air.

FOSSIL FUELS: Fuels formed by biological processes and transformed into solid or fluid minerals over geological time. Fossil fuels include coal, petroleum, and natural gas. Fossil fuels are non-renewable on the timescale of human civilization, because their natural replenishment would take many millions of years.

GREENHOUSE GASES: Gases that cause Earth to retain more thermal energy by absorbing infrared light emitted by Earth's surface. The most important greenhouse gases are water vapor, carbon dioxide, methane, nitrous oxide, and various artificial chemicals such as chlorofluorocarbons. All but the latter are naturally occurring, but human activity over the last several centuries has significantly increased the amounts of carbon dioxide, methane, and nitrous oxide in Earth's atmosphere, causing global warming and global climate change.

INDUSTRIAL REVOLUTION: The period, beginning about the middle of the eighteenth century, during which humans began to use steam engines as a major source of power.

LIMESTONE: A carbonate sedimentary rock composed of more than 50% of the mineral calcium carbonate ($CaCO_3$).

little used, however, from the fall of the Roman Empire to the beginning of the Industrial Revolution in Europe. Inventors in England and France experimented with fast-setting cements and structural concrete in the late 1700s and early 1800s, and concrete reinforced by internal steel rods was first used for building construction in the 1890s. Today, cement-based concrete is the most abundant manufactured material on Earth.

Three parts of the cement-manufacture process release carbon dioxide:

1. *Calcination.* Most cement is made from a mixture of calcined limestone, clay, and chalk. Limestone is a rock that consists mostly of calcium carbonate ($CaCO_3$). Calcined limestone has been heated in a furnace to break $CaCO_3$ down into CaO and CO_2. The CO_2 is released to the atmosphere. In producing 2.2 lb (1 kg) of clinker (dry powder), calcination releases about 1.1 lb (0.5 kg) of carbon dioxide.
2. *Fuel.* Large amounts of fuel are burned in special ovens or kilns to provide heat for calcination. How much carbon dioxide is released by fuel-burning during calcination depends on the kind of fuel used. Coal, natural gas, fuel oil, and other fuels are all used by industry, with coal being the most common fuel.
3. *Electricity.* Electricity is used to run the machines that crush and grind limestone before and after calcination. Smaller amounts of electricity are used for conveyor belts, packing, and other miscellaneous aspects of manufacture. Worldwide, most electricity is produced by burning coal, which releases carbon dioxide. Most of the CO_2 emitted during cement manufacture is due to calcination and fuel-burning, not electricity usage.

■ Impacts and Issues

More than 150 countries produce cement or the baked powder called clinker that is the main ingredient of cement. China produces about 33% of the world CO_2 output from cement-making, the United States about 6%, and India about 5%.

Like all processes that release greenhouse gases and so contribute to climate change, cement manufacturing is hazardous. However, since concrete usage is entwined with all aspects of economic activity, and strong, cheap substitutes do not exist for most applications, cement production is unlikely to be reduced. On the contrary, its use is growing steadily worldwide. Yet, there are other ways in which CO_2 releases during cement manufacture may be reduced.

Much of the heat energy supplied by fuel-burning in conventional kilns during calcination of limestone is wasted: by substituting more efficient machinery and using "dry" rather than "wet" calcination processes, direct fuel usage can be reduced by up to 48%, which would reduce CO_2 emissions by 27%. However, some estimates of how much efficiency can be improved without diminishing profits are as low as 11%, which would yield a CO_2 reduction of only 5%. Substitution of lower-carbon fuels, such as natural gas, would reduce CO_2, as would shifting construction practices to the use of blended cement, which is cement in which some limestone-based clinker is replaced by industrial wastes such as coal fly ash (the ash left over from burning coal in power plants) or volcanic ash.

France, Germany, and the Netherlands committed to a voluntary United Nations-led initiative in the mid 1990s to reduce their CO_2 emissions per ton of cement. A private-sector effort to reduce cement CO_2 intensity, the Cement Sustainability Initiative, was announced in 1999 by the World Business Council for Sustainable Development. As of 2007, the Initiative still consisted primarily of commissioned studies on ways to reduce emissions. Some individual companies have, however, changed actual manufacturing processes. For example, U.S. Concrete claimed to have reduced its CO_2 output in 2006 by 328,000 tons by switching largely to blended-cement manufacture using coal fly ash. As of mid-2007, the governments of China, India, and the

United States had not officially committed to reductions in cement CO_2 emissions.

SEE ALSO *Carbon Credits; Carbon Cycle; Carbon Dioxide (CO_2); Carbon Dioxide Concentrations; Carbon Dioxide Equivalent (CDE); Industry (Private Action and Initiatives).*

BIBLIOGRAPHY

Periodicals

Worrell, Ernst, et al. "Carbon Dioxide Emissions from the Global Cement Industry." *Annual Review of Energy and the Environment* 26 (2001): 303–329.

Web Sites

Hanle, Lisa J. "CO_2 Emissions Profile of the U.S. Cement Industry. U.S. Environmental Protection Agency, 13th International Emission Inventory Conference, Working for Clean Air in Clearwater." *United States Environmental Protection Agency,* June 10, 2004. <www.epa.gov/ttn/chief/conference/ei13/ghg/hanle.pdf> (accessed August 5, 2007).

"The Sustainable Cement Initiative." *World Business Council for Sustainable Development,* 2002. <http://www.wbcsdcement.org/pdf/cement_initiative_arp.pdf> (accessed August 6, 2007).

"Why Cement-Making Produces Carbon Dioxide." Climate Change Fact Sheet 30. *United Nations Environment Programme. Information Unit on Climate Change.* <http://www.cs.ntu.edu.au/homepages/jmitroy/sid101/uncc/fs030.html> (accessed August 5, 2007).

Larry Gilman

Chaos Theory and Meteorological Predictions

■ Introduction

Nonlinear systems can exhibit apparent disorder (randomness) even when future behaviors are well defined by initial conditions and external factors are eliminated. When such situations occur, systems exhibit deterministic chaos or, simply, chaos. For instance, the back-and-forth motion of a pendulum may appear to be steady but, in reality, it is a disordered system guided by chaos theory. Where the ball lands on a roulette wheel may appear to be random but, again, it depends on initial conditions and external influences. It, too, is a chaotic system. Any system that changes a large amount with only a small modification is especially sensitive on its initial conditions. These systems have a basis in chaos theory.

Scientists involved with chaos theory attempt to examine, describe, and quantify complex and unpredictable dynamics of systems that are sensitive to their initial conditions but follow mathematic laws—even though their outward appearance appears random. Meteorology, and the prediction of weather and climate, is a classic example of such an unpredictable (chaotic) system.

Despite an array of advanced technological equipment, weather forecasters still have difficulty predicting an exact forecast several days in advance. *Digital Vision/Royalty Free*

WORDS TO KNOW

DETERMINISTIC: Able to occur in only one way: determined by the laws of nature. In contrast to stochastic, random, or chaotic processes, which are inherently difficult to forecast even when the physical laws governing them are well understood.

DIFFERENTIAL EQUATION: In mathematics, any equation in which one or more derivatives (rates of change) of a variable appear along with the variable itself. Differential equations are needed to describe many physical processes in engineering, physics, and other sciences, and are essential to climate modeling.

EL NIÑO: A warming of the surface waters of the eastern equatorial Pacific that occurs at irregular intervals of 2 to 7 years, usually lasting 1 to 2 years. Along the west coast of South America, southerly winds promote the upwelling of cold, nutrient-rich water that sustains large fish populations, that sustain abundant sea birds, whose droppings support the fertilizer industry. Near the end of each calendar year, a warm current of nutrient-poor tropical water replaces the cold, nutrient-rich surface water. Because this condition often occurs around Christmas, it was named El Niño (Spanish for boy child, referring to the Christ child). In most years the warming lasts only a few weeks or a month, after which the weather patterns return to normal and fishing improves. However, when El Niño conditions last for many months, more extensive ocean warming occurs and economic results can be disastrous. El Niño has been linked to wetter, colder winters in the United States; drier, hotter summers in South America and Europe; and drought in Africa.

ENTROPY: Measure of the disorder of a system.

METEOROLOGY: The science that deals with Earth's atmosphere and its phenomena and with weather and weather forecasting.

NONLINEAR SYSTEM: Physical system in which changes are not always additive or smooth. Climate is a nonlinear system composed of numerous nonlinear subsystems: abrupt climate change occurs at tipping points where the nonlinearity of the system causes drastic change in response to a small additional change in conditions.

■ Historical Background and Scientific Foundations

Chaos in a system was discovered by American mathematician and meteorologist Edward Lorenz (1917-) during research performed at Massachusetts Institute of Technology in the United States. In the late 1950s and early 1960s, Lorenz modeled the weather using twelve differential equations. He wanted to save time on one occasion and started the program in the middle, rather than at its initial conditions, and stored computer data to three decimals rather than the usual six. Instead of getting an expected close approximation to his result, Lorenz got a very different answer. His 1962 paper "Deterministic Nonperiodic Flow" is considered the beginning of chaos theory.

Lorenz rationalized that a small change in the initial conditions can drastically change the long-term behavior of a meteorological system. He called this phenomenon the "butterfly effect." In its extreme case, Lorenz contended it was possible for the flapping of butterfly wings to cause a massive storm a half world away. His 1972 paper "Predictability: Does the Flap of a Butterfly's Wings in Brazil Set off a Tornado in Texas?" originated the term. Based on his results, Lorenz stated that it is impossible to predict the weather accurately.

The meteorological processes and forecasting of weather and climate, along with various other natural systems, are subject to the second law of thermodynamics, which states that entropy (disorder) of an isolated system not in equilibrium will increase in entropy over time. Ultimately, the ability to predict meteorological events is tied with chaos theory.

■ Impacts and Issues

Even though Lorenz contended it was impossible to accurately predict meteorological events, when computers were invented their ability to handle massive amounts of variables changed that impossibility to, at least, a possibility. Meteorologists in the twenty-first century attempt to predict weather and climate using complicated mathematical equations that model the behavior of Earth's atmosphere. They would also like to be able to estimate global climate changes caused by human activities.

The atmosphere itself and individual weather systems are difficult to predict. For instance, because the atmosphere is chaotic, a weather forecast for a 10-day period has little validity by the tenth day and sizeable errors can be noticed after only a few days. However, larger weather systems can be characterized from historical records by knowing initial conditions in the atmosphere. To determine these initial conditions, a large number of initial, but not identical, states are collected. There are about 10 different systems (or regimes) that define weather variability in the northern hemisphere. For instance, the dynamics of ocean water and the atmosphere can raise the water temperature in the Pacific Ocean by about 7°F (4°C), which sometimes causes an El Niño event.

Thus, the meteorologist's goal is not to predict one specific climatic event. Instead, it is to determine the general chance for minor changes in climate (well within average variability of weather from year to year) versus the chance for major changes in climate (which may result in drastic weather changes not experienced in normal climates). Chaos theory allows for the prediction of long-term meteorological events such as global warming and the greenhouse effect.

Modern weather prediction can work when meteorologists gather large amounts of data from accurate sensing devices on Earth and in space about past and current weather and use complicated computer programs to estimate future weather.

However, it is unwise to make premature predictions about meteorological events. Because of chaos theory, it is difficult to calculate the weather with perfect accuracy since meteorology is a chaotic system controlled by an infinite number of variables. Any inaccuracies in the initial conditions, big or small, will have dramatic consequences on the final outcome. These inaccuracies can include defective weather satellites, restrictions of only making approximate measurements, and myriad other reasons.

To succeed over chaos theory, infinitely precise measurements must be done on infinitely accurate computers. In reality, this is impossible. Totally accurate predictions of weather and climate are not possible due to chaos theory. However, scientists will continue to improve on their analyses of weather forecasting with regards to general patterns and forecasts.

SEE ALSO *Climate Change; El Niño and La Niña; Global Warming; Greenhouse Effect; Meteorology.*

BIBLIOGRAPHY

Books

The Chaos Avant-garde: Memories of the Early Days of Chaos Theory, edited by Ralph Abraham and Ueda Yoshisuke. River Edge, NJ: World Scientific, 2000.

Danielson, Eric William. *Meteorology.* Boston: McGraw-Hill, 2003.

Environmental Modelling and Prediction, edited by Gongbing Peng, Lance M. Leslie, and Yaping Shao. New York: Springer, 2002.

Hirsch, Morris W. *Differential Equations, Dynamical Systems, and an Introduction to Chaos.* Boston: Academic Press, 2004.

Peitgen, Heinz-Otto. *Chaos and Fractals: New Frontiers of Science.* New York: Springer, 2004.

China: Climate and Energy Policies

■ Introduction

China, with more than 1.3 billion inhabitants, is the world's most populous country. As of 2008, it was also the largest emitter of greenhouse gases, having just surpassed the United States. China's increased greenhouse emissions are the result of rapid industrialization since the late 1970s and especially since 1990. CO_2 is released in China by cement production, coal burning, underground coal fires, and increased burning of natural gas and petroleum for industrial processes and vehicles.

China intends to continue its rapid economic growth in the following ways: building several new coal-fired electricity generating plants a week; increasing end-use energy efficiency; purchasing a larger share of the world's petroleum supply (China is the second-largest oil importer in the world, after the United States); building windmills and other sources of renewable energy; and creating dozens of nuclear power plants.

In Taiyuan, a factory chimney next to a statue of Mao Zedong (1893–1976), the founder of the People's Republic of China, remains dormant after ceasing operation due to local government regulation in March 2007. In June, China unveiled a climate change plan to reduce greenhouse gases, but rejected mandatory caps as unfair to developing countries. *AP Images.*

In Beijing, China, a man rides by garbage and wood used for fuel. The country's rapid industrialization since the 1970s has led to a great increase in pollution in the nation's cities. *AP Images.*

For more than a decade, China has been in a policy standoff with the United States over whether international legal agreements, such as the Kyoto Protocol, should impose similar greenhouse-gas emission restraints on developing countries such as China and on fully industrialized countries like the United States. China argues that developing countries should not face mandatory greenhouse gas emission caps, so that it can catch up in the industrialization process, while the United States argues that not restraining China and other developing nations equally would give those nations an unfair market advantage.

Historical Background and Scientific Foundations

Modern industrial society, which first appeared in Europe in the late 1700s, depends on abundant, affordable energy. It has obtained most of this energy from fossil fuels, such as coal, petroleum (oil), and natural gas. However, all these fuels release carbon dioxide into the atmosphere, which, in turn, slows the loss of heat energy from Earth into space and acts as the primary cause of global climate change.

Many countries—especially in Asia, Africa, and South America—are not as industrialized as those in North America, Europe, Australia, and Japan. These less-industrialized states, usually called developing countries, hope to achieve the economic success of the states that industrialized long ago and so acquire more wealth and power.

China is the largest of these developing nations. Its economy, already the third largest in the world, is growing faster than that of any other large country (11% in the first quarter of 2007 alone). This rapid growth has been accompanied by pollution problems, greatly increased greenhouse emissions, increased dependence on foreign oil, environmental pollution and destruction, and other problems. China is the world's largest emitter of carbon dioxide, sulfur oxides, and chlorofluorocarbons (which damage the ozone layer). It is also, with Japan, one of the two largest importers of lumber cut from tropical forests, whose destruction is contributing to global climate change.

The world's total greenhouse gas emissions have increased 75% since 1970; China's have increased faster, growing by 80% from 1990 to 2007. China's emissions intensity—the amount of greenhouse gas emitted per unit of national economic production, measured as Gross Domestic Product (GDP)—is among the world's highest. This means that it uses energy less efficiently than most other countries. However, China's energy intensity has actually decreased since 1990, thanks mostly to government-promoted energy efficiency measures.

Scientists predict that China's greenhouse gas emissions will rise between 65% and 80% from 2007 to 2020. Most of this increase will come from burning coal to generate electricity. Mainly because of coal-burning, China contains 16 of the 20 most polluted cities in the world, and about 300,000 people die in China every year as a result of air pollution.

China has officially acknowledged that the problem of global climate change is serious and real, and it has

WORDS TO KNOW

AFFORESTATION: Conversion of unforested land to forested land through planting, seeding, or other human interventions. Unforested land must have been unforested for at least 50 years for such intervention to qualify as afforestation; otherwise, it is termed reforestation.

CHLOROFLUOROCARBONS: Members of the larger group of compounds termed halocarbons. All halocarbons contain carbon and halons (chlorine, fluorine, or bromine). When released into the atmosphere, CFCs and other halocarbons deplete the ozone layer and have high global warming potential.

FOSSIL FUELS: Fuels formed by biological processes and transformed into solid or fluid minerals over geological time. Fossil fuels include coal, petroleum, and natural gas. Fossil fuels are non-renewable on the timescale of human civilization, because their natural replenishment would take many millions of years.

GREENHOUSE GASES: Gases that cause Earth to retain more thermal energy by absorbing infrared light emitted by Earth's surface. The most important greenhouse gases are water vapor, carbon dioxide, methane, nitrous oxide, and various artificial chemicals such as chlorofluorocarbons. All but the latter are naturally occurring, but human activity over the last several centuries has significantly increased the amounts of carbon dioxide, methane, and nitrous oxide in Earth's atmosphere, causing global warming and global climate change.

HYDROELECTRIC POWER: Electric power derived from generators that are driven by hydraulic or water turbine engines.

KYOTO PROTOCOL: Extension in 1997 of the 1992 United Nations Framework Convention on Climate Change (UNFCCC), an international treaty signed by almost all countries with the goal of mitigating climate change. The United States, as of early 2008, was the only industrialized country to have not ratified the Kyoto Protocol, which is due to be replaced by an improved and updated agreement starting in 2012.

OZONE LAYER: The layer of ozone that begins approximately 9.3 mi (15 km) above Earth and thins to an almost negligible amount at about 31 mi (50 km) and shields Earth from harmful ultraviolet radiation from the sun. The highest natural concentration of ozone (approximately 10 parts per million by volume) occurs in the stratosphere at approximately 15.5 mi (25 km) above Earth. The stratospheric ozone concentration changes throughout the year as stratospheric circulation changes with the seasons. Natural events such as volcanoes and solar flares can produce changes in ozone concentration, but man-made changes are of the greatest concern.

RENEWABLE ENERGY: Energy obtained from sources that are renewed at once, or fairly rapidly, by natural or managed processes that can be expected to continue indefinitely. Wind, sun, wood, crops, and waves can all be sources of renewable energy.

announced policies for dealing with the problem. To what extent these announced policies reflect actual intentions will be revealed over the next 10 to 20 years. China, like other nations, is particularly concerned that its anti-greenhouse gas emission commitments do not put it at a disadvantage in the global marketplace.

On June 4, 2007, China released a 62-page document detailing its first-ever national climate change program. China rejected mandatory caps for greenhouse gas emissions and repeated its argument that countries with a long history of high emissions, rather than newcomers to the high-emissions category, such as itself, should take primary responsibility for decreasing greenhouse gas emissions.

Policies for Mitigating Climate Change

In its 2007 plan and other energy directives released piecemeal since 2000, China has described a number of policies for mitigating climate change. The following is a sampling of these announced policies.

1. Close numerous inefficient electric generating plants by 2010, totaling about 8% of China's electric capacity. These plants are to be replaced by new, more efficient plants.
2. Close inefficient factories making cement, aluminum, steel, and other materials; replace their capacity with larger, more efficient plants.
3. Mandate efficiency levels for buildings, industry, and appliances.
4. Produce 16% of China's primary energy from renewable sources by 2020. Today, only 7% of China's primary energy comes from renewables, including hydroelectric dams. China is building electricity-generating wind turbines at a rapid pace, but wind turbines still provided less than 1% of the country's electricity as of 2007.
5. Quadruple nuclear electric capacity by 2020. Nuclear power provided about 2.3% of China's electricity as of 2007. The official goal announced in February 2007 was to generate 9% of China's electricity from nuclear power by 2015.
6. Increase forest area (afforestation) to absorb carbon dioxide.
7. In cooperation with the United States and European Union countries, study technologies for

storing carbon dioxide generated by coal-burning power plants underground rather than releasing it into the air.

Impacts and Issues

China, like other countries, has already been affected by global climate change. For example, the water supply to northern China had decreased by 12% as of 2007. Scientists estimated in 2007 that China's production of its three main crops (rice, wheat, and corn) might decrease to only 40% of its present level by 2050.

China has ratified the Kyoto Protocol and the United Nations Framework Convention on Climate Change (UNFCCC), both international treaties designed to limit greenhouse-gas emissions and so mitigate future climate change and its dangers. As a developing country, China is exempt until at least 2012 from emission limits described in the Kyoto Protocol.

China does, however, participate in the Clean Development Mechanism (CDM), an aspect of the Kyoto Protocol designed to reward developed countries for helping developing countries reduce their greenhouse emissions. Under the CDM, developed countries are allowed to emit more greenhouse gases if they help developing countries emit less. As of 2007, CDM aid to China had accounted for 40% of all emissions credits given to developed countries under the Kyoto Protocol, more than any other country. Most of these credits were earned by destroying stocks of the greenhouse gas trifluoromethane (used in refrigeration and some manufacturing processes) and by making arrangements to capture and burn methane released from landfills. Methane causes 25 times more global warming, ton for ton, than carbon dioxide, while trifluoromethane causes 11,700 times more.

According to some experts, China's ambitious targets for reducing pollution and rapidly building dams and nuclear power plants may be overoptimistic. A nuclear power plant takes approximately 10 years to build, while the number of sites where dams can be built is limited. Moreover, large dams are problematic as greenhouse-gas emission fighters because they can be major emitters of methane, carbon dioxide, and nitrous oxide.

Some of China's other climate-change mitigation goals also may be unrealistic for technical reasons. For example, in 2005 China declared a goal of reducing the energy intensity of its economy (the amount of energy consumed per unit of GDP) in 2010 by 20%, relative to the level in 2005. This included a specific goal of reducing energy intensity by 4% in 2006, but energy intensity actually declined by only 1.23%.

Furthermore, official information in China has historically been unreliable at times, which may make it difficult for international observers to accurately judge China's progress toward energy and climate policy goals. Like other governments, China has been known to distort information for propaganda purposes, and its own State Environmental Protection Administration has warned that local officials often report false numbers to the central government in order to receive financial rewards and promotions.

SEE ALSO *Asia: Climate Change Impacts; Carbon Cycle; China: Total Carbon Dioxide Emissions; Chlorofluorocarbons and Related Compounds; Climate Change; Coal; Economics of Climate Change; Greenhouse Effect; Greenhouse Gases; Intergovernmental Panel on Climate Change (IPCC); United Nations Environment Programme (UNEP); United Nations Framework Convention on Climate Change (UNFCCC).*

BIBLIOGRAPHY

Periodicals

He, Kebin, Hong Huo, and Qiang Zhang. "Urban Air Polution in China: Current Status, Characteristics, and Progress." *Annual Review of Energy and Environment* 27 (2002): 397–431.

Liu, Jianguo, and Jared Diamond. "China's Environment in a Globalizing World." *Nature* 435 (June 30, 2005): 1179–1186.

Ren, Xin, Lei Zeng, and Dadi Zhou. "Sustainable Energy Development and Climate Change in China." *Climate Policy* 5 (2005): 185–198.

Schiermeier, Quirin. "China Struggles to Square Growth and Emissions." *Nature* 446 (April 26, 2007): 954–956.

Streets, David G., et al. "Recent Reductions in China's Greenhouse Gas Emissions." *Science* 294 (November 30, 2001): 1835–1837.

Web Sites

"Beijing's New Thinking on Energy Security." *China Brief*, April 12, 2006. <http://jamestown.org/china_brief/article.php?articleid=2373181> (accessed August 6, 2007).

"Climate Change Mitigation Measures in the People's Republic of China: International Brief No. 1." *Pew Center on Global Climate Change*, April 2007. <http://www.pewclimate.org/docUploads/International%20Brief%20-%20China.pdf> (accessed August 8, 2007).

"Press Conference on National Climate Change Program." *China View*, June 6, 2007. <http://news.xinhuanet.com/english/2007-06/04/content_6197309.htm> (accessed August 8, 2007).

Larry Gilman

China: Total Carbon Dioxide Emissions

Introduction

China is the world's most populous country, with over 1,311,000,000 (1.311 billion) people as of 2006, about a fifth of the world's population. It is also one of the world's largest emitters of carbon dioxide (CO_2). The United States was the world's largest emitter of CO_2 and other greenhouse gases for most of the twentieth century, but in 2007 the Netherlands Environmental Assessment Agency, an arm of the Dutch government, announced that China had become the world's leading CO_2 emitter. Because China's population is more than four times that of the United States, each American is still creating several times more greenhouse gas than is being emitted for each Chinese person.

Historical Background and Scientific Foundations

Greenhouse gas emissions are caused primarily by deforestation and the burning of fossil fuels. Fuel-burning in cars, industrial factories, and electrical power plants is associated with industrialization, which is defined as the changeover of a nation's economy from agriculture to

The air in Beijing is heavy with pollution in this image from February 2001. By 2007, China had surpassed the United States to become the world's biggest emitter of carbon dioxide, the greenhouse gas most attributed to global warming. *AP Images.*

Although China is one of the world's biggest producers of greenhouse gases, it does not emit as much per person as the United States. *AP Images.*

mechanized manufacturing. Industrialization began in Europe in the late 1700s and transformed life in Europe and the United States in the nineteenth century, spreading to Japan and other parts of the world in the late nineteenth and early twentieth centuries.

China was a latecomer to this economic change, with major industrialization beginning only after World War II and proceeding slowly and erratically until about 1990. Since then, China's economy has expanded quickly into car manufacturing, aerospace, electronics, and many other areas. Much of the world's product manufacturing now occurs there, and more Chinese people (still a minority) are driving cars and seeking consumer luxuries. China's economy grew by 11% in the first quarter of 2007 alone. This rapid industrialization has led to a rapid rise in the amount of greenhouse gas, especially carbon dioxide, emitted by China.

In 2005, China's CO_2 emissions were still about 2% below those of the United States. In 2006, the world's total CO_2 emissions increased by about 2.6%. Most of this increase was caused by a 4.5% increase in the amount of coal-burning, about two-thirds of which is attributed to China. China builds, on average, about one coal-fired electric power plant every four days. China's CO_2 emissions increased by 8.7% in 2006 to approximately 6,830 million standard U.S. tons (6,200 million metric tons) per year. The United States's greenhouse gas-emissions decreased by about 1.3% in 2006. As a result, China emitted about 8% more CO_2 than the United States in 2006. European greenhouse gas emissions remained approximately level in 2005 to 2006.

American and Chinese scientists writing in the journal *Science* in 2001 claimed that China's CO_2 emissions actually decreased by 7.3% from 1996 to 2000 because of economic setbacks and reforms in the government-owned coal and energy industries. If this decrease did occur, it has since been swept away by massive increases.

Most of China's CO_2 production is from coal-burning; about 9% is from the manufacture of cement. A significant fraction comes from out-of-control underground coal fires, which burn between 100 and 200 million tons of coal yearly, accounting for 2 to 3% of all the CO_2 released by fossil-fuel burning in the world. It is difficult to measure the extent of such coal fires precisely, but they may have accounted for about a fifth of China's greenhouse gas emissions as of 2006.

■ Impacts and Issues

International disputes over greenhouse emissions by China and other developing countries have made it impossible, so far, to achieve global legal agreement on a strategy for mitigating climate change. China argues that because the United States and European countries were able to develop their economies for generations without restraining their greenhouse gas emissions, those countries should be required to restrain their emissions more stringently than China and other countries that are only now industrializing.

WORDS TO KNOW

DEFORESTATION: Those practices or processes that result in the change of forested lands to non-forest uses. This is often cited as one of the major causes of the enhanced greenhouse effect for two reasons: 1) the burning or decomposition of the wood releases carbon dioxide; and 2) trees that once removed carbon dioxide from the atmosphere in the process of photosynthesis are no longer present and contributing to carbon storage.

FOSSIL FUELS: Fuels formed by biological processes and transformed into solid or fluid minerals over geological time. Fossil fuels include coal, petroleum, and natural gas. Fossil fuels are non-renewable on the timescale of human civilization, because their natural replenishment would take many millions of years.

GREENHOUSE GASES: Gases that cause Earth to retain more thermal energy by absorbing infrared light emitted by Earth's surface. The most important greenhouse gases are water vapor, carbon dioxide, methane, nitrous oxide, and various artificial chemicals such as chlorofluorocarbons. All but the latter are naturally occurring, but human activity over the last several centuries has significantly increased the amounts of carbon dioxide, methane, and nitrous oxide in Earth's atmosphere, causing global warming and global climate change.

INDUSTRIALIZATION: Shift of a large portion of a region or country's economy to mechanized manufacturing and away from agriculture. Industrialization has been an increasingly global process since the beginning of the Industrial Revolution, usually dated to about 1750.

KYOTO PROTOCOL: Extension in 1997 of the 1992 United Nations Framework Convention on Climate Change (UNFCCC), an international treaty signed by almost all countries with the goal of mitigating climate change. The United States, as of early 2008, was the only industrialized country to have not ratified the Kyoto Protocol, which is due to be replaced by an improved and updated agreement starting in 2012.

The United States has replied that China's economy is already competitive and that, as a matter of economic fairness, any greenhouse gas emission agreements must apply equally to all nations. (Most other industrialized countries have signed the Kyoto Protocol, the only international legal instrument so far which seeks to constrain greenhouse gas emissions.) In 1997, the year before the Kyoto Protocol was opened for signatures, the United States Senate unanimously passed the Byrd-Hagel Resolution (S. Res. 98), which stated that the United States would not be signatory to the Kyoto Protocol or any other agreements that "mandate new commitments to limit or reduce greenhouse gas emissions for" the United States and other developed nations "unless the protocol or other agreement also mandates new specific scheduled commitments to limit or reduce greenhouse gas emissions for Developing Country Parties within the same compliance period."

SEE ALSO *Asia: Climate Change Impacts; China: Climate and Energy Policies.*

BIBLIOGRAPHY

Periodicals

Jiang, Wenran. "Beijing's 'New Thinking' on Energy Security." *Jamestown Foundation China Brief* 6 (April 12, 2006). Also available at <http://jamestown. org/china_brief/article.php?articleid=2373181>.

Schiermeier, Quirin. "China Struggles to Square Growth and Emissions." *Nature* 446 (2007): 954–956.

Streets, David G., et al. "Recent Reductions in China's Greenhouse Gas Emissions." *Science* 294 (2001): 1835–1837.

Web Sites

"China Now No. 1 in CO_2 Emissions; USA in Second Position." *Netherlands Environmental Assessment Agency*, June 22, 2007. <http://www.mnp.nl/en/dossiers/Climatechange/moreinfo/Chinanowno1inCO2 emissionsUSAinsecondposition.html> (accessed August 5, 2007).

Larry Gilman

Chlorofluorocarbons and Related Compounds

Introduction

Chlorofluorocarbons (CFCs) are greenhouse gas compounds that contain only carbon, fluorine, and chlorine. Chlorofluorocarbons and related compounds, including halons and hydrobromofluorocarbons, destroy ozone molecules in the stratosphere and cause ozone depletion. The ozone layer is a layer in Earth's stratosphere that absorbs most of the ultraviolet radiation entering Earth's atmosphere.

The greatest concern is that ozone is the only absorber of UVB (medium-wave ultraviolet) radiation, which can cause skin cancer. As ozone is depleted, a greater amount of UVB radiation reaches the surface of Earth and the risk of skin cancer increases. Concern over the depletion of the ozone layer has led to successful international action to reduce the use of CFCs and related compounds.

Historical Background and Scientific Foundations

In 1974, Mexican-American chemist Mario Molina and American chemist F. Sherwood Rowland published a

When scientists learned that CFCs were damaging the ozone layer, many nations around the world signed a treaty to greatly reduce the use of CFCs. Hydrofluorocarbons became one of the new alternatives. Although HFCs have been an effective part of the solution to ozone depletion, they are of increasing concern as they are a major greenhouse gas, which contributes to the melting of the polar ice cap. *AP Images.*

paper in the journal *Nature* stating that CFCs were depleting stratospheric ozone. The hypothesis was initially controversial, but was confirmed by later research, including an alarm sounded by the British scientist Joseph Farman, who had studied atmospheric chemistry near the South Pole since 1957, and reported in 1985 that ozone levels above Antarctica had reduced by more than one third. Satellite observation of the increase in the size of the Antarctic ozone hole was then confirmed by the National Aeronautics and Space Administration (NASA) satellite Nimbus-7.

CFCs were being released into the atmosphere because of their use in refrigeration and air conditioning as well as their use as solvents, fire extinguishers, and as aerosol propellants in spray cans. Molina and Rowland described how CFCs are extremely stable in the lower atmosphere and remain in the atmosphere for about 100 years. They also described how they deplete stratospheric ozone in a chain reaction that results in the production of molecules that destroy additional molecules.

In a paper in *American Scientist*, Rowland stated that one CFC molecule can destroy around 100,000 ozone molecules. For their research into CFCs, Molina and Rowland were awarded a share in the 1995 Nobel Prize in chemistry.

Impacts and Issues

In 1981, action began on developing an international agreement to reduce CFC production and consumption. This led to the Montreal Protocol on Substances that Deplete the Ozone Layer, an international treaty that went into effect on January 1, 1989. The Montreal Protocol originally called for a reduction of CFC production and consumption to 50% of 1986 levels by 1999, then 0% as of 1996. By September 2007, 190 countries had signed the treaty.

In 2007, the EPA reported that the thinning of the ozone layer had slowed and that it could recover to its pre-1980 level between the years 2060 and 2075. It also reported that this could result in more than $4 trillion in societal health benefits in the United States alone, while preventing more than 6 million premature skin cancer deaths within the following century.

Although CFCs do contribute to the greenhouse effect, their properties also cause cooling during ozone destruction. So far, scientists estimate that the cumulative effect of warming versus cooling resulting from CFC release into the atmosphere has been about equal. Ozone destruction, therefore, not climate change itself, remains the largest problem associated with CFCs.

SEE ALSO *Aerosols; Antarctica: Observed Climate Changes; Atmospheric Pollution; Montreal Protocol; Ozone (O_3).*

WORDS TO KNOW

AEROSOL: Particles of liquid or solid dispersed as a suspension in gas.

ATMOSPHERIC CHEMISTRY: Study of the chemistry of planetary atmospheres. The interactions of pollutants, greenhouse gases, Earth's natural atmosphere, solar radiation, and other factors are all studied under the aegis of atmospheric chemistry.

OZONE: An almost colorless, gaseous form of oxygen with an odor similar to weak chlorine. A relatively unstable compound of three atoms of oxygen, ozone constitutes, on average, less than one part per million (ppm) of the gases in the atmosphere. (Peak ozone concentration in the stratosphere can get as high as 10 ppm.) Yet ozone in the stratosphere absorbs nearly all of the biologically damaging solar ultraviolet radiation before it reaches Earth's surface, where it can cause skin cancer, cataracts, and immune deficiencies, and can harm crops and aquatic ecosystems.

STRATOSPHERE: The region of Earth's atmosphere ranging between about 9 and 30 mi (15 and 50 km) above Earth's surface.

ULTRAVIOLET RADIATION: The energy range just beyond the violet end of the visible spectrum. Although ultraviolet radiation constitutes only about 5% of the total energy emitted from the sun, it is the major energy source for the stratosphere and mesosphere, playing a dominant role in both energy balance and chemical composition.

BIBLIOGRAPHY

Books

Parson, E. A. *Protecting the Ozone Layer: Science and Strategy.* New York: Oxford University Press, 2003.

Periodicals

Molina, M. J., and F. S. Rowland. "Stratospheric Sink for Chlorofluoromethanes: Chlorine-atom Catalyzed Distribution of Ozone." *Nature* 249 (1974): 810-812.

Rowland, F. S. "Chlorofluorocarbons and the Depletion of Stratospheric Ozone." *American Scientist* 77 (1989): 36-45.

Web Sites

"Achievements in Stratospheric Ozone Protection Progress Report." *Environmental Protection Agency (EPA)*, 2007. <http://www.epa.gov/ozone/pdffile/spd-annual-report_final_lowres_4-25-07.pdf> (accessed October 26, 2007).

Tony Hawas

Clean Development Mechanism

Introduction

The Clean Development Mechanism (CDM) is an arrangement defined by the Kyoto Protocol to the United Nations Framework Convention on Climate Change (UNFCCC), an international treaty designed to fight greenhouse warming. The CDM allows wealthier nations to earn credits toward their greenhouse-gas reduction goals by funding projects in developing (poorer, less technologically and industrially advanced) countries that will reduce greenhouse-gas emissions overall.

The reasoning behind the CDM is that projects to reduce greenhouse emissions will be cheaper to implement in developing countries, but the money to fund such projects will be found mostly in developed countries. Since it makes no difference where a ton of carbon dioxide (CO_2) or other greenhouse gas is kept out of the atmosphere, the CDM seeks to match funds with reduction opportunities so as to produce a cost-effective greenhouse abatement strategy.

Most CDM projects are in India, China, and Latin America. About 60% of CDM funds went to projects in China in 2006. Critics have argued that some of the claims for emissions reductions under the program are exaggerated; others have stated that Africa is underserved by the program. Still others are opposed to the concept of carbon credits, which, they say, among other faults, allow business as usual to continue in industrialized countries where deep cuts in greenhouse emissions need to be made.

Historical Background and Scientific Foundations

Scientific understanding of global warming began to solidify in the late 1970s and 1980s. In response, member states of the United Nations negotiated the UNFCCC in 1992. This treaty committed its signatories to create national and regional programs containing measures to reduce global warming by decreasing emissions of greenhouse gases and enhancing removal of those gases from the atmosphere by various carbon sinks (for example, forests).

However, the 1992 version of the treaty did not establish any specific arrangements for these programs. This was remedied by negotiations over the next several years. In 1997, the Kyoto Protocol was drafted, an add-on to the UNFCCC that would have to be signed again by each UNFCCC participant wishing to participate.

The UNFCCC divided the world's countries into three groups—Annex I, Annex II, and developing. Developing countries were relatively poor, having only partly industrialized economies. Annex I countries were fully industrialized. Annex II countries were the most prosperous Annex I countries (members of the Organisation for Economic Co-operation and Development), who would contribute funds to pay for emissions reductions in developing countries. Under the protocol, the Annex I countries committed to reducing their greenhouse emissions by 2012 to 5.2% below 1990 levels. A violator would face stiffer reductions requirements under whatever follow-up to the Kyoto Protocol would govern the period after 2012.

Under Article 12 of the Kyoto Protocol, Annex I countries (or companies based in them) can pay for projects in developing countries that reduce greenhouse emissions or enhance carbon sinks. In exchange, the Annex I country that funds such a project receives a certain amount of carbon credits, termed Certified Emission Reductions (CERs). The more carbon that is kept out of the atmosphere, the more CER carbon credits are earned.

Carbon credits are fiscal instruments, abstract objects, like dollars, euros, or stock shares, that have an arbitrarily agreed-upon value. The agreed-upon value of a single carbon credit is 1 metric ton of CO_2 equivalent, that is, either 1 ton of actual CO_2 or an amount of some other greenhouse gas, such as methane, that would cause as much global warming as 1 ton of CO_2. The Kyoto Protocol defined several types of carbon credit, of which the CER is one.

WORDS TO KNOW

BIOMASS: The sum total of living and once-living matter contained within a given geographic area. Plant and animal materials that are used as fuel sources.

CARBON CREDIT: A unit of permission or value, similar to a monetary unit (e.g., dollar, euro, yen) that entitles its owner to emit one metric ton of carbon dioxide into the atmosphere.

CARBON SINK: Any process or collection of processes that is removing more carbon from the atmosphere than it is emitting. A forest, for example, is a carbon sink if more carbon is accumulating in its soil, wood, and other biomass than is being released by fire, forestry, and decay. The opposite of a carbon sink is a carbon source.

FOSSIL FUELS: Fuels formed by biological processes and transformed into solid or fluid minerals over geological time. Fossil fuels include coal, petroleum, and natural gas. Fossil fuels are non-renewable on the timescale of human civilization, because their natural replenishment would take many millions of years.

GREENHOUSE GASES: Gases that cause Earth to retain more thermal energy by absorbing infrared light emitted by Earth's surface. The most important greenhouse gases are water vapor, carbon dioxide, methane, nitrous oxide, and various artificial chemicals such as chlorofluorocarbons. All but the latter are naturally occurring, but human activity over the last several centuries has significantly increased the amounts of carbon dioxide, methane, and nitrous oxide in Earth's atmosphere, causing global warming and global climate change.

KYOTO PROTOCOL: Extension in 1997 of the 1992 United Nations Framework Convention on Climate Change (UNFCCC), an international treaty signed by almost all countries with the goal of mitigating climate change. The United States, as of early 2008, was the only industrialized country to have not ratified the Kyoto Protocol, which is due to be replaced by an improved and updated agreement starting in 2012.

METHANE: A compound of one hydrogen atom combined with four hydrogen atoms, formula CH_4. It is the simplest hydrocarbon compound. Methane is a burnable gas that is found as a fossil fuel (in natural gas) and is given off by rotting excrement.

SUSTAINABLE DEVELOPMENT: Development (i.e., increased or intensified economic activity; sometimes used as a synonym for industrialization) that meets the cultural and physical needs of the present generation of persons without damaging the ability of future generations to meet their own needs.

All carbon credits allow the possessor, whether a country or a private party, to emit 1 ton of carbon dioxide. In addition, the sum of carbon credits and emitted tonnage is fixed. Funding a CDM project that produces, say, 10,000 carbon credits (prevents the release of 10,000 tons of CO_2 or an equivalent amount of some other greenhouse gas) generates 10,000 new credits (CERs).

The funding country or company owns the credits and can apply them toward its own treaty commitment to reduce emissions by 5.2% below 1990 levels by 2012. Ideally, the nation receiving the funding for the CDM project benefits from the project as a form of sustainable development. For example, consider a biomass-energy project in Pagara, India, that registered with the CDM managerial structure on October 23, 2005. The project involved the construction of an electric power plant for Deepak Spinners Limited, a company that manufactures a synthetic blended yarn of polyester, viscose, and acrylic. The power plant would supply power to the yarn factory not from fossil fuel, but from soya crop wastes, bagasse (the biomass remaining after sugar cane processing), and wheat chaff. Biomass of this type does not, ideally, contribute new CO_2 to the atmosphere, but simply returns CO_2 that the plants recently removed. The project was designed to displace the release of 17,424 metric tons of CO_2 per year. Therefore, those who funded the project (in this case, the governments of Germany and Italy) were to earn 17,424 CER carbon credits per year.

Germany or Italy can apply its share of the 17,424 CERs earned from the Pagara project to meeting their Kyoto emissions reductions commitment. It is also possible that they can earn money by selling their credits on the European carbon-credits market. This is permitted under Kyoto if the country acquiring the CER is already meeting its Kyoto emissions limit. The cost of a CER earned from the Pagara project is probably, even counting overhead, far below the 2007 market price of a carbon credit on the European carbon market, namely around 20 euros (US$28). By selling the carbon credit, the Annex I country would, in this case, make a profit.

As of 2007, CERs were already being traded on the European market. However, freely exchanging CERs will not be possible until the computerized system called the International Transaction Log comes online. The ITL will permit CERs to be exchanged on the carbon market along with five other types of Kyoto-defined carbon credits (called "Kyoto units" by the UNFCCC secretariat).

■ Impacts and Issues

According to the UNFCCC secretariat, as of late 2007 over 2,600 CDM projects had entered the validation process. Each CDM project undergoes a searching evaluation that involves site visits, interviews, and audits of the accounts of participants in the developing country where the project is to take place. Follow-up validation visits are also made. If all are approved, these 2,600+ projects are scheduled to generate 2.5 billion CERs by the end of 2012, that is, to reduce by 2.5 billion tons the amount of carbon dioxide that would otherwise be in the atmosphere by that time. (Global human CO_2 emissions are about 27 billion metric tons per year as of 2007.)

Some CDM projects have been criticized for enriching the already rich. For example, one chemical factory in China was outfitted with a $5 million incinerator to destroy HFC-23, a refrigerant chemical that is far more effective at causing global warming, ton for ton, than CO_2. Because the greenhouse mitigation to be realized from the project was so great, the foreign firms paying for the incinerator were actually charged $500 million for the CERs generated by the project. The $495 million difference was split by the factory's owners and by a group of consultants and bankers in London. By comparison, only $150 million of CDM funding went to all of Africa in 2006. Citing such questionable deals, Canada withdrew its share of funding for the CDM aspect of the Kyoto mechanism in September 2006. The United States never ratified the Kyoto Protocol and does not participate in any of its mechanisms.

Despite criticism of the CDM, most of the world's countries continue to participate in the young program (which only began operations after the Kyoto Protocol entered into force in 2005). The CDM program has grown rapidly since its inception, with $4.8 billion in transfer payments made in 2006 alone. Growth in CDM projects continued to be rapid in 2007, with the European market for carbon credits generated by such projects surging in anticipation of the International Transaction Log. If successful on a large scale, the CDM may blunt one of the most common criticisms of the Kyoto Protocol, namely, that it imposed no caps on the emissions of the developing countries, which are growing much more rapidly than those of the Annex I countries. In the summer of 2007, the UNFCCC secretariat, World Bank, and African governments held a conference in Nairobi, Kenya, with the goal of increasing African participation in the CDM mechanism.

■ Primary Source Connection

The following source discusses Clean Development Mechanism (CDM) policy under the United Nations Framework Convention on Climate Change (UNFCCC). Annex I parties and countries are a group consisting of all the developed countries within the Organization of Economic Co-operation and Development, and several economies in transition. At the inception of the UNFCCC, Annex I parties included most of the nations of Europe, Canada, Australia, New Zealand, and the United States. In recent years, more Annex I parties have been added by amendment.

POLICIES, MEASURES, AND INSTRUMENTS

... 6.3.2.2 The Clean Development Mechanism (Article 12)

The purposes of the CDM are to assist non-Annex I Parties to achieve sustainable development and to contribute to the ultimate objective of the Convention while assisting compliance by Annex I Parties.... The CDM allows a project to reduce emissions, or possibly to enhance sinks, in a country without a national commitment to generate certified emission reductions (CERs) equal to the reduction achieved. Annex I Parties can use CERs to meet national emissions limitation commitments. In contrast to JI, for which there is little peer-reviewed literature, the literature is rapidly growing on the CDM.

A process for independent review of the certification of the emission reductions achieved is necessary for the credibility of the CDM. Article 12.4 establishes an executive board for the CDM and Article 12.5 specifies that emission reductions must represent real, measurable, and long-term benefits related to the mitigation of climate change and be certified by designated operational entities. The certification process and the respective roles of the operational entities and the executive board remain to be defined, but they will be critical.

The host government must approve proposed CDM projects. As part of its approval process it will need to assess whether the proposed project contributes to sustainable development. Some Parties have proposed criteria or procedures that the host government be required to follow when determining whether a project contributes to sustainable development of the country.

Investments in CDM projects by Annex I governments could lead to a reduction in their official development assistance (ODA). The effect of government investment in CDM projects on the level of ODA will be difficult to determine since the level of ODA in the absence of CDM projects is unobservable. However, historical figures compiled by the OECD Development Assistance Committee could be used to try to deal with this.

Article 12.8 specifies that a share of the proceeds from CDM projects will be used to cover administrative expenses and to assist developing country Parties that are particularly vulnerable to the adverse effects of climate change to meet the costs of adaptation. Articles 6 and 17 do not impose a comparable levy on JI projects or international transfers of AAUs, although a number of developing countries have proposed that the levy be applied to all three mechanisms.

CDM projects can begin to create CERs upon ratification of the Kyoto Protocol. The advantage is that it supports developing countries obtaining access to cleaner technologies earlier. It means that a supply of CERs should be available prior to the start of the 2008 to 2012 commitment period when they can be used by Annex I Parties. Parkinson et al. (1999) argue that creation of CERs during 2000 to 2007, which are credited towards 2008 to 2012 compliance, increases the emissions trajectories of Annex I countries for 2000 to 2012. They estimate that increased Annex I emissions offset 30-60% of the CERs created during 2000 to 2012.

Some analysts argue that the CDM facilitates the transfer of CERs from low-cost emission reduction actions to Annex I investors when they might subsequently be needed by the host government to meet a future emissions limitation commitment. However, this assumes a fixed stock of emission reduction actions. In practice, the stock of possible emission reduction (or possibly sink enhancement) actions changes over time in response to turnover of the capital stock, technological change, and other developments. Rose et al. (1999) analyzes the optimal strategy for a host government given a dynamic stock of potential projects.

Numerous issues related to implementation of the CDM remain to be negotiated, including:

- host and project eligibility;
- eligibility of sequestration actions;
- demonstrating contribution to sustainable development;
- project financing arrangements;
- monitoring, verification, and reporting requirements;
- baseline establishment;
- CER certification, registry, and trading conditions;
- the share of proceeds for administrative expenses and adaptation assistance;
- adaptation assistance fund administration;
- supplementarity provisions;
- executive board composition and responsibilities;
- process for designation of operational entities; and
- penalties for non-compliance.

IPCC. "POLICIES, MEASURES, AND INSTRUMENTS" *CLIMATE CHANGE 2001: MITIGATION*, CH. 6. <HTTP://WWW.GRIDA.NO/CLIMATE/IPCC_TAR/WG3/INDEX.HTM> (ACCESSED NOVEMBER 30, 2007).

SEE ALSO *Carbon Credits; Emissions Trading; Kyoto Protocol.*

BIBLIOGRAPHY

Periodicals

Bradsher, Keith. "Clean Power that Reaps a Whirlwind." *The New York Times* (May 9, 2007).

Bradsher, Keith. "Outsize Profits, and Questions, in Effort to Cut Warming Gases." *The New York Times* (December 21, 2006).

Yamaguchi, Mitsutsune. "CDM Potential in the Power-Generation and Energy-Intensive Industries of China." *Climate Policy* 5 (2005): 167–184.

Zegras, P. Christopher. "As If Kyoto Mattered: The Clean Development Mechanism and Transportation." *Energy Policy* 35 (2007): 5136–5150.

Web Sites

"Clean Development Mechanism." *United Nations Framework Convention on Climate Change Secretariat*, June 22, 2007. <http://unfccc.int/kyoto_protocol/mechanisms/clean_development_mechanism/items/2718.php> (accessed November 2, 2007).

Larry Gilman

Climate Change

■ Introduction

Climate is the average weather of a region over time. Temperature, winds, heat waves and cold snaps, rainfall, when seasons begin and end, and other weather conditions are all aspects of climate. Climates are shaped by a global machinery of ocean currents, winds, forests, ice caps, mountain ranges, bacteria, planetary orbital motions, and many other factors.

Earth's climate is changing. In a way, this is nothing new: global and regional climates have been changing since Earth formed about 4.5 billion years ago. Ice ages and heat waves have gripped the planet for millions of years on end, sudden warmings and coolings have happened in as little as a decade, ice caps and jungles have spread and shrunk, and seas have risen and fallen by many meters.

But the changes in global climate seen in the last half-century are more drastic than any seen for over a thousand years, and in one way are unique: according to the scientific data, the changes are caused by human activity. Since the Industrial Revolution began in the latter half of the eighteenth century, humans have been digging, pumping, and burning ever-increasing amounts of coal, oil, and natural gas. These fossil fuels contain carbon. When carbon (C) burns, it combines with oxygen (O_2) to form carbon dioxide (CO_2), a clear, odorless gas. In the atmosphere, CO_2 acts like an invisible blanket that warms the planet. Not all gases have this warming effect; the main ingredients of Earth's atmosphere, oxygen (21%) and nitrogen (N) (78%), do not. Those that do have it are called greenhouse gases by analogy to the glass ceilings that keep greenhouses warm inside.

As CO_2 and the other greenhouse gases—including methane (CH_4), nitrous oxide (N_2O), and water vapor (H_2O)—increase, Earth's invisible, gaseous greenhouse roof traps more energy and the planet gets warmer. Since 1750, human activities have increased the atmospheric concentration of CO_2 by over a third, to levels not seen on Earth for at least 800,000 years. From 1906 to 2005, Earth's global average surface-air temperature rose by about 1.3°F (.74°C). Most of the CO_2 increase and most of the warming have happened since 1970, in step with increased burning of fossil fuels.

Natural influences on climate continue to operate, but they have lately been overwhelmed by human greenhouse emissions from fuel burning, agriculture, deforestation, and other activities. These are now the dominant drivers of climate change.

In large part, the greenhouse effect is natural and necessary. Without it, the oceans would freeze over and most life would die. However, humans have greatly increased the greenhouse effect in a mere century or two, giving Earth's delicately balanced climate machine a sudden, unintentional push. The consequences of that push are only beginning to be seen. Warming, the primary or basic effect, triggers a host of other changes: more rain and snow in some places and less in others, more floods and droughts, melting of mountain glaciers and ice caps, rising seas, and extinctions of plants and animals. This is why scientists prefer the phrase "global warming" when speaking of the temperature increase as such and "global climate change" when speaking of global warming plus all the climate changes that warming causes. A few regions, such as parts of Antarctica, may even become cooler or see more snow as the world gets warmer, but such local exceptions do not contradict global warming. There is also the haunting possibility of abrupt, tipping-point climate changes that are difficult to predict but potentially drastic.

In the early twenty-first century, there is global concern about climate change. The well-being or survival of hundreds of millions of people may soon be threatened by rising sea levels, disrupted food production, extreme weather, and emergent diseases. Estimates of the costs of climate change over the next century range in the scores of trillions of dollars, while irreversible losses such as species extinctions and lost lives cannot even be reckoned in terms of money.

Scientists have a high certainty that global climate change is happening, and are almost certain that it is

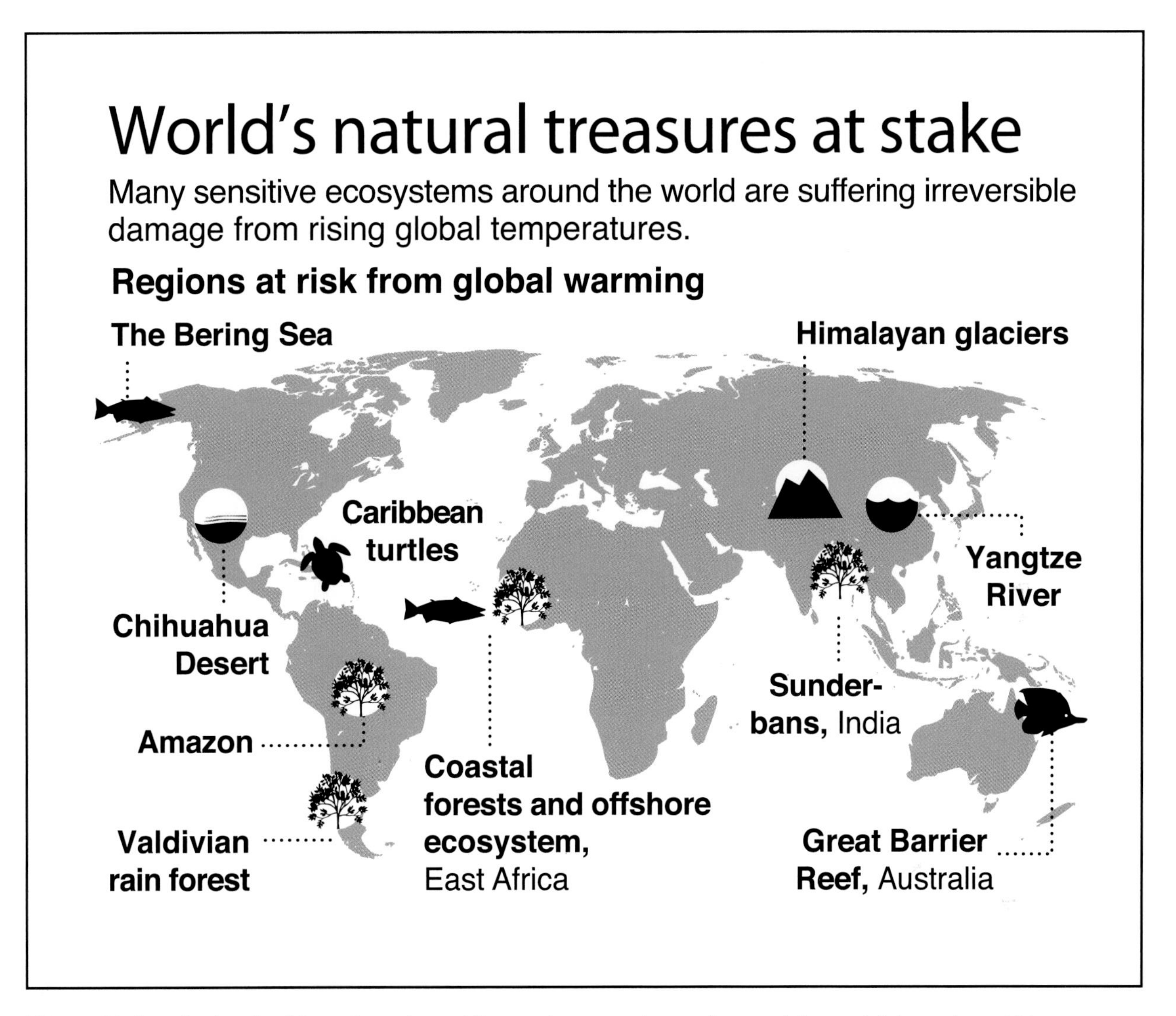

This graphic from the Associated Press shows the world's natural treasures that are threatened due to global warming. *AP Images.*

mostly anthropogenic (caused by humans). Since the 1990s, public debate has turned from the question of whether climate change is real to the question of what can be done to mitigate (lessen) its impact and to prevent further damage to Earth's environment.

Historical Background and Scientific Foundations

The Greenhouse Effect

The sun is hot, and therefore radiates most of its energy at high frequencies (rates of vibration), about half in the range that we perceive as visible light. This high-frequency light from the sun passes easily through Earth's atmosphere, which is transparent to it (does not absorb it). Earth reflects about a third of the sunlight it receives back out into space at once. Another third is absorbed by the atmosphere and the last third is absorbed by Earth's surface. When light is absorbed by air and land, its energy is converted mostly to heat. Air, water, and land, warmed by the sun, re-radiate or give back this energy as a form of low-frequency light called infrared light or radiation.

Infrared light radiated by Earth's surface or lower atmosphere cannot simply go straight out into space: it has to go through the atmosphere, and in doing so often strikes molecules of greenhouse gas, especially water vapor (H_2O), CO_2, methane (CH_4), nitrous oxide (N_2O), and artificial chemicals such as chlorofluorocarbons (CFCs). These gases are not transparent to infrared light, but absorb it and are warmed by doing so. This causes the gas molecules themselves to radiate infrared light. Some of this re-radiated infrared light escapes into space, but some shines back down at Earth, warming it, or is absorbed by other greenhouse molecules in the atmosphere, warming them.

Harbor seals rest on an ice floe in Alaskan waters. According to the National Wildlife Federation, climate change has been a contributing factor in the decline of harbor seals, stellar sea lions, and a variety of seabirds since the 1970s. *Image copyright Vera Bogaerts, 2007. Used under license from Shutterstock.com.*

Eventually, almost all the energy Earth receives from the sun will be radiated back out into space, but greenhouse gases slow down this energy loss. The slower Earth loses energy, the warmer it gets. In sum, greenhouse gases warm Earth by slowing the rate at which it loses heat. Without the greenhouse effect, the average temperature of Earth's surface-level atmosphere would be about 0°F (−18°C), well below the freezing point of water. The oceans would freeze over and life would thrive only near hot-water vents on the ocean floor. Thanks to the natural greenhouse effect, Earth's average surface temperature is actually 59°F (15°C).

Human beings, by adding billions of tons of CO_2 and other gases to the atmosphere, have strengthened Earth's greenhouse effect in just a couple of centuries. A given amount of greenhouse gas added to the atmosphere causes what climatologists call a radiative imbalance: that is, it causes Earth to radiate, for a while, less energy than it receives from the sun. This adds energy to the land, sea, and air, making them warmer. As Earth warms, its infrared glow eventually brightens to the point where it again loses energy to space as fast as the sun supplies it—but this new balance may not be achieved for thousands of years.

Radiative imbalance—the loss or gain over a given area at the top of Earth's atmosphere of more or less energy than that area receives from the sun—is measured in watts per square meter (a square meter is 1.2 square yards). The watt (W) is the standard unit of power: a 100-watt light bulb dissipates energy at the rate of 100 watts. The standard unit of radiative imbalance, also called radiative forcing, is thus watts per square meter (W/m^2). As of 2007, greenhouse gases and tiny airborne particles (aerosols) added to the atmosphere by humans were causing about 1.6 W/m^2 of radiative forcing.

The amount of energy released directly by burning fuels is too small to have any measurable effect on climate. Climate change is caused by chemicals added to the atmosphere, not by the heat released by power plants, heating systems, automobiles, and other machines.

Early Theories of Global Warming

The idea that Earth's atmosphere might act as a one-way valve for solar energy, letting light in but not letting heat out, was first suggested in the early 1800s. In 1824, French scientist Joseph Fourier (1768–1830) described the greenhouse effect accurately, using the scientific language of his day, when he wrote that "the temperature [of Earth] can be augmented by the interposition of the atmosphere, because heat in the state of light finds less resistance in penetrating the air, than in repassing into the air when converted into non-luminous heat."

In 1895, Swedish chemist Svante Arrhenius (1859–1927) suggested that changes in atmospheric CO_2 concentrations could change Earth's climate. He estimated that doubling CO_2 would increase average global temperature by 9°F (5°C). This was not far wrong, by today's standards: In 2007, the United Nations' Intergovernmental Panel on Climate Change (IPCC) said that the result of doubling CO_2 would most likely be a 5.4°F (3°C) increase in global temperature. In 1908,

Arrhenius was the first to suggest the possibility of an anthropogenic (human-caused) greenhouse effect. Human beings, he suggested, by burning fossil fuels such as coal and so increasing the amount of CO_2 in the atmosphere, might warm Earth's climate. Unable to foretell the huge increase in fossil-fuel use that was about to occur in the twentieth century, Arrhenius suggested that a greenhouse effect might become noticeable in 3,000 years; in fact, it was detectable by the 1990s, less than 100 years later.

In the 1930s, British inventor Guy Stewart Callendar (1898–1964) estimated that doubling CO_2 would cause 3.6°F (2°C) of global warming and theorized correctly that warming would be greater in the polar regions. He and some other researchers of that time were, however, mistaken in their belief that anthropogenic global warming was already detectable in the climate record.

The Science Matures: 1950s–1990s

By the mid-1950s, understanding of the physics and chemistry of Earth's climate system was advancing rapidly and the possibility of anthropogenic climate change was widely discussed among scientists. However, nobody had yet found a way to make the precise measurements of atmospheric greenhouse gases that would show whether humans were actually increasing such gases in the atmosphere: perhaps, some scientists theorized, the oceans were absorbing CO_2 as fast as we were releasing it. But in 1958, American scientist Charles David Keeling (1928–2005) developed sensitive new instruments to measure atmospheric CO_2 and installed them on the summit of the Mauna Loa volcano in Hawaii. After just a few years, his data showed a clear result: although CO_2 fell each summer as green plants grew in the Northern Hemisphere (where most of the world's land is), it rose again in the fall and winter, and it always rose farther than it fell. The result was an upward-tilted zig-zag line. Keeling's measurements showed that atmospheric CO_2 was indeed increasing: a greenhouse effect might, therefore, be occurring.

Keeling's work was a turning point in climate science, and his chart of rising CO_2, now known as the Keeling Curve, has become an icon of global warming. Human activities have raised atmospheric CO_2 from about 280 parts per million in 1750 to 383 parts per million as of November 2007, a 36.8% increase. (Parts per million refers to the number of molecules in a mixture; for example, 280 parts per million CO_2 in air means that of every one million molecules of air, 280 are CO_2 molecules.) The Keeling Curve shows that CO_2 is still increasing.

Through the 1970s, however, despite the knowledge that atmospheric CO_2 was increasing, scientists were still uncertain about whether Earth was about to experience global cooling or global warming. Aerosols—small solid or liquid particles—tended to cool Earth while greenhouse gases tended to warm it, and both are added to the atmosphere by fuel-burning. Perhaps, scientists also speculated, the amount of energy from the sun was changing, or as-yet-unknown natural processes were changing climate. Although a few hasty articles in news magazines proclaimed that we were on the verge of a new Ice Age, the scientific consensus (the opinion held by the great majority of scientists) was that we did not yet know enough to say what was going to happen to Earth's climate in the near future.

Meanwhile, the Keeling Curve continued to climb. Surface temperature measurements from thousands of weather stations showed warming trends in most parts of the world, and data from hundreds of tide gauges showed that sea levels were rising worldwide. Layered cylinders of muck from ocean bottoms and of ancient snow layers from deep inside the ice caps of Greenland and Antarctica began to reveal more about paleoclimate (ancient climate), hinting that relatively small changes to radiative forcing, such as slight changes in Earth's orbit, might trigger large changes in climate. The first mathematical descriptions of the climate machine, called climate models, were developed in the 1960s. Although crude by today's standards, they predicted several degrees of warming over the next century, a reasonably

WORDS TO KNOW

ANTHROPOGENIC: Made by people or resulting from human activities. Usually used in the context of emissions that are produced as a result of human activities.

GREENHOUSE GAS: A gaseous component of the atmosphere contributing to the greenhouse effect. Greenhouse gases are transparent to certain wavelengths of the sun's radiant energy, allowing them to penetrate deep into the atmosphere or all the way into Earth's surface. Greenhouse gases and clouds prevent some infrared radiation from escaping, trapping the heat near Earth's surface where it warms the lower atmosphere. Alteration of this natural barrier of atmospheric gases can raise or lower the mean global temperature of Earth.

RADIATIVE FORCING: A change in the balance between incoming solar radiation and outgoing infrared radiation. Without any radiative forcing, solar radiation coming to Earth would continue to be approximately equal to the infrared radiation emitted from Earth. The addition of greenhouse gases traps an increased fraction of the infrared radiation, reradiating it back toward the surface and creating a warming influence (i.e., positive radiative forcing because incoming solar radiation will exceed outgoing infrared radiation).

correct result. Also, starting in the 1970s, satellites began to monitor Earth's temperature from outer space, supplying a flood of new, independent information about climate.

By the mid-1980s, the doubts of most scientists had been answered: the world was indeed warming. What was more, it would continue to warm, and human beings were the major cause. Scientists warned that global warming would cause a host of other climate changes, many dangerous or costly, from rising seas to shifting rainfall patterns. It became increasingly clear that the question of global warming was more than a matter of abstract curiosity: all human life ultimately depends on farming, and all farming ultimately depends on climate. Also, hundreds of millions of people live within three feet (a meter or so) of sea level: rising waters might force their resettlement, if that were possible.

In 1985, a United Nations scientific conference in Austria agreed that significant human-caused global warming was probably about to occur. In 1987, the tenth congress of the U.N.'s World Meteorological Organization recommended ongoing, long-term assessment of climate change by an international group of scientists. The new group, the IPCC, was formed in 1988 and given the job of reporting on the scientific community's understanding of climate so that decision-makers could make informed decisions. The IPCC's reports, issued every two to five years, have been highly influential in the global discussion of climate change; in 2007, the organization shared a Nobel Peace Prize with former U.S. vice president and climate-change activist Al Gore (1948–). The first IPCC Assessment Report was issued in 1990 and the fourth in 2007. A fifth report is due around 2012.

The IPCC's 1990 report advised that global warming was probably happening and might cause many problems. Mildly alarmed—there were still many doubts and uncertainties—almost all the world's nations sent representatives to a climate summit in Rio de Janeiro, Brazil, in 1992. There a treaty addressing the problem of global climate change was negotiated. This treaty, the United Nations Framework Convention on Climate Change (UNFCCC), did not place binding obligations on any countries but did acknowledge the reality of anthropogenic global climate change. Under the UNFCCC, industrialized countries made a non-binding commitment to reduce their greenhouse-gas emissions and to help developing (poorer) countries reduce theirs as well.

Throughout the 1990s, despite efforts by a few scientists and many political commentators to cast doubt on the reality, dangerousness, or human-caused nature of climate change, a scientific consensus on climate change emerged. Computerized climate models became more complex and realistic every year, the Keeling Curve continued to climb (as well as similar records charting increases in other greenhouse gases), global average temperature continued to rise, and paleoclimate studies showed that, thanks to human activity, there was more CO_2 in the air than there had been for at least 600,000 years. Later, data from Antarctic ice cores pushed this back to 800,000 years. Earth, climatologists confirmed, was probably warmer by the late 1990s than it had been for at least 1,100 years, maybe much longer.

The countries that had signed the UNFCCC in 1992, including the world's then-largest greenhouse polluter, the United States, held regular meetings in the following years to discuss the changing science of climate change and to plan counter-action. These meetings issued a protocol or add-on to the UNFCCC, the Kyoto Protocol of 1997. The Kyoto Protocol was rejected by U.S. leaders and those of a few other countries—Australia did not sign the protocol until December 2007—but it was affirmed by all other signatories of the UNFCCC. Under Kyoto, industrialized countries made binding promises to reduce their own greenhouse emissions and to help developing countries do the same.

Kyoto was controversial. The United States refused to commit to the protocol because it did not require rapidly developing nations such as China to reduce their emissions. In any case, Kyoto did serve as a beginning for international action on climate change, establishing mechanisms for carbon emissions trading and formalizing the commitment of most industrial nations to reduce their emissions. Kyoto was not meant to be the final word on climate action; signers agreed from the beginning to replace Kyoto with an updated agreement starting in 2012.

In 2001 and in 2007, the IPCC released its third and fourth Assessment Reports on climate change. The 2007 report, prepared by over 2,500 scientists and economists appointed by scores of governments, declared that global warming was "unequivocal" (certain) and that there was at least a 90% probability that human beings were the cause. The report had an unprecedented impact on world opinion on climate change, creating a heightened sense of urgency. It also appeared to have greatly reduced the U.S. news media's longstanding practice of false balance, that is, framing stories as if there were a balanced climate "debate" between two more-or-less equally authoritative groups of scientists, one alleging the reality of climate change and the other denying it. After the 2007 IPCC report, it was clear that such a balanced debate did not exist, despite the continued existence of a small group of dissident scientists and a large number of climate denialists, that is, persons (rarely having scientific training) who insist that global climate change is a delusion or hoax. Such voices were more drastically marginalized than ever after 2007, when the world climate-science community came as close to speaking with a single voice as any scientific community had ever spoken; and what it said was that climate change is real, accelerating, dangerous, and human-caused. Importantly, it also said that climate change could be mitigated (made less severe) at a fairly low cost if prompt action were taken to reduce greenhouse emissions. As of the end of 2007, however, the United States and China, the world's largest greenhouse polluters, both remained unfriendly to the idea of binding emissions limits, making the future of efforts to control global climate change uncertain.

Impacts and Issues

Global climate change is global but not uniform. That is, it affects different regions differently. For example, the region at and around the North Pole, the Arctic, is experiencing more drastic changes than most other parts of the world. It is warming at twice the global average rate and, by 2007, faster melting of both the floating sea-ice cap over the North Pole and the massive ice sheet on the island of Greenland was occurring than scientists had predicted as recently as 2001. In the summer of 2007, the floating north-polar ice cap shrank to the smallest size ever observed, opening up a clear-water channel around the northern edge of North America from the Pacific to the Atlantic—the fabled Northwest Passage—for the first time in recorded history. Arctic permafrost (permanently frozen soil) has been melting over larger areas, an event that may soon release large quantities of the greenhouse gas methane, which would cause still more rapid global warming.

Water from the melting of mountain glaciers, Greenland's ice cap, and ice on the West Antarctic Peninsula runs downhill to the sea, increasing the amount of water in the oceans. Global warming also heats up the topmost layer of the ocean, causing it to expand. The combined effects of thermal expansion and added water have caused sea levels to rise about 6 in (15 cm) since 1900, with sea level rising more quickly in the last few decades. The habitability of low-lying coastal cities, where hundreds of millions of people live, is threatened by rising oceans. Many small islands in the Pacific, the Indian Ocean, the Bahamas, and elsewhere may be rendered completely uninhabitable by rising seas. Even if greenhouse-gas emissions leveled off in the near future, sea-level rise and other climate changes will probably continue for centuries, though more slowly and to less extreme conclusions, because it will take that long for Earth's massive oceans to get into balance with the radiative forcing from greenhouse gases already added to the atmosphere.

Most types of extreme weather are predicted to become more common with global climate change. Extreme weather includes heat waves, downpours, droughts, and powerful storms. The only form of extreme weather likely to become less common is cold waves. In August 2007, the World Meteorological Organization announced that during the first half of 2007, Earth showed significant increases above long-term global averages in the frequency of extreme weather events, including heavy rainfalls, tropical cyclones, heat waves, and wind storms. Global average temperatures for January and April of 2007 were the highest recorded for those two months since records began in the 1800s.

In some areas, such as the western United States, water shortages may eventually become severe and chronic, interfering with agriculture. Increased flooding and drought may interfere with agriculture in other parts of the world, such as Asia and Africa. Deterioration of coral reefs and other changes to the seas, including acidification from high CO_2 in the atmosphere dissolving in the oceans, may cause fisheries to decline. Fisheries presently provide 2.6 billion people with 20% or more of their dietary protein. Wind patterns, ranges inhabited by plants and animals, and rainfall patterns are also projected to change. Hurricanes may grow more frequent and severe, destroying property along coasts.

All continents will see effects of climate change, but so far the continental United States and most of Europe have seen less drastic effects than many other regions and are likely to continue to do so. The IPCC has predicted that the most severe effects of climate change will afflict

IN CONTEXT: A TIME FOR ACTION

The National Academy of Sciences counsels: "Despite remaining unanswered questions, the scientific understanding of climate change is now sufficiently clear to justify taking steps to reduce the amount of greenhouse gases in the atmosphere. Because carbon dioxide and some other greenhouse gases can remain in the atmosphere for many decades, centuries, or longer, the climate change impacts from concentrations today will likely continue well beyond the 21st century and could potentially accelerate. Failure to implement significant reductions in net greenhouse gas emissions will make the job much harder in the future—both in terms of stabilizing their atmospheric abundances and in terms of experiencing more significant impacts."

"Governments have proven they can work together successfully to reduce or reverse negative human impacts on nature. A classic example is the successful international effort to phase out of the use of chlorofluorocarbons (CFCs) in aerosol sprays and refrigerants that were destroying the Earth's protective ozone layer. At the present time there is no single solution that can eliminate future warming."

SOURCE: *Staudt, Amanda, Nancy Huddleston, and Sandi Rudenstein.* Understanding and Responding to Climate Change. *National Academy of Sciences, 2006.*

Climate change activists often target power plants as relics of the past that need to be replaced with renewable energy sources. Their concern is fueled by reports from experts stating that greenhouse gases such as carbon dioxide and methane are more prevalent in the atmosphere today than at any time in the past 650,000 years. *Christopher Furlong/Getty Images.*

the world's poorest countries, which are mostly in the tropics. Not only do these countries happen to be located where they will be more severely affected by climate change, but they have fewer resources—less money, less technological flexibility—to help them adapt to climate change. Many people will likely be driven from their homes in poorer countries by stresses associated with climate change. In 2005, the United Nations University Institute for Environment and Human Security estimated that the number of environmental refugees might rise to 150 million by 2050, mostly as a result of climate change.

Animals and plants will also be endangered by changing climate. In 2004, scientists calculated that, out of a random sample, 1,103 animal and plant species, 15–37% would, by 2050, be committed to extinction as a result of climate change. Extinction is the death of all members of a species, the irreversible death of the species itself. The scientists estimated that in many if not most ecological regions, climate change will become the greatest threat to biodiversity—the most likely cause of most extinctions—by 2050.

Efforts to mitigate climate change by reducing greenhouse-gas emissions are being made, such as the Kyoto Protocol, but so far these efforts have not reached a level where they are likely to greatly affect the future course of climate change. Some nations have decreased their emissions over the last decade, but others, notably China and India, have greatly increased theirs along with increasing economic activity, including increased manufacturing for export to Western markets. As of 2008, global CO_2 emissions were accelerating, not slowing, according to the International Energy Agency.

Choices by individual persons that reduce greenhouse emissions can do much to reduce global climate change. Such choices include the more efficient use of energy and materials in all departments of life. Purchase of carefully chosen carbon offsets may also be helpful. However, greatly reducing greenhouse-gas emissions will require the restructuring of the world's present energy economy, which has been founded on cheap fossil fuels for two centuries. Only a new energy economy, one founded on efficiency, thrift, and non-emitting or low-emitting energy technologies, can produce the 50-to-90% reductions in emissions that some scientists argue must occur by about 2050 to lessen the chance that Earth will experience uncontrolled greenhouse warming and climate change. However, it is unlikely that such a global change will happen in time without global cooperation to reduce greenhouse emissions by all the world's major governments, perhaps acting on an improved Kyoto model.

■ Primary Source Connection

This statement from the Joint Science Academies states their position that global climate change is an important issue that will have severe impacts on the world population.

The declaration acknowledges that global climate change is primarily due to human activity and that it is the responsibility of all countries to address global climate change.

The Joint Science Academies is an association of the heads of the national science academies of the G8+5 countries (Canada, France, Germany, Italy, Japan, Russia, the United Kingdom, and the United States plus Brazil, China, India, Mexico, and South Africa.)

JOINT SCIENCE ACADEMIES' STATEMENT: GLOBAL RESPONSE TO CLIMATE CHANGE

Climate change is real

There will always be uncertainty in understanding a system as complex as the world's climate. However there is now strong evidence that significant global warming is occurring. The evidence comes from direct measurements of rising surface air temperatures and subsurface ocean temperatures and from phenomena such as increases in average global sea levels, retreating glaciers, and changes to many physical and biological systems. It is likely that most of the warming in recent decades can be attributed to human activities. This warming has already led to changes in the Earth's climate.

The existence of greenhouse gases in the atmosphere is vital to life on Earth—in their absence average temperatures would be about 30 centigrade degrees lower than they are today. But human activities are now causing atmospheric concentrations of greenhouse gases—including carbon dioxide, methane, tropospheric ozone, and nitrous oxide—to rise well above pre-industrial levels. Carbon dioxide levels have increased from 280 ppm in 1750 to over 375 ppm today—higher than any previous levels that can be reliably measured (i.e., in the last 420,000 years). Increasing greenhouse gases are causing temperatures to rise; the Earth's surface warmed by approximately 0.6 centigrade degrees over the twentieth century. The Intergovernmental Panel on Climate Change (IPCC) projected that the average global surface temperatures will continue to increase between 1.4 centigrade degrees and 5.8 centigrade degrees above 1990 levels, by 2100.

Reduce the causes of climate change

The scientific understanding of climate change is now sufficiently clear to justify nations taking prompt action. It is vital that all nations identify cost-effective steps that they can take now, to contribute to substantial and long-term reduction in net global greenhouse gas emissions.

Action taken now to reduce significantly the build-up of greenhouse gases in the atmosphere will lessen the magnitude and rate of climate change. As the United Nations Framework Convention on Climate Change (UNFCCC) recognises, a lack of full scientific certainty about some aspects of climate change is not a reason for delaying an immediate response that will, at a reasonable cost, prevent dangerous anthropogenic interference with the climate system.

As nations and economies develop over the next 25 years, world primary energy demand is estimated to increase by almost 60%. Fossil fuels, which are responsible for the majority of carbon dioxide emissions produced by human activities, provide valuable resources for many nations and are projected to provide 85% of this demand. Minimising the amount of this carbon dioxide reaching the atmosphere presents a huge challenge. There are many potentially cost-effective technological options that could contribute to stabilising greenhouse gas concentrations. These are at various stages of research and development. However barriers to their broad deployment still need to be overcome.

Carbon dioxide can remain in the atmosphere for many decades. Even with possible lowered emission rates we will be experiencing the impacts of climate change throughout the 21st century and beyond. Failure to implement significant reductions in net greenhouse gas emissions now, will make the job much harder in the future.

Prepare for the consequences of climate change

Major parts of the climate system respond slowly to changes in greenhouse gas concentrations. Even if greenhouse gas emissions were stabilised instantly at today's levels, the climate would still continue to change as it adapts to the increased emissions of recent decades. Further changes in climate are therefore unavoidable. Nations must prepare for them.

The projected changes in climate will have both beneficial and adverse effects at the regional level, for example on water sources, agriculture, natural ecosystems and human health. The larger and faster the changes in climate, the more likely it is that adverse effects will dominate. Increasing temperatures are likely to increase the frequency and severity of weather events such as heat waves and heavy rainfall. Increasing temperatures could lead to large-scale effects such as melting of large ice sheets (with major impacts on low-lying regions throughout the world). The IPCC estimates that the combined effects of ice melting and sea water expansion from ocean warming are projected to cause the global mean sea-level to rise between 0.1 and 0.9 metres between 1990 and 2100. In Bangladesh alone, a 0.5 metre sea-level rise would place about 6 million people at risk from flooding.

Developing nations that lack the infrastructure or resources to respond to the impacts of climate change will be particularly affected. It is clear that many of the world's poorest people are likely to suffer the most from climate change. Long-term global efforts to create a more healthy, prosperous and sustainable world may be severely hindered by changes in the climate.

The task of devising and implementing strategies to adapt to the consequences of climate change will require worldwide collaborative inputs from a wide range of experts, including physical and natural scientists, engineers, social scientists, medical scientists, those in the humanities, business leaders and economists.

Conclusion

We urge all nations, in line with the UNFCCC principles, to take prompt action to reduce the causes of climate change, adapt to its impacts and ensure that the issue is included in all relevant national and international strategies. As national science academies, we commit to working with governments to help develop and implement the national and international response to the challenge of climate change.

G8 nations have been responsible for much of the past greenhouse gas emissions. As parties to the UNFCCC, G8 nations are committed to showing leadership in addressing climate change and assisting developing nations to meet the challenges of adaptation and mitigation.

We call on world leaders, including those meeting at the Gleneagles G8 Summit in July 2005, to:

- Acknowledge that the threat of climate change is clear and increasing.
- Launch an international study to explore scientifically-informed targets for atmospheric greenhouse gas concentrations, and their associated emissions scenarios, that will enable nations to avoid impacts deemed unacceptable.
- Identify cost-effective steps that can be taken now to contribute to substantial and long-term reduction in net global greenhouse gas emissions. Recognise that delayed action will increase the risk of adverse environmental effects and will likely incur a greater cost.
- Work with developing nations to build a scientific and technological capacity best suited to their circumstances, enabling them to develop innovative solutions to mitigate and adapt to the adverse effects of climate change, while explicitly recognising their legitimate development rights.
- Show leadership in developing and deploying clean energy technologies and approaches to energy efficiency, and share this knowledge with all other nations.
- Mobilise the science and technology community to enhance research and development efforts, which can better inform climate change decisions.

ROYAL SOCIETY. "JOINT SCIENCE ACADEMIES' STATEMENT: GLOBAL RESPONSE TO CLIMATE CHANGE." JUNE 2005. <HTTP://WWW.ROYALSOCIETY.ORG/DISPLAYPAGEDOC.ASP?ID=20742> (ACCESSED NOVEMBER 30, 2007).

SEE ALSO *Feedback Factors; Global Warming; Greenhouse Effect; Greenhouse Gases.*

BIBLIOGRAPHY

Books

Gore, Al. *An Inconvenient Truth: The Planetary Emergency of Global Warming and What We Can Do About It.* Emmaus, PA: Rodale Press, 2006.

McCaffrey, Paul. *Global Climate Change.* Minneapolis, MN: H. W. Wilson, 2006.

Metz, B., et al, eds. *Climate Change 2007: Mitigation of Climate Change: Contribution of Working Group III to the Fourth Assessment Report of the Intergovernmental Panel on Climate Change.* New York: Cambridge University Press, 2007.

Parry, M. L., et al, eds. *Climate Change 2007: Impacts, Adaptation and Vulnerability: Contribution of Working Group II to the Fourth Assessment Report of the Intergovernmental Panel on Climate Change.* New York: Cambridge University Press, 2007.

Solomon, S., et al, eds. *Climate Change 2007: The Physical Science Basis: Contribution of Working Group I to the Fourth Assessment Report of the Intergovernmental Panel on Climate Change.* New York: Cambridge University Press, 2007.

Weart, Spencer. *The Discovery of Global Warming.* Cambridge, MA: Harvard University Press, 2004.

Periodicals

Alley, Richard B. "Abrupt Climate Change." *Scientific American* (November 2004): 62–69.

Collins, William, et al. "The Physical Science Behind Climate Change." *Scientific American* (August 2007).

Oreskes, Naomi. "The Scientific Consensus on Climate Change." *Science* 306 (2004): 1686.

Thomas, Chris D. "Extinction Risk from Climate Change." *Nature* 427 (2004): 145–148.

Thomas, Karl, and Kevin Trenberth. "Modern Global Climate Change." *Science* 302 (2003): 1719–1723.

Web Sites

"Climate Change." *U.S. Environmental Protection Agency*, November 19, 2007. <http://epa.gov/climatechange/index.html> (accessed December 9, 2007).

Intergovernmental Panel on Climate Change. <http://www.ipcc.ch> (accessed December 9, 2007).

United Nations Framework Convention on Climate Change. <http://unfccc.int/2860.php> (accessed December 9, 2006).

Larry Gilman

Climate Change Science Program

Introduction

The Climate Change Science Program (CCSP) is a U.S. collaborative interagency research program that manages government study into climate change. The CCSP oversees and integrates the research activities of 13 federal agencies including the Department of Energy, Environmental Protection Agency (EPA), Department of Agriculture, National Aeronautics and Space Administration (NASA), National Science Foundation, and the Smithsonian Institution. In total, the CCSP manages over $1 billion of climate change research annually.

WORDS TO KNOW

CLIMATE CHANGE: Sometimes used to refer to all forms of climatic inconsistency, but because Earth's climate is never static, the term is more properly used to imply a significant change from one climatic condition to another. In some cases, climate change has been used synonymously with the term, global warming; scientists, however, tend to use the term in the wider sense to also include natural changes in climate.

Historical Background and Scientific Foundations

In 1990, the Global Change Research Act was passed in the United States. This established the U.S. Global Change Research Program (USGCRP), a coordinated interagency program designed to study changes in the global environment and the implication of such changes. In 2001, President George W. Bush established the Climate Change Research Initiative (CCRI). The focus of the CCRI was conducting short-term research that would allow for improved decision-making and policy decisions.

In February 2002, the Climate Change Science Program (CCSP) was launched to oversee and integrate both the USGCRP and the CCRI. In "Preview of Our Changing Planet: The U.S. Climate Change Science Program for Fiscal Year 2008," the CCSP's mission is stated as follows: to "facilitate the creation and application of knowledge of the Earth's global environment through research, observations, decision support, and communication."

Impacts and Issues

In July 2003, the CCSP released a 10-year strategic plan outlining future research activities. The plan identified five research goals:

1. To improve knowledge of Earth's climate and its variability;
2. To improve quantification of climate change forces;
3. To reduce uncertainty in climate change predictions;
4. To understand the adaptability of ecosystems and human systems to climate change;
5. To identify the limits of knowledge to manage risks and opportunities related to climate change.

The funding for research is provided by the government agencies that are part of the program. The 2007 budget request was $240.6 million for the first goal, $304.8 million for goal two, $283.2 million for goal three, $160.1 million for goal four, and $150.7 million for goal five. This equates to a total of $1.1 billion.

SEE ALSO *Global Change Research Program; United States: Climate Policy.*

BIBLIOGRAPHY

Books

National Research Council of the National Academies. *Implementing Climate and Global Change Research:*

A Review of the Final U.S. Climate Change Science Program Strategic Plan. Washington, DC: National Academies Press, 2004.

Web Sites

"Our Changing Planet: The U.S. Climate Change Science Program for Fiscal Year 2008." *U.S. Climate Change Science Program*, December 15, 2006. <http://www.usgcrp.gov/usgcrp/Library/ocp2008preview/default.htm> (accessed November 6, 2007).

"Strategic Plan for the Climate Change Science Program Final Report, July 2003." *U.S. Climate Change Science Program*, May 19, 2007. <http://www.climatescience.gov/Library/stratplan2003/final/default.htm> (accessed November 6, 2007).

Tony Hawas

Climate Change Skeptics

Introduction

Climate change skeptics are those who doubt or disbelieve some aspect of the mainstream scientific view of global climate change or, in some cases, discount that view entirely. The mainstream or consensus view is that Earth is getting warmer; that human-released (anthropogenic) greenhouse gases, such as carbon dioxide and methane, are the main cause of today's warming; that global warming's consequences will be mostly negative; and that the future course of climate change can be influenced by human actions, such as burning less fossil fuel.

In science, skepticism is the principle that one should proportion one's belief in any claim to the available evidence—the weaker the evidence, the weaker the belief. In this sense, all scientists consider themselves skeptics, only some disagree about exactly what the evidence is and how it should be weighed. In this positive sense, some scientists are skeptical of certain aspects of the mainstream view of climate change. These scientific skeptics are a minority of the scientific community, but the work of dissidents or skeptics is often an essential part of the scientific process.

A much larger group of climate change skeptics consists of individuals with little or no scientific expertise. These persons, who have decided that global warming is a hoax, exaggeration, or delusion, tend to be interested only in those fragments of scientific literature that appear, in isolation, to support their predetermined views. These persons are often termed deniers rather than skeptics, as they tend to ignore or deny scientific knowledge.

Historical Background and Scientific Foundations

Scientific disputes about changing climate have gone on since the 1970s, when theories of a new ice age (global cooling) were competing with theories of global warming. In the 1980s, scientific opinion solidified around the concept of global warming. By the early 1990s, warming was considered a present fact or future

Chief NASA climate scientist Dr. Jim Hansen, director of the Goddard Institute for Space Studies, has tried to raise awareness of the causes and possible effects of climate change since the 1980s. During that time, he claims that he has felt pressure from his employers and various political administrations not to speak out about potential dire consequences of global warming. *AP Images.*

WORDS TO KNOW

ANTHROPOGENIC: Made by people or resulting from human activities. Usually used in the context of emissions that are produced as a result of human activities.

FOSSIL FUEL: Fuel formed by biological processes and transformed into solid or fluid minerals over geological time. Fossil fuels include coal, petroleum, and natural gas. Fossil fuels are non-renewable on the timescale of human civilization, because their natural replenishment would take many millions of years.

GREENHOUSE GASES: Gases that cause Earth to retain more thermal energy by absorbing infrared light emitted by Earth's surface. The most important greenhouse gases are water vapor, carbon dioxide, methane, nitrous oxide, and various artificial chemicals such as chlorofluorocarbons. All but the latter are naturally occurring, but human activity over the last several centuries has significantly increased the amounts of carbon dioxide, methane, and nitrous oxide in Earth's atmosphere, causing global warming and global climate change.

HOCKEY STICK CONTROVERSY: Controversy from 1998 to approximately 2006 over the validity of the hockey-stick graph, a chart of global average air temperatures indicating that recent climate warming is unprecedented for at least the last 1,100 years or so. Critics argued the graph was flawed and misleading, but in 2006, after a careful review of the evidence, the National Academy of Sciences affirmed that the graph is essentially accurate, though its shape is less certain for earlier dates.

ICE AGE: Period of glacial advance.

INTERGOVERNMENTAL PANEL ON CLIMATE CHANGE (IPCC): Panel of scientists established by the World Meteorological Organization (WMO) and the United Nations Environment Programme (UNEP) in 1988 to assess the science, technology, and socioeconomic information needed to understand the risk of human-induced climate change.

SKEPTICISM: Doubt about the truth of a claim. Skepticism, as opposed to denialism, may have a reasonable basis: in fact, skepticism is essential to the scientific process of discovering new knowledge, in which claims are carefully tested before being accepted as correct.

URBAN HEAT ISLAND: Area of warm weather in and immediately around a built-up area. Pavement and buildings absorb solar energy while being little cooled by evaporation compared to vegetation-covered ground. Skeptics of global climate change at one time argued that the expansion of urban heat islands near and around weather stations has caused an illusion of global warming by biasing temperature measurements. Although urban heat islands do exist, the argument that they produce an illusion of global warming has been discredited.

possibility by most climate and weather scientists. By the late 1990s, the great majority of scientists with relevant qualifications agreed that global climate change is real, is caused primarily by human activity, and might be mitigated (made less serious) by reducing emissions of greenhouse gases, especially carbon dioxide.

Disbelief or doubt about climate change occurs on a spectrum, from authentic scientific doubt to irrational denial or misuse of scientific evidence. Today, a small minority of scientists with expertise in climate or weather science maintain that global warming is not happening or that the observed warming is not human-caused. Many scientists disagree about climate-science details such as the range of uncertainty that should be assigned to predictions of future sea-level rise. A large number of nonscientific voices can be found on the Internet, in opinion columns, on radio and television, and in political circles who claim that climate change is a hoax, fantasy, or scare tactic invented by environmentalists. These nonscientific voices have had almost no influence on the scientific world, but have been successful in confusing public perceptions of global-warming science.

Climate-Change Skepticism and Denial among Non-Scientists

Public Opinion. Climate change was not widely discussed by nonscientists until the early 1970s, when the environmental movement was raising general concern about global harm caused by technology (pollution, extinction, etc.). The public was confused later in the 1970s by scientific debate over whether Earth might, in fact, be entering a new ice age rather than facing global warming. The ice-age theory received significant media coverage, but by the end of the decade had been rejected by scientists in favor of global warming. By the 1990s, polls found that about half of Americans thought that global warming was already happening or would soon happen; only about one in eight thought it would never happen. Levels of climate-change concern actually dipped in the early 2000s, with only about one-third of polled Americans saying they worried "a great deal" about climate change.

After 2004, concern rose again. Concern about global warming appears to be associated with political beliefs, with political conservatives being less likely to credit mainstream scientific views. In 2007, a Gallup poll found that about 59% of independents were worried about global warming, 75% of Democrats, and 34% of Republicans. European public opinion generally matched U.S. opinion until the late 1990s, but since then there has been an increase of belief in climate change outside the United States. A 2003 survey of 19 non-U.S. countries by the Pew Research Center found 81% of respondents very concerned about global climate change.

Nonscientist Commentators. Statements that global warming is unreal, is not caused by humans, or is not dangerous even if it is real are often made by nonscientists on radio, television, and the Internet or in print. The most common misuse of science by such commentators is selective citation of studies that appear to support the speaker's predetermined opinion, coupled with silence about the much greater quantities of data supporting mainstream scientific views. Definitely incorrect claims are also sometimes made, such as the statement that computerized climate models ignore water vapor. This claim was made, for example, by columnist Alexander Cockburn in *The Nation* on May 14, 2007: "Water is exactly that component of the Earth's heat balance that the global warming computer models fail to account for." Because of the access to public attention that some of these speakers enjoy, their opinions often reach a larger public than do views expressed by qualified scientists.

Claims that global warming is a hoax, plot, or illusion come from voices across the political spectrum. Alexander Cockburn, the left-wing journalist already mentioned, states: "There is still zero empirical evidence that anthropogenic production of carbon dioxide is making any measurable contribution to the world's present warming trend. The greenhouse fearmongers rely on unverified, crudely oversimplified models to finger mankind's sinful contribution." In reality, there is mounting evidence from many different sources that anthropogenic climate change is occurring. This evidence is described in detail in the Intergovernmental Panel on Climate Change (IPCC)'s report *Climate Change 2007: The Physical Science Basis.* Also, computer climate models are not "unverified, [and] crudely oversimplified" but increasingly accurate and constantly tested against real-world data.

The majority of anti-climate-change political commentators are political conservatives. The list of prominent climate skeptics or deniers who self-identify as conservative includes, for example, William Buckley, George Will, and Anne Coulter. Will has recalled media excitement over the possibility of global cooling in the 1970s and mocked recent scientific work showing wildlife species ranges are being shifted by global warming (*Washington Post,* December 22, 2004; April 2, 2006). Often, such commentators will claim that mainstream scientific opinion is actually on their side. Coulter claimed in 2007 that "[t]here are more reputable scientists defending astrology than defending 'global warming'" (*Human Events,* February 28, 2007). Similarly, Bonner Cohen, senior fellow at the National Center for Public Policy Research, stated on C-SPAN in 2006 that "if you go to climate scientists, climatologists, the people who look at this, as opposed to the scientific community at large, you will find absolutely no consensus.... The vast majority of them are somewhat agnostic on the whole thing" (video and transcript at ThinkProgress.org; see Bibliography).

There is, contrary to these statements, a strong consensus among climate scientists in favor of the reality and human-caused nature of climate change. An article by Naomi Oreskes in the December 3, 2004 issue of *Science* noted that every major scientific body in the United States whose members' specialization is relevant to climate change, such as the American Meteorological Association, has issued a statement affirming the consensus scientific view. Oreskes also surveyed 928 articles discussing climate change published from 1993 to 2003—all those in the Institute for Scientific Information database mentioning "climate change" in their abstracts. Seventy-five percent of these articles explicitly or implicitly accepted the consensus position and not one disagreed with it. Other surveys of scientific opinion have also found large majorities in support of the consensus position.

Another prominent conservative opponent of climate change is Rush Limbaugh, a syndicated radio, television, and print columnist with a weekly radio audience of more than 10 million listeners. Limbaugh denies that human activity has damaged the ozone layer and that present-day climate change is human-caused. For example, he has written that a "fact you never hear the environmentalist wacko crowd acknowledge is that 96 percent of the so-called 'greenhouse' gases are not created by man, but by nature" (Limbaugh, 1993). Yet the IPCC's regular reports on climate change since 1990, along with scores of other books and articles about climate change, do make it clear that most of the carbon dioxide, water vapor, methane, and other greenhouse gases in the atmosphere are natural in origin—indeed, essential to keeping Earth warm enough to support human life. The issue has never been whether human beings are responsible for *most* of the greenhouse gases put into the atmosphere, but whether they are adding enough to cause significant change in the global climate. Most climate and weather scientists agree that they are.

Fiction writer Michael Crichton has also reached a wide public with his anti-climate change message. In Crichton's bestselling 2004 novel *State of Fear,* evil environmentalists conspire to create apparently natural disasters and other evidence to cause panic over global warming. Crichton's book, which claimed to be based on science, was heavily criticized by climate scientists, but did receive a journalism award from the American Society of Petroleum Geologists, the only scientific society in the world to have adopted an official statement in denial of anthropogenic global climate change. Most of the society's members are employed by the oil industry, which has opposed the scientific consensus view of climate change.

Industry Funding of Climate Change Denial. Corporations seeking to influence public opinion against climate change have funded nonscientific groups that deny climate change. In 2006, the United Kingdom's national scientific academy, the Royal Society (the equivalent of the National Academy of Sciences) announced that the oil company ExxonMobil had distributed several million dollars to 36 groups in the United States alone. These groups, the society said in a letter to ExxonMobil, "have been misinforming the public about the science of climate change." The Union of Concerned Scientists reported in 2007 that ExxonMobil had spent a total of $16 million to fund denial groups and create public confusion about climate change.

As a result of this negative publicity, ExxonMobil softened its anti-climate change stance in 2007. In February, the company's vice president for public affairs, Kenneth P. Cohen, said ExxonMobil had never denied climate change, that "the global ecosystem is showing signs of warming, particularly in polar areas," and that "the appropriate debate isn't on whether the climate is changing but rather should be on what we should be doing about it." The company reportedly cut off funding in 2006 to at least one denial group, the American Enterprise Institute, which received over $1.8 million from ExxonMobil from 1998 to 2006. In early 2007, the American Enterprise Institute offered $10,000 to any scientist who would produce a paper undermining the IPCC's report *Climate Change 2007: The Physical Science Basis.* It was not yet known, as of late 2007, if there would be any takers.

Climate-Change Skepticism from Scientists

A few scientists have disputed the reality or anthropogenic nature of global climate change. Such skeptics, including economist Ross McKitrick and minerals consultant Steven McIntyre, have raised a number of arguments against global warming. For example, they have argued that the famous "hockey stick" graph showing that recent global temperatures are much greater than those for the last 1,000 years, which featured prominently in Al Gore's hit documentary *An Inconvenient Truth*, is flawed. They have also pointed to disagreement between surface measurements of Earth's temperature and satellite measurements, which seemed to show that Earth's atmosphere had actually cooled. Some have argued that changes in the sun's heat output, not anthropogenic greenhouse gases, are responsible for global warming, and that warming measurements have been exaggerated because they are made in the vicinity of hot spots created by the pavement and buildings of large cities, "urban heat islands."

In the early 2000s, however, most of these scientific uncertainties collapsed under the weight of new scientific evidence. The U.S. Congress charged the U.S. National Research Council with investigating the merits of the hockey stick graph, which had also appeared in expert testimony to Congress. In 2006, the council affirmed the basic validity of the famous graph. Although the statistical mathematical methods used in producing the graph were found to be inadequate, replacing them with correct methods did not change the graph's shape. The data show that today's global warming is unprecedented in the last 1,000 years.

A 2005 study commissioned specifically to look into the puzzle of the satellite data found that the data had originally been analyzed incorrectly. When the analytical errors were fixed, the apparent contradiction disappeared. Again, it was verified that Earth is indeed warming. Urban heat islands and solar variation have also both been ruled out as explanations of global warming by peer-reviewed scientific studies conducted in recent years.

Some scientists continue to disagree with estimates of the magnitude of future climate change. John Christy, a climatologist at the University of Alabama, said in 2007, "don't sign me up for that catastrophic view of climate change" (quoted in *Nature*, February 8, 2007). Another example is Christopher Landsea, a meteorologist at the U.S. National Hurricane Center, who announced in 2005 that he was withdrawing from the IPCC on the grounds that he had "come to view the part of the IPCC to which my expertise is relevant as having become politicized." In particular, he claimed that fellow lead author Kevin Trenberth used his IPCC position to "promulgate to the media and general public his own opinion that the busy 2004 hurricane season was caused by global warming, which is in direct opposition to research written in the field and is counter to conclusions" in the IPCC's 2001 Assessment Report. Landsea has argued that global warming has not yet affected hurricane activity detectably and that "studies that any impact in the future from global warming upon hurricane[s] will likely be quite small."

The 2007 IPCC report did not, ultimately, assert that anthropogenic global warming is definitely responsible for increased hurricane frequency or severity. Since the report came out, however, scientists have published data showing that hurricane intensity and frequency may indeed have increased over the last century as a result of anthropogenic ocean warming. The relationship between hurricanes and global warming is a point on which authentic scientific disagreement continues.

■ Impacts and Issues

Studies have been conducted to resolve questions raised by scientific skeptics of climate change—a direct benefit

IN CONTEXT: FACING THE FACTS

According to the National Academy of Sciences: "The fact is that Earth's climate is always changing. A key question is how much of the observed warming is due to human activities and how much is due to natural variability in the climate. In the judgment of most climate scientists, Earth's warming in recent decades has been caused primarily by human activities that have increased the amount of greenhouse gases in the atmosphere [figure omitted]. Greenhouse gases have increased significantly since the Industrial Revolution, mostly from the burning of fossil fuels for energy, industrial processes, and transportation. Greenhouse gases are at their highest levels in at least 400,000 years and continue to rise."

"Global warming could bring good news for some parts of the world, such as longer growing seasons and milder winters. Unfortunately, it could bring bad news for a much higher percentage of the world's people. Those in coastal communities, many in developing nations, will likely experience increased flooding due to sea-level rise and more severe storms and surges. In the Arctic regions, where temperatures have increased almost twice as much as the global average, the landscape and ecosystems are rapidly changing."

"Although the potential effects of climate change are widely acknowledged, there is still legitimate debate regarding how large, how fast, and where these effects will be. Climate science is just beginning to project how climate change might affect regional weather. Estimating climate change impacts also requires projecting society's future actions, particularly in the areas of population growth, economic growth, and energy use."

SOURCE: *Staudt, Amanda, Nancy Huddleston, and Sandi Rudenstein.* Understanding and Responding to Climate Change. *National Academy of Sciences, 2006.*

to the scientific process. Many disputes have now been resolved in favor of the mainstream or consensus scientific view of climate change. As of 2007, scientific skepticism about the main elements of the climate-change consensus appeared to be decreasing. Some critics were shifting their focus from arguing that climate change is not real to arguing that it would be too expensive to take action to mitigate it. However, with the release of the third part of the IPCC's latest climate report *Climate Change 2007: Mitigation of Climate Change*, these critics again found themselves on the outside of a widely supported scientific consensus. The IPCC report contends that greenhouse gas concentrations in the atmosphere could be stabilized at low cost or possibly even at a profit.

Internet-based attacks on the consensus scientific view continued unabated even after the release of the 2007 IPCC report, as did mockery by nonscientific commentators, such as Cockburn, Coulter, and Limbaugh. The response of public opinion to the climate-change issue will depend on the quality of media and educational coverage of mainstream science and on how apparent the effects of climate change are to people in daily life.

■ Primary Source Connection

An overwhelming majority of the scientific community accepts global climate change theories. The following article notes that while researchers continue to debate—typically within a consensus range—about smaller aspects of climate change theory, scientific skepticism of human-influenced climate change is increasingly rare.

ERRORS CITED IN ASSESSING CLIMATE DATA

Some scientists who question whether human-caused global warming poses a threat have long pointed to records that showed the atmosphere's lowest layer, the troposphere, had not warmed over the last two decades and had cooled in the tropics.

Now two independent studies have found errors in the complicated calculations used to generate the old temperature records, which involved stitching together data from thousands of weather balloons lofted around the world and a series of short-lived weather satellites.

A third study shows that when the errors are taken into account, the troposphere actually got warmer. Moreover, that warming trend largely agrees with the warmer surface temperatures that have been recorded and conforms to predictions in recent computer models.

The three papers were published yesterday in the online edition of the journal *Science*.

The scientists who developed the original troposphere temperature records from satellite data, John R. Christy and Roy W. Spencer of the University of Alabama in Huntsville, conceded yesterday that they had made a mistake but said that their revised calculations still produced a warming rate too small to be a concern.

"Our view hasn't changed," Dr. Christy said. "We still have this modest warming."

Other climate experts, however, said that the new studies were very significant, effectively resolving a puzzle that had been used by opponents of curbs on heat-trapping greenhouse gases.

"These papers should lay to rest once and for all the claims by John Christy and other global warming skeptics that a disagreement between tropospheric and surface temperature trends means that there are problems with surface temperature records or with climate models," said Alan Robock, a meteorologist at Rutgers University.

The findings will be featured in a report on temperature trends in the lower atmosphere that is the first product to emerge from the Bush administration's 10-year program intended to resolve uncertainties in climate science.

Several scientists involved in the new studies said that the government climate program, by forcing everyone involved to meet five times, had helped generate the new findings.

"It felt like a boxing ring on occasion," said Peter W. Thorne, an expert on the weather balloon data at the Hadley Center for Climate Prediction and Research in Britain and an author of one of the studies.

Temperatures at thousands of places across the surface of the earth have been measured for generations. But far fewer measurements have been made of temperatures in the air from the surface through the troposphere, which extends up about five miles.

Until recently Dr. Christy and Dr. Spencer were the only scientists who had plowed through vast volumes of data from weather satellites to see if they could indirectly deduce the temperature of several layers within the troposphere.

They and other scientists have also tried to analyze temperature readings gathered by some 700 weather balloons lofted twice a day around the world.

But each of those efforts has been fraught with complexities and uncertainties.

The satellites' orbits shift and sink over time, their instruments are affected by sunlight and darkness, and data from a succession of satellites has to be calibrated to account for eccentricities of sensitive instruments.

Starting around 2001, the satellite data and methods of Dr. Christy and Dr. Spencer were re-examined by Carl A. Mears and Frank J. Wentz, scientists at Remote Sensing Systems, a company in Santa Rosa, Calif., that does satellite data analysis for NASA.

They and several other teams have since found more significant warming trends than the original estimate.

But the new paper, by Dr. Mears and Dr. Wentz, identifies a fresh error in the original calculations that, more firmly than ever, showed warming in the troposphere, particularly in the tropics.

The error, in a calculation used to adjust for the drift of the satellites, was disclosed to the University of Alabama scientists at one of the government-run meetings this year, Dr. Christy said.

The new analysis of data from weather balloons examined just one possible source of error, the direct heating of the instruments by the sun.

It found that when data were examined in a way that accounted for that effect, the temperature record produced a warming, particularly in the tropics, again putting the data in line with theory.

"Things being debated now are details about the models," said Steven Sherwood, the lead author of the paper on the balloon data and an atmospheric physicist at Yale. "Nobody is debating any more that significant climate changes are coming."

Andrew C. Revkin

REVKIN, ANDREW C. "ERRORS CITED IN ASSESSING CLIMATE DATA." *THE NEW YORK TIMES* AUGUST 12, 2005.

SEE ALSO *Hockey Stick Controversy; IPCC Climate Change 2007 Report: Criticism; IPCC Climate Change 2007 Report: Physical Science Basis; Media Influences: False Balance; Public Opinion; Urban Heat Islands.*

BIBLIOGRAPHY

Books

Crichton, Michael. *State of Fear: A Novel.* New York: HarperCollins, 2004.

Limbaugh III, Rush H. *See, I Told You So.* New York: Pocket Books, 1993.

Metz, B., et al, eds. *Climate Change 2007: Mitigation of Climate Change: Contribution of Working Group III to the Fourth Assessment Report of the Intergovernmental Panel on Climate Change.* New York: Cambridge University Press, 2007.

Solomon, S., et al, eds. *Climate Change 2007: The Physical Science Basis: Contribution of Working Group I to the Fourth Assessment Report of the Intergovernmental Panel on Climate Change.* New York: Cambridge University Press, 2007.

Weart, Spencer R. *The Discovery of Global Warming.* Cambridge, MA: Harvard University Press, 2003.

Periodicals

Begley, Sharon. "Global-Warming Deniers: A Well-Funded Machine." *Newsweek* (March 18, 2007).

Cockburn, Alexander. "Is Global Warming a Sin?" *The Nation* (May 14, 2007).

Hopking, Michael. "Climate Skeptics Switch Focus to Economics." *Nature* 445 (February 8, 2007): 582–583.

Mahlman, J. D. "Science and Nonscience Concerning Human-Caused Climate Warming." *Annual*

Review of Energy and the Environment 23 (1998): 83–105.

Mufson, Steven. "ExxonMobil Warming Up to Global Climate Issue." *Washington Post* (February 10, 2007).

Oreskes, Naomi. "The Scientific Consensus on Climate Change." *Science* 306 (December 3, 2004): 1686.

Revkin, Andrew C. "Errors Cited in Assessing Climate Data." *The New York Times* (August 12, 2005).

Sample, Ian. "Scientists Offered Cash to Dispute Climate Study." *The Guardian* (February 2, 2007).

Will, George. "Global Warming? Hot Air." *The Washington Post* (December 23, 2004).

Will, George. "Let Cooler Heads Prevail." *The Washington Post* (April 2, 2006).

Web Sites

Coulter, Anne. "Let Them Eat Tofu!" *Human Events*, February 28, 2007. <http://www.humanevents.com/article.php?id=19625> (accessed December 5, 2007).

"Industry-Backed Author: The 'Vast Majority' of Climatologists Don't Believe in Global Warming." *ThinkProgress.org*, August 9, 2006. <http://thinkprogress.org/2006/08/09/vast-majority-climatologists/> (accessed December 5, 2007).

Landsea, Christopher. "An Open Letter to the Community from Chris Landsea," January 17, 2005. <http://www.lavoisier.com.au/papers/articles/landsea.html> (accessed August 13, 2007).

Mooney, Kevin. "UN Climate Summary Designed to Dupe, Critics Say." *Cybercast News Service*, February 2, 2007. <http://www.cnsnews.com/ViewCulture.asp?Page=/Culture/archive/200702/CUL20070202a.html> (accessed August 13, 2007).

Larry Gilman

Climate Engineering

Introduction

Climate engineering refers to the deliberate manipulation of the atmosphere to alter the climate in ways that are desirable. A well-known example is cloud seeding, or the addition of particles to the atmosphere in an effort to encourage precipitation. Cloud seeding is climate engineering on a small scale. Other atmospheric alterations are global in scale.

Climate engineering is an attempt to correct things that are going wrong in the atmosphere. If, as many climate scientists accept, the undesirable atmospheric changes have been at least partly the result of human activities that have polluted the atmosphere with compounds that increase the ability of the atmosphere to retain heat, then climate engineering is one attempt to repair the problems that humans have caused.

Engineered change in climate can also involve strategies here on the ground. Examples include diverting the course of rivers, altering wildlife migration patterns, and planting trees to increase the trapping of carbon dioxide (CO_2).

Two scientists examine an ice core drilled near the Rothera Research Station in Antarctica. Chemical analysis of ice core samples can reveal changes in climate over several hundred years. Studying the samples helps scientists learn how the atmosphere changed, which may lead to better models for predicting how it will change in the future. Such models can also help scientists in their work in climate engineering. *British Antarctic Survey/Photo Researchers, Inc.*

This computer rendition shows a warship using its guns to fire reflective particles high into the atmosphere to reduce global warming and control Earth's climate. The particles, which would reflect the sunlight back toward the sun and away from Earth, would have to be continuously replaced. As a result, the color of the sky would change from blue to white. *Victor Habbick Visions/Photo Researchers, Inc.*

Historical Background and Scientific Foundations

In 1945, a gathering of scientists at Princeton University in New Jersey agreed that manipulating climate was possible. Their motivation was a desire to control the weather as a means of decreasing crop production, particularly in the Soviet Union. This was seen as a way of dealing with the perceived military threat posed by the Soviet Union to the United States. Indeed, during the Cold War (1945–1991) that took place between the two nations, the United States conducted research on what was called climatological warfare. This research was actually tested in the field, when cloud seeding was tried unsuccessfully in the early 1970s during the Vietnam war (1954–1975) to bog down Vietcong troop movements.

In cloud seeding, particles of silver iodide (AgI) are scattered into clouds from an airplane. The particles act as surfaces, on which water vapor precipitates to form

WORDS TO KNOW

CLIMATOLOGICAL WARFARE: Speculative use of deliberately engineered climate or weather changes as weapons of war: studied intensively in the United States from 1945 to the 1970s, after which such research was abandoned.

CLOUD SEEDING: The introduction of particles of (usually) dry ice or silver iodide with the hope of increasing precipitation from the cloud.

WATER VAPOR: The most abundant greenhouse gas, it is the water present in the atmosphere in gaseous form. Water vapor is an important part of the natural greenhouse effect. Although humans are not significantly increasing its concentration, it contributes to the enhanced greenhouse effect because the warming influence of greenhouse gases leads to a positive water vapor feedback. In addition to its role as a natural greenhouse gas, water vapor plays an important role in regulating the temperature of the planet because clouds form when excess water vapor in the atmosphere condenses to form ice and water droplets and precipitation.

IN CONTEXT: GEO-ENGINEERING OPTIONS

"Geo-engineering options, such as ocean fertilization to remove [CO_2] directly from the atmosphere, or blocking sunlight by bringing material into the upper atmosphere, remain largely speculative and unproven, and with the risk of unknown side-effects. Reliable cost estimates for these options have not been published *(medium agreement, limited evidence)*."

[Editor's note: The assessment of certainty "*(medium agreement, limited evidence)*" is according to the IPCC, "based on the expert judgment of the authors on the level of concurrence in the literature on a particular finding (level of agreement), and the number and quality of independent sources qualifying under the IPCC rules upon which the finding is based.]

SOURCE: *Metz, B., et al, eds.* Climate Change 2007: Mitigation of Climate Change: Contribution of Working Group III to the Fourth Assessment Report of the Intergovernmental Panel on Climate Change. *New York: Cambridge University Press, 2007.*

raindrops. If the drops are heavy enough, they will fall as rain or snow.

Although cloud seeding can sometimes induce rainfall, it is still too unpredictable for routine use. In addition, as argued in court cases, the deliberate production of rainfall in one region robs those downwind of the moisture.

As of 2007, there is concern over the warming of the Arctic and the melting of the polar ice cap. However, in the early 1960s, a few scientists proposed to melt the northern ice cap by spreading dust or soot, the aim being to provide open water for polar navigation (primarily military). However, even then the majority of scientists cautioned against such actions, arguing that so little was known of climatic interactions that an unforeseen and undesirable consequence could occur.

By the 1970s, the warming of the atmosphere had been discovered and the notion that human activities were the cause was gaining some support. Schemes to spread reflective material on the ocean surface or seeding the atmosphere to promote the formation of sunlight-reflecting clouds had been proposed.

Other climate engineering schemes were intended to trap CO_2 so that the gas would not accumulate in the atmosphere. The massive planting of trees would produce forests capable of soaking up carbon. Adding nutrients to the ocean could fuel the growth of carbon-absorbing plankton, which would transport the carbon to the ocean floor when they died. An equally exotic but feasible plan was to trap CO_2 being emitted from the furnaces of coal burning facilities, compress the gas to turn the CO_2 into a liquid, and pump the liquid deep in the earth or to the bottom of the ocean (an undersea experiment was actually done in 1999, and research on the technique is ongoing as of 2007 in countries including Japan and the United States).

In 2006, scientists proposed that the increased temperature of the atmosphere that has occurred over the past 150 years could be reversed by pumping tiny particles into the upper atmosphere to deflect the sun's rays away from Earth. In this scheme, the particles would be pumped from Earth's surface to the atmosphere via a hose, 15.5-mi (25-km) in length, held aloft by a blimp. However, once begun, the process could never be stopped, since the rebounded warming of the atmosphere could increase global temperatures by almost 10°F (6°C) each decade. Within a half century, life on Earth would be threatened. Understandably, the project is controversial.

■ Impacts and Issues

As the aforementioned example illustrates, the deliberate manipulation of the atmosphere is a very contentious issue. The climate engineering actions of one nation could affect all nations. Whether a nation has the legal or moral right to alter climate is questionable, especially since the precise details of how the atmosphere functions are still not clear. An action intended to change one aspect of the atmosphere could have other, unforeseen impacts or, in the case of the pumping of particles into the atmosphere courtesy of a blimp-supported hose, very undesirable known impacts.

Another aspect of climate engineering concerns the reason for the action. History shows that climate engineering has been contemplated for economic or military reasons rather than humanitarian. Whether a climate engineering strategy for global warming would be done for the common good or just to benefit a select group is unknown.

SEE ALSO *Atmospheric Circulation; Gaia; Global Warming; Mitigation Strategies.*

BIBLIOGRAPHY

Books

Hillman, Mayer, Tina Fawcett, and Sudhir Chella Rajan. *The Suicidal Planet: How to Prevent Global Climate Catastrophe.* New York: Thomas Dunne Books, 2007.

Lovelock, James. *The Revenge of Gaia: Earth's Climate Crisis and the Fate of Humanity.* New York: Perseus Books, 2007.

Periodicals

Flemming, James R. "The Climate Engineers." *The Wilson Quarterly* (Spring 2007).

Web Sites

Kirby, Alex. "Blue-Sky Thinking about Climate." *BBC News*, January 7, 2004. <http://news.bbc.co.uk/2/hi/science/nature/3370211.stm> (accessed November 14, 2007).

Smalley, Eric. "Climate Engineering Is Doable, as Long as We Never Stop." *Wired*, July 25, 2007. <http://www.wired.com/science/planetearth/news/2007/07/geoengineering> (accessed November 14, 2007).

Climate Lag

Introduction

Climate lag is defined as a delay that can occur in a change of some aspect of climate due to the influence of a factor(s) that is slow-acting. An example of climate lag is the full effect of the release of a particular amount of carbon dioxide into the atmosphere. Following its release, some of the gas is absorbed by ocean water to be released into the atmosphere later on as part of the global carbon cycle. Thus, the full effect of carbon dioxide on the warming of the atmosphere will not be apparent until the ocean-bound gas has been released.

Climate lag is an important concept in climate modeling, and in forming policies to deal with climate change. The climate change that is apparent at a certain point in time may not be an accurate indication of the eventual change. Basing an emissions reduction strategy on current data may not completely address the problem.

A Cambodian child sits on one of countless logs, which have been cleared from the land. Scientists warn that deforestation, the release of greenhouse gases, and other environmentally unfriendly practices need to be mitigated now in order to limit catastrophic global warming in the future. However, they also caution that global warming is likely to continue to some degree regardless of any countermeasures because carbon dioxide absorbed by ocean waters and forests (carbon sinks) can slowly make its way into the atmosphere. *AP Images.*

WORDS TO KNOW

CARBON CYCLE: All parts (reservoirs) and fluxes of carbon. The cycle is usually thought of as four main reservoirs of carbon interconnected by pathways of exchange. The reservoirs are the atmosphere, terrestrial biosphere (usually includes fresh-water systems), oceans, and sediments (includes fossil fuels). The annual movements of carbon, the carbon exchanges between reservoirs, occur because of various chemical, physical, geological, and biological processes. The ocean contains the largest pool of carbon near the surface of Earth, but most of that pool is not involved with rapid exchange with the atmosphere.

CLIMATE MODEL: A quantitative way of representing the interactions of the atmosphere, oceans, land surface, and ice. Models can range from relatively simple to quite comprehensive.

GREENHOUSE GASES: Gases that cause Earth to retain more thermal energy by absorbing infrared light emitted by Earth's surface. The most important greenhouse gases are water vapor, carbon dioxide, methane, nitrous oxide, and various artificial chemicals such as chlorofluorocarbons. All but the latter are naturally occurring, but human activity over the last several centuries has significantly increased the amounts of carbon dioxide, methane, and nitrous oxide in Earth's atmosphere, causing global warming and global climate change.

KYOTO PROTOCOL: Extension in 1997 of the 1992 United Nations Framework Convention on Climate Change (UNFCCC), an international treaty signed by almost all countries with the goal of mitigating climate change. The United States, as of early 2008, was the only industrialized country to have not ratified the Kyoto Protocol, which is due to be replaced by an improved and updated agreement starting in 2012.

■ Historical Background and Scientific Foundations

Climate lag is a function of scale. The volume of water in the ocean is huge—over 322 million cubic mi (1.34 billion cubic km). As a result, changes in the ocean chemistry will occur slowly and will be apparent as only slight changes. An example is the pH of the ocean (pH is a measure of the quantity of acidic and basic compounds). Measurements done throughout the twentieth century have revealed a pH decrease in the ocean. Although the decrease is slight, this means that the ocean is becoming more acidic. The scientific agreement is that this is due to the absorption of atmospheric carbon dioxide by the water.

Because elements such as carbon dioxide cycle between the land, water, and air on a global scale, the carbon dioxide in the ocean will ultimately cycle into the atmosphere. Although not all scientists agree, the majority of climate scientists view the changing ocean pH as a consequence of climate lag. As carbon dioxide has accumulated in the atmosphere, its increased absorption by the ocean has followed. According to an article published in the journal *Science* in 2004, almost half of the carbon dioxide produced by human-related activities in the past two centuries has been absorbed by the ocean. In the future, this carbon dioxide will be liberated from the ocean, providing a long-term source of the greenhouse gas.

Similarly, there is a climate lag with respect to ocean temperature. The rising atmospheric temperature does not immediately produce warming of the ocean. The vast volume of ocean water requires centuries to warm even a few degrees.

■ Impacts and Issues

Climate lag has important implications for efforts to slow the warming of Earth's atmosphere. Because carbon dioxide is absorbed by ocean waters and slowly released to the atmosphere, there will continue to be a source of the greenhouse gas even if the release of carbon dioxide by human-related industrial and other activities were to stop.

Scientists at the U.S. National Center for Atmospheric Research have modeled the contribution of ocean-bound carbon dioxide to global warming. They determined that even if greenhouse gas production remained constant, the temperature of the atmosphere will continue to rise for another 100 to 200 years. As well, continued melting of polar ice over the next several centuries will continue to drive an increase in sea level.

Some critics of the Kyoto Protocol have used climate lag to argue that the emission targets are not nearly stringent enough to achieve the desired climate change, since the reduction in produced greenhouse gases will be small in comparison to the gases still to be released from the ocean. Others invoke climate lag to argue against the protocol for the opposite reason, claiming that to realistically deal with greenhouse-gas production, the emission reduction targets would have to be so great as to be unachievable. The latter view was part of Canada's withdrawal of support for the Kyoto Protocol in 2007.

SEE ALSO *Abrupt Climate Change; Carbon Dioxide (CO_2); Greenhouse Effect; Temperature Record.*

BIBLIOGRAPHY

Books

DiMento, Joseph F.C., and Pamela M. Doughman. *Climate Change: What It Means for Us, Our Children, and Our Grandchildren*. Boston: MIT Press, 2007.

Gore, Al. *An Inconvenient Truth: The Planetary Emergency of Global Warming and What We Can Do About It*. New York: Rodale Books, 2006.

Seinfeld, John H., and Spyros N. Pandis. *Atmospheric Chemistry and Physics: From Air Pollution to Climate Change*. New York: Wiley Interscience, 2006.

Periodicals

Feely, R. A., C. L. Sabine, K. Lee, et al. "Impact of Anthropogenic CO_2 on the $CaCO_3$ System in the Oceans." *Science* 305 (2004): 362-366.

Clouds and Reflectance

Introduction

Clouds are composed of frozen droplets or crystals of water vapor. When gathered together in a cloud formation, the billions of objects can efficiently reflect incoming sunlight back into space. This natural shielding provided by clouds can create cooler surface temperatures.

Scientists are still trying to understand how the increasing warming of Earth's atmosphere will affect the types of clouds that form and the extent of cloud cover. The lack of scientific consensus is hampering efforts to develop climate models to understand the climate changes that could occur in the twenty-first century.

Historical Background and Scientific Foundations

The ice crystals that make up a cloud form a structure that has plate-like sides. Each side acts like a mirror. Depending on the position of the crystals in the cloud, incoming sunlight will be reflected straight back if a surface is horizontal, or at an angle if the encountered surfaces are not quite horizontally positioned.

Nephologists (the term given to scientists who study clouds) have determined that on a global scale, clouds reflect about 17% of incoming sunlight back into space. The efficiency of reflectance depends on the depth of the cloud. For example, deep clouds such as the stratocumulus type can reflect up to 90% of the incoming sunlight. The

A mother and her children swim in the Missouri River as sunlight reflects from the surface of the water in Great Falls, Montana. The sunlight is colored orange by smoke from various large wildfires burning in Montana and Idaho in August 2007. *AP Images.*

WORDS TO KNOW

CIRRUS CLOUD: Thin clouds of tiny ice crystals that form at 20,000 ft (6 km) or higher. Cirrus clouds cover 20–25% of the globe, including up to 70% of tropical regions. Because they both reflect sunlight from Earth and reflect infrared (heat) radiation back at the ground, they can influence climate.

CLIMATE MODEL: A quantitative way of representing the interactions of the atmosphere, oceans, land surface, and ice. Models can range from relatively simple to quite comprehensive.

GREENHOUSE GASES: Gases that cause Earth to retain more thermal energy by absorbing infrared light emitted by Earth's surface. The most important greenhouse gases are water vapor, carbon dioxide, methane, nitrous oxide, and various artificial chemicals such as chlorofluorocarbons. All but the latter are naturally occurring, but human activity over the last several centuries has significantly increased the amounts of carbon dioxide, methane, and nitrous oxide in Earth's atmosphere, causing global warming and global climate change.

NEPHOLOGISTS: Scientists who specialize in the study of clouds.

STRATOCUMULUS CLOUDS: Type of cloud that has rounded shapes and forms at low altitudes, i.e., below about 8,000 ft (2,400 m).

WATER VAPOR: The most abundant greenhouse gas, it is the water present in the atmosphere in gaseous form. Water vapor is an important part of the natural greenhouse effect. Although humans are not significantly increasing its concentration, it contributes to the enhanced greenhouse effect because the warming influence of greenhouse gases leads to a positive water vapor feedback. In addition to its role as a natural greenhouse gas, water vapor plays an important role in regulating the temperature of the planet because clouds form when excess water vapor in the atmosphere condenses to form ice and water droplets and precipitation.

base of these types of clouds is dark because of the low level of light that has penetrated fully through the cloud. In contrast, cirrus clouds, which are thin and whispy, reflect less sunlight.

Although it is clear that the global atmosphere is warming, climatologists do not agree on how this will influence cloud formation, and, therefore, the reflectance of sunlight.

It has been predicted that warmer temperatures will increase the evaporation of water from the surface and, therefore, will cause more numerous and thicker clouds to form. The result would be increased reflectance of incoming sunlight, which would cool Earth's surface. However, studies of various areas around the globe where the atmosphere has been warming have revealed that clouds tend to be thinner, since the bottom of the cloud rises higher while the top of the cloud stays at the same altitude. So, it may be that a warming atmosphere will produce clouds that are less capable of reflecting sunlight. In that case, the increased heat would be absorbed by the accumulated greenhouse gases, warming the atmosphere still further.

■ Impacts and Issues

According to the 2007 Assessment Report from the Intergovernmental Panel on Climate Change (IPCC), the years 1995 to 2006 ranked among the 12 warmest years since people began keeping records in 1850. Although the IPCC has stated that "warming of the climate system is unequivocal," the lack of agreement on the behavior and reflectance capability of clouds in a warming atmosphere hampers efforts to model the climate change in the twenty-first century.

The IPCC report reaffirmed this lack of consensus. A warming atmosphere likely means that rainfall and snowfall will be more frequent in northern Europe, Canada, United States, and the far north. But, increased droughts forecast for equatorial regions points to fewer clouds in those regions of the globe. But current climate models do not all offer the same long-term outlook.

Indeed, if the observations that warmer temperatures produce thinner and less light reflective clouds holds true, then the current estimates for atmospheric warming may be low. In equatorial regions, the increased intensity of sunlight could add to the predicted hardship due to water shortages. The IPCC has predicted that African countries could experience 50% less rainfall by 2020, with water shortages affecting up to 250 million people. The consequences for agriculture, economies, and survival could be extreme.

SEE ALSO *Atmospheric Structure; Greenhouse Effect; Insolation.*

BIBLIOGRAPHY

Books

Day, John A. *The Book of Clouds.* New York: Sterling, 2005.

Houghton, John. *Global Warming: The Complete Briefing.* Cambridge: Cambridge University Press, 2004.

Trefil, Calvo. *Earth's Atmosphere.* Geneva IL: McDougal Littell, 2005.

Coal

Introduction

Most of the world's electricity is produced by burning coal, a black, rocky substance consisting mostly of carbon. Coal is also used to fire cement-production kilns and to produce coke (a reduced form of coal) burned in steel manufacture. Coal is destructive to mine, and burning it releases pollutants that cause acid rain and lung cancer, including sulfur dioxide (SO_2) and mercury.

However, coal burning is also the cheapest source of baseline electric generation and its use is increasing around the world. Many proposals have been made for technologies to collect the carbon dioxide (CO_2) emitted from coal burning and inject it underground or into the ocean. As of 2007, a number of test projects for this technique, known as carbon capture and storage or sequestration, were under way, but no coal-fired power plant was actually sequestering its CO_2 output. Some liquid fuels were also being produced from coal, although without sequestration this technology actually increases the amount of CO_2 released per unit of energy released.

Proposals have been made to generate clean-burning hydrogen from coal by sequestering the CO_2 released by this process. Despite high hopes for such technologies, especially carbon capture and storage, as of early 2008 the burning of coal remained the largest single source of greenhouse-gas emissions and was a major source of air pollution as well, causing several hundred thousand deaths annually.

Historical Background and Scientific Foundations

Formation and History of Coal

Coal is a flammable black rock consisting mostly of carbon. It is composed of the compressed, chemically transformed remnants of plants that grew in vast swamps that flourished from about 286 to 300 million years ago, a period named the Carboniferous because of its association with coal. Because coal was formed in swamps, it tends to occur in extensive horizontal layers called seams or beds. In the millions of years after its formation, coal was covered by sediments that eventually hardened into sedimentary rock.

Today, to obtain coal humans must either tunnel under this rock or, if it is not too thick, strip it away—along with the overlying landscape—to expose the underlying coal seam. This practice, known as strip mining, includes the practice of mountaintop removal, in which the upper portions of entire mountains are removed and cast down into adjacent valleys. Strip mining destroys the original ecology and landscape of the mined area, but because of its cheapness is increasingly used worldwide.

The several grades or qualities of coal are classified according to their hardness and composition. Softer, lower-carbon coals burn more poorly and pollute more, but are in greater supply. The hardest, highest-carbon coal is anthracite, a black rock that is over 90% carbon by weight. Bituminous coal is from 70% to 90% carbon, and tends to be black or dark brown; lignite, the lowest grade, is 60% to 70% carbon. Although estimates of world coal resources are based partly on guesswork, most experts agree that many billions of tons of coal remain in underground deposits.

Small quantities of coal also appear on Earth's surface as outcroppings. This outcrop coal has always been accessible to human beings, and has been used as a fuel for several thousand years. However, for most of the last 800,000 years, wood has been humankind's primary fuel because it is more widely distributed. Coal mining began in earnest in the thirteenth century in Britain, with the rest of Europe soon following suit. Coal became an important fuel at this time because the forests of Europe had been mostly cut down for fuel and timber, creating the world's first energy crisis. Coal mining solved the

WORDS TO KNOW

ACID RAIN: A form of precipitation that is significantly more acidic than neutral water, often produced as the result of industrial processes.

ANTHROPOGENIC: Made by people or resulting from human activities. Usually used in the context of emissions that are produced as a result of human activities.

BASELINE GENERATION: Generation of electricity for baseline demand, that is, demand that is steady around the clock. Demand for electricity rises above baseline during the daytime and during heat waves, when power demand for air conditioning peaks.

CARBON SEQUESTRATION: The uptake and storage of carbon. Trees and plants, for example, absorb carbon dioxide, release the oxygen, and store the carbon. Fossil fuels were at one time biomass and continue to store the carbon until burned.

FOSSIL FUELS: Fuels formed by biological processes and transformed into solid or fluid minerals over geological time. Fossil fuels include coal, petroleum, and natural gas. Fossil fuels are non-renewable on the timescale of human civilization, because their natural replenishment would take many millions of years.

GREENHOUSE GASES: Gases that cause Earth to retain more thermal energy by absorbing infrared light emitted by Earth's surface. The most important greenhouse gases are water vapor, carbon dioxide, methane, nitrous oxide, and various artificial chemicals such as chlorofluorocarbons. All but the latter are naturally occurring, but human activity over the last several centuries has significantly increased the amounts of carbon dioxide, methane, and nitrous oxide in Earth's atmosphere, causing global warming and global climate change.

INDUSTRIAL REVOLUTION: The period, beginning about the middle of the eighteenth century, during which humans began to use steam engines as a major source of power.

OUTCROPPING: Any rock formation that is accessible from the surface without digging: a protruding mass of rock, usually connected to a larger, buried mass of similar rock.

RENEWABLE ENERGY: Energy obtained from sources that are renewed at once, or fairly rapidly, by natural or managed processes that can be expected to continue indefinitely. Wind, sun, wood, crops, and waves can all be sources of renewable energy.

crisis and began the large-scale emission by human beings of CO_2 to Earth's atmosphere.

Decaying or burning wood also releases CO_2, but does not increase total atmospheric CO_2 because the carbon in the wood was extracted by the growing tree from CO_2 in the air originally. Although the carbon found in coal and other fossil fuels was also originally extracted from the air by green plants, this was done so many millions of years ago that the reappearance of that carbon in today's atmosphere is equivalent to adding brand-new carbon to the environment.

From the Middle Ages to the early eighteenth century, coal burning remained small in scale by today's standards and did not release enough CO_2 to affect Earth's climate. Coal was burned in furnaces or fireplaces to heat buildings and to cook, brew, and the like, but machinery that could be powered by coal had not yet been invented. In 1712, British inventor Thomas Newcomen (1663-1729) demonstrated the world's first coal-powered steam engine for pumping water. The Newcomen engine proved a commercial success and served as a forerunner for more efficient coal-powered steam engines in factories, boats, and trains. It allowed the water in coal mines to be pumped out more cheaply than did the horse-powered pumps that had been used until that time. For the first time, a fossil fuel was both expanding demand for itself and making its own extraction more economical.

With the invention of commercially viable and popular steam engines, the switch from muscle and wind power to mechanical power based on fossil fuels had begun. By 1800 it was far advanced in the textile industry and other sectors, and CO_2 began to be released in quantities large enough to eventually change Earth's climate. The concentration of CO_2 in the atmosphere in 1750, about 280 parts per million, is used today as the pre-industrial standard against which subsequent anthropogenic (human-made) increases in greenhouse gases are measured.

By 2007, atmospheric CO_2 stood at about 383 parts per million, a 36.8% increase from 1750. Most of this increase had come from coal, with the second-largest share of total anthropogenic CO_2 coming from oil. As of 2007, humans had released a total of about 170 billion tons of carbon into the atmosphere from burning coal, about half of which had been absorbed by the oceans and other carbon sinks.

Coal Usage and Resources

Anthropogenic greenhouse-gas emissions have increased ever since the beginning of the Industrial Revolution, with the most rapid growth occurring most recently. Despite added dependence on oil, which became essential to transportation in the late nineteenth and early twentieth centuries, and on natural gas, which was first exploited on a mass scale in the 1950s and 1960s, coal has remained important. As of 2006, the world was using more coal than ever, namely about 6.3 billion tons (5.7 billion metric tons), 7.6% more than in 2005 and about twice as much as in 1990. As of 2005, 35% of world primary energy (that is, heat energy from fuel before it is transformed into electricity or other forms, as well as electricity produced directly, as by hydroelectric dams, solar cells, or windmills) was obtained from oil, 25% from coal, 21% from natural gas, 12% from renewables (including hydroelectricity), and 6% from nuclear power. About 90% of coal was used for electricity generation, with most of the rest being used to provide heat in the steel and concrete industries.

Coal's role was particularly important in making electricity. Heat from burning coal is used to make pressurized steam, which turns turbines that turn generators that produce electricity. In 2006, 40% of world electricity came from coal, 20% from natural gas, 16% from hydroelectric dams, 15% from nuclear power, 7% from oil, and 2% from miscellaneous sources, including non-hydro renewables. In specific countries, the mix varied: in Poland, for example, 93% of electricity was coal-generated in 2006, while in Germany only 47% was. In the United States, the world's largest energy consumer, 56% of electricity was coal-generated.

Although wind power had become the cheapest form of new electric generating capacity by 2007, it could not be used for baseline generation. Baseline generation is the making of electricity to meet steady, around-the-clock demand. Wind is not suitable for baseline because wind turbines only make electricity when the wind is blowing, so coal remains the cheapest source of baseline electricity. In 2007, coal cost about one sixth as much per unit of energy released as did oil or natural gas.

Estimates of how much coal remains in the ground worldwide vary widely. Energy experts distinguish between coal reserves and coal resources. Reserves are quantities of undug coal that are proven to exist and to be recoverable. Resources are coal deposits whose existence is assumed based on past patterns of discovery, but which have not been actually found. Estimates of world coal resources (undiscovered coal) are unreliable, and shrank from 1980 to 2005 by about 50%, to a little less than 1 trillion tons. World coal reserves are considerably smaller, about half a trillion tons.

Eighty-five percent of proven coal reserves are found in six countries, namely the United States, Russia, India, China, Australia, and South Africa. The United States holds about 30% of world reserves, some 120 billion tons. However, figures even for supposedly proven reserves can be inaccurate: in 2004, Germany decreased its official estimate of its hard coal reserves by 99%, from 23 billion tons (21 billion metric) to 183 million tons (166 million metric). The change, according to the World Energy Council, was due to the fact that earlier estimates of reserves—supposedly proven—were in fact mostly speculative.

In 2005, China was the world's largest coal producer and consumer, producing about 1.1 billion tons (998 million metric) a year, with reserves of 59 billion tons (54 billion metric), giving an annual depletion rate of 1.9% per year. The United States was second, producing 576 billion tons (523 billion metric) a year; Australia was third, producing 202 billion tons (183 billion metric) a year; and India was fourth, producing 200 billions tons (183 billion metric) a year. Although the United States was producing more coal by weight in 2005 than ever before, more of the coal being produced was low-carbon sub-bituminous coal. Thus, in terms of energy production, the country's coal industry had actually peaked in 2000. Although global coal production was rising, a 2007 study by the Energy Watch Group predicted that world coal production was likely to peak around 2030 at about 30% above 2007 levels, then decline slowly thereafter to 1990 levels by about 2100.

■ Impacts and Issues

Coal and the Environment

Coal is the most destructive form of obtaining primary energy in routine use. (The byproducts of nuclear power generation can be exploited to produce nuclear weapons, which have essentially unlimited destructive capacity, but this is not routine.) Coal's damage to the environment begins during mining. Strip mining, which supplies about 70% of coal from the United States, annihilates existing landscapes above the area mined for coal and destroys additional area that is buried under the removed overburden, as the layer of soil, rock, and forest covering the coal seam is termed. In both shaft and strip mines, sulfur associated with coal dissolves in water, forming sulfuric acid (H_2SO_4) that runs off into streams. Toxic heavy metals dissolve in the acidic water and accumulate in aquatic food chains. Solid mine wastes are bulky, toxic, and often flammable. Coal mines release methane (CH_4), another greenhouse gas.

When coal is burned, it releases scores of pollutants, including mercury, sulfur dioxide (which returns to Earth as acid rain), nitrogen oxides, soot particles, mercury, cadmium, uranium, and lead. According to a 2007 World Bank report, air pollution from coal causes on the order of 400,000 deaths annually in China. A 2004 study commissioned by environmental groups but carried out by a firm often employed by the U.S. Environmental Protection Agency (EPA) found that coal burning causes about 24,000 deaths a year in the United States, including over 8,000 from lung cancer. The accuracy of the latter study was disputed by representatives of the coal industry.

A Beijing, China, resident moves coal that is used for heating homes and cooking. Although China wants to reduce its dependence on coal, the country's rapid industrialization has led to the building of at least one new coal plant each week. *AP Images.*

Coal and Climate Change

Coal is a major contributor to anthropogenic climate change. As of 2004, about 84% of global anthropogenic greenhouse-gas emissions were from energy production, with 95% of these emissions (that is, about 80% of all greenhouse-gas emissions) consisting of CO_2 released from burning fossil fuels. The other 5% of greenhouse-gas emissions from energy production consisted of methane from coal mining and hydroelectric dams and nitrogen-oxygen (NO_x) compounds from fuel burning. Because coal is more carbon-intensive than oil or natural gas, releasing almost all its energy from combustion of carbon rather than of hydrogen, it produces a larger share of CO_2 emissions than it does of primary energy. Coal produced only 25% of global primary energy, but 40% of global CO_2 emissions. Coal is about twice as carbon-intensive as natural gas.

In the early 2000s, global CO_2 emissions were rapidly increasing. From 2003 to 2004 alone, annual emissions increased by over 1.2 billion tons (1.1 billion metric) of CO_2, with 86% of this rise caused by increasing energy demand in developing countries. Increased coal usage accounted for 60% of the 2003 to 2004 increase in global CO_2 emissions.

Large underground coal fires, especially in China, are a significant source of greenhouse gases, although estimates of their magnitude vary widely. Estimates of CO_2 output from Chinese coal fires alone range from 150 to 450 million tons (136 to 408 million metric) of CO_2 per year (0.33-1% of annual global CO_2 emissions).

Alternative Coal Technologies

If global warming is to be stabilized at a 3.6°F (2°C) change, widely cited by scientists as the approximate limit for non-catastrophic climate change, low-carbon power sources will have to dominate by mid-century. A 2003 study published in *Science* projected that by about 2050, 75% to 100% of total power demand would have to be met by non-CO_2-releasing sources to stabilize warming at this level. At the same time, however, energy demand was projected to grow. Although many experts were urging a rapid expansion of nuclear power, and by 2007 renewable energy resources such as wind were already expanding rapidly (over 30% per year for wind power), most expansion in primary energy was expected to come from coal.

By 2007, concern was mounting about the impact of coal on climate. In October of that year, the state of Kansas became the first governmental body in the United States to cite CO_2 emissions as a reason for refusing a construction permit for a new coal-fired electric generation plant. A legal basis for the action had been supplied in April 2007, by a U.S. Supreme Court decision ruling that greenhouse gases, including CO_2, are pollutants under the terms of the federal Clean Air Act.

The only way to use steady or increasing amounts of coal while decreasing CO_2 emissions is to employ carbon capture and storage (CCS) technologies, also known as carbon sequestration or clean coal technology. There are a number of CCS technology concepts, but all involve pumping CO_2 from power plants into ocean waters or deep underground reservoirs rather than allowing it to escape into the air. As of 2007, several industrial-scale demonstration projects were in operation. For example, in Ketzin, Germany, a pilot project was injecting CO_2 into sandstone 2,600 ft (800 m) underground. Over two years, the project planned to sequester 66,000 tons (60,000 metric tons) of CO_2, about the same amount emitted in a year by 40,000 cars. The European Union was considering requiring all new coal plants in Europe to include CCS technology from 2020 onward, with some experts urging a moratorium on all new coal-plant construction until CCS technology could be applied.

Concerns about CCS involved its cost, whether marine sequestration would enhance acidification of the oceans, and whether gas pumped into underground reservoirs would remain there. Also, some critics of the technology were concerned that making coal less harmful in greenhouse terms will encourage reliance on the fuel, whose extraction is environmentally destructive regardless of how it is burned.

SEE ALSO *Carbon Sequestration Issues; Energy Contributions; Nuclear Power; Renewable Energy; Wind Power.*

BIBLIOGRAPHY

Books

Miller, Bruce G. *Coal Energy Systems.* San Diego, CA: Academic Press, 2004.

Periodicals

Berstein, Lenny, et al. "Carbon Dioxide Capture and Storage: A Status Report" *Climate Policy* 6 (2006): 241-246.

Caldeira, Ken. "Climate Sensitivity Uncertainty and the Need for Energy Without CO_2 Emission." *Science* 299 (2003): 2052-2054.

"Coal Use Grows Despite Warming Worries." *The New York Times* (October 28, 2007).

Holloway, Sam. "Storage of Fossil Fuel-Derived Carbon Dioxide Beneath the Surface of the Earth." *Annual Review of Energy and the Environment* 26 (2001): 145-166.

Kintisch, Eli. "Report Backs More Projects to Sequester CO_2 from Coal." *Science* 315 (2007): 1481.

Mufson, Steven. "Democrats Push Coal-to-Liquids Energy Plan." *The Washington Post* (June 13, 2007).

Mufson, Steven. "Power Plant Rejected Over Carbon Dioxide for First Time." *The Washington Post* (October 19, 2007).

Quadrelli, Roberta. "The Energy-Climate Challenge: Recent Trends in CO_2 Emissions from Fuel Combustion." *Energy Policy* 35 (2007): 5938-5952.

Sanderson, Katharine. "King Coal Constrained." *Nature* 3449 (2007): 14-15.

Stauffer, Hoff. "New Sources Will Drive Global Emissions." *Energy Policy* 35 (2007): 5433-5435.

Web Sites

"Emissions of Greenhouse Gases in the United States 2005." *U.S. Energy Information Administration*, November 2006. <http://www.eia.doe.gov/oiaf/1605/ggrpt/carbon.html> (accessed November 5, 2007).

Metz, Bert H., et al. "IPCC Special Report on Carbon Dioxide Capture and Storage." *Intergovernmental Panel on Climate Change*, 2005. <http://www.ipcc.ch/activity/srccs/SRCCS.pdf> (accessed November 5, 2007).

Sharp, Philip, et al. "The Future of Coal." *Massachusetts Institute of Technology*, 2007. <http://web.mit.edu/coal/The_Future_of_Coal.pdf> (accessed November 5, 2007).

Larry Gilman

Coastal Populations

Introduction

As human activities—such as building along coastlines, mining of beach sand, cutting of mangrove forests, and damming of rivers—place stresses on coastal environments and sea levels rise due to climate change, the ecosystems and outlines of many coasts around the world are changing. Such changes will probably accelerate in coming decades, affecting human populations of many coastal areas. In many areas, there is no option for reducing impacts to ecosystems by sea-level rise except mitigating climate change itself by emitting fewer greenhouse gases.

In developed or wealthier nations, some of the coastal effects of climate change and other human activities may be prevented or adapted to. This will cost money—but then, so will doing nothing—and experts have warned that the costs of inaction will greatly outweigh the costs of action. In poorer areas, such as many parts of Latin America, Southeast Asia, and Africa, and the small island countries of the Pacific Ocean, there will not be enough money to adapt effectively to coastal changes. In these areas, as well as in parts of the developed world, there may be no choice but for populations to retreat inland from rising seas and more dangerous coastal storms.

The threat of rising sea levels stands to impact people who make their living on coastal farms like this seaweed farm in Zanzibar, Tanzania. *Image copyright Vera Bogaerts, 2007. Used under license from Shutterstock.com.*

WORDS TO KNOW

ANTHROPOGENIC: Made by people or resulting from human activities. Usually used in the context of emissions that are produced as a result of human activities.

DELTA: Triangular–shaped area where a river flows into an ocean or lake, depositing sand, mud, and other sediment it has carried along its flow.

EROSION: Processes (mechanical and chemical) responsible for the wearing away, loosening, and dissolving of materials of Earth's crust.

ESTUARY: Lower end of a river where ocean tides meet the river's current.

GREENHOUSE GASES: Gases that cause Earth to retain more thermal energy by absorbing infrared light emitted by Earth's surface. The most important greenhouse gases are water vapor, carbon dioxide, methane, nitrous oxide, and various artificial chemicals such as chlorofluorocarbons. All but the latter are naturally occurring, but human activity over the last several centuries has significantly increased the amounts of carbon dioxide, methane, and nitrous oxide in Earth's atmosphere, causing global warming and global climate change.

MANGROVE FOREST: Coastal ecosystem type based on mangrove trees standing in shallow ocean water: also termed mangrove swamp. Mangrove forests support shrimp fisheries and are threatened by rising sea levels due to climate change.

PERMAFROST: Perennially frozen ground that occurs wherever the temperature remains below 32°F (0°C) for several years.

RIVER DELTA: Flat area of fine-grained sediments that forms where a river meets a larger, stiller body of water such as the ocean. Rivers carry particles in their turbulent waters that settle out (sink) when the water mixes with quieter water and slows down; these particles build the delta. Deltas are named after the Greek letter delta, which looks like a triangle. Very large deltas are termed megadeltas and are often thickly settled by human beings. Rising sea levels threaten settlements on megadeltas.

STORM SURGE: Local, temporary rise in sea level (above what would be expected due to tidal variation alone) as the result of winds and low pressures associated with a large storm system. Storm surges can cause coastal flooding, if severe.

WETLANDS: Areas that are wet or covered with water for at least part of the year.

Historical Background and Scientific Foundations

Global climate change places several stresses on coastal ecosystems and human populations. Increased glacial melting, as in the Himalayas and elsewhere, is increasing river flow (until the glaciers are gone); hurricanes and monsoons are increasing in severity; permafrost is melting; and sea levels are rising. All of these changes affect coastlines and therefore the people who live near them. Sea-level rise is perhaps the most important single aspect of climate change as far as coastal populations are concerned, because it affects almost all coastlines at the same time and does so in several ways at once. Rising sea level moves coastlines inland, bringing waves and therefore erosion to new areas. Flat coastal lands and wetlands may be covered with rising water, or rendered vulnerable to exceptional high-water events such as storm surges (brief increases in local sea level caused by large storms such as hurricanes). They may also be eroded at their edges, forcing natural and human communities to move inland—if they can.

Sea levels have changed little for the last 7,000 years. They have been particularly stable for the last 2,000 to 3,000 years. In the late nineteenth century, however, at which time anthropogenic (human-caused) climate change began to swell the upper layer of the oceans by warming them and to deepen the oceans by melting glaciers, the oceans began to rise at an average global rate of about 0.07 in per year (1.7 mm per year). Sea level rose at this rate for most of the twentieth century. From 1993 to 2007, it has been rising at 0.12 in per year (3 mm per year), almost twice the average rate of the first 90 years of the twentieth century.

The affects of climate change on coastlines threaten property and lives in many locations around the world. More than 100 million people live no more than 3 ft (about 1 m) above sea level, while about 400 million live no more than 66 ft (20 m) above sea level and within 12 mi (20 km) of a coast. The number of people per square mile in coastal areas is, on global average, about three times greater than elsewhere. All these populations are at increased risk from erosion or storm flooding due to ongoing climate change.

Eleven of the 15 largest cities in the world are located on coasts or estuaries. (An estuary is a partly enclosed body of water connected to the ocean on one side and a river on the other.) For example, New York, London, and Jakarta, the capital of Indonesia, are all located on estuaries. About 53% of the population of the United States lives on a coast, and further development of coastal regions is continuing rapidly. In developed nations, people are increasingly attracted to coastline developments because such locations are often thought to be more pleasant or desirable. Waterfront property tends to bring a high price on the real-estate market. All these developments will eventually be at risk from rising sea levels.

IN CONTEXT: PLANNING FOR SEA LEVEL RISE

Within the United States, Maine is one of the only states whose laws and building codes specifically acknowledge the possible impacts of global climate change. Collating data from observed and recorded tidal changes on Maine's coastlines since 1912, the Intergovernmental Panel on Climate Change 2001 report, and recommendations from the U.S. Geological Survey, the State of Maine adopted an official policy of planning for a 2-3 ft (0.61-0.91 m) rise in sea level over the next century.

Maine's state and municipal laws try to accommodate the diverse interests of coastal preservation, long-term planning, and current development. Existing coastal structures are permitted to remain at the water's edge. However, the state prohibits rebuilding of some storm-damaged structures, especially those closest to the water. For example, if a structure is at least 50% damaged, it cannot be rebuilt on the same site unless the owner can demonstrate with "clear and convincing evidence" that the site will remain stable and intact after a 3-ft (0.91-m) rise in sea level.

In poorer countries, populations are settled by the ocean because they draw their living from it or are attracted to the job markets in cities that were originally sited by the ocean for trade and fishing. The possible impacts on densely settled populations in the deltas—the very large, flat areas of land built up at the outflows of large rivers, such as the Ganges, Nile, Indus, and Mekong—are catastrophic. As of 2006, about 300 million people lived on some 40 large river deltas around the world. These deltas are large, flat deposits of sediment fanning out from river outlets. They are often fertile, easily built on, and attractive to settlement because they are in contact with both a major river and the ocean.

Deltas are experiencing faster effective sea-level rise than actual sea-level rise, because human activities are causing many of them to subside (become lower). Diverting rivers upstream for irrigation or damming them for hydroelectric generation reduces the amount of sediment—fine-grained mineral material, such as clay and sand—that is carried to the delta system. When the decrease is great enough, erosion can cause the delta to subside and shrink rather than to grow. Alternatively, as with the Sundarbans delta in India and Pakistan, at the outflow of the Ganges River, increased flow from accelerated melting of glaciers upstream in the Himalayas can raise water levels in the delta, increasing effective sea-level rise by raising river levels rather than by lowering the delta by starving it of sediment, as is more commonly the case.

Even some large settlements in the developed world are profoundly vulnerable to rising seas and increasingly severe storms. For example, much of the city of New Orleans, Louisiana, in the United States is below sea level and is protected from instant flooding only by an elaborate system of artificial barriers called levees. The devastation in this city brought by Hurricane Katrina in 2005 exemplifies the impact of a severe storm on such a heavily populated coastal area.

Impacts and Issues

The Arctic Climate Impact Assessment, written by an international scientific group, reported in 2004 that a 1.5-ft (50-cm) rise in sea level by 2100, which is within the range pronounced likely in 2007 by the Intergovernmental Panel on Climate Change (IPCC), would flood low-lying parts of the Florida and Louisiana coasts, pushing the coastline inland by about 150 ft (45 m).

The IPCC, although careful to note that no single hurricane event can ever be attributed to climate change, says that such events illustrate the consequences of climate change as storms become more intense and possibly more common. These consequences include the loss of what are called ecosystem services, that is, features of natural systems like coastal wetlands upon which human communities depend. Hurricane Katrina, which resulted in more than 1,800 deaths along the U.S. Gulf Coast and caused over $100 billion of damage in 2005, also destroyed 150 sq mi (388 sq km) of coastal wetlands, islands, and levees that previously protected New Orleans from storm surge. Thus, storm damage can render a coastal settlement more vulnerable to future storm damage. Sea-level rise combined with other human activities, such as damming of rivers, is rendering many large cities built on river deltas, including New Orleans, increasingly vulnerable to such effects.

The long-term effects of climate change on coastal populations will depend largely on how much the sea level rises and how rapidly. There is scientific disagreement about how much sea-level rise is likely to occur in this century and beyond. The IPCC predicted in 2001 that sea-level rise in the twenty-first century would probably be between 0.3 and 2.9 ft (0.09–0.9 m); in 2007, it narrowed the likely range to 0.6–1.9 ft (0.18–0.6 m), an apparently reassuring change. However, of all the statements in the IPCC's 2007 Assessment Report on climate change, this has probably been more vigorously criticized by scientists than any other. Many experts have argued that the IPCC underestimates future sea-level rise by leaving out of account possible accelerated melting of the Greenland and West Antarctic ice sheets. Accelerated melting in Greenland has been confirmed by satellite observations since December 2005, the cut-off date for research to be included in the IPCC's 2007 report. Greenland's ice sheet contains enough water to raise sea level by about 23 ft (7 m), although complete melting could not take place within a single century.

Primary Source Connection

Much of the world's population and economic development are located along coastlines. Thus, anticipated sea-level rise has sparked intensive study of vulnerable coastal populations. Here, Vivien Gornitz, a lead scientist at Columbia University's Center for Climate Systems Research, describes methods and uncertainties involved in estimating coastal risk from climate change.

COASTAL POPULATIONS, TOPOGRAPHY, AND SEA LEVEL RISE

A frequently cited consequence of global climate change is the potential impact of sea level rise (SLR) on coastal populations. Eleven of the world's 15 largest cities lie along the coast or on estuaries. In the United States, around 53% of the population lives near the coast. In spite of this widespread concern over sea level rise, no accurate worldwide estimate of the number of people likely to be affected by coastal inundation or flooding has been published.

In a recent study with Christopher Small (Columbia Univ. Lamont-Doherty Earth Observatory) and Joel E. Cohen (Rockefeller Univ.), I attempted to provide an improved assessment of human vulnerability to sea level rise by integrating the best currently available information about global population distributions, elevation, and sea level. Population data are compiled from a 1997 study by Tobler and others; topography comes from the EROS Data Center (Sioux Fall[s], SD) 30 arc second gridded elevations, and the sea level records from the Permanent Service for Mean Sea Level, Bidston Observatory.... We considered the following scenarios:

- Extrapolation of current trends. No climate change occurs and current rates of sea level rise are extrapolated from 1990 to 2100.
- Goddard Institute for Space Studies (GISS) coupled ocean-atmosphere General Circulation Model (GCM) (Russell et al. 1995) at 4° by 5° horizontal resolution. In the GS run, CO2 increases by 1%/yr and sulfate aerosols increase annually up to 2050, with a slight decrease thereafter.
- The IPCC IS92a "best estimate" includes effects of sulfate aerosols.

The two model scenarios show increases of around 40-45 cm in sea level over the next 100 years, which is about three times that of extrapolating the current trend....

In addition to permanent inundation due to sea level rise, the coast is at risk to flooding by storms. Average country-wide surge levels for given return periods are taken from a 1993 Delft Hydraulics study. Total flood levels comprise storm surges, tides, and local sea level change, including vertical land motions.... Most sea level stations show significantly less than 1m subsidence over the next 100 years. Areas experiencing uplift include Scandinavia and Alaska, due to ongoing glacial rebound, and Japan due to tectonic uplift.

Analysis of the spatial distribution of population with respect to topography reveals that the number of people and population density diminishes rapidly with increasing elevation and increasing distance from the shoreline. Approximately 400 million people live within 20 m of sea level and within 20 km of a coast, worldwide. However, this figure is not very precise—the spatial distribution of coastal populations is not known to better than 20-30 km. In fact, a major conclusion of our study is that the available data are still inadequate to permit quantitatively precise global estimates of the number of people likely to be affected by plausible levels of sea level rise or storm surges in the coastal zone. In the near future, airborne and satellite-based radar and laser altimeters can map coastal topography and its changes at much higher resolutions than those used in this study. Satellite monitoring of coastal land cover transformations will provide means of quantifying habitation patterns, and thus indirectly, population trends.

Vivien Gornitz

GORNITZ, VIVIEN. "COASTAL POPULATIONS, TOPOGRAPHY, AND SEA LEVEL RISE." NASA: GODDARD INSTITUTE FOR SPACE STUDIES. MARCH 2000. <HTTP://WWW.GISS.NASA.GOV/RESEARCH/BRIEFS/GORNITZ_04/> (ACCESSED NOVEMBER 30, 2007).

SEE ALSO *Coastlines, Changing; Hurricanes; Sea Level Rise.*

BIBLIOGRAPHY

Books

Committee on Mitigating Shore Erosion Along Sheltered Coasts, National Research Council. *Mitigating Shore Erosion Along Sheltered Coasts.* Washington, DC: National Academies Press, 2007.

Parry, M. L., et al, eds. *Climate Change 2007: Impacts, Adaptation and Vulnerability: Contribution of Working Group II to the Fourth Assessment Report of the Intergovernmental Panel on Climate Change.* New York: Cambridge University Press, 2007.

Periodicals

Ericson, Jason P. "Effective Sea-Level Rise and Deltas: Causes of Change and Human Dimension Implications." *Global and Planetary Change* 50 (February 2006): 63–82.

Foley, Jonathan A. "Tipping Points in the Tundra." *Science* 310 (October 28, 2005): 627–628.

Web Sites

Evans, Rob L. "Rising Sea Levels and Moving Shorelines." *Woods Hole Oceanographic Institution*, April 4, 2007. <http://www.whoi.edu/page.do?pid=12457&tid=282&cid=2484> (accessed September 29, 2007).

Gornitz, Vivien. "Coastal Populations, Topography, and Sea Level Rise." *NASA Goddard Institute for Space Studies*, March 2000. <http://www.giss.nasa.gov/research/briefs/gornitz_04/> (accessed November 30, 2007).

Park, Sung Bin. "A Sea of People to Show How Climate Change Will Alter New York City." *NYC Indymedia*, April 4, 2007. <http://www.transalt.org/press/media/2007/890.html> (accessed September 29, 2007).

Larry Gilman

Coastlines, Changing

■ Introduction

Coasts are places where land meets ocean; a coast's shape, as seen from above, is its coastline. Coastlines change when either the land or the ocean changes. Land changes include erosion, deposition (increase of land by the arrival of solid material, often small particles brought to the coast by rivers), or rising or falling of the land itself due to geological forces. The ocean may change by shifting its current and wave patterns or by rising or falling in level. Rising or falling sea level affects coastlines simultaneously all over the world and so is the most important determinant of changes to coastlines.

Sea level has risen and fallen hundreds of times over the 4.5-billion-year history of Earth, and has been rising from about 22,000 years ago to the present, though not at a steady rate. Recently, sea-level rise has been accelerated by anthropogenic (human-caused) climate change, which has caused the top layer of the ocean to warm (and so expand) and ice to begin melting more quickly in Greenland and Antarctica. As a result, coastlines are changing worldwide, and are projected to change much more in coming decades and centuries as global warming continues.

■ Historical Background and Scientific Foundations

Coastlines change greatly over geological time as continents drift together and apart, sea levels fall and rise during ice ages and interglacial periods, and other geological processes rework the world. For example, the whole interior portion of North America from what is now the Arctic Ocean to the Gulf of Mexico was a shallow sea during the Cretaceous period, about 100 million years ago. The coastlines of that sea, the Western Interior Seaway, no longer exist. And as sea levels rise and fall many meters over geological time, entire islands appear and disappear or are greatly altered in coastal outline.

Coastlines have changed even in recent human prehistory. For example, about 12,000 years ago, during the withdrawal of the glaciers of the most recent Ice Age, the removal of the ice's weight on what is now northern Germany and the Baltic Sea allowed the underlying rock to tip like a see-saw. Inhabited coastal areas sank beneath the sea as the coastline of northern Europe shifted southward, while inland areas gained altitude.

The amount of water locked up in ice sitting on land in Greenland, Antarctica, and elsewhere has a large effect on sea levels and thus coastlines worldwide. Since the end of the last major Ice Age about 22,000 years ago, the melting of glaciers has liberated enough water to raise sea level about 400 ft (120 m). All but a few feet of that increase took place before 7,000 years ago. Sea level stabilized about 2,000–3,000 years ago and did not rise significantly again until the late nineteenth century, when it began rising steadily again. During the twentieth century, it rose at an average rate of 0.07 in/year (1.8 mm/year). From 1993 to the present, it has been rising at 0.12 in/year (3 mm/year), almost double the average rate from 1900 to 1990.

Steep coastlines are least affected by sea-level rise, because each inch of increased water depth only climbs a little way up the shore, affecting a relatively narrow band of ground. The effect is much greater for gently sloping or nearly flat shores, such as river deltas (flat areas of fine particles carried to the sea by rivers and dumped at their outlets), swamps, and beaches, where each inch of sea-level rise may cause the sea to advance several inches—sometimes many yards—inland. Rising seas also shrink small, flat islands and seep into their underground freshwater supplies.

The effects of sea-level rise on coastal lands can be destructive. Along the East Coast of the United States, for example, three quarters of the coast that is not protected from the ocean by natural or artificial barriers is experiencing erosion because of sea-level rise over the last 100 to 150 years. The impacts of storms have also increased along the East Coast because of sea level rise.

WORDS TO KNOW

CRETACEOUS PERIOD: Geological period from 145 million years ago to 65 million years ago. During the Cretaceous, the supercontinent that geologists call Gondwana broke up, radically altering regional climates. The Cretaceous ended in the Cretaceous-Tertiary (K-T) extinction event, probably caused at least partly by an asteroid impact, which included the extinction of dinosaurs and about 85% of all other species.

DEPOSITION: Process by which water changes phase directly from vapor into a solid without first becoming a liquid.

EROSION: Processes (mechanical and chemical) responsible for the wearing away, loosening, and dissolving of materials of Earth's crust.

ICE AGE: Period of glacial advance.

MANGROVE FOREST: Coastal ecosystem type based on mangrove trees standing in shallow ocean water; also termed mangrove swamp. Mangrove forests support shrimp fisheries and are threatened by rising sea levels due to climate change.

PERMAFROST: Perennially frozen ground that occurs wherever the temperature remains below 32°F (0°C) for several years.

RIVER DELTA: Flat area of fine-grained sediments that forms where a river meets a larger, stiller body of water such as the ocean. Rivers carry particles in their turbulent waters that settle out (sink) when the water mixes with quieter water and slows down; these particles build the delta. Deltas are named after the Greek letter delta, which looks like a triangle. Very large deltas are termed megadeltas and are often thickly settled by human beings. Rising sea levels threaten settlements on megadeltas.

WETLANDS: Areas that are wet or covered with water for at least part of the year.

However, rising sea level is not the only force causing shores to erode around the world. Other climate-related coastal changes, including the melting of permafrost, have contributed to the fast retreat of some Arctic coastlines. Reduced winter sea-ice cover along some shores, such as those of the gulf of the St. Lawrence River in Canada, allows waves to attack open shore during more of the year and thus speed erosion. Increasingly severe storms, a consequence of global climate change, also speed coastal erosion.

Human activities not directly related to global climate change that affect coastlines include: the damming of rivers, which starves river deltas of silt (small particles carried downstream); the removal of mangrove forests whose roots anchor mud and sand along some shallow tropical coasts; the mining of beaches for sand; and the pumping of gas, oil, and water from beneath the land, which causes the land to subside (sink). Hurricane Katrina, which inundated the U.S. city of New Orleans Louisiana, in August 2005, was particularly destructive because land subsidence placed much of the city below sea level. Rising sea level can have more impact on coastal wetlands, which include the flattest types of coastline, when combined with other human pressures on the environment.

Sea-level rise is not uniform all over the world. Although the world's oceans are in effect a single body of water, they are connected by a complex system of deep and shallow circulations, with regional differences in saltiness and temperature that affect sea level. Sea level is actually falling in a few places. In most places, however, the trend is toward steadily increasing sea levels and retreating coastlines.

Impacts and Issues

Erosion, changes to wetlands, and changes to coastal vegetation have already been observed in many parts of the world due to climate change and other forces. A few of these changes are as follows:

- Shoreline erosion is affecting 75% of the unprotected coastline of the eastern United States. Louisiana's shoreline retreated by 3.3 ft/year (1 m/year) from 1988 to 2002. Nineteen percent of the shoreline recently examined in the Manitounuk Strait in Canada is retreating due to melting of permafrost.
- Wetland changes include the loss and degradation of marshes in Chesapeake Bay in the United States; loss of land area in Venice, Italy; wetland losses in the Thames estuary in England; decreases in salt-marsh area in Long Island and Connecticut in the United States; and changes in other parts of the world. In some places, where rates of sediment accumulation have increased, salt marshes can keep up with sea-level rise, as along the coast of Normandy, France.
- Coastal vegetation changes include the replacement of grassy marshes by mangroves in Florida in the United States and similar changes in southeastern Australia.

In the large delta at the outflow of the Ganges river in India, rising sea level combined with increased river flow from melting glaciers in the Himalayas (an effect of global warming) have recently destroyed some of the Sundarbans islands. About 31 square mi (80 square km) of land have vanished over the last 30 years due to rising water, forcing about 600 families to relocate, often with an increase in poverty. Flooding and erosion are expected to continue in such areas, and will probably worsen in coming decades.

Sea level may rise anywhere from about a foot to more than a yard (0.3-1 m) in the coming century. Forecasts are rendered uncertain by many variables, especially the future melting rate of some of the Greenland ice cap and the West Antarctic ice sheet (the rest of the Antarctic ice sheet is thought to be very stable). Coastal wetland ecosystems, particularly where there is no opportunity to migrate inland, will be degraded.

As with other effects of climate change, developed countries with more monetary resources and better technology will find it easier to adapt to some coastal changes than will less developed nations. The processes causing sea-level rise are difficult to reverse and will continue to operate for centuries, even if human beings stabilize the amount of greenhouse gases they are putting into the atmosphere.

Impacts on natural systems are likely to be severe even in the next century. By the year 2080, scientists estimate that up to 20% of the world's coastal wetlands could vanish due to sea-level rise.

IN CONTEXT: WINNERS AND LOSERS

According to the National Academy of Sciences: "There will be winners and losers from the impacts of climate change, even within a single region, but globally the losses are expected to far outweigh the benefits. The larger and faster the changes in climate, the more difficult it will be for human and natural systems to adapt without adverse effects."

"Unfortunately, the regions that will be most severely affected are often the regions that are the least able to adapt. Bangladesh, one of the poorest nations in the world, is projected to lose 17.5% of its land if sea level rises about 40 inches (1 meter), displacing millions of people. Several islands throughout the South Pacific and Indian oceans will be at similar risk of increased flooding and vulnerability to storm surges. Coastal flooding will likely threaten animals, plants, and fresh water supplies. Tourism and local agriculture could be severely challenged."

SOURCE: *Staudt, Amanda, Nancy Huddleston, and Sandi Rudenstein.* Understanding and Responding to Climate Change. *National Academy of Sciences, 2006.*

■ Primary Source Connection

An overwhelming majority of researchers agree that climate change will alter coastlines worldwide. However, there remains some uncertainty about the possible extent of sea-level rise and how it will reshape coastlines. The following article discusses the threat of rising sea levels to coastal property development in the United States and the emerging technology researchers are using to predict and map coastline change.

R. L. Evans is a marine geophysicist and an associate scientist at Woods Hole Oceanographic Institution, the world's largest nonprofit marine research organization.

RISING SEA LEVELS AND MOVING SHORELINES: NEW TOOLS AND TECHNIQUES SHOW PROMISE FOR BETTER PREDICTIONS AND DECISIONS ABOUT COASTLINE CHANGE

Nae man can tether time or tide.

—Robert Burns

For the past century, the pace and density of development near the ocean has been unprecedented, and much of it is incompatible with the dynamic nature of the shoreline. More than $3 trillion are invested in dwellings, resorts, infrastructure, and other real estate along the Atlantic and Gulf coasts of the United States, and more than 155 million people live in coastal counties. The coastal population is estimated to rise by 3,500 people per day.

Yet, as the devastating hurricane season of 2004 showed, there is a price to be paid for living at sea level and building on sand. Even without extreme storms, the shoreline naturally advances and retreats on scales ranging from seconds to millennia.

As a growing population hugs the coast, understanding the complex processes by which coastlines change has never been more relevant and more important to our well-being.

A rising tide

Changes to the shoreline are inevitable and inescapable. Shoals and sandbars become islands and then sandbars again. Ice sheets grow and shrink, causing sea level to fall and rise as water moves from the oceans to the ice caps and back to the oceans. Barrier islands rise from the seafloor, are chopped by inlets, and retreat toward the mainland. Even the calmest of seas are constantly moving water, sand, and mud toward and away from the shore, and establishing new shorelines.

Coastal changes have accelerated in the past century. Although sea level has been rising since the end of the last glaciation (nearly 11,000 years), the rate of sea-level rise has increased over the past 200 years as average temperatures have increased. Global warming has added water to the oceans by melting ice in the polar regions. But the greater contributor is thought to be thermal expansion of the oceans—a rise in sea level due to rising water temperature. Sea level has risen 10 to 25 centimeters in the past 100 years, and it is predicted to rise another 50 centimeters over the next century (with some

Aerial view of Martha's Vineyard

Computer-enhanced view

In an effort to warn people about the possible effects of coastal erosion due to global warming, the National Environmental Trust presented this aerial view of Martha's Vineyard in Massachusetts from 1998 (top) and a computer-enhanced version of what the area could look like in 100 years (bottom) if nothing is done to reduce greenhouse gases. *AP Images.*

estimates as high as 90 centimeters). Whether or not human activities have contributed to the change, the sea is definitely rising, and it jeopardizes our rapidly growing coastal communities.

Coastal erosion accelerates as sea level rises. Erosion decreases the value of coastal properties because it decreases "the expected number of years away from the shoreline," as researchers and underwriters put it. This quiet loss of U.S. property value amounts to $3 to $5 billion per year. Then there is the actual loss of property, including structures, which amounts to as much as $500 million a year.

Eroding coastlines are also at greater risk from storm damage. Property damage from hurricanes along the eastern U.S. is estimated to average $5 billion per year, with the cost in 2004 alone estimated at more than $21 billion. Such calculations rarely account for the long-term costs of flooding and erosion, damage to natural landforms or ecosystems, and lost recreation and tourism opportunities.

There is significant debate about how to best manage coastal resources to cope with the changing shoreline. When and where will the coast change? And what, if anything, should we do about it?

Billions of tax dollars are being spent to restore and protect our wetlands, maintain our beaches and waterways, and rebuild coastal infrastructure. For example, the State of Louisiana is proposing to spend $14 billion over the next 40 years to restore coastal barriers along the Mississippi River delta. Despite these vast sums of money, very little is being invested in basic research that can improve our ability to predict shoreline change, inform managers in their decision-making, or provide more accurate risk assessment.

More than just a beach problem

The coast is an incredibly complex system, of which beaches are only one part. All aspects of the system—rivers, estuaries, dunes, marshes, beaches, headlands, the surf zone, and the seafloor—influence and respond to the others. But many parts of the system have yet to be studied in sufficient detail to fully understand their role in shoreline change.

Beach erosion threatens property near the shoreline, but it also profoundly influences a critical part of our coastal ecosystem: the marshes. Tidal marshes in estuaries and behind barrier islands are the dominant habitat along the Atlantic Coast of the U.S., and they are particularly vulnerable to rising sea level.

Marshes are ecologically and economically important because they regulate the exchange of water, nutrients, and waste between dry land and the open ocean. They filter and absorb nutrients and pollutants, and buffer coastlines from wave stress and erosion. And tidal marshes provide nursery grounds for countless species of fish and invertebrates. They are among the most biologically productive ecosystems in the world, producing more biomass per area than most other ecosystems.

Whereas researchers have been studying the fertility and biologic productivity of marshes for many years, they have only recently started to determine how these coastal wetlands grow and erode. As sea level rises, we need to know the threshold at which marshes can no longer grow fast enough to keep pace with rising waters. If the rate of sea-level rise doubles over the next 100 years—or quadruples, as some more extreme models project—tidal marshes and coastal ecosystems will likely experience unprecedented changes. Some may disappear altogether. Our coast may return to its condition at the end of the last glaciation, 11,000 years ago, when sea level was rising too fast for marshes to be established....

How is the shoreline changing with time and geography?

Many studies of nearshore processes have been conducted on long, straight shorelines, and scientists have made some progress in understanding how waves, sandbars, and currents interact in simplified situations. But the mechanisms driving shoreline change are not well understood in regions where the nearshore region has complicated seafloor topography, inlets, or headlands—which means most beaches.

Waves traveling across the continental shelf are reflected, refracted, amplified, and scattered by underwater topography, and research has suggested that erosional hotspots along the coast are often the result of these seafloor formations. Banks, shoals, canyons, and even different types of sediment cause waves to decay and break differently. Wave-induced currents cause sediments to erode and accrete and reshape the seafloor near the coast, changing how future waves will evolve.

The complex dynamics between waves and seafloor evolution need to be unraveled before we can make predictions about changes to the shoreline. We need to build a network of wave-measuring instruments along different coastlines and feed those measurements into computer models of how the shoreline reacts to waves and currents. These models will help us make predictions about how water might circulate and how sediment might move in response to those different underwater formations.

How will barrier islands respond to sea level rise?

Barrier islands account for approximately 15 percent of the world's shoreline, and they dominate the Atlantic and Gulf coasts of the United States. Built by the action of waves and currents, these narrow ridges of sand usually run parallel to the mainland, protecting the coast from erosion. These natural barriers are bisected by tidal inlets and channels, and they shelter back-barrier salt marshes, tidal flats and deltas, and mangroves. Though usually no more than a few meters above sea level, these islands are often covered with human developments.

The long-term fate of today's barrier islands is dependent on future sea-level rise. The latest report of the Intergovernmental Panel on Climate Change predicts that global warming will cause sea level to rise by 50 to 90 centimeters in the next 100 years. At the higher end of these estimates, many back-barrier marshes will struggle to keep up with the inundation.

Sand will move from barrier beaches to the nearshore underwater regions in order to re-establish equilibrium between the slope of the beach and the higher tides and waves. The water levels and topography behind these barriers could gradually or catastrophically change. Inlets will become more dynamic, while deltas will enlarge. Whole marshlands might disappear, being converted to tidal lagoons or bays. Catastrophic amounts of sand could be lost from some beaches.

To properly protect barrier beaches—or learn when to abandon them—we need to map and monitor them regularly. We also need to dig into the sediments of the coast to piece together the history of past changes. Such efforts will allow us to model how tidal systems are likely to respond to rising ocean waters.

What is the impact of storms?

Intense storms such as hurricanes, nor'easters, and typhoons often result in substantial loss of life and resources, yet we know little about the processes that govern their formation, intensity, and movement. Nor do we know much about their history, due to the relatively short history of reliable weather observations. With little data on how coastal systems have responded to storms in the past, we have been ill-equipped to model and project how climate and sea-level change will affect future storm trends.

Geological investigations of coastal environments can provide long-term records of environmental change. Evidence of past storms can be found in back-barrier sediments: When a storm washes sand over the dunes and into back bays and marshes, it forms dateable layers in the muddy sediments. Mapping regional occurrences of these "overwash" deposits can allow researchers to estimate the storminess of years past and help improve models of the probability of future storm strikes....

What can the past tell us about the future of the shoreline?

Natural records from a variety of sources—deep-sea sediments, ice sheets, corals, calcium carbonate formations in caves—show that abrupt environmental changes are common in Earth's history. Sea level rise rates during the past 11,000 years have been uncharacteristically steady, and may be ripe for change. That our coastlines have developed such remarkable diversity during these stable times (environmental stress usually promotes diversity; calm promotes homogeneity) suggests the shape of the shore is affected by a lot more than sea level.

Coasts are complex, transitional environments that respond to changes in both continental and deep ocean processes. The sediments on- and offshore are great recorders of this variability, yet these archives have yet to be systematically studied and compared with what we have learned from inland and deep-sea environmental proxies for climate.

The high stakes of high water

Resource managers and civic leaders have a great responsibility for managing the coast and human use of it, but they have not always had the best information available to make scientifically sound decisions. The link between sea-level rise and shoreline change, while undoubtedly present, remains controversial.

For this reason, coastal managers want more reliable data on sea-level rise. They need studies that apply our knowledge of basic processes to more complex, human-altered shorelines (seawalls, bulkheads, jetties, groins). They need scientific analyses of the effects of adding and removing sediments from the shoreline.

There is no doubt that sea level is rising. It's not the first time, and the rate at which it is changing may or may not be unusual. What is different this time is that humans have congregated along the shoreline without much awareness of how much or how soon the sands might shift. We have the ability to make better decisions about our lives along the coast. We just have to start making the measurements that can provide the right answers.

R. L. Evans

EVANS, R. L. "RISING SEA LEVELS AND MOVING SHORELINES." *OCEANUS*. WOODS HOLE OCEANOGRAPHIC INSTITUTION, NOVEMBER 16, 2004. <HTTP://WWW.WHOI.EDU/CMS/FILES/DFINO/2005/4/V43N1-EVANS_2388.PDF> (ACCESSED NOVEMBER 23, 2007).

SEE ALSO *Beach and Shoreline; Coastal Populations; Sea Level Rise; Small Islands: Climate Change Impacts.*

BIBLIOGRAPHY

Books

Committee on Mitigating Shore Erosion Along Sheltered Coasts, National Research Council. *Mitigating Shore Erosion Along Sheltered Coasts.* Washington, DC: National Academies Press, 2007.

Parry, Martin, et al, editors. *Climate Change 2007: Impacts, Adaptation, and Vulnerability. Contribution of Working Group II to the Fourth Assessment Report of the Intergovernmental Panel on Climate Change.* New York: Cambridge University Press, 2007.

Periodicals

Dean, Cornelia. "Expert Federal Panel Urges New Look at Land Use Along Coasts in Effort to Reduce Erosion." *The New York Times* (October 13, 2006).

Web Sites

Gornitz, Vivien. "Coastal Populations, Topography, and Sea Level Rise." *NASA, Goddard Institute for Space Studies,* March 2000. <http://www.giss. nasa.gov/research/briefs/gornitz_04/> (accessed October 25, 2007).

Park, Sung Bin. "A Sea of People to Show How Climate Change Will Alter New York City." *NYC Indymedia,* April 4, 2007. <http://www.transalt.org/press/media/2007/890.html> (accessed October 25, 2007).

Larry Gilman

Compact Fluorescent Light Bulbs

Introduction

A compact fluorescent light bulb is a device that creates light using about one fourth as much power as a conventional, incandescent light bulb for a given amount of light. Large amounts of electricity are used to power light bulbs in industrial countries. Because most electricity worldwide is generated by burning coal, which releases the greenhouse gas carbon dioxide (CO_2), replacing incandescent bulbs with compact fluorescent light bulbs (CFLs) can reduce the amount of greenhouse gases emitted, especially in warmer climates, and have an impact on the amount of global climate change. CFLs contain small amounts of the toxic metal mercury and are more expensive than incandescent light bulbs. They last longer than incandescents and, averaged over the lifetime of the device, cost less to run.

Compact fluorescent light bulbs use about one fourth as much power as conventional, incandescent light bulbs. Although they are more expensive to buy, they pay for themselves through the power they save. *Image copyright R. Mackay, 2007. Used under license from Shutterstock.com.*

WORDS TO KNOW

GREENHOUSE GASES: Gases that cause Earth to retain more thermal energy by absorbing infrared light emitted by Earth's surface. The most important greenhouse gases are water vapor, carbon dioxide, methane, nitrous oxide, and various artificial chemicals such as chlorofluorocarbons. All but the latter are naturally occurring, but human activity over the last several centuries has significantly increased the amounts of carbon dioxide, methane, and nitrous oxide in Earth's atmosphere, causing global warming and global climate change.

ULTRAVIOLET: Light that vibrates or oscillates at a frequency of between 7.5×10^{14} and 3×10^{16} Hz (oscillations per second), more rapid than the highest-frequency color visible to the human eye, which is violet (hence the term "ultraviolet," literally above-violet). Ultraviolet light is absorbed by ozone (O_3) in Earth's stratosphere. This absorption serves both to shield the surface from this biologically harmful form of radiation and to heat the stratosphere, with important consequences for the global climate system.

WATT: Unit of power or rate of expenditure of energy. One watt equals 1 joule of energy per second. A 100-watt light bulb dissipates 100 joules of energy every second, i.e., uses 100 watts of power. Earth receives power from the sun at a rate of approximately 1.75×10^{17} watts.

Historical Background and Scientific Foundations

Conventional light bulbs operate on the principle of heating a small wire or filament until it glows brightly. Most of the energy consumed by an incandescent bulb is turned into heat, not light. Fluorescent light bulbs operate on the principle that certain gas mixtures, such as mercury vapor mixed with xenon or argon, emit ultraviolet radiation (a form of light invisible to the human eye) when excited by an electric current. A coating on the inside of a glass tube filled with such a gas can absorb the ultraviolet radiation and re-radiate it as visible light.

Incandescent bulbs convert about 90% of the electricity they consume into heat, whereas fluorescent light bulbs convert only about 30% into heat. The result is that a fluorescent bulb uses much less electricity to provide a given amount of light. Heat from light bulbs is often undesirable. In air-conditioned buildings, for example, electricity must be purchased to remove the heat produced by interior lighting, so owners pay twice, once to make the unwanted heat and once to remove it.

Scientists first noticed the production of electromagnetic radiation by electrified gases in the late nineteenth century. The invention of the commercial fluorescent light bulb is credited to German inventor Edmund Germer (1901–1987), who in 1926 patented a fluorescent bulb that used an inner bulb coating to convert ultraviolet light to relatively pleasing white light.

Fluorescent lights have traditionally been designed as long tubes, either straight or looping, because lower electric currents (which are easier to produce and safer for the consumer) are needed to produce a given amount of light from a longer tube. Small or "compact" fluorescent light bulbs that could be screwed into a conventional light socket would require either complex, maze-like glassware to pack long gas paths into small volumes or high currents that would waste power.

In the 1970s, a number of inventors sought solutions to these design barriers. Several designs that worked in the laboratory were produced, but no commercially viable design was put forward until the idea of bending a tube into a double-spiral shape was invented by Edward Hammer at the General Electric Corporation in 1976. Although it was more difficult and expensive to make such tubes than to make conventional bulbs, gradual improvements in technique made it possible for spiral-bulb compact fluorescents to be marketed starting in 1995.

Impacts and Issues

CFLs cost much more per unit than incandescent light bulbs, but last longer: about 7,500 hours versus only 1,000 for an incandescent bulb. Because they use less power, burden air conditioning less, and last longer, they end up costing less despite their higher up-front cost. A savings of $30 or more per bulb is cited by the U.S. Environmental Protection Agency (EPA) when CFLs are used instead of traditional incandescent bulbs. Because most electricity is generated by burning coal and CFLs save electricity, CFLs tend to cause less carbon dioxide to be emitted, which helps mitigate global climate change.

Globally, electric lighting causes carbon dioxide emissions equivalent to 70% of those from passenger vehicles. Thus, if every American home replaced just one incandescent bulb with a CFL, the carbon dioxide savings would be roughly equivalent to taking 800,000 cars off the road. Because of the cost and other advantages, some governments have considered, or have taken, action to speed the replacement of incandescent with fluorescent bulbs. In 2007, Australia became the first nation to announce that it would phase out incandescent bulbs entirely by 2012. Also in 2007, California was considering legislation that would ban the sale of incandescent light bulbs between 25 watts and 150 watts.

American consumers have been slow to adopt CFLs: only 2% to 5% of the 2 billion light bulbs sold in the United States each year are CFLs. Critics have pointed out that CFLs, like all fluorescent bulbs, contain the highly toxic metal mercury—about 5 milligrams (mg) per bulb. About 600 million fluorescent bulbs containing a total of 13,600 kg (30,000 lb) of mercury are thrown into U.S. landfills every year. However, because coal-burning also releases mercury and because CFLs prevent the burning of so much coal, an incandescent bulb causes the release, on average, of about 3.7 times more mercury per hour of lighting provided than does a CFL.

A more efficient, less toxic, and longer-lasting lighting technology—light-emitting diodes (LEDs)—is currently under development. LEDs remain relatively expensive, however, and are unlikely to displace CFLs in most applications in the near future.

Because of their mercury content, broken CFLs should not be touched with bare hands. They should also be recycled as toxic waste, not dumped in ordinary trash. Such dumping is illegal in California and several other U.S. states.

SEE ALSO *Energy Efficiency; Solar Illumination.*

BIBLIOGRAPHY

Periodicals

Kleiner, Kurt. "Shades of Success." *Nature* 447, no. 7146 (June 14, 2007): 766–767.

Yi, Matthewe. "Lawmaker Takes on Light Bulbs." *San Francisco Chronicle* (February 9, 2007): B1.

Web Sites

"Compact Fluorescent Bulbs and Mercury: Reality Check." *Popular Mechanics*, June 11, 2007. <http://www.popularmechanics.com/blogs/home_journal_news/4217864.html> (accessed August 6, 2007).

"Compact Fluorescent Light Bulbs." *U.S. Environmental Protection Agency (EPA)*. <http://www.energystar.gov/index.cfm?c=cfls.pr_cfls> (accessed August 6, 2007).

Larry Gilman

Continental Drift

Introduction

Continental drift is the slow movement of continents over the surface of Earth. The geological theory explaining continental drift is plate tectonics. The large, more or less rigid rafts of rock on which the continents float are called plates. Tectonics are any large-scale processes that shape Earth's crust, from the Greek *tekton* for "builder." Earth-shaping processes that involve plates are therefore termed "plate tectonics."

Over geological time—millions or billions of years—continental drift has a strong effect on climate, both local and global. Rearrangement of the layout of oceans and continents changes the ocean circulation pattern, leading to warming or cooling. In addition, the regional climate of a land mass changes gradually as the land moves toward or away from the equator. When continents temporarily stick together in larger masses, climate is changed over large areas by shifted rainfall patterns (the interiors of large continents tend to be dry). Volcanoes, earthquakes, and the creation of mountain ranges are all caused by plate tectonics, and these can also affect regional or global climate. Volcanoes, for example, can affect climate by adding greenhouse gases to the atmosphere.

Historical Background and Scientific Foundations

During the nineteenth century, it seemed obvious to geologists that continents are too big and stable to move. German scientist Alfred Wegener (1880–1930) proposed the concept of continental drift in his 1915 book *The Origins of Continents and Oceans.* He suggested that the puzzle-piece match between the east and west sides of the Atlantic Ocean, as well as other coastal coincidences, was too close to be a mere accident, and explained this correspondence by proposing that the continents had once been joined together in a single huge landmass. He proposed that continental drift could explain the mystery of paleoclimate—the fact that some parts of Earth, as shown by their fossils, were once much warmer or colder than they are today. If they were once located near the equator or poles and had since drifted to other latitudes, paleoclimate would be explained.

Wegener's theory was rejected by American scientists, who condemned it for trying to explain too much in a single framework—fossils, paleoclimate, mountain-building, and more. Wegener had no convincing explanation for why the continents might move about, which counted against his view. However, European scientists tended to be more receptive. Scientific research in the 1920s through the 1940s helped make Wegener's theory more plausible by showing that low-level radioactivity is present throughout Earth's rocks, which could supply heat to make the interior of Earth soft and mobile. British geologist Arthur Holmes (1890–1965) proposed that slow, gigantic convection currents in the earth might drive continental drift.

Convection currents occur in any liquid that is heated from below. The heated liquid expands and rises to the surface, where it gives up some of its heat. It becomes denser as it cools, then sinks. A heated fluid is thus often self-organized into cells or rotating masses of fluid—convection currents. The Earth's interior, heated from within by radioactive elements and cooled on the surface by radiation of heat into space, acts like such a fluid, although a slow-moving one.

In Holmes's view, which is now widely accepted, the continents float on convection currents in Earth's mantle like patches of scum in a boiling pot—drifting about, bumping against each other, separating again in continuous movement. In the 1950s, studies of rock magnetism showed that either the continents or the poles had moved in the deep past. (In fact, both have.) It was theorized that fresh crust wells up along the mid-ocean mountain ridges, flowing away in twin sheets in opposite directions from a central line. So, for example, the Old World on one side of the Atlantic Ocean is being pushed eastward by a sheet of ocean floor

WORDS TO KNOW

BIOGEOCHEMISTRY: The study of how substances and energy are exchanged between living things and the nonliving environment.

CONVECTION CURRENT: Circular movement of a fluid in response to alternating heating and cooling.

CRUST: The hard, outer shell of Earth that floats upon the softer, denser mantle.

JURASSIC PERIOD: Unit of geological time from 200 million years ago to 145 million years ago, famous in popular culture for its large dinosaurs. Global average temperature and atmospheric carbon dioxide concentrations were both much higher during the Jurassic than today.

MANTLE: Thick, dense layer of rock that underlies Earth's crust and overlies the core.

MILANKOVITCH CYCLES: Regularly repeating variations in Earth's climate caused by shifts in its orbit around the sun and its orientation (i.e., tilt) with respect to the sun. Named after Serbian scientist Milutin Milankovitch (1879–1958), though he was not the first to propose such cycles.

PALEOCLIMATE: The climate of a given period of time in the geologic past.

PANGEA: A supercontinent that was assembled from Gondwana, Euramerica, Siberia, and the Cathaysian and Cimmerian terranes. The assembly of Pangea lasted from the Late Carboniferous to the Middle Triassic, while the break up of this supercontinent began in the Jurassic and has continued to the present. The term was first used by Alfred Wegener to refer to the supercontinent of the Mesozoic. It can also be spelled "Pangaea" and comes from the Greek, meaning "all lands."

PERMIAN PERIOD: Geological period from 299 to 250 million years ago. During the Permian, all of Earth's continental plates were united in the supercontinent Pangea.

PLATE TECTONICS: Geological theory holding that Earth's surface is composed of rigid plates or sections that move about the surface in response to internal pressure, creating the major geographical features such as mountains.

THERMOHALINE CIRCULATION: Large-scale circulation of the world ocean that exchanges warm, low-density surface waters with cooler, higher-density deep waters. Driven by differences in temperature and saltiness (halinity) as well as, to a lesser degree, winds and tides. Also termed meridional overturning circulation.

TRIASSIC PERIOD: Geological period from 251 million years ago to 199 million years ago. Global climate was particularly warm during the Triassic; the poles were ice-free and all Earth's continental plates were clumped into the supercontinent Pangea.

generated at the mid-Atlantic ridge, while the New World is being pushed the other way by a twin sheet. If this is so, then as Earth's magnetic field reverses every few hundred thousand years, the magnetism in the rocks of the ocean floor should record these reversals in alternating stripes parallel to the central ridge.

In 1966, data showed that the predicted magnetic striping of the Atlantic Ocean floor does exist. From that time on, the theory of plate tectonics became a fundamental principle of modern geology.

As Wegener had believed, continental drift explains some aspects of ancient climate change (paleoclimate). About 225 million years ago, all the continental plates were collected in a single giant mass, a supercontinent that geologists have named Pangea (Greek for "all lands"). Because atmospheric moisture comes primarily from ocean evaporation, the central parts of large continents tend to be dry, and this was the case with Pangea. It was surrounded by a single giant world-ocean, Panthalassa (Greek for "all-sea").

The formation of this supercontinent—the latest in a long series of supercontinents that have formed and broken up over the 4.5-billion year history of Earth—caused a global ice age. Mountains were pushed up as continental plates crammed together. This caused an increase in erosion from streams running downhill, which changed ocean biogeochemistry so that carbon dioxide was removed from the atmosphere. Removing carbon dioxide reduced the natural greenhouse effect, chilling the planet. Glaciers spread from mountainous areas and covered large parts of Pangea.

The location of land and sea areas is a major influence on global climate, because water (being darker) is a more effective absorber of solar energy than is land. Therefore, the more of the global ocean area that happens to be in the tropics, the warmer the world will tend to be. Another effect of continental layout is that the continents determine where ocean currents can flow. The ocean's largest currents are thermohaline, that is, governed by temperature (thermo-) and saltiness (-haline). Tropical waters gain heat, then eventually flow northward as surface currents to the poles, where they lose heat and gain salt that has been pushed out of freezing sea-ice. Denser because it is cooler and saltier, this northern water sinks and flows back to the tropics through the deep ocean, eventually to rise again in the tropics and complete the loop. If continents shift so that north-south thermohaline circulations are blocked, heat transfer from the tropics to the poles will be slowed, making the polar regions colder.

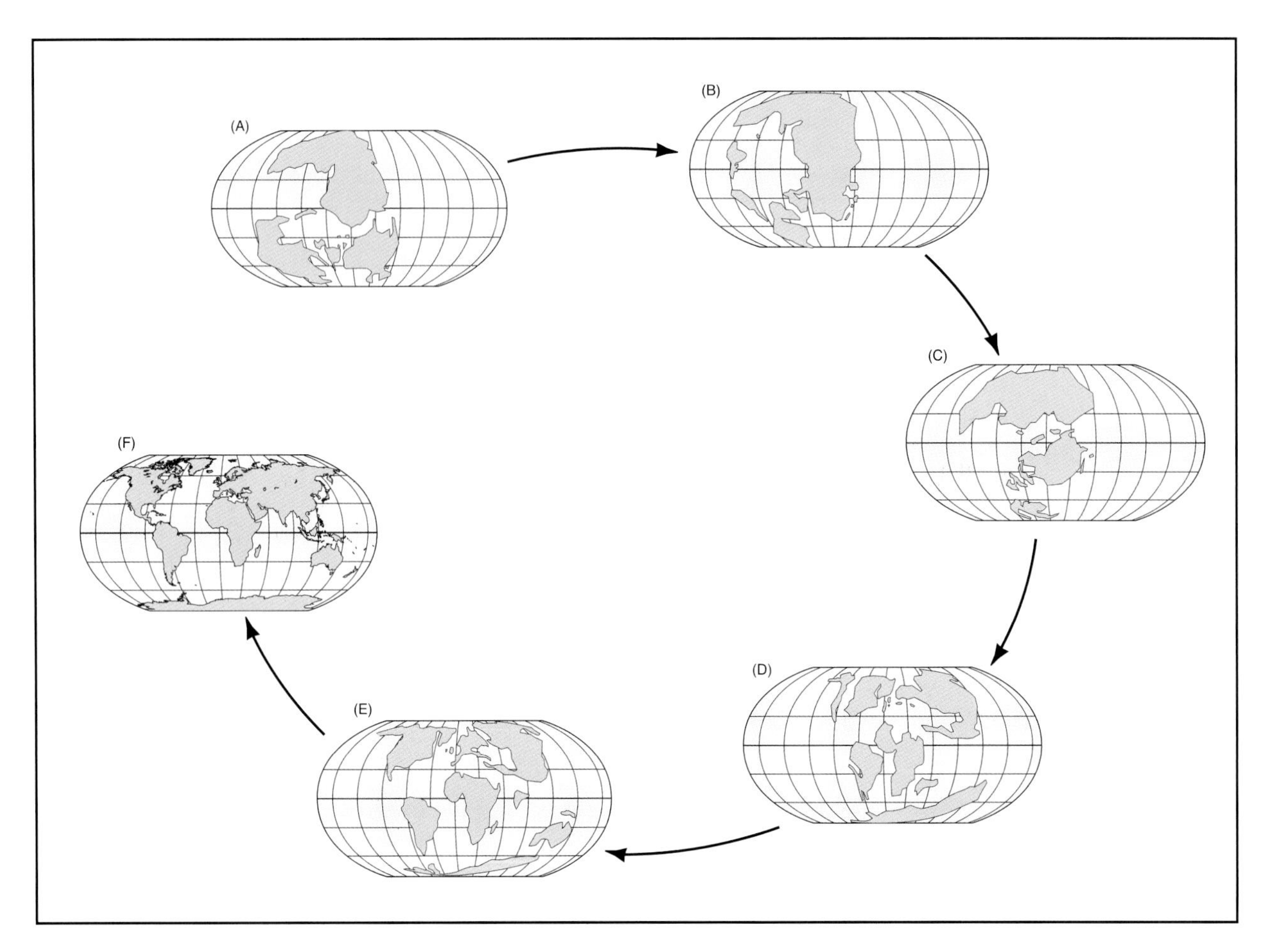

The distribution of landmasses at various points in Earth's history illustrate the theory of continental drift: (A) 320 million years ago; (B) 250 million years ago; (C) 135 million years ago; (D) 100 million years ago; (E) 45 million years ago; and (F) present. *The Gale Group.*

During the Triassic period, about 200 million years ago, Pangea began to break up. Ocean circulations changed so that tropical waters could now circle the globe around the equator without running into a supercontinent. This allowed the ocean water to spend more time in the tropics and grow warmer. It is not understood how these changes affected the poles, but it is known that from 100 to 70 million years ago, during the Jurassic period, the polar regions were warm enough to support forests. As the continents continued to drift apart, the world's present system of ocean currents arose, establishing today's climate system.

Impacts and Issues

Whether land or sea dominates in the tropical areas is what geophysicists call a first-order control; that is, like the geometry of the solar system and the energy output of the sun, it has a strong effect on Earth's climate. Changes in land-sea layout due to continental drift can cause regional average temperature changes greater than 45°F (25°C) even over geologically brief time-spans, that is, over a few million years. Changes in ocean circulation caused by rearrangement of the continents are second-order controls, that is, they produce changes up to 27°F (15°C). Milankovitch cycles—which are determined by regular, recurrent shifts in Earth's orbit happening over many thousands of years—are third-order controls, causing temperature changes no greater than 18°F (10°C). Volcanic eruptions, El Niño, La Niña, and atmospheric carbon-dioxide concentrations are fourth-order controls, producing temperature fluctuations usually less than 9°F (5°C).

Continental drift is too slow to be an influence on the global climate changes that are now being seen. The distance between North America and Europe, for example, is increasing at only about 1 in (2.54 cm) per year.

Primary Source Connection

Incoming solar radiation can either be reflected, absorbed, or reradiated. The amount of solar radiation absorbed compared to that reflected or reradiated is referred to as the global heat budget. The following excerpt from an article in the journal *Geografiska Annaler* discusses the potential effects of plate tectonics on landmass area, ocean currents, and Earth's heat budget. Dramatic changes in each could affect climate change.

GLOBAL HEAT BUDGET, PLATE TECTONICS, AND CLIMATE CHANGE

Introduction

Knowledge of the controlling factors of the heat budget of the Earth is critical to the understanding of the past, present and future climate of our environment. This includes the nature and origin of ice ages, causes of climatic changes and the potential effects of the works of man, e.g., the increasing CO_2 content of the atmosphere.... Many theories have been suggested, ranging from the effects of massive clouds of volcanic dust, ... the presence of an open Arctic Ocean,... the rise of the Tibetan Plateau,... to the Milankovitch theory of orbital cycles, but none satisfactorily explains how the heat budget works.

Fortunately, recent advances in several fields have yielded additional information that suggests a solution to the problem. Since the Earth has suffered periodic, widespread glaciations with intervening warmer periods for at least 2000 Ma,... it must be in some form of thermal equilibrium, with the climate fluctuating in response to changes in certain critical controls. This paper will show that the new evidence concerning the difference in heat absorption by land and water, the transport of excess heat polewards from the tropics, and the changes in distribution of land and sea resulting from plate tectonics, appear to explain the major fluctuations in the temperatures recorded in the geological record during the last 350 Ma. However, no sequence of climatic change in one place can be used to determine the variations in mean global heat budget during this time period.

Plate tectonics and the heat budget of the Earth

The original theory of continental drift of Wegener (1924) has evolved into the concept of plate tectonics, which regards the surface of the Earth as consisting of plates, most of which are in motion. This implies that the distribution of land and water is always changing. The area between the two tropics is the primary heat-absorbing zone: the larger the percentage of ocean in the primary heat-absorbing zone, the higher the absorbed heat and the sea-surface temperatures will be there, and the greater the potential for heat transfer to the poles. At present, about 30% of the heat absorbed in the primary heat-absorbing zone is transferred polewards.

The actual distribution of land and sea will also be critical to the efficient transfer of heat. If land areas block the movement of the warm and cold ocean currents, heat transfer polewards will be greatly reduced. Conversely, if the ocean currents can travel freely into and out of the polar regions, the heat transfer will be very efficient, producing warm temperature conditions in the adjacent polar lands.

Stuart A. Harris

HARRIS, STUART A. "GLOBAL HEAT BUDGET, PLATE TECTONICS, AND CLIMATE CHANGE." *GEOGRAFISKA ANNALER* 84A (2002): 1–9.

SEE ALSO *Great Conveyor Belt; Milankovitch Cycles; Snowball Earth.*

BIBLIOGRAPHY

Periodicals

Harris, Stuart A. "Global Heat Budget, Plate Tectonics and Climatic Change." *Geografiska Annaler* 84A (2002): 1–9.

Web Sites

Sleep, Norman H. "Plate Tectonics and the Evolution of Climate." *Review of Geophysics*, Vol. 33 Suppl. (1995). <http://www.agu.org/revgeophys/sleep00/sleep00.html> (accessed August 6, 2007).

Larry Gilman

Coral Reefs and Corals

Introduction

Corals are colonial animals, that is, collections of separate creatures that combine to form a single organism. The individual coral organisms—polyps, tiny creatures a fraction of an inch across that resemble a tube crowned with tentacles—inhabit cells or pores in a rocklike skeleton made of calcium carbonate (limestone, $CaCO_3$). To survive, corals harbor single-celled green plants called zooxanthellae in their body tissues. The relationship between coral polyps and zooxanthellae is symbiotic, or mutually beneficial: the coral polyps provide a home and nutrients for the algae, and the algae photosynthesize about 60% of the food the polyps need.

Coral polyps, by building new limestone structures or coral heads on top of older ones, can create reefs or submerged masses of coral many yards thick and wide and, in some cases, hundreds of miles long. These reefs support highly diverse populations of fish and other creatures and protect shorelines from storms and waves. When water temperatures are high, coral polyps lose their algae, which can cause the coral's death. Along with pollution or other pressures caused by human activity, increased acidity of the oceans caused by anthropogenic carbon dioxide in the atmosphere and global warming caused mostly by that same carbon dioxide are threatening coral reefs worldwide.

Historical Background and Scientific Foundations

Coral reefs are built by coral polyps, tiny animals that live in large colonies and which are themselves inhabited by thousands of microscopic golden-brown algae termed zooxanthellae. The tissues of the coral animals are mostly transparent and the limestone reef skeleton is white, so the coloration of a living coral reef comes primarily from the zooxanthellae. The zooxanthellae contain chlorophyll, the same chemical that colors green plants and enables them to exploit the energy of sunlight.

Using their chlorophyll, the zooxanthellae produce oxygen and food for the coral polyps. The zooxanthellae meet about 60% of the polyps' food needs, with the other 40% consisting of plankton caught by the polyps with their tentacles. The polyps in turn provide the algae with shelter and nutrients such as phosphorus and nitrogen. The polyps lodge in pores or cavities in the surface of a limestone skeleton and spread a living, skinlike covering over the skeleton's entire surface. A skinlike layer of their tissues is occupied by zooxanthellae, giving living coral its characteristic dark color.

Over centuries, successive generations of coral polyps manufacture the vast rocky heaps or shoals of limestone that are known as coral reefs. Most of a coral reef, its great hidden bulk, is dead: only the surface layer hosts living organisms. Since the zooxanthellae algae need sunlight, like any green plants, reefs grow only in shallow water; however, as sea bottoms subside slowly (or sea levels rise), dead coral can build up in thick layers below the living coral, which manages to stay up in the sunlit layer of the ocean. When a ring of thick coral builds up around an island as it sinks or as the ocean rises, it may become thousands of feet thick over millions of years and remain visible when the island itself is far below water. The origin of such structures, called coral atolls, was first explained by English naturalist Charles Darwin (1809-1882) in 1842.

Corals are often compared to underwater forests, manufacturing food and providing a complex environment in which other organisms can shelter. They are the most species-rich ecosystem in the oceans. About a million different species of plant and animal are harbored in a major reef, only about 10% of which have been described by science. Large fish often school along a reef while smaller ones swim in its crevices. Because of their beauty and accessibility, reefs are common destinations for tourists. For example, tourism at the Great Barrier

Coral reefs are especially popular with tourists. Some countries fear that global warming will continue to damage the coral, which in turn will impact their tourism business and economy. *Image copyright Dennis Sabo, 2007. Used under license from Shutterstock.com.*

Reef generated about $4.5 billion in income for Queensland, Australia, from 2004 to 2005.

Corals can survive only in water that is 64°F (18°C) or warmer, and so are found only in the tropical and semitropical seas of the world. At the same time, as was only recently discovered in the 1990s, they cannot tolerate excessively warm water. A few days of water temperatures that are a few degrees above ordinary summertime maximum cause a coral reef to give up most of its zooxanthellae. Since the zooxanthellae give the coral its color, heat-stressed coral takes on a whitish appearance, a phenomenon called coral bleaching. Bleached coral is not necessarily dead, but its growth is greatly slowed and the coral may take years to recover. When bleaching continues too long, coral dies.

Large bleaching events have occurred several times since the 1980s, notably in association with the El Niños of 1982-83, 1987-88, and 1997-98. Severe bleaching also struck in 2002 and 2005. Coral bleaching events occur naturally due to the recurring weather cycles known as El Niño and La Niña, so cycles of damage and recovery are normal for corals. Although El Niños have increased in number, intensity, and duration since the 1970s, there is not yet any clear link between this trend and global warming.

Impacts and Issues

Coral reefs are stressed by a number of human activities other than climate change, but climate change is likely to increase the damage caused by these other stresses. For example, over-harvesting of fish encourages the marine organisms on which many fish feed, such as seaweeds, to compete with corals for space. Encouraging coral's competitors injures the coral. Nutrient loading occurs when nitrogen and phosphorus from agriculture or sewage are mixed with the ocean, encouraging phytoplankton and seaweed to grow in the shallow waters required by corals. Phytoplankton darken the water, reducing photosynthesis in the zooxanthellae, and seaweeds compete with corals for space. Sediment loading occurs when small mineral particles washed to the sea by erosion interfere with polyp feeding and force the polyps to use up energy cleaning out the particles. Rapid burial under even a thin layer of sediment can kill a coral reef outright. By making bleaching events more common, global warming will probably make corals more vulnerable to these other forms of damage.

There is scientific dispute over whether corals will be able to evolve some resistance to increasing temperature. According to the adaptive bleaching hypothesis, corals shift the type of zooxanthellae on which they depend when repeatedly exposed to bleaching, thus adapting to higher temperatures. There is some evidence supporting this view. However, it is very uncertain whether, if such adaptation occurs, it could continue for more than a couple of degrees of warming in water temperature.

Although warming of the world's seas will make some presently cool regions suitable for coral, the area that will be made newly suitable for coral is forecast to be small compared to that which will be lost.

WORDS TO KNOW

ANTHROPOGENIC: Made by people or resulting from human activities. Usually used in the context of emissions that are produced as a result of human activities.

CORAL ATOLL: Low tropical island, often roughly ring-shaped, formed by coral reefs growing on top of a subsiding island. The rocky base of the atoll may be hundreds of feet below present-day sea level. Atolls, like other low-lying islands, are threatened with submergence by rapid sea-level rise caused by anthropogenic climate change.

CORAL BLEACHING: Decoloration or whitening of coral from the loss, temporary or permanent, of symbiotic algae (zooxanthellae) living in the coral. The algae give corals their living color and, through photosynthesis, supply most of their food needs. High sea surface temperatures can cause coral bleaching.

CORAL POLYP: Living organism that, as part of a colony, builds the rocky calcium carbonate ($CaCO_3$) skeleton that forms the physical structure of a coral reef.

EL NIÑO: A warming of the surface waters of the eastern equatorial Pacific that occurs at irregular intervals of 2 to 7 years, usually lasting 1 to 2 years. Along the west coast of South America, southerly winds promote the upwelling of cold, nutrient-rich water that sustains large fish populations, that sustain abundant sea birds, whose droppings support the fertilizer industry. Near the end of each calendar year, a warm current of nutrient-poor tropical water replaces the cold, nutrient-rich surface water. Because this condition often occurs around Christmas, it was named El Niño (Spanish for boy child, referring to the Christ child). In most years the warming lasts only a few weeks or a month, after which the weather patterns return to normal and fishing improves. However, when El Niño conditions last for many months, more extensive ocean warming occurs and economic results can be disastrous. El Niño has been linked to wetter, colder winters in the United States; drier, hotter summers in South America and Europe; and drought in Africa.

GREAT BARRIER REEF: World's largest coral reef system, located along the northeastern coast of Australia. The reef system, which is approximately 1,600 mi (2,600 km) long, contains 3,000 individual reefs and 900 islands and is threatened by sea-level rise and ocean warming caused by anthropogenic climate change.

LA NIÑA: A period of stronger-than-normal trade winds and unusually low sea-surface temperatures in the central and eastern tropical Pacific Ocean; the opposite of El Niño.

PH: Measures the acidity of a solution. It is the negative log of the concentration of the hydrogen ions in a substance.

SEDIMENT LOADING: Presence of moving mineral particles in rivers or streams. Faster-moving water can carry more sediment. Erosion increases sediment loading, limited by stream capacity for increased sediment loading.

SYMBIOTIC: A relationship or pattern of exchange that is mutually beneficial to two or more creatures. Algae are symbiotic with lichens and corals; intestinal bacteria are symbiotic with mammals.

ZOOXANTHELLAE: Algae that live in the tissues of coral polyps and, through photosynthesis, supply them with most of their food. The relationship is symbiotic, as the polyps supply the zooxanthellae with a hospitable environment. When water temperatures are too high, corals lose their zooxanthellae. If the loss is too pronounced for too long, the coral dies.

The greatest threat to corals from climate change, most experts agree, is not warming but the increasing concentration of carbon dioxide (CO_2) in the atmosphere. As of 2007, the oceans were absorbing about a third of all anthropogenic (human-released) CO_2. This was changing the chemistry of the oceans and thus affecting the ability of various organisms to produce calcium carbonate, which is the material not only of coral skeletons but of the shells of clams, whelks, mussels, and other shelly animals. When a CO_2 molecule dissolves in water, it combines with the water in such a way as to release hydrogen ions. The hydrogen ions make the water more acidic. Fewer carbonate ions (CO_3^{2-}) can exist stably in acidic water. It is carbonate ions, along with calcium ions, that shelly organisms, including coral polyps, use to produce their calcium carbonate shells. Fewer carbonate ions makes shell-building more difficult. There are far more calcium ions than carbonate ions, so carbonate ions are what limit the ability of shelly organisms to make shells in today's oceans.

Over the last 200 years, the average pH (a measure of acidity) of the ocean's waters has fallen by 0.1, a 30% increase in the number of hydrogen ions. This change has reduced the number of carbonate ions available to shelly organisms. At a certain acidity, the concentration of carbonate ions falls so low that corals have difficulty making their skeletons at all; this concentration may occur if today's atmospheric concentration of CO_2, about 375 parts per million, rises to over 500 parts per million. This much CO_2 will probably exist in the atmosphere by the end of the twenty-first century.

■ Primary Source Connection

Human impact on Earth's coral reefs is substantial. Global climate change is likely to exacerbate coral disease and reef death worldwide, with a negative impact on marine species and human economies dependent on healthy reefs. The following excerpt discusses the impacts of climate change on the health and development of coral reefs, highlighting problems such as bleaching and reef death.

The Pew Center on Global Climate Change is a United States-based think-tank that gathers scientific and economic data to inform policymakers on global climate change issues. The report's authors are researchers in atmospheric and marine sciences.

CORAL REEFS & GLOBAL CLIMATE CHANGE: POTENTIAL CONTRIBUTIONS OF CLIMATE CHANGE TO STRESSES ON CORAL REEF ECOSYSTEMS

The "Coral Reef Crisis"

Coral reefs have declined over the course of human history, culminating in the dramatic increase in coral mortality and reef degradation of the past 20–50 years. This "coral reef crisis" is well-documented and has stimulated numerous publications on the future of coral reefs and their vulnerability to environmental change. The causes of this crisis are a complex mixture of direct human-imposed and climate-related stresses, and include factors such as outbreaks of disease, which have suspected but unproven connections to both human activities and climate factors. By 1998, an estimated 11 percent of the world's reefs had been destroyed by human activity, and an additional 16 percent were extensively damaged in 1997–98 by coral beaching. Widespread coral bleaching, unknown before the 1980s, has brought recognition that reefs are threatened by global-scale climate factors as well as by more localized threats, and that different types of stress may interact in complex ways.

Although the crisis is widespread, individual reefs and even whole regions exhibit considerable variation in both health and responses to stress. The Caribbean region has been particularly hard-hit by problems, many of which are well-studied. Caribbean case studies and inter-ocean contrasts help to illustrate both the consistencies and the variations in coral reef responses to complex environmental changes.

Climate and Environmental Change

Over the past one to two centuries, human population growth and development have greatly altered not only local environments, but also the global environment as a whole. Major systematic changes include rising atmospheric concentrations of greenhouse gases (GHGs) that influence the earth's energy budget and climate. In addition, the global phosphorus and nitrogen cycles have accelerated because of artificial fertilizer use and massive changes in land use, the hydrologic cycle has been altered by river damming and water diversion as well as climate change, major natural ecosystems have been altered by fishing, forestry, and agriculture, and the ecological and biogeochemical implications of increased atmospheric CO_2 levels go well beyond the effects on global temperature.

Because coral reefs occur near the junction of land, sea, and atmosphere, their natural habitats experience both the marine and terrestrial results of any climatic change and are vulnerable to human activities....

Climatic Change Stresses to Coral Reefs

Global climate change imposes interactive chronic and acute stresses, occurring at scales ranging from global to local, on coral reef ecosystems.... Gas bubbles preserved in polar ice caps show that atmospheric CO_2 concentrations over the past 400,000 years have oscillated between about 180 and 310 parts per million volume, or ppmv; past temperature and sea-level variations mimic the CO_2 fluctuations, with relatively constant minimum (glacial period) and maximum (interglacial) values. Accompanying this CO_2 increase is an observed increase in temperature, and a decrease in pH of the surface ocean. IPCC projections show an even greater departure from geologically recent climates by the end of the present century.

Coral Bleaching

The atmosphere and the ocean have warmed since the end of the 19th century and will continue to warm into the foreseeable future, largely as a result of increasing greenhouse gas concentrations. El Niño-Southern Oscillation (ENSO) events have increased in frequency and intensity over the last few decades. This combination (warming and intense El Niño events) has resulted in a dramatic increase in coral bleaching.

"Bleaching" describes the loss of symbiotic algae by the coral or other host. Most of the pigments in the usually colorful corals depend on the presence of these plant cells. The living tissue of coral animals without algae is translucent, so the white calcium carbonate skeleton shows through, producing a bleached appearance. Bleaching is a general stress response that can be induced in both the field and the laboratory by high or low temperatures, intense light, changes in salinity, or by other physical or chemical stresses. Bleaching is the extreme case of natural variation in algal population density that occurs in many corals.

Three types of bleaching mechanisms are associated with high temperature and/or light: "animal-stress bleaching," "algal-stress bleaching," and "physiological bleaching." Although all are important to understanding climate-coral interactions, two are particularly relevant to present concerns: algal-stress bleaching, an acute response to impairment of photosynthesis by high temperature coupled with high light levels; and physiological bleaching, which reflects depleted reserves, reduced tissue biomass, and less capacity to house algae as a result of the added energy demands of sustained above-normal

White coral syndrome is shown in the Great Barrier Reef near Australia in 2007. Studies indicate the coral is being impacted by climate change, disease, and coastal development. AP Images.

temperatures. A rising baseline in warm-season sea-surface temperatures on coral reefs suggests that physiological bleaching is at least partly to blame in some bleaching events (e.g., in the Caribbean in 1987 and on the Great Barrier Reef in 2001). Such chronic temperature stress may also underlie some less obvious causes of reef decline, such as low rates of sexual reproduction.

The temperature threshold for bleaching is not an absolute value, but is relative to other environmental variables (especially light) and to the duration and severity of the departure from the normal temperature conditions of a reef. Bleaching due to thermal stress is not, therefore, limited to areas of normally high water temperature. However, regions where higher temperatures are the norm seem likely to be more vulnerable to increased physiological bleaching.

Coral bleaching events of greatest concern are acute episodes of high mortality and protracted debilitation of survivors in the form of diminished growth and reproductive rates. Corals with branching growth forms, rapid growth rates, and thin tissue layers appear to be most sensitive to bleaching, and usually die if seriously bleached. Slow-growing, thick-tissued, massive corals appear to be less sensitive and commonly recover from all but the most extreme episodes. Bleaching thus selectively removes certain species from reefs and can lead to major changes in the geographic distribution of coral species and reef community structures.

Global Warming and Reef Distribution

The global distribution of reef-building corals is limited by annual minimum temperatures of ~18°C (64°F). Although global warming might extend the range of corals into areas that are now too cold, the new area made available by warming will be small, and the countervailing effects of other changes suggest that any geographic expansion of coral reefs will be very minor.

Coral reefs require shallow, clear water with at least some hard seafloor, and their propagation depends primarily on ocean currents. The west coasts of North and South America, Europe, and Africa experience cool water flowing toward the equator and are thus "upstream" from potential sources, causing restricted distributions of coral reefs. In areas such as the southeastern United States and near the Amazon River, reef expansion along the coast is blocked by muddy coastal shelves, river deltas, and turbid water. Only southern China, Japan, Australia, and southern Africa present geographically realistic opportunities for reef expansion. Additionally, sea-surface temperature (SST) gradients are very steep in the vicinity of 18°C (the annual minimum temperature threshold for coral reef growth), and ocean model projections suggest that SST warming associated with doubled CO_2 will only move the 18°C contour by a few hundred kilometers, especially in the critical western boundary areas. The overall positive effects of warming on habitat availability and ecosystem distribution will be very minor compared to the overall negative effects.

Reduced Calcification Potential

The oceans currently absorb about a third of the anthropogenic CO_2 inputs to the atmosphere, resulting in significant changes in seawater chemistry that affect the ability of reef organisms to calcify. Photosynthesis and respiration by marine organisms also affect seawater CO_2 concentration, but the overwhelming driver of CO_2 concentrations in shallow seawater is the concentration of CO_2 in the overlying atmosphere. Changes in the CO_2 concentration of seawater through well-known processes of air-sea gas exchange alter the pH (an index of acidity) and the concentrations of carbonate and bicarbonate ions. Surface seawater chemistry adjusts to changes in atmospheric CO_2 concentrations on a time scale of about a year. Projected increases in atmospheric CO_2 may drive a reduction in ocean pH to levels not seen for millions of years.

Many marine organisms use calcium (Ca^{2+}) and carbonate (CO_3^{2-}) ions from seawater to secrete $CaCO_3$ skeletons. Reducing the concentration of either ion can affect the rate of skeletal deposition, but the carbonate ion is much less abundant than calcium, and appears to play a key role in coral calcification. The carbonate ion concentration in surface water will decrease substantially in response to future atmospheric CO_2 increases, reducing the calcification rates of some of the most important $CaCO_3$ producers. These include corals and calcareous algae on coral reefs and planktonic organisms such as coccolithophores and foraminifera in the open ocean. . . .

Sea Level

The predicted rise of sea level due to the combined effects of thermal expansion of ocean water and the addition of water from melting icecaps and glaciers is between 0.1 and 0.9 meter (4–36 inches) by the end of this century. Sea level has remained fairly stable for the last few thousand years, and many reefs have grown to the point where they are sea-level-limited, with restricted water circulation and little or no potential for upward growth. A modest sea-level rise would therefore be beneficial to such reefs. Although sea-level rise might "drown" reefs that are near their lower depth limit by decreasing available light, the projected rate and magnitude of sea-level rise are well within the ability of most reefs to keep up. A more likely source of stress from sea-level rise would be sedimentation due to increased erosion of shorelines.

Robert W. Buddemeier et al.

BUDDEMEIER, ROBERT W., ET AL. *CORAL REEFS & GLOBAL CLIMATE CHANGE: POTENTIAL CONTRIBUTIONS OF CLIMATE CHANGE TO STRESSES ON CORAL REEF ECOSYSTEMS*, PEW CENTER ON GLOBAL CLIMATE CHANGE. FEBRUARY 2004.

SEE ALSO *El Niño and La Niña; Extinction; Great Barrier Reef; Sea Temperatures and Storm Intensity.*

BIBLIOGRAPHY

Books

Parry, M. L., et al, eds. *Climate Change 2007: Impacts, Adaptation and Vulnerability: Contribution of Working Group II to the Fourth Assessment Report of the Intergovernmental Panel on Climate Change.* New York: Cambridge University Press, 2007.

Periodicals

Brown, Barbara E., and John C. Ogden. "Coral Bleaching." *Scientific American* (January 1993): 64-70.

Ellperin, Jullet. "Yes, the Water's Warm . . . Too Warm." *The Washington Post* (July 15, 2007).

Web Sites

Buddemeier, Robert W., et al. "Coral Reefs & Global Climate Change: Potential Contributions of Climate Change to Stresses on Coral Reef Ecosystems." *Pew Center on Global Climate Change*, February 2004. <http://www.pewclimate.org/docUploads/Coral_Reefs.pdf> (accessed October 26, 2007).

Pomerance, Rafe. "Coral Bleaching, Coral Mortality, and Global Climate Change." *Bureau of Oceans and International Environmental and Scientific Affairs, U.S. Department of State*, March 5, 1999. <http://www.state.gov/www/global/global_issues/coral_reefs/990305_coralreef_rpt.html> (accessed October 26, 2007).

"Reef 'At Risk in Climate Change.'" *Australian Research Council Centre of Excellence, Coral Reef Studies*, April 10, 2007. <http://www.coralcoe.org.au/news_stories/climatechange.html> (accessed October 26, 2007).

Larry Gilman

Cosmic Rays

Introduction

Cosmic rays are naturally occurring high-energy particles—protons, helium nuclei, and electrons—that travel near the speed of light. Some scientists have argued that cosmic rays may cause cloud droplets to form in Earth's atmosphere. If so, the cloudiness of Earth's atmosphere might increase when the sun is emitting more cosmic rays or when the solar system is passing through a part of the galaxy where cosmic rays are more abundant. Increased clouds could affect climate.

As of 2007, most climate scientists did not agree that cosmic rays are a significant influence on Earth's climate, but a small number of scientists disagreed. An experimental device under construction at the European Organization for Nuclear Research (CERN) particle accelerator laboratory in Switzerland, due to be completed in 2010, could decide the question.

Georges-Henri Lemaitre (1894–1966, right) proposed a theory concerning cosmic rays in the early 1930s that brought him wide scientific acknowledgement. A Catholic priest and astronomer from Belgium, he is shown here meeting with Albert Einstein (1879–1955, left) in Pasadena, California, in 1932. *AP Images.*

Historical Background and Scientific Foundations

Scientists have long known that slight changes in the energy output of the sun might affect the climate of Earth. American astronomer Jack Eddy pointed out in the 1970s that a cold period in Earth's climate during the seventeenth and eighteenth centuries, the Little Ice Age, coincided with a historic low point in the number of sunspot numbers known as the Maunder minimum. In 1997, Danish scientists Henrick Svensmark and Eigil Friis-Christensen noted that Earth's global cloudiness had decreased by 3% during the period 1987–1990, at the same time that cosmic rays had decreased by 3.5% due to the sun's regular cycle of activity.

They proposed that cosmic rays might increase cloud cover by the following mechanism: Because cosmic rays have high energy, they can strip electrons from many numbers of atoms when they strike Earth's atmosphere. These atoms, now bearing positive electric charges, might cause water to start condensing out of humid air into cloud droplets. This, in turn, might increase the number of clouds, their density, or both.

The cosmic-ray theory was controversial because most scientists agree that the global warming occurring rapidly today is caused by an increase in the amount of carbon dioxide in Earth's atmosphere, not by changing levels of solar activity. In 2003, two other scientists, Nir J. Shaviv and Ján Veizer, took the dispute to a new level by arguing that over the last 545 million years, about two thirds of Earth's changes in climate could be attributed to rises and falls in the number of cosmic rays striking Earth from outside the solar system. Independently of changes in cosmic ray flow from our own star, rises and falls in cosmic ray abundance also occur as the solar system's orbital path around the center of the galaxy takes it into and out of our galaxy's spiral arms over millions of years.

Impacts and Issues

There is evidence from several independent sources that changing energy output from the sun has affected Earth's climate in the past. Moreover, the solar hypothesis and the carbon dioxide hypothesis do not necessarily exclude each other: Earth's climate might be influenced by multiple factors. As of 2008, however, most scientists had not been convinced that galactic and solar cosmic rays ever had a significant influence on climate.

In 2004, German scientist Stefan Rahmstorf and a group of 10 other scientists from the United States, Switzerland, and elsewhere published a paper replying to the 2003 paper by Shaviv and Veizer arguing that cosmic rays dominate Earth's climate. The new paper was widely reported in the scientific press as a refutation of the cosmic-ray idea. In 2007, British scientists Mike Lockwood and Claus Fröhlich published a study finding that "the observed rapid rise in global mean temperatures seen after 1985 cannot be ascribed to solar variability," no matter what mechanism for solar influence on Earthly climate is invoked or how that influence might be amplified, as for example by cosmic rays. Present-day climate change, the authors concluded, is due to human activity, not solar influence.

Nevertheless, the possibility of a cosmic-ray connection to climate during other periods, whether slight or great, remains open. At the CERN particle accelerator laboratory in Geneva, Switzerland, an experiment called CLOUD (Cosmics Leaving Outdoor Droplets) is under construction. Due to be completed in 2010, CLOUD will enable scientists to observe the effects of artificial cosmic rays on air and water vapor and may settle the question of whether cosmic rays can create clouds. Even if they can, some scientists assert, it is unlikely that solar cosmic rays could explain the recent warming of Earth.

WORDS TO KNOW

COSMICS LEAVING OUTDOOR DROPLETS (CLOUD): Experiment at the CERN particle physics laboratory in Geneva, Switzerland, designed to investigate whether cosmic rays (high-energy particles from outer space) can influence Earthly weather and climate by triggering the formation of cloud particles. Began collecting data in 2006.

SUNSPOTS: Comparatively dark, cool patches that appear on the sun's surface in synchrony with increased solar activity every 11 years. By interacting with stratospheric ozone, sunspot activity affects Earth's climate, mostly at high altitudes but subtly at the surface (perhaps a few tenths of a degree of warming in the Northern Hemisphere).

WATER VAPOR: The most abundant greenhouse gas, it is the water present in the atmosphere in gaseous form. Water vapor is an important part of the natural greenhouse effect. Although humans are not significantly increasing its concentration, it contributes to the enhanced greenhouse effect because the warming influence of greenhouse gases leads to a positive water vapor feedback. In addition to its role as a natural greenhouse gas, water vapor plays an important role in regulating the temperature of the planet because clouds form when excess water vapor in the atmosphere condenses to form ice and water droplets and precipitation.

SEE ALSO *Atmospheric Chemistry.*

BIBLIOGRAPHY

Periodicals

Kanipe, Jeff. "A Cosmic Connection." *Nature* 443 (2006): 141–143.

Lockwood, Mike, and Claus Fröhlich. "Recent Opposite Directed Trends in Solar Climate Forcings and the Global Mean Surface Air Temperature." *Proceedings of the Royal Society A*, May 2007. Also available online at http://www.pubs.royalsoc.ac.uk/media/proceedings_a/rspa20071880.pdf (accessed August 8, 2007).

Rahmstorf, Stefan, et al. "Cosmic Rays, Carbon Dioxide, and Climate." *Eos* 85 (2004): 38–40.

Schiermeier, Quirin. "No Solar Hiding Place for Greenhouse Skeptics." *Nature* 448 (2007): 8–9.

Shaviv, Nir J., and Ján Veizer. "Celestial Driver of Phanerozoic Climate?" *GSA [Geological Society of America] Today* (July, 2003): 4–10.

Web Sites

"Cosmic Rays Are Not the Cause of Climate Change, Scientists Say." *American Geophysical Union*, January 21, 2004. <http://www.agu.org/sci_soc/prrl/prrl0405.html> (accessed August 8, 2007).

Larry Gilman

Cow Power

■ Introduction

Cow power refers to the generation of relatively clean, renewable electricity from the anaerobic decomposition of livestock manure. In a typical system, manure is fed into an anaerobic digester, basically a covered tank where bacteria break down organic materials in the absence of oxygen and produce biogas, which can then be burned in a modified natural gas engine that spins an electric generator. One cow's manure—about 30 gallons (114 l) a day—can produce enough electricity to light two 100-watt light bulbs 24 hours a day. Biogas can also be produced from the anaerobic decomposition of sewage sludge, landfill waste, or other organic materials.

Biogas is typically 50–70% methane (CH_4), a high-energy fuel, and 30–50% carbon dioxide (CO_2). Methane is about 21 times more effective at heating Earth's atmosphere than CO_2 and may account for 23% of human-caused global warming. About 8% of methane emissions generated from human activities comes from the anaerobic

Agriculture produces significant amounts of methane, a greenhouse gas. Scientists, researchers, and others have looked into ways to reduce the emission of methane gas into the atmosphere. Here, a rancher in California stands near a trough of flowing manure, which is on its way to a methane digester to be processed and turned into electricity. *AP Images.*

Juehnde, Germany, became the first village in the country to become energy self-sufficient through the creation of a bioenergy electrical plant. Using cow manure, wood chips, and plant remains, the plant creates heat and electricity. *Sean Gallup/Getty Images.*

decomposition of livestock manure. Cow power and other biogas collecting systems keep methane out of the atmosphere by burning the gas, which converts it into CO_2 and so reduces its impact on global climate.

Historical Background and Scientific Foundations

Biogas from anaerobic digestion was first used to heat bathwater in Assyria in the tenth century BC and in Persia in the sixteenth century. Through the 1700s, scientists determined that the decomposition of organic material could produce flammable gas, and in the early 1800s isolated methane—the primary component of natural gas, a fossil fuel often found in association with petroleum—as one of the gases produced by the anaerobic digestion of cattle manure. The first modern biodigester was built at a leper colony in Bombay, India, in 1859, and biogas from a sewage plant was used to light the streets of Exeter, England, after 1895.

Fuel shortages during World War II (1939–1945) and the energy crisis of the 1970s greatly increased interest in biogas energy in the United States and Europe, but early efforts in the United States to operate digesters were often plagued by design and construction problems. Over the 1990s and 2000s, increasing recognition of global climate change, the emergence of carbon regulations and carbon markets, climbing fossil fuel prices, and a growing worldwide push for renewable power sources have reinvigorated broad interest in harnessing biogas for energy.

The countries of the European Union were producing 32,122 gigawatt hours' worth of biogas by 2002, a 6% increase over 2001. In the United States, there were 125 operational anaerobic digesters by 2006. The U.S. Environmental Protection Agency (EPA) estimates that these systems produced 275 gigawatt hours of electricity and kept approximately 88,134 tons (80,000 metric tons) of methane out of the atmosphere in 2007 alone.

Biogas is also popular for cooking, heating, and producing electricity in developing countries, where small-scale systems better serve dispersed, rural populations than do centralized gas and power plants. In China, 7.5 million household biogas systems and 750 industrial-scale plants had been installed as of 2002. About 3.4 million family-sized biogas plants had been installed in India as of 2002.

Impacts and Issues

In addition to climate benefits, manure-based biogas-to-electricity systems can increase the income and long-term viability of agricultural operations like dairies, where commodity prices are often low. Such systems also control odors and help clean pathogens and weed seeds from manure by essentially cooking it with the concentrated heat from decay. The solid byproduct of the process can be used as animal bedding, saving farmers money, or sold as compost, generating cash income.

Closed biodigesters may also help farmers better comply with regulations protecting water from manure runoff.

Large systems can be prohibitively expensive, however, with traditional digesters costing as much as $3 million. Farms generally must be large, with a large number of animals contained in a small area, to collect enough manure for electricity sales to offset the infrastructure costs of generating power. To overcome this hurdle, some governments and utilities offer grants and cost-sharing programs to smaller farmers. Green pricing programs, where utilities purchase power from farmers at a premium and sell it at elevated cost to customers who choose to pay more because of its added environmental value, also assist deployment of biogas-to-electricity systems. The sale of carbon credits, which represent a biogas system's climatic benefits, may also help offset costs.

The Blue Spruce Farm in Bridport, Vermont, is an example of a success story. In 2005, the 2,000 cow dairy—the first to participate in Central Vermont Public Service (CVPS) Company's "Cow Power" program—produces enough electricity to power 300 to 400 average homes. CVPS provided incentives to offset the initial cost of system installation, and now purchases and sells the electricity at a premium. Overall, the Blue Spruce Farm has invested about $1.3 million in its system, but expects to recoup the cost in electricity sales over seven years and eventually to realize a profit.

SEE ALSO *Biofuel Impacts; Biomass; Methane; Renewable Energy.*

WORDS TO KNOW

ANAEROBIC: Lacking free molecular oxygen (O_2). Anaerobic environments lack O_2; anaerobic bacteria digest organic matter such as dead plants in anaerobic environments such as deep water and the digestive systems of cattle. Anaerobic digestion releases methane, a greenhouse gas.

BIOGAS: Methane produced by rotting excrement or other biological sources. It can be burned as a fuel.

CARBON CREDIT: A unit of permission or value, similar to a monetary unit (e.g., dollar, euro, yen) that entitles its owner to emit one metric ton of carbon dioxide into the atmosphere.

FOSSIL FUELS: Fuels formed by biological processes and transformed into solid or fluid minerals over geological time. Fossil fuels include coal, petroleum, and natural gas. Fossil fuels are non-renewable on the timescale of human civilization, because their natural replenishment would take many millions of years.

METHANE: A compound of one hydrogen atom combined with four hydrogen atoms, formula CH_4. It is the simplest hydrocarbon compound. Methane is a burnable gas that is found as a fossil fuel (in natural gas) and is given off by rotting excrement.

RENEWABLE ENERGY: Energy obtained from sources that are renewed at once, or fairly rapidly, by natural or managed processes that can be expected to continue indefinitely. Wind, sun, wood, crops, and waves can all be sources of renewable energy.

BIBLIOGRAPHY

Books

Smith, P., et al. "Agriculture." In *Climate Change 2007: Mitigation of Climate Change: Contribution of Working Group III to the Fourth Assessment Report of the Intergovernmental Panel on Climate Change*, edited by B. Metz et al. New York: Cambridge University Press, 2007.

Periodicals

Blank, Michelle. "Cow Power." *High Country News* (May 14, 2007).

Demirbas, M. F., and Mustafa Balat. "Recent Advances on the Production and Utilization Trends of Bio-fuels: A Global Perspective." *Energy Conversion and Management* 47 (2006): 2371-2381.

Dunn, David. "Utility Turns Biomass Into Renewable Energy." *Biocycle* (September 2004).

Martinot, Eric, et al. "Renewable Energy Markets in Developing Countries." *Annual Review of Energy and the Environment* 27 (2002): 309-348.

Mazza, Patrick. "Harvesting Clean Energy for Rural Development: Biogas." *Climate Solutions Special Report* (February 2002).

Raloff, Janet. "Cow Power." *Science News* (November 18, 2006).

Web Sites

"How Energy Happens." *CVPS Cow Power.* <http://www.cvps.com/cowpower/How%20It%20Works.html> (accessed November 4, 2007).

"Methane: Science." *U.S. Environmental Protection Agency (EPA)*, October 19, 2006. <http://www.epa.gov/methane/scientific.html> (accessed November 4, 2007).

"Methane: Sources and Emissions." *U.S. Environmental Protection Agency (EPA)*, October 19, 2006. <http://www.epa.gov/methane/sources.html> (accessed November 4, 2007).

"Non-CO_2 Gases Economic Analysis and Inventory." *U.S. Environmental Protection Agency (EPA)*, March, 6, 2007. <http://www.epa.gov/nonco2/econ-inv/international.html> (accessed November 4, 2007).

Sarah Gilman

Cryosphere

Introduction

The cryosphere refers to the areas of Earth that are covered by ice and snow. These are primarily located near the poles and at high altitudes. The cryosphere includes ice and snow that is seasonal and short-term, such as frozen freshwater lakes and snow cover. It also includes areas of long-lasting ice such as the polar ice caps, glaciers, and permafrost. The word cryosphere comes from the Greek word *kryos,* meaning cold.

The role of the cryosphere in climate change is not yet clear, as complex feedback mechanisms make predicting impacts on the relationship between climate change and changes in the cryosphere difficult to achieve. However, the cryosphere is thought to play a role in enhancing climate change, as well as acting as an early indicator of climate change. Most scientists agree on two observations: mean global temperatures have risen at an increasing rate over the past several decades, and the amount of Earth's snow and ice cover has decreased over the same period.

WORDS TO KNOW

ALBEDO: A numerical expression describing the ability of an object or planet to reflect light.

INTERGOVERNMENTAL PANEL ON CLIMATE CHANGE (IPCC): Panel of scientists established by the World Meteorological Organization (WMO) and the United Nations Environment Programme (UNEP) in 1988 to assess the science, technology, and socioeconomic information needed to understand the risk of human-induced climate change.

MEAN: A measure of central tendency (average) found by adding all the terms in a set and dividing by the number of terms.

PERMAFROST: Perennially frozen ground that occurs wherever the temperature remains below 32°F (0°C) for several years.

POLAR ICE CAP: Ice mass located over one of the poles of a planet not otherwise covered with ice. In our solar system, only Mars and Earth have polar ice caps. Earth's north polar ice cap has two parts, a skin of floating ice over the actual pole and the Greenland ice cap, which does not overlay the pole. Earth's south polar ice cap is the Antarctic ice sheet.

SOLAR RADIATION: Energy received from the sun is solar radiation. The energy comes in many forms, such as visible light (that which we can see with our eyes). Other forms of radiation include radio waves, heat (infrared), ultraviolet waves, and x-rays. These forms are categorized within the electromagnetic spectrum.

Historical Background and Scientific Foundations

The Intergovernmental Panel on Climate Change (IPCC) noted in its report *Climate Change 2001: The Scientific Basis* that the inclusion of the cryosphere in global climate models is rudimentary. However, the cryosphere is also considered to be an early indicator of climate change, with melting glaciers, melting ice caps, and reduced snowfalls being signs of the impact of global warming.

Impacts and Issues

A major issue with the cryosphere is the high albedo of snow compared to land, and sea ice compared to the sea. On land, the high albedo (reflective fraction) of snow causes more solar radiation to be reflected from Earth. Without snow cover, more solar radiation is absorbed by land, and this enhances the warming effect. The same occurs for sea ice, where sea ice reflects more solar radiation than seawater. Without frozen sea ice present, a greater amount of radiation is absorbed and the warming effect is enhanced. This suggests that the cryosphere could play a role in accelerating global warming.

Sea ice also impacts ocean circulation because it contains less salt than seawater, and its formation increases the salinity and density of surface water. This density difference between surface water and deep water drives ocean circulation. With less sea ice forming, this ocean circulation could be reduced, which would then also influence the redistribution of heat.

The cryosphere also plays a role in providing freshwater to regions throughout the world. Melting snow and melting glaciers are a major source of freshwater throughout Europe, North America, South America, and Asia. Reductions in snowfall and changes in the melting patterns of snow and glaciers could have a significant impact on global freshwater supplies.

SEE ALSO *Albedo; Antarctica: Melting; Antarctica: Observed Climate Changes; Arctic Melting: Greenland Ice Cap; Arctic Melting: Polar Ice Cap; Arctic People: Climate Change Impacts; Glacier; Glacier Retreat; Global Warming; Greenland: Global Implications of Accelerated Melting; Icebergs; Melting; Ocean Circulation and Currents; Oceans and Seas; Permafrost; Polar Ice; Sea Level Rise.*

BIBLIOGRAPHY

Books

Houghton, J. T., et al, eds. *Climate Change 2001: The Scientific Basis.* Cambridge: Cambridge University Press, 2001.

Web Sites

"1999 EOS Science Plan: Cryospheric Systems." *National Aeronautics and Space Administration (NASA)*, November 8, 2007. <http://eospso.gsfc.nasa.gov/science_plan/Ch6.pdf> (accessed November 8, 2007).

"Special Report on the Regional Impacts of Climate Change: An Assessment of Vulnerability." *Intergovernmental Panel on Climate Change (IPCC).* <http://www.grida.no/climate/ipcc/regional/index.htm> (accessed October 15, 2007).

"Our Planet." *United Nations Environment Programme (UNEP)*, May 2007. <http://www.unep.org/pdf/Ourplanet/2007/may/en/OP–2007-05-en-FULLVERSION.pdf> (accessed November 6, 2007).

Tony Hawas

Cultural Impacts: Venice in Peril, a Case Study

Introduction

The sea continuously laps at the foundations of Venice, Italy. The initial trade-off for peril by the sea for the city founded in the fifth century, was found in great profits reaped from its strategically important port. By the Renaissance, Venice was one of the world's richest cities, a vital link in trade between the East and West. In the modern era, Venetian wealth is preserved and conveyed through beauty in art and architecture.

The sea is relentless and has time and tide as its allies. Numbers tell a tale of a battle slowly lost. With some degree of flooding along the lagoon now evident as many as 200 days per year, Venice endures almost thirty times the number of flood days experienced just a century ago. San Marco Square regularly floods and when flooding is anticipated, elevated wooden walkways are hurriedly set up in the square. The sight of local residents carrying galoshes on the vaporettos (waterbuses) is now as common as Londoners carrying umbrellas.

Climatologists assert that global warming will result in sea level increases, and such sea level rise will exacerbate Venice's flooding problems.

The city of Venice, Italy, is in danger due to rising waters associated with climate change. *Brand X Pictures/ Royalty Free.*

Historical Background and Scientific Foundations

Natural Factors

Fluctuations of absolute global sea level result from climate change—including the normal cyclical growth and decay of Earth's polar ice caps (eustasy). Such cycles do not alone determine sea level relative to a specific coastal segment. Rates of sediment supply and transport along with patterns of deposition, erosion, crustal subsidence, and uplift also influence elevation at a particular location.

On a much larger geological scale, the Venetian region of Italy is very slowly dipping downward, at a rate that ensures that Venice will continue to sink at least a few centimeters a century.

Human Influence

The *Consorzio Venezia Nuova*, the public authority responsible for coordinating efforts to protect Venice from flooding, has inherited the task from a long line of agencies that date back as far as the fourteenth century Venetian Magistry of the Waters. For now the *Consorzio* uses stop-gap measures such as raising sidewalks and extending temporary flood walkways to protect the city, These measures, however, are merely life-support designed to keep Venice going while longer-term remedies are found and implemented.

Some prior stop-gap measures may have done long-term harm. For example, excess pumping of groundwater resulted in soil compaction and sinking. The practice was stopped in the 1980s and the city stabilized with regard to the rate of subsoil subsidence. Many hydrogeologists assert that the acceleration in the sink rate of the city during the twentieth century is a result of the combination of the over-extraction of groundwater by industry combined with ill-tested digging projects in the canal and lagoon. In addition to being a treasured cultural attraction, Venice still remains an industrially important city. Oil tankers course through deep-water channels, many cut in the 1950s, to supply nearby refineries. The effect of such digging also made Venice more vulnerable to flooding.

WORDS TO KNOW

CARBON SEQUESTRATION: Storage or fixation of carbon in such a way that it is isolated from the atmosphere and cannot contribute to climate change. Sequestration may occur naturally (e.g., forest growth) or artificially (e.g., injection of CO_2 into underground reservoirs).

INTERGOVERNMENTAL PANEL ON CLIMATE CHANGE (IPCC): Panel of scientists established by the World Meteorological Organization (WMO) and the United Nations Environment Programme (UNEP) in 1988 to assess the science, technology, and socioeconomic information needed to understand the risk of human-induced climate change.

SEA LEVEL: The datum against which land elevation and sea depth are measured. Mean sea level is the average of high and low tides.

Impacts and Issues

Forecasts call for Venice to continue to sink.

To slow the sinking, construction is underway on an extended mobile flood barrier (MOSE) that will feature closable gates to protect Venice from sea surges during extreme high tides or other flooding conditions. MOSE is projected to be ready for operation by 2011, and its initial costs are projected to approach $4 billion dollars. Given the complexity of the engineering tasks and fragility of the local ecosystem that will require many downstream project modifications, many political officials are bracing for the potential that the ultimate costs of the project will soar well beyond initial projections.

If sea levels continue to rise, however, not even MOSE is projected to be able to hold back the seas for more than a century. Proponents of other countermeasures contend that this is all the more reason to re-examine other measures to counter subsidence (sinking).

For example, in a recent issue of *Nature* a group of geomechanical engineers from the University of Padua proposed that a restoration of the underground fluids via modern injection technology could actually raise the level of Venice. Optimistic projections indicate that Venice could be raised by 8 to 12 in (20 to 30 cm) by a large-scale injection project, a level that would counter the settling and subsidence of more than 8 in (20 cm) since 1950.

The group's plan called upon technology already used in the oil and gas industry to counter Venice's subsidence. Fluid pumping was initially considered in the 1970s, but the proposal was set aside because the technology to accomplish the task at that time was not as economically feasible—and because some engineers feared that uneven rising could do more harm than good.

The newest proposals for underground injections call for injections of seawater (the cheapest solution) and/or carbon dioxide (a byproduct of local power generation) into sandy layers located approximately 2,300 ft (700 m) below the lagoon. This would be one of the first potential practical engineering applications for carbon sequestration.

Layers of sand at a depth of 2,300 ft (700 m) are geologically squeezed between impermeable layers of clay and rock. The depth is significant because attempts

IN CONTEXT: SEA LEVEL

Because tides raise and lower the actual sea level daily, and by different amounts in different parts of the world, scientists refer not to the actual level of water at any given time, but to the sea level datum plane, a reference height used in measuring land elevation and water depths. It refers to the vertical distance from the surface of the ocean to some fixed point on land, or a reference point defined by people. Sea level became a standardized measure in 1929. Mean sea level is the average of the changes in the level of the ocean over time, and it is to this measure that we refer when we use the term sea level.

Constant motion of water in the oceans causes sea levels to vary. Sea level in Maine is about 10 in (25 cm) higher than it is in Florida. The Pacific coasts' sea level is approximately 20 in (50 cm) higher than the Atlantic.

Mean sea level can also be influenced by air pressure. If the air pressure is high in one area of the ocean and low in another, water will flow to the low pressure area. Higher pressure exerts more force against the water, causing the surface level to be lower than it is under low pressure. That is why a storm surge (sea level rise) occurs when a hurricane reaches land. Air pressure is unusually low in the eye of a hurricane, and so water is forced toward the eye, creating coastal flooding.

Increases in temperature can cause sea level to rise. Warmer air will increase the water temperature, which causes water molecules to expand and increase the volume of the water. The increase in volume causes the water level to become higher.

Mean sea level has risen about 4 in (10 cm) during the last hundred years. Several studies indicate that this is due to an average increase of 1.8°F (1°C) in worldwide surface temperatures. Most climate scientists believe that rising sea levels will create environmental, social, and economic problems, including the submerging of coastal lands, higher water tables, salt water invasion of freshwater supplies, and increased rates of coastal erosion.

Sea level can be raised or lowered by tectonic processes, which are movements of Earth's crustal plates. Major changes in sea level can occur over geologic time due to land movements, ice loading from glaciers, or an increase and decrease in the volume of water trapped in ice caps.

About 30,000 years ago, sea level was nearly the same as it is today. During the Ice Age 15,000 years ago, it dropped and has been rising ever since.

---

to raise Venice from this level should be more uniform than the projections associated with earlier generations' injection proposals that called for injections into layers only about 160 ft (50 m) below the floor of the lagoon. Engineers feared that injections at the more superficial depth might result in uneven leveling that might prove disastrous to building integrity. Costs of such an injection project are now comparable to the proposed budgets of other proposed countermeasures that are technologically more ambitious, but also more risky in terms of outcome.

The idea to pump fluids underground is not intended to replace the construction of an extended mobile flood barrier (MOSE), but given that MOSE may offer protection for only another 100 to 150 years, any elevation increase resulting from other anti-subsidence countermeasures such as injection would increase MOSE's effectiveness and extend its projected useful life and help partially, but not fully, offset the anticipated accelerated sea level rise due to global warming.

The degree of peril to Venice caused by global warming can be generalized, but precise projections are difficult, especially given the controversy of the expected rise on sea levels associated with climate change. If by 2100, sea levels rise between 7 to 23 in (18 to 59 cm) as predicted by the Intergovernmental Panel on Climate Change (IPCC), projections of the protections imposed by MOSE will prove insufficient. If more dire scenarios are realized, such as might be caused by increased contributions to sea-level rise from accelerated glacial movement in the Greenland and West Antarctic ice sheets, the peril for Venice, and the economic cost of protection, must likewise escalate.

SEE ALSO *Europe: Climate Change Impacts; IPCC Climate Change 2007 Report: Impacts, Adaptation and Vulnerability; Sea Level Rise.*

BIBLIOGRAPHY

Web Sites

"Case Studies on Climate Change and World Heritage." *United Nations Educational, Scientific, and Cultural Organization. World Heritage Centre*, 2007. <http://whc.unesco.org/documents/publi_climatechange.pdf> (accessed November 16, 2007).

"Save Venice" *Media Center for Art History, Columbia University*. <http://www.savevenice.org/site/pp.asp?c=9eIHKWMHF&b=67611> (accessed December 7, 2007).

"Venice in Peril." *British Committee for the Preservation of Venice*, 2003. <http://www.veniceinperil.org/news_articles/newsarticles.htm> (accessed December 7, 2007).

"Venice's 1,500-year Battle with the Waves." *BBC News*, July 17, 2003. <http://news.bbc.co.uk/1/hi/world/europe/3069305.stm> (accessed December 07, 2007).

K. Lee Lerner

Dams

Introduction

Dams are structures built across streams or rivers, usually to block water flow and cause the formation of a lake upstream of the dam. The water accumulated or impounded behind a dam can be used for drinking, irrigation of crops, or generating electricity. Electricity generated by dams, termed hydroelectricity, is made by allowing water to run downhill through fan-like turbines, a modern application of the ancient waterwheel principle for extracting work from falling water. The turbines turn generators that produce electricity. About 19% of the world's electricity is hydroelectricity.

Dams have long been criticized for flooding landscapes, destroying ecosystems, and in some cases, forcing the evacuation of millions of people. However, for many years it was assumed that dams are a climate-neutral way to produce electricity since they burn no chemical fuel and produce no carbon dioxide (CO_2) emissions in daily operation. In the 1990s, researchers realized that some dams, especially in the tropics, are significant sources of methane, a potent greenhouse gas. Research published in 2006 reported that dams may account for more than 4% of anthropogenic warming.

Historical Background and Scientific Foundations

Overview

The first large dam was built by the Egyptians around 2800 BC. It was a massive barrier 37 ft (11 m) tall and capable of impounding 20 million cubic ft (566,000 cubic m) of water. The dam failed after a few years when water overflowed its crest, eroding the downstream side of the structure and weakening it until the dam collapsed.

Since that time, most civilizations have built dams of various sizes. However, the building of truly large dams did not occur until the twentieth century. During the 1930s, engineers began building dams large enough to block whole river basins. The Hoover Dam was completed across the Colorado River in the United States in 1936, becoming the world's largest at that time. When the lake behind the dam had filled, it backed 110 mi (177 km) upstream and covered 247 square mi (639 square km). Hoover is still the second-tallest dam in the United States.

Since Hoover, thousands of dams have been built worldwide. Over half (172 out of 292) of the world's large rivers are now dammed at one point or more. All of the world's largest 21 rivers are dammed. Globally, there are more than 45,000 large dams, that is, dams over 49 ft (15 m) high, which altogether store more than 1,557 cubic mi (6,500 cubic km) of water, an amount equal to 15% of the world's annual freshwater runoff. There are over 300 very large dams, namely those over 492 ft (150 m) high or impounding over 6 cubic mi (25 cubic km) of water. The latest of these giant dams, and the most controversial and environmentally destructive to date, is the Three Gorges dam across the Yangtze River in China. It will impound over 9 cubic mi (39 cubic km of water) after filling and is forcing the evacuation of over a million people.

As of 2004, 2.2% of the world's primary energy was obtained from hydroelectric dams, or about 19% of the world's electricity. Irrigation water from dams is used to grow 12–16% of the world's food.

Dams and Climate Change

There are several connections between dams and climate change. First, hydroelectricity was long assumed to have less global-warming impact than electricity generated using fossil fuels. Although CO_2 is emitted during the manufacture of the cement, concrete, and steel that go into making a modern dam, these emissions, for a typical dam, amount to less than 10% of the emissions from generating the same amount of electricity by burning fossil fuels. However, starting in 1993, researchers began

Arizona's Glen Canyon Dam, which dams the Colorado River and creates Lake Powell, has been controversial since its inception in the 1960s. In recent years, drought has plagued the Colorado. Evidence of the low water levels is found on the Navajo sandstone of the canyon walls. The white bath tub ring found on the rock is calcium carbonate, which indicates how high the water level once was. Continued drought can reduce water and energy benefits from such dams. *Image copyright Tom Grundy, 2007. Used under license from Shutterstock.com.*

noting that some large dams might be significant sources of the greenhouse gas methane (CH_4), which has about 23 times the global-warming effect of CO_2, ton for ton. As of 2008, this issue continued to be hotly debated, with some scientists challenging the objectivity of scientists working for (or partly funded by) the hydropower industry, who have done most of the research on dam methane to date.

Methane is produced after the dam is constructed. When a typical tropical dam is built, large areas of upstream forest are drowned. Wood and other types of organic (carbon-containing) matter that decay under water are mostly broken down by anaerobic bacteria—bacteria that thrive in the absence of oxygen—using digestive chemistry that produces CH_4 as a waste product; in contrast, microorganisms that digest organic matter where oxygen is present produce CO_2. The anaerobic rotting of wood and other organic matter under water produces methane.

An initial pulse of methane is released as the trees, soil carbon, and forest litter covered by the filling of a new dam lake rot. However, this is not the major source of dam methane, which continues to be produced as long as the dam stands. In most tropical areas, rainy seasons alternate with dry seasons. During the dry season, water continues to flow through the dam and the lake water level lowers. This creates large mud flats and shallow zones, because tropical river valleys are generally wide and gently sloping rather than narrow and steep. Water weeds and other soft, quick-growing vegetation flourish in these flats and shallows.

When the rainy season comes, the water level rises again, drowning these plants, which quickly decay underwater, releasing methane. In effect, the rising and falling lake is a vast, solar-powered machine for harvesting carbon from CO_2 in the air and re-releasing it as CH_4 with greatly increased greenhouse warming effect.

In 2005, Philip Fearnside, one of the most-cited researchers on methane production by dams, found that in 1990, 13 years after its lake filled, the Curuá-Una Dam in Brazil emitted 3.6 times more greenhouse gas than oil-burning plants generating the same amount of electricity would have emitted. That is, the dam released enough methane to cause 3.5 times as much global warming as the CO_2 released by the oil plants would have.

Ivan Lima and colleagues, among other scientists, have suggested that at least some of this methane could be collected and burned to produce electricity. This would both extract useful energy and convert the methane (CH_4)

WORDS TO KNOW

ANAEROBIC BACTERIA: Single-celled creatures that thrive in anaerobic environments, that is, environments lacking free molecular oxygen (O_2). They digest organic matter and release methane, a greenhouse gas.

ANTHROPOGENIC: Made by people or resulting from human activities. Usually used in the context of emissions that are produced as a result of human activities.

CLEAN DEVELOPMENT MECHANISM: One of the three mechanisms set up by the Kyoto Protocol to, in theory, allow reductions in greenhouse-gas emissions to be implemented where they are most economical. Under the Clean Development Mechanism, polluters in wealthy countries can obtain carbon credits (greenhouse pollution rights) by funding reductions in greenhouse emissions in developing countries.

FOSSIL FUELS: Fuels formed by biological processes and transformed into solid or fluid minerals over geological time. Fossil fuels include coal, petroleum, and natural gas. Fossil fuels are non-renewable on the timescale of human civilization, because their natural replenishment would take many millions of years.

GREENHOUSE GASES: Gases that cause Earth to retain more thermal energy by absorbing infrared light emitted by Earth's surface. The most important greenhouse gases are water vapor, carbon dioxide, methane, nitrous oxide, and various artificial chemicals such as chlorofluorocarbons. All but the latter are naturally occurring, but human activity over the last several centuries has significantly increased the amounts of carbon dioxide, methane, and nitrous oxide in Earth's atmosphere, causing global warming and global climate change.

HYDROELECTRICITY: Electricity generated by causing water to flow downhill through turbines, fan-like devices that turn when fluid flows through them. The rotary mechanical motion of each turbine is used to turn an electrical generator.

METHANE: A compound of one hydrogen atom combined with four hydrogen atoms, formula CH_4. It is the simplest hydrocarbon compound. Methane is a burnable gas that is found as a fossil fuel (in natural gas) and is given off by rotting excrement.

WORLD BANK: International bank formed in 1944 to aid in reconstruction of Europe after World War II, now officially devoted to the eradication of world poverty through funding of development projects. Although ostensibly independent, the bank is always headed by a person appointed by the President of the United States.

to CO_2, greatly reducing its greenhouse potential. Most dams allow deep, high-pressure, methane-rich water to run through their turbines. This causes the release of much of the methane in the water and may account for up to 70% of methane emissions from dam lakes (the rest is emitted from the lake surface). Some or most of this methane might be collected for burning. As of 2007, however, such technologies had not yet been demonstrated on a practical scale.

There are other relationships between dams and climate change. Dams in high latitudes (far from the equator) do not produce large amounts of methane, and so can indeed produce electricity with less greenhouse impact than comparable fuel-burning plants would. This does not reduce their other environmental impacts, however. Counting all dams worldwide, the greenhouse impact of dams is less than half that of the least-polluting fuel-burning alternatives.

Dams are also linked to climate change because they may serve as storage reservoirs in places where precipitation patterns and flood risk may be altered by climate change. Thus, some dams may help with mitigation of and adaptation to the effects of climate change. Finally, some dams may have to contend with decreased rainfall and water flow from changing climate, which might reduce water and energy benefits from those dams.

Impacts and Issues

Reports in the 1990s of significant methane emissions from dams led to intense debate on the climate-change mitigation merits of large dams. In 1997, a workshop was convened by the International Union for the Conservation of Nature and Natural Resources (an 83-state organization headquartered in Geneva, Switzerland) and the World Bank to discuss the question. The two groups established the World Commission on Dams to investigate the effectiveness and costs of dams, including their relationship to climate change.

In November 2000, the commission released its report. Although stating that "dams have made an important and significant contribution to human development, and benefits derived from them have been considerable," the report also said that "in too many cases an unacceptable and often unnecessary price has been paid to secure those benefits, especially in social and environmental terms, by people displaced, by communities downstream, by taxpayers and by the natural environment." The commission estimated that 40–80 million people had been displaced by dams and many species driven to extinction; it also stated that most of the benefits from 1% to 28% of anthropogenic (human-caused) greenhouse emissions might come from dams. A study by Ivan Lima and colleagues stated that 4% of global emissions are from dams.

The World Commission on Dams made a number of recommendations for how proposed dam projects should be judged, including whether people who would be displaced consent to the project. Its findings have been controversial because one of the commission's founders, the World Bank, is the world's largest funder of dam projects, having provided a total of $75 billion (1998 dollars) for 538 large dams in 92 countries since its inception in 1945. The World Bank has not accepted the commission's recommended international standards for dam-building. Europe's two largest public banks, the European Investment Bank and the European Bank for Reconstruction and Development, announced in 2005 that they would abide by the commission's standards. A directive of the European Union mandates that carbon-trading hydropower projects under the Clean Development Mechanism must abide by the commission's standards.

SEE ALSO *Clean Development Mechanism; Energy Contributions; Methane.*

BIBLIOGRAPHY

Periodicals

Fearnside, Philip M. "Do Hydroelectric Dams Mitigate Global Warming? The Case of Brazil's Curuá-Una Dam." *Mitigation and Adaptation Strategies for Global Change* 10 (2005): 675–691.

Lima, Ivan B. T., et al. "Methane Emissions from Large Dams as Renewable Energy Resources: A Developing Nation Perspective." *Mitigation and Adaptation Strategies for Global Change* (February 2007).

Yardley, William. "Climate Change Adds Twist to Debate Over Dams." *New York Times* (April 23, 2007).

Web Sites

"Citizen's Guide to the World Commission on Dams." *International Rivers Network*, 2002. <http://www.irn.org/wcd/wcdguide.pdf> (accessed November 13, 2007).

"Climate Change and Dams: An Analysis of the Linkages Between the UNFCCC Legal Regime and Dams." *United Nations Environment Programme*, November 2000. <http://www.dams.org/docs/kbase/contrib/env253.pdf> (accessed November 13, 2007).

"Dams and Development: A New Framework for Decision-Making." *World Commission on Dams*, June 22, 2007. <http://www.dams.org//docs/report/wcdreport.pdf> (accessed November 13, 2007).

Grahame-Rowe, Duncan. "Hydroelectric Power's Dirty Secret Revealed." *New Scientist*, February 24, 2005. <http://www.newscientist.com/article/dn7046.html> (accessed November 13, 2007).

Larry Gilman

Delhi Declaration

Introduction

In November 2002, in Delhi, India, representatives of some 185 nations adopted the Delhi Declaration. The process of negotiation that led to its final draft highlighted substantial differences in national attitudes toward global climate change, especially those of developing nations as opposed to developed nations.

Historical Background and Scientific Foundations

Staged in Delhi between October 22 and November 1, 2002, the Eighth Session of the Conference of the Parties (COP) to the Climate Change Convention—the annual meeting of the signatories of the United Nations Framework Convention on Climate Change (UNFCCC)—was intended primarily as a discussion on the implementation of the Kyoto Protocol. However, the conference was dominated by tensions between developed and less developed countries, as represented by the Group of 77 (G77). G77 nations sought cash aid from wealthier countries to help them adapt cleaner technologies to reduce greenhouse-gas emissions and rejected limits on their emissions. Many G77 nations sought the support of the United States, an opponent of the Kyoto Protocol, to remove mentions of Kyoto from the final declaration. The European Union (EU) strongly objected.

At one stage, India's environment minister, who was chairing the conference, threatened to end the meeting without a resolution if parties could not reach a consensus. Under continued pressure from many industrialized nations, a reference to the Kyoto Protocol appeared in the final wording of the Delhi Declaration. However, the declaration lacked specifics on how developing countries would regulate and reduce their greenhouse-gas emissions.

WORDS TO KNOW

CONFERENCE OF THE PARTIES (COP) TO THE CLIMATE CHANGE CONVENTION: Annual meeting of representatives from nations that are signatories of (parties to) the United Nations Framework Convention on Climate Change, a treaty drafted in 1992. The treaty entered into force in 1994, and COPs have been held ever since. It was at COPs that the Kyoto Protocol was drafted.

GREENHOUSE GASES: Gases that cause Earth to retain more thermal energy by absorbing infrared light emitted by Earth's surface. The most important greenhouse gases are water vapor, carbon dioxide, methane, nitrous oxide, and various artificial chemicals such as chlorofluorocarbons. All but the latter are naturally occurring, but human activity over the last several centuries has significantly increased the amounts of carbon dioxide, methane, and nitrous oxide in Earth's atmosphere, causing global warming and global climate change.

GROUP OF 77 (G77): Group of developing countries, originally founded in 1964 with 77 members but having 130 members as of 2007. The group was formed as a counterpoint to the Group of 7 (G7) nations, a group of the world's wealthiest nations formed to advance their economic interests.

KYOTO PROTOCOL: Extension in 1997 of the 1992 United Nations Framework Convention on Climate Change (UNFCCC), an international treaty signed by almost all countries with the goal of mitigating climate change. The United States, as of early 2008, was the only industrialized country to have not ratified the Kyoto Protocol, which is due to be replaced by an improved and updated agreement starting in 2012.

Impacts and Issues

The Delhi Conference exposed some of the dilemmas facing developing nations when it came to climate change. Rapid urbanization and industrialization are aiding the economies of developing nations, but industrialization and development often negatively impact the environment. The cleanest technologies are often the most expensive to implement. Thus, adopting greener solutions for energy production and industrial emissions management is more of a burden on less-wealthy nations.

Although per capita emissions in the developing world are substantially smaller than those in the most developed nations, growing populations and rapid industrialization have created a growing emissions problem. While the United States remains the world's largest per capita emitter of greenhouse gases, in 2007 China overtook the United States as the world's largest overall polluter by total volume.

Under the Kyoto Protocol, developing countries are exempt from the emissions reduction targets to which developed nations are subject. At the Delhi Conference, the most developed nations, led by the European Union, asserted that many developing nations begin negotiations toward establishing targets and restricting greenhouse-gas emissions after 2012. Delegates from developing countries repeatedly rejected such demands. This schism continues to burden ongoing international efforts toward reducing emissions worldwide.

SEE ALSO *Europe: Climate Policy; Kyoto Protocol; United Nations Framework Convention on Climate Change (UNFCCC).*

BIBLIOGRAPHY

Web Sites

"Conference of the Parties 8 (COP 8)—Climate Talks in New Delhi." *Pew Center on Global Climate Change.* <http://www.pewclimate.org/what_s_being_done/in_the_world/cop_8_india/> (accessed November 21, 2007).

"Delhi Declaration." *United Nations*, 2003. <http://unfccc.int/cop8/latest/1_cpl6rev1.pdf> (accessed November 21, 2007).

"India Rejects Climate Change Pressure." *BBC News*, October 30, 2002. <http://news.bbc.co.uk/1/hi/world/south_asia/2374551.stm> (accessed November 21, 2007).

James Corbett

Desert and Desertification

Introduction

Deserts are very dry regions that have seasonal high temperatures, low sporadic rainfall, and a high evaporation rate. Areas that receive less than 10 in (25.4 cm) of rain a year are generally classified as deserts. Dry (arid) regions are usually found in area of high pressure (subtropical highs, leeward sides of mountains, etc.) associated with descending divergent air masses that are common between 30 degrees N and 30 degrees S latitude. Arid lands cover 25% of Earth's land surface.

Desert temperatures and rainfall patterns are influenced by climate changes. Interrelated ocean-atmosphere systems are making deserts drier and reducing the diversity of their ecosystems. Wind and dry conditions in deserts create dust that can affect regions far removed from a desert.

Desertification—the degradation of productive land to desert—is an environmental crisis that is affecting the land where an estimated 100–200 million people live. It is a complex process that, like climate change, is being accelerated by human activities.

Drought can lead to desertification, as shown in this image of a river bed in Marrakech, Morocco. The dried-up river bed is home to few plants and other vegetation. *AP Images.*

Historical Background and Scientific Foundations

Climatic processes driven by the interaction of ocean currents and the atmosphere create deserts. The intensity of solar radiation near the equator creates dry winds and deserts in subtropical regions, particularly in Africa. Most of the world's desert ecosystems are located in two belts near the tropics at 30 degrees north and 30 degrees south of the equator. At the coast, dry winds induce upwelling of deep cold ocean waters that are then driven by ocean coastal currents. Variations occur in rainfall in African deserts due to global weather cycles associated with changes in ocean-atmospheric conditions between the South Pacific and Indian Oceans.

WORDS TO KNOW

ALBEDO: A numerical expression describing the ability of an object or planet to reflect light.

BIODIVERSITY: Literally, "life diversity": the number of different kinds of living things. The wide range of organisms—plants and animals—that exist within any given geographical region.

BIOMASS: The sum total of living and once-living matter contained within a given geographic area. Plant and animal materials that are used as fuel sources.

CONFERENCE OF THE PARTIES (COP) TO THE CLIMATE CHANGE CONVENTION: Annual meeting of representatives from nations that are signatories of (parties to) the United Nations Framework Convention on Climate Change, a treaty drafted in 1992. The treaty entered into force in 1994, and COPs have been held ever since. It was at COPs that the Kyoto Protocol was drafted.

EL NIÑO: A warming of the surface waters of the eastern equatorial Pacific that occurs at irregular intervals of 2 to 7 years, usually lasting 1 to 2 years. Along the west coast of South America, southerly winds promote the upwelling of cold, nutrient-rich water that sustains large fish populations, that sustain abundant sea birds, whose droppings support the fertilizer industry. Near the end of each calendar year, a warm current of nutrient-poor tropical water replaces the cold, nutrient-rich surface water. Because this condition often occurs around Christmas, it was named El Niño (Spanish for boy child, referring to the Christ child). In most years the warming lasts only a few weeks or a month, after which the weather patterns return to normal and fishing improves. However, when El Niño conditions last for many months, more extensive ocean warming occurs and economic results can be disastrous. El Niño has been linked to wetter, colder winters in the United States; drier, hotter summers in South America and Europe; and drought in Africa.

LA NIÑA: A period of stronger-than-normal trade winds and unusually low sea-surface temperatures in the central and eastern tropical Pacific Ocean; the opposite of El Niño.

LEEWARD: Downwind of an object: opposite of windward.

The reduction of upwelling cold water in the Pacific along the coastline of North and South America, called El Niño, and the reverse process called La Niña, cause extremes in weather conditions that create desert regions near the coasts. Also, the leeward sides of coastal mountains are deserts because they do not receive ocean-generated moisture.

Normal variations in climate are being affected by climate change that has warmed deserts at an average rate of 0.4–1.4°F (0.2–0.8°C) per decade between 1976 and 2000—a rate that is higher than the average global temperature increase. Global warming is expected to increase rainfall in higher latitudes, but most deserts in the subtropical areas are expected to continue to get drier.

■ Impacts and Issues

As deserts become drier, the ecosystem changes. Higher temperatures result in higher evaporation rates that, in turn, increase soil acidity and reduce desert vegetation. Grasses are replaced with desert shrubs that provide lower nutrition. Biodiversity is impacted as animals move away and plant species start to disappear.

As a result of human activities, there has been a 36% increase in atmospheric carbon dioxide—a significant greenhouse gas—since the mid-1800s. Plants use carbon dioxide in photosynthesis, thus removing it from the atmosphere. The uptake of atmospheric carbon dioxide stored in desert plants is the lowest of any biomass on Earth, and getting lower.

The reduction of plant cover in deserts is increasing the albedo, the reflection of sunlight from Earth's surface back into the atmosphere. This is expected to actually cool areas away from the deserts, but to increase the degradation of lands adjacent to the deserts, a process called desertification. Desertification is also induced by such human activities as overgrazing, clearing woody vegetation, poor management of the land in farming, and water contamination from overpopulation of fragile semi-arid regions.

Desertification is spreading largely at the expense of grasslands and croplands. In Africa, deserts are expanding north and south from the Sahara. Desertification also is expanding deserts in the Middle East, Central Asia, and western and northern China.

Desertification produces deserts, and deserts produce dust which can travel great distances. Dust from the Asian deserts has been carried by winds over the North Pacific Ocean to North America. Dust from the Sahara in Africa has been carried to North and Central America. Climate change is increasing dryness in deserts and increasing wind speeds. The combination of dryness and increased wind speeds is increasing dust emissions to non-desert regions.

Wind blown dust particles—largely made of an aluminum-silicate mineral—are usually less than 2 micrometers in size. Significant amounts of wind-blown desert dust have settled onto coral reefs and contributed to their decline. Desert dust affects the atmosphere as it partly reflects and partly absorbs solar radiation. It affects rainfall, visibility, and human health in regions far from the deserts.

IN CONTEXT: U.N. EFFORTS TO COMBAT DESERTIFICATION

The United Nations first recognized desertification as a global problem in 1977. That year, representatives from several U.N. agencies, national governments, and international organizations met to devise and adopt the Plan of Action to Combat Desertification. This first plan focused on encouraging desertification prevention and water conservation work by national governments and international conservation agencies. Projects under the plan continued for over a decade, but a conference of the U.N. Environment Program in 1991 asserted that desertification was accelerating, severely affecting some of the most underdeveloped parts of the world. Furthermore, conference participants noted that issues of desertification, soil and freshwater salination, water shortages, agricultural land degradation, and political and economic turmoil often compound each other.

The following year, the U.N. Conference on Environment and Development proposed a new strategy of combating desertification at the local and community level. The new approach integrated economic and environmental concerns. For example, anti-desertification projects in sub-Saharan Africa targeted communities in need of improved access to freshwater and agricultural land-management assistance. Communities were taught agricultural and conservation practices that prevented soil erosion and nutrient depletion. New, deeper wells were dug to provide access to water.

In June 1994, the Convention to Combat Desertification was adopted. Nearly 180 signatory parties met two years later at the first Conference of the Parties. The eighth session of the Conference of the Parties took place in Madrid, Spain, in September 2007. Although the Convention to Combat Desertification still sponsors community-level programs to alleviate the effects of desertification, increasing attention is now paid to the effects of global climate change on desertification issues.

The U.N. designated June 17 for annual observance of World Day to Combat Desertification.

SEE ALSO *Africa: Climate Change Impacts; Albedo; Coral Reefs and Corals; Ocean Circulation and Currents.*

BIBLIOGRAPHY

Web Sites

"Desertification." *U.S. Geological Survey,* October 29, 1997. <http://pubs.usgs.gov/gip/deserts/desertification/> (accessed August 20, 2007).

"Global Deserts Outlook." *United Nations Environment Programme,* 2006. <http://www.unep.org/geo/GDOutlook/> (accessed August 20, 2007).

"UN Issues Desertification Warning." *BBC News,* June 28, 2007. <http://news.bbc.co.uk/2/hi/africa/6247802.stm> (accessed August 20, 2007).

Drought

Introduction

Drought does not happen suddenly. It evolves in a region when weather conditions produce declining rainfall amounts that cause vegetation to dry, soil moisture to be reduced, and surface water sources to decline and ultimately to dry up. The length of a drought depends entirely on weather conditions. It can be only a few weeks in duration or it can last years to become a serious weather-related catastrophe. Satellite images of vegetation in a region can provide evidence of long-term drought conditions where they exist. As vegetation dries up, wildfires become more frequent and intense.

About 10% to 15% of global land areas were considered to be undergoing moderate to severe drought conditions in the 1970s. That number rose to around 30% by 2002. Changing climate conditions that have increased global temperatures are considered by atmospheric scientists to be the cause for at least half of the increase in droughts around the world. All climate models used by the scientists indicate that global warming will increase. Scientists also are predicting more severe drying conditions if the concentration of greenhouse gases increases in the atmosphere.

In February 2006, a woman walks past carcasses in a drought-stricken region in Kenya, Africa. The drought, which had lasted 14 months at that time, had claimed several dozen human lives as well as 70% of the region's cattle. *Tony Karumba/AFP/Getty Images.*

WORDS TO KNOW

DIPOLE: A system that has two equal but opposite characteristics separated by a distance.

EL NIÑO: A warming of the surface waters of the eastern equatorial Pacific that occurs at irregular intervals of 2 to 7 years, usually lasting 1 to 2 years. Along the west coast of South America, southerly winds promote the upwelling of cold, nutrient-rich water that sustains large fish populations, that sustain abundant sea birds, whose droppings support the fertilizer industry. Near the end of each calendar year, a warm current of nutrient-poor tropical water replaces the cold, nutrient-rich surface water. Because this condition often occurs around Christmas, it was named El Niño (Spanish for boy child, referring to the Christ child). In most years the warming lasts only a few weeks or a month, after which the weather patterns return to normal and fishing improves. However, when El Niño conditions last for many months, more extensive ocean warming occurs and economic results can be disastrous. El Niño has been linked to wetter, colder winters in the United States; drier, hotter summers in South America and Europe; and drought in Africa.

GREENHOUSE GASES: Gases that cause Earth to retain more thermal energy by absorbing infrared light emitted by Earth's surface. The most important greenhouse gases are water vapor, carbon dioxide, methane, nitrous oxide, and various artificial chemicals such as chlorofluorocarbons. All but the latter are naturally occurring, but human activity over the last several centuries has significantly increased the amounts of carbon dioxide, methane, and nitrous oxide in Earth's atmosphere, causing global warming and global climate change.

LA NIÑA: A period of stronger-than-normal trade winds and unusually low sea-surface temperatures in the central and eastern tropical Pacific Ocean; the opposite of El Niño.

SAHEL: The transition zone in Africa between the Sahara Desert to the north and tropical forests to the south. This dry land belt stretches across Africa and is under stress from land use and climate variability.

SOUTHERN OSCILLATION: A large-scale atmospheric and hydrospheric fluctuation centered in the equatorial Pacific Ocean. It exhibits a nearly annual pressure anomaly, alternatively high over the Indian Ocean and over the South Pacific. Its period is slightly variable, averaging 2.33 years. The variation in pressure is accompanied by variations in wind strengths, ocean currents, sea-surface temperatures, and precipitation in the surrounding areas. El Niño and La Niña occurrences are associated with the phenomenon.

TREE RINGS: Marks left in the trunks of woody plants by the annual growth of a new coat or sheath of material. Tree rings provide a straightforward way of dating organic material stored in a tree trunk. Tree-ring thickness provides proxy data about climate conditions: most trees put on thicker rings in warm, wet conditions than in cool, dry conditions.

Historical Background and Scientific Foundations

Historical records of weather events have been developed from studies of tree rings, ice core samples from glaciers, and studies of sediment layers in lakes. However, systematic instrumental records of weather events go back only a little over 100 years. Over the past 50 years, climatologists have discovered the relationship between regional climate events, such as droughts, and much larger interactions between the atmosphere and oceans.

The occurrence of major droughts is believed to be related to ocean-atmosphere oscillation cycles, such as the El Niño/Southern Oscillation (ENSO) cycle. It was not until the 1980s that climatologists saw that El Niño, (an upwelling of warm water in the Pacific Ocean) and La Niña (an upwelling of cold water in the Pacific Ocean) were related to the see-saw atmospheric pressure changes known as the Southern Oscillation. The ENSO cycle affects climate across North America. Similar oscillation patterns in other atmosphere-ocean systems are now considered to contribute to weather conditions that cause major droughts around the world.

Global warming is impacting the oscillation cycles and affecting the location and severity of droughts. Warming temperatures will increase evaporation from ocean surfaces. The extra atmospheric moisture is expected to increase average rainfall overall but not uniformly around the world. Warmer temperatures also will cause the land to dry more quickly where rainfall is less plentiful. Although some regions will experience greater rainfall, other regions will suffer more severe droughts. The percent of land area that suffers severe droughts is expected to continue to rise.

The Standard Precipitation Index (SPI) was developed in 1993 to provide early warning of drought conditions and to assess the extent of a drought. SPI values are determined from data collected on precipitation in a specific region for a specified length of time. The SPI index is keyed to reflect local drought conditions. If low precipitation conditions persist for three to six months, agriculture is impacted. Longer low precipitation periods can result in reduced groundwater supplies and increased potential for severe wildfires. The severity of wildfires depends on the terrain of the region and the type of vegetation.

IN CONTEXT: U.S. SCIENTISTS CALL FOR NATIONAL DROUGHT POLICY

In a February 2007 meeting, members of the American Association for the Advancement of Science (AAAS) called upon the United States to adopt a national drought policy that plans for global climate change. The researchers asserted that parts of the United States, including some of the nation's most densely populated urban centers and agriculturally productive areas, needed long-term solutions for water management, desertification prevention, and drought adaptation. The group of scientists focused on the western United States, citing the region's recent history of drought and its decades-long battle over freshwater resources, especially those from the Colorado River.

Scientists note that the western United States has long experienced periods of drought, but that global climate change may increase the frequency, length, and severity of future droughts. Past droughts have highlighted the region's precarious balance of high water-needs but vulnerability to water scarcity. Water shortages during regional droughts have affected farms, fisheries, and residential households. Increased urban development in arid areas has increased demand for freshwater in areas where it is naturally scarce, requiring complex water-sharing agreements with more water-rich neighboring states. Furthermore, most of the population of the western United States is on or near the Pacific Coast, making large population centers vulnerable to sea-level rise and freshwater and soil salination, compounding the anticipated long-term effects of climate change in the region.

The AAAS researchers, calling for a national drought policy, noted that droughts are currently responded to on an ad-hoc basis, addressing issues only as they arise. The group asserted that current drought assistance programs, agricultural regulations, and land and water-use laws may prove insufficient to cope with the effects of climate change. Although some states have addressed the effects of climate change through land and resource management laws, federal leaders have been hesitant to adopt policies that acknowledge specific threats of global climate change. As of November 2007, no national policy taking into account the AAAS recommendations had been proposed.

An older, more complex assessment tool for drought conditions is the Palmer Drought Severity Index, developed in the United States in 1968. It is based on a supply-and-demand concept for water use in addition to precipitation measurements. The Palmer index is used by relief agencies to determine where emergency assistance may be needed due to extreme drought conditions. The Palmer index also classifies excess rainfall conditions.

Impacts and Issues

Severe droughts have social and economic consequences. In the 1930s, a seven-year drought in the Great Plains of the United States created such desperate conditions that millions of people migrated to other regions of the country, such as California, to find better living conditions. A writer for the *Washington Evening Star* coined the term "Dust Bowl" to describe the effects of this severe Great Plains drought. North America is one of the regions that has historically had severe droughts.

Climate change is expected to intensify droughts in all the regions that already experience periodic droughts. In addition to North America, parts of Africa and Australia are already experiencing exceptionally dry conditions. The Australian Commonwealth Scientific and Industrial Research Organisation issued a report in January 2007 warning the Sydney, Australia, region to expect temperatures to rise 9°F (5°C) above the global average in the next 20 years.

Sydney, Australia, could experience severe droughts nine out of every ten years by the second half of the century. The region is already dry and bush fires are frequent, but that scenario is expected to worsen, since scientists predict rainfall will decrease by 40% by 2070. Periodic extreme storms are predicted for the Australian coast to add to the country's inhospitable climate, if climate change continues as current studies predict.

Australia's climate is under the influence of three ocean-atmospheric systems: the El Niño/Southern Oscillation in the Pacific Ocean, the Indian Ocean Dipole, and the Antarctic Circumpolar Wave for the Southern Ocean. The Indian Ocean Dipole is a year-to-year climate pattern that involves alternating warmer and colder ocean surface temperatures in the Indian Ocean. The three systems create a complex weather model for Australia.

Africa is another region already suffering from droughts that are thought to be tied to global warming. The Sahel region across Africa, from just below the Sahara desert in the east to the Atlantic coast on the west, and bordered by the wet tropical regions to the south, is being impacted by climate change. That region of Africa is influenced by the temperatures in the Indian Ocean. The Sahel region has had periods of adequate rainfall, but has suffered from drought in more recent years. The African droughts have created serious human crises that are frequently publicized in United Nations articles.

Southern Africa also is a drought-prone region likely to suffer more droughts if present climate changes continue. The gloomy outlook for droughts in southern Africa was highlighted by Lloyd's of London, the world's oldest insurer. In May 2007, Lloyd's cited global warming as the cause for drought that is increasing in many parts of the world, and particularly in southern Africa. Lloyd's has formed a partnership with other major groups, including Harvard University's Center for Health and Global Environment and the Insurance Information Institute, to study the severity and consequences of natural disasters that may arise in the future in many regions of the globe as a result of climate change.

SEE ALSO *Desert and Desertification; Dust Storms; El Niño and La Niña; Ocean Circulation and Currents; Wildfires.*

BIBLIOGRAPHY

Periodicals

Baines, Peter. "Australia's Climate Cerberus: The Puzzle of Three Oceans." *Ecos* 97 (1998): 22–25.

Web Sites

"Climate Change Warning for Sydney." *BBC News*, January 31, 2007. <http://news.bbc.co.uk/2/hi/asia-pacific/6315885.stm> (accessed September 1, 2007).

"Climate Change 'Will Dry Africa.'" *BBC News*, November 29, 2005. <http://news.bbc.co.uk/1/hi/sci/tech/4479640.stm> (accessed September 1, 2007).

"Drought and Wildland Fire." *UCAR: The University Corporation for Atmospheric Research.* <http://www.ucar.edu/research/climate/drought.jsp> (accessed September 1, 2007).

Dust Storms

Introduction

Dust storms occur when sustained high winds blow near Earth's surface in arid regions, pick up large quantities of fine sand or dust, and transport the materials over great distances. More than 40% of the world's land surface is arid, either already desert land or in the process of desertification. These desert margins undergoing desertification are prone to dust storms. Human activities—particularly mismanagement of land uses in marginal arid regions—and climate change that is causing desert areas to get even drier are contributing to an increase in dust storms. Dust storms arise most frequently in four regions: Central Asia, North America, Africa, and Australia.

Airborne particles in dust storms are a health hazard and can cause considerable damage because of their abrasive effect on any surface in the storm's path. When dust storms pass over industrial areas, emissions from the combustion of fossil fuels can be incorporated in the storm causing even more serious health hazards.

Historical Background and Scientific Foundations

Core samples taken from ocean floors and studies of glaciers indicate that dust storms have occurred for at least 70 million years. The term dust storm also includes sand storms, the difference being the size of the soil particles involved. Sand particles are about 0.02 in (0.5 mm) in diameter, while dust particles are smaller. The size of the particles transported and the wind speed determine the character of a dust storm.

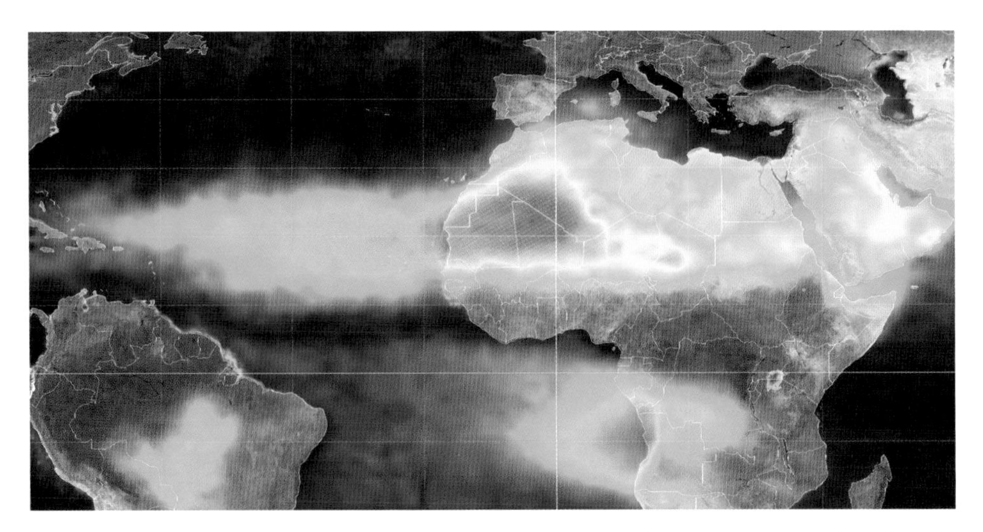

This ultraviolet satellite image of an African dust storm shows aerosols from airborne dust and smoke (red, orange, yellow, and green). The Sahara desert and the Arabian Peninsula are the source of the dust that is spreading westward across the Atlantic Ocean toward Central and North America. Smoke from biomass burning in Southern Africa and South America also spreads west. *Dr. Jay Herman/NASA/Photo Researchers, Inc.*

During the Great Depression in the United States in the 1930s, dust storms rolled across the plains, prompting some people to move to California in search of work and shelter. *Courtesy of the Library of Congress.*

Blown sand generally reaches a height up to about 49 ft (15 m) in winds of 10 mph (16 km/h). Heavy sand particles tend only to creep along the surface, moving sand dunes from one location to another. At higher wind speeds, sand particles are lifted and then bounce along the surface, a process called saltation. This process can loosen finer particles and trigger a more extensive dust storm. Very fine dust particles can be lifted by turbulent air hundreds or thousands of miles into the atmosphere and be carried many thousands of miles with the prevailing winds. Dust from extreme storms has been documented as high as 35,000 ft (10,668 m). Fine clay dust particles that are extremely small (0.00004 in/ 0.001 mm in size) can remain in the atmosphere for years.

Dust clouds prevent solar radiation from reaching the ground temporarily, causing a cooling effect at Earth's surface. However, dust clouds also absorb solar radiation thereby heating the cloud itself. Dust storms are most likely to occur in the spring when weather systems are most turbulent. Once a dust storm begins, the dust cloud can travel thousands of miles across the globe. Dust from the Sahara region of Africa is the source of much of the wind blown dust around the world, although some dust storms that originated on the Mongolian-China border have reached the United States and Canada.

WORDS TO KNOW

DESERT: A land area so dry that little or no plant or animal life can survive.

DESERTIFICATION: Transformation of arid or semiarid productive land into desert.

SALTATION: Any sudden or jumping change. Particles of snow or sand lofted by wind undergo saltation, building dunes. Climate change may encourage saltation by drying out soils.

■ Impacts and Issues

Dust from the Sahara region can travel across the Atlantic Ocean. The change in solar heating associated with a dust cloud temporarily alters Earth's climate. Cooling at the surface can occur far downwind from the dust cloud. The U.S. National Aeronautics and Space Administration (NASA) Goddard Institute of Space Studies has researched the effect of major dust storms on climate. The cooling effect from a major Arabian dust storm was observed to extend to regions from northern Asia to the

Pacific and North America. The average surface temperature was reduced by 1.8°F (1°C).

The relationship between climate change and dust storms continues to be a focus of research. Recent studies using data from the Advanced Very High Resolution Radiometer on U.S. National Oceanic and Atmospheric Administration (NOAA) polar-orbiting satellites indicate there is a correlation between increased dust storm activity in the Sahara region of North Africa and a reduction of Atlantic Ocean hurricanes. Dust storms and sand storms are seen easily by instruments on NOAA satellites. For the past 25 years, NOAA has used the data gathered by these satellites to monitor the sources of dust storms and sand storms and track their movements.

SEE ALSO *Desert and Desertification; Drought; Hurricanes; Wind Power.*

BIBLIOGRAPHY

Web Sites

"Desertification." *U.S. Geological Survey,* October 29, 1997. <http://pubs.usgs.gov/gip/deserts/desertification/> (accessed August 20, 2007).

"Dust and Sandstorms from the World's Drylands." *United Nations Convention to Combat Desertification,* 2007. <http://www.unccd.int/publicinfo/duststorms/menu.php> (accessed August 25, 2007).

"Dust Storms, Sand Storms and Related NOAA Activities in the Middle East." *NOAA (National Oceanic and Atmospheric Administration) Magazine,* April 2003. <http://www.magazine.noaa.gov/stories/mag86.htm> (accessed August 25, 2007).

Earthquakes

Introduction

Partial melting of glaciers around Greenland's southeastern and southwestern shores, where ice from the cap flows into the sea, has apparently lubricated the rock beneath the glaciers and caused their motion to accelerate. So massive are the glaciers that their jerky flow causes measurable earthquakes. The quakes are not strong enough to be a danger to human beings, but they are one indication that the Greenland ice cap is changing more quickly in response to global warming than scientists had predicted. The number of Greenland glacial quakes increased greatly starting in 2002. Greenland's contribution to sea-level rise in the next century and beyond may, therefore, be greater than scientists had previously predicted.

Historical Background and Scientific Foundations

Glacial earthquakes in Alaska and Greenland were first reported in 2003 by Göran Ekström and colleagues. Such quakes had escaped detection in the past because the vibrations they produce are mostly slow-moving (low-frequency), lacking the vast-moving (high-frequency) vibrations also produced by ordinary earthquakes that occur as a result of rock motions. Glacial earthquakes happen as glaciers—large, slow-moving rivers of ice—make sudden shifts forward. One glacial earthquake in Alaska was produced by ice sliding over a period of 30 seconds to 1 minute, which is 15–30 times longer than the rock slippages that produce typical earthquakes. This long, drawn-out motion explains the lack of high-frequency vibrations in glacial quakes.

Ekström and colleagues applied new mathematical techniques to already-recorded seismic data (records of vibrations in the ground) to detect earthquakes that earlier researchers had missed. They found 46 quakes with magnitudes between 4.6 and 5.0 on the Richter scale that had occurred in 1999 to 2001. Forty-two of these quakes were in Greenland, one in Alaska, and three were along the Antarctic coast, all places where glaciers are active.

The Greenland quakes tended to be seasonal, with more happening in the summer, when surface melting allows water to penetrate to the bottom of glaciers and decrease friction with surface rock. The area of the Greenland ice cap over which seasonal surface melting is observed has been increasing rapidly in recent years. Surface meltwater pours into holes in the glacier called moulins, penetrating eventually to the interface of rock and glacier and acting as a lubricant.

In 2006, after observing the glacial quakes for several more years and analyzing older data, Ekström and other colleagues reported that 182 glacial earthquakes had occurred in Greenland from January 1993 to October 2005. The record showed that starting in 2002, the number of quakes had increased steadily. By 2005, twice as many events were being detected per year as were detected in any previous year.

Impacts and Issues

More than two and half times as much Greenland ice melted in 2004 to 2006 as did in 2002 to 2004. The higher melt rate released 59 cubic mi (248 cubic km) of water a year, enough to raise sea level by 0.02 in (0.5 mm per year).

Almost all melting through 2004 occurred near the coasts of southern Greenland, but shortly thereafter the melting began to accelerate along Greenland's northwest coast as well. In September 2007, scientists observing the Ilulissat glacier in northwest Greenland reported that the glacier had accelerated to the point where its movement—about two yards (1.8 m) per hour—was visible to the naked eye.

Historic treasures, such as the Parthenon in the center of the Acropolis in Athens, Greece, may face increased dangers from natural disasters as the climate changes. Scientists are studying how climate change could impact the materials used in the construction of such buildings and how likely they are to withstand such events. *Image copyright Nick Koumaris, 2007. Used under license from Shutterstock.com.*

Finnish scientist Veli Kallio stated that glacial earthquakes in north-west Greenland had not existed until just three years earlier. Polar expert Robert Correll, as quoted in the September 8, 2007 issue of *The Independent*, said that when he first visited the Ilulissat glacier in 1968, moulins were rare: now, he said, they were "phenomenal. Now they are like rivers 10 or 15 meters in diameter and there are thousands of them."

The glacial lurches that cause Greenland's glacial quakes are both a side-effect and a contributor to its accelerated melting. They are a side-effect because they are enabled by the lubricating effect of surface meltwaters flowing down through moulins to the ground. They are a contributor because they are part of the glaciers' accelerated flow to the sea. When they meet the sea, glaciers crumble, breaking up into large chunks of ice that float away as icebergs and eventually melt. As soon as an iceberg enters the ocean, it raises sea level just as an ice cube put into a drink raises the level of the drink. (Melting of the iceberg or ice cube does not change the level further.)

The accelerated melting of Greenland's coastal ice impacts sea-level rise. The Intergovernmental Panel on Climate Change (IPCC), which issued its Fourth Assessment Report on climate change in early 2007, used no data collected later than the end of 2005. As such, it did not take into account the accelerated melting documented largely in 2006. The IPCC predicted likely sea-level rise by 2100 of 8–24 in (20–60 cm), but many scientists now studying Greenland's accelerated melting say that the upper end of the IPCC range is actually the lower limit of what should be expected, with a likelier upper limit being 6.6 ft (2 m). In all scenarios, larger rises would probably occur beyond 2100, which is an arbitrary cutoff date.

■ Primary Source Connection

Researchers measuring changes in the world's ice sheets and glaciers have noted an increase in glacial earthquakes, low-frequency waves similar to typical earthquakes but which can be traced to the movement of glaciers and ice shelves. The increase in seismic activity associated with Earth's large ice masses indicates that the planet's climate is warming.

Science is the journal of the American Association for the Advancement of Science (AAAS).

SEASONALITY AND INCREASING FREQUENCY OF GREENLAND GLACIAL EARTHQUAKES

Some glaciers and ice streams periodically lurch forward with sufficient force to generate emissions of elastic waves that are recorded on seismometers worldwide. Such glacial earthquakes on Greenland show a strong seasonality as well as a doubling of their rate of occurrence over the past 5 years. These temporal patterns suggest a link to the hydrological cycle and are indicative of a dynamic glacial response to changing climate conditions.

WORDS TO KNOW

GLACIER: A multi-year surplus accumulation of snowfall in excess of snowmelt on land and resulting in a mass of ice at least 0.04 mi^2 (0.1 km^2) in area that shows some evidence of movement in response to gravity. A glacier may terminate on land or in water. Glacier ice is the largest reservoir of freshwater on Earth and is second only to the oceans as the largest reservoir of total water. Glaciers are found on every continent except Australia.

ICE CAP: Ice mass located over one of the poles of a planet not otherwise covered with ice. In our solar system, only Mars and Earth have polar ice caps. Earth's north polar ice cap has two parts, a skin of floating ice over the actual pole and the Greenland ice cap, which does not overlay the pole. Earth's south polar ice cap is the Antarctic ice sheet.

INTERGOVERNMENTAL PANEL ON CLIMATE CHANGE (IPCC): Panel of scientists established by the World Meteorological Organization (WMO) and the United Nations Environment Programme (UNEP) in 1988 to assess the science, technology, and socioeconomic information needed to understand the risk of human-induced climate change.

MELTWATER: Melted ice in a glacier's bottom layer, caused by heat that develops as a result of friction with Earth's surface.

MOULIN: Vertical shaft or crevice in a glacier into which meltwater from the glacier's surface flows. Moulins allow liquid water to penetrate to the bottom of a glacier, lubricating its contact with the ground and accelerating its flow.

RICHTER SCALE: A scale used to compare earthquakes based on the energy released by the earthquake.

SEISMIC: Related to earthquakes.

Continuous monitoring of seismic waves recorded at globally distributed stations has led to the detection and identification of a new class of earthquakes associated with glaciers. These "glacial earthquakes" are characterized by emissions of globally observable low-frequency waves that are incompatible with standard earthquake models for tectonic stress release but can be successfully modeled as large and sudden glacial-sliding motions. Seismic waves are generated in the solid earth by the forces exerted by the sliding ice mass as it accelerates down slope and subsequently decelerates. The observed duration of sliding is typically 30 to 60 s. All detected events of this type are associated with mountain glaciers in Alaska or with glaciers and ice streams along the edges of the Antarctic and Greenland ice sheets. The Greenland events are most numerous, and we present new data indicating a strong seasonality and an increasing frequency of occurrence for these events since at least 2002....

Summer melting of the Greenland Ice Sheet has become more widespread during the past decade, and many outlet glaciers have thinned, retreated, and accelerated during the same time period. We investigate temporal changes in the frequency of Greenland glacial earthquakes by counting events for each year since 1993. Detections for 2005 are for January to October and are based on the subset of seismic data available in near-real time. A clear increase in the number of events is seen starting in 2002. To date in 2005, twice as many events have been detected as in any year before 2002. The control group of detected events was used to determine whether an improvement in the detection capability of the seismic network could explain the observed increase. No clear trend is seen in the number of control events detected, indicating that the observed increase in the number of glacial earthquakes is real.

Recent evidence suggests that ice sheets and their outlet glaciers can respond very quickly to changes in climate, primarily through dynamic mechanisms affecting glacier flow. The seasonal signal and temporal increase apparent in our results are consistent with a dynamic response to climate warming driven by an increase in surface melting and the supply of meltwater to the glacier base. The number of events detected at each outlet glacier using the global seismic network is relatively small, and it is therefore difficult to draw robust conclusions about behavior at any single glacier. However, both the seasonal and temporal patterns reported here are observed for independent subsets of the data corresponding to east and west Greenland. The increase in number of glacial earthquakes over time thus appears to be a response to large-scale processes affecting the entire ice sheet. We note also that a part of the increase in the number of glacial earthquakes in west Greenland is due to the occurrence of more than two dozen of these earthquakes in 2000 to 2005 at the northwest Greenland glaciers, where only one event (in 1995) had previously been observed.

Understanding the mechanisms of the dynamic response of ice sheets to climate change is important in part because ice-sheet behavior itself affects global climate, through, for example, the modulation of freshwater input to the oceans. Glacial earthquakes represent one mechanism for the dynamic thinning of outlet glaciers, providing for the transport of a large mass of ice a distance of several meters (e.g., 10 km^3 by 10 m) over a duration of 30 to 60 s. Although the mechanics of sudden sliding motions at the glacier base are not known, the seasonal and temporal patterns reported here suggest that the glacial earthquakes may serve as a marker of ice-sheet response to external forcing. Continuous monitoring of ice velocity at outlet glaciers, along with regional seismic monitoring, would provide important insight into the nature of the dynamic response of ice sheets to changes in climate.

Göran Ekström, Meredith Nettles, and Victor C. Tsai.

EKSTRÖM, GÖRAN, MEREDITH NETTLES, AND VICTOR C. TSAI. "SEASONALITY AND INCREASING FREQUENCY OF GREENLAND GLACIAL EARTHQUAKES." *SCIENCE* 311 (2006): 1756–1758.

SEE ALSO *Arctic Melting: Greenland Ice Cap; Greenland: Global Implications of Accelerated Melting.*

BIBLIOGRAPHY

Periodicals

Dowdeswell, Julian. "The Greenland Ice Sheet and Global Sea-Level Rise." *Science* 311 (2006): 963–964.

Ekström, Göran, et al. "Glacial Earthquakes." *Science* 302 (2003): 622.

Howden, Daniel. "Shockwaves from Melting Icecaps Are Triggering Earthquakes, Say Scientists." *The Independent*, September 8, 2007.

Joughlin, Ian. "Greenland Rumbles Louder as Glaciers Accelerate." *Science* 311 (2006): 1719–1720.

Murray, Tavi. "Greenland's Ice on the Scales." *Nature* 443 (2006): 277–278.

Larry Gilman

Economics of Climate Change

■ Introduction

The economics of climate change refers to the study of the economic costs and benefits of climate change, along with the economic impact of actions aimed at limiting its effects.

Participants in the climate change debate—from government to non-governmental organizations (NGOs) to academia—have increasingly used economic assessments to determine the costs of addressing climate change.

In recent years, influential economic analyses, such as the 2006 *Stern Review*, have influenced the climate debate and shifted climate change policy by countering assertions that greener technologies are too expensive. Increasing support among economists for action to mitigate possible causes of climate change is especially significant given the public perception—particularly in the United States and many developing nations—that the main arguments against action are economic (i.e., that the costs of effective policies would be too large, or that delay is economically justifiable).

■ Historical Background and Scientific Foundations

The field of climate change economics has evolved in tandem with increased scientific knowledge of climate

German minister for construction and housing, Wolfgang Tiefensee, unveils a new climate change campaign in March 2007. The campaign's slogan reads: "Don't burn your money: Dress your house warm. Renovation saves heating costs, protects the climate and creates jobs." *AP Images.*

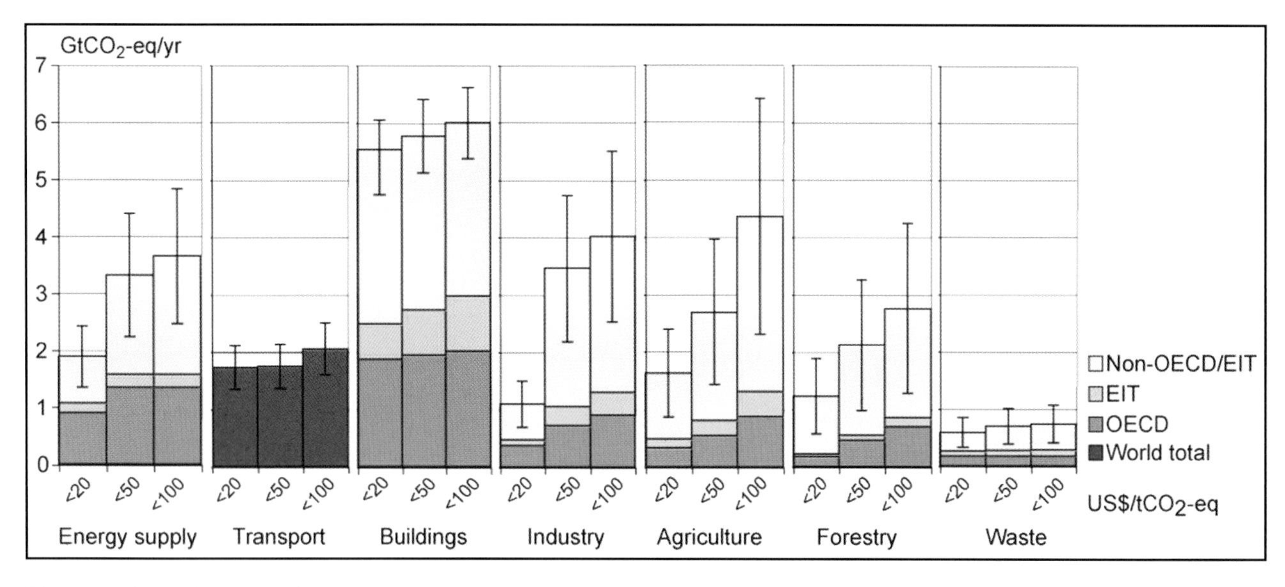

The image shows the estimated economic mitigation potential (ability to offset the projected growth of global emissions or reduce emissions below current levels) for different regions in each sector for 2030. Uncertainties in the estimates are shown as ranges in the tables above to reflect the ranges of baselines, rates of technological change, and other factors that are specific to the different approaches. Furthermore, uncertainties also arise from the limited information for global coverage of countries, sectors and gases. EIT stands for Economies in Transition and refers to countries of the former Soviet bloc which are transitioning to a market economy. OECD stands for the 30 member countries of the Organisation for Economic Co-operation and Development. *Climate Change 2007: Mitigation of Climate Change, Summary for Policymakers, Intergovernmental Panel on Climate Change.*

change itself. In February 1997, 2,500 U.S. economists, including eight Nobel laureates, signed the "Economists' Statement on Climate Change"—a short, three-point document acknowledging human impact on Earth's climate, and the "environmental, economic, social, and geopolitical risks" justifying preventative measures. There were many counteractive policies, it said, "for which the total benefits outweigh the total costs." And that "the most efficient approach to slowing climate change is through market-based policies." This can be the starting point for the modern study of the economics of climate change: the majority of studies and reports published since then have concerned themselves with finding the most monetarily efficient way of mitigating global warming.

Not all scientists and economists are agreed that positive mitigation action is the best way forward. However, in 2004, Danish academic Bjørn Lomborg, who later wrote *Cool It: The Skeptical Environmentalist's Guide to Global Warming* (2007), founded the Copenhagen Consensus, a project in which a panel of economists established priorities for advancing global welfare using the theory of welfare economics. Welfare economics is the study of how best to distribute economic product among competing claimants. The experts ranked 17 global challenges—from climate change to water sanitation—in order of priority. The panel ranked lowest the three projects addressing climate change.

Opponents of the Kyoto Protocol have used economic arguments to bolster claims that Kyoto implementation would be too expensive or harm economic growth. At a July 2002 hearing on U.S. climate change policy under the President George W. Bush administration, James L. Connaughton, the chairman of the Council on Environmental Quality, testified that implementing the Kyoto Protocol would reduce U.S. economic output by "up to $400 billion" in 2010. His estimate was similar to a $397 billion estimate that appeared in a 1998 report by the Energy Information Administration (EIA), an independent statistical and analytical agency within the U.S. Department of Energy. The figures projected conflicted with the results of a number of other studies, however.

Initially most economics research on climate change employed simple cost-benefit analyses framed by two key questions: What are the potential costs of cutting greenhouse-gas emissions? Can such reductions be achieved without sacrificing economic output? In other words: do the costs of climate change prevention outweigh its possible benefits, or vice-versa? Published in October 2006 by the British government, the *Stern Review on the Economics of Climate Change*—a 700-page report authored by economist and Cambridge University Fellow Sir Nicholas Stern—also asked a third question: what are the economic costs if we do nothing? The report's answers, which mentioned that inaction on climate change issues could lead to environmental devastation and long-term economic crisis, attracted international attention for the report.

The *Stern Review* estimated that without international action to counteract climate change "the overall costs and risks of climate change will be equivalent to

WORDS TO KNOW

CARBON TAX: Mandatory fee charged for the emission of a given quantity of CO_2 or some other greenhouse gas. Under a carbon taxation scheme, polluters who emit greenhouse gases must pay costs that are directly proportional to their emissions. The purpose of a carbon tax is to reduce greenhouse emissions. Carbon taxation is the main alternative to emissions trading.

CARBON TRADING: Buying and selling of carbon credits, abstract instruments (like money) that each represent the right to emit 1 ton of carbon dioxide or an equivalent amount of other greenhouse gases. Carbon trading presently takes place under the European Union Emission Trading Scheme and the Chicago Climate Exchange.

ECONOMISTS' STATEMENT ON CLIMATE CHANGE: A statement endorsed by over 2,500 economists including eight Nobel Prize winners: the economists stated that "global climate change carries with it significant environmental, economic, social, and geopolitical risks, and that preventive steps are justified." They also empasized that preventive steps would not necessarily be harmful to economies and might in fact "improve productivity in the long run."

GREENHOUSE GASES: Gases that cause Earth to retain more thermal energy by absorbing infrared light emitted by Earth's surface. The most important greenhouse gases are water vapor, carbon dioxide, methane, nitrous oxide, and various artificial chemicals such as chlorofluorocarbons. All but the latter are naturally occurring, but human activity over the last several centuries has significantly increased the amounts of carbon dioxide, methane, and nitrous oxide in Earth's atmosphere, causing global warming and global climate change.

***STERN REVIEW*:** *The Stern Review on the Economics of Climate Change*, a controversial report commissioned by the government of the United Kingdom and written by British economist Nicholas Stern (1946–). The *Stern Review* predicted that climate change, if not mitigated, will eventually severely damage world economic growth, causing disruptions comparable to those of the world wars and Great Depression of the twentieth century.

losing at least 5% of global GDP each year.... If a wider range of risks and impacts is taken into account, the estimates of damage could rise to 20% of GDP or more." Stern contrasted this with the economic cost of taking action to reduce climate change—around 1% of global GDP. He wrote of the necessity of taking prompt action:

> The investment that takes place in the next 10–20 years will have a profound effect on the climate in the second half of this century and in the next. Our actions now and over the coming decades could create risks of major disruption to economic and social activity, on a scale similar to those associated with the great wars and the economic depression of the first half of the twentieth century. And it will be difficult or impossible to reverse these changes.

At the core of Stern's recommendations was the necessity to stabilize greenhouse-gas levels in the atmosphere between 450 and 550 parts per million (ppm) CO_2 equivalent (CO_2e). The current level is 430ppm CO_2e and according to Stern is rising at more than 2ppm each year. In order to achieve this, emissions would need to be reduced by at least 25% from current levels by 2050. In the long term, Stern recommended that annual emissions be brought down to more than 80% below current levels, while acknowledging that such action represented a major challenge.

Stern put forward a three-point plan for action. First, implement a regulatory system through which carbon could be taxed and traded. Second, foster government support for the development and the deployment of low-carbon technologies. Third, educate the public and policymakers about climate change and increase individual responsibility. Stern further asserted that action by individual countries was insufficient and that addressing climate change requires a sustained, cooperative international response.

The environmental costs of climate change, Stern argued, would be felt first by the least-developed nations and developing countries, many of whom had contributed least to greenhouse-gas levels. Stern argued that developing countries should not be required to bear the full costs of stabilizing greenhouse-gas levels alone.

Stern acknowledged that climate change prevention could cause some mild to moderate market losses. However, he noted that some of those initial losses would be offset by growth in the development and production of green technologies. Entire new markets will be created in low-carbon energy technologies and other low-carbon goods and services worth hundreds of billions of dollars each year. "Changes in energy technologies and in the structure of economies have created opportunities to decouple growth from greenhouse gas emissions," argued Stern, adding that "tackling climate change is the pro-growth strategy for the longer term."

Stern marked the onset of a new consensus amongst economists regarding climate change. The International Monetary Fund's October 2007 *World Economic Outlook* concluded that the economic impacts of climate change were potentially substantial, and could include a number of wide-ranging effects. These ranged from a

IN CONTEXT: INVESTMENT BENEFITS

"Government support through financial contributions, tax credits, standard setting and market creation is important for effective technology development, innovation and deployment. Transfer of technology to developing countries depends on enabling conditions and financing *(high agreement, much evidence)*."

"Public benefits of RD&D investments are bigger than the benefits captured by the private sector, justifying government support of RD&D."

"Government funding in real absolute terms for most energy research programmes has been flat or declining for nearly two decades (even after the UNFCCC came into force) and is now about half of the 1980 level."

"Governments have a crucial supportive role in providing appropriate enabling environment, such as, institutional, policy, legal and regulatory frameworks, to sustain investment flows and for effective technology transfer without which it may be difficult to achieve emission reductions at a significant scale. Mobilizing financing of incremental costs of low-carbon technologies is important. International technology agreements could strengthen the knowledge infrastructure."

SOURCE: *Metz, B., et al, eds.* Climate Change 2007: Mitigation of Climate Change: Contribution of Working Group III to the Fourth Assessment Report of the Intergovernmental Panel on Climate Change. *New York: Cambridge University Press, 2007.*

direct negative impact on output and productivity from long-term temperature change, particularly for agriculture, fisheries, and tourism; to costs from sea-level rise and increased flooding. It also highlighted many of the concerns mentioned by Stern, such as the increased risk of widespread migration and conflict; weakened tax bases; and increasing national debt burdens. Positive effects were also mentioned, such as long-term environmental health, natural resource protection, and the "double dividend" promised by mitigation schemes.

■ Impacts and Issues

No economic paper can predict the future. Behind each analysis is an economic model with its own set of assumptions, its own definitions of how the economy works, and its own data sets. As such, it remains endlessly open to interpretation and counter argument.

The United States has been accused by a majority of environmental organizations of using worst-case scenarios to support its decision to reject the Kyoto Protocol. An oft-quoted figure—provided by the White House Council on Environmental Quality—of a $400 billion cost for Kyoto implementation is at odds with figures cited in other studies. A 1997 study by the Council of Economic Advisors (CEA) found that the U.S. costs of implementing the Kyoto Protocol could be as little as $7–12 billion, depending on the extent of international emissions trading allowed and the participation of developing countries. The estimate of the White House Council on Environmental Quality assumed that all reductions would be achieved domestically, while the CEA estimate allowed for the purchase of emissions reductions from other nations. Both reports used different growth rates. Neither methodology used was incorrect, rather its subjective elements left room for interpretation. Therefore, the different types of models produced different types of cost estimates for implementing the Kyoto Protocol.

The methodology of the *Stern Report* came under scrutiny when it was published in 2006. One of the most renowned experts on the economics of climate change, Richard Tol, produced a detailed critique of the report, which he concluded was "alarmist and incompetent." Tol wrote that Stern used only the most pessimistic impact studies and had no real cost-benefit analysis. Tol concluded that "unsound analyses like the *Stern Report* only provide fodder for those skeptical of climate change and climate policy." This skepticism was mirrored by several other interest groups and individuals. For example, OPEC, the Organization of Petroleum Exporting Countries, called the *Stern Report* misguided and stated that the scenarios used by Stern had little scientific or economic basis.

Nevertheless, the *Stern Report* attained widespread support of an unlikely coalition of climatologists, scientific bodies, NGOs, and economists. "It provides the vital missing link between global economics and the emerging and overwhelming evidence of human influence on climate change," said Chris Huntingford of the Centre for Ecology and Hydrology. Martin Rees, president of the Royal Society, put the impact he hoped it would have in the following terms: "this should be a turning point in a debate which has pitted short term economic interests against long-term costs to the environment, society and the economy."

The overwhelming message of the *Stern Review* was that it was time to move on from arguing about the existence of climate change to taking preventative action. The report's conclusions were that a rise in temperatures of 9–11°F (5–6°C) could trigger a global loss of economic wealth of up to 20%. What arguably separated the *Stern Review* from other major studies and articles on climate change, was not its basic hypothesis, but rather

California Governor Arnold Schwarzenegger (left) and British Prime Minister Tony Blair met with a coalition of business leaders in 2007 to discuss what actions could be taken to create a low-carbon economy as well as other issues pertaining to global climate change. *AP Images.*

its highly controversial appeal to the self-interest of developed countries. Citing financial disarray and possible economic collapse as well as the possibility of the developed world being flooded with climate change refugees, Stern put forth a financial choice to the world: spend 1% of your GDP now and avoid a potential loss of 20% later.

Upon the publication of the *Stern Review,* the British Chancellor of the Exchequer (prime minister as of June 2007) Gordon Brown, promised to expand carbon trading and legislate on carbon reductions, with an independent body to monitor progress and to push to reduce European-wide emissions by 30% by 2020 and over 60% by 2050. Although such pledges were welcomed by a majority of British environmentalists, Brown immediately also received criticism from the same sector for not going far enough. For example, many environmental advocates sought regulation of private car and airline emissions, higher taxes on large polluters, and increased investment in clean technologies.

European leaders reached a consensus to reduce greenhouse emissions by 60% by 2050. But the *Stern Report* also sought to persuade skeptical governments, particularly that of the United States, to undertake reforms. Stern traveled to Washington, D.C., to discuss his findings. He met with President George W. Bush, who called the report a "contribution to the body of knowledge on climate change," but did not endorse a shift in current U.S. climate change policy. Eileen Claussen, the president of the Pew Center on Global Climate Change, suggested that the *Stern Report* would only belatedly have an impact in the United States. "We are at the beginning of a serious debate about what to do about climate change.... [m]uch of the opposition will be over the costs, and Stern has been able to talk to the costs of inaction."

SEE ALSO *Environmental Policy; Stern Review; United States: Climate Policy.*

BIBLIOGRAPHY

Periodicals

De Leo, Giulio, L. Rizzi, A. Caizzi, and M. Gatto. "The Economic Benefits of the Kyoto Protocol." *Nature* 413 (October 4, 2001): 478–79.

Giles, Jim. "How Much Will It Cost to Save the World?" *Nature* 444 (November 2, 2006): 6–7.

"A Global Call to Arms." *Nature* 444 (November 2, 2006): 2.

"The Heat Is On: Survey—Climate Change." *Economist* (September 7, 2006).

Sachs, Jeffrey. "Averting Disaster: At What Cost?" *Nature Reports: Climate Change* (June 7, 2007).

Spash, Clive. "Economics, Ethics, and Long Term Environmental Damages." *Environmental Ethics* 15 (1993).

Web Sites

"The Economists' Statement on Climate Change." *Redefining Progress: The Nature of Economics,* 2001. <http://www.rprogress.org/publications/2001/econstatement.htm> (accessed December 7, 2007).

Shogren, Jason, and Michael Toman. "How Much Climate Change Is Too Much? An Economics Perspective." *Climate Change Issues Brief, No. 25,* September 2005. <http://www.rff.org/Documents/RFF-CCIB-25.pdf> (accessed November 26, 2007).

Stern, Nicholas, et al. "The Stern Review of the Economics of Climate Change: Final Report." *HM Treasury* (British government). <http://www.hm-treasury.gov.uk/independent_reviews/stern_review_economics_climate_change/stern_review_report.cfm> (accessed November 26, 2007).

Tol, Richard S. J. "The Stern Review of the Economics of Climate Change: A Comment." <http://www.fnu.zmaw.de/fileadmin/fnu-files/reports/sternreview.pdf> (accessed November 26, 2007).

James Corbett

Economies in Transition

Introduction

A transitional economy is one in the process of changing from a centrally planned economy to a free market. Economists generally regard the process to take ten years.

Historical Background and Scientific Foundations

Since the collapse of communism from 1989 to 1990, dozens of countries have seen their economies shift from centrally planned fixed economies to a free market. Transition economies undergo a period that can roughly be dissected into three often-overlapping parts: economic liberalization; macroeconomic stabilization; and privatization of state-run enterprises. Other reforms also take place, such as land or property ownership changes as well as the creation of stock exchanges. Such changes often lead to increased income inequality, dramatic inflation, and a fall of the gross domestic product (GDP).

Impacts and Issues

In the context of climate change and the terms of the Kyoto Protocol, transitional economies assume considerable significance. This is because targets of the Kyoto treaty are marked against 1990 emission levels. This was the last year of the Union of Soviet Socialist Republics (USSR) and a time when the Soviet client economies of Eastern Europe were functioning almost at full capacity. Peace also still reigned in Yugoslavia. As such, emissions levels were high.

The subsequent break up of the USSR led to the collapse of most former Soviet republics' economies. In

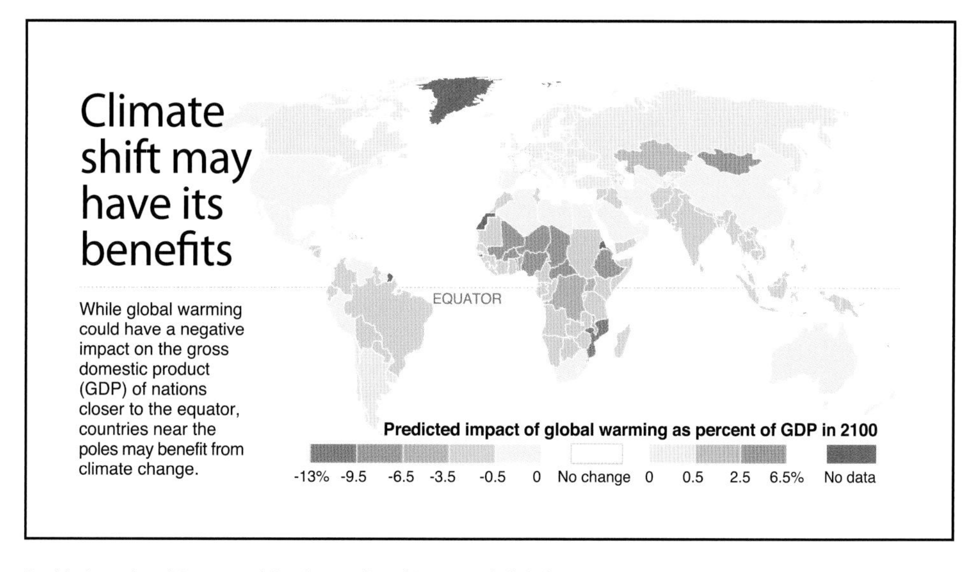

In this Associated Press graphic, the predicted impact of global warming as a percent of gross domestic product (GDP) is presented. [Source: Yale University School of Forestry and Environmental Studies.] *AP Images.*

WORDS TO KNOW

ECONOMIC LIBERALIZATION: Reduced government regulation of corporate behavior and state ownership of resources (e.g., oil, mines). Economic liberalization has been presented by many of its advocates as the only possible route to economic prosperity and is blamed by critics for such disasters as Chile's slide into greater unemployment and poverty starting in 1979, when military dictator Augusto Pinochet imposed economic liberalization measures. Those who criticize economic liberalization consider the term to be a euphemism (a nice-sounding name for something that is not nice). Liberalization in the form of "structural adjustment programs" has often been imposed on countries as a condition for receiving loans from the World Bank.

GROSS DOMESTIC PRODUCT (GDP): A measure of total economic activity, whether of a nation, group of nations, or the world: slightly different from Gross National Product (GNP, used by economists until the early 1990s). Defined as the total monetary value of all goods and services produced over a given period of time (usually one year). GDP's limitations have been pointed out by many economists. GDP is a bulk or aggregate statistic and does not take into account inequity: thus, a country's GDP might increase while 99% of its population got poorer, as long as the richest 1% grew sufficiently richer during the same period.

MACROECONOMIC STABILIZATION: Stabilization of an economy at the level of a whole nation (i.e., at the macroeconomic level). In many countries, governments manipulate certain financial variables to seek macroeconomic stabilization: for example, in the United States, the Federal Reserve Bank raises and lowers interest rates.

PRIVATIZATION: Transfer of state-owned resources to private owners, whether by sale or outright gift.

the former USSR, wealth was only 63% of its 1990 level ten years later. In client states, such as Poland where so-called economic "shock therapy" was applied, the transition to market economies was painful and beset by factory closures and industrial decline. In Yugoslavia, civil war broke out in 1991, grinding its economy to a halt almost for the rest of the decade.

In all of these countries, economic recession and turmoil fed industrial decline and a huge fall in carbon emissions. Russia, for example, saw a 23.9% fall between 1992 and 1998 and may have experienced a drop as high as 40% from its Soviet-era high. Albania, a particularly extreme example, experienced a fall in emissions of 72% from 1990 to 1991 alone as its economy went into meltdown after the collapse of its Soviet bankrollers.

Even when these economies recovered, the building of cleaner, more efficient industrial plants, the use of new technologies, or population decrease (another side-effect of transitory economies) have meant that carbon emissions have seldom reached their 1990 levels. As a result, most former Soviet-bloc countries have, in a swoop, met their Kyoto pledges and acquired a considerable surplus for trading on the emergent carbon market. In Russia's case, with carbon credits worth up to $60 billion, this was the key to its ratification of the Kyoto Protocol.

The other side of the case is China. With a malfunctioning and inefficient centrally planned economy and a majority of its population living in the countryside, in 1990 it boasted comparatively low carbon emissions. However, market reforms from the early 1990s onward have transformed it into an economic superpower. A majority of its people now lives in cities, and Chinese factories supply the world with its manufactured goods. China's CO_2 emissions increased by 109% between 1990 and 2004 and may double again by 2012. A nation whose lack of development meant it was not party to Kyoto targets in 1997 overtook the United States as of 2007 as the world's leading producer of CO_2 emissions.

SEE ALSO *China: Total Carbon Dioxide Emissions; Emissions Trading; Kyoto Protocol.*

BIBLIOGRAPHY

Periodicals

Baumert, Kevin A., Elana Petkova, Diana Barbu. "Capacity for Climate: Economies in Transition after Kyoto." *World Resources Institute* (1999).

Kramer, Andrew E. "Russian Energy Giant to Bundle Carbon Credits with Gas Sales." *New York Times* (April 25, 2007).

"World Bank; Transition the First Ten Years: Analysis and Lessons for Eastern Europe and the Former Soviet Union." *World Bank* (2002).

James Corbett

El Niño and La Niña

Introduction

El Niño refers to an unusually warm, persistent current of surface water found in the tropical eastern Pacific Ocean. An El Niño event generally lasts one to two years and appears every two to seven years, usually between the end of one year and the beginning of the next year. The event is located along the western coast of Ecuador and Peru in South America. Similar events usually occur annually, but are not classified as El Niño because of their smaller nature.

La Niña, also sometimes called El Viejo (anti-El Niño), refers to an unusually cold, enduring current of surface water. Generally, these colder surface-water temperatures are found, like El Niño, in the tropical eastern Pacific Ocean. La Niña events occur about half as often, and are much less severe than El Niño events.

Meteorologists define El Niño and La Niña events as any sustained surface temperature change in the tropical eastern Pacific Ocean of over 0.9°F (0.5°C), whether it is a large temperature increase (El Niño) or decrease (La Niña). Together, these two ocean temperature fluctuations are known as the El Niño/Southern Oscillation.

Historical Background and Scientific Foundations

The effects of El Niño, which is Spanish for male child, were first noted by British physicist Gilbert Thomas Walker (1868–1958). The first mention of an unusually warm current being called an "El Niño" supposedly occurred when Peruvian sailors and fishermen noticed warm waters around Christmas-time and named it after the infant Jesus.

Normally, eastern trade winds move tropical waters in the Pacific Ocean westward toward southeastern Asia, such as Indonesia, Papua New Guinea, and Australia. These waters are warmer because they are heated longer by the sun due to their equatorial proximity.

However, periodically, an El Niño occurs when trade winds weaken, or even reverse, and allow the warm western Pacific waters to drift back eastward toward the western coast of South America. The cool, nutrient-rich water is replaced by warmer, but nutrient-deficient water, which causes reductions in marine fish and plant life if the condition remains over many months. Such a change in ocean patterns also causes global atmospheric changes, such as storms and heavy rains to South America, and unusual weather to North America. It also causes droughts in southeastern Asia and southern Africa.

Early signs indicating an El Niño include rising air pressures and drier than normal air in the western Pacific (causing arid conditions) and declining air pressures and more moist air in the western Pacific and Indian Oceans (causing wetter conditions).

A La Niña, which is Spanish for female child, follows an El Niño, especially if the El Niño is a major one. A La Niña generally brings atmospheric and oceanic changes that are the opposite of those of a El Niño. In particular, the Pacific Northwest in North America becomes much wetter than

WORDS TO KNOW

EQUATORIAL: Concerning or related to Earth's equator, the line dividing the Northern and Southern Hemispheres.

PERUVIAN ANCHOVETA: A fish species related to anchovies, found in the southeastern Pacific, especially off the coasts of Chile and Peru. Populations vary with climate changes such as those produced by El Niño.

TRADE WINDS: Surface air from the horse latitudes (subtropical regions) that moves back toward the equator and is deflected by the Coriolis Force, causing the winds to blow from the Northeast in the Northern Hemisphere and from the Southeast in the Southern Hemisphere. These steady winds are called trade winds because they provided trade ships with an ocean route to the New World.

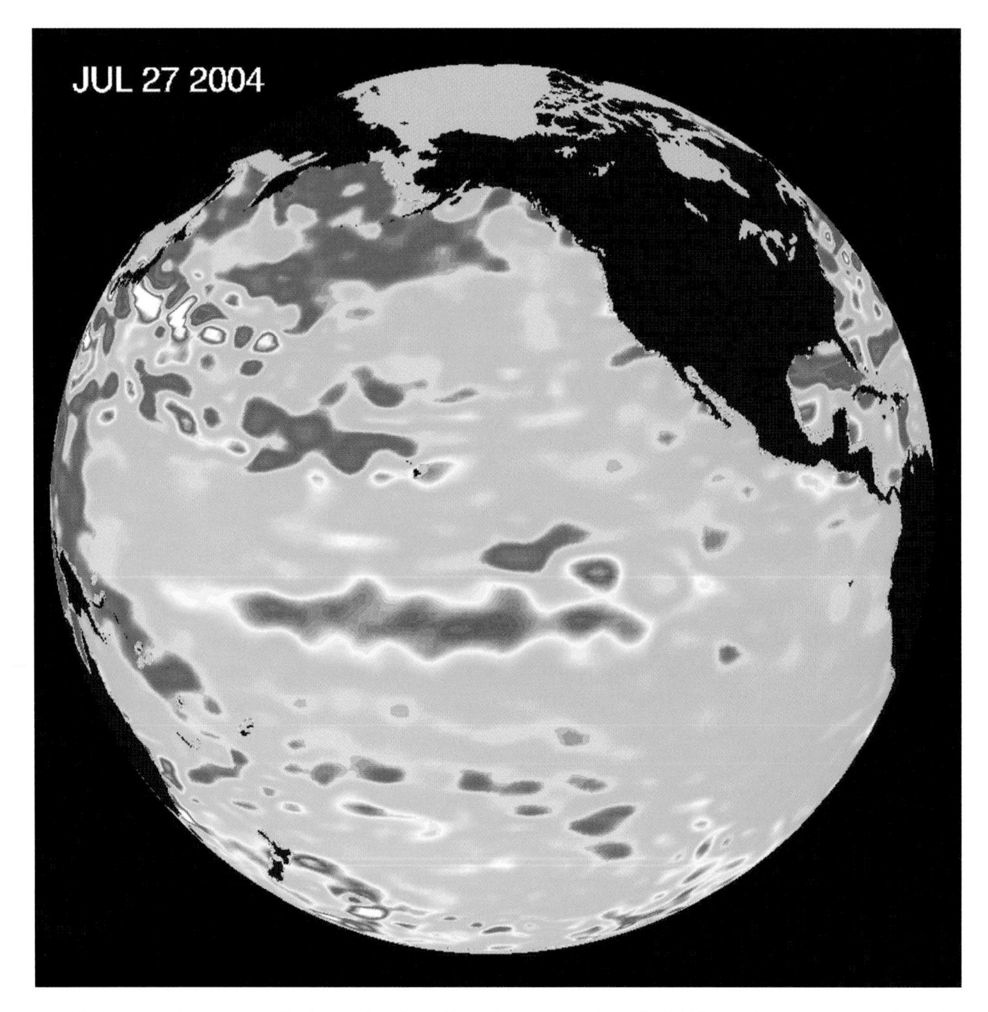

In this image taken over a 10-day cycle of satellite observations in July 2004, weaker-than-normal trade winds are shown in the western and central equatorial Pacific during June, triggered by a Kelvin wave, moving eastward. The image, produced by the Jason satellite, shows warmth as a red area in the central Pacific that has a sea-level height about 4 in (10 cm) above normal. Sea levels depicted in blue indicate areas where the heights are 2 to 5 in (5 to 13 cm) below normal. Scientists pay close attention to Kelvin waves as they may signal the start of an El Niño event. *AP Images.*

normal, while the southwestern United States is much dryer and warmer.

Impacts and Issues

Major El Niño and La Niña events can cause devastating global economic problems, along with dramatic changes in the climate and weather, especially in the Southern Hemisphere. In fact, a major El Niño event is considered a primary cause of global climate variability, especially for areas adjoining the Atlantic, Indian, and Pacific Oceans.

When a prolonged El Niño occurs, its warm waters cause severe storms and usually produce abnormally high rainfall in the east-central and eastern regions of the Pacific Ocean. Warmer and wetter weather usually result in the Americas, especially in Peru and Ecuador in South America. Other parts of South America usually receive abnormal weather, such as milder winters in central Chile.

As a result of El Niño, the midwestern portions of Canada and the United States often receive warmer than normal temperatures in winter, while the southwestern part of North America receives wetter but colder temperatures. All the while, the Pacific Northeast receives dry and foggy winters and warm and sunny springs. By contrast, as a result of La Niña, the midwestern region of North America is drier than normal.

El Niño events can destroy vast numbers of marine life. Large fish populations are fed from the nutrients contained in cold ocean waters. When warmer waters replace colder ones, nutrients are not readily available. Large losses of fish and marine life can result during prolonged warming periods. For instance, the number of Peruvian anchoveta (*Engraulis ringens*) off the coast of Peru were sharply reduced in 1972 due to El Niño. Similarly, sea lions off the Peruvian coast once numbered in the hundreds of thousands. However, because of El

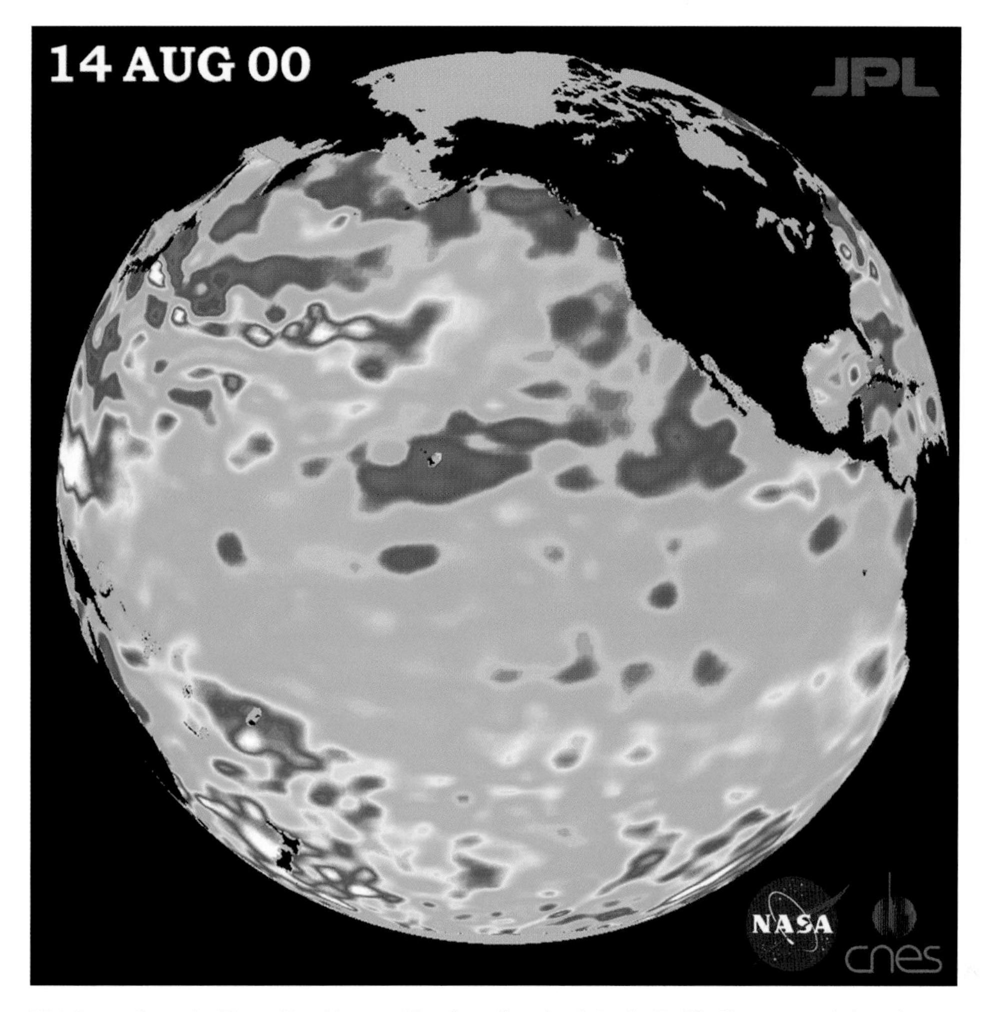

This image from the Topex-Poseidon satellite shows heat levels in the Pacific Ocean recorded on August 14, 2000. The image shows that for the first time in three years, the tropic Pacific Ocean wasn't running unusually hot or cold. Climatologists joked that the current condition should be called "No Niña." The red areas mark the warmer waters and the blue areas mark the cold. *AP Images.*

Niño, their numbers have dropped to the tens of thousands. Many have died or migrated to other areas.

The scientific community has speculated that global warming may have a cycle similar to El Niño and La Niña cycles. They are investigating the possible relationship. In addition, scientists are still debating whether global warming increases the intensity and/or frequency of El Niño and La Niña, events.

As the exact cause(s) of El Niño and La Niña events are not known, such cycle similarities are being studied seriously by scientists. Many theories have been proposed; however, so far, none have been validated. In fact, it is still difficult for scientists to predict the next event. Recent El Niño events have occurred in 1997 to 1998; 2002 to 2003; 2004 to 2005; and late-2006 to mid-2007. The last strong La Niña event occurred from 1988 to 1989. In fact, the 1997-to-1998 event first brought attention to the worldwide consequences—massive food shortages for sea creatures and humans, and unusual weather patterns—that a major El Niño event can cause.

SEE ALSO *Abrupt Climate Change; Climate Change; Global Warming; Ocean Circulation and Currents; Oceans and Seas; Oscillation; Sea Temperatures and Storm Intensity; Solar Radiation; South America: Climate Change Impacts.*

BIBLIOGRAPHY

Books

Glantz, Michael H. *Currents of Change: Impacts of El Niño and La Niña on Climate and Society.* New York: Cambridge University Press, 2001.

Philander, S. George. *Our Affair with El Niño: How We Transformed an Enchanting Peruvian Current into a Global Climate Hazard.* Princeton, NJ: Princeton University Press, 2004.

Pittock, A. Barrie. *Climate Change: Turning Up the Heat*. Sterling, VA: Earthscan, 2007.

Voituriez, Bruno. *El Niño: Fact and Fiction*. Paris, France: United Nations Educational, Scientific and Cultural Organization, 2000.

Web Sites

"NOAA El Niño Page." *National Oceanic and Atmospheric Administration (NOAA)*. <http://www.elnino.noaa.gov/> (accessed November 27, 2007).

Electromagnetic Spectrum

Introduction

Light is electromagnetic radiation. Climate change studies make frequent reference to solar radiation, insolation, etc., and various types of light (e.g., ultraviolet, infrared, etc.) that are part of the electromagnetic spectrum of light.

The electromagnetic spectrum is a continuous range of frequencies or wavelengths (each determines the other) of electromagnetic radiation. The spectrum ranges from long-wavelength, low frequency radio waves to short-wavelength, high frequency gamma rays. The electromagnetic spectrum is traditionally divided into radio waves, microwaves, infrared radiation, visible light, ultraviolet rays, x rays, and gamma rays. The divisions between these types of rays are invented, not physical.

Historical Background and Scientific Foundations

In a vacuum such as in space, light travels at 186,000 miles per second (300,000 kilometers per second), or what is called the speed of light. The nature of light has been the subject of controversy for thousands of years.

The Greeks were the first to theorize about the nature of light. Led by the scientists Euclid and Hero (first century AD), they came to recognize that light traveled in a straight line. However, they believed that vision worked by intromission—that is, that light rays originated at the eye and traveled to the object being seen. Despite this erroneous hypothesis, the Greeks were able to successfully study the phenomena of reflection and refraction and derive the laws governing them. In reflection, they learned that the angles of incidence and reflection were approximately equal; in refraction, they saw that a beam of light would bend as it entered a denser medium (such as water or glass) and bend back the same amount as it exited.

Another contributor to the embryonic science of optics was Arab mathematician and physicist Alhazen (965–1039). Experimenting around the year 1000, Alhazen showed that light comes from a source (the sun) and reflects from an object to the eyes, thus allowing the object to be seen. He also studied mirrors and lenses and further refined the laws of reflection and refraction.

By the twelfth century, scientists felt they had solved the riddles of light and color. English philosopher Francis Bacon (1561–1626) contended that light was a disturbance in an invisible medium which could be detected by the eye; subsequently, color was caused by objects "staining" the light as it passed. More productive research into the behavior of light was sparked by the new class of realistic painters, who strove to better understand perspective and shading by studying light and its properties.

In the early 1600s, the refracting telescope was perfected by Galileo and Johannes Kepler, providing a reliable example of the laws of refraction. These laws were further refined by Willebrord Snell, whose name is most often associated with the equations for determining the refraction of light. By the mid-1600s, enough was known about the behavior of light to allow for the formulation of a wide-range of theories.

English physicist and mathematician Sir Isaac Newton (1642–1727) was intrigued by the so-called "phenomenon of colors"—the ability of a prism to produce colors from white light. It had been generally accepted that white was a single color, and that a prism could somehow combine white light with others to form a multicolored mixture. Newton, however, doubted this assumption. He used a second prism to recombine the rainbow spectrum back into a beam of white light; this showed that white light must be a combination of colors, not the other way around.

Newton performed his experiments in 1666 and announced them shortly thereafter, subscribing to the corpuscular (or particulate) theory of light. According to this theory, light travels as a stream of particles that originate from a bright source and are absorbed by the eye. Aided by Newton's reputation, the corpuscular theory soon became accepted throughout Great Britain and in parts of Europe.

In the European scientific community, many scientists believed that light, like sound, traveled in waves. This group of scientists was most successfully represented by Dutch physicist Christiaan Huygens (1629–1695), who challenged Newton's corpuscular theory. He argued that a wave theory could best explain the appearance of a spectrum as well as the phenomena of reflection and refraction.

Newton immediately attacked the wave theory. Using some complex calculations, he showed that particles, too, would obey the laws of reflection and refraction. He also pointed out that, if truly a wave form, light should be able to bend around corners, just as sound does; instead it cast a sharp shadow, further supporting the corpuscular theory.

In 1660, however, Italian mathematician and physicist Francesco Maria Grimaldi (1618–1663) examined a beam of light passing through a narrow slit. As it exited and was projected upon a screen, faint fringes could be seen near the edge. This seemed to indicate that light did bend slightly around corners; the effect, called diffraction, was adopted by Huygens and other theorists as further proof of the wave nature of light.

One piece of the wave theory remained unexplained. At that time, all known waves moved through some kind of medium—for example, sound waves moved through air and kinetic waves moved through water. Huygens and his allies had not been able to show just what medium light waves moved through; instead, they contended that an invisible substance called ether filled the universe and allowed the passage of light. This unproven explanation did not earn further support for the wave theory, and the Newtonian view of light prevailed for more than a century.

The first real challenge to Newton's corpuscular theory came in 1801, when British physicist and physician Thomas Young (1773–1829) discovered interference in light. He passed a beam of light through two closely spaced pinholes and onto a screen. If light were truly particulate, Young argued, the holes would emit two distinct streams that would appear on the screen as two bright points. What was projected on the screen instead was a series of bright and dark lines—an interference pattern typical of how waves would behave under similar conditions.

If light is a wave, then every point on that wave is potentially a new wave source. As the light passes through the pinhole, it exits as two new wave fronts, which spread out as they travel. Because the holes are placed close together, the two waves interact. In some places the two waves combine (constructive interference), whereas in others they cancel each other out (destructive interference), thus producing the pattern of bright and dark lines. Such interference had previously been observed in both water waves and sound waves and seemed to indicate that light, too, moved in waves.

The corpuscular view did not die easily. Many scientists had allied themselves with the Newtonian theory and they were unwilling to risk their reputations to support an antiquated wave theory. Also, English scientists were not pleased to see one of their countrymen challenge the theories of Newton; Young, therefore, earned little favor in his homeland.

Throughout Europe, however, support for the wave nature of light continued to grow. In France, Etienne-Louis Malus (1775–1826) and Augustin Jean Fresnel (1788–1827) experimented with polarized light, an effect that could only occur if light acted as a transverse wave (a wave which oscillated at right angles to its path of travel). In Germany, Joseph von Fraunhofer (1787–1826) was constructing instruments to better examine the phenomenon of diffraction and succeeded in identifying within the sun's spectrum 574 dark lines corresponding to different wavelengths.

In 1850 two French scientists, Jéan Foucault and Armand Fizeau, independently conducted an experiment that would strike a serious blow to the corpuscular theory of light. One of their instructors, Dominique-Françios Arago, had suggested that they attempt to measure the speed of light as it traveled through both air and water. If light were particulate, it should move faster in water; if, on the other hand, it were a wave, it should move faster in air. The two scientists performed their experiments, and each came to the same conclusion: light traveled more quickly through air and was slowed by water.

Even as more and more scientists subscribed to the wave theory, one question remained unanswered: through what medium did light travel? The existence of ether had never been proven—in fact, the very idea of it seemed ridiculous to most scientists. In 1872, Scottish physicist James Clerk Maxwell (1831–1879) suggested that waves composed of electric and magnetic fields could propagate in a vacuum, independent of any medium. Maxwell developed a set of equations that accurately described electromagnetic phenomena and allowed the mathematical and theoretical unification of electrical and magnetic phenomena. When Maxwell's calculated speed of light fit well with experimental determinations of the speed of light, Maxwell and other physicists realized that visible light should be a part of a broader electromagnetic spectrum containing forms of electromagnetic radiation that varied from visible light only in terms of wavelength and wave frequency. Frequency is defined as the number of wave cycles that pass a particular point per unit time, and is commonly measured in Hertz (cycles per second). Wavelength defines the distance between adjacent points of the electromagnetic wave that are in equal phase (e.g., wavecrests).

Maxwell's hypothesis was later proven by German physicist Heinrich Rudolph Hertz (1857–1894), who showed that such waves would also obey all the laws of reflection, refraction, and diffraction. It became generally accepted that light acted as an electromagnetic wave.

Hertz, however, had also discovered the photoelectric effect, by which certain metals would produce an electrical potential when exposed to light. As scientists studied the photoelectric effect, it became clear that a wave theory could not account for this behavior; in fact, the effect seemed to indicate the presence of particles. For the first time in more than a century, there was new support for Newton's corpuscular theory of light.

At the beginning of the twentieth century, German physicist Maxwell Planck proposed that atoms absorb or emit electromagnetic radiation only in certain bundles termed quanta. In his work on the photoelectric effect, German-born American physicist Albert Einstein used the term photon to describe these electromagnetic quanta. Planck determined that energy of light was proportional to its frequency (i.e., as the frequency of light increases, so does the energy of the light). Planck's constant, $h = 6.626 \times 10^{-34}$ joule-second in the meter-kilogram-second system (4.136×10^{-15} eV-sec), relates the energy of a photon to the frequency of the electromagnetic wave and allows a precise calculation of the energy of electromagnetic radiation in all portions of the electromagnetic spectrum.

By employing the quantum theories of Planck and Einstein, American physicist Arthur Holly Compton (1892–1962), who showed that the bundles of light—which he called photons—would sometimes strike electrons during scattering, causing their wavelengths to change, was able to describe light as both a particle and a wave, depending upon the way it was tested. Although this may seem paradoxical, it remains an acceptable model for explaining the phenomena associated with light and is the dominant theory of modern physics.

Although electromagnetic radiation is now understood as having both photon (particle) and wavelike properties, descriptions of the electromagnetic spectrum generally utilize traditional wave-related terminology (i.e., frequency and wavelength).

Exploration of the electromagnetic spectrum quickly resulted practical advances. Hertz regarded Maxwell's equations as a path to a "kingdom" or "great domain" of electromagnetic waves. Based on this insight, in 1888, Hertz demonstrated the existence of radio waves. A decade later, Wilhelm Röentgen's discovery of high-energy electromagnetic radiation in the form of X-rays quickly found practical medical use.

WORDS TO KNOW

ELECTROMAGNETIC SPECTRUM: The entire range of radiant energies or wave frequencies from the longest to the shortest wavelengths—the categorization of solar radiation. Satellite sensors collect this energy, but what the detectors capture is only a small portion of the entire electromagnetic spectrum. The spectrum usually is divided into seven sections: radio, microwave, infrared, visible, ultraviolet, x-ray, and gamma-ray radiation.

FREQUENCY: The rate at which vibrations take place (number of times per second the motion is repeated), given in cycles per second or in hertz (Hz). Also, the number of waves that pass a given point in a given period of time.

WAVELENGTH: Distance between the peaks or troughs of a cyclic wave. The character and effects of electromagnetic radiation are determined by its wavelength: very short-wavelength rays (e.g., X rays) are biologically harmful, somewhat longer-wavelength rays are classified as ultraviolet light, rays of intermediate wavelength are visible light, and longer wavelengths are infrared radiation and radio waves.

Basic Physics

Electromagnetic fields and photons exert forces that can excite electrons. As electrons transition between allowed orbitals, energy must be conserved. This conservation is achieved by the emission of photons when an electron moves from a higher potential orbital energy to a lower potential orbital energy. Accordingly, light is emitted only at certain frequencies characteristic of every atom and molecule. Correspondingly, atoms and molecules absorb only a limited range of frequencies and wavelengths of the electromagnetic spectrum, and reflect all the other frequencies and wavelengths of light. These reflected frequencies and wavelengths are often the actual observed light or colors associated with an object.

In addition to light visible to the human eye (e.g., light in the visible spectrum), other regions in the electromagnetic spectrum have distinct and important components. Radio waves, with wavelengths that range from hundreds of meters to less than a centimeter, transmit radio and television signals. Within the radio band, FM radio waves have a shorter wavelength and higher frequency than AM radio waves. Still higher frequency radio waves with wavelengths of a few centimeters can be utilized for radar imaging.

Microwaves range from approximately 1 ft (30 cm) in length to the thickness of a piece of paper. The atoms in food placed in a microwave oven become agitated (heated) by exposure to microwave radiation. Infrared radiation comprises the region of the electromagnetic spectrum where the wavelength of light is measured from one millimeter (in wavelength) down to 400 nm. Infrared waves are discernible to humans as thermal radiation

IN CONTEXT: THE COLORS OF THE RAINBOW

The region of the electromagnetic spectrum that contains light at frequencies and wavelengths that stimulate the rod and cones in the human eye is termed the visible region of the electromagnetic spectrum. Color is the association that the eye and brain make with various frequencies in the visible region; that is, particular colors are associated with specific wavelengths of visible light. Mixed wavelengths produce more complex color sensations. A nanometer (10^9 m) is the most common unit used for characterizing the wavelength of visible light. Using this unit, the visible portion of the electromagnetic spectrum is located between 380 nm–750 nm and the component color regions of the visible spectrum are Red (670–770 nm), Orange (592–620 nm), Yellow (578–592 nm), Green (500–578 nm), Blue (464–500 nm), Indigo (444–464 nm), and Violet (400–446 nm). Because the energy of electromagnetic radiation (i.e., the photon) is inversely proportional to the wavelength, red light (longest in wavelength) is the lowest in energy. As wavelengths contract toward the blue end of the visible region of the electromagnetic spectrum, the frequencies and energies of colors steadily increase.

A common way to remember the order of colors of the electromagentic spectum (from longest to shortest wavelength) is to use the mnemonic name (a name created from the first letters of the colors) ROY G. BIV (from ***R***ed, ***O***range, ***Y***ellow, ***G***reen, ***B***lue, ***I***ndigo, and ***V***iolet).

(heat). Just above the visible spectrum in terms of higher energy, higher frequency, and shorter wavelengths is the ultraviolet region of the spectrum with light ranging in wavelength from 400 to 10 billionths of a meter. Ultraviolet radiation is a common cause of sunburn even when visible light is obscured or blocked by clouds. X-rays are a highly energetic region of electromagnetic radiation with wavelengths ranging from about ten billionths of a meter to 10 trillionths of a meter. The ability of X-rays to penetrate skin and other substances renders them useful in both medical and industrial radiography. Gamma rays, the most energetic form of electromagnetic radiation, are light with wavelengths of less than about ten trillionths of a meter and include waves with wavelengths smaller than the radius of an atomic nucleus (10^{15} m). Gamma rays are generated by nuclear reactions (e.g., radioactive decay and nuclear explosions).

Cosmic rays are not a part of the electromagnetic spectrum because they are not a form of electromagnetic radiation. Rather, they are high-energy charged particles with energies similar to, or higher than, observed gamma electromagnetic radiation energies.

Wavelength, Frequency, and Energy

The wavelength of radiation is sometimes given in units with which we are familiar, such as inches or centimeters, but for very small wavelengths, they are often given in angstroms (abbreviated Å). There are 10,000,000,000 angstroms in 3.3 ft (1 m).

An alternative way of describing a wave is by its frequency, or the number of peaks which pass a particular point in one second. Frequencies are normally given in cycles per second, or hertz (abbreviation Hz), after Hertz. Other common units are kilohertz (kHz, or thousands of cycles per second), megahertz (MHz, millions of cycles per second), and gigahertz (GHz, billions of cycles per second). The frequency and wavelength, when multiplied together, give the speed of the wave. For electromagnetic waves in empty space, that speed is the speed of light, which is approximately 186,000 miles per second (300,000 km per sec).

In addition to the wavelike properties of electromagnetic radiation, it also can behave as a particle. The energy of a particle of light, or photon, can be calculated from its frequency by multiplying by Planck's constant. Thus, higher frequencies (and lower wavelengths) have higher energy. A common unit used to describe the energy of a photon is the electron volt (eV). Multiples of this unit, such as keV (1,000 electron volts) and MeV (1,000,000 eV), are also used.

Properties of waves in different regions of the spectrum are commonly described by different notation. Visible radiation is usually described by its wavelength, for example, while X-rays are described by their energy. All of these schemes are equivalent, however; they are just different ways of describing the same properties.

Wavelength Regions

The electromagnetic spectrum is typically divided into wavelength or energy regions, based on the characteristics of the waves in each region. Because the properties vary on a continuum, the boundaries are not sharp, but rather loosely defined.

Radio waves are familiar to us due to their use in communications. The standard AM radio band is at 540–1650 kHz, and the FM band is 88–108 MHz. This region also includes shortwave radio transmissions and television broadcasts.

Microwaves used in microwave ovens (which heat food by causing water molecules to rapidly vibrate and rotate) at a frequency of 2.45 GHz. In astronomy, radiation emitted at a wavelength of 8.2 inches (21 cm) has been used to map neutral hydrogen throughout the galaxy. Radar is also included in this wave region.

The infrared region of the spectrum lies just beyond the visible wavelengths. It was discovered by William Herschel in 1800 by measuring the dispersing sunlight with a prism, and measuring the temperature increase just beyond the red end of the spectrum.

The visible wavelength range is the range of frequencies with which we are most familiar. These are the wavelengths to which the human eye is sensitive, and which most easily pass through Earth's atmosphere. This region is further broken down into the familiar colors of the rainbow, which fall into a characteristic wavelength (and inversely, frequency) gradient.

The ultraviolet range lies at wavelengths just short of the visible. Although humans do not use UV to see, it has many other important effects on Earth. The ozone layer high in Earth's atmosphere absorbs much of the UV radiation from the sun, but that which reaches the surface can cause suntans and sunburns.

Higher frequency (and thus higher energy), shorter wavelength X-rays have broad application in medicine. X-ray wavelength electromagnetic radiation can pass through the soft tissues of the body, allowing doctors to examine bones and teeth. As with other forms of high energy radiation, Earth's upper atmosphere acts as a shield and so most X-rays do not penetrate Earth's atmosphere (astronomers must place X-ray telescopes in space).

Gamma rays, the most energetic of all electromagnetic radiation (highest frequency, shortest wavelength photons), are produced by nuclear processes, for example, during radioactive decay or in nuclear reactions in stars or in space.

■ Impacts and Issues

The small percentage of high energy electromagnetic radiation that does penetrate Earth's atmosphere can cause genetic mutations and is a driving force in evolutionary biology (biologic change over time).

The role of variation in the amount of electromagnetic radiation emitted by the sun on climate change is questionable. Although some scientists assert that the natural variations or cycles account for long-term climatic changes, Intergovernmental Panel on Climate Change (IPCC) scientists conclude that the variations do not account for recent or anticipated climate change and that the current changes are driven by human activity.

SEE ALSO *Solar Energy; Solar Illumination; Solar Radiation.*

BIBLIOGRAPHY

Books

Ohanaian, Hans C. *Classical Electrodynamics.* Hingham, MA: Infinity Science Press, 2006.

Robinson, Keith. *Spectroscopy: The Key to the Stars.* New York: Springer, 2007.

Web Sites

"Imagine the Universe. The Electromagnetic Spectrum." *High Energy Astrophysics Science Archive Research Center, NASA.* <http://imagine.gsfc.nasa.gov/docs/science/know_l1/emspectrum.html> (accessed December 7, 2007).

Emissions Trading

Introduction

Emissions trading is the buying and selling of units of credits that entitle their owners to emit a certain amount of a given kind of pollution. The total number of credits in circulation is fixed so that total emissions of the pollutant will be kept at some chosen level. The term "cap-and-trade" is often used to describe emissions trading: a cap on emissions of a certain pollutant is set, and polluters are allowed to trade credits that entitle them to pollute, with total emissions not to exceed the cap. The largest emissions trading schemes today are carbon trading schemes, which are being implemented by some state governments of the United States, by the European Union, and globally under the rules of the Kyoto Protocol.

Historical Background and Scientific Foundations

Pollution Trading Theory

The idea of pollution trading can be traced to the work of American economist Ronald Coase (1910–). In 1960, Coase published an influential article, "The Problem of Social Cost," in which he argued that pollution is not simply, as he put it, "something bad," but a combination of good and bad: the good aspect is that it is beneficial to the polluter (a cheap way of getting business done) and the bad aspect is the harm the pollution does to everybody, also called the social cost of the pollution. Coase argued that in a perfect market—that is, a world of buyers and sellers in which it costs nothing to have all needed information and to negotiate with other players—pollution will always be regulated by market forces to an optimum level that will maximize total product. A perfect market would, Coase argued, always produce the correct amount of pollution—neither too much nor too little.

Coase acknowledged, however, that no real-world market could never be perfect in the required sense, and thought that governments would have to intervene to give pollution rights to those who could make the most profit from them (i.e., who could maximize the good that emitting the pollution would do). Over the next decade, other economists, such as J. H. Dales, argued that government should not allocate pollution rights but should decide what the best overall amount of pollution is, then allow polluters to trade pollution rights with each other under a predetermined limit or cap. This was the origin of the cap-and-trade emissions trading concept.

The U.S. Clean Air Act Amendments of 1990 set up a national cap-and-trade program for pollution rights for sulfur dioxide (SO_2). The program, which is projected to reduce SO_2 in the United States by 50% from 1980 to 2010 at only 20% of the cost of direct regulation, has been widely hailed as proof that the cap-and-trade concept can work.

Kyoto Protocol

The 1992 treaty known as the United Nations Framework Convention on Climate Change (UNFCCC) commits its signatories (signing countries) to reduce their greenhouse emissions to stabilize global climate. The original treaty did not say exactly how this was to happen, but in 1997, most countries signed the Kyoto Protocol to the UNFCCC. The protocol requires signatories to monitor and report their greenhouse emissions and requires developed or industrialized countries to cap and trade emissions of six greenhouses gases, including CO_2. As is customary in climate studies, emissions of the other five gases are counted in terms of how many tons of CO_2 would be required to produce the same amount of climate change. For example, since methane (CH_4) is 21 times as powerful a greenhouse gas as CO_2, 1/21 of a ton of methane is rated as 1 ton CO_2e (CO_2 equivalent).

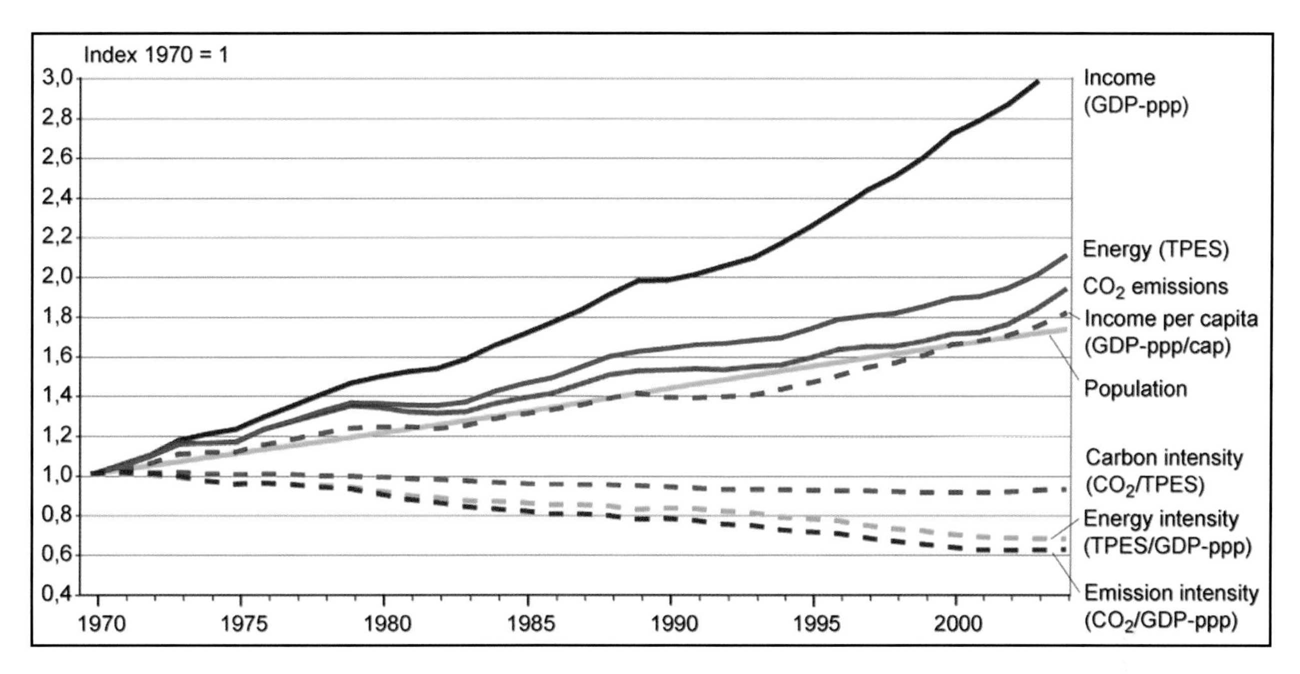

The image shows the relative development of income, population, energy and CO_2 emissions. As the world population (Pop), Gross Domestic Product (GDP-ppp), and Income per capita (GDP-ppp/Pop) rose from 1970–2004, so did the Total Primary Energy Supply (TPES) and CO_2 emissions (from fossil fuel burning, gas flaring and cement manufacturing). This was even as Energy Intensity (TPES/GDP-ppp), Carbon Intensity of energy supply (CO_2/TPES), and Emission Intensity of the economic production process (CO_2/GDP-ppp) for the period dropped slightly. *Climate Change 2007: Mitigation of Climate Change, Summary for Policymakers, Intergovernmental Panel on Climate Change.*

Industrialized countries that signed the Kyoto Protocol agreed to reduce their greenhouse emissions to 5.2% below 1990 levels by 2012. The treaty entered into force on February 16, 2005. A point of major dissatisfaction among critics of the treaty is that it does not cap the greenhouse emissions of large, developing countries such as India and China. By 2007, China, according to some analysts, edged out the United States as the world's largest single emitter of greenhouse gases.

Two Kinds of Carbon Trading

The carbon trading system established by the Kyoto Protocol, like many other carbon markets, is actually a blend of two kinds of carbon trading. The first is basic emissions trading, the cap-and-trade mechanism described earlier; the second is trade in project-based credits.

In basic emissions trading, the number of emissions credits in existence is fixed. Polluters trade this limited stock of credits directly with each other. Consider, for example, a country with only two power plants, Plant A and Plant B, both presently emitting 110 tons of carbon per year. Both plants are allotted 100 carbon credits for next year, which entitle each plant to emit 100 tons of carbon. Both plants must therefore either emit 10 tons less per year or buy credits entitling them to emit 110 tons. Managers at Plant A discover that they can reduce emissions at a cost of $.50 per ton. The market price of carbon credits is $1 per ton. Thus, it would make sense for Plant A to cut its emissions not by a mere 10 tons but by 20 tons, at a cost of $10. Now emitting only 90 tons, Plant A possesses 10 unused carbon credits, which it can sell for a total of $10. By selling the credits, it saves $5 (the cost of simply reducing its emissions by 10 tons).

Plant B, meanwhile, finds that for technical reasons—it generates power from a different fuel, perhaps—to make emissions cuts would cost it $1.50 per ton. Plant B's cost for getting emissions down from 110 to 100 tons would therefore be $15. It would be cheaper to buy 10 carbon credits than to cut 10 tons of emissions: $10 to buy the credits versus $15 to cut the emissions. Plant B will therefore save $5 by buying the credits.

The emissions credits are traded. Plant B buys the 10 extra credits that Plant A is selling. It now owns 110 credits and can emit 110 tons of carbon. Both Plant A and Plant B have saved $5 over what the 100-ton cap would have cost them, and the total amount of carbon emitted has been reduced from 110 to 100 tons, as mandated by the government.

The other form of carbon trading built into the Kyoto Protocol is trading in project-based credits. In this system, a polluter can invest in projects that absorb carbon or reduce emissions of carbon in some other country, generally a poorer one where such measures are cheaper. In exchange for reducing carbon in the poorer country, the wealthier polluter is given credits that allow it to keep polluting proportionally.

WORDS TO KNOW

CAP AND TRADE: The practice, in pollution-control or climate-mitigation schemes, of mandating an upper limit or cap for the total amount of some substance to be emitted (e.g., CO_2) and then assigning allowances or credits to polluters that correspond to fixed shares of the total amount. These allowances or credits can then be bought and sold by polluters, in theory allowing emission cuts to be bought where they are most economically rational.

CARBON CREDITS: Units of permission or value, similar to monetary units (e.g., dollars, euros) that entitle their owner to emit one metric ton of carbon dioxide into the atmosphere per credit.

FREE MARKET: Economic system in which price-setting and other behaviors are not constrained by special laws or regulations. Perfectly free markets are a philosophical ideal, not found in the real world.

GREENHOUSE GASES: Gases that cause Earth to retain more thermal energy by absorbing infrared light emitted by Earth's surface. The most important greenhouse gases are water vapor, carbon dioxide, methane, nitrous oxide, and various artificial chemicals such as chlorofluorocarbons. All but the latter are naturally occurring, but human activity over the last several centuries has significantly increased the amounts of carbon dioxide, methane, and nitrous oxide in Earth's atmosphere, causing global warming and global climate change.

KYOTO PROTOCOL: Extension in 1997 of the 1992 United Nations Framework Convention on Climate Change (UNFCCC), an international treaty signed by almost all countries with the goal of mitigating climate change. The United States, as of early 2008, was the only industrialized country to have not ratified the Kyoto Protocol, which is due to be replaced by an improved and updated agreement starting in 2012.

SOCIAL COST: The cost of emitting a given quantity of a pollutant, translated into monetary terms. Usually used in reference to the social cost of carbon, the total future economic cost of emitting a ton of carbon (or an equivalent amount of some other greenhouse gas) at a given time.

Thus, under Kyoto, industrialized countries can meet their carbon-reduction goals either by reducing their own emissions, by trading emissions, or by acquiring carbon credits earned by funding reductions elsewhere. The system for acquiring project-based credits under Kyoto is called the Clean Development Mechanism.

The Kyoto system is not the only carbon market. Nine states in the northeast region of the United States have joined to form the Regional Greenhouse Gas Initiative (RGGI), which proposes to start a cap-and-trade emissions system for power plants starting in 2009. Some corporations in the United States, totaling about 4% of the country's emissions, have been participating since 2003 in a voluntary carbon market called the Chicago Climate Exchange. Here, the cap is not government-mandated but is set by a contractual agreement by the participating companies to reduce their group emissions by 6% by 2010. The world's largest emissions market is the European Union Trading Scheme, which began operating in 2005 and includes the 27 states of the European Union (EU). The EU scheme covers emissions not only from power plants but other energy-intensive industries, with coverage of about 46% of total EU CO_2 emissions.

Impacts and Issues

Compulsory emissions trading, as in the European Union, has been criticized as an ineffectual system for reducing greenhouse emissions by free-market conservatives and libertarians. These critics argue that voluntary schemes, including source reductions and voluntary trading such as the Chicago Climate Exchange, would reduce emissions more efficiently.

Many environmental organizations support emissions trading, but others criticize the practice. For example, the World Wide Fund for Nature, which supported the creation of the European Union Emissions Trading Scheme, announced in 2006 that European governments had been "guilty of allowing their industries to produce as much carbon dioxide as they wish at no cost" by setting caps above actual emissions levels. Others have argued that since carbon credits are a commodity with monetary value, simply allocating them to industries constitutes a multibillion-dollar giveaway to polluters.

The integrity and realism of Kyoto's Clean Development Mechanism projects has also been challenged. For example, the consultancy companies that design and monitor such projects are sometimes clients or subdivisions of the same energy companies that are paid to carry them out. When such conflicts of interest exist, projects generating credits for polluters in developed countries may be abating carbon emissions less than is reported.

Defenders of carbon markets argue that large-scale emissions trading mechanisms are still in their infancy, and will become more effective as governments learn from experience.

Primary Source Connection

On January 1, 2005, the member states of the European Union commenced a trading scheme for emissions that capped the amount of emissions from large emitters of greenhouse gases. The following article evaluates the effectiveness of emissions trading schemes in Europe and compares emissions trading to other regulatory devices such as fixed emissions quotas.

Author Claudia Kemfert is a member of the German Institute for Economic Research and Faculty of Economics at the University of Humboldt, Germany.

THE ENVIRONMENTAL AND ECONOMIC EFFECTS OF EUROPEAN EMISSIONS TRADING

Abstract

In this article, we analyse the effects of emissions trading in Europe, with special reference to Germany. We look at the value of the flexibility gained by trading compared to fixed quotas. The analysis is undertaken with a modified version of the GTAP-E model using the latest GTAP version 6 database. It is based on the national allocation plans (NAP) as submitted to and approved by the EU. We find that, in a regional emissions trading scheme, Germany, Great Britain and the Czech Republic are the main sellers of emissions permits, while Belgium, Denmark, Finland and Sweden are the main buyers. The welfare gains from regional emissions trading—for the trading sectors only—are largest for Belgium, Denmark and Great Britain; smaller for Finland and Sweden, and smallest for Germany and other regions. When we take into account the economy-wide and terms-of-trade effects of emissions trading, however, (negative) terms-of-trade effects can offset the (positive) allocative efficiency gains for the cases of the Netherlands and Italy, while all other regions end up with positive net welfare gains. All regions, however, experienced increases in real GDP as a result of regional emissions trading.

1. Introduction

The European Union considers climate change as 'one of the greatest environmental, social and economic threats facing the planet.' It therefore took a leading role in the negotiations for international action against climate change, in particular the Kyoto Protocol. In order to set an example, it accepted relatively ambitious targets. Whereas all Annex B countries were to reduce the emissions of greenhouse gases by about 5%, the EU has committed to an 8% reduction. . . .

Therefore, in 2000 the EU Commission launched the European Climate Change Programme (ECCP), a continuous multi-stakeholder consultative process which serves to identify cost-effective ways for the EU to meet its Kyoto commitments, to set priorities for action, and to implement concrete measures. One of the main elements of this programme was the establishment of a European CO_2 emissions trading scheme (EU ETS). The EU considers this as 'a cornerstone in the fight against climate change', which will help its Member States to achieve compliance with their commitments under the Kyoto Protocol and the EU burden-sharing at lower costs. The basic idea of emissions trading is to limit the amount of emissions by creating rights to emissions and to make these rights—which are called allowances—tradable. The scarcity of emission allowances gives them a market value and those emitters whose avoidance costs are lower than the market value of allowances will reduce their emissions and buy fewer certificates or sell excess emissions rights, and vice versa for other emitters.

There is a fundamental difference between the EU ETS and the emissions trading scheme as envisaged under the Kyoto Protocol. In the latter case, emissions trading is to occur between the Parties to the protocol at the level of the States. Under the EU ETS, however, trading is to occur between individual emitters, which comprise 11,428 installations in 25 Member States. There have been other studies which look at the effects of emissions trading in Europe. Böhringer et al. (2004), for example, used a set of 'reduced form' equations which represent marginal abatement costs derived from a general equilibrium model to conduct simulation experiments to analyse the efficiency and equity aspects of different allocation rules for the EU ETS (European Commission, 2001, 2004). In these studies, the approach adopted is often 'partial equilibrium' in nature, which implies that important market interactions (including terms-of-trade effects) are not taken into account. . . .

2. The European emissions trading scheme

The EU ETS started on 1 January 2005. The first trading period—which has been nicknamed the 'warming-up phase' or 'learning phase'—covers the years 2005-2007. The second phase corresponds to the Kyoto period 2008–2012.

The framework for EU ETS has been defined by a Directive in October 2003 which outlines the basic features of the scheme, but leaves substantial scope for the Member States to decide on important aspects of the implementation. The most important features set by the EU are the following:

- The European ETS is a cap-and-trade system; i.e., the absolute quantity of emission rights (rather than relative or specific emissions) is fixed at the beginning.
- Only one of the six greenhouse gases of the Kyoto Protocol (CO_2) is subject to the ETS, at least during the first period from 2005 to 2007. The main reason for this is that CO_2 is the greenhouse gas which is easiest to monitor, since the emissions are directly related to the use of fossil fuels for which most countries have already established a monitoring system in order to levy energy taxes.
- The EU ETS is implemented as a downstream system; i.e. the users (rather than the producers and

New vehicles are ready to be shipped from a wharf in Sydney, Australia, in June 2007. Emissions from vehicles are one of the leading causes of greenhouse gases in the atmosphere. Then-Australian Prime Minister John Howard announced that carbon trading might be an essential tool to curb global warming. *AP Images.*

importers of fossil fuels) will be obliged to hold emission allowances.

This has some fundamental consequences. All users of fossil fuels which are covered by the ETS have to be monitored and can participate actively in the trading system. In order to limit the administrative costs of the ETS, the system is restricted to large installations. Therefore, only installations belonging to one of four broad sectors, which are listed in the Directive and which exceed a sector-specific threshold, are subjected to emissions trading. The four sectors are:

- Energy activities (such as electric power, direct emissions from oil refineries)
- Production and processing of ferrous metals (iron and steel)
- Mineral industry (such as cement, glass and ceramic production)
- Pulp and paper.

The thresholds refer to the production capacity of the installation, e.g., in the case of combustion installations these are installations with a rated thermal input exceeding 20 MW. The emissions trading scheme will cover around 45% of the EU's total CO_2 emissions, or about 30% of its overall greenhouse gas emissions. This partial coverage of the ETS is likely to produce inefficiencies which can only be avoided if the total quantities of allowances are set at a level which equalizes the marginal avoidance costs between the emissions trading sector and other emitters. This, however, is unlikely to be the case because the marginal avoidance costs of these emitters are not known.

- Allowances are issued by each Member State, but trading can take place between any EU participants.
- The so-called 'linking Directive' will allow participants in emissions trading to count credits from Clean Development Mechanism and Joint Implementation emission reduction projects around the world toward their obligations under the European Union's emissions trading scheme, even if the Kyoto Protocol did not enter into force.

Within this framework, the Member States have three important tasks. First, they have to decide which quantity of emissions should be allocated to the installations participating in the ETS. This decision must take into consideration the burden-sharing target of the country and must list the policies and measures which are to be applied in the sectors which are not part of the ETS. However, in almost all countries, business representatives have made strong lobbying efforts to ensure that emissions trading will not impair their competitive position. This has led to very generous allocations in some cases. Second, they have to draw up a list of all installations which are subject to emissions trading. Third, they have to decide how to allocate the total quantity to individual installations. The Directive sets some general rules according to which the allocation has to be made, but there is substantial scope for national priorities. These decisions have to be set down in a national allocation plan (NAP)....

5. Conclusions

Our study has shown that emissions trading is an important policy instrument to achieve a particular climate policy objective such as the fulfilment of the Kyoto obligations by the EU at minimum costs. The use of this 'flexible' policy instrument is seen to result in significant efficiency gains, measured either in terms of the reduction in marginal abatement costs or in terms of the efficiency gains for both (permit) buying and selling sectors. For buying sectors (those with high MACs without trading), the efficiency gains represent reductions in overall compliance costs. For selling sectors (those with low MACs without trading), increases in income from emission trading overcompensate additional abatement costs. As a result, real GDP is seen to increase for all regions. However, the efficiency gains in some cases may not be sufficient to offset the losses in revenue due to emissions trading (emissions permit purchasing); hence some regions may still experience a net welfare loss. For these regions, a net welfare loss implies a negative change in net national income even if there is a positive change in gross domestic product. This uneven distribution of the total welfare gains (income from emissions trading) across regions may warrant some attention being given to the initial distribution of the burden of emissions reductions across regions.

Claudia Kemfert, et al.

CLAUDIA KEMFERT, ET. AL. "THE ENVIRONMENTAL AND ECONOMIC EFFECTS OF EUROPEAN EMISSIONS TRADING," *CLIMATE POLICY* 6 (2006) 441–455.

SEE ALSO *Clean Development Mechanism; Kyoto Protocol; Offsetting; Social Cost of Carbon (SCC).*

BIBLIOGRAPHY

Books

Lohmann, Larry, et al. *Carbon Trading—A Critical Conversation on Climate Change, Privatisation and Power.* Uppsala, Sweden: Dag Hammarskjöld Foundation, 2006.

Periodicals

Betz, Regina, and Misato Sato. "Emissions Trading: Lessons Learnt from the 1st Phase of the EU ETS and Prospects for the 2nd Phase." *Climate Policy* 6 (2006): 351–359.

Chameides, William, and Michael Oppenheimer. "Carbon Trading Over Taxes." *Science* 315 (2007): 1670.

Coase, Ronald. "The Problem of Social Cost." *Journal of Law and Economics* 3 (1960): 1–44.

Hopkins, Michael. "The Carbon Game." *Nature* 432 (2004): 268–340.

Kanter, James. "Carbon Trading: Where Greed Is Green." *International Herald Tribune* (June 20, 2007).

Kemfert, Claudia, et al. "The Environmental and Economic Effects of European Emissions Trading." *Climate Policy* 6 (2006): 441–455.

Web Sites

"About IETA." *International Emissions Trading Association (IETA)*, 2007. <http://www.ieta.org/ieta/www/pages/index.php?IdSiteTree=2> (accessed November 6, 2007).

"Emission Impossible: Access to JI/CDM Credits in Phase II of the EU Emissions Trading Scheme." *World Wildlife Fund UK*, June 2007. <http://assets.panda.org/downloads/emission_impossible__final_.pdf> (accessed November 6, 2007).

"Emissions Trading." *United Nations Framework Convention on Climate Change Secretariat*, June 22, 2007. <http://unfccc.int/kyoto_protocol/mechanisms/emissions_trading/items/2731.php> (accessed November 6, 2007).

Vallette, Jim, et al. "A Wrong Turn from Rio: The World Bank's Road to Climate Catastrophe." *Sustainable Energy and Economy Network*, December 2004. <http://www.seen.org/PDFs/Wrong_turn_Rio.pdf> (accessed November 6, 2007).

Larry Gilman

Endangered Species

Introduction

A species whose numbers have declined so drastically that it is in danger of extinction throughout all or a significant part of its habitat is considered endangered. Extinct species no longer occur anywhere on Earth, and once they have disappeared, they are gone forever. Extirpated species have disappeared locally or regionally, but still survive in other regions or in captivity. Threatened species are at risk of becoming endangered in the foreseeable future. Drastic reductions in the population of a species or its complete disappearance diminish the biodiversity of the ecosystems it inhabits and the planet as a whole.

Historical Background and Scientific Foundations

There are many reasons that a species may become threatened or endangered. Species have become extinct throughout the history of life on Earth due to selective pressures, natural climate changes, and catastrophic geologic events. However, the pace of extinction has accelerated dramatically in modern times due to human activities, and scientists estimate that present extinction rates are 1,000 to 10,000 times higher than the average natural extinction rate. Among the human activities that contribute to species endangerment are: land use practices, such as urban development, agricultural conversion, and deforestation; unsustainable levels of hunting or harvesting; habitat damage due to water, air, or soil pollution; intentional or accidental introduction of predators or diseases; and global climate change caused by humans.

An accurate assessment of the number of endangered species worldwide does not yet exist. This is due, in part, to the fact that the smallest, yet most multitudinous, species of plants, insects, and microorganisms have not been tallied. In addition, scientists believe that large numbers of as yet unknown (to science) and

An emperor penguin tends to his newly hatched chick in captivity. Out in the wild, such penguins are suffering from the effects of climate change, especially in the western Antarctic Peninsula. *AP Images.*

unnamed species can be found in tropical rainforests. Since these forests are disappearing rapidly due to timber extraction and human population encroachments, all the species that inhabit these forests are likely to be endangered.

Since the number of species evaluated in the compilation of lists of endangered species is limited, the large numbers of plants and animals recorded on these lists is particularly alarming. IUCN-The World Conservation Union through its Species Survival Commission (SSC) has been assessing the status of species, subspecies, varieties, and selected subpopulations on a global scale for more than 40 years to highlight those at risk and promote their conservation. The IUCN Red Lists are intended to provide a comprehensive, apolitical, global approach for evaluating the conservation status of plant and animal species. The 2007 Red List of Threatened Species, released in September 2007, assessed 41,415 species. Of these species, 16,306 are threatened with extinction to a greater or lesser degree, an increase of nearly 200 species over the previous year. In addition, 65 species are found only in captivity or cultivation. According to the 2007 IUCN Red List, 25% of mammals, 12.5% of birds, 33% of amphibians, and 70% of the world's assessed plants are in jeopardy.

In the United States, species must meet a stringent set of criteria to be listed as endangered or threatened under the Endangered Species Act (ESA). In 2007, the U.S. Fish and Wildlife Service (FWS) listed 1,174 animal species and 1,921 plant species as threatened or endangered around the world. A listing under the ESA provides a species with a number of conservation benefits including protection from harm due to federal activities, restrictions on hunting and trade, development and implementation of recovery plans by the FWS (for species under U.S. jurisdiction), and protection of critical habitat.

Studies of species are often grouped by habitat. Freshwater ecosystems are among the most threatened. Marine species have historically been poorly covered in studies of endangered species due to the difficulties of studying these species in the world's vast oceans. Arid and semi-arid regions urgently need further study particularly since severe degradation of arid and semi-arid lands around the world is being accelerated by global warming. Global warming has also recently focused attention on polar regions, since one of the effects of global climate change appears to be accelerated melting of glaciers and Arctic ice. The plight of polar bears has attracted significant media attention.

Polar bears and other cold-loving species have very specialized adaptations for the harsh environments at the poles, and they have low tolerance for environmental change. Reductions in sea ice due to global warming influence not only the availability of polar bear hunting and denning habitat, but may impact the overall distribution and abundance of polar bears. Scientists have found evidence of polar bear drownings because some bears must swim longer distances (up to 60 mi/96 km) over open ocean in their search for food. In 2007, scientists studying the effects of climate and ice changes to help determine whether polar bears should be protected under the ESA estimated that two-thirds of the world's polar bears will disappear by 2050. These scientists concluded that polar bears are unlikely to be driven to extinction, but their habitat will contract significantly to the Arctic archipelago of Canada and areas off the northern coast of Greenland where sea ice persists even in warm summers.

Arid regions are getting hotter and drier, threatening the survival of plant and animal species already at their heat-tolerance limits. In marginal drylands, biodiversity is threatened by over cultivation. In the United States, for example, more than 50 species are now endangered in the Arizona desert, including some cactus varieties.

Marine and coastal waterways and freshwater systems also are vulnerable to climate changes. Freshwater fish have been seriously impacted with an estimated 20% of these fish species now extinct. Spawning and feeding habitats have changed. A number of salmon varieties are on the endangered species list due to habitat disruption, water pollution, and overfishing.

Plants have received less attention than animals, but they have not escaped the effects of climate changes. One quarter of all the wild potato species are predicted to be extinct within 50 years. Changes in rainfall patterns and increasing temperatures cause soil erosion and an increased leaching of soil nutrients. Wildfires are increasing in drier regions. In general, there are three main categories of potential effects of climate change on plant species. For plants existing in areas of climatic extremes,

WORDS TO KNOW

BIODIVERSITY: Literally, "life diversity": the number of different kinds of living things. The wide range of organisms—plants and animals—that exist within any given geographical region.

EXTINCTION: The total disappearance of a species or the disappearance of a species from a given area.

EXTIRPATED: The condition in which a species is eliminated from a specific geographic area of its habitat.

HABITAT: The area or region where a particular type of plant or animal lives and grows.

THREATENED: When a species is capable of becoming endangered in the near future.

IN CONTEXT: POLAR BEARS AND THE ENDANGERED SPECIES ACT

Initial studies suggest that global climate change may already have taken a toll on Arctic polar bear populations. Polar bears require summer sea ice for hunting. Typically, polar bears swim between large and stable patches of sea ice, but the animals cannot swim long distances. Already researchers are noting thinner polar bears, decreasing birth rates, and declining cub survival rates in Arctic bear populations. The National Center for Atmospheric Research (NCAR) stated that, under a worst-case scenario, summer sea ice could disappear by 2040.

The U.S. Minerals Management Service released a report that documented the drowning deaths of at least 4 polar bears off of the Alaskan coast in September 2004. The report asserted that warmer than average temperatures prompted record melting of sea ice, stranding hunting polar bears. The report prompted several environmental organizations—including the Center for Biological Diversity, the Natural Resources Defense Council (NRDC), and Greenpeace—to sue the U.S. government for failing to take action on an earlier petition to have polar bears evaluated for inclusion as a threatened species under the Endangered Species Act. Environmental groups and researchers asserted that the U.S. government failed to address the question of adding polar bears to the threatened species list because of a policy of downplaying the possible effects of global climate change. Under the act, the secretary of the interior should have responded within 90 days and opened the issue for public comment. Government officials claimed that they were actively studying polar bear populations for over two years.

Under a court settlement, the U.S. Department of the Interior announced that it was formally proposing to designate polar bears as a threatened species. Identifying polar bears as threatened could give environmental groups more power to challenge current standards governing carbon dioxide emissions and other leading contributors to climate change.

global warming may have drastic impacts, for example, forcing plants that require cold temperatures out of habitats like mountain peaks. If this is their only habitat, extinction will occur. Shifts in global biome distribution are another likely impact of global warming. Coniferous forests will shift farther north and grasslands and deserts will expand. Global warming also may alter how plants function in their existing environment by lengthening the growing season in some areas, especially in northern latitudes.

■ Impacts and Issues

Climate changes are affecting the environments inhabited by endangered species in many ways, including altering the timing of ecological events, such as the flowering of plants and the breeding of animals. Such changes are impacting the relationships of interdependent species whose interactions may be disrupted by differing rates of change. The habitats of species are shifting with the changing temperatures, but some habitats are disappearing rather than shifting. The species that inhabit these habitats may become extirpated from that region or may be driven to extinction.

Climate change impacts biodiversity by changing the distribution of species and threatening their survival as warm seasons grow longer. Global warming is reducing sea ice in polar regions and this melting is, in turn, accelerating global warming. Snow and ice reflect sunlight, but exposed darker land and ocean surfaces increase absorption of solar radiation. The melting ice caps in Greenland and Antarctica raise the sea level and add freshwater to the ocean. The added freshwater alters salinity and may cause shifts in ocean circulation.

Studies of threatened and endangered species are not consistent around the globe. The status of vertebrates is best documented with about 40% assessed. Species-rich environments, such as tropical forests and the oceans, are not well studied. What has become evident, however, is that species are declining fastest in tropical and subtropical regions and in the poorer regions of the world where conservation responses are often limited. Brazil, Indonesia, and the Philippines are among those regions.

A number of conservation organizations and governmental initiatives have been created to address the threats to Earth's biodiversity, including the World Summit on Sustainable Development, the international Convention on Biodiversity, and the United Nations Framework Convention on Climate Change and its Kyoto Protocol. More than 170 countries have signed the Convention on International Trade in Endangered Species of Wild Flora and Fauna (CITES), which aims to ensure that international trade in specimens of wild animals and plants does not threaten their survival. Many individual countries have passed legislation to protect endangered species within their borders and to provide funding for recovery efforts. The World Wildlife Fund, Conservation International, and other similar nongovernmental organizations promote field research, sponsor species recovery efforts, and increase the public visibility of endangered species issues and their impact.

Although extremely limited when compared to the enormity of the problem, there have been some success stories. The California condor has been successfully returned to the wild as a result of a concerted captive breeding and release program, and the black-footed ferret is making a comeback in the western prairies of the

United States due to a similar program. A wide range of recovery initiatives extending over several decades resulted in the removal of the bald eagle from the U.S. endangered species list in 2007. In Canada, a red squirrel population seems to be adapting successfully to changing climate conditions. Scientists studying these squirrels have found that they now produce offspring on average 18 days earlier in the year than their great-grandmothers did, probably due to warmer spring temperatures. These squirrels provide some of the first hard evidence that mammals can adapt to a warmer world in a few generations.

Primary Source Connection

This article discusses how climate change may affect the production of baseball bats. For decades, ash from the Northeastern and Midwestern United States has been the wood of choice for baseball bats. Warmer temperatures could cause ash wood to become softer, and some fear the species could be wiped out.

BALMY WEATHER MAY BENCH A BASEBALL STAPLE

RUSSELL, Pa. —Careers at stake with each swing, baseball players leave little to sport when it comes to their bats. They weigh them. They count their grains. They talk to them.

But in towns like this one, in the heart of the mountain forests that supply the nation's finest baseball bats, the future of the ash tree is in doubt because of a killer beetle and a warming climate, and with it, the complicated relationship of the baseball player to his bat.

"No more ash?" said Juan Uribe, a Chicago White Sox shortstop, whose batting coach says he speaks to his ash bats every day. Uribe is so finicky about his bats, teammates say, that he stores them separately in the team's dugout and complains bitterly if anyone else touches them.

At a baseball bat factory tucked into the lush tree country here in northwestern Pennsylvania, the operators have drawn up a three-to-five-year emergency plan if the white ash tree, which has been used for decades to make the bat of choice, is compromised.

In Michigan, the authorities have begun collecting the seeds of ash trees for storage in case the species is wiped out, a possibility some experts now consider inevitable.

As early as this summer, federal officials hope to set loose Asian wasps never seen in this country with the purpose of attacking the emerald ash borer, an Asian beetle accused of killing 25 million ash trees in Michigan, Illinois, Indiana, Ohio and Maryland since it was spotted in the United States five years ago.

In late June, officials found signs of the ash borer's arrival in Pennsylvania, setting off a new alarm for the makers of baseball bats, most of which come from this rocky, cool range on the New York border.

Along with the ash borer beetle, a warming of the local climate could also affect the ash used for bats, some scientists say. As temperatures rise, the ash wood that now makes an ideally dense but flexible bat might turn softer because of a longer growing season. Eventually, some scientists predict, the ash tree could vanish from the region.

A warmer climate could also aid the emerald ash borer's invasion, some scientists contend, although others disagree, by creating stressed trees and the possibility of a quicker reproduction cycle in the beetle.

"We're watching all this very closely," said Brian Boltz, the general manager of the Larimer & Norton company, whose Russell mill each day saws, grades and dries scores of billets destined to become Louisville Slugger bats. "Maybe it means more maple bats. Or it may be a matter of using a different species for our bats altogether."

Such uncertainty does not sit well with professional players, some of whom shun (or break) bats that have failed them and worship those that have sent balls out of the park. (Some widely suspect that the well-known players get the best-quality wood, and the rookies, something softer.) Baseball, after all, is a game of routine, of instinct, of superstition.

The magic in a perfect bat is not easy to define. "You can't describe it—it's a feel," said Scott Podsednik, an outfielder for the White Sox. "When you pick it up and take a couple of swings with it, you just know."

After batting practice one morning, Podsednik's teammate Uribe sheepishly confirmed his lectures to his bats (his beloved "Hoosier HB 23" models). "I tell them: 'do your job and if you don't do your job, I'm going to have to go back to the Dominican Republic,'" Uribe said in Spanish. "Sometimes they listen; sometimes they don't."

For much of a century, ash was the wood that ruled the realm of baseball bats, but it has faced threats before: First, competition from aluminum and composite bats (which whisked away much of the youth and amateur market but are barred from professional baseball) and then, in the past decade, from the sugar maple.

When it became known that Barry Bonds of the San Francisco Giants, who is closing in on baseball's career home-run record, was using maple bats, change swept through baseball's clubhouses.

Some bat makers say professional players are now about evenly divided between ash and maple, which is more expensive and which some players (catchers, especially) say tends to explode more violently when a bat breaks.

"Maple is all the rage with the young players coming up now," said Tom Hellman, the clubhouse manager for the Chicago Cubs, whose responsibilities include ordering bats and keeping track of them. "But the older players still want their ash."

Science has never definitively established whether ash makes the optimal bat. Terry Bahill, an engineer at the University of Arizona and a co-author of "Keep Your Eye on the Ball: Curveball, Knuckleballs and Fallacies of Baseball," said researchers could measure how much energy was dissipated when a bat struck a baseball and how much force was required to bend a bat.

"But in the end," Mr. Bahill said, "we can't tell you which bat is going to be more effective because a human being is going to be swinging this bat. So the players making decisions about bats are making them on feelings, not scientific data."

Some scientists, however, do see a threat to the quality of the northern white ash posed by rising temperatures over a period of decades. Ash that grows in the warmer Southeastern States is held to be softer, in part because of the longer growing season, said Ron Vander Groef, who runs a factory in Dolgeville, N.Y., which make Rawlings bats.

There are also some concerns that the numbers of white ash trees in the North could significantly decline. Louis R. Iverson, a research landscape ecologist with the United States Forest Service, has helped map how habitat changes could affect 134 tree species by the end of the century. In a worst-case scenario, the white ash (and the sugar maple) diminish in numbers and shift farther north.

Still, the emerald ash borer, or Agrilus planipennis Fairmaire, is the most immediate threat. Discovered in the United States near Detroit in 2002, the beetles, which are shiny green, will destroy a tree in two to three years. The larvae tunnel inside the trees, cutting off water and food.

The ash borer is native to Asia, where the trees are naturally resistant to it.

"It just doesn't look good," Dan Herms, an associate professor of entomology at Ohio State University, said of the prospect of stopping the beetle in this country. "The current technology won't be able to stop it."

Dr. Herms strongly disputes any link between the ash borer and climate change, saying that the beetle has survived in a wide range of temperatures in Asia.

For now, the baseball bat makers are bracing for the worst. At the mill in Russell, even as machines cranked and hummed with ash billets last month, state investigators were barring the movement of wood from four Western Pennsylvania counties after adult beetles were discovered.

Some suppliers say they are harvesting trees years earlier than planned because of the ash borer's arrival.

In the end, baseball players may be faced with switching to, and holding conversations with, bats made of maple or some new wood yet untested by the hardball.

Monica Davey

DAVEY, MONICA. "BALMY WEATHER MAY BENCH A BASEBALL STAPLE." *THE NEW YORK TIMES*. JULY 11, 2007.

See Also *Desert and Desertification; Extinction; Forests and Deforestation; United Nations Environment Programme (UNEP).*

BIBLIOGRAPHY

Web Sites

Biever, Celeste. "Red Squirrels Evolving with Global Warming." *New Scientist,* February 12, 2003. <http://environment.newscientist.com/channel/earth/climate-change/dn3382> (accessed September 21, 2007).

Black, Richard. "Gorillas Head Race to Extinction." *BBC News,* September 12, 2007. <http://news.bbc.co.uk/go/pr/fr/-/2/hi/science/nature/6990095.stm> (accessed September 14, 2007).

Carlton, Jim. "Is Global Warming Killing the Polar Bears?" *The Wall Street Journal Online,* December 14, 2005. <http://online.wsj.com/public/article_print/SB113452435089621905-vnekw47PQGtDyf3iv5XEN71_o5I_20061214.html> (accessed September 14, 2007).

"Climate Change and Biodiversity." *United Nations Environmental Programme. Convention on Biological Diversity.* <http://www.cbd.int/climate/default.shtml> (accessed August 27, 2007).

"Endangered Species." *World Wildlife Fund.* <http://www.worldwildlife.org/endangered/> (accessed August 27, 2007).

"The Endangered Species Act and What We Do." *U.S. Fish and Wildlife Service.* <http://www.fws.gov/endangered/whatwedo.html> (accessed August 27, 2007).

IUCN-The World Conservation Union. <http://www.iucn.org> (accessed August 27, 2007).

"New Polar Bear Findings." *U.S. Geological Survey.* <http://www.usgs.gov/newsroom/special/polar_bears/default.asp> (accessed September 21, 2007).

Miriam C. Nagel

Energy Contributions

Introduction

There are two main sources of human-caused greenhouse-gas emissions, namely energy production and land-use change. Today, most energy is produced by burning fossil fuels, especially coal, oil, and natural gas. Burning these fuels—whether to run vehicles, heat buildings or industrial processes, or generate electricity—releases carbon dioxide (CO_2), the most important greenhouse gas. The two other large-scale sources of energy are nuclear power and hydroelectric dams. Other energy sources, such as biofuels, thermal and photovoltaic solar power, geothermal energy, and windmills, produce a small but rapidly growing fraction of world energy supply.

All forms of energy production have environmental impacts, but these impacts vary widely. In some cases, such as nuclear power, their nature is disputed by supporters and opponents of the technology. With regard only to the release of greenhouse gases, not other possible environmental harms, fossil fuels are by far the worst offenders. However, large hydroelectric dams emit significant methane (CH_4), biofuels may or may not reduce greenhouse emissions significantly depending on how they are produced, and windmills, solar cells, and nuclear power plants do not emit greenhouse gases while in operation because they do not emit waste gases, but do emit greenhouse gases at least during manufacture.

Today, debate swirls around the question of whether a massive new round of investment in nuclear power is essential to mitigating climate change, or would set back mitigation by absorbing funds that could be more effectively spent fighting greenhouse emissions by increasing energy efficiency. All energy generation schemes, not only nuclear power, can be seen as diverting funds from efficiency measures that can reduce energy demand more cheaply than that demand can be met by any primary source.

Historical Background and Scientific Foundations

Human beings have been burning fuel, usually wood, for almost 800,000 years. Fossil fuels—most of which consist of the transformed remnants of ancient forests (coal) or tiny oceanic organisms (petroleum and natural gas)—were first widely used in the Middle Ages in Europe, after the forests had been depleted for fuel and timber. In Britain, where the forests were smaller originally, the crisis arrived sooner and the practice of mining for coal underground began first, in the 1200s. Coal was heaped up in furnaces or fireplaces to heat spaces or boil liquids; it was not used to make steam or run machinery. Europe's increased reliance on coal was the beginning of large-scale human emissions of greenhouse gases.

Although burning or decaying wood also releases CO_2, this does not increase the total amount of carbon in the atmosphere. Growing trees extract carbon from the air, and burning wood returns that carbon to the air. The carbon in fossil fuels was also originally obtained from the air by plants, but this happened so many millions of years ago that its reappearance in the atmosphere today has the effect of adding brand new carbon to Earth's environment.

For centuries after coal-burning became important, the amount of coal burned was small by modern standards. Mining techniques were crude and could not reach the largest, deepest deposits, and demand was relatively low. This began to change in the 1700s, when the switch to more complex, energy-intensive technologies began, the phase of history called the Industrial Revolution. In 1712, British inventor Thomas Newcomen (1663–1729) was the first to demonstrate a coal-powered steam engine for pumping water, allowing coal mines to be kept open more cheaply. Though Newcomen's steam engine was inefficient it was still far cheaper than the horse-powered pumps that had been standard up to that time.

WORDS TO KNOW

BIOFUEL: A fuel derived directly by human effort from living things, such as plants or bacteria. A biofuel can be burned or oxidized in a fuel cell to release useful energy.

CARBON SEQUESTRATION: The uptake and storage of carbon. Trees and plants, for example, absorb carbon dioxide, release the oxygen, and store the carbon. Fossil fuels were at one time biomass and continue to store the carbon until burned.

DEFORESTATION: Those practices or processes that result in the change of forested lands to non-forest uses. This is often cited as one of the major causes of the enhanced greenhouse effect for two reasons: 1) the burning or decomposition of the wood releases carbon dioxide; and 2) trees that once removed carbon dioxide from the atmosphere in the process of photosynthesis are no longer present and contributing to carbon storage.

FOSSIL FUELS: Fuels formed by biological processes and transformed into solid or fluid minerals over geological time. Fossil fuels include coal, petroleum, and natural gas. Fossil fuels are non-renewable on the timescale of human civilization, because their natural replenishment would take many millions of years.

GEOTHERMAL ENERGY: Energy obtained from Earth's internal heat, which is maintained by the breakdown of radioactive elements. Geothermal means, literally, Earth-heat. Geothermal energy may be used either directly as heat (e.g., to heat buildings or industrial processes) or to generate electricity.

GREENHOUSE GASES: Gases that cause Earth to retain more thermal energy by absorbing infrared light emitted by Earth's surface. The most important greenhouse gases are water vapor, carbon dioxide, methane, nitrous oxide, and various artificial chemicals such as chlorofluorocarbons. All but the latter are naturally occurring, but human activity over the last several centuries has significantly increased the amounts of carbon dioxide, methane, and nitrous oxide in Earth's atmosphere, causing global warming and global climate change.

INDUSTRIAL REVOLUTION: The period, beginning about the middle of the eighteenth century, during which humans began to use steam engines as a major source of power.

INTERGOVERNMENTAL PANEL ON CLIMATE CHANGE (IPCC): Panel of scientists established by the World Meteorological Organization (WMO) and the United Nations Environment Programme (UNEP) in 1988 to assess the science, technology, and socioeconomic information needed to understand the risk of human-induced climate change.

METHANE: A compound of one hydrogen atom combined with four hydrogen atoms, formula CH_4. It is the simplest hydrocarbon compound. Methane is a burnable gas that is found as a fossil fuel (in natural gas) and is given off by rotting excrement.

PHOTOVOLTAIC SOLAR POWER: Electricity produced by photovoltaic cells, which are semiconductor devices that produce electricity when exposed to light. Depending on the cell design, between 6% and (in laboratory experiments) 42% of the energy falling on a photovoltaic cell is turned into electricity.

RADIATIVE FORCING: A change in the balance between incoming solar radiation and outgoing infrared radiation. Without any radiative forcing, solar radiation coming to Earth would continue to be approximately equal to the infrared radiation emitted from Earth. The addition of greenhouse gases traps an increased fraction of the infrared radiation, reradiating it back toward the surface and creating a warming influence (i.e., positive radiative forcing because incoming solar radiation will exceed outgoing infrared radiation).

RENEWABLE ENERGY: Energy obtained from sources that are renewed at once, or fairly rapidly, by natural or managed processes that can be expected to continue indefinitely. Wind, sun, wood, crops, and waves can all be sources of renewable energy.

By the end of the 1700s, more sophisticated steam engines were being used to run factory machinery, a practice that became universal during the 19th century in the industrializing countries of Europe and North America. The beginning of the modern or post-industrial increase in greenhouse gas concentrations in the atmosphere, especially of CO_2, is usually dated to about 1750. Greenhouse-gas emissions have increased ever since, with the most recent growth being the most rapid. In the 1800s, coal began to be used not only for heat and factory work but to fuel locomotives and steamships. In the early 1900s, trucks and cars became essential to industrialized economies and were dependent on gasoline refined from petroleum. At about the same time, coal and other fuels began to be used to generate large amounts of electricity. Finally, from the 1950s onward, large quantities of natural gas also became essential to modern civilization.

None of the old fuels, since the replacement of wood with coal, have ever become truly obsolete. In the early 2000s, the world is using more coal than ever, more oil than ever, and more natural gas than ever. Although nuclear power and renewable energy sources such as wind and solar have contributed some energy in recent decades and renewable energy production is growing rapidly, only a relatively small slice of world primary energy comes from these sources.

Mining trucks are shown carrying loads of oil-laden sand in Alberta, Canada, in 2005. Although Canada ratified the Kyoto Protocol, it became the first country to announce that it would not be able to meet its Kyoto target of a 6% emissions reduction over the years 2008–2009. Instead, Canada's emissions were increasing. *AP Images.*

The Energy Usage Picture: Primary Supply and Usage by Sector

Primary energy is the energy content of a fuel or other source. As of 2004, world primary energy supply was obtained from the following sources: oil (petroleum), 34.3%; coal, 25.2%; natural gas, 20.9%; renewables, 13.1%; nuclear power, 6.5%. The 13.1% of supply from renewables could be further broken down into combustibles, 10.6%; hydroelectric dams, 2.2%; and tidal, wind, solar, and geothermal energy, 0.5%.

Most primary energy is wasted, not used. How much of a primary energy fuel is wasted depends on how the fuel is used: for example, burning one gallon (3.8 l) of gas in an efficient car produces more useful work than burning it in an SUV, but the gallon of gas contains the same amount of primary energy in either case, namely about 124,000 British thermal units. (In metric units, gasoline's primary energy density is 34.6 million joules per liter.) Since 70% of primary fuel energy is wasted in standard centralized electric-generating plants, whether coal-fired or nuclear—about 30% is turned into electricity, and the rest is thrown away as waste heat—the energy sources that produce electricity directly, without producing and discarding heat (wind, solar, hydroelectric dams), actually supply a greater share of useful energy than their primary energy share.

Most coal—in the United States, about 92%—is burned to generate electricity, although a significant fraction is burned in kilns to make cement and a growing amount is used to make liquid fuel. Petroleum is refined to produce a variety of specialty fuels, including gasoline, kerosene, diesel fuel, aviation fuel, heating oil, and liquefied petroleum gas (mostly propane). Almost the entire modern transportation system depends on petroleum fuels, but very little oil is burned to generate electricity—only about 2.4% of oil usage in the United States as of 2005. Natural gas, which consists mostly of methane, is burned as-is, after the addition of a scenting agent to alert users to leaks. It is used to generate about 20% of the nation's electricity. Most of the remainder is used for residential space heat and industrial processes and space heat.

Greenhouse Gas Emissions

As of 2007, atmospheric CO_2 was the most important greenhouse gas, accounting for 63% of the radiative forcing involved in anthropogenic (human-caused) climate change. Radiative forcing is the amount of energy retained by Earth, rather than radiated to space, as a result of a greenhouse gas's presence in the atmosphere. About 84% of CO_2 emissions were from fossil-fuel combustion, with the remainder coming from land-use changes, mostly deforestation. CO_2 from fossil-fuel energy was the largest single contributor to both historic and ongoing accumulation of greenhouse gases in the atmosphere. In the United States, CO_2 emissions were 86% of greenhouse pollution in 2006.

Globally, coal and oil contribute about equally to annual global emissions of anthropogenic CO_2: oil's share was 40% in 2004 and coal's share was 39%. The remainder of CO_2 emissions, about 20%, was from deforestation and natural gas. However, coal's contribution was expected to climb. The U.S. Energy Information Administration (EIA) predicted in 2007 that by 2030, oil's share would be down to 36% and coal's up to 43%, with natural gas' share remaining about the same.

Historically, the first populations to industrialize—Europe and the United States—have been the greatest producers of greenhouse gases from energy fuels. As of 2004, the developing and least-developed countries, containing about 80% of the world's population, had produced only 23% of all emissions released since the beginning of the Industrial Revolution. The emissions share of the developing nations was growing rapidly; they accounted for 73% of emissions growth in 2004. However, this still only amounted to 41% of global emissions. Most of these emissions were from fuel-burning.

The EIA predicted that total amounts of CO_2 would continue to rise, despite statements by NASA scientist James Hansen and some other climatologists that global emissions of CO_2 and other greenhouse gases must be reduced by 80% below present-day levels by 2050 in order to prevent potentially catastrophic climate change. So far, the EIA's prediction of rising emissions has the virtue of describing actual trends; as of 2004, CO_2 emissions were accelerating, not slowing or decreasing. (The EIA's results were released in 2007, but referred to 2004 because energy analysis figures often lag reality by several years due to the time it takes to collect and analyze data.)

CO_2 emissions grew at 1.1% per year from 1990 to 1999, but at more than 3% per year from 2000 to 2004. This post-2000 growth rate was greater than in the most pessimistic, fossil-fuel-intensive scenario posited by the Intergovernmental Panel on Climate Change (IPCC) in the late 1990s and was driven mostly by increases in fossil-fuel combustion, including industrial process heat, transportation, space heat, and flaring of gas from wells.

CO_2 emissions in the United States fell in 2006 by about 1.3%. The drop was unusual: in only three out of the 16 years from 1990 to 2006 had CO_2 emissions dropped rather than risen, with typical changes being about +1% per year. Long-term growth (or shrinkage) of CO_2 emissions is influenced by overall economic growth, the energy intensity of the economy (how much energy is used per unit of business done), and the carbon intensity of the energy supply (how much CO_2 is emitted from the mix of primary energy sources). The drop in CO_2 emissions in the United States for 2006 illustrates the interaction of all these factors with weather and climate across energy-consumption sectors: residential CO_2 emission fell 3.7%, mostly because of warm weather, commercial and industrial emissions fell by about 1%, and transportation emissions stayed about constant. Warm weather and increased use of natural gas, the least carbon-intensive fossil fuel, accounted for the 2006 CO_2 dip.

Greenhouse gas emissions can be described by fuel, as discussed earlier. They can also be described by economic sector. Global CO_2 emissions in 2000 by energy sector were as follows:

Electricity generation: 29.5%

Industrial processes: 20.6%

Transportation: 19.2%

Residential and commercial use: 12.9%

Processing of fossil fuels: 8.4%

Biomass burning and deforestation: 9.1%

■ Impacts and Issues

Nonzero Emissions from Zero Emissions Sources?

Greenhouse gas emissions are produced directly by fossil-fuel burning, but not by renewables, hydroelectric power, or nuclear power. However, these other forms of energy do not have zero environmental or even climate impact, partly because energy must be invested to build the systems that produce the renewable energy. For example, large amounts of concrete, steel, and electricity must be consumed to build and fuel a nuclear power plant, and windmills consist mostly of steel, which requires energy to manufacture. Since most of the existing energy supply comes from fossil fuels, building non-emitting energy sources therefore releases greenhouse gases. This makes it incorrect to characterize windmills, nuclear power plants, and similar systems as zero-emissions energy sources, at least as long as most primary energy still comes from fossil fuels. Over the lifetime of a unit, however, the reduction in emissions versus generating the same energy from fossil fuels may be great. Net lifetime emission gains vary widely between technologies. In cases where a complex life-cycle of mining, refinement, construction, safeguards, waste disposal, and decommissioning is involved, as in the case of nuclear power, the subject is highly contentious, with expert estimates varying sharply.

Emissions from Dams

Hydroelectric dams do not use fuel: their energy is indirectly solar. Sun-driven evaporation lifts water above sea level as vapor and drops it as snow or rain inland, where it runs downhill in streams and rivers and gives up some of its gravitational potential energy to the turbines that produce hydroelectricity. Dams appear, therefore, to be perfect non-emitters of greenhouse gases.

Recent research, however, has shown that many hydroelectric dams are actually large greenhouse-gas emitters. When a dam is built, it causes water to back up and fill an artificial lake. This lake generally drowns a large area of forest. In areas where trees are only killed by shallow water, not submerged, their decay releases CO_2. Submerged trees, which must decay under oxygen-free (anaerobic) conditions, release methane. Also, vegetation flourishes in the vast mud flats that are exposed during low-water or "drawdown" periods; this vegetation is submerged when the water rises again and decays, also releasing methane. Methane release from many dams, especially in tropical and subtropical areas, is thus an ongoing process, not a one-time pulse from drowned trees after construction.

A study published in 2004 found that in 1990 the Curuá-Una Dam in Brazil, 13 years after its lake filled with water, was producing 3.6 times as much global greenhouse warming, because of its methane emissions, as if the same amount of electricity were being produced by burning oil. Some researchers have suggested that at least some of this methane could be collected and burned to produce electricity, both extracting useful energy and converting the methane to CO_2. Burning methane reduces its greenhouse potential by about 95%, since methane is about 21 times as powerful a greenhouse gas as CO_2. However, dam-methane recovery technology had not yet been demonstrated as of 2007.

Reducing Emissions

If emissions of greenhouse gases are to be stabilized or reduced, energy must be either produced with fewer emissions, used more efficiently and sparingly, or both. Lower emissions are claimed for natural gas, nuclear power, renewable energy sources, and advanced coal technologies that sequester CO_2 underground rather than spewing it into the air.

All of these technologies have their supporters and critics. For example, although it is a fact that natural gas releases less carbon at the point of combustion for each unit of energy produced, a Carnegie Mellon University team claimed in 2007 that importing liquefied natural gas (LNG) to the United States from other countries could produce 35% more greenhouse gas emissions than domestic coal burned in projected (but not yet built) facilities that use carbon sequestration. The reason is that imported LNG is a complex and roundabout technology, which requires that gas be first extracted, chilled to −264°F (−163°C) to make it a liquid, shipped in large tankers over thousands of miles, heated again to a gaseous state, and finally distributed through pipelines.

The coal technologies to which LNG was compared by the Carnegie Mellon analysts, however, were still speculative: no large-scale coal plant with carbon sequestration had yet been built in 2007. In October 2007, the U.S. Department of Energy announced permits for three regional-scale sequestration demonstration projects. It remains to be seen whether the technology will prove both effective and affordable.

In 2007, the United Nations released an analysis of the global potential for energy efficiency as a way of abating global climate change. The report said that to reduce rising global greenhouse-gas emissions to 2007 levels by 2030 would require only 0.3–0.5% of the projected 2030 global gross domestic product. Greater savings could be realized at higher cost. In general, it is cheaper to save a unit of energy than to generate it by any means. The UN Climate Change Secretariat, Yvo de Boer, stated: "Energy efficiency is the most promising means to reduce greenhouse gases in the short term."

SEE ALSO *Automobile Emissions; Aviation Emissions and Contrails; Biofuel Impacts; Carbon Sequestration Issues; Energy Efficiency; Nuclear Power; Petroleum; Renewable Energy; Wind Power.*

BIBLIOGRAPHY

Books

Boyle, Godfrey. *Renewable Energy.* New York: Oxford University Press, 2004.

Periodicals

Caldeira, Ken, et al. "Climate Sensitivity Uncertainty and the Need for Energy Without CO_2 Emission." *Science* 299 (2003): 2052-2054.

Green, Chris, et al. "Challenges to a Climate Stabilizing Energy Future." *Energy Policy* 35 (2007): 616-626.

Hoffert, Martin I., et al. "Energy Implications of Future Stabilization of Atmospheric CO_2 Content." *Nature* 395 (1998): 881-884.

Quadrelli, Roberta, and Sierra Peterson. "The Energy-Climate Challenge: Recent Trends in CO_2 Emissions from Fuel Combustion." *Energy Policy* 35 (2007): 5938-5952.

Raupach, Michael R. "Global and Regional Drivers of Accelerating CO_2 Emissions." *Proceedings of the National Academy of Sciences* (104) 2007: 10288-10293.

Stauffer, Hoff. "New Sources Will Drive Global Emissions." *Energy Policy* 35 (2007): 5433-5435.

Web Sites

Carnegie Mellon University. "Natural Gas Imported to US for Electricity Generation May Be Environmentally Worse than Coal." *Sciendaily,* August 23, 2007. <http://sciendaily.com/releases/2007/08/070822132122.htm> (accessed November 4, 2007).

"Emissions of Greenhouse Gases in the United States 2005." *U.S. Energy Information Administration,*

November 2006. <http://www.eia.doe.gov/oiaf/1605/ggrpt/carbon.html> (accessed November 4, 2007).

"Global Climate Change: The Role for Energy Efficiency." *United Nations Framework Convention on Climate Change Secretariat*, August 2007. <http://www.NCSEonline.org/nle/crsreports/climate/clim-23.cfm> (accessed November 4, 2007).

Spadaro, Joseph R., et al. "Assessing the Difference: Greenhouse Gas Emissions of Electricity Generation Chains." *IAEA [International Atomic Energy Agency] Bulletin*, 2000. <http://f40.iaea.org/worldatom/Periodicals/Bulletin/Bull422/article4.pdf> (accessed November 4, 2007).

Larry Gilman

Energy Efficiency

Introduction

Energy efficiency is proportion of energy used, rather than wasted, during the production or consumption of energy. Higher efficiency equals less waste; lower efficiency equals more waste. Efficiency can be further divided into conversion efficiency, distribution (or transmission) efficiency, and end-use efficiency. Conversion efficiency is the fraction of useful energy obtained during conversion of energy from one form to another, such as from heat to mechanical motion or electricity. Distribution efficiency is the fraction of a given amount of energy that is successfully sent through a transmission system such as a steam pipe or electric transmission line. End-use efficiency is the fraction of a given amount of energy that is used to accomplish a desired task once it has been delivered to a device, such as heating a home, running a computer, or moving a vehicle.

Most of the energy produced by burning fuels is lost because of low efficiencies in conversion, transmission, and end use. In general, conversion efficiency is very low, transmission efficiency rather high, and end-use efficiency fair to low. Greenhouse-gas emissions from fuel-burning could be greatly reduced by changing technology and human behavior to increase efficiency. Such changes would also have other benefits, such as reducing toxic pollution and environmental destruction associated with mining and drilling. Because energy is so expensive, large efficiency savings can often be realized at low cost or even at a profit.

Historical Background and Scientific Foundations

Energy efficiency was of urgent concern during World War I (1914–1918) and World War II (1939–1945), when many countries dependent on fossil fuels found themselves cut off from easy access to petroleum. For example, gasoline was rationed in all major combatant nations during World War II, including the United States, then the world's largest oil producer.

However, efficiency was not a permanent peacetime concern until fairly late in the twentieth century. In the 1970s, oil embargoes by the Organization of the Petroleum Exporting Countries (OPEC) triggered widespread public awareness that fossil fuels might run out, and not just in some remote future, but relatively soon. It became obvious that dependence on oil imports weakened national security all the time, not just during wars. Also, the destructiveness of coal mining, oil drilling, and the possible environmental side-effects of nuclear power increased public awareness that energy use always exacts an environmental cost.

In the 1980s and early 1990s, scientists became reasonably certain of the reality of global climate change, and in the late 1990s and early 2000s, developing economies such as those of India and China rapidly increased their demands for oil and natural gas, raising fuel prices to historically unprecedented levels.

All these concerns—limited supply, import insecurity, pollution, global warming, and high fuel costs—have conspired to make energy efficiency an abiding interest of citizens, engineers, corporate managers, and politicians in the early twenty-first century. Increased efficiency gets the same job done with less energy; less energy used means less fuel (or other primary energy) purchased, which means longer-lasting supplies, lessened import dependence, lessened pollution, lessened global warming, and lower costs. Although exotic, expensive forms of energy efficiency can be invented—affordable or even profitable opportunities for efficiency exist in almost all departments of energy use: buildings, vehicles, electronics, manufacturing, and more.

There is broad agreement among scientists who study energy usage that increasing energy efficiency is not only one of the most cost-effective ways to combat human-caused global climate change, but that there is no hope of significantly mitigating climate change unless energy efficiency is greatly increased worldwide.

WORDS TO KNOW

FOSSIL FUELS: Fuels formed by biological processes and transformed into solid or fluid minerals over geological time. Fossil fuels include coal, petroleum, and natural gas. Fossil fuels are non-renewable on the timescale of human civilization, because their natural replenishment would take many millions of years.

GREENHOUSE GASES: Gases that cause Earth to retain more thermal energy by absorbing infrared light emitted by Earth's surface. The most important greenhouse gases are water vapor, carbon dioxide, methane, nitrous oxide, and various artificial chemicals such as chlorofluorocarbons. All but the latter are naturally occurring, but human activity over the last several centuries has significantly increased the amounts of carbon dioxide, methane, and nitrous oxide in Earth's atmosphere, causing global warming and global climate change.

INTERGOVERNMENTAL PANEL ON CLIMATE CHANGE (IPCC): Panel of scientists established by the World Meteorological Organization (WMO) and the United Nations Environment Programme (UNEP) in 1988 to assess the science, technology, and socioeconomic information needed to understand the risk of human-induced climate change.

RADIO WAVES: Electromagnetic waves that oscillate or vibrate between 3 and 300 billion times per second. Radio waves are physically identical to light waves, except that they do not vibrate as rapidly.

RATIONING: Mandatory distribution of fixed amounts of food, fuel, or other goods in order to conserve a scarce resource.

SOLAR ENERGY: Any form of electromagnetic radiation that is emitted by the sun.

According to a 2000 report to the U.S. Congress from the Congressional Research Service, "Increased energy efficiency is generally thought to be the primary way to reduce the nation's growth in CO_2 emissions."

Technical Definitions of Efficiency

The energy efficiency of any system is defined as the ratio of useful energy output to total energy input. For example, if a fifth of the energy contained in the gasoline burned by a vehicle is turned into mechanical energy—ends up actually moving the vehicle—the vehicle's efficiency is 20%. Waste energy ends up as heat that is ejected to the environment from the radiator or out the tailpipe.

The three basic types of efficiency—conversion, distribution, and end-use—refer to different types of energy systems. Conversion systems include power plants, which convert the chemical energy in fuels into electricity and heat, and devices such as batteries, which convert electricity into chemical potential energy when they are charged and back into electricity when they are used to supply current. Drive trains, electrical transformers, and other devices are also conversion systems.

The conversion efficiency of a typical coal-fired power plant, defined as the ratio of the energy in the coal burned to the amount of electricity sent out from the plant, is about 30%. The efficiency of a typical lead-acid automobile battery, defined as the ratio of electricity used in charging the battery to the electricity obtained when discharging it, is 75–85%.

Distribution efficiency is the ratio of the useful energy extracted from a distribution system to the energy put into that system. For example, electricity from a large power plant is fed into a distribution system of high-voltage transmission lines—the grid—but not all that energy comes back out again in the form of useful electricity. Some energy is lost in heating the wires, and some is radiated from the system as radio waves. (So much energy is radiated in this way, in fact, that a fluorescent light bulb will glow in the dark underneath a typical high-voltage transmission line.) Distribution losses consume 8–9% of the electricity that is produced by power plants.

End-use efficiency is the ratio of energy that devices use to do useful work to the energy delivered to those devices. For example, the end-use efficiency of an incandescent (standard) light bulb is about 2%; that is, it turns only about 2% of the electricity it consumes into visible light. The rest is turned into heat. A compact fluorescent light (CFL) bulb has an end-use efficiency of 7–8%. Thus, CFL bulbs make about four times as much visible light for each watt-hour of electricity they consume as do incandescent bulbs (a 16-watt CFL bulb replaces a 60-watt incandescent bulb); their end-use efficiency is four times greater. Yet they still have low end-use efficiency, overall.

Many energy analysts define two other kinds of efficiency in addition to these three classic kinds. The first is extractive efficiency, the efficiency of converting primary energy—coal, oil, or uranium in the ground, for example—into available fuel energy. All energy spent extracting the fuel from the ground, processing it, and transporting it to where it will be burned must be counted as loss in calculating extractive efficiency. Extractive efficiency comes before conversion efficiency in the energy stream.

The other non-classical type of efficiency is hedonic efficiency, named from the Greek word *hedone* for "pleasure." Hedonic efficiency is the rate at which energy services are converted into human welfare, and comes at the very end of the energy stream. For example, a CFL bulb burning in a room that is occupied 10% of the time has a hedonic efficiency of 10%. If the room is never visited by human beings and does not need to be illuminated at all, the bulb has a hedonic efficiency of 0%. A computer that is left on overnight has a hedonic of efficiency of 0% during that time.

Interactions between Efficiencies

Although conversion and distribution efficiencies are straightforward to calculate, end-use efficiency and hedonic efficiencies are more complex because they depend on how "use" is defined. This, in turn, depends on what human energy users desire in terms of energy services. In the case of a light bulb, energy turned by the bulb directly into heat, not light, is generally considered waste, because light bulbs are generally purchased to produce light, not heat.

Sometimes, however, heat is desired, as in buildings during the winter. In this case, the heat emitted by the light bulb is not, strictly speaking, wasted. Yet most of the money spent on that electricity is wasted, because heating indoor spaces with electricity is generally much more expensive than other common heating options (heating oil, wood stoves, propane heaters, passive solar heat, etc). Waste heat from light bulbs is a very expensive contribution to a building's heat budget.

Further, many buildings are air conditioned to get rid of unwanted heat. To continue the light-bulb example, waste heat from lights adds to the cooling burden of an air-conditioned building: the building owner must not only purchase the energy that the light bulb turns into unwanted heat, but must purchase still more energy to remove that unwanted heat from the building. Typically, an air conditioner consumes about 0.5 units of electrical energy for each unit of heat energy it removes. An air conditioner must therefore consume half a watt of power for each watt of waste heat produced by a light bulb. So, for a 100-watt bulb generating 98 watts of heat, an additional 98/2 = 49 watts of electricity must be purchased to keep the building cool. The actual efficiency of the light bulb is, in this context, found by dividing the 2 watts of visible light (useful, desired function) by the 100 watts purchased to run the bulb plus the 49 watts purchased to get rid of the bulb's waste heat: 2/(100 + 49) = .013. This is a mere 1.3% efficiency, significantly worse than the solitary bulb's 2% efficiency.

This example shows how efficiencies can be connected with each other. Waste in one place (an inefficient light bulb) can lead to further waste in another (the air-conditioning system). By the same token, increased efficiency in one place can lead to increased efficiency in others. Replacing an incandescent bulb with a CFL bulb in an air-conditioned building leads to a total efficiency gain not of 4 times (comparing just the bulbs), but of at least 9 times (taking air conditioning into account).

Energy Conversion along Chains

Consider a situation where a single energy unit of visible light is produced by a light bulb. To make that unit of visible light energy, 100 units of electrical energy must be delivered to the bulb. Assuming that there is 9% loss of electric power between the power plant and the bulb, this means that 109.9 units must be sent out from the power plant to produce the 1 unit of visible light. Because there is 70% loss at the power plant from fuel to electricity, 366 units of fuel energy must be burned to produce those 109.9 units of electrical energy. The efficiency of the total system, fuel to light, is only 1/366 = 0.27%.

However, if a more efficient light bulb is installed, say a CFL bulb that uses 25 units of electrical energy instead of 100 to produce 1 unit of light energy, then only about 27 units of electricity need be delivered to the home. In this case, only 27.5 units need to be transmitted from the power plant, and only 91.5 units of fuel energy need to be produced. The amount of energy saved at the light bulb is 75 units, but the amount saved at the power-plant is 274.5 units. In short, downstream savings (at the user's end) multiply upstream savings (at the production end). Reductions in mining damage, air pollution, and climate change are correspondingly large.

Consider this whole story for a typical improvement in end-use efficiency. If enough end-users install enough efficient appliances, a smaller power plant or fewer power plants can be built, as well as smaller or fewer transmission lines, saving money and land as well as fuel costs. Some of the money saved from these measures may be invested in buying further efficiency improvements.

■ Impacts and Issues

Because of the extreme cheapness of energy during the early post-World War II era, consumer and industrial concerns about cost focused mostly on capital costs—how much a light bulb, refrigerator, automobile, heating system, or other energy-using object cost, not how much energy it used. This changed permanently with the oil shocks of the 1970s. Since that time, appliances, buildings, and other energy-use sectors have become steadily and significantly more efficient. Automobiles are an exception, due to the vogue for the large, inefficient private vehicles known as sport utility vehicles (SUVs); from 1985 to 2002, the average mileage of U.S. cars declined from 26 to 24 miles per gallon (10.8 to 10.2 km/L).

Widespread increases in efficiency show up as decreases in energy intensity, that is, fuel or primary energy consumption per dollar of gross domestic product (a measure of total economic activity). The energy intensity of an economy is how much energy it uses to get a given amount of business done. As efficiency rises, the economy gets each unit of business done using less primary energy, and energy intensity decreases. Energy intensity may also decrease if an economy shifts the kind of business it is doing to lower-energy activities. For instance, if an aluminum smelting plant is replaced by a telemarketing center, the same amount of business may get done as measured in dollars, but energy intensity will be less.

IN CONTEXT: ENERGY INFRASTRUCTURE INVESTMENTS

"New energy infrastructure investments in developing countries, upgrades of energy infrastructure in industrialized countries, and policies that promote energy security, can, in many cases, create opportunities to achieve GHG [greenhouse gas] emission reductions compared to baseline scenarios. Additional co-benefits are country-specific but often include air pollution abatement, balance of trade improvement, provision of modern energy services to rural areas and employment (high agreement, much evidence)."

SOURCE: *Metz, B., et al. "IPCC, 2007: Summary for Policymakers." In:* Climate Change 2007: Mitigation of Climate Change. Contribution of Working Group III to the Fourth Assessment Report of the Intergovernmental Panel on Climate Change. *New York: Cambridge University Press. 2007.*

Historical Savings

Energy intensity has declined in much of the industrialized world over the last several decades. In the United States, energy intensity declined by 46% from 1975 to 2005. This was mostly due to efficiency increases rather than to shifting to services that replaced manufacturing. In the early 2000s, U.S. energy intensity was declining by about 2.5% per year.

Efficiency improvements have appeared in lighting, refrigeration, air conditioning, and heating. Buildings, for example, have realized higher efficiency through using better insulation, sealing air leaks, orienting windows and shading to harvest (or exclude) solar energy as appropriate to the seasons and to local climate, and other measures. Industries have become more efficient by using techniques such as cogeneration, where waste heat from one process is used for another, such as space heating, that requires a lower intensity of heat, rather than being vented directly to the environment. Yet despite the dramatic gains in efficiency that have already occurred, only a fraction of the possible efficiency gains have yet been realized.

Potential Savings

An analysis of the global potential for energy efficiency as a way of mitigating global climate change was released by the United Nations in 2007. The report said that greenhouse emissions could be reduced even more. To reduce rising global greenhouse-gas emissions to 2007 levels by 2030, the report said, would require the world to spend only 0.3–0.5% of projected 2030 global gross domestic product. According to the U.N. Climate Change Secretariat, Yvo de Boer, "Energy efficiency is the most promising means to reduce greenhouse gases in the short term." The secretariat also claimed that most of the cost-effective efficiency opportunities are in developing countries.

However, the report was criticized as being conservative. The UN's own Intergovernmental Panel on Climate Change (IPCC) has stated that to avoid possibly catastrophic global warming, emissions must be reduced by 80% from today's levels: simply holding them steady at 2007 levels will not mitigate enough climate change.

Further, even if it is true that the opportunities for efficiency improvements are greater in developing countries, they are still very great in the developed countries. Energy expert Amory B. Lovins, who was a prominent and early champion of end-use efficiency in the 1970s, claimed in 2005 that efficiency improvements in vehicles, electrical usage, and other areas could save half of U.S. gas and oil usage and 75% of U.S. electricity usage for less than it would cost to supply the energy itself (i.e., at a profit). Globally, the potential for efficiency savings was even greater, in agreement with the UN's statements, because many countries still had higher energy intensities than the United States.

■ Primary Source Connection

Energy efficiency refers to technologies or programs that are designed to use less energy to perform the same task or work. By reducing energy consumption, energy efficient devices or systems also reduce the production of greenhouse gases, particularly carbon dioxide (CO_2), which is involved in energy production. One example of energy efficiency is the use of technology to produce an automobile that is more fuel efficient while delivering the same performance. Energy efficiency and CO_2 reduction have been the primary goals of the energy policy of the United States since the 1970s.

This article, by the National Council for Science and the Environment (NCSE), details the use of federal energy efficiency standards over the last several decades. NCSE is a non-profit group that is dedicated to improving the use of science in environmental policy making.

GLOBAL CLIMATE CHANGE: THE ROLE FOR ENERGY EFFICIENCY

Energy efficiency is increased when an energy conversion device, such as a household appliance, automobile engine, or steam turbine, undergoes a technical change that enables it to provide the same service (lighting, heating, motor drive) while using less energy. Energy efficiency is often viewed as a resource option like coal, oil or natural gas. It provides additional economic value by preserving the resource base and reducing pollution.

Energy security, a major driver of federal energy efficiency programs in the past, is now somewhat less of an issue. On the other hand, worldwide emphasis on

High-energy grade garbage is ready to be packed at a garbage processing center in Nuemuenster, Germany, in 2007. The Mechanical-Biological Waste Processing Center converts 55% of the garbage it receives into high-energy fuel, which it then sends to the city's converted power plant. The burning of garbage has helped the plant reduce its reliance on coal, which has lessened its emissions of CO_2. *Sean Gallup/Getty Images.*

environmental problems of air and water pollution and global climate change have emerged as important drivers of support for energy efficiency policies and programs. Also, energy efficiency is seen as a technology strategy to improve the competitiveness of U.S.-made appliances, cars, and other energy-using equipment in world markets. The Clinton Administration views energy efficiency as the flagship of its energy policy for global climate change and other environmental reasons.

From 1975 through 1985, high energy prices served as a strong catalyst to improved energy efficiency. However, the sharp drop in oil and other energy prices that began in 1986 has dampened the impact of prices on energy efficiency improvements.

Federal policies and programs have also made a significant contribution to improved energy efficiency. One such program is DOE's energy efficiency R&D program, which employs a "technology-push" strategy. That is, it produces new, ever-more efficient technologies that form a basis for new products and services in the private sector. In contrast, EPA's energy star programs employs a "market-pull" strategy wherein businesses, institutions, and consumers are encouraged to buy more energy-efficient equipment.

The role of energy prices and the environmental benefits of energy efficiency often lead to a discussion about barriers and market failures. However, the resultant debate over the effectiveness of market forces to stimulate energy efficiency and the merit of federal policies and programs that support energy efficiency is not the focus of this report. Instead, this paper is focused on the projected contribution of energy efficiency to reducing CO2 emissions.

Energy efficiency is proposed as a cost-effective and reliable means for reducing the nation's growth in CO2 emissions due to fossil fuel use. Recognition of that potential has led to high expectations for the control of future CO2 emissions through even more energy efficiency improvements than have occurred through past programs, regulation, and price effects. Thus, in a recent context of low energy prices and rising fossil fuel use, the Clinton Administration has proposed increased government support for energy efficiency programs as its primary initiative to reduce emissions of CO2 and other "greenhouse gases" that may cause global climate change.

However, there is a debate over [projected] estimates of the future potential for energy efficiency to curb the growth of CO2 emissions through 2010. This paper discusses this debate, which is centered on differences between key reports by the Department of Energy (DOE) and the Energy Information Administration (EIA). A DOE report by five of its research laboratories projects that further gains in energy efficiency could be the largest future contributor to CO2 emissions reduction. However, EIA has criticized the DOE report's assumptions about the character of future energy efficiency measures, economic growth rates, future government R&D policies, and market adoption of energy efficiency measures.

The paper also describes a debate over the analysis of actual CO_2 emission reductions from past energy efficiency measures. In this case, methodological issues are at the core of disagreements between the General Accounting Office (GAO) and the Environmental Protection Agency (EPA) about the best way to assess emission savings from EPA's various energy efficiency programs.

Finally, the paper notes that federal efforts to curb global climate change through increased energy efficiency may be affected by a number of issues being debated by Congress, including program appropriations, new tax incentives, and legislation on electricity restructuring.

Energy Use Impact on Global Climate Change

Wherever energy efficiency and conservation measures reduce fossil fuel use, they will reduce carbon dioxide (CO_2) emissions, as well as pollutants that contribute to water pollution, acid rain, and urban smog. Human activities, particularly burning of fossil fuels, have increased atmospheric CO_2 and other trace gases. If these gases continue to accumulate in the atmosphere at current rates, many experts believe global warming could occur through intensification of the natural "greenhouse effect," that otherwise moderates Earth's climate. Excess CO_2 is the major contributor to this effect. The influence of human-induced emissions on the "greenhouse effect" is a subject of continuing research and controversy.

U.S. use of fossil energy (coal, oil, natural gas) currently produces about one-fourth of the world's CO_2 emissions. Since 1988, the federal government has accelerated programs that study the science of global climate change and created programs aimed at mitigating fossil fuel-generated carbon dioxide (CO_2) and other human-generated emissions. The federal government has funded programs for energy efficiency as a CO_2 mitigation measure at DOE, EPA, the Agency for International Development (AID), and the World Bank. The latter two agencies have received funding for energy efficiency-related climate actions through foreign operations appropriations bills.

Efforts to study greenhouse gas emissions and to devise programs to reduce them accelerated after the 1992 United Nations Conference on Environment and Development (UNCED) concluded with the signing of the Rio Declaration, Agenda 21 (an action program), and the Framework Convention on Climate Change (UNFCCC). Agenda 21 promotes the development, transfer, and use of improved energy-efficient technologies, the application of economic and regulatory means that account for environmental and other social costs, and other energy efficiency-related measures. The United States ratified the UNFCCC in 1992, and the Convention entered into force in 1994. The UNFCCC calls for each nation to develop a strategy for emissions reduction, inventory emissions, and promotion of energy and other technologies that reduce emissions.

Energy Efficiency and Energy Use

Increased energy efficiency of combustion and other fuel-using equipment has a long record of reducing the rate of growth in fossil fuel use and, thereby, reducing carbon emissions. This improvement is reflected in the ratio of U.S. energy use to Gross Domestic Product (GDP), which fell from 19,750 British thermal units (Btu's) per dollar in 1971 to 14,040 Btu's per dollar in 1986. This represents an average annual reduction of 1.81% in the energy/GDP ratio. For the period from 1972 to 1986, energy efficiency improvements cut energy use by 30% or 32 quadrillion Btu's per year. By 1988, recognition of this accomplishment had led to a focus on energy efficiency programs as a key strategy for future control of CO_2 emissions.

However, from 1986 to 1998, the rate of energy efficiency improvement slowed. The energy/GDP ratio declined from 14,040 Btu's per dollar in 1986 to 12,480 Btu's per dollar in 1998, but this represents an average annual reduction of 0.85%, which is less than half the rate for 1972 to 1986. Further, the decline in oil prices since the mid–1980s has led to historically low gasoline prices which, in turn, encouraged motorists to buy less fuel-efficient automobiles, such as sport utility vehicles, and to increase travel by about 24%. Overall, national petroleum use for transportation grew 21%, or 4.3 Q during this period. Also, since 1994, electric utility industry restructuring at the state level caused utility spending for energy efficiency to fall 48% by 1998 and the resultant rate of energy savings fell 20% from 1996 to 1998. Meanwhile, coal use for electricity production grew 33% from 1986 to 1998.

Thus, despite the increase in efficiency as measured by Btu/$, total fossil fuel use, has been rising steadily due to low energy prices, economic growth, and population growth. This growth includes oil and coal, which are the most intense emitters of carbon dioxide (CO_2). As a result, CO_2 emissions have been rising, eclipsing the 1993 Clinton Administration Climate Change Action Plan (CCAP) goal of reducing emissions to the 1990 level by 2000. In fact, Energy Information Administration (EIA) projections show fossil energy use and emissions increases continuing through 2010.

Fred Sissine

SISSINE, FRED. "GLOBAL CLIMATE CHANGE: THE ROLE FOR ENERGY EFFICIENCY." NATIONAL COUNCIL FOR SCIENCE AND

THE ENVIRONMENT. CONGRESSIONAL RESEARCH SERVICE, FEBRUARY 3, 2000.

See Also *Adaptation; Energy Contributions; Industry (Private Action and Initiatives); Lifestyle Changes.*

BIBLIOGRAPHY

Books

Casten, Thomas R. *Turning Off the Heat: Why America Must Double Energy Efficiency to Save Money and Reduce Global Warming.* Amherst, NY: Prometheus Books, 1998.

Hordeski, Michael Frank. *New Technologies for Energy Efficiency.* Lilburn, GA: Fairmont Press, 2002.

Periodicals

Wing, Ian Sue, and Richard S. Eckaus. "The Implications of the Historical Decline in U.S. Energy Intensity for Long-Run CO_2 Emission Projections." *Energy Policy* 35, no. 11 (2007): 5267–5268.

Web Sites

Lovins, Amory B. "Energy End-Use Efficiency." *Rocky Mountain Institute*, September 19, 2005. <http://www.rmi.org/images/PDFs/Energy/E05-16_EnergyEndUseEff.pdf> (accessed October 26, 2007).

Sissine, Fred. "Energy Efficiency: Budget, Oil Conservation, and Electricity Conservation Issues." *Congressional Research Agency*, May 25, 2006. <http://www.ncseonline.org/NLE/CRSreports/06jun/IB10020.pdf> (accessed October 26, 2007).

Sissine, Fred. "Global Climate Change: The Role for Energy Efficiency." *United Nations*, February 3, 2000. <http://unfccc.int/cooperation_and_support/financial_mechanism/items/4053.php> (accessed October 26, 2007).

Larry Gilman

Energy Industry Activism

Introduction

The human activity that contributes most to global climate change is the burning of fossil fuels. These fuels—oil, coal, and natural gas—are extracted by a relatively small group of large companies, often run by governments (for example, the Kuwait Petroleum Corporation). For much of the last 20 years, many oil and coal companies have resisted the scientific mainstream view that global climate change is real and caused in large part by fossil fuels.

Since the 1980s, these companies have often seen concern about global warming as a direct threat to their business: if fewer fossil fuels are burned, they reason, demand for their product will go down and they will make less money. As a result, some energy companies have funded advertising campaigns, scientific research, front groups, and political lobbying efforts in an attempt to discredit the scientific view of global warming, or to at least create a sense that climate science is too uncertain for any action to be taken.

Historical Background and Scientific Foundations

In the mid 1970s, scientific opinion wavered briefly between theories of imminent global cooling and global warming. By the end of the decade, it had shifted decisively toward global warming, and scientific concerns had begun to influence public opinion. By 1980, more than a third of Americans had heard of the greenhouse effect. The energy industries and some other business interests became worried that concerns about global warming might lead to profit-injuring government regulations.

Starting in the early 1990s, some energy companies developed a variety of front groups and funding strategies to spread disbelief or doubt about the science of global climate change. A front group is an organization that is created by another organization, such as a business or government, to evoke a false appearance of independence. Front groups established by the energy industry have been given names to make them sound like scientific or environmentalist groups, such as the Global Climate Coalition (GCC), Information Council on the Environment, and the Greening Earth Society. A few of these groups and their relationships to the energy industry are outlined below.

The Information Council on the Environment

The Information Council on the Environment (ICE) was created in 1991 by utility and coal industry groups, including the National Coal Association and the Western Fuels Association. ICE at once began a $500,000 ad campaign to discredit the idea of global warming. According to a strategy advice document from Cambridge Reports, a polling firm hired to help design the ICE campaign, the goal of the ads was "to reposition global warming as theory rather than fact." Cambridge Reports also created the name of the group to match the acronym ICE, which would suggest the opposite of warming; the name "Informed Citizens for the Environment" was also considered, but rejected because it did not sound as authoritative.

Print and radio ads were tested in several U.S. states. The ad that appeared in Minnesota featured a cartoon horse wearing a scarf and earmuffs saying, "If the Earth is getting warmer, why is Minneapolis getting colder?" (In fact, it had warmed.) Another print ad read, "Some say the Earth is warming. Some also said the Earth was flat." Cambridge Reports also advised that the ads should tell people that "some members of the media scare the public about global warming to increase their audience and influence."

These and other features of the ICE ad campaign were described in a July 1991 article in the *New York*

Times and in other media outlets, producing negative publicity that led to a stoppage of the campaign.

The George C. Marshall Institute

In early 1992, two months before the Earth Summit in Rio de Janeiro, Brazil, where the United Nations Framework Convention on Climate Change (UNFCCC) treaty was drafted, the George C. Marshall Institute, a conservative think-tank founded in 1984, issued a study claiming that global warming might well be caused by increasing radiation from the sun. Since 1989, the institute had been engaged in what its Web site describes as a "critical examination of the scientific basis for climate change."

In 1992, the institute was funded entirely by private conservative foundations, but eventually it began to accept funding directly from the energy industry. For example, in 1999 it accepted a grant from the oil company Exxon (just before Exxon's merger with Mobil to form the world's most profitable company, ExxonMobil, in November 1999). The chief executive officer of the institute as of 2007, William O'Keefe, is a former vice-president of the American Petroleum Institute (an oil-industry advocacy group) and in 2002 registered as a lobbyist for ExxonMobil. As of 2007, the Marshall Institute continued to oppose mainstream science on climate change.

ExxonMobil Activism

ExxonMobil has been the most publicly visible of energy companies seeking to influence public and political opinion about climate change. In 1989, Exxon and other U.S. firms founded the Global Climate Coalition (GCC) to counter reports from the Intergovernmental Panel on Climate Change (IPCC) indicating that global climate change is real and human-caused. The basic strategy of the GCC, like similar groups, was to create an exaggerated impression of uncertainty in climate science, implying that too little is known to justify any action—especially any action that might impinge on the well-being of large energy companies. During the 1990s, the GCC spent over $63 million on ads and lobbying to prevent action on climate change, including a 1997 ad campaign against ratification of the Kyoto Protocol that cost $13 million.

In 1996, U.S. Undersecretary for Global Affairs Timothy Wirth told a climate convention in Geneva, Switzerland, that groups such as the GCC were "naysayers and special interests bent on belittling, attacking and obfuscating climate change science" (Masood, 1996). In 1997, the GCC suffered a setback as British Petroleum (BP) and the Arizona Public Service Company pulled out of the group. The CEO of the latter company explained, "I was concerned that to continue to attack the science—which the GCC is basically doing—is not the way forward" (Masood, 1996). Other energy companies left in succeeding years, and despite president William O'Keefe's 1996 claim that "We're adding members, not losing them," the GCC closed down in 2001.

In 1998, Exxon, the oil company Chevron, and the American Petroleum Institute considered setting up a $5 million Global Climate Science Data Center as what they called a "sound scientific alternative" to the IPCC. The proposed organization was never created, apparently due to the ongoing crumbling of the GCC (Shell Oil quit in 1998). Also in 1998 the American Petroleum Institute drafted a memo, later leaked, detailing a Global Climate Science Communications Plan for manufacturing public uncertainty about global warming.

Exxon's primary public-relations efforts from the late 1990s through the early 2000s have been through contributing funds to groups that oppose the scientific

WORDS TO KNOW

DENIALISTS: People who insistently refuse to accept a scientific or historical fact in the face of overwhelming physical evidence and near-universal expert agreement.

EARTH SUMMIT: Alternative name for the United Nations Conference on Environment and Development, Rio de Janeiro, Brazil, June 3–14, 1992, the meeting at which the United Nations Framework Convention on Climate Change (UNFCCC) was developed.

FOSSIL FUELS: Fuels formed by biological processes and transformed into solid or fluid minerals over geological time. Fossil fuels include coal, petroleum, and natural gas. Fossil fuels are non-renewable on the timescale of human civilization, because their natural replenishment would take many millions of years.

GREENPEACE: Nonprofit environmental group formed in 1971, originally to protest nuclear testing and whaling. The group remains active, now addressing a wide range of issues, including climate change.

INTERGOVERNMENTAL PANEL ON CLIMATE CHANGE (IPCC): Panel of scientists established by the World Meteorological Organization (WMO) and the United Nations Environment Programme (UNEP) in 1988 to assess the science, technology, and socioeconomic information needed to understand the risk of human-induced climate change.

KYOTO PROTOCOL: Extension in 1997 of the 1992 United Nations Framework Convention on Climate Change (UNFCCC), an international treaty signed by almost all countries with the goal of mitigating climate change. The United States, as of early 2008, was the only industrialized country to have not ratified the Kyoto Protocol, which is due to be replaced by an improved and updated agreement starting in 2012.

consensus view of climate change. These groups include the American Enterprise Institute, the Competitive Enterprise Institute, the George C. Marshall Institute, the Cato Institute, and many more.

In 2006, the Royal Society (the United Kingdom's governmental science-advisory group, similar to the National Academy of Sciences) wrote to ExxonMobil to ask the company to stop funding such groups. The Royal Society found that ExxonMobil had given $2.9 million in 2005 alone to 39 U.S. groups that, according to the society, "misinformed the public about climate change through their websites."

In 2007, news media reported that ExxonMobil had softened its stance on the climate issue. According to Kenneth Cohen, Exxon's vice president for public affairs, ExxonMobil already cut funding for the Competitive Enterprise Institute and other denialist organizations in 2006. A Greenpeace report in 2007 stated, however, that ExxonMobil had continued to fund the majority of the groups it had assisted in the past, giving $2.1 million in 2006 to 41 denialist groups, including $421,000 to groups listed in the American Petroleum Institute's 1998 Global Climate Science Communications Plan.

■ Impacts and Issues

There is no way to precisely measure the effectiveness of public-relations campaigns such as that conducted over the last 15 to 20 years by portions of the energy industry, especially some oil and coal companies. However, it appears that these efforts may have been successful in confusing the issue. As of 2001, only 15% of Americans could identify fossil-fuel burning as the main cause of global warming, compared to 26% of Mexicans.

As of 2007, the number of Americans who knew that the majority of scientists agree that anthropogenic carbon dioxide is the main cause of global warming was between 33% and 60%, depending on how the question was worded. However, as of 2006, 87% saw the threat of global warming as either critical or important—less than in some countries, more than in others.

Whatever its past successes, and apart from continued corporate contributions to denialist groups, energy-industry efforts to undermine the scientific consensus on climate change have largely faltered or even reversed. In 2001, the CEO of BP acknowledged publicly that human-released greenhouse gases cause global warming and urged that action should be taken. BP has collaborated in climate-change initiatives with Royal Dutch/Shell, making a commitment to reduce greenhouse emissions with specific deadlines and investing in alternative energy technologies.

SEE ALSO *Climate Change Skeptics; Environmental Protection Agency (EPA); IPCC Climate Change 2007 Report: Criticism; Media Influences: False Balance; Public Opinion.*

BIBLIOGRAPHY

Books

Weart, Spencer. *The Discovery of Global Warming.* Cambridge, MA: Harvard University Press, 2004.

Periodicals

Adam, David. "Royal Society Tells Exxon: Stop Funding Climate Change Denial." *The Guardian* (September 20, 2006).

Eilperin, Juliet. "AEI Critiques of Warming Questioned." *The Washington Post* (February 5, 2007).

Krauss, Clifford, and Jad Mouawad. "Exxon Chief Cautions Against Rapid Action to Cut Carbon Emissions." *The New York Times* (February 14, 2007).

Masood, Ehsan. "Companies Cool to Tactics of Global Warming Lobby." *Nature* 383 (1996): 470.

"Oil Industry Lobby Plans Rival to UN Climate Science Panel." *Nature* 392 (1998): 856.

Schrope, Mark. "A Change of Climate for Big Oil." *Nature* 411 (2001): 516–517.

Wald, Matthew L. "Pro-Coal Ad Campaign Disputes Warming Idea." *The New York Times* (July 8, 1991).

Web Sites

"ExxonMobil's Continued Funding of Global Warming Denial Industry: Analysis by Greenpeace USA Research Department." *Greenpeace USA.* <http://www.greenpeace.org/usa/assets/binaries/exxon-secrets-analysis-of-fun.pdf> (accessed November 11, 2007).

George C. Marshall Institute: Better Science for Public Policy. <http://www.marshall.org/> (accessed November 11, 2007).

"Global Climate Science Communications Action Plan" (leaked oil-company documents). *Denial and Deception: A Chronicle of ExxonMobil's Efforts to Corrupt the Debate on Global Warming,* Greenpeace, 1998. <http://www.greenpeace.org/usa/assets/binaries/leaked-api-comms-plan-1998> (accessed November 11, 2007).

Jiang, Wenran. "Smoke, Mirrors, and Hot Air: How ExxonMobil Uses Big Tobacco's Tactics to Manufacture Uncertainty on Climate Science." *Union of Concerned Scientists,* January, 2007. <http://www.ucsusa.org/assets/documents/global_warming/exxon_report.pdf> (accessed November 11, 2007).

Ward, Bob. "Letter to ExxonMobil." *Royal Society,* September 4, 2006. <http://www.royalsoc.ac.uk/displaypagedoc.asp?id=23780> (accessed November 11, 2007).

Larry Gilman

Enhanced Greenhouse Effect

■ Introduction

The enhanced greenhouse effect refers to human activities that are adding to the warming of the atmosphere due to the greenhouse effect—the presence of gases that increases the atmosphere's retention of the heat energy of the sun.

The burning of fossil fuels including coal, oil, and natural gas, along with the clearing of land for agricultural use and urban development, are increasing the amount of the heat-retaining greenhouse gases in the atmosphere.

Although the reality of the enhanced greenhouse effect was debated by nations including the United States even to the end of the twentieth century, the evidence that human-related activities since the mid-eighteenth century have enhanced atmospheric warming is now considerable. Organizations including the United Nations' Intergovernmental Panel on Climate Change (IPCC) no longer debate the reality of the enhanced greenhouse effect. In May 2007, U.S. President George W. Bush acknowledged that human activities are affecting the atmosphere and called for a reduction in emissions of carbon dioxide (CO_2). Despite this, the United States remains opposed to the greenhouse gas reduction targets of the Kyoto Protocol, which was adopted in 1997.

A farmer carries firewood as he walks along the banks of the Basantar River at an industrial area in India. In June 2007, Indian Prime Minister Manmohan Singh told officials at the G-8 Summit in Germany that efforts to combat global warming and reduce greenhouse gases must not hinder the development of poorer countries. *AP Images.*

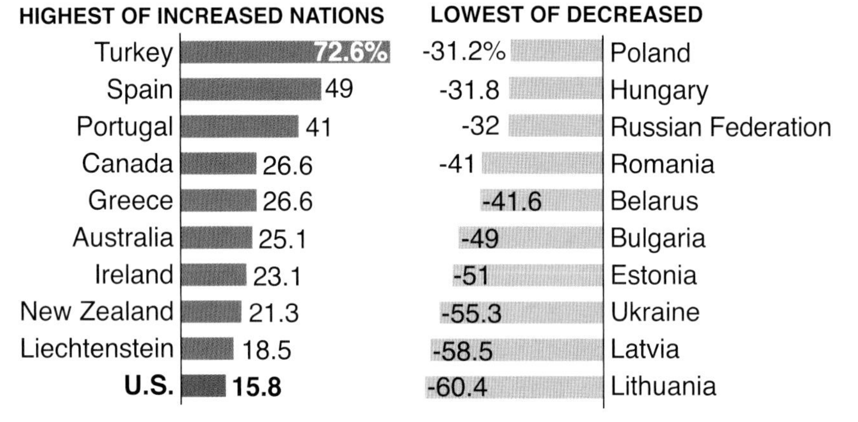

This graphic from the Associated Press shows the percentage change of greenhouse-gas emissions in various countries between 1990 and 2004. [Source: United Nations.] *AP Images.*

Historical Background and Scientific Foundations

The atmosphere naturally contains gases that retain some of the heat energy of the sun. This reduces the amount of heat escaping from the atmosphere to space, and it is vital for maintaining life on Earth. If the atmosphere did not retain heat, Earth would become frozen and lifeless.

The capacity of the atmosphere to retain heat is due to the presence of gases including water vapor, carbon dioxide (CO_2), methane (CH_4), and nitrous oxide (N_2O). These greenhouse gases allow sunlight to pass through the atmosphere to the surface of Earth. Some of this radiation is reflected back toward space as infrared radiation (heat). But, infrared radiation can also be absorbed by the greenhouse gases, which retain the heat in the atmosphere.

Before the mid-eighteenth century, the amounts of greenhouse gases remained relatively constant and allowed the atmosphere to retain a suitable amount of solar energy to power life. But, as nations such as Britain and later the United States began to shift from being predominantly agricultural to industrial, the amount of fossil fuels that were burned increased. The burning of fuels including coal, oil, and natural gas releases greenhouse gases (in particular CO_2) into the atmosphere.

Increasingly, as forests have been cleared to make way for agricultural land, factories, and suburban housing, the amount of CO_2-trapping vegetation has decreased. Additionally, wood-burning, either for disposal or fuel, releases CO_2 to the atmosphere.

As the quantities of CO_2 and the other greenhouse gases in the atmosphere rose, the temperature of the atmosphere began to increase beginning around the 1850s and has accelerated since 1950. It is this human-caused contribution of greenhouse gases and resulting global warming that is the enhanced greenhouse effect.

Compared to 200 years ago, the 2007 levels of greenhouse gases are much higher. As two examples, the levels of CO_2 and CH_4 are about 30% and approximately 145% higher in 2007 than at the beginning of the nineteenth century, according to a 2006 report from the U.S. Department of Energy (DOE).

The United States' contribution to the enhanced greenhouse effect is mainly through energy use. In 1990, CO_2 emissions from human sources were about 6,150 million metric tons. By 2006, the DOE report chronicled, this amount had risen to about 7,100 million metric tons, a 15% increase. Carbon dioxide consistently accounts for

WORDS TO KNOW

AGROFORESTRY: The practice of sustainably combining forestry with agriculture by combining trees with shrubs, crops, or livestock. Two forms of agroforestry are alley cropping (strips of crops alternating with strips of woodland) and silvopasture (pasturing grazing livestock under widely spaced trees). Benefits include greater soil retention, biodiversity encouragement, higher monetary returns per acre, and more diverse product marketing.

CLIMATE MODELS: Mathematical representations of climate processes. Climate models are computer programs that describe the structure of Earth's land, ocean, atmospheric, and biological systems and the laws of nature that govern the behavior of those systems. Detail and accuracy of models are limited by scientific understanding of the climate system and by computer power. Climate models are essential to understanding paleoclimate, present-day climate, and future climate.

FOSSIL FUELS: Fuels formed by biological processes and transformed into solid or fluid minerals over geological time. Fossil fuels include coal, petroleum, and natural gas. Fossil fuels are non-renewable on the timescale of human civilization, because their natural replenishment would take many millions of years.

GREENHOUSE GASES: Gases that cause Earth to retain more thermal energy by absorbing infrared light emitted by Earth's surface. The most important greenhouse gases are water vapor, carbon dioxide, methane, nitrous oxide, and various artificial chemicals such as chlorofluorocarbons. All but the latter are naturally occurring, but human activity over the last several centuries has significantly increased the amounts of carbon dioxide, methane, and nitrous oxide in Earth's atmosphere, causing global warming and global climate change.

INDUSTRIAL REVOLUTION: The period, beginning about the middle of the eighteenth century, during which humans began to use steam engines as a major source of power.

INFRARED RADIATION: Electromagnetic radiation of a wavelength shorter than radio waves but longer than visible light that takes the form of heat.

KYOTO PROTOCOL: Extension in 1997 of the 1992 United Nations Framework Convention on Climate Change (UNFCCC), an international treaty signed by almost all countries with the goal of mitigating climate change. The United States, as of early 2008, was the only industrialized country to have not ratified the Kyoto Protocol, which is due to be replaced by an improved and updated agreement starting in 2012.

WATER VAPOR: The most abundant greenhouse gas, it is the water present in the atmosphere in gaseous form. Water vapor is an important part of the natural greenhouse effect. Although humans are not significantly increasing its concentration, it contributes to the enhanced greenhouse effect because the warming influence of greenhouse gases leads to a positive water vapor feedback. In addition to its role as a natural greenhouse gas, water vapor plays an important role in regulating the temperature of the planet because clouds form when excess water vapor in the atmosphere condenses to form ice and water droplets and precipitation.

approximately 82% of greenhouse-gas emissions in the United States, and the country contributes approximately 22% of global CO_2 emissions.

When the level of greenhouse-gas emissions are related to population, Canada is one of the world's largest polluters per capita, with emissions of 22 tons (20 tonnes) per person per year.

Impacts and Issues

In its 2007 Assessment Report, the IPCC stated that the atmospheric concentrations of CO_2, CH_4, and N_2O have substantially grown "as a result of human activities since 1750." The IPCC noted that the concentrations of these greenhouse gases is now far greater than the amounts of the gases that were present in the atmosphere prior to the Industrial Revolution, as determined from measurements taken from ice cores. Since the ice core measurements enable atmospheric data to be gathered going back thousands of years, the large increase in greenhouse gases since 1750 is compelling evidence for the enhanced greenhouse effect.

The need to lessen the enhanced greenhouse effect has been recognized by the global community. Lessening the amount of greenhouse gases generated by reducing the use of fossil fuels, recycling materials, and using alternate, greenhouse-gas-free sources of energy is one approach. Other alternative energy sources such as nuclear power are more controversial.

A second approach seeks to reduce greenhouse gases by directly adding compounds into the atmosphere or altering Earth's surface. Agroforestry is an example of a surface altering strategy. The planting of trees can generate forests that soak up CO_2 from the air. As of 2007, research is ongoing concerning the feasibility of trapping CO_2 released from coal-burning plants, converting the gas to a liquid, and transferring it to reservoirs underground or at the ocean floor.

Altering Earth's atmosphere to try to control warming, such as by pumping light-reflective material into the atmosphere, is still conceptual as of late 2007. This approach is also controversial. Some critics argue that the strategy does not address the problem—the contribution of human activities to atmospheric warming—but rather is a stop-gap measure that allows the environmental damage to continue.

Despite present efforts, more needs to be done if the enhanced greenhouse effect is to be halted or reversed, according to the IPCC. In December 2007, delegates from 191 countries and organizations attending the United Nations Climate Change Conference held in Bali, Indonesia, began negotiations on climate change efforts after 2012, when the Kyoto Protocol on the reduction of greenhouse gas emissions expires.

SEE ALSO *Abrupt Climate Change; Aerosols; Bali Conference; Carbon Dioxide (CO_2); Greenhouse Effect.*

BIBLIOGRAPHY

Books

DiMento, Joseph F. C., and Pamela M. Doughman. *Climate Change: What It Means for Us, Our Children, and Our Grandchildren.* Boston: MIT Press, 2007.

Gore, Al. *An Inconvenient Truth: The Planetary Emergency of Global Warming and What We Can Do About It.* New York: Rodale Books, 2006.

Lovelock, James. *The Revenge of Gaia: Earth's Climate Crisis and the Fate of Humanity.* New York: Perseus Books, 2007.

Solomon, S., et al, eds. *Climate Change 2007: The Physical Science Basis: Contribution of Working Group I to the Fourth Assessment Report of the Intergovernmental Panel on Climate Change.* New York: Cambridge University Press, 2007.

Environmental Impact Statement (EIS)

Introduction

An Environmental Impact Statement (EIS) is a structured analytical report (a document that examines various issues relevant to a particular topic). It is prepared when a development may significantly impact the quality of the human environment—the area's air, water, or land environments.

The official definition of an environmental impact statement according to the International Association for Impact Assessment is "the process of identifying, predicting, evaluating and mitigating the biophysical, social and other relevant effects of development proposals prior to major decisions being taken and commitments made." Put another way, an environmental impact statement is intended to indicate how the project will likely affect the environment. If it is realized that the proposed development could damage the environment, the statement should present alternatives.

The purpose of an environmental impact statement is to allow agencies that are responsible for approving the particular project to have a clear understanding of how the project will influence the environment. With this information in hand, they then make their decision to approve the

Protesters demonstrate in front of the governor's mansion in Carson City, Nevada, in February 2007, demanding that an environmental impact statement be prepared and public hearings held regarding the U.S. government's plans to detonate a 700-ton explosion in the Nevada desert. *AP Images.*

WORDS TO KNOW

EXTINCTION: The total disappearance of a species or the disappearance of a species from a given area.

MITIGATION: To make changes to alleviate a situation or condition. For example, to implement changes that will reduce greenhouse gases.

project, reject it, or call for modifications that will help safeguard the environment. Human development changes the environment. A well-prepared environmental impact statement helps guide the development so that damage is minimal. Development in the absence of this reasoned consideration of the adverse effects can be disastrous.

Historical Background and Scientific Foundations

In the United States, environmental impact statements were created with the passage of the National Environmental Policy Act (NEPA) by Congress in 1969. The need for the statements in the approval of federal government developments became law a year later.

Then, as now, an environmental impact statement is necessary in the approval process for developments that may significantly affect "the human environment," quoting the legislation. The human environment is a broad concept. The downtown core of a bustling metropolis like New York City or Los Angeles is definitely a human environment. But so is relatively unpopulated northern Alberta, Canada, where the development of a vast region to extract oil from the sandy ground (the Tar Sands) generates noxious byproducts that pollute the water and air. It can be argued that any development that affects Earth affects the human environment.

A typical environmental impact statement has four sections. The first section is a justification for the proposed development, and it explains the need and rationale for the development. The second section provides a description of the environment that could be affected by the development. The third section presents alternatives to the proposed development. This section is helpful to those evaluating the statement, since it provides them with options to consider and shows that the development's proponents have considered all the options. Finally, the fourth section analyzes how each of the approaches could affect the environment.

Development projects can be approved even though the environmental impact statement has recognized that environmental harm will be done. As long as the harm is known and understood in advance, plans can be made to minimize or confine the damage.

Not all projects require an environmental impact statement prepared as part of their approval. If the activity is not likely to be harmful to the environment, then the approval process requires that a less detailed document—called an environmental assessment—be prepared.

Part of the power of an environmental impact statement is its public nature. Even before a statement is prepared, anyone can notify officials of points of interest that they think should be addressed. The first version of the statement is a public document, and anyone can comment on it. Even when the final environmental impact statement has been issued, discussion can continue and contentious parts of the statement can be revised or an explanation provided of why a recommendation from earlier discussions was not adopted. In addition, the final statement can be contested in court in an effort to halt the development.

Impacts and Issues

The NEPA style of environmental impact statement has been the model for similar programs at the state and municipal levels, and in other countries including Canada and the European Community.

A recent example of the power of an environmental impact statement is the reaction to a statement filed in July 2007 by EnCana Corporation, a world leader in the discovery and recovery of natural gas based in Calgary, Alberta. The company prepared an environmental impact statement for the drilling of more than 1,200 wells in the Suffield National Wildlife Area, a 177-square-mi (458-square-km) protected region in the southern part of the province. Opponents of the development pointed out hundreds of omissions or lack of information that they say makes the claims of environmental safety in the developed region dubious. The opposition has prompted Environment Canada to investigate and coordinate public hearings on the development. As combative as this process can be, it helps ensure protection of the developed region and its wildlife.

Another well-known example of the influence of an environmental impact statement occurred in the 1980s in the northwestern region of the United States. An environmental impact statement for a proposed logging operation in an old growth forest on federal government lands was contested by the Sierra Club on the basis that the development would threaten species already at risk of extinction, in particular the northern spotted owl (*Strix occidentalis caurina*). The opposition prompted the U.S. Fish and Wildlife Service to designate the owl a threatened species. Because logging in national forests would threaten the habitat of the owl, this activity was stopped in 1991.

Because an environmental impact statement can potentially delay and even stop a development, some developers may see the statement as a necessary hurdle to be overcome to gain project approval. This can produce a statement that is so loaded with detail that it is hard to understand and, thus, to criticize.

SEE ALSO *Carbon Footprint; Energy Efficiency; Sustainable Energy Policy Network; Terraforming; Wind Power.*

BIBLIOGRAPHY

Books

DiMento, Joseph F. C., and Pamela M. Doughman. *Climate Change: What It Means for Us, Our Children, and Our Grandchildren.* Boston: MIT Press, 2007.

Gore, Al. *An Inconvenient Truth: The Planetary Emergency of Global Warming and What We Can Do About It.* New York: Rodale Books, 2006.

Seinfeld, John H., and Spyros N. Pandis. *Atmospheric Chemistry and Physics: From Air Pollution to Climate Change.* New York: Wiley Interscience, 2006.

Web Sites

"EnCana's Environmental Impact Statement (EIS) Severely Lacking." *Federation of Alberta Naturalists,* July 30, 2007 <http://fanweb.ca/issues/suffield/news-releases/encanas-environmental-impact-statement-eis-severely-lacking> (accessed November 13, 2007).

"NEPA: Past, Present, and Future." *U.S. Environmental Protection Agency,* September 21, 2007 <http://www.epa.gov/history/topics/nepa/01.htm> (accessed November 13, 2007).

Environmental Policy

Introduction

Environmental policy is a statement by a governmental body or other organization of its intentions and principles toward the environment. It commonly refers to a government's use and creation of laws and regulatory mechanisms concerning environmental issues and sustainability. Environmental policy is an increasingly central tenet of an international, national, regional, or local governmental body's principles. In recent years, many businesses have adopted environmental policies, too.

Historical Background and Scientific Foundations

As early as 1647, Massachusetts forbade the pollution of Boston Harbor as a public health measure. In Great Britain, policy measures extend at least as far back as the thirteenth century, when smoke-heavy sea coal was banned from London. Many governments have long taken an interest in the environment, but environmental policy as a cogent statement of its intent toward the natural environment, governed by systematic law and overseen by its agencies, is a modern idea.

The term environmental policy was coined by professor Lynton Caldwell of Indiana University in 1963. Writing in *Public Administration Review*, under the title "Environment: A New Focus for Public Policy?," Caldwell focused not on individual issues, such as pollution or conservation, but on the problem of governing environmental issues on the whole. He asserted that environmental problems required a comprehensive, ecological approach, focusing on interrelated problems and creating a central, integrated public policy.

The National Environmental Policy Act (NEPA) was signed into law in the United States on January 1, 1970, by President Richard M. Nixon, who declared the 1970s to be the "decade of the environment." That same year the Nixon administration passed the Clean Air Act, which established national minimum standards for air pollution and drew on a raft of legislation to cut pollution by the middle of the decade; and the Water Quality Improvement Bill, which authorized the federal government to clean up oil spills and bill polluters. It also formed the Council on Environmental Quality (CEQ), the Environmental Protection Agency (EPA), and the National Oceanic and Atmospheric Administration (NOAA).

In the United Kingdom in 1970, Prime Minister Harold Wilson set up the Royal Commission on Environmental Pollution "to advise on matters, both national and international concerning the pollution of

The Clean Air Act, first enacted in 1970, regulates how much pollution can be emitted into the air. *Image copyright Victor Soares, 2007. Used under license from Shutterstock.com.*

the environment; on the adequacy of research in this field; and the future possibilities of danger to the environment." By the year's end, this had given way to the creation of the Department of the Environment. This development was followed by Canada in 1971 and gradually through Western Europe and parts of Asia during the 1970s and 1980s.

Environmental policy and regulation was further unified by the creation of intergovernmental agencies. In 1972, the United Nations Environment Programme (UNEP) was established following the U.N. Conference on the Human Environment. Its intention was to reflect the environmental conscience of the U.N. and coordinate the development of environmental policy on both global and regional levels.

One year later, in 1973, the European Community (the precursor of the European Union) launched its first Environmental Action Program. Sharing many of UNEP's ideas on sustainable development, its most important objectives included: the prevention, reduction and containment of environmental damage; the conservation of an ecological equilibrium; and the rational use of natural resources. As the EU gained in size and influence, its mandate over environmental policy also increased. In 1993, the EU Commission established the office of the Environment Directorate General, which was afforded responsibility for initiating and drafting new environmental legislation.

Some disparity exists between national and international definitions of environmental policy. According to the U.S. NEPA, the role of environmental policy is to "encourage productive and enjoyable harmony between man and his environment; to promote efforts which will prevent or eliminate damage to the environment and biosphere and stimulate the health and welfare of man; to enrich the understanding of the ecological systems and natural resources important to the nation."

The Polish Government frames it in less idealistic terms: "[environmental policy] aims at correcting market and regulatory failures to improve environmental quality. Ideally, environmental policy should be designed to maximize the net benefits to society by achieving the optimal level of environmental quality."

By contrast, the U.N. definition in the context of climate change is somewhat more opaque: policies "are actions that can be taken and/or mandated by a government—often in conjunction with business and industry within its own country, as well as with other countries—to accelerate the application and use of measures to curb greenhouse gas emissions."

WORDS TO KNOW

BIOSPHERE: The sum total of all life-forms on Earth and the interaction among those life-forms.

CLEAN AIR ACT: Any law that seeks to control air pollution may be referred to as a "clean air act." In climate literature, the U.S.Clean Air Act of 1970 is often spoken of as "the" Clean Air Act.

KYOTO PROTOCOL: Extension in 1997 of the 1992 United Nations Framework Convention on Climate Change (UNFCCC), an international treaty signed by almost all countries with the goal of mitigating climate change. The United States, as of early 2008, was the only industrialized country to have not ratified the Kyoto Protocol, which is due to be replaced by an improved and updated agreement starting in 2012.

NATIONAL ENVIRONMENTAL POLICY ACT (NEPA): U.S. federal law signed by President Richard Nixon on January 1, 1970. The act requires federal agencies to produce Environmental Impact Statements describing the impact on the environment of projects they propose to carry out.

SUSTAINABILITY: The quality, in any human activity (farming, energy generation, or even the maintenance of a society as a whole), of being sustainable for an indefinite period without exhausting necessary resources or otherwise self-destructing.

■ Impacts and Issues

When viewed in the context of climate change, national environmental policy increasingly tends to follow or be dictated by international agreements. There exists a consensus among a majority of industrialized countries that because climate change is an internationalized problem, cooperative solutions must be sought. However, a few industrialized nations have opted out of large international agreements. The United States and Australia failed to ratify the 1997 Kyoto Protocol (although Australia announced intentions to ratify in December 2007). Large emerging economies like China and India have also refused to participate in some international environmental policies, despite both nations' rapid growth of pollution problems and substantially increased greenhouse-gas emissions.

At the same time, environmental policy at the United Nations' level largely lacks enforcement ability. UNEP, the U.N.'s designated authority in environmental issues, is mandated to coordinate the development of environmental policy, rather than proscribe a global environmental agenda. Its policy reflects a global consensus, balancing the disparate needs of less-developed and most-developed nations. However, scientific environmental research published under the auspices of the U.N., such as that of the Intergovernmental Panel on Climate Change (IPCC), often advocates more dramatic and comprehensive solutions—particularly with regard to climatic change.

Regional intergovernmental agencies are also an important part of modern environmental policy. The European Union dictates up to 80% of its member state's environmental policy. The EU and its 27-member states have exerted considerable force at the international negotiating table. The EU led efforts to revive the Kyoto Protocol and continues to advocate a more widely encompassing international policy on climate change with stricter emissions reduction targets after 2012. Some of its member states already are committed to 60% emissions reductions by 2050.

SEE ALSO *Environmental Pollution; United Nations Environment Programme (UNEP); United States: Climate Policy.*

BIBLIOGRAPHY

Books

Andrews, Richard N. L. *Managing the Environment, Managing Ourselves: A New History of American Environmental Policy.* New Haven, CT: Yale University Press, 1999

Dowey, Scott Hamilton. *Don't Breath the Air: Air Pollution and U.S. Environmental Politics, 1945–1970.* College Station, TX: Texas A&M University Press, 2000.

Gareis, Sven Bernhard, and Johannes Varwick. *The United Nations: An Introduction.* New York: Palgrave Macmillan, 2005.

McCormick, John. *Environmental Policy in the European Union.* New York: Palgrave Macmillan, 2001.

Weiss, Thomas G., and Sam Daws, eds. *The Oxford Handbook on the United Nations.* New York: Oxford University Press, 2007.

Periodicals

Caldwell, Lynton K. "Environment: A New Focus for Public Policy?" *Public Administration Review* 23:3 (September 1963): 132–139.

Web Sites

Hay, Christian. "EU Environmental Policies: A Short History of the Policy Strategies." *European Environmental Bureau, EU.* <http://www.eeb.org/publication/chapter-3.pdf> (accessed November 21, 2007).

James Corbett

Environmental Pollution

■ Introduction

Environmental pollution is the discharge of material, in any physical state, that is dangerous to the environment or human health. Most industrialized and developing countries, and many intergovernmental organizations, have developed maximum exposure values for pollutants. Many regulate atmospheric emissions to keep their levels compatible with environmental equilibrium and human health.

Although pollution was regarded primarily as a localized problem until the last third of the twentieth century, environmental pollution is a significant part of the contemporary international political agenda. This is especially true of atmospheric and marine pollution, since neither is contained within national borders. With regard to climate change, environmental pollution is the defining issue: limiting anthropogenic greenhouse-gas emissions is acknowledged by scientists as fundamental to mitigating the effects of global climate change.

Greenpeace protesters erect a mock flood dam in Bonn, Germany, in 1997 at the end of the United Nations Climate Conference. Their banner reads: "Climate protection made in USA. Before the deluge," claiming that the U.S. position will delay action on dangerous climate change for another decade. *AP Images.*

WORDS TO KNOW

CLEAN AIR ACT: Any law that seeks to control air pollution may be referred to as a "clean air act." In climate literature, the U.S.Clean Air Act of 1970 is often spoken of as "the" Clean Air Act.

NATIONAL ENVIRONMENTAL POLICY ACT (NEPA): U.S. federal law signed by President Richard Nixon on January 1, 1970. The act requires federal agencies to produce Environmental Impact Statements describing the impact on the environment of projects they propose to carry out.

OZONE LAYER: The layer of ozone that begins approximately 9.3 mi (15 km) above Earth and thins to an almost negligible amount at about 31 mi (50 km) and shields Earth from harmful ultraviolet radiation from the sun. The highest natural concentration of ozone (approximately 10 parts per million by volume) occurs in the stratosphere at approximately 15.5 mi (25 km) above Earth. The stratospheric ozone concentration changes throughout the year as stratospheric circulation changes with the seasons. Natural events such as volcanoes and solar flares can produce changes in ozone concentration, but man-made changes are of the greatest concern.

PEA SOUPER: Nineteenth-century term for a thick smog episode in London. London continued to suffer extreme smog episodes through the 1950s.

Historical Background and Scientific Foundations

Awareness of the wider problems caused by environmental pollution is only a comparatively recent phenomenon. Until at least the middle of the twentieth century, it was regarded primarily as a localized problem. However, local action was rarely sufficient to address problems that were not confinable to one location. Centralized research and legislation on far-reaching environmental issues—such as polluted rivers or smog—were scant.

This was especially true for outdoor air pollution—a principal contributory factor to climate change. The picture started to change after a lethal episode of air pollution in Donora, Pennsylvania, in 1948, left 20 dead and hundreds seriously ill; and severe episodes of smog were observed in London in 1952. By 1957, the U.S. Public Health Service had organized an air pollution division in the Bureau of State Services and started a program of health effects research and training programs in universities. At this time, research focused on human health impacts in urban centers, but it formed a starting point for debate about alternative air pollution control strategies.

The birth of the modern environmental movement in the latter–1960s gave rise to greater awareness of environmental pollution, and also to political pressure to implement measures to counteract it. In the United States, the passage of the National Environmental Policy Act (NEPA) in 1970 transformed pollution from a localized or statewide concern into a federal one. A raft of federal laws legislating all manner of pollution—airborne (Clean Air Act 1970); water (Clean Water Act 1972); oil (Water Quality Improvement Bill 1970); and noise (Noise Control Act 1979)—was passed during the 1970s.

By the 1970s, a link between respiratory disease and particulate air pollution had been well established. Links were also established between annual respiratory mortality and exposure to air pollution from coal burning and other atmospheric emissions from heavy industry. Partly as a consequence of growing public concern about air pollution, atmospheric scientists were able to secure additional funding for new research into human impacts on the atmosphere. Within a matter of years, it became clear that sulfate aerosols, CFCs, and other anthropogenic emissions, including CO_2, played a significant role in the depletion of the planet's atmospheric ozone layer.

Environmental pollution became an international concern. In the late 1980s, a series of international conferences explored the problem of climate change. The period also witnessed the founding of scientific research bodies such as the Intergovernmental Panel on Climate Change (IPCC). In 1987, some 50 countries adopted the Montreal Protocol, which specified a 50% reduction in fully halogenated CFCs by the century's end. In 1990, the Montreal Protocol was amended to include a total ban on CFCs. This was followed by the U.N. Framework Convention on Climate Change (UNFCCC) in 1992, and the Kyoto Protocol in 1997.

The IPCC and other international scientific bodies studying climate change have attempted to divorce the scientific study of environmental pollution from the social, political, and even physical contexts of its production. The IPCC's otherwise comprehensive glossary does not provide a definition for the term environmental pollution. As such, defining what constitutes environmental pollution remains the domain of national governments and specific international conferences and agreements. However, there is some consensus.

According to the U.S. National Aeronautics and Space Administration (NASA), environmental pollution is: "alterations of the natural environment that are harmful to life, normally as produced by human activities.... [This includes] atmospheric, water, soil, noise, and thermal pollution."

According to the British government (as per its Pollution and Prevention Control Act 1999): "Environmental pollution means pollution of the air, water or land which may give rise to any harm." The act further states that "'harm' means—(a) harm to the health of human beings or other living organisms; (b) harm to the quality of the environment, … (c) offence to the senses of human beings." Environmental harm includes: "(i) harm to the quality of the environment taken as a whole, (ii) harm to the quality of the air, water or land, and (iii) other impairment of, or interference with, the ecological systems of which any living organisms form part."

According to the EU (as per its 1996 Pollution Directive): "'Pollution' shall mean the direct or indirect introduction as a result of human activity, of substances, vibrations, heat or noise into the air, water or land which may be harmful to human health or the quality of the environment, result in damage to materiel property, or impair or interfere with amenities and other legitimate uses of the environment."

In the context of climate change, the IPCC also publishes comprehensive guidelines for national greenhouse gas inventories. These provide methodologies to the UNFCCC's signatories for estimating anthropogenic emissions by sources. While resisting use of the term "pollution," the IPCC guidelines form the internationally recognized definition of humanmade atmospheric pollutants that impact global warming. They are: Carbon dioxide (CO_2); methane (CH_4); nitrous oxide (N_2O); hydrofluorocarbons (HFCs); perfluorocarbons (PFCs); sulfur hexafluoride (SF_6); nitrogen trifluoride (NF_3); trifluoromethyl sulfur pentafluoride (SF_5CF_3); halogenated ethers; and other halocarbons not covered by the Montreal Protocol.

Impacts and Issues

The defining of environmental pollution as a national and international problem—rather than a local one—helped shape environmental policy. Today, clear conceptions of what forms of pollution influence global warming serve as the principal agents for mitigating climate change. Efforts to cancel emissions of such pollutants may well become the defining issue of the age.

Already, efforts to counteract environmental pollution have had substantial impacts. Landmark legislation, such as U.S. and U.K. Clean Air Acts, has improved urban and rural environments and the lives of millions of its people. For example, "pea soupers," the term once given to London's smog-ridden mornings, are long a part of the past. For the first time in centuries, salmon are once again found in the city's river Thames. Global efforts to counteract pollution have had a positive impact on human lives.

Regulation of ozone-depleting substances and greenhouse-gas emissions is helping combat atmospheric environmental pollution. The terms of the amended Montreal Protocol—which outlawed fully halogenated CFCs—has aided the decrease or leveling off of atmospheric concentrations of the most important chlorofluorocarbons and related chlorinated hydrocarbons. Recognition of other pollutants that are harmful to the atmosphere and moves to limit their emission has not yet stabilized greenhouse gas-levels, although it may have decreased the rate at which atmospheric concentrations are increasing.

SEE ALSO *Atmospheric Pollution; Chlorofluorocarbons and Related Compounds; Environmental Policy.*

BIBLIOGRAPHY

Books

Cherni, Judith A. *Economic Growth Versus the Environment: The Politics of Wealth, Health, and Air Pollution.* New York: Palgrave, 2002.

Dowey, Scott Hamilton. *Don't Breath the Air: Air Pollution and U.S. Environmental Politics, 1945–1970.* College Station, TX: Texas A&M University Press, 2000.

Hill, Marquita K. *Understanding Environmental Pollution.* Cambridge: Cambridge University Press. 2004.

Trivedy, R. K. *Encyclopaedia of Environmental Pollution and Control.* Karad: Enviro-media, 1995.

Periodicals

Pope III, C. Arden, et al. "Health Effects of Particulate Air Pollution: Time for Reassessment." *Environmental Health Perspectives* 103 (1995): 472–80.

Whittenberger, James L. "Health Effects of Air Pollution: Some Historical Notes." *Environmental Health Perspectives* 91 (1989): 129–130.

James Corbett

Environmental Protection Agency (EPA)

Introduction

The Environmental Protection Agency (EPA) is an agency of the U.S. government that employs over 17,000 people, more than half of whom are engineers, scientists, and policy analysts. The agency is headquartered in Washington, D.C., and maintains 10 regional offices and over a dozen laboratories. Established in 1970 by President Richard Nixon (1913–1994) with the approval of the U.S. Senate and U.S. House of Representatives, it was charged with reducing pollution and protecting the environment. In recent years it has been embroiled in controversy as the presidential administration of George W. Bush (1946–), which has opposed both domestic and international legal limits on the emission of greenhouse gases, has used its authority to weaken the EPA's official position on global climate change.

Historical Background and Scientific Foundations

The EPA was established by President Richard Nixon's Reorganization Plan Number 3 (July 9, 1970) in response

Massachusetts Attorney General Tom Reilly speaks to reporters outside the U.S. Supreme Court in November 2006. In the first global warming case heard by the Supreme Court, 12 states and 13 environmental groups sued the EPA stating that the agency's inaction on greenhouse-gas emissions was harming the environment. The court later ruled against the EPA. *Brendan Smialowski/Getty Images.*

to the rapidly growing environmental movement. The first Earth Day, on April 22, 1970, drew some 20 million participants and has been credited with influencing the passage of the Clean Air Act of 1970 and Nixon's decision to create the EPA. Nixon's Reorganization Plan required the new agency to establish and enforce environmental protection standards, conduct research on pollution and on methods for controlling it, and provide grants and technical assistance to help others control pollution of the environment. The EPA was officially created on December 2, 1970, after Nixon's Reorganization Plan was cleared by hearings in the U.S. Senate and U.S. House of Representatives.

Since that time, the EPA has defined numerous regulations, funded research, and given grants to university groups to study pollution and other aspects of the environment. As of 2007, its administrator was Stephen L. Johnson (1951–), appointed by President George W. Bush on January 26, 2005.

In 2003 and 2007, the White House Office of Management and Budget (OMB) produced reports analyzing the costs and benefits of federal environmental regulations. Both reports found that the benefits, measured in dollars, exceeded the costs. In its 2007 report, the OMB found that the costs of the previous 10 years of EPA regulation had been \$39–\$46 billion while the benefits had been \$98–\$480 billion. Costs include more expensive cars and other machinery including pollution-reduction features, costs of waste containment and proper disposal, and other pollution-reducing measures. Benefits include reduced medical costs, reduced damage to buildings and forests from acid rain, and more.

The EPA has funded research on climate change, has collected basic data on greenhouse-gas emissions (published, for example, in its *Inventory of U.S. Greenhouse Gas Emissions and Sinks 1990–2002*), and places information about global climate change on its Web site, which refers the reader to the latest (2007) report of the Intergovernmental Panel on Climate Change (IPCC). It is, therefore, officially aligned with the mainstream scientific view of present-day climate change. However, the EPA is subject to political control from the White House, whose Office of Management and Budget must approve all new federal regulations and standards. Under the Bush administration, the EPA has been accused of foot-dragging on the regulation of greenhouse-gas emissions and altering official documents to downplay the dangers of global climate change.

IN CONTEXT: ENERGY STAR

In 1992, the U.S. Environmental Protection Agency (EPA) introduced the Energy Star program to promote energy efficiency and reduce greenhouse-gas emissions. Energy Star encourages companies to produce more efficient home appliances, from dishwashers to lighting. The Energy Star program asserts that households that use less energy help avoid greenhouse-gas emissions associated with electrical power generation. The voluntary labeling program awards an Energy Star designation to products with energy-saving features. To further aid consumers in choosing the most efficient and cost-effective appliances, new appliances are labeled with information on energy use and the average annual cost of operation.

Energy Star has promoted the use of many common energy-saving features such as stand-by settings on computers and better insulation on water heaters and refrigerators. Energy Star also promotes community-wide action, advocating the use of energy-efficient technologies like LED traffic lights and compact fluorescent light bulbs.

In 2001, the European Union adopted the Energy Star program. Proponents of Energy Star assert that better consumer labeling has raised awareness of home energy use and prompted manufacturers to develop and market more energy-efficient products. Critics claim that the qualifications to receive an Energy Star rating should be stricter and better inform consumers of low-energy alternatives. In 2006, Energy Star helped consumers save over \$14 billion dollars in utility bills and reduce greenhouse-gas emissions through energy conservation.

Impacts and Issues

Weakening of Climate Change Reports

In 2001, the Bush White House appointed Philip A. Cooney as its chief of staff for the White House Council on Environmental Quality, the group that defines White House policies on environmental issues. Before receiving the appointment, Cooney held the position of "climate team leader" at the American Petroleum Institute, an oil-industry group opposing caps on greenhouse-gas emissions. Cooney was also a lobbyist for the group.

In his new position, Cooney had authority to edit reports written by EPA scientists before they were released to the public. Cooney made hundreds of changes to EPA documents, nearly all introducing doubt about the reality and danger of human-caused climate change. He began by adding phrases like "the weakest links in our knowledge," "a lack of understanding," "uncertainties," "considerable uncertainty," and "perhaps even greater uncertainty" to the Administration's 2002 Climate Action Report to the United Nations. An investigation by the House Oversight Committee later found that Cooney—a lawyer with no scientific training—was given veto power over the statements of working scientists such as the head of the federal Climate Change Science Program Office.

Cooney made more than 100 edits to the EPA document "Our Changing Planet," all directed toward downplaying the threat of climate change. In April 2003, he edited the EPA's Draft Report on the

WORDS TO KNOW

ACID RAIN: A form of precipitation that is significantly more acidic than neutral water, often produced as the result of industrial processes.

CLEAN AIR ACT OF 1970: Extension of the 1963 U.S. Clean Air Act that tasked the newly established Environmental Protection Agency with developing and enforcing regulations to reduce air pollution.

GREENHOUSE GASES: Gases that cause Earth to retain more thermal energy by absorbing infrared light emitted by Earth's surface. The most important greenhouse gases are water vapor, carbon dioxide, methane, nitrous oxide, and various artificial chemicals such as chlorofluorocarbons. All but the latter are naturally occurring, but human activity over the last several centuries has significantly increased the amounts of carbon dioxide, methane, and nitrous oxide in Earth's atmosphere, causing global warming and global climate change.

Environment, deleting and adding claims. He deleted the sentences "Climate change has global consequences for human health and the environment" and "Greenhouse gases are accumulating in the atmosphere as the result of human activities, causing surface air temperatures and subsurface ocean temperatures to rise." He replaced the latter sentence with: "Some activities emit greenhouse gases and other substances that directly or indirectly may affect the balance of incoming and outgoing radiation, thereby potentially affecting climate on regional and global scales." In an internal memo, senior EPA scientists complained that the Draft Report on the Environment as altered by Cooney "no longer accurately represents scientific consensus on climate change."

In 2005, the *New York Times* reported on the changes Cooney had been making to EPA and other documents. A senior EPA scientist told the *New York Times* that such manipulation of EPA science "has somewhat of a chilling effect and has created a sense of frustration." Two days after the *New York Times* report, Cooney resigned. The following week, he was hired by the public affairs department of the oil company ExxonMobil.

Automobile Emissions

In 2005, a coalition of state governments, environmental groups, and the territory of American Samoa brought a case to federal court in the United States, *Massachusetts, et al. v. Environmental Protection Agency, et al.*. They claimed that the EPA was refusing to regulate carbon dioxide and other greenhouse gases known to be changing Earth's climate. The EPA, the plaintiffs argued, was required to act to regulate air pollution and was breaking the law by refusing to regulate greenhouse gases.

The case was appealed to the U.S. Supreme Court. In April 2007, the court ruled against the EPA. "Under the clear terms of the Clean Air Act," Justice John Stevens wrote in the majority opinion, "EPA can avoid taking further action only if it determines that greenhouse gases do not contribute to climate change or if it provides some reasonable explanation as to why it cannot or will not exercise its discretion to determine whether they do." The decision means that the EPA must now evaluate the danger posed by greenhouse gases. If it finds that these gases are a form of air pollution that "may reasonably be anticipated to endanger public health or welfare" (in the words of the federal Clean Air Act), it must regulate them.

However, the U.S. Supreme Court did not set any timeline for the EPA to evaluate or regulate greenhouse gases. Therefore, the EPA could remain inactive for an indefinite time without suffering any penalty. Some observers, such as Emma Marris writing in the journal *Nature* (in April 2007), predicted that no action will be taken until a new presidential administration is inaugurated in January 2009.

In October 2007, the EPA announced that it began developing regulations concerning the geologic sequestration of carbon dioxide (CO_2). Geologic sequestration involves capturing the carbon dioxide that results from energy production or industrial activity and isolating it from the atmosphere by permanently storing it deep underground in rock layers. The captured CO_2 is injected more than 2,625 ft (800 m) underground, where it becomes super-critical and behaves like a liquid, taking up less space and over time, dissolving in water present in the pore spaces of the rock. The final regulations, to be released in late 2008, are designed to promote and standardize geologic sequestration, and protect underground drinking water reservoirs.

See Also *United States: Climate Policy.*

BIBLIOGRAPHY

Books

Collin, Robert W. *The Environmental Protection Agency: Cleaning Up America's Act.* Westport, CT: Greenwood Press, 2005.

Periodicals

Greenhouse, Linda. "Justices Say E.P.A. Has Power to Act on Harmful Gases." *New York Times* (April 3, 2007).

Hakim, Danny. "E.P.A. Holds Back Report on Car Fuel Efficiency." *New York Times* (July 28, 2005).

Marris, Emma. "Car Emissions Are EPA's Problem." *Nature* 446 (April 5, 2007): 589.

Revkin, Andrew C. "Bush Aide Softened Greenhouse Gas Links to Global Warming." *New York Times* (June 8, 2005).

Revkin, Andrew C., and Matthew L. Wald. "Material Shows Weakening of Climate Change Reports." *New York Times* (March 20, 2007).

Web Sites

"Climate Change." *U.S. Environmental Protection Agency.* <http://www.epa.gov/climatechange/index.html> (accessed October 2, 2007).

"EPA History." *U.S. Environmental Protection Agency.* <http://www.epa.gov/history/index.htm> (accessed October 2, 2007).

Larry Gilman

Environmental Protests

Introduction

Environmental protests are demonstrations seeking to bring recognition of or action against human-made degradation of a natural environment. Protests predate the modern environmental or green movement.

Environmental protests take on a diverse array of forms against the multitude of scientific, social, conservational, and political consequences of environmental degradation. They are among the foremost reasons that green issues have risen in public consciousness and political discourse.

Historical Background and Scientific Foundations

On a localized basis, environmental protests occurred throughout history. For example, in 1739, Benjamin Franklin (1706–1790) and his neighbors successfully petitioned the Pennsylvania Assembly to stop waste dumping and ban tanneries from Philadelphia's commercial district. In 1858, public outrage about pollution on London's River Thames—the year when the smell of raw sewage and industrial effluent led to the so-called

Environmental protesters form a blockade to prevent access to Appalachian Power Company's Clinch River Plant in Carbo, Virginia, in 2006. The activists demonstrated against the use of coal to generate electricity. *AP Images.*

"Great Stink"—resulted in the building of a vast public sewerage system that is still used to this day.

Nevertheless, it was not until the 1960s and early 1970s, when environmentalism emerged as a broad social movement, that such protests became less sporadic. In the United States, there were grassroots protests against the uncontrolled use of pesticides, particularly DDT, that followed publication of the Rachel Carson 1962 surprise bestseller, *Silent Spring*. In this period, anti nuclear organizations, such as the U.S. National Committee for a Sane Nuclear Policy (SANE) and Britain's Campaign for Nuclear Disarmament (CND) were primarily concerned with arms control, but sometimes exploited environmental concerns to further their causes. For example, SANE protested against the prevalence of the isotope, Stronium 90, in the human food chain as a result of nuclear power production and missiles testing. High levels of Stronium 90 have been linked to bone cancer and leukemia.

The first large-scale nationwide environmental demonstration in the United States was the first Earth Day. It was organized by Wisconsin Senator Gaylord Nelson who was concerned about the absence of environmentalism on the public agenda. His intention was to "tap into the environmental concerns of the general public and infuse the student anti-war energy into the environmental cause [in order to] generate a demonstration that would force this issue onto the political agenda," according to Envirolink.org. On April 22, 1970, 20 million Americans took to the streets, parks, and auditoriums to demonstrate for a healthy environment.

At the same time, environmental protests started to take on a more radical complexion. From the early 1970s, Greenpeace—initially a local Canadian enviro-protest group, but by the decade's end an international organization—began a series of high profile and daring sea raids against U.S. and French atmospheric nuclear tests in the Pacific Ocean. The publicity garnered from such protests allowed Greenpeace to expand its international membership base. Its other great cause was to stop whaling. From the mid–1970s, the group set out on several missions to sabotage whale hunting expeditions. Greenpeace has since been involved in expeditions to sabotage and draw attention to illegal logging in Brazilian and Southeast Asian rainforests as well as protests against the introduction of genetically modified crops.

The late 1980s and early 1990s marked a great surge in environmental protest. In Great Britain, an unlikely coalition of seasoned environmental campaigners and concerned (mostly middle class and socially conservative) residents formed to protest against motorway building in the south of England. In France, urban and industrial pollution and the nuclear power industry were favorite causes of environmental protesters. In Italy, more than 4 million people took part in environmental protests in 1990 alone, mainly against urban air pollution. Even in

WORDS TO KNOW

ACID RAIN: A form of precipitation that is significantly more acidic than neutral water, often produced as the result of industrial processes.

EARTH DAY: An annual global commemoration (every April 22) of concerns about the environment, first celebrated in 1970.

EARTH LIBERATION FRONT (ELF): An underground group in North America that describes itself as "an international underground organization that uses direct action in the form of economic sabotage to stop the destruction of the natural environment." The group claims to have destroyed $100 million worth of property. The U.S. Federal Bureau of Investigation declared it the number one domestic terror threat as of March 2001.

ECO-TERRORISM: Criminal violence against persons or property carried out by an environmentally-oriented group for symbolic purposes. As of 2007 no eco-terrorist act had entailed harm to persons, only to property (for example, sport utility vehicles), so the use of the word "terrorism" in this context is, critics of the term say, inflammatory and inaccurate.

GREENPEACE: Nonprofit environmental group formed in 1971, originally to protest nuclear testing and whaling. The group remains active, now addressing a wide range of issues, including climate change.

KYOTO PROTOCOL: Extension in 1997 of the 1992 United Nations Framework Convention on Climate Change (UNFCCC), an international treaty signed by almost all countries with the goal of mitigating climate change. The United States, as of early 2008, was the only industrialized country to have not ratified the Kyoto Protocol, which is due to be replaced by an improved and updated agreement starting in 2012.

LIVE EARTH: A popular music concert that occurred at 11 locations simultaneously around the world on July 7, 2007. The event was organized by former U.S. Vice President Al Gore to publicize the problem of global climate change and to raise funds for projects to combat it.

OZONE: An almost colorless, gaseous form of oxygen with an odor similar to weak chlorine. A relatively unstable compound of three atoms of oxygen, ozone constitutes, on average, less than one part per million (ppm) of the gases in the atmosphere. (Peak ozone concentration in the stratosphere can get as high as 10 ppm.) Yet ozone in the stratosphere absorbs nearly all of the biologically damaging solar ultraviolet radiation before it reaches Earth's surface, where it can cause skin cancer, cataracts, and immune deficiencies, and can harm crops and aquatic ecosystems.

developing countries, where environmental degradation was a less recognized sociopolitical concern, grassroots campaigns—such as those against the damming of India's Narmada River—garnered international attention. This was augmented by international conservation campaigns, such as Save the Whale, and growing public awareness of issues such as ozone deletion and acid rain. At the Earth Summit, staged in the Brazilian city of Rio de Janeiro in 1992, a United Nations conference had become a jamboree of global environmental interests at which tens of thousands of campaigners traveled to articulate their causes.

Environmentalism has been termed "the great survivor" of the social and protest movements that first arose in the late 1960s. However, in the first years of the twenty-first century, protests have also taken new forms. Protests are less about waving banners or marching on government buildings than educating, raising awareness, and prompting individual action. For example, former U.S. Vice President Al Gore's 2006 documentary film, *An Inconvenient Truth*, which warns of the dangers of climate change, is representative of this new medium of protest. Another example were the Live Earth concerts in July 2007, staged in 11 locations and broadcast worldwide. During the concerts, people were asked to support a seven-point pledge, including volunteering to become carbon neutral and demanding better environmental policies from their governments.

On the world stage, the United States's refusal to acknowledge the Kyoto Protocol prompted protest. In April 2007, more than 1,300 events were organized in every state under the banner Step It Up 2007 to pressure Congress into legislating for an 80% cut in carbon dioxide emissions by 2050. Unable to exact federal support for Kyoto, the U.S. Conference of Mayors has produced its own Climate Protection Agreement, in which mayors commit to reduce emissions in their cities to 7% below 1990 levels by 2012—the same cuts called upon at Kyoto. As of November 2007, more than 500 mayors have committed to this goal.

■ Impacts and Issues

Along with local, national, and international green organizations—and advances made in both the quality and breadth of scientific environmental research—the environmental protest movement has been central in publicizing green issues.

The original super-protest, Earth Day, is an annual global event, with its organizers claiming the participation of 500 million people in 175 countries each April 22. Earth Day's founders claimed that the show of support in the early 1970s paved the way for such landmark legislation as the Clean Air Act and the creation of the U.S. Environmental Protection Agency (EPA). Early critics claimed that Earth Day was communist or socialist in origin and theme, noting the coincidence that the first Earth Day occurred on the centenary of former USSR leader Vladimir Illyich Lenin's birth. Most Earth Day celebrations are now considered mainstream environmental awareness events.

Some protests have been met by legal action from corporations targeted by campaigners. A famous example occurred when a London-based group waged a campaign against McDonalds in the mid-1980s, accusing the restaurant chain of a number of environmental violations, including the destruction of the Amazonian rainforest, cruelty to animals, and unfair labor policies. McDonalds sued the activists for defamation, leading to the so-called McLibel trial, the longest in British history. After years of deliberations, a judge found in McDonalds favor, but only token damages were awarded after the judge found elements of the activists' case to be correct. Other companies to utilize legal action against protesters include the biotech giant Monsanto; the Australian logging company, Gunns; and the oil corporation ExxonMobil.

A minority of green activists have become frustrated by the lack of radicalization within the environmental protest movement. A small minority has used violence to further its aims. Extremist protests include those carried out by the Earth Liberation Front (ELF), active in Britain, Canada and the United States. The group's protests or direct actions have involved sabotaging large sport-utility vehicles (SUVs), infiltrating and sabotaging companies involved in logging and energy production, and destroying suburban construction projects. ELF is described by the FBI as "one of the most active extremist elements in the United States." Most members of the mainstream environmental movement disapprove of the radical direct-actions of the green movement's extremist fringe, opting instead for peaceful demonstration or targeted acts of civil disobedience that do not endanger lives or vandalize property.

Some environmental protests have had a wider impact. In September 1988, 45,000 Soviet citizens lined the coast of Latvia to protest against the pollution they said was destroying the Baltic Sea. This is now seen as a pivotal moment in Baltic history, marking the starting point for a series of demonstrations against the Soviet regime, which would lead to Baltic independence three years later.

SEE ALSO *Environmental Policy; Green Movement; Sustainability.*

BIBLIOGRAPHY

Books

Burchell, Jon. *The Evolution of Green Politics: Development and Change within European Green Parties.* London: Earthscan, 2002.

Dryzek, John S., et al. *Green States and Social Movements: Environmentalism in the U.S., U.K.,*

Germany and Norway. Oxford: Oxford University Press, 2003

Liddick, Donald R. *Eco Terrorism: Radical Environmental and Animal Liberation Movements*. Westport, CT: Praeger, 2006.

Rootes, Christopher, ed. *Environmental Protest in Western Europe*. Oxford: Oxford University Press, 2003

Rubin, Charles T. *The Green Movement: Rethinking the Roots of Environmentalism*. New York: Free Press, 1994.

James Corbett

Erosion

Introduction

Erosion is the process of transportation of weathered rock and soil material by the processes of water, wind, or ice movement and/or by gravity or the actions of some organisms. For erosion to begin, rock or soil must first be subjected to the processes of weathering, including disintegration (breaking apart) and decomposition (chemical breakdown). Erosion ends with deposition, which is the process of eroded material (rock fragments, sediments, or dissolved chemical elements) coming to rest in a place that is apart from where erosion began. Erosion may be natural, human-induced, or the result of a combination of these two factors.

Historical Background and Scientific Foundations

Erosion has played a huge part in the history of planet Earth. Mountain building through geological time has raised vast mountain ranges, which in turn have been reduced by weathering and erosion to low plains. Tectonic forces within Earth, which have routinely affected

Several people are seen on a stretch of eroded soil next to the Tonala River near the city of Villahermosa, Mexico, in April 2007. Scientists predict that due to rising global temperatures, Latin America's diverse ecosystems will struggle with intense droughts and flooding. A United Nations report estimated that up to 70 million people in the region will be left without enough water. *AP Images.*

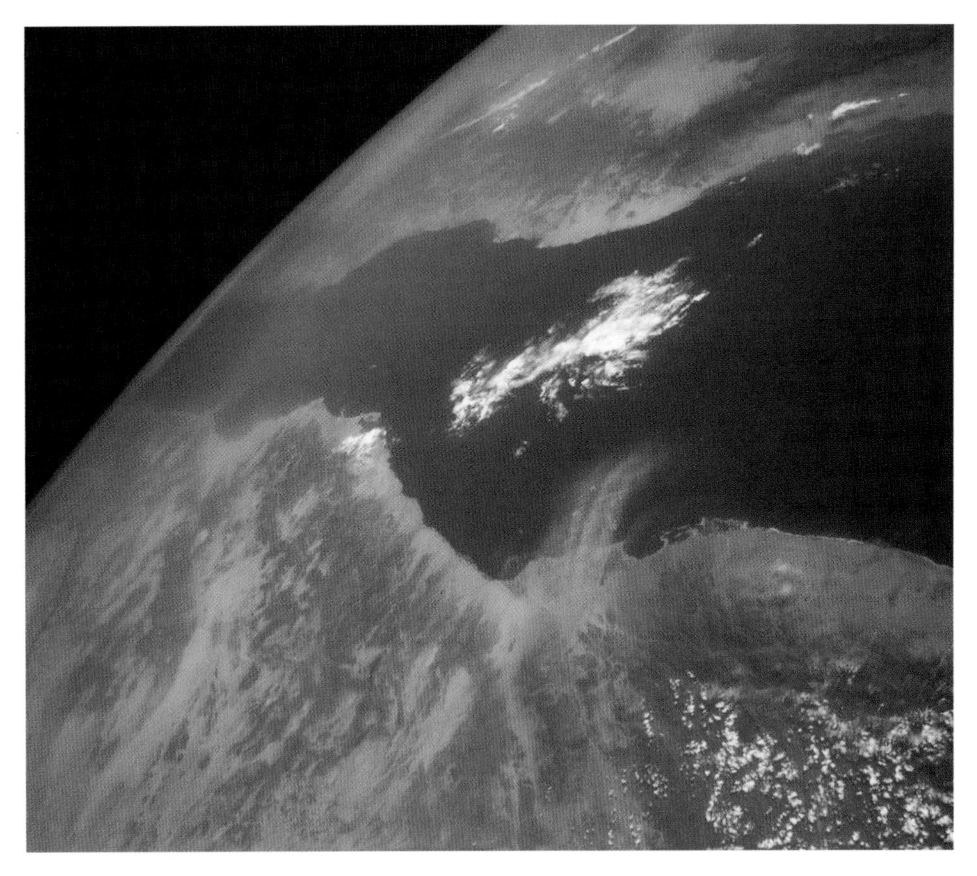

A dust plume blows over the Red Sea. Winds pick up eroded material from the land and carry it out over the sea. *Photodisc/Royalty Free.*

the crust over time, move rock to higher elevations. Yet, these materials are moved to lower elevations by erosion. In fact, if mountains were not continually built at a faster rate than erosion, the continental surface of Earth would have been reduced to a nearly featureless landscape long ago.

Eroded material can be picked up and moved by water, wind, and ice. Each of these transportation media (water, wind, and ice) requires a minimum velocity of movement to pick up (or entrain) eroded particles. For water and wind, these velocities are relatively high compared to ice, which can move materials of any size at almost any velocity. As the velocity of water and wind increases, progressively larger particles can be moved. As velocity decreases, particles are deposited or laid to rest.

During transportation, water and wind move particles by rolling, bouncing, and in suspension (floating). Particles interact during transportation and thus they are abraded (worn) or comminuted (broken into smaller pieces) during transportation. Transportation of particles moves them from the point of origin (for example, an eroded hillside) to the point of deposition (for example, a river delta or sand dune). The point of deposition may be a temporary location from which the particles are moved again at another time. An example of a temporary location is a gravel or sand bar in a river, which is a temporary location for river sediment. During a flood on the river, such bars may be eroded and the sediment moved downstream to another bar or ultimately downstream to the river's delta.

Gravity pulls on all natural and human-made materials. The effect of this is the natural tendency for materials that can slide or flow to move downslope at varying rates. The movement of water, wind, and ice noted earlier is in response to gravity. However, gravity acts directly upon rocks, sediment, and soil to move these materials downslope as well. If the affected rock, sediment, or soil moves as a single large mass, it is called a slump. If the rock, sediment, or soil moves as a layer where the upper part is moving faster than the lower parts, it is called a flow. An example of a common type of flow is soil creep, which is the slow, progressive flow of soil down the side of a hill or other sloping surface. Gravity effects like slump and flow work in combination with water, wind, and ice to move large quantities of materials all the time.

The actions of some organisms may affect erosion in some situations. The trampling effect of animals walking on sloping surfaces may weaken the rock, sediment, or soil in that area and thus contribute to erosion. Trampling may remove vegetation as well, which weakens

WORDS TO KNOW

DECOMPOSITION: The breakdown of matter by bacteria and fungi. It changes the chemical makeup and physical appearance of materials.

DISINTEGRATION: Spontaneous nuclear transformation characterized by the emission of energy and/or mass from the nucleus.

SEDIMENTS: Disintegration and decomposition products of natural materials

TECTONIC FORCES: Forces that shape Earth's crust over geological time. The largest tectonic force is supplied by the slow convection of Earth's interior, driven by radiation of heat to space through the crust, much like the roiling of liquid in a boiling pot.

natural materials or allows water, wind, or ice to more effectively transport such materials. Some organisms in the seas actively eat rock and sediment and thus disintegrate (by chewing) and decompose (by digesting) rock fragments and sediment. Some agricultural, construction, timbering, and related practices affect erosion in some situations. Like the actions of organisms mentioned earlier, these practices can assist gravity and the flow of water and wind in the process of erosion of landscapes.

Impacts and Issues

Erosion may be human-induced, but much of natural erosion may be induced by climate change and its attendant effects. For example, increased erosion due to increased rainfall in an area may be climatically controlled. Climate control necessarily implies that the change is long-term and part of a long-term trend, as opposed to short-term effects from weather systems that come and go over shorter time spans.

SEE ALSO *Dust Storms; Floods; Glaciation.*

BIBLIOGRAPHY

Books

Montgomery, David R. *Dirt: The Erosion of Civilization.* Berkeley: University of California Press, 2007.

Periodicals

U.S. Department of Agriculture. "Predicting Soil Erosion by Water: A Guide to Conservation Planning with the Revised Universal Soil Loss Equation (RUSLE)." *Agriculture Handbook* 703 (1996).

Web Sites

Pidwirny, Michael. "Introduction to the Lithosphere: Erosion and Deposition." *PhysicalGeography.net*, 2007. <http://www.physicalgeography.net/fundamentals/10w.html> (accessed December 3, 2007).

David T. King Jr.

Ethanol

Introduction

Ethanol—also known as drinking, ethyl, or grain alcohol—is a compound of carbon, hydrogen, and oxygen (C_2H_5OH) that can be burned as a fuel. Ethanol can be refined from petroleum, but its interest as an alternative energy source depends on the fact that it can be produced biologically. Yeast cells ingest sugars and excrete ethanol. The sugars can be extracted from kernels of grain, grapes, sugarcane, or other plant materials. Ethanol can be added to gasoline to substitute for some of it or to raise its octane or make it burn more cleanly.

Ethanol can also be burned by itself in specially designed engines. In the United States and Brazil, ethanol production is subsidized by the government in an effort to reduce national reliance on imported oil. The U.S. ethanol program, which relies on corn, is particularly controversial. There is a wide scientific consensus that little more energy, if any, is extracted from burning corn ethanol than is required to produce it. Defenders of corn ethanol argue that ethanol's energy balance is at least positive, and will be much better when methods are perfected for producing ethanol from cellulose, a more abundant plant material. There is also disagreement about whether ethanol contributes less to climate change than do equal amounts of fossil fuel.

Historical Background and Scientific Foundations

The use of ethanol as a fuel goes back to the early days of the internal combustion engine. Henry Ford (1863–1947), the first person to apply assembly-line techniques to manufacturing cars, stated that ethanol was "the fuel of the future." Ford even had the Model T, the first car to be marketed to millions of middle-class consumers, designed so that it could run either on gasoline or ethanol. Germany and France had few oil wells and hoped to increase their energy independence by promoting ethanol as a motor fuel. There was debate about which fuel was superior, gasoline or ethanol. In Europe in the late 1890s and early 1900s, a number of road races were held between vehicles burning pure gasoline, pure ethanol, or various blends of the two. In 1906, a third of the heavy locomotives built by the Deutz Gas works in Germany burned pure ethanol and a tenth of the automobile engines built by Otto Gas Engine Works in Philadelphia, Pennsylvania, burned pure ethanol.

Ethanol was eclipsed, however, by gasoline. One reason was that gasoline has 1.5 times as much energy per gallon as ethanol, so a gasoline-powered car can go farther on a tank of fuel. Efforts were made to revive ethanol in the 1920s and 1930s, with a blend of 90% gasoline and 10% ethanol being offered at many gas stations. At that time, ethanol was proposed not primarily as a substitute for gasoline, but as an antiknock additive. Knocking is the annoying sound caused by premature ignition of fuel in an engine's cylinders, which lowers efficiency.

However, the use of a highly toxic (but profitable) lead compound, tetraethyl lead, was successfully promoted by the companies Du Pont and General Motors starting in the early 1920s. About 7 million tons (6.3 million metric tons) of lead were added to the environment in the United States in the twentieth century as the result of using tetraethyl lead instead of ethanol as an antiknock additive. Lead was phased out by law in the United States from 1973 to 1996. Phaseout was completed in Japan in 1980 and in most European countries by the late 1990s. Leaded gasoline is still used in much of the developing world.

The Arab oil embargo of 1973 created awareness in many countries of their dependence on imported oil, and ethanol was advocated as a solution. The U.S. Energy Tax Act of 1978 lifted the $.04/gallon federal excise tax on gasoline blends containing at least 10% ethanol, and in the following decades other tax incentives were created to encourage the U.S. ethanol industry, including a $.54/gallon tariff on imported ethanol, still in effect as of late 2007. By 2005, 1.6 billion bushels of corn, about 14% of

Some municipalities have turned to using ethanol- or biodiesel-powered vehicles. Here, an ethanol-powered snowplow is shown in Minnesota. *U.S. Department of Energy.*

the U.S. crop, was being used to produce some 3 billion gallons (11.3 billion liters) of ethanol per year.

The U.S. ethanol industry began especially rapid growth after the Energy Policy Act of 2005, which mandated that 7.5 billion gallons (28.4 billion liters) of U.S. gasoline consumption, about 5%, come from renewable sources by 2012. This refers primarily to ethanol, with comparatively tiny contributions from biodiesel.

Brazil began an ambitious ethanol program in 1975 in response to the oil shock of 1973. By 2007, most automotive fuel sold in Brazil contained 70% gasoline and 30% ethanol; about 4 million cars were burning pure ethanol; and most new vehicles contained "flexible-fuel" engines that could burn pure gasoline, pure ethanol, or any blend. Brazil's ethanol is produced not from corn but from sugarcane, which contains more sugar per ton and is therefore a more efficient feedstock for ethanol production. Large volumes of sugarcane are not grown in the United States.

One of the most commonly voiced criticisms of the U.S. corn-ethanol program is that as much, or almost as much, fossil-fuel energy is needed to produce a gallon of ethanol as can be obtained by burning it. Most recent scientific studies support this view. One reason for the poor energy balance of corn ethanol is that only a tiny fraction of the corn plant—the sugar in the kernels—is turned into alcohol. The rest consists of the woody material lignocellulose (cellulose bound with lignin), which cannot be digested by yeast to produce ethanol.

Research on ways to ferment cellulose to produce ethanol has been conducted for decades. Methods proposed include genetically engineered yeast or bacteria, enzymes, or chemicals to liberate sugars from cellulose so that yeast can digest them. As of 2007, very little ethanol was produced from cellulosic fermentation (also called lignocellulosic fermentation), as no process yet affordable compared to either fossil fuels or sugar ethanol. Many experts state that if the process is perfected, it will significantly improve the energy balance of ethanol. Others disagree, claiming that the energy balance even of cellulosic fermentation is negative. The different claims are based on different listings or accountings of all the energy inputs required for the process.

■ Impacts and Issues

Brazilian Program

Brazil's large ethanol program has been deemed a success by many observers. But it also has its critics, who argue that it has depended on cheap labor, in some cases even slave labor, and that the conversion of large areas of land to single-crop agriculture to support ethanol production is damaging society. For example, in 2007, more than a thousand workers kept as virtual slaves by Para Pastoril e Agricola SA, one of Brazil's largest ethanol producers, were liberated by a Brazilian government raid. Many Brazilian cane workers, harvesting with 8-14 tons (7.2-12.7 metric tons) of sugarcane apiece each day with machetes, are kept in permanent debt and are not free to leave. As some harvesting machines are introduced, supervisors demand even larger harvests from the hand laborers. Seasonal jobs, concentration of land ownership in the hands of large corporations, deforestation to make room for plantations, and other problems are also cited.

Corn Ethanol

As mentioned earlier, corn ethanol has been accused of producing no significant energy. The U.S. Department of Agriculture claims an energy ratio of 1.34, that is, it says that for 1 unit of fossil-fuel energy invested, corn ethanol yields 1.34 units of energy. Most studies, however, agree that the gain is smaller, or that energy is actually lost. Although any plant, when burned, simply returns carbon to the atmosphere that it extracted from the atmosphere, and so makes no net contribution to climate change, in practice the use of fossil fuels to raise and process corn does add carbon dioxide to the atmosphere. Corn ethanol, gallon for gallon, presently contributes about as much to climate change as gasoline.

Critics also point to the fact that modern agribusiness is a destructive industry, leading to the rapid erosion of irreplaceable soil. Even if ethanol yielded an energy profit, these critics say, it would only be a means of mining the soil, which is a non-renewable resource. Some agricultural scientists have also argued that the removal of roots, husks, and stalks from fields to make cellulosic ethanol would accelerate erosion and bleed soil fertility. Such materials slow erosion and aid the penetration of the soil by rainwater.

Critics of ethanol also claim that it puts arable land at the service of the automobile, against which the poor and hungry of the world must then compete for food. In 2007 the *New York Times* reported that soaring food prices, driven partly by sharply rising demand for corn ethanol, had caused U.S. foreign food-aid dollars to buy less than half the food they bought in 2000.

The official government position is that ethanol will help U.S. farmers, produce cleaner air, and reduce U.S. dependence on foreign oil. As of 2007, this position is supported by leaders of both the Republican and Democratic parties.

■ Primary Source Connection

The following journal article outlines biomass ethanol's potential to aid in reduction of greenhouse gas emissions, but does not discuss many of the potential drawbacks and limitations of ethanol use. While supporters of ethanol note its clean-burning potential and renewable source, critics assert that current production methods (including those of industrialized farming) require large amounts of petroleum, reducing biomass ethanol's "green" potential. The author here notes that researchers are working to develop greener ethanol production technologies, and that biomass ethanol faces several challenges before becoming a significant petroleum alternative.

WORDS TO KNOW

BIODIESEL: A fuel made from a combination of plant and animal fat. It can be safely mixed with petro diesel.

CELLULOSIC FERMENTATION: Digestion of high-cellulose plant materials (e.g., wood chips, grasses) by bacteria that have been bred or genetically engineered for that purpose. The useful product is ethanol, which can be burned as a fuel. Cellulosic fermentation is one way of producing cellulosic ethanol.

ENERGY POLICY ACT OF 2005: U.S. federal law passed in 2005 that offers tens of billions of dollars of subsidies and loan guarantees for energy technologies it categorizes as "clean," including renewables, some forms of coal-burning, and nuclear power. Most commentators believed that the majority of the loan guarantees authorized by the act would go to the nuclear industry.

FOSSIL FUEL: Fuel formed by biological processes and transformed into solid or fluid minerals over geological time. Fossil fuels include coal, petroleum, and natural gas. Fossil fuels are non-renewable on the timescale of human civilization, because their natural replenishment would take many millions of years.

LIGNOCELLULOSE: Any of several substances making up woody cell walls in plants, consisting of cellulose mixed with lignin (both organic polymers). Digestion of lignocellulose to produce ethanol is a sought-after technology for producing biofuel.

BIOMASS ETHANOL: TECHNICAL PROGRESS, OPPORTUNITIES, AND COMMERCIAL CHALLENGES

No other sustainable option for production of transportation fuels can match ethanol made from lignocellulosic biomass with respect to its dramatic environmental, economic, strategic, and infrastructure advantages. Substantial progress has been made in advancing biomass ethanol (bioethanol) production technology to the point that it now has commercial potential, and several firms are engaged in the demanding task of introducing first-of-a-kind technology into the marketplace to make bioethanol a reality in existing fuel-blending markets. Opportunities have also been defined to further reduce the cost of bioethanol production so it is competitive without tax incentives.

This chapter provides a brief review of the key factors that drive interest in producing ethanol from biomass sources such as agricultural (e.g., sugar cane bagasse) and forestry (e.g., wood trimmings) residues, significant fractions of municipal solid waste (e.g., waste paper and yard waste), and herbaceous (e.g., switchgrass) and woody (e.g., poplar) crops. Next, a state-of-the-art bioethanol process is outlined, followed by an economic pro forma analysis to provide a sense of the important cost drivers. Against this backdrop, progress made in advancing bioethanol technology is reviewed to define the key accomplishments made possible through sustained research and development. Then two important areas meriting much greater emphasis are outlined. The first is in developing a solid technical foundation built on fundamental principles to help overcome the barriers that impede introduction of first-of-a-kind technology into the marketplace. The second is in aggressively funding research to advance bioethanol technology to the point at which it can be competitive as a pure fuel in the open marketplace. Hopefully, this chapter will provide a better appreciation of how bioethanol production technology has been improved and the vast potential it has for continued advancements and large-scale benefits. . . .

Greenhouse Gas Reductions

Perhaps the most unique attribute of bioethanol is very low greenhouse gas emissions, particularly when compared with the emissions from other liquid transportation fuel options. Because nonfermentable and unconverted solids left after making ethanol can be burned or gasified to provide all of the heat and power to run the process, no fossil fuel is projected to be required to operate the conversion plant for mature technology. In addition, many lignocellulosic crops require low levels of fertilizer and cultivation, thereby minimizing energy inputs for biomass production. The result is that most of the carbon dioxide released for ethanol production and use in a cradle-to-grave (often called a full-fuel-cycle) analysis is recaptured to grow new biomass to replace that harvested, and the net release of carbon dioxide is low. If credit is taken for export of excess electricity produced by the bioethanol plant and that electricity is assumed to displace generation by fossil fuels such as coal, it can be shown that more carbon dioxide can be taken up than is produced.

The impact of bioethanol on greenhouse gas emissions can be particularly significant because the transportation sector is a major contributor to greenhouse gas emissions, accounting for about one-third of the total. As part of a Presidential Advisory Committee on reducing greenhouse gas emissions from personal vehicles, a survey of experts in the field clearly showed that most alternatives to petroleum (e.g., hydrogen production from solar energy) required significant changes in the transportation infrastructure to be implemented, whereas others that could be more readily used (e.g., methanol production from coal or natural gas) would have little impact on reduction of greenhouse gas emissions. On the other hand, ethanol is a versatile liquid fuel, currently produced from corn and other starch crops, that is blended with ~10% of the gasoline in the United States and is widely accepted by vehicle manufacturers and users. Vehicles that use high-level ethanol blends (e.g., in E85, a blend of 85% ethanol in gasoline) are now being introduced throughout the United States. In addition, bioethanol production technology could be commercialized in a few years and would not require extended time frames to be applied. Overall, the evidence suggests that the best choice from the coupled perspectives of greenhouse gas reduction, integration into the existing infrastructure, and rapid implementation is the production of ethanol from lignocellulosic biomass.

Although surveys show that Americans are concerned about the prospects of global climate change, the issue has not received broad political support, perhaps owing to the influence of special-interest groups. On the other hand, much of Europe, Canada, and other countries are actively seeking to reduce greenhouse gas emissions. Ironically, much more attention has been focused on developing bioethanol technology in the United States, whereas other countries have only recently shown interest in the area. Thus, there is tremendous potential for application of U.S. technology in many other regions of the world, benefiting all concerned. . . .

Conclusions

Biomass ethanol is a versatile fuel and fuel additive that can provide exceptional environmental, economic, and strategic benefits of global proportions. Bioethanol can play a particularly powerful role in the quest to reduce greenhouse gas emissions that will be difficult for any other transportation fuel options to match. Because of the widespread abundance of biomass, bioethanol can also be invaluable for meeting the growing international demand for fuels by developing nations as well as enhancing the energy security of developed countries. Furthermore, conversion of waste materials to ethanol provides an important disposal option as new regulations restrict historical approaches. It also is important to note that bioethanol is among the few options available for sustainable production of liquid fuels. Finally, although gasoline is continually being reformulated to reduce its environmental impact, ethanol has favorable properties that can provide air and water quality attributes comparable, if not superior, to gasoline and can provide particular benefits when used as a pure fuel in properly optimized engines and ultimately fuel cells.

Tremendous progress has been made in reducing the cost of enzymatic-based technology for bioethanol production, with current estimated costs showing the technology to be potentially competitive now, particularly for niche markets. A key to these advances has been in achieving higher yields, faster rates, and greater concentrations of ethanol through improved pretreatment technology, development of better cellulase enzymes, and synergistic combination of cellulose hydrolysis and

Steam billows from an ethanol production plant in South Dakota. *Image copyright Jim Parkin, 2007. Used under license from Shutterstock.com.*

fermentation steps that make progress in overcoming the natural recalcitrance of biomass. Genetic engineering of bacteria so that they ferment the diverse range of sugars in lignocellulosic materials to ethanol with high yields is a milestone achievement essential to economic success.

Although progress has been impressive, the cost of bioethanol production must be reduced further if it is to be competitive without special tax incentives on a large scale for the fuel market. Because enzyme-based systems can build off the emerging achievements of biotechnology, they show particular promise for further cost reductions, and sensitivity studies, process modeling, and macroscopic economic analyses reveal that there are no fundamental barriers to advancing the technology. Cost estimates reveal that pretreatment is a particularly expensive step, both directly and indirectly. From a technology perspective, the sensitivity studies clearly show that ethanol yield is a strong economic driver, and there are significant gains from improving the yields of all process steps. It is important that even greater cost reductions can result from improving pretreatment and biological-conversion process configurations. In fact, specific advanced pretreatment and bioprocessing configurations based on continued progress in overcoming the recalcitrance of biomass have been identified that would reduce the cost of bioethanol production to levels that it can compete in a nonsubsidized market. However, even though the advanced pretreatment configuration chosen significantly reduces cost, it would represent about two-thirds of an overall advanced design scenario, suggesting that further improvements beyond those envisioned should be sought, with tremendous impact. This result also implies that emphasis on novel pretreatment technology with extremely low-cost potential is badly needed instead of pursuing relatively minor improvements over dilute sulfuric-acid approaches, and such advances will probably best come through improving our knowledge of how pretreatment works. Interestingly, although feedstock cost reductions are constrained to levels that will have moderate impact for large-scale bioethanol production, more productive and less expensive biomass would make it feasible to feed larger plants that realize significant economies of scale.

It is just as important to take the next step and commercialize bioethanol technology so that its tremendous benefits can be realized. However, because bioethanol plants must typically be large to be profitable, substantial capital outlay is required, and risk management is essential to attract investors to finance the introduction of first-of-a-kind technology. Although large pilot and perhaps even semi-works demonstration projects may be required to provide an adequate level of comfort, significantly more emphasis on developing solid fundamental principles for design of biomass processing operations would greatly reduce the tremendous costs and delays associated with technology scale-up. Building expert teams to work cooperatively to understand key bioethanol-processing steps in the context of applying and advancing the technology is the most effective approach to realize the low-cost potential of bioethanol and realize its benefits on a large scale. In the final analysis, researchers, research managers, program leaders, and funding authorities who have had the vision and courage to advance bioethanol technology to the point that it now has commercial potential need to facilitate advancing and applying the technology in the face of even greater challenges to achieve widespread impact. In addition, entrepreneurs, financiers, engineers, and contractors with equal vision and courage are needed to take the technology to its first commercial applications.

Charles E. Wyman

WYMAN, CHARLES E. "BIOMASS ETHANOL: TECHNICAL PROGRESS, OPPORTUNITIES, AND COMMERCIAL CHALLENGES," *ANNUAL REVIEW OF ENERGY AND THE ENVIRONMENT* 24 (1999): 189–226.

SEE ALSO *Biofuel Impacts.*

BIBLIOGRAPHY

Books

Worldwatch Institute. *Biofuels for Transport: Global Potential and Implications for Energy and Agriculture.* London: Earthscan Publications, Ltd., 2007.

Periodicals

Dugger, Celia. "As Prices Soar, U.S. Food Aid Buys Less." *The New York Times* (September 29, 2007).

Farrell, Alexander et al. "Ethanol Can Contribute to Energy and Environmental Goals." *Science* 311 (2006): 506-508.

Krauss, Clifford. "Sudden Surplus Arises as Threat to Ethanol Boom." *The New York Times* (September 30, 2007).

Pimentel, David. "Ethanol Fuels: Energy Balance, Economics, and Environmental Impacts Are Negative." *Natural Resources Research* 12 (2007): 127-134.

Rosner, Hillary. "Cooking Up More Uses for the Leftovers of Biofuel Production." *The New York Times* (August 8, 2007).

Sanderson, Katharine. "A Field in Ferment." *Nature* 444 (2007).

Wald, Matthew L. "Is Ethanol for the Long Haul?" *Scientific American* (January 2007): 42-49.

Wyman, Charles E. "Biomass Ethanol: Technical Progress, Opportunities, and Commercial Challenges." *Annual Review of Energy and the Environment* 24 (1999): 189-226.

Web Sites

Al-Kais, Mahdi, and Jose Guzman. "'How Residue Removal Affects Nutrient Cycling." *Department of Agronomy, Iowa State University,* May 22, 2007. <http://www.ipm.iastate.edu/ipm/icm/2007/5-21/cycling.html> (accessed October 26, 2007).

Friedemann, Alice. "Peak Soil: Why Cellulosic Ethanol, Biofuels are Unsustainable and a Threat to America." *CultureChange.org,* April 10, 2007. <http://www.culturechange.org/cms/index.php?option=com_content&task=view&id=107&Itemid=1> (accessed October 26, 2007).

Kenfield, Isabella. "Brazil's Ethanol Plan Breeds Rural Poverty, Environmental Degradation." *Americas Program, Center for International Policy,* March 6, 2007. <http://americas.irc-online.org/am/4049> (accessed October 26, 2007).

"'Slave' Labourers Freed in Brazil." *BBC News,* July 3, 2007. <http://news.bbc.co.uk/2/hi/americas/6266712.stm> (accessed October 26, 2007).

Larry Gilman

Europe: Climate Change Impacts

Introduction

Europe, inhabited by more than 700 million people, is the westernmost portion of the Eurasian continent but is traditionally treated as a continent due to its cultural distinctness. The 27 member states of the European Union include 60% of the population of Europe and most of its economic output (though only about 17% of its land area). These 27 states have established a European Climate Change Program to identify the most environmentally effective and affordable policies that can be enacted in Europe to cut the greenhouse-gas emissions that, most scientists agree, are causing the global climate to change.

Europe, like most other regions of the world, is already experiencing observable effects from climate change. These include shifting rainfall patterns, shifting ranges of wild animal and plant species, impacts on agriculture, rising sea levels, and increased heat waves.

Historical Background and Scientific Foundations

Europe has been inhabited by human beings or closely related species for hundreds of thousands of years. The Industrial Revolution that made anthropogenic (human-

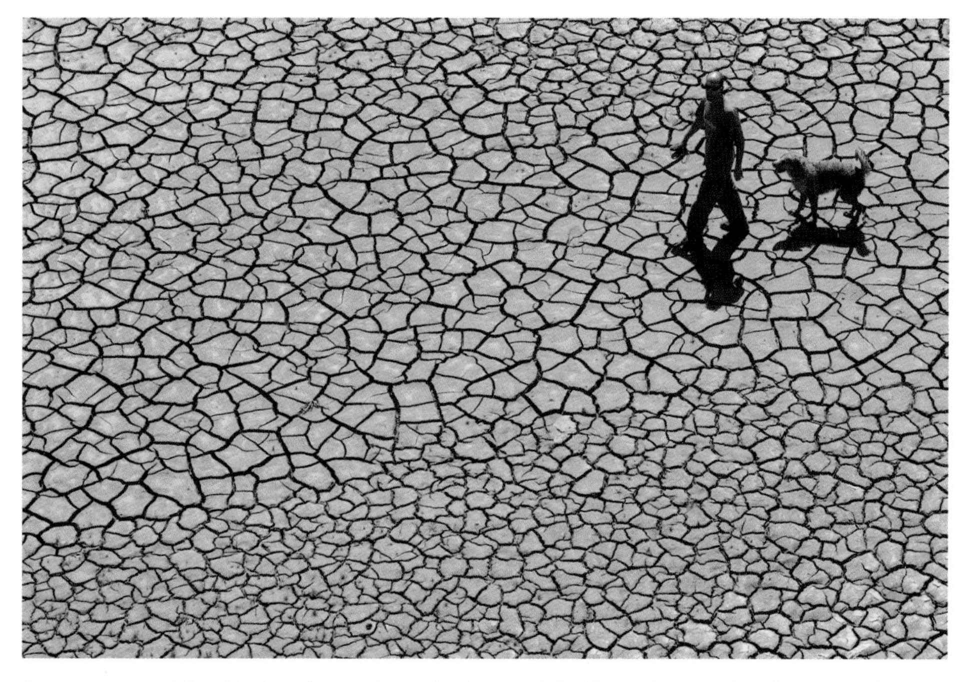

A man is seen walking his dog along a dry cracked reservoir in Alcora, Spain, in 2005. Scientists believe that climate change could greatly impact Europe's Mediterranean region. Changes could lead to sapping electric power generation, reversing long-standing tourism trends, raising sea levels in coastal areas, and leaving millions of people without sufficient water. *AP Images.*

The Thames Barrier is composed of a series of 10 movable parts. It is positioned across the Thames river to keep flood waters from reaching London. With predictions of sea-level rising due to global warming, some residents are wondering if the Thames Barrier will be enough. *Image copyright Pres Panayotov, 2007. Used under license from Shutterstock.com.*

caused) climate change possible began in Europe in the latter half of the eighteenth century and has since spread to most of the world. Today, Europe remains both an industrial and agricultural area, responsible for about 21% of global meat production and 20% of global cereal grain production. It is the world's most densely populated continent (about 156 persons per square mile, or 60 per square kilometer). Due to its wealth and technological sophistication, it is comparatively well-equipped to adapt to changes in climate, and is implementing an array of policies designed to reduce carbon dioxide and other greenhouse-gas emissions by reducing fuel consumption.

There is no doubt that Europe has grown warmer: there has been a well-established warming trend in Europe for over a century, namely about 1.6°F (.9°C) per decade averaged over the period 1901 to 2005. Moreover, this trend has accelerated, being about .74°F (.41°C) per decade from 1979 to 2005. Daily temperature variability has increased due to a rise in the number of warm extremes (for example, heat waves and exceptionally warm days).

In 2007, the United Nations' Intergovernmental Panel on Climate Change (IPCC) released the report of its Working Group II, *Climate Change 2007: Impacts, Adaptation and Vulnerability*. The report noted the following points, among others, about the impacts of climate change on Europe:

- Impacts of climate change are already being seen in Europe. These include the retreat of glaciers, decreased permafrost area, lengthening of the growing season, shifts of species ranges, the heat wave of 2003, and more. These changes are consistent with past projections of impacts of climate change: that is, scientists successfully predicted that these climate changes would occur (although the exact timing of individual weather events such as the 2003 heat wave cannot be predicted, and scientists have not attempted to do so).
- Climate-related hazards will increase in Europe in the twenty-first century. These include floods, coastal flooding from increased storminess and higher sea levels, more frequent and prolonged droughts, higher wildfire risk (including catastrophic fires on peatlands in central Europe), rock avalanches in the mountains, and more. Health risks from heat, flooding, and some infectious diseases may increase unless adaptive measures are successful. Some impacts, the IPCC notes, may be positive, including decreased risk from extreme cold; however, the overall health risks are expected to increase.
- Europe's natural ecosystems will be affected, and most organisms and ecosystems will have difficulty adapting to climate change. Inland migration of beaches from rising sea levels will destroy up to 20% of coastal wetlands, which will stress the species that rely on those ecosystems. Many areas of Arctic permafrost are likely to disappear. Forests will expand northward; tundra area will decrease. Populations in North

WORDS TO KNOW

ANTHROPOGENIC: Made by people or resulting from human activities. Usually used in the context of emissions that are produced as a result of human activities.

GREENHOUSE GASES: Gases that cause Earth to retain more thermal energy by absorbing infrared light emitted by Earth's surface. The most important greenhouse gases are water vapor, carbon dioxide, methane, nitrous oxide, and various artificial chemicals such as chlorofluorocarbons. All but the latter are naturally occurring, but human activity over the last several centuries has significantly increased the amounts of carbon dioxide, methane, and nitrous oxide in Earth's atmosphere, causing global warming and global climate change.

INDUSTRIAL REVOLUTION: The period, beginning about the middle of the eighteenth century, during which humans began to use steam engines as a major source of power.

KYOTO PROTOCOL: Extension in 1997 of the 1992 United Nations Framework Convention on Climate Change (UNFCCC), an international treaty signed by almost all countries with the goal of mitigating climate change. The United States, as of early 2008, was the only industrialized country to have not ratified the Kyoto Protocol, which is due to be replaced by an improved and updated agreement starting in 2012.

PEATLAND: An area where peat is forming or has formed. Peat is a dense, moist substance formed of compacted ground-cover plants, mainly moss and grasses. Over geologic time, peat can turn into coal.

PERMAFROST: Perennially frozen ground that occurs wherever the temperature remains below 32°F (0°C) for several years.

TUNDRA: A type of ecosystem dominated by lichens, mosses, grasses, and woody plants. It is found at high latitudes (arctic tundra) and high altitudes (alpine tundra). Arctic tundra is underlain by permafrost and usually very wet.

WETLANDS: Areas that are wet or covered with water for at least part of the year.

Atlantic fisheries are likely to increase. Under some possible scenarios for future emissions of greenhouse gases, up to 60% of alpine (high-mountain environment) plant species may become extinct.

■ Impacts and Issues

The 27 member states of the European Union (EU) have committed to a wide range of policies to decrease their contribution to the greenhouse-gas emissions that are causing most climate change. For example, all members of the EU are signatories of the Kyoto Protocol of the United Nations Framework Convention on Climate Change, a treaty committing its member states to the timely stabilization of greenhouse-gas concentrations in the atmosphere. Building on the provisions of Kyoto, the EU has instituted its Emissions Trading Scheme, which now governs carbon dioxide emissions from about 11,500 large emitters of carbon dioxide in manufacturing and power generation.

Some critics have objected that the European emissions trading scheme is ineffective or inadequate. Market-oriented critics object that the scheme is less effective than emissions taxes, or constitutes unwarranted government interference in the market. Environmentally minded critics object that it allows excessively high caps on emissions and treats life-threatening sources of pollution as a tradable commodity, with severe impacts on some local communities. Other experts defend the system as workable.

The Kyoto Protocol's targets expire in 2012. The EU officially favors an early start to negotiations on a post-Kyoto international climate treaty. Such a regime, the EU says, should see participation by "all major emitters." This is likely a reference to the United States, the world's all-time largest emitter of greenhouse gases, which is not a participant in the Kyoto process, and to China, which is unconstrained by Kyoto and probably now the world's largest annual emitter. The EU also hopes to see innovations in emissions-saving technologies and the pursuit of strategies to adapt to future global warming.

See Also *Africa: Climate Change Impacts; Arctic People: Climate Change Impacts; Asia: Climate Change Impacts; Australia: Climate Change Impacts; North America: Climate Change Impacts; Small Islands: Climate Change Impacts; South America: Climate Change Impacts.*

BIBLIOGRAPHY

Books

Parry, M. L., et al, eds. *Climate Change 2007: Impacts, Adaptation and Vulnerability: Contribution of Working Group II to the Fourth Assessment Report of the Intergovernmental Panel on Climate Change.* New York: Cambridge University Press, 2007.

Web Sites

"Impacts of Europe's Changing Climate." EEA Report No. 2. *European Environment Agency.* 2004. <http://reports.eea.europa.eu/climate_report_2_2004/en/impacts_of_europes_changing_climate.pdf> (accessed September 16, 2007).

Larry Gilman

Europe: Climate Policy

Introduction

The 48 countries of Europe, although they occupy only 2% of the world's land area and contain only 11% of its population, are economically, politically, and culturally influential. Although Europe contains poor countries, it is one of the world's wealthiest and most technologically advanced regions, producing almost a third of global gross domestic product. It is also responsible for 15% of global greenhouse emissions. Eighty percent of its emissions are from its energy and transport sectors. Per person, Europe emits about half as much greenhouse gases as does the United States.

Many European countries are united under the banner of the European Union (EU), though individual countries retain full sovereignty. The climate policy of Europe is formed at several levels. First, individual nations produce their own internal climate policies. Second, climate policy is formed at the level of the EU. Third, nations are individually committed to global-scale climate efforts such as the United Nations Environment Programme (UNEP), the United Nations Framework Convention on Climate Change (UNFCCC, 1992), and the Kyoto Protocol (1997), which is an elaboration and extension of the UNFCCC. The UNFCCC and Kyoto are the basic legal framework of the EU's climate policies.

Since its formation in 1993, the EU has consistently accepted the reality of climate change and argued for action to reduce the situation through binding caps on greenhouse emissions. As of 2007, results of the EU's various climate efforts had been mixed, with some European countries' greenhouse-gas emissions increasing and others decreasing since 1990 (the year most often used for comparison). Europe continues to revise and adjust its climate policy, with negotiations for a 2012 treaty to succeed Kyoto beginning in late 2007.

This article discusses climate policy at the level of the EU, not at the level of individual countries.

Historical Background and Scientific Foundations

The European Union, with 27 member states as of late 2007, was legally established by the Maastricht Treaty of 1993, which revised the pre-existing structure of the European Economic Community (formed 1957). It is concerned primarily with economic affairs, and does not set a united European foreign or military policy as does the federal government of the United States. The executive branch of the EU is the European Commission, headquartered in Brussels, Belgium.

In 1990, in response to the first Assessment Report on climate change issued by the United Nations' Intergovernmental Panel on Climate Change (IPCC), the EU committed to stabilizing its overall emissions of carbon dioxide (CO_2), the most important greenhouse gas, at 1990 levels by 2002. This goal was achieved. The European Commission's first climate initiative was in 1991 (while it was still the executive of the European Economic Community, not yet the EU), a directive to reduce emissions of CO_2 through renewable energy and voluntary commitments by auto manufacturers.

In 1992, all of the member states of the EU, along with the United States and almost all other countries, signed the UNFCCC, a nonbinding treaty acknowledging the reality of climate change and the need to reduce emissions of greenhouse gases. The UNFCCC named no specific goals, but called for further international negotiation to determine such goals. These negotiations produced the Kyoto Protocol in 1997, an extension of (protocol to) the UNFCCC. The protocol specified binding caps or upper limits on greenhouse-gas emissions from developed (industrialized) countries. The countries of the EU, for example—more specifically, the 15 countries that were EU members before 2004—are committed under Kyoto to reduce their greenhouse-gas emissions to 8% below 1990 levels by 2012. The

WORDS TO KNOW

CAP AND TRADE: The practice, in pollution-control or climate-mitigation schemes, of mandating an upper limit or cap for the total amount of some substance to be emitted (e.g., CO_2) and then assigning allowances or credits to polluters that correspond to fixed shares of the total amount. These allowances or credits can then be bought and sold by polluters, in theory allowing emission cuts to be bought where they are most economically rational.

CARBON CREDIT: A unit of permission or value, similar to a monetary unit (e.g., dollar, euro, yen) that entitles its owner to emit one metric ton of carbon dioxide into the atmosphere.

CARBON TAX: Mandatory fee charged for the emission of a given quantity of CO_2 or some other greenhouse gas. Under a carbon taxation scheme, polluters who emit greenhouse gases must pay costs that are directly proportional to their emissions. The purpose of a carbon tax is to reduce greenhouse emissions. Carbon taxation is the main alternative to emissions trading.

CLEAN DEVELOPMENT MECHANISM: One of the three mechanisms set up by the Kyoto Protocol to, in theory, allow reductions in greenhouse-gas emissions to be implemented where they are most economical. Under the Clean Development Mechanism, polluters in wealthy countries can obtain carbon credits (greenhouse pollution rights) by funding reductions in greenhouse emissions in developing countries.

EMISSIONS CAP: Government-mandated upper limit on total amount to be emitted of some pollutant (e.g., carbon dioxide) by all polluters in a country, region, or class.

GREENHOUSE GASES: Gases that cause Earth to retain more thermal energy by absorbing infrared light emitted by Earth's surface. The most important greenhouse gases are water vapor, carbon dioxide, methane, nitrous oxide, and various artificial chemicals such as chlorofluorocarbons. All but the latter are naturally occurring, but human activity over the last several centuries has significantly increased the amounts of carbon dioxide, methane, and nitrous oxide in Earth's atmosphere, causing global warming and global climate change.

KYOTO PROTOCOL: Extension in 1997 of the 1992 United Nations Framework Convention on Climate Change (UNFCCC), an international treaty signed by almost all countries with the goal of mitigating climate change. The United States, as of early 2008, was the only industrialized country to have not ratified the Kyoto Protocol, which is due to be replaced by an improved and updated agreement starting in 2012.

RENEWABLE ENERGY: Energy obtained from sources that are renewed at once, or fairly rapidly, by natural or managed processes that can be expected to continue indefinitely. Wind, sun, wood, crops, and waves can all be sources of renewable energy.

goal was to stabilize global warming at a low enough level to avoid dangerous climate change. Under the terms of the UNFCCC, the Kyoto Protocol is defined as a first step, not a finished product.

At the time of Kyoto, a historic division in climate policy occurred, with Europe affirming Kyoto and the United States—the world's largest national economy and at that time the world's largest emitter of greenhouse gases—refusing to do so. The United States declined to ratify Kyoto asserting that it unfairly allowed rapidly developing countries such as China to increase their emissions without restriction, which would put Kyoto-capped countries at an economic disadvantage. Over a decade later, the divide between U.S. and European climate policy continues, with the U.S. government continuing to oppose Kyoto's central principle of action, binding emissions caps.

In 2000, the European Commission set up the first European Climate Change Programme (ECCP), a consultative body tasked with pulling together expert knowledge on possible actions and policies relating to the mitigation of climate change (that is, reducing the severity of its consequences). EU climate change policy has been based on the recommendations of the ECCP since its formation.

In the first phase of the first ECCP (2000–2001), six expert working groups were set up to identify cost-effective measures to reduce greenhouse emissions that could be pursued at the EU level. Five were devoted to research, industry, transport, energy consumption, and energy supply. The sixth was devoted to the three market-based schemes for reducing greenhouse emissions specified by the Kyoto Protocol, namely: 1) the clean development mechanism, 2) emissions trading, and 3) joint implementation. All three of these flexible mechanisms, as they are called, involve the exchange of carbon credits under a legally defined cap or upper limit. A carbon credit is a symbolic tool, similar to a unit of money, that has a certain agreed-upon value, namely, a single ton of CO_2 or a quantity of some other greenhouse gas, such as methane (CH_4), that produces as much greenhouse warming as would a single ton of CO_2. The goal of all three of Kyoto's flexible mechanisms is to allow for reductions in emissions to be made where they can be purchased most cheaply, whether in industrialized countries or poorer parts of the world. Since there is only one atmosphere, it does not, in theory, matter where a ton of CO_2 is removed from the air, as long as it is removed.

The first ECCP produced recommendations for action by the EU in 2001. These recommendations were then weighed and acted upon by the European Commission and the European Parliament (which, together with the Council of the European Union, forms the legislative branch of the EU). The first ECCP was succeeded in October 2005 by the second ECCP, which formed a new set of working groups to examine various aspects of climate policy and produce further recommendations. As of November 2007, the second ECCP was still preparing its reports.

Impacts and Issues

In response to the recommendations of the first ECCP, the EU took several steps. The most significant was the establishment of the EU Emissions Trading Scheme, which began operation in January 2005. This is a cap-and-trade scheme in which carbon credits are allotted to individual greenhouse polluters by national governments. A given number of credits allows the polluter (say, a power plant) to emit an equal number of tons of CO_2 or an equivalent amount of some other gas. Polluters may buy and sell the credits or, under the Clean Development Mechanism, obtain new credits by funding greenhouse mitigation projects in developing countries. The system embraces only countries in the EU and those developing countries that agree to participate in the Clean Development Mechanism.

As of 2006, the system embraced about 11,500 European greenhouse polluters in the power, heat, and industrial sectors. In 2004, the European Commission proposed extending the Emissions Trading Scheme to include airlines operating out of EU airports. Implementation of an Aviation Emissions Trading Scheme was scheduled to begin in 2008.

Europe's climate policy has been criticized as ineffective. Although European emissions overall appeared on track as of late 2007 to meet the Kyoto goal of 8% reductions below 1990 levels by 2012, some countries had accomplished significant reductions while others had seen significant increases. For example, Germany's 2007 emissions were 19% below its 1990 levels, a decrease achieved despite brisk economic growth, but France's emissions grew 9% in the same period. (For comparison, U.S. emissions rose 16.3% from 1990 to 2005.) The European Emissions Trading Scheme has been accused of being fruitless because caps were set too high, allowing polluters to make little or no changes in practice while endowing them with a new, valuable commodity to trade, namely carbon credits.

In November 2006, responding to criticism that emissions quotas had been set too high to be meaningful, the European Commission announced that EU member states must decrease emission-credits quotas for their largest polluters. In March 2007, the 27 EU heads of state convened a climate summit and decided to commit the EU to mandatory reduction of its overall greenhouse-gas emissions of 20–30% below 1990 levels by 2020. Talks for the successor treaty to Kyoto commenced in late 2007. It seemed likely that European climate policy's emphasis on binding emissions caps and direct carbon taxes (in some countries) would remain unacceptable to U.S. business interests and U.S. politicians.

Primary Source Connection

This article discusses efforts by the European Union (EU) to reduce greenhouse gas emissions in Europe. The EU's European Climate Change Programme (ECCP) has been instrumental in devising measures and standards for EU member states. ECCP's efforts, along with additional measures taken by individual member states, have reduced overall greenhouse gas emissions in Europe over the last several years.

EUROPEAN CLIMATE CHANGE PROGRAMME: EU ACTION AGAINST CLIMATE CHANGE

EU policies on climate change

The European Union has long been committed to international efforts to tackle climate change and felt the duty to set an example through robust policy-making at home. At European level a comprehensive package of policy measures to reduce greenhouse gas emissions has been initiated through the European Climate Change Programme (ECCP). Each of the 25 EU Member States has also put in place its own domestic actions that build on the ECCP measures or complement them.

The First European Climate Change Programme (2000–2004)

The European Commission established the ECCP in 2000 to help identify the most environmentally effective and most cost-effective policies and measures that can be taken at European level to cut greenhouse gas emissions. The immediate goal is to help ensure that the EU meets its target for reducing emissions under the Kyoto Protocol. This requires the 15 countries that were EU members before 2004 to cut their combined emissions of greenhouse gases to 8% below the 1990 level by 2012.

The ECCP builds on existing emissions-related activities at EU level, for instance in the field of renewable energy and energy demand management. It also dovetails with the EU's Sixth Environmental Action Programme (2002–2012), which forms the strategic framework for EU environmental action and includes climate change among its four top priorities, as well as the EU's Sustainable Development Strategy.

Among various protests staged in Europe was this one outside of the European Union (EU) summit in Brussels, Belgium. Here, people are urged to cut energy waste by choosing renewables. *AP Images.*

The ECCP is a multi-stakeholder consultative process that has brought together all relevant players, such as the Commission, national experts, industry and the NGO community. Stakeholder involvement is an essential element of the ECCP because it enables the programme to draw on a broad spectrum of expertise and helps to build consensus, thereby facilitating the implementation of the resulting policies and measures.

The first ECCP examined an extensive range of policy sectors and instruments with potential for reducing greenhouse gas emissions. Coordinated by an ECCP Steering Committee, 11 working groups were established covering the following areas:

- Flexible mechanisms: emissions trading
- Flexible mechanisms: Joint Implementation and Clean Development Mechanism
- Energy supply
- Energy demand
- Energy efficiency in end-use equipment and industrial processes
- Transport
- Industry (sub-groups were established on fluorinated gases, renewable raw materials and voluntary agreements)
- Research
- Agriculture
- Sinks in agricultural soils
- Forest-related sinks

Each working group identified options and potential for reducing emissions based on cost-effectiveness. The impact on other policy areas was also taken into account, including ancillary benefits, for instance in terms of energy security and air quality. . . .

Progress towards meeting the EU's Kyoto commitment

The EU's greenhouse gas emissions have been falling thanks to the combined impact of policies and measures resulting from the first ECCP, domestic action taken by Member States and the restructuring of European industry, particularly in central and eastern Europe.

By 2003, combined emissions from today's 25 Member States (EU-25) were down 8% compared to their levels in the respective base years (mostly 1990). Emissions from the 15 'old' Member States (EU–15) had fallen by 1.7%, or 2.9% averaged over 1999–2003.

Latest projections show that additional domestic policies and measures planned by Member States but not yet implemented will take the reduction in EU–15 emissions to 6.8% below 1990 levels by 2010. In addition, use of the Kyoto Protocol's flexible mechanisms Joint Implementation and the Clean Development Mechanism will reduce emissions by a further 2.5%, taking EU–15 emissions in 2010 to 9.3% below 1990 levels—more than enough to meet the EU–15's reduction commitment of 8%.

For the EU-25 latest projections indicate that the implementation of additional policies and measures planned by Member States will reduce emissions to 9.3% below base year levels by 2010. Use of the Kyoto mechanisms will achieve an additional 2% reduction.

Future perspectives

The EU is convinced that strong global action to combat climate change will continue to be needed after 2012, when the Kyoto Protocol's targets are due to have been met. It therefore favours an early start to negotiations on an international climate regime for the post–2012 period.

In a policy document published in early 2005 the Commission outlined the key elements that a post–2012 regime will need to incorporate. The Commission wants to see participation by all major emitters and economic sectors, greater innovation in emissions-saving technologies, continued use of cost-effective market-based instruments and the implementation of strategies for adapting to the level of climate change that is already unavoidable.

Many of the EU policies and measures now in place will be important for cutting greenhouse gas emissions beyond 2012. It is already foreseen, for example, that the EU Emissions Trading Scheme will continue in five-year periods after 2012.

But it is also clear that deeper emissions cuts will be needed after 2012 if the international community is to win the battle against climate change, and further EU policies and measures will be required to achieve these. Consequently the Commission has initiated the Second European Climate Change Programme (ECCP II).

The Second European Climate Change Programme (2005–)

ECCP II was launched in October 2005 at a major stakeholder conference in Brussels. It will explore further cost-effective options for reducing greenhouse gas emissions in synergy with the EU's 'Lisbon strategy' for increasing economic growth and job creation.

New working groups have been established, covering carbon capture and geological storage, CO_2 emissions from light-duty vehicles, emissions from aviation, and adaptation to the effects of climate change.

The aviation group will focus on the technical aspects of bringing aircraft emissions into the Emissions Trading Scheme, which the Commission considers the most promising way to tackle the rapid growth in emissions from this sector.

One working group will also assess the implementation of the ECCP I policies and measures in the Member States and their effects in terms of emission reductions. This will feed into a broader ECCP I review process and give guidance to the Commission and the Member States on any supplementary efforts that may be needed to meet the EU's Kyoto commitment.

EUROPEAN CLIMATE CHANGE PROGRAMME: EU ACTION AGAINST CLIMATE CHANGE. EUROPEAN COMMUNITIES, 2006.

SEE ALSO *Carbon Credits; Clean Development Mechanism; Emissions Trading; Kyoto Protocol; United Nations Environment Programme (UNEP); United Nations Framework Convention on Climate Change (UNFCCC); United States: Climate Policy.*

BIBLIOGRAPHY

Books

Harris, Paul G. *Europe and Global Climate Change: Politics, Foreign Policy and Regional Cooperation.* Northampton, MA: Edward Elgar Publishing, 2007.

Periodicals

Biermann, Frank. "Between the USA and the South: Strategic Choices for European Climate Policy." *Climate Policy* 5 (2005): 273–290.

Hepburn, Cameron. "Carbon Trading: A Review of the Kyoto Mechanisms." *Annual Review of Environment and Resources* 32 (2007): 375–393.

Schiermeier, Quirin. "Europe Moves to Secure Its Future Energy Supply." *Nature* 445 (January 18, 2007): 234–235.

Web Sites

Dohmen, Frank, Alexander Neubacher, Sebastian Knauer, and Wolfgang Reuter. "A Europe Divided Over Climate Policy." *Der Spiegel* (Germany), February 17, 2007. <http://www.spiegel.de/international/spiegel/0,1518,467367,00.html> (accessed November 20, 2007).

"European Climate Change Programme: EU Action Against Climate Change." *European Communities*, 2006. <http://ec.europa.eu/environment/climat/pdf/eu_climate_change progr.pdf> (accessed November 20, 2007).

Larry Gilman

Extinction

Introduction

Extinction is the complete disappearance of a species of plant or animal. Extinction can be natural; over 99% of all species that have existed in the history of Earth are now extinct. When many species become extinct at nearly the same time, the event is called a mass extinction. Earth is now experiencing a mass extinction caused by human beings, perhaps the first mass extinction ever caused by a single species rather than by natural forces like meteor impacts, volcanoes, or continental drift. After all past mass extinctions, biodiversity (number of species) recovered as new species evolved to fill the ecological niches left vacant by the dead species; however, this took millions of years. On a human time-scale, that is, over hundreds or even tens of thousands of years, extinction causes irreversible loss of biodiversity.

Climate change is only one means by which human beings are causing extinction, and is not yet the dominant one. Destruction of habitat, such as the cutting down of rainforests, has been the means of most human-caused extinctions to date. However, as global average temperatures increase by 3.6–5.4°F (2–3°C) above pre-industrial levels, which may well occur in this century, some 20 to 30% of all known species are likely to be put at higher risk of extinction. In some regions, the extinction-risk rate will be as low as 1%; in others, as high as 80%.

Historical Background and Scientific Foundations

Without extinction, human beings would not exist, as humans likely would not have evolved unless the dinosaurs had died. Even mass extinctions—the sudden or relatively sudden extinction of many species—have happened repeatedly. For example, 251 million years ago, the Permian-Triassic extinction event killed about 96% of all marine (sea-living) species and about 70% of all land-dwelling species. About 65 million years ago, the Cretaceous-Tertiary extinction event killed about half of all species, including the dinosaurs, which created the opportunity for larger mammals and, eventually, humans to evolve.

Today, most biologists accept that humans are in the midst of the first mass extinction since the Cretaceous-Tertiary event. This event is termed the Holocene extinction. (The Holocene is the most recent geographic epoch, stretching from about 10,000 years ago to the present.) This new wave of extinctions has been caused by human activities including hunting, agriculture, pollution, and habitat destruction. A century ago, some scientists advanced the theory that the mass extinction of large mammals in North America (giant ground sloths, giant bears, mammoths, camels, and dozens of other creatures) occurring about 10,500 years ago was caused by hunting done by early Americans. This view, called the overkill hypothesis, is supported by the fact that the extinction of animals on islands often occurs after colonization by humans. However, the hypothesis has been recently challenged by scientists who point to climate change at the end of the Ice Age as a more likely cause of the extinctions.

Ten thousand years ago, all climate changes were caused by natural forces. In the last several decades, however, climate change has been supplemented by anthropogenic (human-caused) greenhouse gases. Although anthropogenic global climate change has only recently become large enough to distinguish definitely from natural changes, it is happening rapidly. In the next century, it will probably become a major cause of extinctions. The United Nations' Intergovernmental Panel on Climate Change (IPCC) stated in 2007 that about 20–30% of all plant and animal species are likely to be at increasingly high extinction risk as the average global temperature warms to 3.6–5.4°F (2–3°C) above the level it held before the Industrial Revolution, when human beings first began to burn large amounts of fossil fuels. This much warming by 2100 is likely even for

WORDS TO KNOW

BIODIVERSITY: Literally, "life diversity": the number of different kinds of living things. The wide range of organisms—plants and animals—that exist within any given geographical region.

CONTINENTAL DRIFT: A theory that explains the relative positions and shapes of the continents, and other geologic phenomena, by lateral movement of the continents. This was the precursor to plate tectonic theory.

EPOCH: Unit of geological time. From longest to shortest, the geological system of time units is eon, era, period, epoch, and stage. Epochs are generally about 500 million years in length.

HOLOCENE EXTINCTION EVENT: The Holocene is the geological period from 10,000 years ago to the present; the Holocene extinction is the worldwide mass extinction of animal and plant species being caused by human activity. Global warming may accelerate the ongoing Holocene extinction event, possibly driving a fourth of all terrestrial plant and animal species to extinction.

ICE AGE: Period of glacial advance.

INDUSTRIAL REVOLUTION: The period, beginning about the middle of the eighteenth century, during which humans began to use steam engines as a major source of power.

SPECIES-AREA RELATIONSHIP: In biology, the relationship between the size of an area and the number of species of plants and animals that can live in that area: the smaller the area, the fewer the species. Expressed algebraically as $S = cA^z$, where S stands for number of species, A for area, and c and z for fixed numbers (constants) that depend on the ecological setting.

conservative scenarios for global warming, in the absence of effective actions to reduce greenhouse-gas emissions and other causes of global climate change; warming by up to 8.1°F (4.5°C) by 2100 is possible.

A 2004 study by Chris D. Thomas and colleagues calculated that, out of a sample of 1,103 animal and plant species, 15–37% would be committed to extinction by 2050 as a result of climate change—that is, that the number of living individuals of each species would be so small that the species' extinction some time after 2050 would become highly likely. Thomas and his colleagues estimated that in many if not most ecological regions, climate change would become the greatest threat to biodiversity by 2050.

One tool that scientists use to estimate extinction risk from climate change is the climate envelope. A species' climate envelope is the range of climate conditions under which populations of that species tend to survive in spite of natural enemies and competitors. As climate changes, the areas where a species' climate envelope exists will shift or, in some cases, shrink. For example, as climate warms, the highest altitude at which trees can grow on mountains, the treeline, will ascend. As there is a limited amount of land on a mountain range, as the treeline ascends, the area above the treeline will shrink. Species that depend on the mountain-tundra environment above the treeline will therefore, have less room in which to live; the area where their climate envelope exists will decrease.

A well-known relationship in biology exists between area and the number of species of plants and animals that can live in that area: the smaller the area, the fewer the species. This is often expressed as a mathematical equation called the species-area relationship. Using S to stand for the number of species, A for area, and c and z for some fixed numbers or constants that depend on the ecological setting, the species-area relationship is $S = cA^z$. For example, if $c = 1$ and $z = 2$, then $S = A^2$. Halving area, A, will in this case reduce the number of species, S, to one-fourth its previous value. This type of rule is usually a good predictor of how many species will go extinct when habitat is lost.

The rule does not work in reverse—that is, increasing area does not bring more biodiversity (more species) into being. Extinction can happen overnight, but the evolution of new species is usually, on the timescale of habitat destruction, too slow to detect.

By sampling species habitats around the world, using projections of likely climate change to estimate how much various habitats will shrink or grow, and using the species-area relationship to predict about how many species will become extinct when habitats shrink, scientists can make an educated guess about how many extinctions are likely to be caused by a given amount of global warming.

■ Impacts and Issues

It is usually estimated that there are 5 to 15 million species of creatures on Earth. If 20% of these are committed to extinction by climate change, then from 1 million to 3 million species will eventually become extinct as a result of global climate change. (Most of these would be species of insects.)

However, there are many uncertainties in guessing how many extinctions will occur because of anthropogenic climate change. First, there is no certainty about how much climate change is going to occur. This depends on natural feedbacks, tipping points, and human choices. Second, the range of uncertainty around climate-change extinction estimates, both in timing and quantity, is large. The species-area relationship is only an approximation (other methods are also used), and it is not known how many ecosystems will respond to climate change.

The 2004 study by Thomas and colleagues mentioned earlier was widely misreported in the media as saying that a million species would be extinct within 50 years. This was not, however, what the scientists said: they emphasized the uncertainties in their predictions and said that many species would be "committed to extinction," not extinct, by 2050. They also said that reducing greenhouse-gas emissions would change their projections, and that reducing global temperatures would eventually bring back some species from the edge of extinction. The study's basic message—that anthropogenic global climate change may very well threaten thousands or millions of species with extinction within the next century or more—was viewed as sufficiently alarming, even without exaggeration.

Climate-related extinctions have already begun. A 2006 study found that global warming almost certainly played a key role in the recent extinction of about 67% of the 110 or so species of the Monteverde harlequin tree frog of the mountains of Costa Rica, making the frogs more vulnerable to a deadly fungal infection.

IN CONTEXT: IMPACTS ON ANIMAL POPULATIONS

According to the National Academy of Sciences: "Even within a single regional ecosystem, there will be winners and losers. For example, the population of Adélie penguins has decreased 22% during the last 25 years, while the Chinstrap penguin population increased by 400%. The two species depend on different habitats for survival: Adélies inhabit the winter ice pack, whereas Chinstraps remain in close association with open water. A 7-9° F rise in midwinter temperatures on the western Antarctic Peninsula during the past 50 years and associated receding sea-ice pack is reflected in their changing populations."

SOURCE: *Staudt, Amanda, Nancy Huddleston, and Sandi Rudenstein.* Understanding and Responding to Climate Change. *National Academy of Sciences, 2006.*

Primary Source Connection

Scientist are in the midst of an argument over the cause of the mass extinction of large mammals in North America over 10,000 years ago. Since the 1960s the overkill theory has been the prevailing theory on this mass extinction. The overkill theory holds that the arrival of humans in North America about 11,000 years ago led to the extinction of these mammals through hunting. A more recent theory, though, suggests that climate change was responsible for the mass extinction. This article from *National Geographic News* presents views from proponents of both theories.

CLIMATE CHANGE CAUSED EXTINCTION OF BIG ICE AGE MAMMALS, SCIENTIST SAYS

A renewed assault is being made on the popular idea that the mass extinction of large mammals in North America around 10,500 years ago was the result of human hunting.

The overkill hypothesis was first put forward more than a century ago and has been widely accepted for the past 30 years. But it does not square with the known facts and has become more a faith-based credo than good science, said Donald Grayson, an archaeologist at the University of Washington.

Understanding what caused the extinction has implications for conservation biology.

Grayson, who specializes in vertebrate paleontology and archaeology, argues that a call by some environmentalists to return Ice Age mammals—elephants, camels, llamas, and other large herbivores—to the southwestern United States is based on bad science.

"Overkill proponents have argued that these animals would still be around if people hadn't killed them and that ecological niches still exist for them," said Grayson. "Those niches do not exist. Otherwise the herbivores would still be there."

Grayson points to climate shifts during the late Pleistocene and related changes in weather and vegetation patterns as the likely culprits in the demise of North America's megafauna.

Islands and Continents

The number of large mammals that became extinct in North America at the end of the late Pleistocene, about 10,500 years ago, is staggering. Among some 35 different kinds of animals that disappeared from the fossil record were mammoths, mastodons, camels, horses, giant ground sloths, and bears.

A leading proponent of the overkill theory, Paul S. Martin, believes the Ice Age megafauna disappeared not because they lost their food supply but because of human hunting.

The extinction of animals as a result of human colonization in island settings has been well documented and the causes widely agreed on. New Zealand is offered as a classic example of human impacts on island animal populations. Before it was colonized, New Zealand was home to 11 species of moas, a large flightless bird species weighing from 45 to more than 400 pounds (20 to 200 kilograms). Within a few hundred years after human settlement, moas were extinct.

The wave of extinctions that followed human colonization of New Zealand occurred as a result of several factors. Humans arriving on the islands brought with them rats, dogs, and other non-indigenous animals that competed for food or preyed on native species. Humans

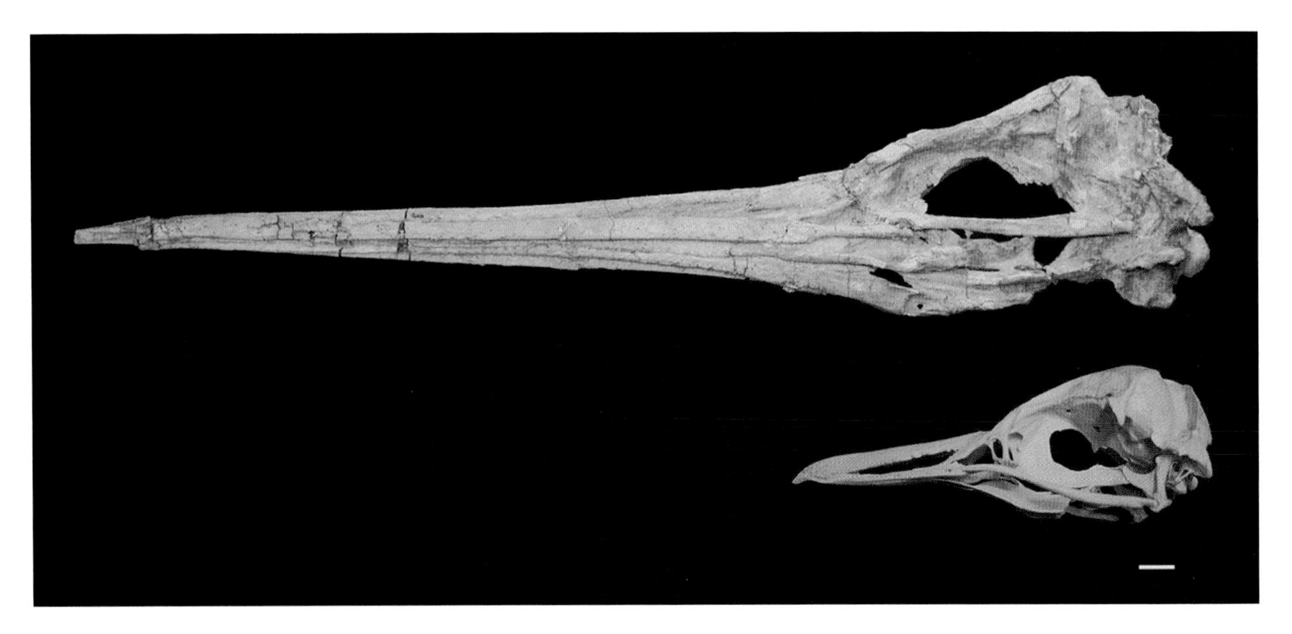

In Peru, two recently discovered fossils reveal that early penguins responded differently to natural climate change than scientists once thought. The smaller fossil shows a penguin that was comparable in size to today's king penguin. The larger fossil shows a penguin that was over 5 ft (1.5 m) tall with a 7 in (18 cm) beak. *AP Images.*

hunted the native animals extensively and destroyed much habitat by burning down forests.

A similar pattern of events occurred on other islands around the world. Proponents of the overkill theory argue that the same process—human colonization followed by massive extinction—also occurred in North America.

But Grayson rejects that idea. "The fossil record simply doesn't stand up to the theory, and comparing continents to islands is simply inappropriate," he said.

Arguing Against Overkill

The overkill hypothesis, Grayson says, rests on five tenets: human colonization can lead to the extinction of island species; the Clovis people were the first humans to arrive in North America, around 11,000 years ago; the Clovis people hunted a wide range of large mammals; the extinction of many species of North American megafauna occurred 11,000 years ago; and therefore, Clovis hunting caused those extinctions.

Grayson disputes several of these tenets.

There is no proof, he said, that the late Pleistocene extinctions occurred in conjunction with the arrival of the Clovis people. "Of the 35 genera to have become extinct beginning around 20,000 years ago, only 15 can be shown to have survived beyond 12,000 years ago," Grayson said. "The Clovis peoples didn't arrive until shortly before 11,000 years ago. That leaves 20 'genera' unaccounted for."

There is also no evidence that the Clovis people hunted anything other than mammoths, he said. Although numerous sites where large numbers of mammoths were killed have been uncovered, no similar sites for any other large mammals have been found in North America.

And while there is no evidence of widespread human-caused environmental change similar to that seen on island settings, there is evidence that animal populations in Siberia and Western Europe, as well as North America, were affected during the same period by climate changes and glacial retreat.

Martin, who in 1967 wrote the seminal book proposing the overkill hypothesis, disagrees that climate change could have caused the extensive extinctions.

"The climate had been changing over a million-year period, with swings from cold to warm, and back again—some much more severe than the one that occurred at the end of the Pleistocene," he said. "It doesn't make sense that just one more [climate shift swing] would make all the difference in the world."

Martin holds that the "dreadful syncopation"—humans arrive, animals disappear—seen in the islands of Oceania, Australia, New Zealand, Madagascar, and other islands fits with what happened in North America.

He suggests that because humans were responsible for the demise of large animals in the desert southwest, these animals should be reintroduced.

Grayson thinks this is dangerous thinking.

"One of the reasons people have glommed on to the overkill hypothesis is 'green' politics," said Grayson. "It plays to the Judeo-Christian theme that human beings are all-powerful and responsible for negative impacts on the environment."

"The hypothesis made a lot of sense 30 years ago," Grayson said, "but now it can be compared to the empirical record, and it just doesn't hold up."

His work appears in the current issue of the *Journal of World Prehistory* and an upcoming issue of the *Bulletin of the Florida Museum of Natural History.*

MAYELL, HILLARY. "CLIMATE CHANGE CAUSED EXTINCTION OF BIG ICE AGE MAMMALS, SCIENTIST SAYS." *NATIONAL GEOGRAPHIC NEWS*, NOVEMBER 12, 2001.

SEE ALSO *Biodiversity; Polar Bears.*

BIBLIOGRAPHY

Books

Parry, M. L., et al, eds. *Climate Change 2007: Impacts, Adaptation and Vulnerability: Contribution of Working Group II to the Fourth Assessment Report of the Intergovernmental Panel on Climate Change.* New York: Cambridge University Press, 2007.

Periodicals

Eilperin, Juliet. "188 More Species Listed as Near Extinction." *The Washington Post* (September 13, 2007).

Higgins, Paul A. T. "Biodiversity Loss Under Existing Land Use and Climate Change: An Illustration Using Northern South America." *Global Ecology and Biogeography* 16 (March 2007): 197–204.

Pounds, J. Alan, et al. "Widespread Amphibian Extinctions from Epidemic Disease Driven by Global Warming." *Nature* 439 (January 12, 2006): 161–167.

Thomas, Chris D. "Extinction Risk from Climate Change." *Nature* 427 (January 8, 2004): 145–148.

Web Sites

Ladle, Richard J., et al. "Crying Wolf on Climate Change and Extinction." *Oxford University, School of Geography*, 2004. <http://www.geog.ox.ac.uk/research/biodiversity/crywolf.pdf > (accessed October 10, 2007).

Mayell, Hillary. "Climate Change Caused Extinction of Big Ice Age Mammals, Scientist Says." *National Geographic News*, November 12, 2001. <http://news.nationalgeographic.com/news/2001/11/1112_overkill.html> (accessed October 10, 2007).

Larry Gilman

Extraterrestrial Climate Change

Introduction

Venus, Earth, and Mars are the second, third, and fourth planets in the solar system counting outward from the sun. Of all the planets in the solar system, Mars and Venus are most similar to Earth. Venus is almost exactly the same size and has a thick, cloudy atmosphere. Mars, although having a mass only about one-tenth that of Earth, has an atmosphere, clouds, frosts, and dust storms, with polar caps of frozen carbon dioxide and water and large amounts of subsurface permafrost. In fact, since both planets have atmospheres, they have climates, and since they have climates, they can experience climate change. Study of Martian and Venusian climate in the 1960s and 1970s helped spark scientific thought about climate change on Earth, and the validity of mathematical principles used to model climate change on Earth is sometimes tested today against observations of those planets.

Historical Background and Scientific Foundations

Since the nineteenth century, many writers, including some astronomers, had fantasized that Mars might have a climate that was extremely similar to Earth's, with intelligent beings building canals across its surface to channel water from the ice caps to the supposedly thirsty tropical zones. Gradually such ideas were corrected in the first half of the twentieth century when better telescopes were built.

In the 1940s, long before it was yet possible to launch robot space probes to the planets, scientists were speculating that data about atmospheres elsewhere in the solar system might improve the understanding of Earth's atmosphere. In 1948, the head of the U.S. Weather Bureau, Harry Wexler (1911–1962), suggested the formation of a "Project on Planetary Atmospheres." However, the idea was premature, since earthbound telescopes could not tell enough about what was happening on Venus and Mars to yield interesting comparisons between their atmospheres and Earth's.

By the mid-1950s, scientists knew from telescopic observations that the Martian atmosphere was almost entirely carbon dioxide (CO_2). They also knew that the atmosphere was very thin compared to Earth's, with a surface pressure of less than one one-hundredth of the pressure at terrestrial sea level. Mariner 4, which in 1965 became the first space probe to reach Mars, radioed back pictures of craters that looked more like the surface of the moon than of an Earth-like world.

In the 1970s, 1990s, and 2000s, many probes successfully orbited and landed on Mars, making on-site measurements of its weather and climate. Proof of the prior existence of large amounts of liquid water on the surface of Mars was discovered by NASA's twin Mars Exploration Rovers, which landed on the planet in late 2003 and early 2004. Since liquid water does not appear on the Martian surface today, Mars had clearly undergone massive climate change at some time in its deep past, several billion years ago.

Meanwhile, knowledge of Venus also was growing. Early fantasies had concluded that it might be an Earth-like jungle world covered by a perpetual deck of clouds, but this turned out to be very far from the truth. From telescopic observations, scientists already knew in the 1940s that Venus had an atmosphere consisting largely of CO_2. In the late 1950s, scientists learned that the surface of Venus was astonishingly hot—about 860°F (460°C)—by bouncing radio waves off the planet's surface. The Venusian surface temperature is hotter than any household oven and hot enough to melt lead (which melts at 621°F or 327°C). Soviet and American space probes visited the planet in the 1960s and 1970s, verifying this fact. In addition, the U.S. Magellan probe orbited Venus from 1990 to 1994, making a detailed radar map of its surface and gathering data on the Venusian atmosphere as well. The European Space Agency's

Scientists also study climate change on other planets. This image from the Hubble Space Telescope on May 4, 2006, shows a second red spot emerging on Jupiter. This was the first time that astronomers witnessed the birth of a new red spot on the planet. The storm is roughly one-half the diameter of its bigger and legendary cousin, the Great Red Spot. Scientists suggest that the new spot may be related to a possible major climate change in Jupiter's atmosphere. *AP Images.*

Venus Express probe, devoted mostly to observing the Venusian atmosphere, began orbiting Venus in April 2006 and was still functioning as of late 2007.

Scientific knowledge of the two planets' climates is now extensive. Their past history of climate change, or possible ongoing climate changes, is a subject that continues to be investigated. Scientists presently believe that both planets once had climates that were more similar to Earth's, but underwent radical climate changes that made one into a burning desert and the other into a frozen desert.

Mars

Of all the bodies and moons in the solar system, Mars has the climate that is least unlike Earth's. Its day, by coincidence, is almost exactly the length of a terrestrial day (24 hours). The tilt of the Martian axis with respect to the plane of the ecliptic (the imaginary flat plane in which all the orbits of the major planets lie, like rings on a disk), is 25°, similar to Earth's tilt of about 23°. The northern and southern hemispheres experience alternating winter and summer seasons, as do those of Earth, with increased precipitation in the winter hemisphere and increased evaporation in the summer hemisphere.

But the dissimilarities are also great. Mars's year is 1.8 times longer than Earth's; its atmosphere is about one-hundredth as thick; and its surface is bone-dry compared even to the driest Earthly conditions. Snow and rain never fall on its surface, although frosts form and clouds sometimes form. It is also far colder, with minimum temperatures of -225°F (-143°C) and an average surface temperature of -69°F (-56°C). Mars's orbit is not as neatly centered on the sun as Earth's, and its orbital shifts over thousands of years produce cyclic climate changes.

Although Mars is blanketed with a greenhouse gas, CO_2 (with traces of another, methane), it is cold because

This computer rendition shows the area that is part of the carbon dioxide ice cap at the Martian South Pole. The large pits are where the carbon dioxide ice cap has sublimed to reveal ice layers below the surface. Much colder than Earth, Mars has a thin atmosphere of carbon dioxide. Observations by the Mars Global Surveyor have shown that this ice cap is gradually shrinking as these pits are enlarged each summer, which scientists suspect may be evidence that the Martian climate is slowly warming. *Chris Butler/Photo Researchers, Inc.*

it is farther from the sun and its atmosphere is so much thinner than Earth's. Once, several billion years ago, Mars had a much thicker, warmer, and moister atmosphere. Oceans briefly washed its surface. The gases of its atmosphere were supplied by large volcanoes. However, because of its low gravity, once the volcanoes became extinct, the atmosphere began to leak away to space. The water froze and Mars became the planet we see it today.

Venus

The climate of Venus is dominated by its thick atmosphere, which has a surface pressure 92 times that of Earth and is composed mostly of CO_2. This CO_2, along with deep cloud layers of sulfur-compound particles, creates an intense greenhouse effect. In 1940, German-American astronomer Rupert Wildt (1905–1976) performed calculations showing that Venus's carbon-dioxide atmosphere must create a hot surface, about 260°F (126°C).

In 1960, American astronomer Carl Sagan (1934–1996) studied the problem and concluded that Venus was at least as hot as Wildt had calculated. He also speculated that Venus had once had a surface ocean like Earth and Mars, but had suffered a runaway greenhouse effect, where CO_2 caused heating, which evaporated the water of the oceans, which enhanced the greenhouse effect. Although Sagan was mistaken about water in the atmosphere of Venus—it is now known that its clouds are particles of sulfur compounds, not water particles like Earthly clouds—his idea that Venus once had an Earth-like climate, early in its history, is now commonly accepted.

In 1979, astrophysicist Michael Hart (1932–) proposed that Earth is luckily located in a very narrow orbital zone between too much energy from the sun and too little. If so, it might be easily tipped over into a Venus-like or Mars-like condition by human-caused global climate change. Scientists have since confirmed that Earth's climate is not as precarious as Hart thought. It is not plausible that Earth will come to resemble either of its neighbors. Rather, Earth is vulnerable to human-caused climate change in a narrower range—enough to cause problems, perhaps severe ones, but far from the sort of change that would threaten to sterilize the planet.

Recent research has suggested that the greenhouse effect on Venus is very intense. When massive volcanoes spew large quantities of greenhouse gases into the Venusian atmosphere, these additional gases may make the greenhouse effect even hotter for millions or hundreds of millions of years. This extra heat can then work its way down into the planet's crust from the atmosphere, causing stresses that deform and wrinkle the crust. These wrinkles have been mapped from orbit by spacecraft using radar to see through the thick haze layer that completely enshrouds the planet.

Impacts and Issues

Efforts to understand the climate on Mars and Venus aid efforts to understand Earth's climate. David Grinspoon, a scientist with the Denver Museum of Nature and Science and a member of the Venus Express team, has urged scientists to test their mathematical models of Earth's climate by applying them to the simpler systems of Venus and Mars. According to Grinspoon in 2007, the planets's systems contain invaluable information for the study of Earth's climate.

In the early 2000s, it was reported that Mars had begun undergoing global warming. Some skeptics or doubters of the reality of anthropogenic (human caused) climate change claimed that there must be some common, non-human cause for simultaneous global warming on Earth and Mars—perhaps increased heat from the sun. However, measurements made by spacecraft orbiting at Mars show that slightly less, not more, solar energy is reaching Mars. The real reason for Mars's recent warming is its shifting orbit around the sun, which causes repetitive or cyclic changes of climate called Milankovitch cycles. Earth also experiences Milankovitch cycles, but these well-understood changes are not the cause of recent global climate change on Earth.

WORDS TO KNOW

ANTHROPOGENIC: Made by people or resulting from human activities. Usually used in the context of emissions that are produced as a result of human activities.

METHANE: A compound of one hydrogen atom combined with four hydrogen atoms, formula CH_4. It is the simplest hydrocarbon compound. Methane is a burnable gas that is found as a fossil fuel (in natural gas) and is given off by rotting excrement.

MILANKOVITCH CYCLES: Regularly repeating variations in Earth's climate caused by shifts in its orbit around the sun and its orientation (i.e., tilt) with respect to the sun. Named after Serbian scientist Milutin Milankovitch (1879–1958), though he was not the first to propose such cycles.

Primary Source Connection

Although climate change is often thought of only in the context of Earth, climate change can occur on other planets, often with severe impacts. On Venus, climate change influences, and is influenced by, the planet's volcanic and tectonic processes. This article from the journal *Science* describes the process by which large volcanic eruptions cause climate change by releasing greenhouse gases, primarily water vapor (H_2O) and sulfur dioxide (SO_2), into the atmosphere. The atmospheric warming caused by these volcanic events then increases surface temperatures to such a degree that it ultimately leads to tectonic deformations.

CLIMATE CHANGE AS A REGULATOR OF TECTONICS ON VENUS

Tectonics, volcanism, and climate on Venus may be strongly coupled. Large excursions in surface temperature predicted to follow a global or near-global volcanic event diffuse into the interior and introduce thermal stresses of a magnitude sufficient to influence widespread tectonic deformation. This sequence of events accounts for the timing and many of the characteristics of deformation in the ridged plains of Venus, the most widely preserved volcanic terrain on the planet.

Venus has had a volcanic and tectonic history differing in important respects from that of Earth. The distribution and states of preservation of impact craters indicate that much of the surface is indistinguishable in age from a mean value of 300 to 700 million years (My). Venus lacks evidence for global plate tectonics, but widespread volcanic plains record globally coherent episodes of deformation that appear to have occurred over short intervals of geological history. These characteristics have been difficult to reconcile with interior dynamical models for the thermal and mechanical evolution of the planet.

It has recently been recognized that global or near-global volcanic events, such as those called upon to account for the history of crater preservation and plains emplacement, can have a significant influence on the climate of Venus. In particular, the injection into the atmosphere from erupting lavas of such volatile species as H_2O and SO_2 can lead to large excursions of surface temperature over time scales ranging from millions to hundreds of millions of years. Here, we explore the implications of these temperature excursions for the state of stress in the Venus interior. In particular, we examine possible coupling between the temporal variations in climate and the nearly global tectonic deformation that would have followed the largest distinct episode of widespread volcanism known to have occurred on the planet.

The most abundant geological terrain on the surface of Venus consists of ridged plains, volcanic plains material deformed by wrinkle ridges subsequent to emplacement. Ridged plains are concentrated in areas of low elevation and make up 60 to 65% of the present surface. Because they are overlain by younger plains units in some areas, the ridged plains must have occupied a still larger fraction of the surface area of Venus at the time of their emplacement. On the basis of stratigraphic relations, impact crater densities, and the small fraction of impact craters embayed by ridged plains lavas, these plains appear to have been emplaced over a narrow interval of geological time, at most a few percent of the average crater retention age for the surface or a few tens of millions of years. Widespread formation of wrinkle ridges evidently occurred shortly after ridged plains emplacement, on the grounds that few impact craters on these plains have been deformed by the wrinkle ridges. An interval no more than 100 My between ridged plains emplacement and the formation of most wrinkle ridges is implied.

Wrinkle ridges, by analogy with similar structures on the other terrestrial planets, are inferred to be the products of horizontal shortening of the lithosphere, the mechanically strong outer layer of the planet. The consistent orientation of most wrinkle ridges over areas thousands of kilometers in extent and the strong tendency for the ridges to encircle broad topographic and geoid highs implies that much of ridge formation was a response to variations in the lithospheric stress field over comparable spatial scales. Although models for gravitational stresses that arise from long-wavelength variations in topography and lithospheric density can match the distribution of many wrinkle ridges, such models do not account for a limited time interval for most ridge formation. Some researchers have suggested that wrinkle ridge formation was but one of a series of widespread episodes in the geological history of Venus during which a single tectonic style dominated deformation on a global scale, although others have questioned the basis for global synchroneity of regional deformational episodes.

The volume of lavas that formed the ridged plains can be estimated from the exposed surface area and estimates of typical thicknesses of such plains material. The exposed surface area of ridged plains is about 3×10^8 km^2. On the basis of crater embayment relations, statistics on volcanic shield burial, and relations between wrinkle ridge width or spacing and plains thickness, the fraction of ridged plains material exceeding 500 m in thickness has been estimated to be 20 to 40%. Given that the greatest thicknesses of plains material may be as large as 2 to 4 km if such plains bury surfaces as rough as the oldest terrain on the planet, that additional plains units may have been nearly contemporaneous with the ridged plains, and that the surface area of ridged plains after emplacement was greater than the presently observed area, the total volume of lavas that erupted to form the ridged plains was probably at least 1 to 2×10^8 km^3. This volume exceeds by as much as an order of magnitude that of even the largest of the major igneous provinces on Earth. The emplacement of ridged plains on Venus represents the largest distinct volcanic episode in the preserved geological history of Venus, as measured either by total volume or by volcanic flux.

The widespread emplacement of the ridged plains should have released large quantities of water and sulfur gases into the Venus atmosphere. These gases affect both the atmospheric greenhouse and the albedo and opacity of the global cloud cover. Perturbations to the atmospheric SO_2 content are modulated by surface-atmosphere chemical reactions, which are limited by kinetics and diffusion into the uppermost crust. Photo dissociation and upper atmospheric hydrogen loss affect the atmospheric H_2O inventory. Climate evolution models incorporating all of these processes predict significant excursions of surface temperature after a large volcanic event. . . .

There is reason to expect a strong coupling between the evolution of climate on Venus, the history of large volcanic eruptions, and the state of stress and large-scale deformation of the surface. Climate-induced changes in the stress field act on a planetary scale, so synchroneity of widespread deformational events is to be expected if this mechanism plays an important role in tectonics. Such planet-scale coherence of tectonic episodes distinguishes the geological history of Venus from that of Earth.

Sean C. Solomon, et al.

SOLOMON, SEAN C., ET AL. "CLIMATE CHANGE AS A REGULATOR OF TECTONICS ON VENUS," *SCIENCE* 286 (1999): 87–90.

SEE ALSO *Global Warming.*

BIBLIOGRAPHY

Books

Weart, Spencer R. *The Discovery of Global Warming.* Cambridge, MA: Harvard University Press, 2004.

Periodicals

Fenton, Lori I., et al. "Global Warming and Climate Forcing by Recent Albedo Changes on Mars." *Nature* 446 (April 5, 2007): 646–649.

Schorghofer, Norbert. "Dynamics of Ice Ages on Mars." *Nature* 449 (September 13, 2007): 192–194.

Solomon, Sean C., et al. "Climate Change as a Regulator of Tectonics on Venus." *Science* 286 (October 1, 1999): 87–89.

Web Sites

"Climate Catastrophes in the Solar System." *European Space Agency,* April 26, 2007. <http://www.esa.int/esaSC/SEM2EHMJC0F_index_0.html> (accessed September 28, 2007).

Larry Gilman

Extreme Weather

■ Introduction

Extreme weather is a weather event such as snow, rain, drought, flood, or storm that is rare for the place where it occurs. For example, normal temperatures at the equator would constitute a heat wave if they occurred at the North Pole. The Intergovernmental Panel on Climate Change (IPCC) suggests that "rare" means in the bottom 10% or top 10% of severity for a given event type in a given location. Because extreme weather is by definition rare, it is difficult to assess the risk of such events, including changes in risk with global warming.

Also, since extreme weather events have always occurred, even before anthropogenic (human-caused) climate change began to be unequivocally present starting in about 1980, it is impossible to attribute any one extreme event to climate change. Climate change is likely to increase the frequency of heat waves or other extreme weather events, but it will never be possible to point to one such event and say that it was caused by climate change.

Nevertheless, it is possible to estimate the effects of climate change on the frequency and average magnitude (strength) of extreme weather events. There is evidence that some weather extremes have already shifted: cold nights have decreased globally, for example, while warm nights have increased (associated with heat waves). Droughts, storm intensity, and heat waves have increased and will continue to do so. Most categories of extreme weather events, with the exception of cold waves, are predicted to continue to increase with global warming.

■ Historical Background and Scientific Foundations

Historical data enable us to see whether global warming has already changed the frequency of various extreme weather events such as heat waves, droughts, floods, and hurricanes. Weather observations of uniform quality for the whole globe have only been available since about 1970, when satellite data were first gathered. Data from earlier decades on tropical cyclones in some parts of the world are spotty, and even after 1970 data on cyclone intensity are not always of good quality.

However, during the twentieth century, a global network of weather stations gradually came into being, and adequate data are available for most regions for the last half-century or more. Since 2000, global data collection has improved greatly, with the collection of continent-scale daily data sets, the placement of more closely spaced instruments, and the recovery of data from national archives.

These data show that from 1950 to the early 2000s, the number of heat waves has increased, along with the incidence of warm nights. In most places the number of heavy precipitation events (unusually heavy rainfalls) has increased, along with flood frequency. The intensity of tropical cyclones (called hurricanes when they occur in the Atlantic) has probably increased since 1995, but the frequency of such storms has probably not increased.

The observational record for several important categories of extreme weather events follows. In the final section, projections for changes in these types of extreme weather events during the twenty-first century are described.

Temperature

On the scale of days rather than of seasons, there is evidence that cold nights have become less common and warm nights more common. This is a global pattern, but specific regions show various patterns. In Central America and northern South America, for example, extremes of both cold and warmth have become more common; in southern South America, cold nights have become rarer and warm nights more common, but there has been no widening of extremes. That is, the more-frequent warm nights are not any warmer, on average, than warm nights used to be in that region.

WORDS TO KNOW

ANTHROPOGENIC: Made by people or resulting from human activities. Usually used in the context of emissions that are produced as a result of human activities.

CLIMATE MODEL: A quantitative way of representing the interactions of the atmosphere, oceans, land surface, and ice. Models can range from relatively simple to quite comprehensive.

CLIMATOLOGIST: Scientist who specializes in the study of climate.

DROUGHT: A prolonged and abnormal shortage of rain.

PRECIPITATION: Moisture that falls from clouds. Although clouds appear to float in the sky, they are always falling, the water droplets slowly being pulled down by gravity. Because their water droplets are so small and light, it can take 21 days to fall 1,000 ft (305 m) and wind currents can easily interrupt their descent. Liquid water falls as rain or drizzle. All raindrops form around particles of salt or dust. (Some of this dust comes from tiny meteorites and even the tails of comets.) Water or ice droplets stick to these particles, then the drops attract more water and continue getting bigger until they are large enough to fall out of the cloud. Drizzle drops are smaller than raindrops. In many clouds, raindrops actually begin as tiny ice crystals that form when part or all of a cloud is below freezing. As the ice crystals fall inside the cloud, they may collide with water droplets that freeze onto them. The ice crystals continue to grow larger, until large enough to fall from the cloud. They pass through warm air, melt, and fall as raindrops.

SAHEL: The transition zone in Africa between the Sahara Desert to the north and tropical forests to the south. This dry land belt stretches across Africa and is under stress from land use and climate variability.

TROPICAL CYCLONES: Large rotating storm systems characterized by a clear, low-pressure center surrounded by spiral arms of thunderstorms. Such storms form in the tropics because they are powered by the thermal energy of warm surface ocean waters. In the Atlantic, tropical cyclones are termed hurricanes.

Beginning in the second half of the twentieth century, heat waves have increased in duration. Although climate scientists frequently caution that no particular weather event can be said to be caused by climate change (or purely natural causes), individual events can illustrate what sorts of event are likely to become more common because of climate change. The 2003 heat wave in Europe is one such event. That June-July-August period was the hottest in Europe since instrumental record-keeping in the region began in 1780, beating the previous record-holder, the summer of 1807, by 2.52°F (1.4°C). Heat waves have become more frequent and longer-lasting in Europe and other areas of the world since the beginning of the twentieth century.

Drought

Warming speeds up the drying of soil and so tends to increase the frequency and severity of droughts even apart from decreases in rainfall. Soil moisture has been found to be decreasing over the Northern Hemisphere since the mid–1950s, especially in Eurasia, northern Africa, Canada, and Alaska. Trends have been smaller and more erratic in the Southern Hemisphere. Only extreme decreases in soil moisture correspond to drought. Drought is not strictly defined: agricultural drought refers to low moisture in the topmost yard (meter) or so of soil, which affects crops, while meteorological drought refers to a long period of low precipitation, and hydrologic drought refers to below-normal levels in streams, lakes, and groundwater.

A widely accepted scientific measure of drought is the Palmer Drought Severity Index (PDSI), which assesses soil moisture by combining data on precipitation, temperature, and locally available water. Global PDSI has varied greatly from year-to-year over the last century, but the overall trend has been upward. Most of the world has seen significant increase in PDSI, that is, more drought. Africa, particularly in the Sahel (the east-west band just south of the Sahara desert), has shown particularly strong increases in drought. The only large areas with moistening trends are central South America, western Russia, Scandinavia, and the east-central United States. This pattern is caused by a combination of slightly lower precipitation with higher temperatures due to global warming. According to one study cited by the IPCC, "very dry" zones worldwide have more than doubled in area since the 1970s due to a combination of natural (El Niño–Southern Oscillation) forcings and anthropogenic global warming.

Precipitation

One result of global warming is increased evaporation. Because evaporation acts to cool land, areas where moisture has increased, such as eastern North and South America, have warmed less than other parts of the world. One result of increased evaporation is that there is more moisture available for extreme precipitation events—unusually massive downpours or snowstorms. The former can be especially destructive, causing flooding. The IPCC finds that there has been a shift toward more precipitation coming from very wet days (upper 5% of wet-day rainfall intensities) in recent decades. There will likely be a 2–4% increase in the number of extreme precipitation vents in middle and high latitudes in the coming century. One study (Palmer and Räisänen, 2002) has found that the frequency of extreme precipitation events will increase by about a factor of five over parts of the United Kingdom by 2100. Data are still not adequate to form a consistent account of such changes in the tropics and subtropics.

A 2002 study by P. C. D. Milly and colleagues investigated changes in great floods, that is, floods above 100-year levels from 29 drainage basins larger than 77,220 square mi (200,000 square km). The scientists found that there has been a substantial increase in such floods since the beginning of the twentieth century, that this increase is consistent with computer climate models, and that the models predict such floods will continue to increase in frequency with global warming.

Tropical Cyclones

Tropical cyclones (called hurricanes when they occur in the Atlantic Ocean) have probably increased in intensity and duration since about 1995. They have probably not increased in frequency (about 60 occur each year worldwide, for reasons that are not yet understood). Warming of sea surface temperatures is expected by some scientists to increase hurricane intensity and duration because hurricanes draw their energy from warm surface waters. Climate scientists are divided over the question of whether the observed and predicted increases in intensity and duration of tropical cyclones are real.

■ Impacts and Issues

Climate scientists project that continued warming of Earth's atmosphere will lead to increased summer drying of soils with increased risk of drought over most of the world's land area. In the business-as-usual (most pessimistic) IPCC projection scenario, the percentage of world land area experiencing extreme drought at any one time increases from 1% today to 30% by 2100. Actual changes will depend on whether efforts to mitigate greenhouse-gas emissions are successful and on the uncertainties involved in predicting Earth's behavior as a physical system.

Although it may seem paradoxical, risk of extreme precipitation and flooding increases even as risk of drought increases. Warmer air has greater water-holding capacity; precipitation will occur in more concentrated events with longer dry periods in between. During the dry periods, soils dry; during the intense precipitation events, water runs off rather than soaking in: dry soils absorb water more slowly than moist soils and sudden bursts of water tend to run off faster than they can be absorbed even under the best conditions.

Heat waves will continue to become more common. The European heat wave of 2003 is representative of the type of heat waves that will become more common as the climate warms. Extreme weather around the world in 2007 was unusually common. Omar Baddour, a climatologist employed by the United Nations's World Meteorological Organization, said that "When we observe such extremes in individual years, it means that this fits well with current knowledge from the IPCC report on global trends" (Associated Press, 2007).

Not all forms of extreme weather will increase under global warming. Computer models project a 50–100% decrease in the number of cold waves (also called cold-air outbreaks or cold snaps) in the Northern Hemisphere relative to rates in the early 2000s. There is not enough evidence, as of late 2007, to say whether tornadoes, hail, lightning, and dust storms are yet occurring at greater frequency or intensity because of global warming.

IN CONTEXT: DESERTIFICATION AND DEATH

Desertification claimed major international attention in the 1970s. This resulted from an extended period of severe drought in the Sahel region during 1968 to 1973, affecting six African countries on the southern border of the Sahara Desert. Although international relief measures were undertaken, millions of livestock died during that prolonged drought, and thousands of people suffered or died of starvation.

Arid lands in parts of North America are among those severely affected by desertification; almost 90% of such habitats are considered to be moderately to severely desertified. The arid and semi-arid lands of the western and southwestern United States are highly vulnerable to this kind of damage.

■ Primary Source Connection

This section from a 2007 Intergovernmental Panel on Climate Change (IPCC) report examines climate data from 1950 to present. A review of this scientific evidence indicates that there has been a noticeable and quantifiable change in the global climate over the last 50 years.

The IPCC is a scientific body that was founded by the United Nations in 1988 under the United Nations Environment Programme and the U.N.'s World Meteorological Organization.

OBSERVATIONS: SURFACE AND ATMOSPHERIC CLIMATE CHANGE

Frequently Asked Question 3.3

Has there been a Change in Extreme Events like Heat Waves, Droughts, Floods and Hurricanes?

Since 1950, the number of heat waves has increased and widespread increases have occurred in the numbers of warm nights. The extent of regions affected by droughts has also

During a prolonged drought and weeks of extreme heat in Tennessee, thousands of dead fish washed up on shore. *AP Images.*

increased as precipitation over land has marginally decreased while evaporation has increased due to warmer conditions. Generally, numbers of heavy daily precipitation events that lead to flooding have increased, but not everywhere. Tropical storm and hurricane frequencies vary considerably from year to year, but evidence suggests substantial increases in intensity and duration since the 1970s. In the extratropics, variations in tracks and intensity of storms reflect variations in major features of the atmospheric circulation, such as the North Atlantic Oscillation.

In several regions of the world, indications of changes in various types of extreme climate events have been found. The extremes are commonly considered to be the values exceeded 1, 5 and 10% of the time (at one extreme) or 90, 95 and 99% of the time (at the other extreme). The warm nights or hot days (discussed below) are those exceeding the 90th percentile of temperature, while cold nights or days are those falling below the 10th percentile. Heavy precipitation is defined as daily amounts greater than the 95th (or for 'very heavy', the 99th) percentile.

In the last 50 years for the land areas sampled, there has been a significant decrease in the annual occurrence of cold nights and a significant increase in the annual occurrence of warm nights. Decreases in the occurrence of cold days and increases in hot days, while widespread, are generally less marked. The distributions of minimum and maximum temperatures have not only shifted to higher values, consistent with overall warming, but the cold extremes have warmed more than the warm extremes over the last 50 years .... More warm extremes imply an increased frequency of heat waves. Further supporting indications include the observed trend towards fewer frost days associated with the average warming in most mid-latitude regions.

A prominent indication of a change in extremes is the observed evidence of increases in heavy precipitation events over the mid-latitudes in the last 50 years, even in places where mean precipitation amounts are not increasing. For very heavy precipitation events, increasing trends are reported as well, but results are available for few areas.

Drought is easier to measure because of its long duration. While there are numerous indices and metrics of drought, many studies use monthly precipitation totals and temperature averages combined into a measure called the Palmer Drought Severity Index (PDSI). The PDSI calculated from the middle of the 20th century shows a large drying trend over many Northern Hemisphere land areas since the mid-1950s, with widespread drying over much of southern Eurasia, northern Africa, Canada and Alaska, and an opposite trend in eastern North and South America. In the Southern Hemisphere, land surfaces were wet in the 1970s and relatively dry in the 1960s and 1990s, and there was a drying trend from 1974 to 1998. Longer-duration records for Europe for the whole of the 20th century indicate few significant trends. Decreases in precipitation over land since the 1950s are the likely main cause for the drying trends, although large surface warming during the last two to three decades has also likely contributed to the drying. One study shows that very dry land areas across the globe (defined as areas with a PDSI of less than -3.0) have more than doubled in extent since the

1970s, associated with an initial precipitation decrease over land related to the El Niño–Southern Oscillation and with subsequent increases primarily due to surface warming.

Changes in tropical storm and hurricane frequency and intensity are masked by large natural variability. The El Niño–Southern Oscillation greatly affects the location and activity of tropical storms around the world. Globally, estimates of the potential destructiveness of hurricanes show a substantial upward trend since the mid-1970s, with a trend towards longer storm duration and greater storm intensity, and the activity is strongly correlated with tropical sea surface temperature. These relationships have been reinforced by findings of a large increase in numbers and proportion of strong hurricanes globally since 1970 even as total numbers of cyclones and cyclone days decreased slightly in most basins. Specifically, the number of category 4 and 5 hurricanes increased by about 75% since 1970. The largest increases were in the North Pacific, Indian and Southwest Pacific Oceans. However, numbers of hurricanes in the North Atlantic have also been above normal in 9 of the last 11 years, culminating in the record-breaking 2005 season.

Based on a variety of measures at the surface and in the upper troposphere, it is likely that there has been a poleward shift as well as an increase in Northern Hemisphere winter storm track activity over the second half of the 20th century. These changes are part of variations that have occurred related to the North Atlantic Oscillation. Observations from 1979 to the mid-1990s reveal a tendency towards a stronger December to February circumpolar westerly atmospheric circulation throughout the troposphere and lower stratosphere, together with poleward displacements of jet streams and increased storm track activity. Observational evidence for changes in small-scale severe weather phenomena (such as tornadoes, hail and thunderstorms) is mostly local and too scattered to draw general conclusions; increases in many areas arise because of increased public awareness and improved efforts to collect reports of these phenomena.

"OBSERVATIONS: SURFACE AND ATMOSPHERIC CLIMATE CHANGE." *CLIMATE CHANGE 2007: THE PHYSICAL SCIENCE BASIS*, CHAPTER 3. INTERGOVERNMENTAL PANEL ON CLIMATE CHANGE, 2007.

SEE ALSO *Drought; Dust Storms; Heat Waves; Hurricanes.*

BIBLIOGRAPHY

Books

Solomon, S., et al, eds. *Climate Change 2007: The Physical Science Basis: Contribution of Working Group I to the Fourth Assessment Report of the Intergovernmental Panel on Climate Change.* New York: Cambridge University Press, 2007.

Periodicals

Benniston, Martin. "Linking Extreme Climate Events and Economic Impacts: Examples from the Swiss Alps." *Energy Policy* 35 (2007): 5384–5392.

Kaufman, Marc. "Across Globe, Extremes of Heat and Rain." *The Washington Post* (August 8, 2007).

Milly, P. C. D., et al. "Increasing Risk of Great Floods in a Changing Climate." *Nature* 415 (2002): 514–517.

Palmer, T. N., and J. Räisänen. "Quantifying the Risk of Extreme Seasonal Precipitation Events in a Changing Climate." *Nature* 415 (2002): 512–513.

Web Sites

Associated Press. "UN: Weather Extremes Match Forecast." *USAToday.com*, August 7, 2007. <http://www.usatoday.com/news/topstories/2007-08-07-1773939871_x.htm> (accessed November 11, 2007).

Sharma, Anju. "Assessing, Predicting, and Managing Current and Future Climate Variability and Extreme Events, and Implications for Sustainable Development." *UNFCCC Workshop on Climate Related Risks and Extreme Events under the Nairobi Work Programme on Impacts, Vulnerability, and Adaptation*, June 2007. <http://unfccc.int/files/adaptation/sbsta_agenda_item_adaptation/application/pdf/background_paper_on_climate_related_risks.pdf> (accessed November 11, 2007).

Larry Gilman

Feedback Factors

Introduction

Predictions of the future effects and severity of climate change are complicated by a number of feedback factors. A feedback, in this case, is an effect of global warming that amplifies or dampens the warming influences of anthropogenic (human-caused) greenhouse-gas emissions. These feedbacks arise from a number of different sources, including changes in ocean temperature, changes in cloud cover and altitude, changes in snow and ice cover, changes in vegetation range and growth, changes in soil temperature, changes in wildfire cycles, and others.

Feedback factors typically work in one of two ways: 1) By influencing how much carbon dioxide (CO_2), methane, and other greenhouse gases are released into or absorbed from the atmosphere through processes such as decay and photosynthesis, or 2) directly through albedo changes—where alterations in land, water, and clouds directly influence the amount of sunlight absorbed or reflected by the planet, and on a smaller scale, by its ecosystems. Feedbacks can be both positive, leading to further warming, or negative, moderating warming or leading to cooling.

Because there are many potential feedbacks to global warming from earth, ocean, and atmosphere, and because they work at different magnitudes, in concert or against one another, it is difficult to work these parameters into computer models for consistent long-term predictions. However, an increasingly large body of evidence suggests

The incidence of large wildfires has been increasing in the American West. Such fires release large amounts of CO_2 into the air. *© 2004 Kathleen J. Edgar.*

WORDS TO KNOW

ALBEDO: A numerical expression describing the ability of an object or planet to reflect light.

CARBON CYCLE: All parts (reservoirs) and fluxes of carbon. The cycle is usually thought of as four main reservoirs of carbon interconnected by pathways of exchange. The reservoirs are the atmosphere, terrestrial biosphere (usually includes freshwater systems), oceans, and sediments (includes fossil fuels). The annual movements of carbon, the carbon exchanges between reservoirs, occur because of various chemical, physical, geological, and biological processes. The ocean contains the largest pool of carbon near the surface of Earth, but most of that pool is not involved with rapid exchange with the atmosphere.

DEFORESTATION: Those practices or processes that result in the change of forested lands to non-forest uses. This is often cited as one of the major causes of the enhanced greenhouse effect for two reasons: 1) the burning or decomposition of the wood releases carbon dioxide; and 2) trees that once removed carbon dioxide from the atmosphere in the process of photosynthesis are no longer present and contributing to carbon storage.

GREENHOUSE GASES: Gases that cause Earth to retain more thermal energy by absorbing infrared light emitted by Earth's surface. The most important greenhouse gases are water vapor, carbon dioxide, methane, nitrous oxide, and various artificial chemicals such as chlorofluorocarbons. All but the latter are naturally occurring, but human activity over the last several centuries has significantly increased the amounts of carbon dioxide, methane, and nitrous oxide in Earth's atmosphere, causing global warming and global climate change.

ICE CORE: A cylindrical section of ice removed from a glacier or an ice sheet in order to study climate patterns of the past. By performing chemical analyses on the air trapped in the ice, scientists can estimate the percentage of carbon dioxide and other trace gases in the atmosphere at that time.

PERMAFROST: Perennially frozen ground that occurs wherever the temperature remains below 32°F (0°C) for several years.

PHOTOSYNTHESIS: The process by which green plants use light to synthesize organic compounds from carbon dioxide and water. In the process, oxygen and water are released. Increased levels of carbon dioxide can increase net photosynthesis in some plants. Plants create a very important reservoir for carbon dioxide.

THERMOHALINE CIRCULATION: Large-scale circulation of the world ocean that exchanges warm, low-density surface waters with cooler, higher-density deep waters. Driven by differences in temperature and saltiness (halinity) as well as, to a lesser degree, winds and tides. Also termed meridional overturning circulation.

WATER VAPOR: The most abundant greenhouse gas, it is the water present in the atmosphere in gaseous form. Water vapor is an important part of the natural greenhouse effect. Although humans are not significantly increasing its concentration, it contributes to the enhanced greenhouse effect because the warming influence of greenhouse gases leads to a positive water vapor feedback. In addition to its role as a natural greenhouse gas, water vapor plays an important role in regulating the temperature of the planet because clouds form when excess water vapor in the atmosphere condenses to form ice and water droplets and precipitation.

that the net effects of various feedback mechanisms on human-caused global warming will ultimately be positive, further increasing warming.

■ Historical Background and Scientific Foundations

Antarctica's Vostok ice cores have captured evidence of climatic variations and changing atmospheric gas concentrations during glacial and interglacial cycles over the last 420,000 years. They show that past warming has typically resulted in increased levels of CO_2 and methane, which in turn contributed to more warming—a positive feedback. The same may be true today.

Scientists have identified several feedback mechanisms that may exacerbate current warming in localized and global ways. Several of these involve water in various stages. For example, warmer air can hold more water vapor, a strong greenhouse gas. Currently, water vapor may exert a warming effect 1 to 2 times greater than that exerted by greenhouse gases released by human activity. Global warming may also affect the oceans' ability to store and redistribute excess heat. Decreasing snow and ice cover at high latitudes and altitudes, meanwhile, reduces the reflectivity of Earth's surface, resulting in more localized warming and more snow and ice melt, which in turn spurs more warming. The effect of clouds, which can cool or warm Earth's surface depending on their thickness and altitude, is also important, but poorly understood.

About half of human greenhouse-gas emissions are currently absorbed by the ocean, vegetation, and soils, but this absorption is sensitive to climate change. For example, organic materials preserved in soils—which contain much of the world's stored carbon—may provide an important positive feedback as they warm and decay more rapidly in a warming climate, cycling more

IN CONTEXT: RADIATIVE FORCING

"The Earth's global mean climate is determined by incoming energy from the Sun and by the properties of the Earth and its atmosphere, namely the reflection, absorption and emission of energy within the atmosphere and at the surface. Although changes in received solar energy (e.g., caused by variations in the Earth's orbit around the Sun) inevitably affect the Earth's energy budget, the properties of the atmosphere and surface are also important and these may be affected by climate feedbacks. The importance of climate feedbacks is evident in the nature of past climate changes as recorded in ice cores up to 650,000 years old."

"Changes have occurred in several aspects of the atmosphere and surface that alter the global energy budget of the Earth and can therefore cause the climate to change. Among these are increases in greenhouse gas concentrations that act primarily to increase the atmospheric absorption of outgoing radiation, and increases in aerosols (microscopic airborne particles or droplets) that act to reflect and absorb incoming solar radiation and change cloud radiative properties. Such changes cause a radiative forcing of the climate system . . ."

"Radiative forcing is a measure of the influence a factor has in altering the balance of incoming and outgoing energy in the Earth-atmosphere system and is an index of the importance of the factor as a potential climate change mechanism. Positive forcing tends to warm the surface while negative forcing tends to cool it."

SOURCE: *Solomon, S., et al., eds.* Climate Change 2007: The Physical Science Basis. Contribution of Working Group I to the Fourth Assessment Report of the Intergovernmental Panel on Climate Change. *New York: Cambridge University Press. 2007.*

CO_2 and methane into the atmosphere. The melting of permafrost, a type of frozen soil that can contain millennia worth of frozen organic materials like root fragments and leaf litter, may release massive amounts of carbon into the atmosphere that have been absent from the carbon cycle for thousands of years, providing a particularly strong warming effect.

■ Impacts and Issues

Many of these positive feedbacks will likely be moderated by negative feedbacks. An influx of freshwater from melting Arctic ice sheets, as well as increased precipitation, may weaken the global thermohaline ocean current (which has a warming effect) and slow Arctic and Nordic warming. Increasing concentrations of CO_2, which plants convert into sugars for food, may stimulate more photosynthesis, and thus more plant growth. Such growth would, in turn, suck more CO_2 from the air.

Melting permafrost and warmer conditions will facilitate the advance of forests and shrubs in latitude and altitude, potentially providing another growing sink for CO_2. But plant growth may ultimately be limited by the amount of available nutrients in the soil, which in turn may be influenced by climate change. Meanwhile, the increasing incidence of large wildfires, which release massive pulses of greenhouse gases into the air, as well as warming resulting from the lower albedo of growing forests and shrublands, may dampen or obscure the benefits of these negative feedbacks.

Based on evidence of past warming from Vostok ice core data, the combined effects of feedback factors may shift predicted average warming from 2.7–8.1°F (1.5–4.5°C) under a doubling of atmospheric CO_2 to a range of 2.9–11°F (1.6–6°C). However, few computer models take a full range of feedbacks into consideration and several of the positive feedbacks listed earlier are still poorly understood in a global context, leading to widely variable predictions of their effects over time. Human land-use practices further complicate the picture, and may enhance or limit ecosystems' abilities to respond to or moderate climate change. More than 30% of Earth's land surface is used by humans for farming and grazing, and increases or decreases in the deforestation associated with these practices may play a significant role in the future effects of climate change.

SEE ALSO *Agriculture: Contribution to Climate Change; Albedo; Anthropogenic Change; Arctic Melting: Greenland Ice Cap; Arctic Melting: Polar Ice Cap; Carbon Cycle; Carbon Dioxide (CO_2); Climate Change; Clouds and Reflectance; Forests and Deforestation; Great Conveyor Belt; Ice Core Research; Permafrost; Wildfires.*

BIBLIOGRAPHY

Periodicals

Betts, Richard A. "Offset of the Potential Carbon Sink from Boreal Forestation by Decreases in Surface Albedo." *Nature* 408 (November 9, 2000): 187–190.

Cox, Peter, et al. "Acceleration of Global Warming Due to Carbon-Cycle Feedbacks in a Coupled Climate Model." *Nature* 408 (November 9, 2000): 184–187.

Cramer, Wolfgang, et al. "Global Response of Terrestrial Ecosystem Structure and Function to CO_2 and Climate Change: Results from Six Dynamic Global Vegetation Models." *Global Change Biology* 7 (April 2001): 357–373.

Field, Christopher B., et al. "Feedbacks of Terrestrial Ecosystems to Climate Changes." *Annual Review of Environment and Resources* 32 (2007): 1–29.

Karl, Thomas R., and Kevin E. Trenberth. "Modern Global Climate Change." *Science* 302 (December 5, 2003): 1719–1723.

Oechel, Walter C., and George L. Vourlitis. "The Effects of Climate Change on Land-Atmosphere Feedbacks in Arctic Tundra Regions." *TREE* 9 (1994): 324–329.

Overpeck, J. T., et al. "Arctic Ecosystems on Trajectory to New, Seasonally Ice-Free State." *EOS* 86 (2005): 309–316.

Powlson, David. "Will Soil Amplify Climate Change?" *Nature* 433 (January 20, 2005): 204–205.

Running, Steven W. "Is Global Warming Causing More, Larger Wildfires?" *Science* 313 (August, 18, 2006): 927–928.

Sturm, Matthew, and Tom Douglas "Changing Snow and Shrub Conditions Affect Albedo with Global Implications." *Journal of Geophysical Research* 110 (September 2005): G01004.

Torn, Margaret S., and John Harte "Missing Feedbacks, Asymmetric Uncertainties, and the Underestimation of Future Warming." *Geophysical Research Letters* 33 (May 2006): L10703.

Walker, Gabrielle. "The World Melting from the Top Down." *Nature* 446 (April 12, 2007): 718–721.

Web Sites

"Feedback Loops in Global Climate Change Point to a Very Hot 21st Century." *ScienceDaily*, May 22, 2006. <http://www.sciencedaily.com/releases/2006/05/060522151248.htm> (accessed November 17, 2007).

Sarah Gilman

Fisheries

Introduction

Fisheries are areas where fish are caught. Fisheries supply a large fraction of the world's food, but are under severe stress today from over fishing. Climate change may add to the stress on ocean fisheries by changing the planktonic food web that is the basis of the marine food chain. Changes in fisheries are associated with damage to coral reefs by water heating and ocean acidification, and other means.

Fish populations are also shifting to different ranges because of warming water, with resulting changes in marine ecology and fishery populations. Climate warming affects freshwater fisheries by reducing dissolved oxygen, increasing stratification (layering of deep water), and shifting the patterns of rainfall that supply lakes and rivers with water.

Historical Background and Scientific Foundations

Humans have harvested fish for thousands of years. However, not enough fish could be harvested to injure large fish populations until the twentieth century, when mechanized ships in large fleets began to ply the seas. Most fish are captured at sea: in 2005, 7.6% of all fish were caught in freshwater, 9.5% were raised on freshwater fish farms (aquacultured), 7.1% were aquacultured at sea, and 75.7% were captured wild at sea. Sixty-nine percent of the total fisheries output was used for human consumption, the remainder for other purposes (for example, pet food). Fish provide 2.6 billion people with 20% or more of their dietary protein. About 15% of the world animal protein supply is from fish.

The world wild-fish catch increased 4.5-fold from 1950 (19 million metric tons) to 2000 (90 million metric tons). However, the global catch leveled off in the early 2000s and has declined slightly since, probably passing its all-time historic peak. There are few fish stocks that are not already being intensively exploited, and many fish populations are declining. Some fishing fleets are subsidized by governments, which encourage over fishing, and in most fisheries there is either a struggle to fill catch quotas before others do or to simply catch every fish possible.

Such practices lead to what fisheries experts call "the race for fish." The race for fish is causing fish abundance to decline in many fisheries, with consequent structural changes to ecosystems, loss of jobs, and the breakdown of fishing communities. Aquaculture can also stress wild fisheries, as the food supplied to captive fish (e.g., shrimp) often consists of small, wild fish caught in indiscriminate fine-meshed driftnets that reduce populations of commercially valuable species by harvesting their young along with adults of small, noncommercial species.

Fish populations fluctuate over decadal time scales (10 or more years), in some cases centennial time scales; fish populations in areas that are geographically separated often vary at the same time, showing the influence of climate. Freshwater fish populations also vary in response to climate changes, which affect water temperatures, dissolved oxygen content (the warmer the water, the less the oxygen), and the amount, timing, and location of precipitation supplying water to rivers and lakes.

Impacts and Issues

Anthropogenic (human-caused) climate change adds stress to fishery ecosystems that are already under historically unprecedented stress from over fishing. The most general result of modern anthropogenic global climate change is warming, which has several effects. First, temperature affects the species mixture and abundance of plankton. Plankton are tiny animals (zooplankton) and green plants (phytoplankton) that float in the upper layers of bodies of water. Phytoplankton play an ecological role similar to that of green plants on land, capturing the energy of sunlight and passing it up the food chain to larger organisms. Almost all oceanic fish are dependent on plankton either directly (as filter feeders, for example, the tiny shrimp called krill) or indirectly (as predators on smaller fish).

Plankton productivity—the amount of plankton for a given area of water surface—is altered by climate variations. Satellite observations show that plankton productivity decreases as sea surface temperature increases: warming of surface waters decreases upwelling of deeper waters, which contain nutrients that phytoplankton need. Decreased upwelling occurs because the warmer that the surface water is, the less dense it is, and the more stably it floats on deeper, cooler waters. Warming therefore decreases planktonic productivity, which inevitably decreases populations of animals higher up in the food chain, such as fish. Such decreases can happen naturally, as for example during the El Niño/Southern Oscillation climate cycle, when the climate and surface waters of the central Pacific warm. It also occurs as a result of anthropogenic global warming. Some studies forecast a possible shift to permanent El Niño conditions as a result of anthropogenic warming.

A general effect of global warming involves migration toward the Poles by animal and plant species, both in water and on land, as species track the climate zone in which they are adapted by evolution to thrive. In the Northern Hemisphere, both fish and plankton species have already been observed to shift their ranges northward. In the North Sea, for example, where waters warmed by 1.9°F (1.05°C) from 1977 to 2001, a study by A. L. Perry and colleagues published in 2005 showed that the ranges of 15 out of 36 fish species had shifted northward by distances ranging from 30 mi (48 km) to 250 mi (403 km).

Other effects on fisheries may be less direct. For example, general circulation models (computer simulations of global climate) predict that with the doubling of atmospheric CO_2, which is possible within the next century, freshwater discharge from the Mississippi River into the Gulf of Mexico might increase by 20%. The Gulf of Mexico is the site of the world's largest and most severe oxygen-depleted zone, sometimes called a dead zone because no fish can live in it. Agricultural chemicals in the river water are the cause of the dead zone, already up to 7,700 mi^2 (20,000 km^2) in size. Increased river discharge is expected to enlarge the dead zone, affecting food webs in the Gulf and probably decreasing fish populations.

Increased stratification—the stabilization of warmer surface waters, with less mixing of nutrient-rich deeper waters up to sunlit layers where phytoplankton can exploit both nutrients and sunlight—is an important climate impact on some large freshwater fisheries. For example, the world's second-largest lake, Lake Tanganyika in Africa, is home to hundreds of species of fish and is a major freshwater fishery, supplying a harvest of 165,000–200,000 metric tons of fish per year, 15–40% of the protein intake of the inhabitants of the four countries that border the lake, namely Burundi, Tanzania, Zambia, and Congo. Since the 1970s, catches of the two primary species of the fishery (both members of the clupeid family, which includes herrings and sardines) have declined by 30–50%, though the lake is not over fished.

Until the early 2000s, the decline was attributed to some unknown environmental factor. This factor was finally shown to be climate warming. Steady atmospheric warming at Tanganyika since 1950 has caused the surface layer of the lake to warm, decreasing mixing of that layer with cooler, nutrient-rich waters from the depths. As of 2003, plankton density in Lake Tanganyika was less than one third than it had been 25 years earlier. Since plankton are the basis of the marine food chain, fish stocks declined along with plankton, and were 30% smaller in 2003 than they were 80 years before. Over the next 80 years, the climate in Tanganyika's area is predicted to warm by a further 2.3–3°F (1.3–1.7°C). This is likely to further increase stratification, reduce mixing, deplete plankton, and shrink the fishery, with possible consequences for human well-being in the surrounding areas.

■ Primary Source Connection

One potentially catastrophic effect of global climate change is a predicted change in Earth's hydrologic (water) cycle, including an increase in freshwater runoff. This source states that climate models indicate increased freshwater runoff in most of the world's major river systems. Greater discharge of freshwater will result in increased hypoxic zones in oceans, causing major disruptions in coastal ecosystems. Hypoxic zones are coastal waters with depleted oxygen content due to fertilizers and other sediments that are carried by freshwater runoff. Hypoxic zones are sometimes referred to as "dead zones" because the oxygen content is so low that the waters cannot support life. One of the largest hypoxic zones in the world is in the Gulf of Mexico where the Mississippi River discharges. The Gulf of Mexico hypoxic zone covers an approximate area of 22,000 km^2.

CLIMATE, HYPOXIA AND FISHERIES: IMPLICATIONS OF GLOBAL CLIMATE CHANGE FOR THE GULF OF MEXICO HYPOXIC ZONE

Introduction

There is a growing consensus among scientists that human activities, which have increased atmospheric concentrations of carbon dioxide (CO_2) by one-third during the last

WORDS TO KNOW

EL NIÑO/SOUTHERN OSCILLATION: Global climate cycle that arises from interaction of ocean and atmospheric circulations. Every 2 to 7 years, westward-blowing winds over the Pacific subside, allowing warm water to migrate across the Pacific from west to east. This suppresses normal upwelling of cold, nutrient-rich waters in the eastern Pacific, shrinking fish populations and changing weather patterns around the world.

PHYTOPLANKTON: Microscopic marine organisms (mostly algae and diatoms) that are responsible for most of the photosynthetic activity in the oceans.

UPWELLING: The vertical motion of water in the ocean by which subsurface water of lower temperature and greater density moves toward the surface of the ocean. Upwelling occurs most commonly among the western coastlines of continents, but may occur anywhere in the ocean. Upwelling results when winds blowing nearly parallel to a continental coastline transport the light surface water away from the coast. Subsurface water of greater density and lower temperature replaces the surface water and exerts a considerable influence on the weather of coastal regions. Carbon dioxide is transferred to the atmosphere in regions of upwelling.

100 years, may be responsible for an increase in global Earth's temperatures. This so-called "global warming" theory is not without challengers who argue that scientific proof is incomplete or contradictory, and that there remain many uncertainties about the nature of climate variability and climate change. Nevertheless, global temperature averages increased by almost 1°C during the last 150 years, and further temperature increase seems probable. General circulation models (GCMs) forced by enhanced greenhouse gas concentrations have projected a global temperature increase of 2 to 6°C over the next 100 years. An increase in global Earth's temperature of 2 to 6°C would likely produce an enhanced global hydrologic cycle that would be manifested in altered freshwater runoff. This hypothesis is supported by several lines of evidence, including "paleofloods," decadal trends in the freshwater runoff and GCM's scenarios.

In the United States, there is historic evidence suggesting that a change in climate enhances the frequency of extreme flood events. An analysis of a 5000-yr old geological record for the southwestern United States suggested that floods occurred more frequently during transitions from cool to warm climate conditions. Apparently, modest changes in climate were able to produce large changes in the magnitude of floods. Additional evidence in support of the above hypothesis came from a 7000-yr old record of over bank floods for the upper Mississippi River tributaries. Approximately 3300 years ago, an abrupt shift in flood behavior occurred, with frequent floods of a magnitude that now recurs every 500 years or more. Also, an analysis of the data collected by the U.S. Geological Survey indicated statistically significant increasing trends in monthly streamflow during the past five decades across most of the conterminous United States. These results seem to support the hypothesis that enhanced greenhouse forcing produces an enhanced hydrologic cycle. One of the GCM studies has examined the impact of global warming on the annual runoff of the 33 world's largest rivers. For a $2xCO_2$ climate, the runoff increases were detected in all studied rivers in high northern latitudes, with a maximum of +47 %. At low latitudes there were both increases and decreases, ranging from +96% to −43%. Importantly, the model results projected an increase in the annual runoff for 25 of the 33 studied rivers.

The northern Gulf of Mexico, which receives inflows of the Mississippi River—the eighth largest river in the world, is one of the coastal areas that may experience increased freshwater and nutrient inputs in the future. According to a GCM study referenced above, the annual Mississippi River runoff would increase 20% if the concentrations of atmospheric CO_2 doubles. This hydrologic change would be accompanied by an increase in summer and winter temperatures over the Gulf Coast region of 2°C and 4°C, respectively. A higher runoff is expected during the May-August period, with an annual maximum most likely occurring in May. While there are no other GCM estimates of the Mississippi River runoff, this result is in agreement with a projected $2xCO_2$ increase in rainfall over the Mississippi River drainage basin.

Here we review probable implications of climate change for the Gulf of Mexico hypoxic zone, focusing on two areas: (1) coupling between climate variability, freshwater runoff of the Mississippi River, and hypoxia in the coastal northern Gulf of Mexico, and, (2) potential implications of global climate change for coastal fisheries in the hypoxic zone. In this analysis we use our previously published physical biological model and extensive long-term data sets collected at station within the core of the Gulf of Mexico hypoxic zone.

Coupling Between Climate and Hypoxia

Climate change, if manifested by increasing riverine freshwater inflow, may affect coastal and estuarine ecosystems in several ways. First, changes in freshwater inflow will affect the stability of the water column, and this effect may be enhanced due to changes in sea surface temperatures. Vertical density gradients are likely to increase, that could decrease vertical oxygen transport and create conditions in the bottom water favorable for the development of severe hypoxia or anoxia. Second, the concentrations of nitrogen (N), phosphorus (P), and silicon (Si) in riverine freshwater inflows are typically an order of magnitude higher than those in coastal waters. The mass fluxes of riverine nutrients are generally well-correlated with integrated runoff values. Consequently, the nutrient inputs to the coastal ocean are expected to increase as a result of the increasing riverine runoff, which could have an immediate effect on the productivity of coastal phytoplankton. Third, the stoichiometric ratios of riverine nutrients, Si:N, N:P and Si:P, may differ from those in the coastal ocean. Increased freshwater inflow, therefore, may also affect coastal phytoplankton communities by increasing or decreasing a potential for single nutrient limitation and overall nutrient balance. Thus, it appears that there is a plausible link between global climate change and the productivity of river-dominated coastal waters.

Changes in the areal extent of hypoxic (<2 mg O_2 l^{-1}) bottom waters provide a representative example of the riverine influence on coastal productivity processes. The northern Gulf of Mexico is presently the site of the largest (up to 20,000 km^2) and most severe hypoxic zone in the western Atlantic Ocean. Hypoxia normally occurs from March through October in waters below the pycnocline, and extends between 5 and 60 km offshore. During the drought of 1988 (a 52-year low discharge record of the Mississippi River), however, bottom oxygen concentrations were significantly higher than normal, and formation of a continuous hypoxic zone along the coast did not occur in midsummer .... The opposite occurred during the Great Flood of 1993 (a 62-year maximum discharge for August and September), when the areal extent of summertime hypoxia doubled with respect to the average hydrologic year. Hypoxia in the coastal bottom waters of the northern Gulf of Mexico develops as a synergistic product of high surface primary productivity, which is also manifested in a high carbon flux to the sediments, and high stability of the water column. Likewise, the 1993 event was associated with both an increased stability of the water column and nutrient-enhanced primary productivity, as indicated by the greatly increased nutrient concentrations and phytoplankton biomass in the coastal waters influenced by the Mississippi River....

Conclusions

Projections of general circulation models suggest that freshwater discharge from the Mississippi River to the coastal ocean would increase 20% if atmospheric CO_2 concentration doubles. A higher Mississippi River runoff would be accompanied by an increase in winter and summer temperatures over the Gulf Coast region of 4.2°C and 2.2°C, respectively. This is likely to affect the global oxygen cycling of the northern Gulf of Mexico, which is presently the site of the largest (up to 20,000 km^2) and the most severe coastal hypoxic zone (<2 mg O_2 l^{-1}) in the western Atlantic Ocean. Model simulations suggest a close coupling between climate variability and hypoxia, and indicate a potential for future expansion of the Gulf's hypoxic zone as a result of global warming. In simulation experiments, a 20% increase in annual runoff of the Mississippi River, relative to a 1985–1992 average, resulted in a 50% increase in net primary productivity of the upper water column (0–10 m) and a 30-60% decrease in summertime subpycnoclinal (10–20 m) oxygen content within the present day hypoxic zone. Those model projections are in agreement with the observed increase in severity and areal extent of hypoxia during the flood of 1993. Because of large uncertainties in the climate system itself, and also at different levels of biological control, it is difficult to predict how climate change may affect coastal food webs. Nevertheless, future expansion of the coastal hypoxic zones would have important implications for habitat functionality and sustainability of coastal fisheries.

Dubravko Justić, et al.

JUSTIĆ, DUBRAVKO, ET AL. "CLIMATE, HYPOXIA AND FISHERIES: IMPLICATIONS OF GLOBAL CLIMATE CHANGE FOR THE GULF OF MEXICO HYPOXIC ZONE." PROCEEDINGS OF THE 2000 LOUISIANA ENVIRONMENTAL STATE OF THE STATE CONFERENCE, 2006. <HTTP://WWW.EPA.GOV/MSBASIN/TASKFORCE/2006SYMPOSIA/146JUSTICE.PDF> (ACCESSED NOVEMBER 30, 2007).

SEE ALSO *Biodiversity; Oceans and Seas.*

BIBLIOGRAPHY

Books

Cushing, D. H. *Climate and Fisheries.* New York: Academic Press, 1982.

Periodicals

Behrenfeld, Michael J. "Climate-Driven Trends in Contemporary Ocean Productivity." *Nature* 444 (December 7, 2006): 752–755.

Ficke, Ashley D., et al. "Potential Impacts of Global Climate Change on Freshwater Fisheries." *Reviews in Fish Biology and Fisheries* 17 (November 2007): 581–613.

Hilborn, Ray, et al. "State of the World's Fisheries." *Annual Review of Environmental Resources* 28 (2003): 359–399.

Lehodey, P., et al. "Climate Variability, Fish, and Fisheries." *Journal of Climate* 19 (October 2006): 5009–5030.

Perry, Allison J., et al. "Climate Change and Distribution Shifts in Marine Fishes." *Science* 308 (June 24, 2005): 1912–1915.

Roessig, Julie M., et al. "Effects of Global Climate Change on Marine and Estuarine Fishes and Fisheries." *Reviews in Fish Biology and Fisheries* 14 (June 2004): 251–275.

Web Sites

"State of the World's Fisheries and Aquaculture 2006." *Food and Agriculture Organization of the United Nations,* June 22, 2007. <ftp://ftp.fao.org/docrep/fao/009/a0699e/a0699e.pdf> (accessed November 19, 2007).

Larry Gilman

Floods

Introduction

Climate is an important influence of floods. A winter thaw or rainy spring can send amounts of water into a river that are greater than the capacity of the watercourse. The excess water has to go somewhere, and the result can be flooding of adjacent land. The vast majority of scientists have come to accept the reality of global warming—the warming of the atmosphere due to human activities. One consequence of the world's changing climate may be increased rainfall and flooding.

Historical Background and Scientific Foundations

For thousands of years, people have settled in low-lying land next to rivers, lakes, or other watercourses. Although the fertile soil is ideal for growing crops, the land is vulnerable to flooding. In 2007, flooding remained the most damaging of all natural disasters. One-third of all the natural disasters occurring around the world each year are floods, and more than 50% of disaster-related deaths are due to floods.

International experts warn that Africa is particularly vulnerable to climate change. Flooding is one of the potential consequences. Here, many Mozambicanque citizens are shown stranded on a bridge during a flood in 2000. *AP Images.*

Excessive rainfall can cause rivers to jump their banks and flood areas near them. The home on the right (on stilts) suffered less flood damage than the house on the left. Floods can occur with little warning, catching people unaware and unprepared. *Digital Vision/Royalty Free.*

In the wake of Hurricane Katrina in August 2005, for example, nearly every levee in the downtown region of New Orleans, Louisiana, was breached, and the resulting flooding turned 80% of the city into a lake. The hurricane- and flood-related damage in New Orleans topped $80 billion; more than 1,000 people died. Over the entire affected Gulf Coast region, the death toll exceeded 2,200 and damage approached $130 billion. As of early 2008, the region had yet to fully recover.

The devastation of Hurricane Katrina is an example of flooding caused by a storm surge, where strong winds push water onto shore. In the case of Hurricane Katrina, the storm surge was huge; in one documented example, waves that struck a Gulfport, Mississippi, hotel were approximately 10 ft (3 m) high, even though the hotel was 250 yd (230 m) from the shoreline and 19 ft (6 m) above sea level. Calculations revealed that the surging waves were 28 ft (8.5 m) high when they came ashore.

Much smaller flood surges are still capable of causing a great deal of damage. Anyone who has tried to wade into the ocean surf and who has been thrown off balance by waves less than a 1 ft (.3 m) tall can attest to the power packed by a wave. In a flood, waves that are several feet tall can wash away an average-sized car. Powerful floods that last hours or days can throw houses off their foundations.

Other floods are caused by heavy rainfall or a sudden snowmelt, which inundates a watercourse with a greater volume of water than the watercourse can handle. Also, the soil that borders the river, stream, or lake becomes saturated with water. This is especially true for clay-rich soil or soil that has been compacted. When this happens, any additional water cannot be absorbed by the soil, and it flows into the watercourse.

A watercourse with a shoreline that is rocky or close to a parking lot (where the water-absorbing soil has been replaced by asphalt) is more prone to flooding than, for example, one with a forested shoreline capable of absorbing water.

Floods can occur in a fairly predictable pattern. A historic example is the Nile river, which, before the construction of dams to control water flow, tended to flood with the increase of rain in the spring. This type of flooding tends to be less destructive because those living near the river anticipate it. In addition, the nutrients that are washed onto the land by the flood are left when the flood waters recede, making the land very fertile for growing crops.

WORDS TO KNOW

CLIMATE MODEL: A quantitative way of representing the interactions of the atmosphere, oceans, land surface, and ice. Models can range from relatively simple to quite comprehensive.

MONSOON: An annual shift in the direction of the prevailing wind that brings on a rainy season and affects large parts of Asia and Africa.

STORM SURGE: Local, temporary rise in sea level (above what would be expected due to tidal variation alone) as the result of winds and low pressures associated with a large storm system. Storm surges can cause coastal flooding, if severe.

TSUNAMI: Ocean wave caused by a large displacement of mass under the surface of the water, such as an earthquake or volcanic eruption.

Floods that occur with no warning are known as flash floods. These floods typically occur when a large amount of rain falls in a short time, such as in a monsoon. The sudden appearance of the excessive amount of water in a small watercourse can create a wall of water that rushes downstream. Just as quickly, the water level can return to normal. In July 2007, for example, a series of flash floods caused by heavy seasonal rains ravaged eight states in the African country of Sudan, killing more than 100 people

Undersea earthquakes release a great deal of energy, which pushes the surrounding ocean water outward from the origin of the earthquake. The result is a ripple effect, similar to the ripples created when a rock is dropped into a pond. The series of waves is called a tsunami. Moving at hundreds of miles per hour, the waves can devastate coastal regions. On December 26, 2004, a magnitude 9.0 earthquake off the coast of Sumatra spawned a series of waves that obliterated coastal villages in Sri Lanka, Thailand, Myanmar, India, and elsewhere. More than 155,000 people were killed in Sri Lanka alone.

■ Impacts and Issues

The flash floods experienced in Sudan in the summer of 2007 have affected 21 other African nations including Uganda, Ethiopia, Democratic Republic of the Congo, and Ghana, displacing more than 700,000 people and killing at least 300 people as of October 2007. At the same time, regions of the United Kingdom were affected by severe floods triggered by June rains of 5.5 in (14 cm), more than double the historic average for the month. Some regions of Britain experienced the most monthly rainfall ever recorded.

Hurricane Katrina was one of the 28 tropical and subtropical storms, and one of the 15 hurricanes that occurred during the 2005 hurricane season. As of late 2007, the 2005 season stood as the most active Atlantic hurricane season ever recorded.

These examples may have their origin in the changing global climate, which the majority of scientists agree is influenced, if not determined, by the warming of the atmosphere. This warming trend, which has been evident since the middle of the nineteenth century, and which coincides with the increased industrialization in developed countries, is popularly known as global warming. Altered air currents in the changing atmosphere can affect weather patterns, with some regions having more rainfall and other regions experiencing drought.

A pair of 2002 studies published in *Nature* highlighted the influence of climate on the occurrence of floods. The studies assessed data from all over the globe on the occurrence of 100-year-floods—those that are so severe that they are predicted to occur very rarely—and on 19 climate models. They revealed an increased frequency of major floods since the mid-twentieth century, and related the increase to global warming.

The World Water Council, a think-tank of water users and suppliers concerned with water-related social and economic issues, has predicted that by 2025 about half of Earth's population will live in areas that will be at increased risk of severe storms and other extreme weather. One consequence could be increased frequency of floods.

Computer models have been used for decades to help understand weather patterns and predict the weather. Daily weather forecasts rely on computer projections along with the monitoring of data. Similarly, flood modeling software exists. Information on the size, shape, and topography of the area that drains into the particular watercourse; the types of soil present in the area; typical weather patterns; and land use are used to better understand how changing parameters affect the potential for flooding. Such models help scientists better understand how climate changes in the real world affect flooding.

SEE ALSO *Erosion; Forests and Deforestation; Hurricanes; Rainfall.*

BIBLIOGRAPHY

Books

DiMento, Joseph, and Pamela M. Doughman. *Climate Change: What It Means for Us, Our Children, and Our Grandchildren.* Boston: MIT Press, 2007.

Hulme, Mike. *Nature's Revenge?: Hurricanes, Floods and Climate Change.* New York: Hodder & Stoughton, 2003.

Morgan, William W. *The Biblical Flood: Global Warming & Bush's Harvest.* Charleston: BookSurge Publishing, 2007.

Periodicals

Milly, P. C. D., R. T. Wetherald, K. A. Dunne, and T. L. Delworth. "Increasing Risk of Great Floods in a Changing Climate." *Nature* 415 (2002): 514–517.

Palmer, T. N., and J. Räisänen. "Quantifying the Risk of Extreme Seasonal Precipitation Events in a Changing Climate." *Nature* 415 (2002): 512–514.

Web Sites

"Causes of Flooding." *Environment Canada.* <http://www.ec.gc.ca/water/en/manage/floodgen/e_cause.htm> (accessed October 31, 2007).

"In Depth: Forces of Nature—Flooding." *Canadian Broadcasting Corporation (CBC)*, June 20, 2005. <http://www.cbc.ca/news/background/forcesofnature/flooding.html> (accessed October 31, 2007).

Fog

Introduction

In meteorology, fog is defined as a mass of liquid water vapor or solid ice crystals that has condensed and suspended itself in the lower layers of the atmosphere just above the surface of Earth. Fog is different from a stratus cloud only in that the cloud's base has dropped down so that it is in contact or in close proximity with the ground, whether it is on level ground, on the top of a hill, or within a valley.

Because of fog's location near the ground, it often causes reduced visibility of a normally clear sky. Specifically, meteorologists define fog as a cloud that reduces visibility on the surface of Earth to less than 0.6 mi (1 km). They have found that fog forms when the difference between the dew point and the outside temperature is 5°F (3°C) or less. When the relative humidity is 100%, fog is much more likely to form.

Historical Background and Scientific Foundations

Fog is generally classified as four types, depending on how it is formed: advection, precipitation, radiation, and upslope. Advection fog is formed whenever wind or

A bank of fog stands between a group of boaters and the city of San Francisco, California, in 2006. *AP Images.*

Fog rolls into the Scottish Highlands. *Brand X Pictures/Royalty Free.*

current (formally called advection) of warm, moist air travels over a relatively cooler body of land or water. Precipitation fog, also called frontal fog, often develops when snowflakes or raindrops descend through an atmospheric layer that is cooler and drier. It also occurs frequently during the passage of warm and cold fronts, when the lower air near the surface is much different in temperature from the upper air.

Radiation fog, which forms only over land after sunset, is caused by the cooling of Earth—that is, land cools more rapidly than water; as a consequence, fog is formed over land. It often occurs in fall and early winter months. Radiation fog usually develops up to 4 feet (just over 1 meter) in depth, but winds and other atmospheric events can cause it to be larger in depth. Upslope fog is formed when air is uniformly cooled by rising and enlarging wind currents, such as when the wind blows up a mountain slope (what is called orographic lift), causing the moisture within it to condense.

■ Impacts and Issues

Changes in local and global climate make forecasting of fog difficult at best. Further meteorological studies are needed to decide if currently used forecasting tools provide accurate guidance to meteorologists and other professionals in various pursuits throughout the world. For instance, because aviation pilots and air traffic controllers need to be able to forecast the occurrence of fog when dealing with airplane flights, the Federal Aviation Administration (FAA) authorizes studies to find better methods and devices to assist these professionals.

According to the University Corporation for Atmospheric Research, the FAA provides research into soil and air temperature, cloud ceiling, visibility, rainfall, moisture levels, and other critical variables to better understand fog. Some of the research involves weather balloons and automated sensing systems placed in key locations in order to determine the behavior of fog.

The warming and cooling of Earth, locally and globally, is controlled to a large part by fog and clouds. They block the heat brought by the sun in the form of sunlight, reflecting it back into space. Such action helps to cool Earth. However, heat already beneath fog and clouds can become trapped by its water particles, which can quickly heat up the planet. Whether Earth becomes warmer or cooler depends in part by the frequency and amount of fog and clouds. A long-term change in their patterns could have a drastic impact on Earth's weather, either making it much colder or warmer.

Atmospheric physicist Anthony D. Del Genio, who is associated with the U.S. National Aeronautics and

WORDS TO KNOW

BLACK CARBON: A type of aerosol (small, airborne particle) consisting mostly of carbon: includes soot, charcoal, and some other dark organic particles.

DEW POINT: The temperature to which air must be cooled for saturation to occur, exclusive of air pressure or moisture content change. At that temperature, dew begins to form, and water vapor condenses into liquid.

FOSSIL FUELS: Fuels formed by biological processes and transformed into solid or fluid minerals over geological time. Fossil fuels include coal, petroleum, and natural gas. Fossil fuels are nonrenewable on the timescale of human civilization, because their natural replenishment would take many millions of years.

GREENHOUSE GASES: Gases that cause Earth to retain more thermal energy by absorbing infrared light emitted by Earth's surface. The most important greenhouse gases are water vapor, carbon dioxide, methane, nitrous oxide, and various artificial chemicals such as chlorofluorocarbons. All but the latter are naturally occurring, but human activity over the last several centuries has significantly increased the amounts of carbon dioxide, methane, and nitrous oxide in Earth's atmosphere, causing global warming and global climate change.

METEOROLOGY: The science that deals with Earth's atmosphere and its phenomena and with weather and weather forecasting.

OROGRAPHY: Branch of geology that deals with the arrangement and character of land altitude variations (hills and mountains); also, the average land altitude over a given region. In computer models of climate, orography in the second sense defines the lower bound of the atmosphere over land.

STRATUS CLOUD: Extensive, layer-like cloud form ("stratum" means layer in Latin). Stratus clouds occur worldwide but play a particularly important role in Arctic climate, where they are prevalent and affect vertical exchanges of heat, moisture, and momentum; this, in turn, affects global climate.

WATER VAPOR: The most abundant greenhouse gas, it is the water present in the atmosphere in gaseous form. Water vapor is an important part of the natural greenhouse effect. Although humans are not significantly increasing its concentration, it contributes to the enhanced greenhouse effect because the warming influence of greenhouse gases leads to a positive water vapor feedback. In addition to its role as a natural greenhouse gas, water vapor plays an important role in regulating the temperature of the planet because clouds form when excess water vapor in the atmosphere condenses to form ice and water droplets and precipitation.

Space Administration's Goddard Institute for Space Studies, states that even a small change in cloud cover can dramatically change the temperature of Earth's atmosphere. Del Genio studies the relationship between water vapor in the atmosphere and meteorological phenomena. He researches whether a warmer climate will result in cloudier or clearer skies; more humid or drier atmosphere; and more or less frequent and/or violent storms. As the world adds larger concentrations of greenhouse gases to the atmosphere, Del Genio and many other scientists around the world are working on mathematical computer models based on satellite and ground-based observations to determine Earth's changing climate.

Other scientists are working at research to determine how artificially made substances affect fog and clouds. For instance, black carbon is a primary ingredient in soot left over when fossil fuels are not completely combusted (burned). When black carbon gets into the atmosphere, it becomes easily trapped in water vapor. When this action happens, scientists contend that fog and clouds absorb more heat from Earth's surface than normal, causing warmer temperatures. Such research to learn more about how human-made substances affect fog and clouds is important to Earth's evolving climate.

SEE ALSO *Atmospheric Pollution; Climate Change; Clouds and Reflectance; Global Warming; Greenhouse Effect; Meteorology; Soot.*

BIBLIOGRAPHY

Books

Allaby, Michael. *Fog, Smog, and Poisoned Rain.* New York: Facts on File, 2003.

Entertainment Services and Technology Association (ESTA). *Introduction to Modern Atmospheric Effects.* New York: ESTA, 2000.

Lutgens, Frederick K. *The Atmosphere: An Introduction to Meteorology.* Upper Saddle River, NJ: Prentice Hall, 2004.

Pretor-Pinney, Gavin. *The Cloudspotter's Guide: The Science, History, and Culture of Clouds.* New York: Berkley Publishing Group, 2006.

Periodicals

McConnell, J.C., et al. "20th-Century Industrial Black Carbon Emissions Altered Arctic Climate Forcing." *Science* 317 (September 7, 2007): 1381-1384.

Web Sites

"Clouds, Clouds, Clouds." *University Corporation for Atmospheric Research.* <http://www.ucar.edu/

news/features/clouds> (accessed November 4, 2007).

Jacobson, Mark Z. "Attribution of Regional and Global Climate Change: Relative Effects of Fossil-Fuel Soot, Methane, Other Greenhouse Gases and Particles, and Urbanization." *The Smithsonian/NASA Astrophysics Data System.* <http://adsabs.harvard.edu/abs/2006AGUFM.A43A0110J> (accessed November 4, 2007).

Forests and Deforestation

Introduction

Forests cover 30% of Earth's total land area. They are an important part of the climate system in several respects. First, wood stores carbon: all green plants obtain the carbon in their tissues by extracting CO_2 from the air, breaking out the carbon, and releasing the oxygen. Trees, the largest plants, store the most carbon, sequestering it in persistent woody tissues that can keep carbon out of the atmosphere for centuries or millennia (4,000 years, in the case of certain bristlecone pines). Second, large amounts of carbon are stored in forest soils, as dead branches, leaves, needles, trunks, and other tree parts accumulate and partly decay. By keeping CO_2 out of the atmosphere, forests mitigate climate change (make it less severe).

Third, trees are dark and so decrease the planet's surface albedo, especially in snowy northern regions, which tends to increase global warming. Fourth, deforestation to clear land for agriculture and to extract lumber and fuelwood, which has been particularly severe in tropical regions for decades, releases carbon from forest trees and soils, enhancing global climate change. Throughout the 1990s, tropical deforestation was responsible for 20–30% of global anthropogenic (human-caused) greenhouse-gas emissions. Deforestation has slowed only slightly in the

Clear cutting has left this Oregon mountain landscape full of tree stumps and brush. If the area is not replanted, erosion (from both wind and water) can take hold and reshape the landscape as the trees are no longer there to provide a buffer. *© 2007 Nancy A. Edgar.*

IN CONTEXT: FOREST-RELATED MITIGATION

"Forest-related mitigation activities can considerably reduce emissions from sources and increase [CO_2] removals by sinks at low costs, and can be designed to create synergies with adaptation and sustainable development *(high agreement, much evidence)*."

"About 65% of the total mitigation potential (up to 100 US$/ tCO_2-eq) is located in the tropics and about 50% of the total could be achieved by reducing emissions from deforestation."

[Editor's note: As used in IPCC reports "a carbon dioxide equivalent (CO_2-eq) is the amount of CO_2 emission that would cause the same radiative forcing as an emitted amount of a well mixed greenhouse gas or a mixture of well mixed greenhouse gases, all multiplied with their respective GWPs [Global Warming Potential] to take into account the differing times they remain in the atmosphere [WGI AR4 Glossary]." tCO_2 denotes tons carbon dioxide equivalents.]

"Climate change can affect the mitigation potential of the forest sector (i.e., native and planted forests) and is expected to be different for different regions and sub-regions, both in magnitude and direction."

"Forest-related mitigation options can be designed and implemented to be compatible with adaptation, and can have substantial co-benefits in terms of employment, income generation, biodiversity and watershed conservation, renewable energy supply and poverty alleviation."

SOURCE: *Metz, B., et al, eds.* Climate Change 2007: Mitigation of Climate Change. Contribution of Working Group III to the Fourth Assessment Report of the Intergovernmental Panel on Climate Change. *New York: Cambridge University Press, 2007.*

early 2000s. Planting and conservation of forests is a primary goal of schemes to mitigate global climate change, such as the Kyoto Protocol.

Historical Background and Scientific Foundations

Forests have existed since shortly after the evolution of land plants about 450 million years ago. The first forests consisted of tree ferns; gymnosperms (evergreens such as pine and fir) were next to evolve, about 362 million years ago; finally, flowering plants, including most deciduous and present-day tropical tree species, first evolved about 130 million years ago. Contrary to a common misconception, trees and land plants do not supply Earth's atmosphere with significant amounts of oxygen. Land plants do emit oxygen, but the oxygen concentration in Earth's atmosphere was at approximately the modern level, thanks to emissions from single-celled green plants floating in the oceans (phytoplankton), the first multicellular plant and animal life that evolved about 600 million years ago, many millions of years before land plants existed.

Deforestation, although more rapid since the mid-twentieth century than at any earlier time, began long before the modern period. For example, about 70% of Europe was forested in AD 900, but by 1900 only 25% was forested. As much as 90% of all deforestation occurred before 1950. Yet the human relationship to forests has not always been destructive. According to ethnobotanists, much of the Amazon rainforest (the most biologically diverse forest on Earth today, though not as large in extent as the boreal pine forest of Siberia) is partly a cultural artifact produced by native peoples selectively propagating favored tree species.

Since the nineteenth century, steadily increasing population in Asia, Africa, and South America has also added pressure on forests, as tens of thousands of square miles of forest have been cleared for cropland. This has been especially intense since 1950, about which time deforestation pressure shifted from Europe and North America to the tropical regions. From 1920 to 1949, about 907,000 mi^2 (2,350,000 km^2) of tropical forest were destroyed. From 1950, to 2001, boreal and temperate hardwood forests have declined only slightly, with growth almost keeping pace with demand for forest products, while 2,100,000 mi^2 (5,500,000 km^2) of tropical forest have disappeared. This is having important consequences for ecosystems, soil loss, hydrologic cycles, species extinction, and climate.

State of the World's Forests

As of 2005, forests covered about 30% of total land area, just under 15,450,000 mi^2 (40,000,000 km^2). The ten countries most rich in forest held two-thirds of global forest area: the top four, which together had about as much forest as the entire rest of the world put together, were the Russian Federation with 3,100,000 mi^2 (8,000,000 km^2), Brazil with 1,850,000 mi^2 (4,780,000 km^2), Canada with 1,200,000 mi^2 (3,100,000 km^2), and the United States with 1,170,000 mi^2 (3,030,000 km^2).

Globally, counting forest destruction, regrowth, planting, and natural increase, forest area was lost at 50,580 mi^2 (131,000 km^2) per year for 1990–2000 and at 49,800 mi^2 (129,000 km^2) per year for 2000–2005. However, losses were not evenly distributed worldwide; South America and Africa accounted for about 80% of forest loss from 1990 to 2005, with smaller losses in North and

WORDS TO KNOW

AFFORESTATION: Conversion of unforested land to forested land through planting, seeding, or other human interventions. Unforested land must have been unforested for at least 50 years for such intervention to qualify as afforestation; otherwise, it is termed reforestation.

ALBEDO: A numerical expression describing the ability of an object or planet to reflect light.

BIOMASS: The sum total of living and once-living matter contained within a given geographic area. Plant and animal materials that are used as fuel sources.

BOREAL FOREST: Type of forest covering much of northern Europe, Asia, and North America, composed mostly of coniferous evergreen trees. Although low in biodiversity, boreal forest covers more of Earth's land area than any other biome.

CARBON SEQUESTRATION: The uptake and storage of carbon. Trees and plants, for example, absorb carbon dioxide, release the oxygen, and store the carbon. Fossil fuels were at one time biomass and continue to store the carbon until burned.

CARBON SINK: Any process or collection of processes that is removing more carbon from the atmosphere than it is emitting. A forest, for example, is a carbon sink if more carbon is accumulating in its soil, wood, and other biomass than is being released by fire, forestry, and decay. The opposite of a carbon sink is a carbon source.

ETHNOBOTANIST: Scientist who specializes in the study of relationships between human cultures and plants, whether wild or domesticated. Culture is intimately related to forests, herbal medicines, sacred or recreational mind-altering substances derived from various plants, crops, and the like.

EVAPOTRANSPIRATION: Transfer of water to the atmosphere from an area of land, combining water transfer from foliage (transpiration) and evaporation from non-living surfaces such as soils and bodies of water.

FOSSIL FUELS: Fuels formed by biological processes and transformed into solid or fluid minerals over geological time. Fossil fuels include coal, petroleum, and natural gas. Fossil fuels are non-renewable on the timescale of human civilization, because their natural replenishment would take many millions of years.

HYDROLOGIC CYCLE: The process of evaporation, vertical and horizontal transport of vapor, condensation, precipitation, and the flow of water from continents to oceans. It is a major factor in determining climate through its influence on surface vegetation, the clouds, snow and ice, and soil moisture. The hydrologic cycle is responsible for 25 to 30% of the mid-latitudes' heat transport from the equatorial to polar regions.

KYOTO PROTOCOL: Extension in 1997 of the 1992 United Nations Framework Convention on Climate Change (UNFCCC), an international treaty signed by almost all countries with the goal of mitigating climate change. The United States, as of early 2008, was the only industrialized country to have not ratified the Kyoto Protocol, which is due to be replaced by an improved and updated agreement starting in 2012.

Central America and Oceania. Forest declined in Asia during 1990–2000 but increased in 2000–2005, the only region to show a turnaround in forest trends. Forest area increased from 1990 to 2005 in Europe.

Forests and Carbon

Globally, more carbon is stored in standing forest biomass (primarily wood and leaves), dead wood, litter (dead leaves, twigs, etc.), and soil than is present in Earth's atmosphere. Standing forest biomass alone, not counting other forms of forest storage, contains about 283 billion metric tons of carbon.

The ability of forests to sequester carbon—that is, take it out of the atmosphere and keep it there for climatically significant periods of time, i.e., decades or centuries—is the largest component of one of the two major carbon sinks or carbon-absorbing mechanisms of the planet. These two mechanisms are the terrestrial and the marine carbon sinks. The terrestrial carbon sink consists primarily of carbon uptake by forests (standing biomass, dead wood, litter, and soils); the marine carbon sink consists of the dissolving of CO_2 in the oceans. Together, these two sinks remove a little over 50% of the CO_2 being added to the atmosphere by human burning of fossil fuels each year. About half of this amount (25% of fossil-fuel emissions) is sequestered by forests. Carbon storage in forest biomass decreased in Africa, Asia, and South America in 1990–2005, but increased slightly in all other regions.

Since the late 1990s, there has been a high-stakes scientific debate about the nature and location of the terrestrial (land) carbon sink. Human activity releases about 7.1 billion metric tons of carbon to the atmosphere each year; about 2 billion tons are absorbed by the oceans, while 1.1 to 2.2 billion appear to be absorbed by the terrestrial sink, mostly forests. Efforts to discover which continent's forests are doing most of the absorbing have, however, given contradictory results.

There are two basic methods for trying to trace carbon sinking by forests. One is called inversion modeling. This method is often characterized as a top-down approach because it starts not with data about forests but with data about atmospheric CO_2. In inversion modeling, readings of atmospheric CO_2 recorded at ocean-sampling stations from 1988 onward are examined. Although Earth's atmosphere is well-mixed, so that a CO_2 measurement anywhere on the planet gives a good approximation of the value everywhere else, slight regional differences can be measured depending on whether large carbon sources or sinks are located upwind of the measuring station. Regional sourcing and sinking of carbon produces a temporary, slight increase or decrease in CO_2. Measurements of these slight differences are fed into a computer model that calculates where sinks and sources of CO_2 are most likely located.

The other basic method for tracing carbon sinking by forest is the inventory method. This is a bottom-up approach, whereby the acreage and standing timber volume of particular forests are counted to see how much carbon is located in which forests, and whether that amount is increasing or decreasing.

In 1998, a dramatic result based on inversion modeling was announced: North America's forests were sequestering 1.7 billion tons of CO_2 per year, about as much as was being emitted on that continent by fossil-fuel burning. About half of North America's forests are in Canada and about half in the United States, while most of the continent's CO_2 emissions from fossil fuels originate in the United States, which in 1998 was still the world's largest emitter of CO_2. If correct, this result implied that the United States and Canada had, as it were, won the greenhouse mitigation lottery: without taking any significant national measures to abate their releases of CO_2, they were already absorbing as much as they were emitting. Other countries, not the United States, the world's largest consumer of fossil fuels, were responsible for the increasing levels of atmospheric CO_2 that were the primary cause of global warming.

The result was swiftly challenged by studies based on inventory methods, which found forests sequestering much smaller amounts of carbon. Some studies also pointed to non-North American forests as major sinks. For example, in 1998 a British group reported that thickening tree trunks in undisturbed South American forests accounted for 40% of the 1.7 billion tons claimed for North America's forests by the inversion modelers. Both claims could not be true: there was not enough carbon to go around. Other scientists pointed out that many of North America's forests were young and that forests sequester less and less carbon as they mature, so the great North American carbon sink, even if real, could not last for many more decades.

Over the next few years, however, a modified picture emerged. Other inverse-emissions studies agreed that there was a carbon sink of about 2 billion tons per year in the Northern Hemisphere but did not necessarily locate it in North America (Russia has more forest than the United States and Canada put together, and is another obvious candidate for carbon sinking; China has expanding forests). A 1999 analysis of changing U.S. land use concluded that at least during the 1980s, U.S. forests accumulated carbon at a rate equal to about 10–30% of U.S. fossil-fuel emissions, consistent with a Northern Hemispheric forest sink but not with the high value for the North American sink found by the 1998 study.

A 2001 study of the issue published in *Science* stated that the apparent conflict between inventories and inversion models could be resolved by using lower estimates from inversion modeling and from direct CO_2 flow (flux) measurements in North America along with higher estimates from improved inventorying. Traditional forest inventorying does not count litter, soil carbon, wood products in landfills, or shrubs encroaching on grasslands due to suppression of wildfires. More realistic inventorying and the less extreme values from inverse modeling and flux measurements agreed on an annual uptake of CO_2 by U.S. forests between .3 and .7 billion tons per year, a good deal less than U.S. annual emissions of about 1.6 billion tons per year but still equal to 20–40% of global fossil-fuel emissions.

It also became clear that Chinese forests were sequestering large amounts of carbon dioxide, over 20 million tons of CO_2 per year by the late 1990s. This was largely the result of government policies of reforestation (restoring former forests) and afforestation (planting new forests) since the late 1970s. Planted forests, by 2001, accounted for 20% of all organic matter and 80% of sequestered carbon in China. These results tended to show that reforestation and afforestation policies could help significantly in mitigating climate change in coming decades, as well as providing traditional forest services such as wood, flood control, and habitat.

■ Impacts and Issues

Does Deforestation Increase Global Warming?

According to a study by G. Bala and colleagues published in the *Proceedings of the National Academy of Sciences* (U.S.) in 2007, total global deforestation would actually have a net cooling effect on Earth's climate. This is because deforestation has both warming and cooling effects, which vary in intensity depending on which climate region the forest being destroyed is in. Deforestation's warming effects are to release CO_2 into the atmosphere, eliminate possible future uptake by trees of CO_2 due to CO_2 fertilization (increased plant growth because of heightened CO_2 levels), and decrease evapotranspiration; its cooling effect is to decrease albedo, particularly in high latitudes (closer to the poles, especially in the Northern Hemisphere, where most high-latitude forested land is located).

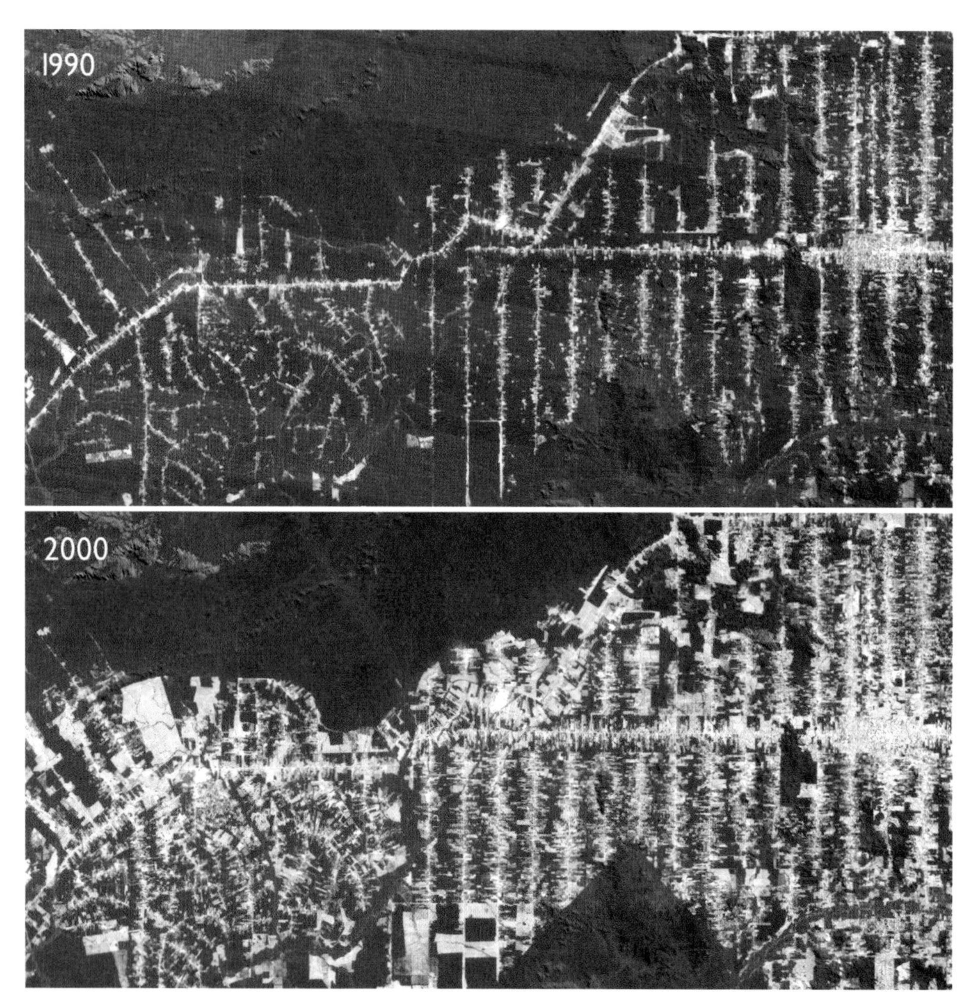

In these satellite images of Brazil's Amazon rainforest taken in 1990 (top) and 2000 (bottom), the extent of deforestation is visible. The images show roads that take loggers into the center of the rainforest, branching off to provide access to trees for logging. Although the rainforests are important areas of biodiversity, nearly one-fifth of the Amazon had been cleared as of 2005, according to the BBC. *M-Sat Ltd/Photo Researchers, Inc.*

The scientists stated that computer experiments in which only certain latitudes (tropics, temperate zones, high-latitude zones) are deforested show that afforestation and prevented deforestation in the tropics would definitely help mitigate global warming, but that afforestation would actually speed global warming if promoted in high-latitude regions such as the expanses of tundra now being rapidly warmed under global climate change. Temperate afforestation (e.g., in the contiguous United States), the researchers said, would "offer only marginal benefits." However, these results seem to contradict the studies cited earlier that show a large carbon sink in the forests of North America, absorbing about 20–40% of global fossil-fuel emissions annually. The scientists cautioned that their results were based on only a single computer study and that, in any case, forests in all parts of the world remain environmentally valuable for many reasons other than mitigation of climate change.

Issues: Tropical Forests

Tropical deforestation accounts for almost a third of annual global CO_2 emissions. Expansion of agricultural lands is a main driver of tropical deforestation. In the early 2000s in Brazil, about 4,000 mi^2 (10,000 km^2) of rainforest were being cleared yearly, mostly for cattle ranching, soy farming, and subsistence farming. This activity is often referred to as slash-and-burn agriculture, because land is cleared by cutting and burning. Burning trees immediately release their stored carbon to the air. The amounts released can be substantial, especially for mature tropical forests: each acre of old-growth Indonesian rainforest contains about 750 tons of CO_2 in standing biomass alone, that is, not counting litter and soil. As of 2005, widespread slash-and-burn agriculture made

Indonesia the world's third largest emitter of greenhouse gas and Brazil its fourth largest.

Because soils contain much of the carbon sequestered by forests, significant carbon releases can follow long after the cutting and burning of the trees themselves. A large-scale demonstration of this fact was given in the 1990s by an Indonesian government plan to clear and drain about 4,000 mi^2 (about 10,000 km^2) of peat-swamp forests. In this case, the soil of the forest region consisted of layers of peat 33–60 ft (10–20 m) thick. Peat is a dense deposit of dead plant matter too water-soaked to fully decompose—essentially young coal. Dried peat has been used as a fuel in some regions (e.g., in Scotland). Exposed, drained, and dried, vast areas of Indonesian peat caught fire in the late 1990s and again in the early 2000s, releasing an estimated 2 or more billion tons of carbon dioxide.

In 2007, the Australian and Indonesian governments announced a joint plan to preserve 270 mi^2 (700 km^2) of remaining Indonesian peat forests, re-flood 770 mi^2 (2,000 km^2) of peatland, and plant 100 million trees. The plan would, if successful, absorb or prevent the emission of more CO_2 per year than is emitted by Australia.

Kyoto and Forestry

The Kyoto Protocol climate treaty, which was created in 1997 and entered into force in 2005 (with the United States and Australia not participating), called for the management of terrestrial carbon sinks, particularly afforestation and reforestation, to reduce CO_2 in the atmosphere. Kyoto is particularly oriented toward the creation of young forest stands: cutting of an old-growth forest followed by replanting is not counted as a source of greenhouse emissions. In a 2000 study published in *Science*, however, Ernst-Detlef Schulze and colleagues suggested that preserving old-growth forests, rather than replacing them with young stands, would sequester far more carbon. Replacing old-growth with young-growth forests replaces a large pool of sequestered carbon in both biomass and soil carbon with a small pool of regrowth carbon and slowing the flow of carbon into soil.

■ Primary Source Connection

Tropical deforestation, the process where tropical forests are cut and usually burned, has been one of the greatest anthropogenic sources of greenhouse gases over the last several decades. In addition to releasing large quantities of carbon dioxide (CO_2) into the atmosphere, tropical deforestation further contributes to a global increase in greenhouse gases due to the decreased ability of tropical forests to remove CO_2 from the air. This article, from the journal *Science*, examines possible policy that could be implemented to reduce tropical deforestation. The reduction of tropical deforestation is one of the most cost-effective and easiest to implement CO_2 reduction plans.

TROPICAL FORESTS AND CLIMATE POLICY: NEW SCIENCE UNDERSCORES THE VALUE OF CLIMATE POLICY INITIATIVE TO REDUCE EMISSIONS FROM TROPICAL DEFORESTATION

Tropical deforestation released ~1.5 billion metric tons of carbon (GtC) to the atmosphere annually throughout the 1990s, accounting for almost 20% of anthropogenic greenhouse gas emissions. Without implementation of effective policies and measures to slow deforestation, clearing of tropical forests will likely release an additional 87 to 130 GtC by 2100, corresponding to the carbon release of more than a decade of global fossil fuel combustion at current rates. Drought-induced tree mortality, logging, and fire may double these emissions, and loss of carbon uptake (i.e., sink capacity) as forest area decreases may further amplify atmospheric CO_2 levels.

A combination of sovereignty and methodological concerns led climate policy-makers to exclude "avoided deforestation" projects from the 2008–12 first commitment period of the Kyoto Protocol's Clean Development Mechanism (CDM). The United Nations Framework Convention on Climate Change (UNFCCC) recently launched a 2-year initiative to assess technical and scientific issues and new "policy approaches and positive incentives" for Reducing Emissions from Deforestation (RED) in developing countries. This process was initiated at the request of several forest-rich developing nations, an indication of willingness to explore approaches to reduce deforestation that do not intrude upon national sovereignty. Recent technical progress in estimating and monitoring carbon emissions from deforestation and diverse climate policy and financing proposals to help developing countries reduce their deforestation emissions are currently being reviewed by the UNFCCC Subsidiary Body on Scientific and Technical Advice.

Whether a successful RED policy process can make an important contribution to global efforts to avoid dangerous climate change depends on two issues. First, are the potential carbon savings from slowing tropical deforestation sufficient to contribute substantially to overall emissions reductions? Second, is it likely that tropical forests (and the forest carbon) protected from deforestation will persist over coming decades and centuries in the face of some unavoidable climate change? The available evidence indicates that the answer to both questions is yes, especially in a future with aggressive efforts to limit atmospheric CO_2. . . .

Reducing deforestation rates 50% by 2050 and then maintaining them at this level until 2100 would avoid the direct release of up to 50 GtC this century (equivalent to nearly 6 years of recent annual fossil fuel emissions, and up to 12% of the total reductions that must be achieved from all sources through 2100 to be consistent with stabilizing atmospheric concentrations of CO_2 at 450 ppm. Emissions reductions from reduced deforestation may be among the least-expensive mitigation options available. The IPCC estimates that reductions equal to or greater than the scale suggested here could be achieved at £U.S.$20 per ton CO_2.

Reducing deforestation not only avoids the release of the carbon stored in the conserved forests, but by reducing atmospheric carbon, it also helps to reduce the impacts of climate change on remaining forests. The experience of the 1997–98 El Niño Southern Oscillation Event (ENSO) demonstrates how climate change can interact with land-use change to put large areas of tropical forests and their carbon at risk. The extended dry conditions triggered by the ENSO across much of the Amazon and Southeast Asia increased tree mortality and forest flammability, particularly in logged or fragmented forests. Globally, increased forest fires during the 1997–98 ENSO released an extra 2.1 $\pm$ 0.8 GtC to the atmosphere.

Even in non-ENSO years, global warming may be putting tropical forest regions at risk of more frequent and severe droughts. Over the last 5 years, a number of Amazon Basin and Southeast Asian droughts have been uncoupled from ENSO events but have coincided with some of the warmest global average temperatures on record.

In recent decades, carbon losses from tropical deforestation have been partly or largely offset by a tropical sink. Forest sinks are, however, unlikely to continue indefinitely, and continued warming will likely diminish and potentially even override any fertilization effects of increasing CO_2. Climate change might also adversely impact tropical forests by reducing precipitation and evapotranspiration, making them drier, more susceptible to fires, and more prone to replacement by shrublands, grasslands, or savanna ecosystems, which store much less carbon. In the Amazon Basin, continued deforestation may disrupt forest water cycling, amplifying the negative impacts of climate change. A new generation of coupled climate-carbon models is being used to explore the prospects for the persistence of tropical forests in a changing climate. A widely discussed early study projected that business-as-usual increases in CO_2 and temperature could lead to dramatic dieback and carbon release from Amazon forests, raising concerns that high sensitivity of tropical forests to climate change might compromise the long-term value of reduced deforestation, with dieback releasing much of the carbon originally conserved. However, of 11 coupled climate-carbon cycle models using the IPCC's mid-to-high range A2 emissions scenario, 10 project that tropical forests continue to act as carbon sinks, albeit declining sinks, throughout the century. The moderate sensitivity indicated by the new results suggests that reducing deforestation can result in longterm carbon storage, even with substantial climate change. Aggressive efforts to reduce industrial and deforestation emissions would likely further reduce the rate of decline and risk of reversal of the tropical sink.

While no single climate policy approach is likely to address the diverse national circumstances faced by forest-rich developing countries seeking to reduce their emissions, there are promising examples of countries with adequate resources and political will that have been able to reduce forest clearing. In some countries, it may be possible at relatively low cost to reduce emissions from deforestation and forest degradation that provide little or no benefit to local and regional economies. For example, reducing accidental fire and eliminating forest clearing on lands that are inappropriate for agriculture are two promising lowcost options for reducing greenhouse gas emissions in Brazil and Indonesia.

Other measures are unlikely to be implemented at large scales without financial incentives that may be feasible only within the framework of comprehensive environmental service payments, such as through carbon-market financing. In forests slated for timber production, for example, moderate carbon prices could support widespread adoption of sustainable forestry practices that both directly reduce emissions and reduce the vulnerability of logged forests to further emissions from fire and drought exacerbated by global warming. On forested lands threatened by agricultural expansion, financing could provide significant incentives for forest retention and enable, for example, more effective implementation of land-use regulations on private property and protected area networks.

Parties to the UNFCCC should consider adopting a range of options, from capacity building supported by traditional development assistance to carbon-market financing to help developing countries meet voluntary national commitments for reductions in forest-sector emissions below historic baselines. Voluntary commitments, which were put forward by several tropical forest nations, would substantially address a concern associated with the project-based approach of the CDM that emissions reductions from a site-specific project might simply be offset by increased deforestation elsewhere.

Key requirements for effective carbon-market approaches to reduce tropical deforestation include strengthened technical and institutional capacity in many developing countries, agreement on a robust system for measuring and monitoring emissions reductions, and commitments to deeper reductions by industrialized countries to create demand for RED carbon credits and to ensure that these reductions are not simply traded off against less emission reductions from fossil fuels.

Beyond protecting the climate, reducing tropical deforestation has the potential to eliminate many negative impacts that may compromise the ability of tropical countries to develop sustainably, including reduction in rainfall, loss of biodiversity, degraded human health from biomass burning pollution, and the unintentional loss of productive forests. Providing economic incentives for the maintenance of forest cover can help tropical countries avoid these negative impacts and meet development goals, while also complementing aggressive efforts to reduce fossil fuel emissions. Industrialized and developing countries urgently need to support the RED policy process and develop effective and equitable compensation schemes to help tropical countries protect their forests, reducing the risk of dangerous climate change and protecting the many other goods and services that these forests contribute to sustainable development.

Raymond E. Gullison, et al.

GULLISON, RAYMOND E., ET AL. "TROPICAL FORESTS AND CLIMATE POLICY: NEW SCIENCE UNDERSCORES THE VALUE OF CLIMATE POLICY INITIATIVE TO REDUCE EMISSIONS FROM TROPICAL DEFORESTATION." *SCIENCE* 316 (MAY 18, 2007): 985–986.

SEE ALSO *Biomass; Biosphere; Carbon Cycle; Carbon Sinks; Feedback Factors; Sequestration.*

BIBLIOGRAPHY

Periodicals

Bala, G., et al. "Combined Climate and Carbon-Cycle Effects of Large-Scale Deforestation." *Proceedings of the National Academy of Sciences* 104 (April 17, 2007): 6550–6555.

Magnani, Federico, et al. "The Human Footprint in the Carbon Cycle of Temperate and Boreal Forests." *Nature* 447 (June 14, 2007): 849–852.

Martin, Philippe H., et al. "Carbon Sinks in Temperate Forests." *Annual Review of Energy and the Environment* 26 (2001): 435–65.

Melillo, J. M., et al. "Tropical Deforestation and the Global Carbon Budget." *Annual Review of Energy and the Environment* 21 (1996): 293–310.

Piao, Shilong, et al. "Changes in Climate and Land Use Have a Larger Direct Impact Than Rising CO_2 on Global River Runoff Trends." *Proceedings of the National Academy of Sciences* 104 (September 25, 2007): 15242–15247.

Randerson, J. T., et al. "The Impact of Boreal Forest Fire on Climate Warming." *Science* 314 (November 17, 2006): 1130–1133.

Saleska, Scott R., et al. "Amazon Forests Green-Up During 2005 Drought." *Science* 318 (October 26, 2007): 612.

Schulze, Ernst-Detlef, et al. "Climate Change: Managing Forests After Kyoto." *Science* 289 (September 22, 2000): 2058–2059.

Williams, Michael. "The History of Deforestation." *History Today* 51 (July 2001): 30–39.

Wofsy, Steven C. "Where Has All the Carbon Gone?" *Science* 292 (June 22, 2001): 2261–2263.

Web Sites

"Global Forest Resources Assessment 2005." *Food and Agriculture Organization of the United Nations.* <http://www.fao.org/docrep/008/a0400e/a0400e00.htm> (accessed November 9, 2007).

Larry Gilman

Gaia

Introduction

The word Gaia incorporates several varied concepts about planet Earth, how it is organized, and how it has evolved. Over the last few decades, research on Gaia theory, once derided by many as unscientific, has been influential in bringing together scientists from disparate fields such as geochemistry, biology, and atmospheric physics in order to promote new understanding of Earth's systems, especially regarding natural and anthropogenic (human-caused) climate change.

All recent scientific definitions of Gaia theory include the assumption that Earth is a self-regulating entity that promotes life. As the creator of the Gaia hypothesis, British chemist, medical researcher, and atmospheric scientist James Lovelock, states that Gaia is a complex system, "organisms and material environment coupled together," that continues to evolve. He adds: "Unless we see the Earth as a planet that behaves as if it were alive, at least to the extent of regulating its climate and chemistry, we will lack the will to change our way of life and to understand that we have made it our greatest enemy."

Historical Background and Scientific Foundations

The Gaia hypothesis was first formulated by Lovelock in the late 1960s, and published in 1972. His neighbor, British author William Golding, suggested the term Gaia (from the Greek personification of Earth, and the goddess that followed Chaos) as an alternative to Lovelock's "cybernetic system with homeostatic tendencies."

The Gaia hypothesis is often oversimplified as merely the idea that Earth (or Gaia) is a living planet. Lovelock actually defined Gaia as "a complex entity involving the Earth's biosphere, atmosphere, oceans, and soil; the totality constituting a feedback or cybernetic system which seeks an optimal physical and chemical environment for life on this planet." In other words, life on Earth and the non-living parts of the planet form a giant system that includes the effects of living things, climate, and the composition of the atmosphere and oceans. These components change to maintain Earth's system in its present state of equilibrium, producing the optimal conditions for life.

Lovelock's initial hypothesis met with scientific skepticism and was criticized for circular reasoning (because Earth is hospitable for life, Gaia exists, and because Gaia exists, there is life on Earth), and its lack of testable hypotheses, although it became popular as an environmental and philosophical concept. In the 1980s and 1990s, however, several conferences, clarifications, and modifications moved the Gaia hypothesis toward its more recent and more scientifically robust position as the Gaia theory. Today, many scientists also categorize different studies using Gaia theory as strong Gaia or weak Gaia, depending on how pervasive they see the importance of life in modifying Earth, and whether purpose or directionality is involved, as some strong Gaian arguments assert.

American microbiologist Lynn Margulis has been one of the scientists engaged in defining, popularizing, and applying Gaia theory. Margulis defines symbiosis as organisms of different species living together in physical contact, and credits her student Greg Hinkle as coming up with the explanation that "Gaia is just symbiosis as seen from space." Margulis goes on to clarify the best-accepted scientific basis for Gaia theory, noting that "Gaia is neither vicious nor nurturing in relation to humanity; it is a convenient name for an Earthwide phenomenon: temperature, acidity/alkalinity, and gas composition regulation. Gaia encompasses the series of interacting ecosystems that compose a single huge ecosystem at the Earth's surface."

Impacts and Issues

An important model demonstrating the feedback mechanisms that are crucial to Gaia theory is Daisyworld, which was introduced in 1983. Daisyworld simulations first illustrate a simple planet containing two species of daisies with different albedo (the amount of light they reflect). As the simulation progresses, quantities of dark and light colored daisies spread, covering the planet, and the climate on Daisyworld changes until it reaches homeostasis. Recent Daisyworld simulations include many different species, diseases, and more environmental factors. All demonstrate that life (or at least a model of life) may alter its environment to an optimal state, as the Gaia theory proposes.

As understanding positive and negative feedback in complicated systems (involving atmosphere, plant and ocean life, soil, and anthropogenic factors) is critical to understanding climate change, Daisyworld is an important teaching tool for contemporary research. As the American physicist Lawrence E. Joseph points out, Gaia is also "a brilliant organizing principle" for promoting the interdisciplinary work that is now essential for understanding climate change and other global environmental issues.

Much of the current research on the systems that make up Earth (or Gaia) is incorporated by the emerging field of Earth System Science (ESS). Many scientists acknowledge that Earth System Science owes a historic and scientific debt to Gaia theory, and argue that the philosophical focus of Gaia theory actually provides a more holistic and humanistic alternative for scientific research than ESS does, which some scientists argue could be better suited for the current climate crisis.

Gaia may continue to prove most important, however, as a vehicle for scientists to convey their work on Earth's complex systems to non-scientists. As anthropologist Mary Catherine Bateson notes, "It may be that many people, possibly all people, can only think about complex wholes—think holistically—by using aesthetic or religious energy.... Gaia is the supersystem. It is intellectually irresistible."

WORDS TO KNOW

ALBEDO: A numerical expression describing the ability of an object or planet to reflect light.

BIOSPHERE: The sum total of all life-forms on Earth and the interaction among those life-forms.

FEEDBACK: Information that tells a system what the results of its actions are and thereby alters future actions.

HOMEOSTASIS: State of being in balance; the tendency of an organism to maintain constant internal conditions despite large changes in the external environment.

HYPOTHESIS: An idea phrased in the form of a statement that can be tested by observation and/or experiment.

SYMBIOSIS: A pattern in which two or more organisms live in close connection with each other, often to the benefit of both or all organisms.

SEE ALSO *Albedo; Atmospheric Chemistry; Atmospheric Structure; Baseline Emissions; Biodiversity; Biogeochemical Cycle; Biomass; Biosphere; Carbon Cycle; Carbon Dioxide (CO_2); Carbon Dioxide Concentrations; Carbon Sinks; Chlorofluorocarbons and Related Compounds; Climate Change; Feedback Factors; Oceans and Seas; Simulations.*

BIBLIOGRAPHY

Books

Harding, Stephan. *Animate Earth: Science, Intuition, and Gaia.* White River Junction, VT: Chelsea Green Publishing, 2006.

Joseph, Lawrence E. *Gaia: The Growth of an Idea.* New York: St. Martin's Press, 1990.

Lovelock, James. *Gaia: A New Look at Life on Earth.* Oxford: Oxford University Press, 1979.

Lovelock, James. *The Revenge of Gaia: Earth's Climate in Crisis and the Fate of Humanity.* New York: Basic Books, 2006.

Margulis, Lynn. *Symbiotic Planet: A New View of Evolution.* New York: Basic Books, 1998.

Margulis, Lynn, and Dorion Sagan. *Slanted Truths: Essays on Gaia, Symbiosis, and Evolution.* New York: Copernicus, 1997.

Schneider, Stephen H. *Scientists Debate Gaia: The Next Century.* Cambridge, MA: MIT Press, 2004.

Periodicals

Kerr, Richard A. "No Longer Willful, Gaia Becomes Respectable." *Science* 240 (1988): 393.

Lovelock, James E. "Gaia as Seen Through the Atmosphere." *Atmospheric Environment* 6, no. 8 (1972): 579–580.

Lovelock, James E. "Hands Up for the Gaia Hypothesis." *Nature.* 344 (1990): 100–102.

Watson, Andrew, and James Lovelock. "Biological Homeostasis of the Global Environment: The Parable of Daisyworld." *Tellus* 35B (1983): 284–289.

Web Sites

Bice, David. "Gaia: Biospheric Control of the Global Climate System." *Carleton College.* <http://www.carleton.edu/departments/geol/DaveSTELLA/Daisyworld/gaia.htm> (accessed November 4, 2007).

"Earth System Science: Gaia and the Human Impact." *Crispin Tickell,* October 18, 2007.<http://www.crispintickell.com/page123.html> (accessed November 7, 2007).

"The Gaia Theory: Model and Metaphor for the 21st Century." *The Gaia Theory Conference 2006,* October 16, 2007. <http://www.gaiatheory.org/> (accessed November 4, 2007).

Sandra L. Dunavan

General Circulation Model (GCM)

Introduction

A general circulation model (GCM) is a complex mathematical model that attempts to provide information about the global climate. GCMs usually include equations that describe the energy changes that occur when regions of different temperature, pressure, chemical composition, velocities, and accelerations interact with each other. Some GCMs focus on modeling the atmosphere (AGCMs) while others model the ocean (OGCMs). More advanced versions couple the atmosphere and ocean (AOGCMs), but become increasingly complex.

Historical Background and Scientific Foundations

In the 1960s, scientists began developing models of circulation patterns and heat transfer in the atmosphere. Led by the work of Syukuro Manabe (1931–) at the Program in Atmospheric and Oceanic Science at Princeton University, a group of theoretical meteorologists developed models of the atmospheric dynamics. In collaboration with Kirk Bryan (1929–), Manabe coupled the AGCM to a model of ocean dynamics. Predictions from this complex system formed the basis for some early reports of the Intergovernmental Panel on Climate Change (IPCC).

AOGCMs are four-dimensional models. Three of the dimensions are spatial: they mathematically slice the ocean and atmosphere into depth layers and further subdivide these layers by latitude and longitude. The models calculate changes that happen in each of these small regions with regard to the fourth dimension, time. Significant research has gone into understanding how changes in resolution and numerical techniques used to calculate changes in the models affect the stability and reliability of the models.

GCMs usually consider variables that affect temperature, in particular energy input and energy dissipation. Examples of dynamics that might affect these variables are velocity, pressure, radiation, albedo, convection, and

WORDS TO KNOW

ALBEDO: A numerical expression describing the ability of an object or planet to reflect light.

CARBON CYCLE: All parts (reservoirs) and fluxes of carbon. The cycle is usually thought of as four main reservoirs of carbon interconnected by pathways of exchange. The reservoirs are the atmosphere, terrestrial biosphere (usually includes freshwater systems), oceans, and sediments (includes fossil fuels). The annual movements of carbon, the carbon exchanges between reservoirs, occur because of various chemical, physical, geological, and biological processes. The ocean contains the largest pool of carbon near the surface of Earth, but most of that pool is not involved with rapid exchange with the atmosphere.

HYDROLOGY: The science that deals with global water (both liquid and solid), its properties, circulation, and distribution.

INTERGOVERNMENTAL PANEL ON CLIMATE CHANGE (IPCC): Panel of scientists established by the World Meteorological Organization (WMO) and the United Nations Environment Programme (UNEP) in 1988 to assess the science, technology, and socioeconomic information needed to understand the risk of human-induced climate change.

MATHEMATICAL MODEL: Used to develop descriptions of systems using numerical data and to make predictions of future states or events based upon current data and known changes in data over time.

METEOROLOGY: The science that deals with Earth's atmosphere and its phenomena and with weather and weather forecasting.

water processes like evaporation of water vapor, precipitation, and hydrology. The next generation of GCMs, called Earth System Models of Intermediate Complexity (EMICs), will specifically incorporate the role of the carbon cycle in climate change into simulations.

■ Impacts and Issues

In 2007, the IPCC released its fourth assessment. The panel placed significant reliance on predictions made by some of the more advanced AOGMs. The major prediction from the AOGMs studied by the IPCC is that Earth will continue to warm at a rate of 0.4°F (0.2°C) per decade.

The panel reports that AOGCMs can reproduce observed measurements of the climate and that they are able to make quantitative estimates of climate change. In particular, confidence in the temperature predictions made by the models is very strong. The models have been shown to be very effective at simulating extreme weather events, like hot and cold spells and the frequency and distribution of tropical cyclones. Some of the variables that are not as well represented in AOGCMs are precipitation and the effects of cloud cover.

SEE ALSO *Albedo; Atmospheric Circulation; Atmospheric Pollution; Atmospheric Structure; Carbon Dioxide (CO_2); Chaos Theory and Meteorological Predictions; Clouds and Reflectance; Extreme Weather; Global Change Research Program; Intergovernmental Panel on Climate Change (IPCC); Meteorology; Simulations; Temperature and Temperature Scales.*

BIBLIOGRAPHY

Books

Solomon, S., et al, eds. *Climate Change 2007: The Physical Science Basis: Contribution of Working Group I to the Fourth Assessment Report of the Intergovernmental Panel on Climate Change.* New York: Cambridge University Press, 2007.

Web Sites

"Estimating the Circulation and Climate of the Ocean." *Massachusetts Institute of Technology,* 2007. <http://www.ecco-group.org/index.htm> (accessed October 28, 2007).

"General Circulation Models." *NASA Goddard Institute for Space Studies.* <http://www.giss.nasa.gov/research/modeling/gcms.html> (accessed October 28, 2007).

"Intergovernmental Panel on Climate Change." *World Meteorological Organization and the United Nations Environmental Programme.* <http://www.ipcc.ch> (accessed October 28, 2007).

"The Next Generation of Climate Models." *HPC Wire,* July 21, 2006. <http://www.hpcwire.com/hpc/732506.html> (accessed October 28, 2007).

Juli Berwald

Geologic Time Scale

Introduction

In climate change studies many reference are made to the geologic record or geologic time. The geologic time scale is an internationally developed and agreed scheme of subdividing the passage of time since the origin of Earth. This time scale is universally used among geologists, paleontologists, and other natural scientists who deal with Earth history and Earth antiquity issues. The original structure of the geologic time scale was developed during the nineteenth and twentieth centuries, mainly as a result of studies of stratigraphy (the study of Earth's rock layers). The finer details of the geologic time scale continue to evolve today. At first, the geologic time scale was a relative time scale, which means that the absolute ages of various intervals within the time scale were not known. With the advent of radiometric age-dating techniques, the ages of the boundaries of these intervals has become established in most instances and the modern geologic time scale became essentially an absolute time scale. Ages of boundaries between subdivisions are usually given in millions of years (abbreviated Ma) before present.

WORDS TO KNOW

INTERNATIONAL UNION OF GEOLOGICAL SCIENCES (IUGS): Non-governmental group of geologists founded in 1961 and headquartered in Trondheim, Norway. The group fosters international cooperation on geological research of a transnational or global nature.

RADIOMETRIC AGE: The age of an object as determined by the levels of certain radioactive substances in its substance. Radioactive substances break down into other substances (some radioactive and some not) over time in a regular way: measurements of their relative quantities can allow an estimate of an object's age (e.g., rock, piece of wood). Many techniques for determining radiometric age are used in the field of radiometric dating.

STRATIGRAPHY: The branch of geology that deals with the layers of rocks or the objects embedded within those layers.

Historical Background and Scientific Foundations

Geologists were unable to accurately measure the dimensions of Earth's history until mass spectrometers (a device that allows the determination of the elements and compounds in a sample) became available in the 1950s. Before that time, inferences were made by comparing the rock record from different parts of the world and estimating how long it would take natural processes to create formations.

Georges Louis Leclerc de Buffon (1707–1788), for example, calculated Earth to be 74,832 years old by figuring how long it would take the planet to cool down to the present temperature. Writing around 1770, he was among the first to suggest that Earth's history can be known about by observing its current state.

James Hutton (1726–1797) did not propose a date for the formation of Earth, but is famous for the statement that Earth contains "no vestige of a beginning—no prospect of an end." The German geologist Abraham Werner (1750–1817) was the first scientist to make use of a stratigraphic column, which is a diagram showing the order of sedimentary layers. The French zoologist and paleontologist Georges Cuvier (1769–1832) observed that specific fossil animals occurred in specific rock layers, forming recognizable groups, or assemblages. William Smith (1769–1768) combined Werner's and Cuvier's approaches, using fossil assemblages to identify identical sequences of layers distant from each other, linking or correlating rocks which were once part of the same rock layer but had been separated by faulting or erosion.

In 1897, the British physicist Lord Kelvin (1824–1907) developed a model that assumed that Earth has been cooling steadily since its formation. Because he did not know that heat is transported by convection currents beneath Earth's surface or that Earth generates its own heat from the decay of radioactive minerals buried inside it, Kelvin proposed that Earth was formed from 20 to 40 million years ago.

In the late eighteenth century, geologists began to name periods of geologic time. In the nineteenth century, geologists such as William Buckland (1784–1856), Adam Sedgwick (1785–1873), Henry de la Beche (1796–1855), and Roderick Murchison (1792–1871) identified widespread rock layers beneath continental Europe, the British Isles, Russia, and America. They named periods of time after the places in which these rocks were first described. For instance, the Cambrian Period was named for Cambria (the Roman name for Wales), and Permian, for the Perm province in Czarist Russia. Pennsylvanian and Mississippian periods, widely used by American geologists, were named for a U.S. state and a region around the upper reaches of a large river, respectively. By the mid-nineteenth century, most of the modern names of the periods of geologic time had been proposed.

Today's Geologic Time Scale

Today, the geologic time scale is hierarchically divided (from largest to smallest) into Eons, Eras, Periods, Epochs, and Ages. The corresponding rocks that represent these subdivisions are referred to as Eonothems, Erathems, Systems, Series, and Stages. At this time, the geologic time scale is administered worldwide by the International Union of Geological Sciences (IUGS) and its International Commission on Stratigraphy. IUGS geological maps display standard colors for the geologic time scale. In the United States, slightly different colors are used on geological maps, which follow the color scheme of the U.S. Geological Survey.

As the precision and accuracy of radiometric age-dating techniques have improved over time, so too are the age dates of geologic time scale boundaries more precise and accurate. Over the past few decades, international working groups have been established under the auspices of the IUGS to refine the definition and ages of geologic time scale boundaries. These working groups have established many reference sites on Earth, called GSSPs (Global Stratotype and Point), where a geological boundary has been established and age-dated. These GSSPs help establish firmly the age of Stage (and thus Age) boundaries in the geologic time scale. These boundaries in turn establish the ages of the other hierarchical groups within the geologic time scale.

The development of the geologic time scale was critically important to the development of the science of geology because it gave a universally accepted framework into which geological events of the past could be fit. This framework permitted correlation of events in Earth history, a key aspect in the evolving science of Earth history (Historical Geology). In addition, the history of life as revealed in the fossil record could be described within this framework. As a communication device, the geologic time scale is essential because the named subdivisions of the geologic time scale are the same in all places over the Earth where they are exposed.

Eons, Eras, Periods, Epochs, and Ages

According to the modern geologic time scale, the Archean was the first eon, which spanned the time from Earth's formation (about 4,570 million years ago) to an arbitrary point in time about 2,500 million years ago. From 2,500 to 542 million years ago is the Proterozoic Eon. All remaining geologic time, up to present, is contained within the Phanerozoic Eon. The term Phanerozoic means visible life, which refers to the fact that fossils are usually quite evident in sedimentary rocks deposited during the Phanerozoic. Within the Archean and Proterozoic eons, Epochs and Stages are not delineated, so the Period is the level of subdivision used by geologists at this time. The ages of these Period boundaries are called GSSAs (Global Standard Stratigraphic Ages), a concept similar to the GSSPs noted earlier.

Within Phanerozoic rocks, there are three Eras. In age order (oldest first), they are Paleozoic, Mesozoic, and Cenozoic. Paleozoic is divided into six Periods. From oldest to youngest, they are: Cambrian, Ordovician, Silurian, Devonian, Carboniferous, and Permian. Mesozoic is similarly divided into three Periods: Triassic, Jurassic, and Cretaceous. Cenozoic is similarly divided into Paleogene, Neogene, and Quaternary Periods.

All Phanerozoic Periods are divided into two or more Epochs. Some Epochs are named, for example, the Miocene and Pliocene Epochs within Neogene, and some are given position names, such as the Lower and Upper Cretaceous. Ages are named subdivisions of Epochs. Some Ages are not formally named within the Cambrian, Ordovician, and Silurian Periods. Stage names, by international agreement, are based mainly on U.K. and European reference locations and thus are named for the British and European sites where they are well known.

There are two Sub-Periods in the geologic time scale, Mississippian and Pennsylvanian. These are vestiges of an interval in the development of the geologic time scale when geologists in the United States advocated splitting the Carboniferous into two Periods. This was never agreed internationally, and the Sub-Periods are retained because of this controversy.

Because of the close association of some dominant fossil groups with some geologic time scale subdivisions, there are informal fossil names for some of the subdivisions. For example, the Mesozoic Era is informally called "the age of the dinosaurs." Likewise, the Cenozoic Era is informally called "the age of mammals."

As radiometric age dates and suitable reference sites become available and are agreed upon, additional GSSPs will be established until the whole of the Phanerozoic has been delineated with GSSPs. Likewise, in older rocks, work will continue to refine GSSAs within Archean and Proterozoic rocks. The status of Quaternary as a Period within the Cenozoic is being examined as is the status of the Holocene as an Epoch. There are many unsettled questions on the organization of Earth's youngest time scale subdivisions and many science groups are interested in this issue, including archeologists, glacial geologists, and others.

Other Geologic Time Scales

As noted, there is one internationally recognized geologic time scale (or International Stratigraphic Chart), which is administered and updated by the IUGS. There are other geologic time scales, which are not as widely embraced as the IUGS effort. For example, the Geological Society of America (GSA) has published a geologic time scale that is widely used, especially in publications of that society. The GSA geologic time scale is not vastly different from the IUGS time scale, but some of the Age names within the older Paleozoic Periods are different as are some of the radiometric age dates at geologic time scale boundaries. The differences in ages may be attributed to the fact that the GSA geologic time scale is a few years older than the IUGS time scale and has not been recently updated. The GSA geologic time scale uses the Tertiary Period for the pre-Quaternary Cenozoic and considers the Paleogene and Neogene (Periods in the IUGS time scale) as Sub-Periods.

Age Determinations

A rock layer may or may not contain evidence that reveals its age. Rock layers for which ages are defined by relationships with the dated rock units around it are examples of relative age determination. That relationship is found by observing the unknown rock layer's stratigraphic relationship with the rock layers for which ages are known. If the known rock layer is on top of the unknown layer, then the lower layer is probably the older of the two. That inference is based on the principle of superposition, which states that when two rock layers are stacked one above the other, the lower one was formed before the overlying one, unless the layers have been overturned.

Every rock and mineral exists in the world as a mixture of elements, and every element exists as a population of atoms. One element's population of atoms will not all have the same number of neutrons, and so two or more kinds of the same element will have different atomic masses or atomic numbers. These different kinds of the same chemical element are called nuclides of that element. A nuclide of a radioactive element is known as a radionuclide.

The nucleus of every radioactive element spontaneously disintegrates over time. This process results in radiation, and is called radioactive decay. Losing high-energy particles from their nuclei transforms the atoms of a radioactive nuclide into the daughter product of that nuclide. A daughter product is either a different element altogether, or is a different nuclide of the same parent element. A daughter product may or may not be radioactive. If it is, it also decays to form its own daughter product. The last radioactive element in a series of these transformations will decay into a stable element, such as lead.

Although there is no way to discern whether an individual atom will decay today or two billion years from today, the behavior of large numbers of the same kind of atom is so predictable that certain nuclides are known as radioactive clocks. The use of these radioactive clocks to calculate the age of a rock is referred to as radiometric age determination. First, an appropriate radioactive clock must be chosen. The sample must contain measurable quantities of the element to be tested for, and its radioactive clock must tell time for the appropriate interval of geologic time. Then, the amount of each nuclide present in the rock sample must be measured.

Each radioactive clock consists of a radioactive nuclide and its daughter product, which accumulate within the atomic framework of a mineral. These radioactive clocks decay at various rates, which govern their usefulness in particular cases. A three-billion-year-old rock needs to have its age determined by a radioactive clock that still has a measurable amount of the parent nuclide decaying into its daughter product. The same radioactive clock would reveal nothing about a two million year old rock, because the rock would not yet have accumulated enough of the daughter product to measure.

The time it takes for half of the parent nuclide to decay into the daughter product is called one half-life. The remaining population of the parent nuclide is halved again, and the population of daughter product doubled, with the passing of every succeeding half-life. The amount of parent nuclide measured in the sample is plotted on a graph of that radioactive clock's known half-life. The absolute age of the rock, within its margin of error, can then be read directly from the time axis of the graph.

When a rock is tested to determine its age, different minerals within the rock are tested using the same radioactive clock. Ages may be determined on the same sample by using different radioactive clocks. When the age of a rock is measured in two different ways, and the results are the same, the results are said to be concordant.

Discordant ages means the radioactive clock showed different absolute ages for a rock sample, or different ages for different minerals within the rock. A discordant age result means that at some time after the rock was formed, something happened to it which reset one of the radioactive clocks.

For example, if a potassium-argon test produces a discordant result, the rock may have been heated to a blocking temperature above which a mineral's atomic framework becomes active enough to allow trapped gaseous argon-40 to escape.

Concordant ages mean that no complex sequence of events-deep burial, metamorphism, and mountain-building, for example, has happened that can be detected by the two methods of age determination that were used.

A form of radiometric dating is used to determine the ages of organic matter (matter from once living things that contain carbon centered molecules and compounds). A short-lived radioisotope, carbon-14, is accumulated by all living things on Earth. At death, the amount of carbon-14 is fixed and then begins to decay into carbon-12 at a known rate (its half-life is 5,730 years). By measuring how much of the carbon-14 is left in the remains, and plotting that amount on a graph showing how fast the carbon-14 decays, the approximate date of the organism's death can be known.

When uranium atoms decay, they emit fast, heavy alpha particles. Inside a zircon crystal, these subatomic particles leave trails of destruction in the zircon's crystal framework. The age of a zircon crystal can be estimated by counting the number of these trails. The rate at which the trails form has been found by determining the age of rocks containing zircon crystals, and noting how torn-up the zircon crystals become over time. This age determination technique is called fission-track dating. This technique has detected the world's oldest rocks, between 3.8 billion and 3.9 billion years old, and yet older crystals, which suggest that Earth had some solid ground on it 4.2 billion years ago.

The age of Earth is deduced from the ages of other materials in the solar system, namely, meteorites. Meteorites are pieces formed from the cloud of dust and debris left behind by a supernova, the explosive death of a star. Through this cloud the infant Earth spun, attracting more and more pieces of matter. The meteorites that fall to Earth today have orbited the sun since that time, unchanged and undisturbed by the processes that have destroyed Earth's first rocks. Radiometric ages for meteorites fall between 4.45 billion and 4.55 billion years.

The radionuclide iodine-129 is formed in nature only inside stars. A piece of solid iodine-129 will almost entirely decay into the gas xenon-129 within a hundred million years. If this decay happens in open space, the xenon-129 gas will float off into space, blown by the solar wind. Alternatively, if the iodine-129 were in a rock within a hundred million years of being formed in a star, then some very old rocks should contain xenon-129 gas. Both meteorites and Earth's oldest rocks contain xenon-129. That means the star that provided the material for the solar system died less than 4.65 billion years ago.

Impacts and Issues

The geologic time scale provides a temporal frame of reference for scientists to communicate about Earth's past. Such communication may involve, among other things, fossil evolution, Earth's tectonic changes, and climatic change. For example, the warm, equitable Cretaceous (144-65 million years ago) climates that supported a tremendous diversity of animal (including dinosaurs) and plant life eventually changed to globally cooler climates during the Neogene period (23 million years ago to the near-present). During the Cretaceous, there was little or no polar ice, and oceans that were much warmer than today covered more of Earth's land area. Climatic changes near the end of the Cretaceous resulting from volcanism, the asteroid impact credited for the Cretaceous-Tertiary mass extinction, and effects of continental drift resulted in more distinct seasons, dropping sea levels, and greater extremes between temperatures at the equator and the poles.

SEE ALSO *Ice Ages.*

BIBLIOGRAPHY

Books

Gradstein, F. M., et al. *A Geologic Time Scale.* Cambridge: Cambridge University Press, 2004.

Tarbuck, E. J., F. K. Lutgens, and D. Tasa. *Earth: An Introduction to Physical Geology.* Upper Saddle River, NJ: Prentice Hall, 2005.

Walker, M. J. C. *Quaternary Dating Methods.* Chichester, UK: Wiley, 2005.

Web Sites

"Geologic Time: Online Edition." *U.S. Geological Survey,* 2007. <http://pubs.usgs.gov/gip/geotime> (accessed December 4, 2007).

"Tour of Geologic Time." *University of California Museum of Paleontology,* 2007.<http://www.ucmp.berkeley.edu/exhibits/geologictime.php> (accessed December 4, 2007).

Geothermal Energy

■ Introduction

Geothermal energy is natural heat produced within the earth that can be removed and used commercially. Reservoirs of hot water and steam under Earth's surface can be accessed by drilling through the rock layers above the reservoir. The naturally heated water can be used to heat buildings, and the steam can be used to generate electricity. Cold water also can be pumped into areas of heated rocks to generate steam.

However, geothermal energy is only a viable power source in areas of geothermal activity, including some areas of the United States, Italy, New Zealand, and Iceland. It is a renewable resource, but underground shifts can affect the reservoirs and alter the supply. Geothermal energy is usually viewed as a renewable, green energy source that produces little, if any, water or air pollution.

■ Historical Background and Scientific Foundations

Geothermal energy originates deep in the earth, where heat from Earth's interior meets bedrock and groundwater resources. This interior heat warms the water, sometimes to temperatures that produce steam when the reservoirs

Long steam pipelines are part of this geothermal energy plant in New Zealand. *Image copyright Joe Gough, 2007. Used under license from Shutterstock.com.*

are tapped. In some areas, the naturally heated water erupts through the surface as hot springs and geysers.

Hot springs have long been used for therapeutic purposes, but the use of geothermal energy for electric power has a much shorter history. The first geothermally generated electricity in the world was produced at Larderello, Italy, in 1904, while the first large-scale geothermal power plant was constructed in New Zealand in the 1950s. The Geysers in California is the largest steam field in the world and has been used to produce electricity since 1960. Today, geothermal power plants are recognized as being clean, sustainable sources of electricity that do not use combustion for energy production.

Geothermal power plants use steam or hot water to power the turbines that generate electricity. The used geothermal water is then pumped down an injection well back into the underground reservoir to sustain the reservoir, to maintain the reservoir's pressure, or to be reheated. There are three primary types of geothermal power plants. Where the underground reservoir contains primarily steam, but very little water, the steam is used directly to turn a turbine generator to produce electricity. In other areas, the underground reservoir does not contain steam, but the underground water is so hot that some of it turns to steam when the pressure is released as the water is pumped to the surface. The steam formed is separated from the hot water and used to turn the turbines. If the water is hot, but not hot enough to generate steam, this hot water is used to heat an organic liquid—that vaporizes at a lower temperature—in a closed system and this vapor is used to power the turbines.

Impacts and Issues

Geothermal power plants emit only a very low level of gases into the atmosphere, since most of the water used to power the turbines to make the electricity is returned to underground reservoirs. Where the hot water from geothermal resources is used for heating, as in greenhouses in Iceland, the water is recycled directly back to the geothermal reservoirs. Geothermal power plants also help reduce the consumption of fossil fuels and the negative environmental impact of the emissions generated by the burning of these fuels.

Although the cost of establishing a geothermal power plant is high when compared to the cost of constructing a fossil fuel power plant, geothermal power plants use a free and natural resource. They are generally clean and nonpolluting when the hot water and steam are pumped back underground rather than being released into the air or surface waters.

WORDS TO KNOW

FOSSIL FUELS: Fuels formed by biological processes and transformed into solid or fluid minerals over geological time. Fossil fuels include coal, petroleum, and natural gas. Fossil fuels are non-renewable on the timescale of human civilization, because their natural replenishment would take many millions of years.

GREEN ENERGY: Energy obtained by any means that causes relatively little harm to the environment. There is no universal agreement on what constitutes a green energy source: for example, whether nuclear power is green is hotly debated, as is the question of whether windmills are by nature green, or only if sited in certain locations.

RENEWABLE ENERGY: Energy obtained from sources that are renewed at once, or fairly rapidly, by natural or managed processes that can be expected to continue indefinitely. Wind, sun, wood, crops, and waves can all be sources of renewable energy.

RESERVOIR: A natural or artificial receptacle that stores a particular substance for a period of time

SUSTAINABLE: Capable of being sustained or continued for an indefinite period without exhausting necessary resources or otherwise self-destructing: often applied to human activities such as farming, energy generation, or even the maintenance of a society as a whole.

TURBINE: An engine that moves in a circular motion when force, such as moving water, is applied to its series of baffles (thin plates or screens) radiating from a central shaft. Turbines convert the energy of a moving fluid into the energy of mechanical rotation.

SEE ALSO *Continental Drift; Earthquakes; Renewable energy.*

BIBLIOGRAPHY

Web Sites

"Environmental Benefits and Impacts of Geothermal Energy." *U.S. Department of Energy. Energy Efficiency and Renewable Energy.* <http://www1.eere.energy.gov/geothermal/environ_impacts.html> (accessed August 14, 2007).

"Geothermal Energy." *Geothermal Education Office.* <http://www.geothermal.marin.org> (accessed August 14, 2007.).

GIS and Climate Change Mapping

■ Introduction

A Geographic Information System (GIS) is a set of computer-based tools that collects, analyzes, and maps spatial data. Scientists use GIS mapping technology that employs methods of statistical analysis, database functions, and visualization benefits to study the impact of climatic changes. On the basis of these insights, researchers create profiles that illustrate the vulnerability of a specific geographical area to changes in its climate.

Greenhouse gas emissions resulting from human activities in the industrialized world have led to a steady increase in average global temperatures. Melting glaciers, increasing numbers of storms, soil erosion, and other natural calamities are the fallout of climate change. A report published in 2006 by *The Pew Center on Global Climate Change* predicts that if greenhouse gas emissions do not subside, temperatures will rise by as much as 10°F (5.6°C) by the end of the twenty-first century.

WORDS TO KNOW

DIGITAL MAPPING: Computerized production of maps, especially in geographic data systems.

GEOCODING: In geographic information systems, the assignment of geological data to particular features on maps or to other data records, such as photographs.

GREENHOUSE GASES: Gases that cause Earth to retain more thermal energy by absorbing infrared light emitted by Earth's surface. The most important greenhouse gases are water vapor, carbon dioxide, methane, nitrous oxide, and various artificial chemicals such as chlorofluorocarbons. All but the latter are naturally occurring, but human activity over the last several centuries has significantly increased the amounts of carbon dioxide, methane, and nitrous oxide in Earth's atmosphere, causing global warming and global climate change.

SPATIAL DATA: Data (individual measurements with numerical values) that are associated with points in space, such as measurement stations dotted over the face of a continent. Spatial data give information about location and are complementary to temporal data, which record changes over time.

VULNERABILITY: The degree to which an ecosystem or human community is susceptible to, or cannot adapt to or cope with, the negative effects of climate change. The type, intensity, and speed of climate change, the adaptive capacity of the system, and the sensitivity of the system to increased climate variability or climate change all determine vulnerability.

■ Historical Background and Scientific Foundations

The GIS mapping method is the result of a steady evolution of digital mapping approaches. The birth of the MIMO (map in–map out) model in 1959 introduced geocoding, data capture and analysis, and display features, which are integral to any GIS mapping software. The U.S. Census Bureau pioneered the large-scale use of digital mapping in the 1960s with the DIME (Dual Independent Map Encoding) data format. In 1966, the Laboratory for Computer Graphics and Spatial Analysis at the Harvard Graduate School of Design developed SYMAP, a grid-based mapping program. Both the methods worked with mainframe computers.

The first attempt at studying climate change with GIS was made by the Canada Land Inventory in the 1960s, leading to the origin of the Canada Geographic Information System. It was used to classify land for agriculture, wildlife sustainability, and forestry.

In 1970, the first GIS symposium was held in Ottawa, Ontario, Canada. In the same decade, GIS software became commercially available to the public. Some of the leading names in this field were M&S Computing (changed to Intergraph Corporation in 1980) and Environmental Systems Research Institute (ESRI). With the advent of high-speed processors and with the increasing affordability of computer hardware, the application of GIS became more feasible. In 1992, ESRI launched ArcView, a mapping system with a graphical user interface. This enhanced the user experience and aided the proliferation of the technology.

By the 1990s, the use of GIS became widespread and also led to the launch of GIS-focused publications and activity groups. The technology made consistent improvements, including the emergence of multimedia GIS. This enabled the integration of GIS information with multimedia resources, such as photographs, animation, and video.

Impacts and Issues

Climate change has far-reaching implications on Earth's ecosystems, including its forest cover, water bodies, human activities, and wildlife. Each life form has a different level of sensitivity to these changes. The level to which the ecosystem will be affected by a climatic change is known as vulnerability. The analytical capabilities of GIS enable a researcher to store large amounts of information about a particular ecosystem and study its vulnerability. Furthermore, data obtained by GIS can help scientists rank areas based on vulnerability, allowing scientists to prioritize their efforts in dealing with vulnerable hot spots.

However, the adoption of GIS technology has also garnered criticism. The technology makes it possible to collect minute details from an identified geography, such as street address and postal codes. Consequently, privacy concerns and the potential for misuse of such data exist. Some government agencies have implemented GIS privacy protocols to regulate the flow of information.

SEE ALSO *Forests and Deforestation; Glacier Retreat; Global Warming; Industry (Private Action and Initiatives).*

BIBLIOGRAPHY

Books

Dow, Kirstin, and Thomas E. Downing. *The Atlas of Climate Change: Mapping the World's Greatest Challenge.* Berkeley: University of California Press, 2006.

Martens, P., and A. J. McMichael. *Environmental Change, Climate and Health: Issues and Research Methods.* New York: Cambridge University Press, 2002.

McMichael, A. J., et al. *Climate Change and Human Health: Risks and Responses.* Geneva: World Health Organization, 2003.

Web Sites

"Climate Change 101: Understanding and Responding to Global Climate Change." *Pew Center on Global Climate Change,* October 12, 2006. <http://www.pewclimate.org/docUploads/Climate101-FULL_121406_065519.pdf> (accessed October 28, 2007).

"GiS Timeline." *Centre for Advanced Spatial Analysis,* May 31, 2000 <http://www.casa.ucl.ac.uk/gistimeline/> (accessed October 28, 2007).

"Vulnerability Mapping: A GIS Based Approach to Identity Vulnerable Regions to Climate Change." *GIS Development,* December, 2005.<http://www.gisdevelopment.net/magazine/years/2005/dec/28_1.htm> (accessed October 28, 2007).

Amit Gupta

Glaciation

Introduction

Glaciation is the ice and snow cover that results when more snow falls in winter than melts in summer. When the snow accumulates over the years, the weight of it turns the mass into ice that flows outward and downward to cover an ever-larger land mass. If the advancing glaciation covers more than 12 million acres (50,000 sq km), it is called an ice sheet. Antarctica and Greenland are each largely covered by an ice sheet.

When the climate changes and less snow falls in winter than melts in summer, glaciation retreats. Left behind from glaciation are tell-tale signs that paleoclimatologists use to trace the history of the fairly regular advances and retreats of glaciation over very large portions of Earth's land mass, especially over a period of 1.8 million years ago to 10,000 years ago called the Pleistocene, an epoch in geologic history that is often referred to as the Ice Ages.

Historical Background and Scientific Foundations

The history of repeating global glaciation is determined from deposits left behind on land, sea floor sediments, and actual ice core samples that are collected by drilling deep into an ice sheet. In 2003, a team of European scientists

Ice falls from a glacier in Alaska. This phenomenon is known as glacial calving. *Image copyright Natalia Bratslavsky, 2007. Used under license from Shutterstock.com.*

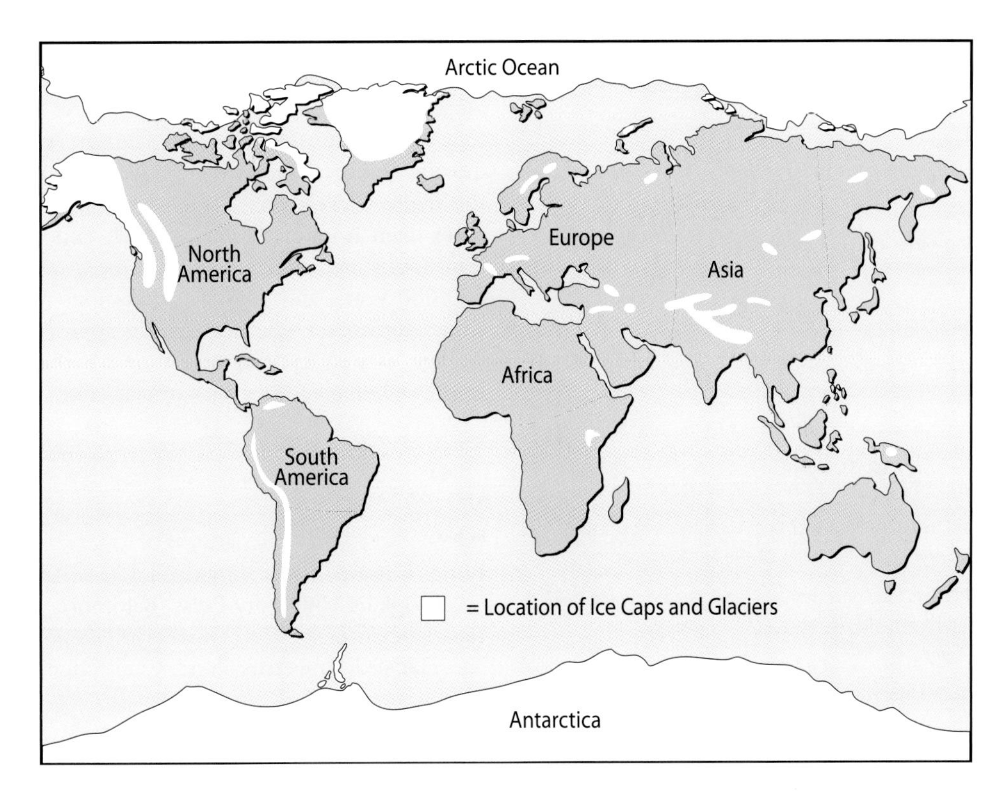

Map depicting the location of ice caps and glaciers throughout the world. *Kopp Illustration, Inc.*

completed drilling of an ice core 9,800 ft (3,000 m) into the North Greenland ice sheet. The effort took seven years. When the core is completely analyzed it will trace the history of the ice sheet back about 120,000 years.

Swiss born geologist Louis Agassiz (1807–1873) first suggested the concept of glaciation over large areas of land when he came to North America and saw evidence of glacial changes in the landscapes of vast areas. He saw boulders of rock types from distant locations that were scarred from being pushed by advancing glacial ice that reminded him of the glacier scoured rocks in valleys in Switzerland but on a much larger scale.

Repeated advances and retreats of ice are also visible in channels carved in the landscape that are now lakes and in glacial deposits of collected sand and rock that form long regular hills called eskers. In some elevated locations, ancient fossil shorelines are evidence of periods when glaciation had retreated and sea levels were much higher than they are now.

Serbian engineer Milutin Milankovitch (1879–1958) developed a mathematical explanation for the evidence of past periods of glaciation based on cyclical variations of Earth's orbit that would change the amount of sunlight reaching Earth and therefore change the climate. At least for the Pleistocene period Milankovitch's theory is in close agreement with glaciation cycles of 20,000, 40,000, and 100,000 years. The variations Milankovitch considered are variations in the shape of Earth's orbit around the sun; changes in the tilt of Earth's axis to the orbit; and the precession effect that changes the direction of Earth's axis of rotation much as a spinning top wobbles.

Periods of glaciation that are more ancient than the Pleistocene Epoch are attributed by scientists to shifting land masses that were associated with Earth's plate movements. The plate movements created polar wandering, mountain ranges where plates converged, and opened and closed ocean basins. Pre-Pleistocene plate movements resulted in changes in the atmosphere and ocean circulation, driving changes in climate.

■ Impacts and Issues

The most recent period of widespread global glaciation began about 70,000 years ago. It reached its peak about 20,000 years ago and ended about 10,000 years ago. During that period about 32% of the total land area was covered in ice. Presently about 10% is covered. The glaciation that remains stores about 70% of Earth's freshwater.

Historically the warm periods have lasted about 10,000 years. Climate is affected by small changes. Some scientists believe in a Chaos Theory that refers to small indeterminate events as the trigger of severe climate changes. There is speculation that adding pollution to the atmosphere is one of those events. The rise of two key greenhouse gases, carbon dioxide and methane, is changing the climate. The combination of adding atmospheric changes to changing oceanic dynamics as glacier ice melts

WORDS TO KNOW

ICE AGE: Period of glacial advance.

ICE CORE: A cylindrical section of ice removed from a glacier or an ice sheet in order to study climate patterns of the past. By performing chemical analyses on the air trapped in the ice, scientists can estimate the percentage of carbon dioxide and other trace gases in the atmosphere at that time.

MILANKOVITCH CYCLES: Regularly repeating variations in Earth's climate caused by shifts in its orbit around the sun and its orientation (i.e., tilt) with respect to the sun. Named after Serbian scientist Milutin Milankovitch (1879–1958), though he was not the first to propose such cycles.

PALEOCLIMATOLOGY: The study of past climates, throughout geological history, and the causes of the variations among.

PLEISTOCENE EPOCH: Geologic period characterized by ice ages in the Northern Hemisphere, from 1.8 million to 10,000 years ago.

PRECESSION: The comparatively slow torquing of the orbital planes of all satellites with respect to Earth's axis, due to the bulge of Earth at the equator, which distorts Earth's gravitational field. Precession is manifest by the slow rotation of the line of nodes of the orbit (westward for inclinations less than 90 degrees and eastward for inclinations greater than 90 degrees).

and adds freshwater to dilute salty ocean water at a record pace, particularly around Greenland, could trigger severe climate changes.

As ice ages in the past have ended, the rising temperatures stimulated changes that increased carbon dioxide and methane levels in the atmosphere. These changes sent temperatures even higher. Current levels of carbon dioxide and methane in the atmosphere are higher than the highest levels ever detected through studies of ice cores from Antarctica. Present CO_2 and methane concentrations are higher than any time in at least the last 650,000 years, and growing. Questions remain among climatologists on whether Earth is approaching another ice age as past cycles suggest or whether the human factor will change the cycles of glaciation.

SEE ALSO *Glacier; Glacier Retreat; Ice Ages.*

BIBLIOGRAPHY

Books

Lurie, Edward. *Louis Agassiz: A Life in Science.* John Hopkins University Press, Baltimore, MD: 1996.

Weart, Spencer R. *The Discovery of Global Warming.* Harvard University Press, Cambridge, MA: 2004.

Web Sites

"All About Glaciers." *National Snow and Ice Data Center*, 2007. <http://www.nsidc.org/glaciers> (accessed August 16, 2007).

"Atmospheric Changes." *U.S. Environmental Protection Agency*, 2007. <http://www.epa.gov/climate change/science/recentac.html> (accessed August 16, 2007).

"Past Climate Cycles: Ice Age Speculations." *American Institute of Physics*, 2007. <http://www.aip.org/climate/cycles.htm> (accessed August 16, 2007).

Glacier

Introduction

A glacier is a mass of ice larger than 247 acres (one tenth of a square kilometer) that starts on land, but can move out over coastal waters in polar regions where calving (breaking apart) produces icebergs. All glaciers move as the result of their weight and gravity. Glaciers require a specific cold climate to exist.

There are three types of glaciers: alpine glaciers in mountain valleys; piedmont glaciers near the base of mountain ranges where glacier masses have combined; and continental glaciers, which are called ice sheets when they are very large. Continental glaciers have covered vast areas of Earth at different times through history.

When glaciers advance they change the landscape. When they retreat, tell-tale signs are left behind. Paleoclimatologists studying the signs left from retreating ancient glaciers have discovered that massive glaciation occurred fairly regularly. Past glaciation cycles appear to be related to subtle changes in solar radiation. Scientists are actively looking for answers to how human activities that are already causing climate changes will influence Earth's glaciation cycles in the future.

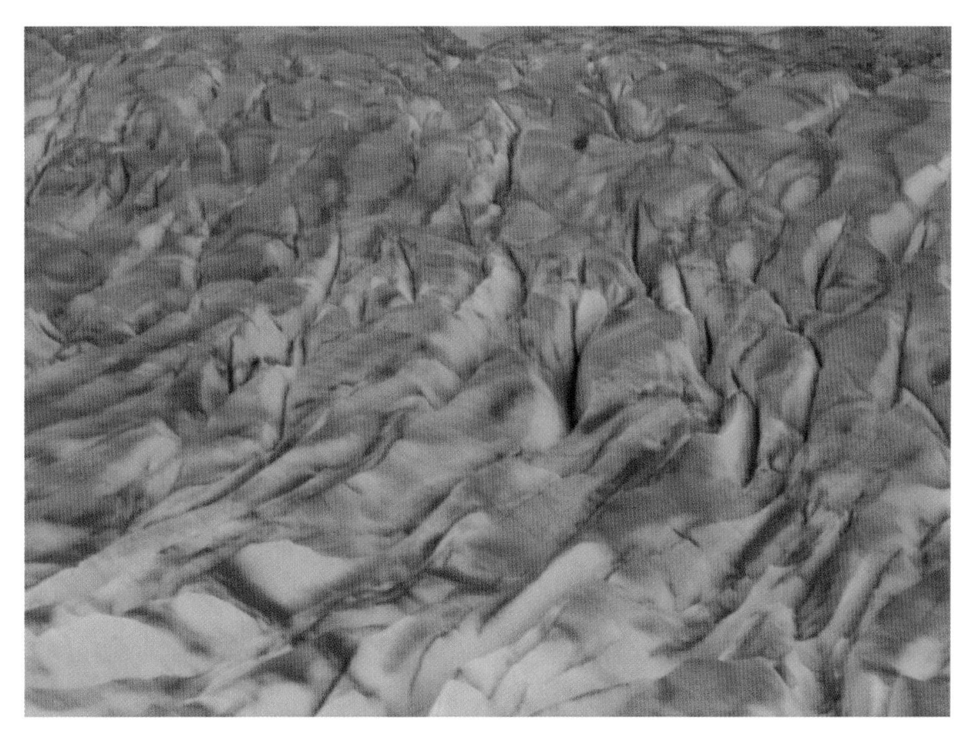

Located in Alaska, the Juneau Icefield is North America's fifth largest icefield and feeds nearly 40 large glaciers. The section shows evidence of soot carried by the wind. *© 1992 Nancy A. Edgar*

WORDS TO KNOW

CALVING: Process of iceberg formation when huge chunks of ice break free from glaciers, ice shelves, or ice sheets due to stress, pressure, or the forces of waves and tides.

GLACIATION: The formation, movement, and recession of glaciers or ice sheets.

ICE AGE: Period of glacial advance.

KETTLE DRUM: Usually termed a kettle: depression or sinkhole left in a mass of rocky debris by a retreating glacier. A large, isolated chunk of ice buried in debris melts, causing a collapse of the overlying debris.

MORAINE: Mass of boulders, stones, and other rock debris carried along and deposited by a glacier.

PALEOCLIMATOLOGY: The study of past climates, throughout geological history, and the causes of the variations among.

PIEDMONT: A region of foothills along the base of a mountain range.

Historical Background and Scientific Foundations

Glaciers start where more snow falls in the winter than melts in the summer. First, a snow field collects. Through successive freezing and thawing the snow becomes grain-sized ice called firn. Over time, the snow field becomes an ice field. When the icy mass gets large enough so it begins to move, it is called a glacier.

Alpine glaciers move down the mountain in channels, scouring the walls to form characteristic U-shaped valleys called cirques. A piedmont glacier forms where several glaciers combine at the foot of mountains. These are common in Alaska.

A glacier picks up surface soil and rocks and carries the load (called till) forward with the ice. When the glacier recedes, till is dropped. Till that is dropped in front of a receding glacier is known as a terminal moraine. The overall landscape left behind a retreating continental glacier is irregular. Glaciated landscape features include hollows that are collectively described as kettle drum topography. Boulders are scarred as they are carried for hundreds of miles or kilometers to new locations. They are called erratics because the rock type is not consistent with the area where the boulders are dropped.

In the mid-1800s, Swiss-born geologist Louis Agassiz (1807–1873) first suggested the concept of continental glaciers when he noticed evidence of glacial changes in the landscapes over large areas of Canada that resembled the glaciated alpine landscapes of Switzerland. Subsequently geologists determined there was not one but a series of periods in ancient history where large glaciers covered up to 32% of the total land area of Earth.

Evidence that there have been eight Ice Age cycles over the past 750,000 years has been collected from deposits left behind on land, and sea floor sediments. Evidence also comes from ice core samples taken from Greenland and Antarctica where continental glaciers that date back to the Ice Ages still cover most of the land area.

Impacts and Issues

Each period of vast continental glaciation was separated by a warmer interglacial period that lasted about 10,000 years. Serbian engineer Milutin Milankovitch (1879–1958) developed a mathematical explanation for the Ice Ages based on variations in solar radiation. The Milankovitch theory is in close agreement with 20,000, 40,000, and 100,000 year cycles of glaciation. The most recent Ice Age began about 70,000 years ago, peaked about 20,000 years ago and ended about 10,000 years ago. Since interglacial periods were generally around 10,000 years, Earth could be approaching another Ice Age.

There is a big question about Earth's future that was not considered by Milankovitch. Where climate was controlled by natural cyclical variations in solar radiation in the past 750,000 years, it may not be controlled only by natural causes in the future. Human activities have changed the climate through greenhouse gases emitted from the widespread use of fossil fuels. Carbon dioxide and methane levels are higher in the atmosphere in the twenty-first century than they have been in any previous interglacial period.

Glaciers along the coast of Greenland are melting at record rates as a result of global warming. Since glaciers contain 70% of Earth's freshwater, when they melt the freshwater changes the dynamics of the ocean water around them. Melting glaciers also cause the sea level to rise. Global warming is adding to sea levels rising as the oceans get warmer and expand. Where glaciers need a very specific cold climate to form, there is no easy answer to the question of when the next Ice Age will start.

SEE ALSO *Glaciation; Glacier Retreat; Ice Ages.*

BIBLIOGRAPHY

Books

Lurie, Edward. *Louis Agassiz: A Life in Science.* John Hopkins University Press, Baltimore, MD: 1996.

Weart, Spencer R. *The Discovery of Global Warming.* Harvard University Press, Cambridge, MA: 2004.

Web Sites

"All About Glaciers." *National Snow and Ice Data Center*, 2007. <http://www.nsidc.org/glaciers> (accessed August 16, 2006).

"Atmospheric Changes." *U.S. Environmental Protection Agency*, 2007. <http://www.epa.gov/climatechange/science/recentac.html> (accessed August 16, 2007).

"Past Climate Cycles: Ice Age Speculations." *American Institute of Physics*, 2007. <http://www.aip.org/climate/cycles.htm> accessed August 16, 2007).

"The Physical Environment: Glacial Systems." *University of Wisconsin Stevens Point*, 2007. <http://www.uwsp.edu/geo/faculty/ritter/geog101/textbook> (accessed August 16, 2007).

Glacier Retreat

Introduction

A glacier is essentially a river of ice. Glaciers form at high latitudes and altitudes where vast quantities of fallen snow accumulate over time and compress into thick masses of ice that flow downhill under their own weight. Glaciers, along with ice caps and ice sheets, cover about 10% of Earth's land mass—mostly in Greenland and Antarctica. Mountain, or alpine, glaciers outside the poles hold only a tiny fraction of this ice, but produce meltwater essential for ecosystem health and for human use in populated areas. Glaciers are sensitive indicators of climate change because their size and mass change relatively quickly in response to long-term climate variations (such as changing temperature, precipitation, and cloud cover) over the course of decades, rather than to short-term weather patterns.

Most of Earth's alpine and non-polar glaciers, which tend to be smaller and more unstable, have shrunken rapidly over the last century. This widespread glacier retreat is most likely the result of climbing global temperatures brought on by human emissions of greenhouse gases such as carbon dioxide (CO_2) combined with a natural warming phase. Changes in humidity, precipitation, cloud cover, and albedo (reflectance) that accompany global warming may themselves have a greater impact on glacier retreat in some tropical mountains. If human greenhouse-gas emissions go unchecked and current warming trends continue, alpine glaciers in many places will likely disappear well before the end of the twenty-first century.

Historical Background and Scientific Foundations

Global ice cover and glacial extent have varied widely over the past 750,000 years, advancing during numerous ice ages, and retreating during intermediate warming periods called interglacials. During the last Ice Age, which ended about 10,000 years ago, glaciers, ice caps, and ice sheets covered 32% of Earth's land mass and 30% of its oceans.

Humans began tracking the length and extent of certain glaciers as early as 1534 in European mountain ranges, but thorough, internationally coordinated glacier monitoring did not start until 1894, when the

In an attempt to prevent the Gurschen Glacier in Switzerland from continuing to melt, crews work to cover about 2,990 sq yards (2,500 sq meters) of the ice with plastic sheeting. Environmental groups protested the action, demanding a change in climate policy instead. *AP Images.*

While he was vice president of the United States, Al Gore (right) visited Glacier National Park in Montana in 1997 to draw attention to the effects of global warming. Here, he and a scientist from the U.S. Geological Survey stand by a sign indicating where glaciers once reached sixty years prior. *Luke Frazza/AFP/Getty Images.*

International Glacier Commission was established in Zurich, Switzerland. Now, long-term data exist for glaciers on every continent. These data generally show that glaciers reached their most recent maximum during the Little Ice Age, a short and relatively mild cooling period that ended sometime in the nineteenth century. General retreat of glacier tongues started in earnest after 1800, and accelerated across the globe after 1850 and through the twentieth century, with a slight lull during the 1970s before escalating rates of loss began in the 1980s and 1990s.

Records for changes in glacier mass balance are not nearly as complete, as they are generally restricted to relatively easy-access glaciers and stretch back only to the mid-twentieth century. Mass balance refers to the equilibrium between a glacier's accumulation of snow and ice through precipitation and its ablation, or loss, of snow and ice by melting and sublimation. Over time, this measure shows whether a glacier's mass is growing (positive), shrinking (negative), or staying the same (zero). Mathematical extrapolations from the data that do exist on mass balance indicate that the global rates of glacier mass loss are increasing rapidly, roughly doubling during 1990/1991–2003/2004 over the mass loss rates from 1960/1961–1989/1990.

Generally, these glacier fluctuations show a strong statistical correlation with air temperatures over the last century, though other factors, such as changing cloud cover, snowfall, humidity, and climbing sea surface temperature, play roles of varying importance in different locations.

Regional Trends

The short, steep, and shallow glaciers in tropical mountain ranges are particularly sensitive to climate change, responding over a scale of a few years. In the tropical Andes of Peru, for example, 10 monitored glaciers retreated 1,935-6,266 ft (590-1,910 m) between 1939 and 1994. The Chacaltaya glacier in Bolivia, meanwhile, lost about 60% of its ice volume between 1940 and 1983, and will likely disappear altogether within a decade.

This trend toward glacier retreat and eventual disappearance in the tropical Andes, which has accelerated markedly since the late 1970s and early 1980s, corresponds with a 0.16-0.27°F (0.09-0.15°C) increase in temperature per decade in the area from 1950 to 1994, with most of the warming taking place after the mid-1970s. But related climate changes, especially increasing humidity and changes in precipitation and cloud cover, may play a larger direct role than temperature in the shrinkage of tropical glaciers. A documented rise in tropical sea surface temperatures, which has forced tropical freeze levels to higher altitudes, is also an important factor in tropical, and potentially worldwide, glacier retreat.

Glaciers in mid- and high-latitude regions of the Northern Hemisphere, such as the northwestern United States, southwestern Canada, and Alaska, have lost mass quickly over the last century due to accelerated summer melting, despite some increases in precipitation. Losses have also been extreme in the European Alps, where glaciers have shrunk 30–40%, and New Zealand, which has lost 20–32% of its glacierized area over the past century.

In Patagonia, massive ice fields and their glaciers have lost about 0.77–3.2 cubic mi (3.2–13.5 cubic km) of ice per year from 1968 to 2000. Patagonia's changes likely stem in part from warmer, drier conditions, but accelerated glacial flow, resulting in the accelerated loss of large chunks of ice at glacier termini through a process known as calving, also plays a role.

Despite trends toward widespread glacier retreat, there are isolated exceptions. Some glaciers in Pakistan's Karakorum have thickened and advanced due to enhanced precipitation, though glaciers in Asian mountain ranges are generally retreating. The tropical Andes also enjoyed periods of glacial advance in the early 1970s and again in

WORDS TO KNOW

ALBEDO: A numerical expression describing the ability of an object or planet to reflect light.

CALVING: Process of iceberg formation when huge chunks of ice break free from glaciers, ice shelves, or ice sheets due to stress, pressure, or the forces of waves and tides.

GREENHOUSE GASES: Gases that cause Earth to retain more thermal energy by absorbing infrared light emitted by Earth's surface. The most important greenhouse gases are water vapor, carbon dioxide, methane, nitrous oxide, and various artificial chemicals such as chlorofluorocarbons. All but the latter are naturally occurring, but human activity over the last several centuries has significantly increased the amounts of carbon dioxide, methane, and nitrous oxide in Earth's atmosphere, causing global warming and global climate change.

ICE AGE: Period of glacial advance.

ICE CAP: Ice mass located over one of the poles of a planet not otherwise covered with ice. In our solar system, only Mars and Earth have polar ice caps. Earth's north polar ice cap has two parts, a skin of floating ice over the actual pole and the Greenland ice cap, which does not overlay the pole. Earth's south polar ice cap is the Antarctic ice sheet.

ICE SHEET: Glacial ice that covers at least 19,500 square mi (50,000 square km) of land and that flows in all directions, covering and obscuring the landscape below it.

LA NIÑA: A period of stronger-than-normal trade winds and unusually low sea-surface temperatures in the central and eastern tropical Pacific Ocean; the opposite of El Niño.

SUBLIMATION: Transformation of a solid to the gaseous state without passing through the liquid state.

1999 corresponding to cold La Niña conditions in the tropical Pacific. New Zealand glaciers thickened and advanced from the late 1970s through the 1990s, as did some Scandinavian glaciers during the 1990s, potentially due to temporarily enhanced precipitation. But glaciers in both areas have resumed retreat since 2000.

Impacts and Issues

If human greenhouse-gas emissions continue to rise unchecked and current warming trends continue, climate models predict the disappearance of many glaciers before the end of the twenty-first century. In the tropical Andes, for example, many glaciers may vanish within 50 years at current rates of retreat. Meanwhile, Montana's Glacier National Park has lost all but 10 of its 27 original glaciers over the past century, and the extent of those remaining has also been vastly reduced. Under a doubling of CO_2 emissions, the park's remaining glaciers will have completely vanished by 2030.

The glaciers and famous snows of Mount Kilimanjaro in Kenya, Africa, will also likely disappear by mid-century for the first time in over 11,000 years if a long-term dry spell, likely part of the climate change picture, persists. However, some studies in the European Alps suggest that many alpine glacier retreat scenarios have been largely overestimated.

The shrinkage and loss of alpine glaciers, which provide important water sources in much of the world, will have direct and severe implications for humans and ecosystems alike. Because they are perennial, glaciers provide an important base flow for streams and rivers beyond more unreliable and now dwindling snowpack melt, sustaining agriculture and drinking water supplies through seasonal variations in many mountain communities from the Andes to the Himalayas. The reduction of the Zongo Glacier in Bolivia, for example, is already creating freshwater supply problems for downstream human communities. The loss of glaciers will also represent a major aesthetic loss for humans who flock to places like Glacier National Park for the scenery, as well as a potential blow to recreational activities such as high-altitude mountaineering.

High-latitude and high-altitude ecosystems will also be affected. Glacier melt provides crucial soil moisture and ensures input of cold water to decrease stream temperatures in glacial basins during warmer seasons, maintaining certain plant communities and cold-adapted aquatic invertebrates and fish. The retreat of glaciers in these areas will thus likely force changes in local vegetation and stream ecology, as well as alter stream sedimentation and open up new ground for plant colonization.

Widespread melting of glaciers will also play a part in climate change-induced sea level rise. Sea level has risen about 5-9 in (13-23 cm) worldwide over the past century, with mountain glacier melt (excluding the glaciers of Greenland's ice sheet) contributing an estimated 1 in (2.7 cm) of this rise between 1865 and 1990. Melting Alaskan glaciers are among the biggest contributors, accounting for about 30% of mountain glacier inputs to sea level rise. The Arctic, the high mountains of Asia, and Patagonian glaciers also play relatively hefty roles. However, these contributions are miniscule compared to the potential sea level rise that could result from the melting of the vast Antarctic and Greenland ice sheets.

SEE ALSO *Albedo; Arctic Melting: Greenland Ice Cap; Arctic Melting: Polar Ice Cap; Glaciation; Glacier.*

BIBLIOGRAPHY

Books

Solomon, S., et al, eds. *Climate Change 2007: The Physical Science Basis: Contribution of Working Group I to the Fourth Assessment Report of the Intergovernmental Panel on Climate Change.*

New York: Cambridge University Press, 2007.

Periodicals

Diaz, Henry F., and Nicholas E. Graham. "Recent Changes in Tropical Freezing Heights and the Role of Sea Surface Temperature." *Nature* 383 (1996): 152–155.

Hall, Myrna H. P., and Daniel P. Fagre. "Modeled Climate-Induced Glacier Change in Glacier National Park, 1850–2100." *BioScience* 53 (2003): 131–140.

Kaser, Georg, et al. "Modern Glacier Retreat on Kilimanjaro as Evidence of Climate Change: Observations and Facts." *Royal Meteorological Society* 24 (2004): 329–339.

Rignot, Eric, et al. "Contribution of the Patagonia Icefields of South America to Sea Level Rise." *Science* 302 (2003): 434–437.

Vuille, Mathias, et al. "20th Century Climate Change in the Tropical Andes: Observations and Results." *Climatic Change* 59 (2003): 75–99.

Zuo, Z., and J. Oerlemans. "Contribution of Glacier Melt to Sea-Level Rise Since AD 1865: A Regionally Differentiated Calculation." *Climate Dynamics* 13 (1997): 835–845.

Web Sites

"All About Glaciers." *National Snow and Ice Data Center.* <http://www.nsidc.org/glaciers> (accessed December 2, 2007).

Sarah Gilman

Global Change Research Program

Introduction

The Global Change Research Program (GCRP) is a project of the U.S. government to coordinate research on all aspects of global change, including climate change. The GCRP began as a presidential initiative under President George H. W. Bush (1924–) in 1989 and was formally established by Congress by the Global Change Research Act of 1990. In 2001, President George W. Bush (1946–) ordered the establishment of a Climate Change Research Initiative and then, in 2002, placed both it and the GCRP under the auspices of a new program called the U.S. Climate Change Science Program (CCSP). The existing GCRP continued to conduct research on aspects of global change other than climate, such as the regular El Niño cycle, but its climate work was integrated with that of the new Climate Change Research Initiative.

Historical Background and Scientific Foundations

In the late 1980s, scientific agreement on the reality of human-caused global climate change was growing. In response, the first President Bush ordered the establishment of a U.S Global Change Research Program in 1989. The program was given an annual budget of $1.87 billion to be divided among eight participating federal agencies, namely the National Aeronautics and Space Administration (NASA), the National Science Foundation, the Department of Energy, the National Oceanic and Atmospheric Administration (NOAA), the Department of the Interior, the Environmental Protection Agency (EPA), the Smithsonian Institution, and the Department of Health and Human Services. About three quarters of the budget went to NASA, which used much of the money to loft and operate Earth-observing satellites such as the EOS AM-1 satellite launched in 1999.

The existence of the GCRP was confirmed by Congress in 1990 through the Global Change Research Act, which declared that the GCRP was "aimed at understanding and responding to global change, including the cumulative effects of human activities and natural processes on the environment, to promote discussions towards international protocols in global change research, and for other purposes." The GCRP had a complex task that involved gathering data, pulling together disparate scientific disciplines, and linking various government bureaucracies. Its successes in basic science included mapping ozone in the atmosphere, studying the effects of tropical forest fires on climate, and building a database on global plankton (which are essential to the oceans' removal of carbon dioxide from the atmosphere).

In February 2002, President George W. Bush established the Climate Change Science Program (CCSP) as a cabinet-level structure to improve government-wide management of climate science. The CCSP, which the administration particularly charged with characterizing uncertainties in climate science, brought together the existing GCRP with the new Climate Change Research Program. The CCSP integrates the efforts of 13 federal agencies. Its $1.7 billion annual budget is provided and managed by the agencies themselves.

Impacts and Issues

From 1989 to 1998, the U.S. government spent approximately $15 billion on the GCRP. The program was always unpopular with political conservatives, who tended to consider global warming unreal and saw the GCRP as a pet of the Democrats. (Senator Al Gore, who later served as vice president from 1993 to 2001, had supported it in the Senate and favored it while in the White House.)

In the late 1990s, the program was also criticized by scientists and members of the Clinton administration, but for different reasons. A panel of 15 scientists convened by the National Academies (the government's official advisory group on scientific questions) criticized the program for being too ambitious in its attempt to understand the whole climate system at once, an attempt that, the

The Smithsonian Institution launched an exhibit, called the "Climate Rollercoaster," on its Web site. The graphic of the rollercoaster showed fluctuations in the Arctic climate over the past 20,000 years. But three small arrows (top right of graphic) were added to indicate that it is unclear how much the Arctic will continue to warm. A former Smithsonian administrator said the arrows were added as a form of self-censorship to avoid any political conflicts with Congress or the Bush administration. *AP Images.*

scientists said, made the program too diffuse and left it vulnerable to budget cutbacks. The panel also said that the program had neglected key questions such as the nature of Earth's water and carbon cycles.

In 2007, another 15-member panel of the National Academies reviewed the new Climate Change Science Program. It found that the program had helped resolve the question of whether Earth's atmosphere is warming significantly—it is—and had achieved other scientific goals. However, it also reported that the CCSP had been hampered by Bush administration polices that reduced the number of Earth-observing satellites and stopped programs to monitor conditions on the surface. In particular, the National Academies panel noted that of the CCSP's $1.7 billion annual budget, only about $25–$30 million was spent each year on studying the impacts of climate change on human beings.

Numerous claims were made during the Bush years that the administration delayed or censored testimony from government scientists about the reality and dangers of climate change. In August 2007, a federal judge ruled in favor of several groups, including Greenpeace and the Center for Biological Diversity, that had sued the Climate Change Science Program for delaying the release of reports on climate change required by the Global Change Research Act of 1990. The judge found that the defendants had violated the law by failing to release a scientific assessment of global change that had been due in November 2004, and a national global-change research plan that had been due in July 2006. The Bush administration had claimed that the 1990 act allowed it to release the reports when it wished. "The defendants are wrong," the judge ruled; "Congress has conferred no discretion upon the defendants as to when they will issue [the reports]."

Primary Source Connection

The Climate Change Science Program (CCSP) was established by the U.S. government in 2002 to marshal the resources of thirteen separate federal agencies in order to

WORDS TO KNOW

CARBON CYCLE: All parts (reservoirs) and fluxes of carbon. The cycle is usually thought of as four main reservoirs of carbon interconnected by pathways of exchange. The reservoirs are the atmosphere, terrestrial biosphere (usually includes freshwater systems), oceans, and sediments (includes fossil fuels). The annual movements of carbon, the carbon exchanges between reservoirs, occur because of various chemical, physical, geological, and biological processes. The ocean contains the largest pool of carbon near the surface of Earth, but most of that pool is not involved with rapid exchange with the atmosphere.

CARBON DIOXIDE: Odorless, colorless, non-poisonous gas, chemical formula CO_2. Carbon dioxide is released by natural processes and by burning fossil fuels. A minor but very important component of the atmosphere, carbon dioxide traps infrared radiation. Atmospheric CO_2 has increased dramatically since the early 1800s. Burning fossil fuels is the leading cause of increased CO_2 levels with deforestation the second major cause. The increased amounts of CO_2 in the atmosphere enhance the greenhouse effect, blocking heat from escaping into space and contributing to the warming of Earth's lower atmosphere.

EL NIÑO: A warming of the surface waters of the eastern equatorial Pacific that occurs at irregular intervals of 2 to 7 years, usually lasting 1 to 2 years. Along the west coast of South America, southerly winds promote the upwelling of cold, nutrient-rich water that sustains large fish populations, that sustain abundant sea birds, whose droppings support the fertilizer industry. Near the end of each calendar year, a warm current of nutrient-poor tropical water replaces the cold, nutrient-rich surface water. Because this condition often occurs around Christmas, it was named El Niño (Spanish for boy child, referring to the Christ child). In most years the warming lasts only a few weeks or a month, after which the weather patterns return to normal and fishing improves. However, when El Niño conditions last for many months, more extensive ocean warming occurs and economic results can be disastrous. El Niño has been linked to wetter, colder winters in the United States; drier, hotter summers in South America and Europe; and drought in Africa.

OZONE: An almost colorless, gaseous form of oxygen with an odor similar to weak chlorine. A relatively unstable compound of three atoms of oxygen, ozone constitutes, on average, less than one part per million (ppm) of the gases in the atmosphere. (Peak ozone concentration in the stratosphere can get as high as 10 ppm.) Yet ozone in the stratosphere absorbs nearly all of the biologically damaging solar ultraviolet radiation before it reaches Earth's surface, where it can cause skin cancer, cataracts, and immune deficiencies, and can harm crops and aquatic ecosystems.

PLANKTON: Floating animal and plant life.

WATER CYCLE: The process by which water is transpired and evaporated from the land and water, condensed in the clouds, and precipitated out onto Earth once again to replenish the water in the bodies of water on Earth.

address global climate change. CCSP's goals include scientific study of global climate change and the use of that scientific data to formulate a strategy to address global climate change on both national and international levels. CCSP issues reports to policymakers discussing the scientific, societal, and economic impacts of global climate change. This source from the United States Global Change Research Program (USGCRP), which has been integrated into CCSP, details the goals of CCSP and the steps that CCSP has taken toward achieving those goals.

PREVIEW OF OUR CHANGING PLANET: THE U.S. CLIMATE CHANGE SCIENCE PROGRAM FOR FISCAL YEAR 2008

Climate plays an important role in shaping the environment, natural resources, infrastructure, economy, and other aspects of life in all countries of the world. Therefore, variations and changes in climate can have substantial environmental and socioeconomic implications. The Climate Change Science Program (CCSP) was established in 2002 to empower the Nation and the global community with the science-based knowledge to manage risks and opportunities of change in the climate and related environmental systems. CCSP incorporates and integrates the U.S. Global Change Research Program (USGCRP) with the Administration's U.S. Climate Change Research Initiative (CCRI). The USGCRP was mandated by Congress in the Global Change Research Act of 1990 (P.L. 101-606, 104 Stat. 3096-3104) to improve understanding of uncertainties in climate science, expand global observing systems, develop science-based resources to support policymaking and resource management, and communicate findings broadly among scientific and stakeholder communities. Thirteen departments and agencies of the U.S. Government participate in CCSP. These departments and agencies are listed in Appendix A of this report....

CCSP Goals and Analysis of Progress Toward These Goals

At the highest conceptual level, five goals have been identified to provide focus and facilitate programmatic integration. These goals encompass the full range of climate-related issues.

The program's detailed objectives, milestones, and products and payoffs complement these overarching goals, and are articulated in the program's *Strategic Plan*. CCSP-participating agencies and departments coordinate their work through discipline-related "research elements," which together support scientific research across a wide range of interconnected issues of climate and global change.

The goals address the most common questions concerning climate change which include:

- To what extent and how is the climate system changing?
- What are the causes of these changes?
- What will the future climate be like and what effects will a changed climate have on ecosystems, society, and the economy?
- How can we best apply knowledge about ongoing and projected changes to decisionmaking? . . .

The primary focus of U.S. climate research has historically been on Goals 1 through 3, which emphasize improvements in fundamental understanding of the climate system, its driving forces, and the tools to make predictions of short-term climate variability and potential long-term climate change more reliable. As the science matures and its societal utility becomes more evident, the importance of Goals 4 and 5 has become more significant. Examples of progress provided under each of the goals are often the result of coordinated research activities from many disciplines conducted or supported across the participating CCSP agencies.

Goal 1: Improve knowledge of the Earth's past and present climate and environment, including its natural variability, and improve understanding of the causes of observed variability and change.

Analyses based on observations provide a solid foundation for the program. These analyses contribute to improved understanding of Earth system processes, help determine the extent of climate variations, and provide true comparisons to test and advance model veracity. In the past year, analyses have enabled several important advances in understanding the nature and variability of the Earth system. The illustrative examples of progress toward CCSP's Goal 1 are drawn from and integrate a variety of different CCSP research elements.

One example of these integrated analyses is illustrated by the progress made in understanding climate change at high latitudes. Temperature and moisture patterns over North America and Europe are experiencing an earlier transition from winter to summer. The warmer spring temperatures produce earlier springtime green-up of vegetation and longer growing seasons. Satellite, airborne, and groundbased observations suggest significant changes occurring in the mass balance of Greenland and Antarctic ice sheets that are inferred to be caused by warming at high latitudes. Climate model simulations suggest that the global pattern of regional temperature and moisture trends is more readily explained by estimated human activity and natural climate forcing than by internal variability alone. These wide-ranging sets of analyses tie together findings from traditionally disparate disciplines including hydrology, glaciology, and ecology.

Progress has been made in understanding changes in atmospheric ozone through observations and comparisons with models of the atmosphere. Satellite observations have shown that the large ozone decreases over Antarctica have been accompanied by significant but smaller summertime ozone increases at higher levels of the Antarctic atmosphere. A chemical-climate model confirms these observations. Other related modeling research has shown that warming of the tropical lower atmosphere due to increased greenhouse gases may accelerate the large-scale motions of the stratosphere and thus alter the global distribution of ozone. . . .

Goal 2: Improve quantification of the forces bringing about changes in the Earth's climate and related systems.

In making long-term climate projections, an understanding of the factors responsible for global environmental change is necessary. These forcing factors include greenhouse gases, land cover changes, tiny airborne particles (aerosols), and solar variability. As in the previous Goal, the following examples of progress toward CCSP Goal 2 result from the integrated focus of multiple CCSP research elements.

Recent climate warming has been particularly intense in boreal and Arctic regions, leading to concern that increasing air temperature in these ecosystems may indirectly increase the incidences of forest fires. Beyond the emission of carbon dioxide (CO_2) and other greenhouse

IN CONTEXT: IMPROVING CLIMATE MODELS

"More than a dozen federal agencies are involved in producing and using climate change data and research. The first efforts at a coordinated government research strategy culminated in the creation of the U.S. Global Change Research Program (GCRP) in 1989. The GCRP coordinated research at these agencies with the aim of 'understanding and responding to global change, including the cumulative effects of human activities and natural processes on the environment.' GCRP made substantial investments in understanding the underlying processes of climate change, documenting past and ongoing global change, improving modeling, and enhancing knowledge of El Niño and the ability to forecast it."

SOURCE: *Staudt, Amanda, Nancy Huddleston, and Sandi Rudenstein.* Understanding and Responding to Climate Change. *National Academy of Sciences, 2006.*

gases from fires, understanding the consequences of large-scale fires on climate is challenging due to the many additional ways in which they influence the atmosphere and surface. A recent study in Alaska found that there was intensification in the climate warming in the first year after a major fire but a slight decrease in the climate warming when averaged over the 80 years of the study. The long-term result, which was primarily due to plant re-growth increasing the summertime reflectivity of the burn-scarred surface, appears to be more significant than the fire-emitted greenhouse gases. The result implies that future increases in boreal fires may not further intensify long-term climate warming.

CCSP's interdisciplinary research on the carbon cycle has produced a set of analyses utilizing long-term observations of several new and mature forests. Results from this work show that forest carbon storage has been increasing in these ecosystems and is not in balance with the carbon lost by respiration and decay. This result is contrary to the contemporary concept of near balance of carbon sources and sinks in mature forests. The gain in forest carbon is typical of findings from the U.S.-based large-scale networks, as well as observations made in mature forests in China. Evidence is therefore mounting that these sinks for atmospheric CO_2 offer significant potential for modulating the rate of atmospheric CO_2 increase.

Incorporation of the sub-surface water table into regional climate models is important since land cover changes produce significant effects on the water table and the hydrologic cycle. Shallow water tables can be either a sink or source of water to the surface soil depending on the relative balance of infiltration versus evaporation. Recent studies using detailed observations and regional climate models have found that the fraction of rainfall that either recharges groundwater or ends up as stream flow tends to decrease when the fraction of land devoted to agriculture increases. This result suggests that intense agriculture can amplify surface water stresses, particularly during drought conditions....

Goal 3: Reduce uncertainty in projections of how the Earth's climate and related systems may change in the future.

The accuracy of estimates of future Earth system conditions at time scales ranging from months to centuries and at spatial scales ranging from regional to global has been significantly improved by CCSP research. The primary tools for Earth system prediction and projection are computer models that reflect the best available knowledge of Earth system processes. The following examples demonstrate the integration of observations and modeling necessary to contribute to the progress being made in CCSP Goal 3.

For a model to produce a realistic climate requires that it include realistic representations of physical processes such as cloudiness, precipitation, and solar energy. Recent innovative studies using newly developed, detailed models of cloud processes that are coupled with a global climate model provide results that are significantly more consistent with observations. The incorporation of improved cloud representation in climate models is expected to reduce the uncertainty in predictions of the global and regional water cycle and surface climate.

Sunlight not reflected back to space provides the driving energy to Earth's weather and climate systems. Clouds are a major component in the global reflectance of sunlight. Year-to-year variability in the global reflectance is dominated by the variability of cloudiness in the tropics. On the other hand, scientists have recently found that the year-to-year variability of reflectance at middle and high latitudes has had little change despite decreases in the highly reflective snow and sea-ice cover. This fascinating result appears to be due to the compensating increase in cloud cover balancing the decreasing surface level reflectance. Clouds continue to provide the largest source of uncertainty in model estimates of climate sensitivity, although a recent study finds evidence that in most climate models used in the Working Group I contribution to the Fourth Assessment Report of the Intergovernmental Panel on Climate Change (IPCC) clouds provide a positive feedback....

Goal 4: Understand the sensitivity and adaptability of different natural and managed ecosystems and human systems to climate and related global changes.

Significant advances have been made in understanding the potential impacts of climate change. One of the characteristics of CCSP research is the use of many different sources of information, including analyses utilizing prehistoric information, direct observations, and model-based projections. Recent research also accounts for the dynamic nature of the response of human and natural systems to climate change. This research encompasses a wide range of potential impacts on societal needs such as water, health, and agriculture, as well as potential impacts on natural terrestrial and marine ecosystems. The integrated approach to developing the understanding sought in CCSP Goal 4 is illustrated in the following examples.

Tools and research resulting from carbon cycle science are highly relevant to carbon management as demonstrated by a recent study that estimated the spatial variability of net primary production and potential biomass accumulation over the conterminous United States. This study's model-based predictions indicate a potential to remove carbon from the atmosphere at a rate of 0.3 GtC yr^{-1} through afforestation of low production crop and rangeland areas. This rate of carbon sequestration would offset about one-fifth of the annual fossil fuel emissions of carbon in the United States.

The changing adaptability of coastal marshes is illustrated by a study of a coastal ecosystem. In a Chesapeake Bay marsh ecosystem, rising sea level, increasing CO_2, and high rainfall were shown to interact and improve the growth of a relatively tall bulrush at the expense of a hay-like cordgrass that grows in thick mats. Such changes in species composition, caused by interacting global change factors, may influence the capacity of coastal marshes to rise in elevation at the pace required to keep abreast of sea-level rise because of species-specific differences in their ability to trap sediment and organic material. . . .

Goal 5: Explore the uses and identify the limits of evolving knowledge to manage risks and opportunities related to climate variability and change.

A substantial investment in basic research focused on global environmental variability and change has provided a significant set of opportunities for applying this knowledge in local and regional planning. To explore and communicate the potential uses and limits of this knowledge, CCSP is taking the following three approaches: the development of scientific syntheses and assessments; exploration of adaptive management and planning capabilities; and development of methods to support climate change policy inquiries . . . .

Appendix A. Climate Change Science Program

The Climate Change Science Program (CCSP) integrates federally supported research on global change and climate change, as conducted by 13 U.S. Government departments and agencies:

- Department of Agriculture (USDA)
- Department of Commerce (DOC)
  —National Oceanic and Atmospheric Administration (NOAA)
  —National Institute of Standards and Technology (NIST)
- Department of Defense (DOD)
- Department of Energy (DOE)
- Department of Health and Human Services (HHS)
- Department of the Interior / U.S. Geological Survey (DOI/USGS)
- Department of State (DOS)
- Department of Transportation (DOT)
- Agency for International Development (USAID)
- Environmental Protection Agency (EPA)
- National Aeronautics and Space Administration (NASA)
- National Science Foundation (NSF)
- Smithsonian Institution (SI).

"PREVIEW OF OUR CHANGING PLANET: THE U.S. CLIMATE CHANGE SCIENCE PROGRAM FOR FISCAL YEAR 2008." CLIMATE CHANGE SCIENCE PROGRAM AND THE SUBCOMMITTEE ON GLOBAL CHANGE RESEARCH. APRIL 2007. <HTTP://WWW.USGCRP.GOV/USGCRP/LIBRARY/OCP2008PREVIEW/OCP08-PREVIEW.PDF> (ACCESSED NOVEMBER 30, 2007).

SEE ALSO *Environmental Protection Agency (EPA); United States: Climate Policy.*

BIBLIOGRAPHY

Periodicals

Lawler, Andrew. "Global Change Fights Off a Chill." *Science* 280, no. 5730 (June 12, 1998): 1682–1684.

Revkin, Andrew C. "Panel Faults Emphasis of U.S. Climate Program." *The New York Times* (September 14, 2007).

Schiermeier, Quirin. "China Struggles to Square Growth and Emissions." *Nature* 446, no. 7139 (April 26, 2007): 954–955.

Streets, David G., et al. "Recent Reductions in China's Greenhouse Gas Emissions." *Science* 294, no. 30 (November 30, 2001): 1835–1837.

Web Sites

"Preview of Our Changing Planet: The U.S. Climate Change Science Program for Fiscal Year 2008." *Climate Change Science Program and the Subcommittee on Global Change Research*, April 2007. <http://www.usgcrp.gov/usgcrp/Library/ocp2008preview/OCP08-preview.pdf> (accessed November 30, 2007).

Sandell, Clayton. "Court Rebukes Bush Administration on Global Warming." *ABC News,* August 21, 2007. <http://abcnews.go.com/Technology/GlobalWarming/story?id=3508197&page=1> (accessed October 26, 2007).

"U.S. Global Change Research Act of 1990." *U.S. Global Change Research Information Office.* <http://www.gcrio.org/gcact1990.html> (accessed October 26, 2007).

U.S. Global Change Research Program (homepage), October 2007. <http://www.usgcrp.gov/usgcrp/default.php> (accessed October 26, 2007).

Larry Gilman

Global Dimming

Introduction

Global dimming is the reduction in the amount of sunshine reaching Earth's surface. Both global dimming and global brightening, the opposite effect, have been observed. Dimming is caused by an increased blockage in the atmosphere of light from the sun. Clouds and aerosols—small particles emitted by burning fuels—can both contribute to dimming. Because dimming reduces the amount of solar energy reaching Earth, it tends to cool the planet, masking or offsetting global warming.

Global dimming has been confirmed for the period of 1960 to 1990, during which time the amount of sunlight reaching Earth's surface decreased about 4%. However, the dimming trend has been reversed from about 1990 to at least 2007 over most of the world. There has been much confusion in the scientific literature over the measurement of global dimming, as well as some sensational media coverage of the subject. Dimming has been hailed both as an imminent threat to human life and as salvation from global warming (and therefore a reason not to curtail the burning of fossil fuels). However, both claims are inaccurate. Although global dimming has reversed, as of 2007 the brightening had not proceeded far enough in all areas to undo the earlier decades of dimming.

Historical Background and Scientific Foundations

The possibility that particles suspended in Earth's atmosphere might cause cooler weather was recognized in the eighteenth century by American statesman and experimenter Benjamin Franklin (1706–1790), who suggested that a large volcanic eruption in Iceland might explain a spell of cold weather observed in Europe. In the 1960s, scientists first became aware of the extent to which tiny aerosol particles suspended in the air as haze could travel hundreds of miles or more from their sources. In the 1970s and 1980s, analysis of ancient ice samples retrieved from the depths of the Greenland ice cap showed traces left on global climate by volcanic eruptions that placed millions of tons of ash and sulfate particles in the atmosphere.

In the middle and late 1970s, early efforts at computer modeling of climate included aerosol particles. Researchers concluded that the tendency of aerosols would be to increase the planet's albedo (reflectivity), both by scattering light back into space and by increasing cloud formation. A cloud droplet or ice crystal is more likely to form around a tiny particle of air pollution than in clean air, other conditions being equal, so aerosols make cloud formation more likely. Aerosols also make denser, more opaque clouds, because smaller, more numerous cloud droplets tend to form when there are many aerosol particles present. However, it was not realized that a decrease in the amount of sunlight reaching the planet's surface—global dimming—would be the result. In 1987, satellite photographs showed that cloudiness was enhanced over major ocean shipping lanes because aerosols from the diesel fuel burned by ships were seeding cloud formation.

In the mid 1980s, Israeli meteorologist Gerry Stanhill reported that the amount of sunlight reaching the ground in Israel had fallen by 22% from the 1950s to the 1980s. (It was Stanhill who coined the term "global dimming.") At about the same time, German meteorologist Beate Liepert noted dimming in the Bavarian Alps. Both scientists began independently to search the scientific literature for measurements of the amount of sunlight reaching Earth's surface over time. They discovered that researchers had independently reported dimmings of 9% in Antarctica, 10% in parts of the United States, 16% in parts of the United Kingdom, and up to 30% in parts of Russia.

These direct measurements of diminished sunlight were confirmed by data on pan evaporation rates. The pan evaporation rate is the amount of water that evaporates over a given period of time from an exposed pan. This rate had decreased in step with the proposed global dimming. Stanhill concluded that the amount of sunlight reaching Earth had, in fact, decreased by 2 to 4% from about 1960 to

WORDS TO KNOW

AEROSOL: Particles of liquid or solid dispersed as a suspension in gas.

ALBEDO: A numerical expression describing the ability of an object or planet to reflect light.

FOSSIL FUELS: Fuels formed by biological processes and transformed into solid or fluid minerals over geological time. Fossil fuels include coal, petroleum, and natural gas. Fossil fuels are non-renewable on the timescale of human civilization, because their natural replenishment would take many millions of years.

GLOBAL BRIGHTENING: Any increase in the amount of sunlight reaching Earth's surface; in particular, any decrease in or reversal of global dimming, the blockage of sunlight by aerosols (fine particles) in Earth's atmosphere.

GREENHOUSE GASES: Gases that cause Earth to retain more thermal energy by absorbing infrared light emitted by Earth's surface. The most important greenhouse gases are water vapor, carbon dioxide, methane, nitrous oxide, and various artificial chemicals such as chlorofluorocarbons. All but the latter are naturally occurring, but human activity over the last several centuries has significantly increased the amounts of carbon dioxide, methane, and nitrous oxide in Earth's atmosphere, causing global warming and global climate change.

PAN EVAPORATION RATE: Rate at which water evaporates from an open pan exposed to the sky. Pan evaporation rate depends on wind, sunlight, humidity, and temperature. Measurements of pan evaporation rates have declined globally over the last century or so: global dimming has been proposed as a possible contributor to this effect.

SAHEL: The transition zone in Africa between the Sahara Desert to the north and tropical forests to the south. This dry land belt stretches across Africa and is under stress from land use and climate variability.

SOLAR ENERGY: Any form of electromagnetic radiation that is emitted by the sun.

1990. Yet this seemed to contradict global warming, which had already been well established by millions of measurements: if aerosols were increasing cloud cover and so darkening Earth, how could the planet be getting warmer?

The role of clouds in global warming and cooling is one of the largest sources of uncertainty in global climate modeling: clouds have a paradoxical double effect on the heating of the planet, reflecting heat radiation back at Earth's surface like a blanket and reflecting solar radiation back out into space like a mirror. The blanket effect tends to warm Earth, the mirror effect to cool it. Thus, confirming enhanced cloud formation in some areas by aerosol pollution did not necessarily confirm the occurrence of a cooling effect, whether regional or global.

To complicate the picture, aerosols can affect climate directly, as well as indirectly (through increased cloud formation). In the 1990s, Project INDOEX (Indian Ocean Experiment) collected data showing that aerosol pollution from India was causing a dark layer of haze some 2 miles (3.2 km) thick over the northern Indian Ocean, and that this brown cloud was reducing the amount of sunlight reaching the ocean by over 10%. The toxic clouds of pollution were also spawning increased cloud cover, which was cooling the ocean by reflecting more solar energy into space.

It is now known that the net effect of sulfate aerosols is cooling. However, some other aerosols cause warming. In particular, the dark, carbon-rich particles called soot or black-carbon particles are excellent absorbers of sunlight. They heat the air around them, and when they sift down to the surface, darkening Arctic snow, they make melting more likely. Melting Arctic snow cover exposes dark land, which accelerates global warming.

■ Impacts and Issues

Wide agreement that global dimming was indeed a global issue was not reached until about 2004, and even since then a large number of scientists have maintained that the phenomenon is not truly global but regional, being especially confined to urban areas, where there are more aerosols in the air. This dispute is partly about words, because if many regional dimming effects occur, then they will produce a dimming of the world as a whole. In any case, most scientists agree that global dimming has reversed itself since about 1990.

In 2005, the journal *Nature* reported that although the amount of solar radiation reaching Earth's surface decreased by 4 to 6% from 1960 to 1990, and that the dimming trend had reversed since that time, brightening had not gone far enough yet to undo all the dimming of earlier decades. Measurements by NASA's Aqua satellite showed global brightening to a degree that agreed with the warming of the oceans, which had been measured and had also been correctly predicted by climate-change computer models.

The main cause of global dimming is sulfate aerosols. Sulfate aerosols have a net cooling effect on Earth, as opposed to black carbon and soot particles, which have a net warming effect. As sulfate aerosol pollution has decreased in the 1990s and beyond, therefore, brightening has replaced dimming. To complicate the picture, the eruption of the volcano Mount Pinatubo in 1991 ejected about 20 million tons of sulfur dioxide into the atmosphere, increasing both darkening and cooling. As the global climate recovered from the Pinatubo eruption through the 1990s and as sulfate aerosols from human sources decreased due to anti-pollution laws, global dimming—whether viewed as a collection of regional phenomena or a single global phenomenon—has decreased.

The effects of global dimming are disputed. A few studies have attributed the occurrence of some droughts to dimming: less solar radiation reaching the surface means less evaporation, which may have decreased rainfall and even caused famine in the Sahel region of northern Africa. However, this connection is far from proven. There is broader scientific agreement that dimming may have partly masked the extent of global warming, and that global brightening is now revealing the full warming effect that human activities have already had. This unmasking may explain why global warming effects lagged model predictions for some years but have exceeded them since the mid-1990s. Other scientists caution that the feedbacks between aerosols, evaporation, humidity, albedo, and cooling are too complex for such conclusions to be drawn, given present knowledge. What is certain is that the world is rapidly warming, regardless of changes in the global dimming effect.

In the early 2000s, some urged that because of global dimming's masking effect on global warming, it would be self-destructive to burn less fossil fuel. In 2005, the Reuters news service ran a headline that read, "Fossil Fuel Curbs May Speed Global Warming." However, concentrations of greenhouse gases are steadily increasing, and aerosol emissions could not increase enough to overcome the warming effects of these gases. Furthermore, aerosols are a direct health hazard, causing many hundreds of thousands of deaths from cancer and respiratory disease around the world each year. One kind of pollution cannot be undone simply by increasing another kind.

Primary Source Connection

This source describes the contents of a new NASA study on the relationship between global dimming and global warming. Global dimming is a process by which less direct sunlight reaches Earth's surface due to aerosol particles, dust, and pollution in the atmosphere that reflect sunlight back into space. Global dimming was observed throughout the second half of the twentieth century. The new NASA study noted that global dimming and global warming could occur simultaneously, with global dimming lessening the effect of global warming.

Stephen E. Cole is on the Earth Science News Team at NASA's Goddard Space Flight Center in Greenbelt, Maryland.

GLOBAL 'SUNSCREEN' HAS LIKELY THINNED, REPORT NASA SCIENTISTS

A new NASA study has found that an important counterbalance to the warming of our planet by greenhouse gases—sunlight blocked by dust, pollution and other aerosol particles—appears to have lost ground.

The thinning of Earth's "sunscreen" of aerosols since the early 1990s could have given an extra push to the rise in global surface temperatures. The finding, published in the March 16 issue of Science, may lead to an improved understanding of recent climate change. In a related study published last week, scientists found that the opposing forces of global warming and the cooling from aerosol-induced "global dimming" can occur at the same time.

"When more sunlight can get through the atmosphere and warm Earth's surface, you're going to have an effect on climate and temperature," said lead author Michael Mishchenko of NASA's Goddard Institute for Space Studies (GISS), New York. "Knowing what aerosols are doing globally gives us an important missing piece of the big picture of the forces at work on climate."

The study uses the longest uninterrupted satellite record of aerosols in the lower atmosphere, a unique set of global estimates funded by NASA. Scientists at GISS created the Global Aerosol Climatology Project by extracting a clear aerosol signal from satellite measurements originally designed to observe clouds and weather systems that date back to 1978. The resulting data show large, short-lived spikes in global aerosols caused by major volcanic eruptions in 1982 and 1991, but a gradual decline since about 1990. By 2005, global aerosols had dropped as much as 20 percent from the relatively stable level between 1986 and 1991.

The NASA study also sheds light on the puzzling observations by other scientists that the amount of sunlight reaching Earth's surface, which had been steadily declining in recent decades, suddenly started to rebound around 1990. This switch from a "global dimming" trend to a "brightening" trend happened just as global aerosol levels started to decline, Mishchenko said.

While the Science paper does not prove that aerosols are behind the recent dimming and brightening trends—changes in cloud cover have not been ruled out—another new research result supports that conclusion. In a paper published March 8 in the American Geophysical Union's Geophysical Research Letters, a research team led by Anastasia Romanou of Columbia University's Department of Applied Physics and Mathematics, New York, also showed that the apparently opposing forces of global warming and global dimming can occur at the same time.

The GISS research team conducted the most comprehensive experiment to date using computer simulations of Earth's 20th-century climate to investigate the dimming trend. The combined results from nine state-of-the-art climate models, including three from GISS, showed that due to increasing greenhouse gases and aerosols, the planet warmed at the same time that direct solar radiation reaching the surface decreased. The dimming in the simulations closely matched actual measurements of sunlight declines recorded from the 1960s to 1990.

Further simulations using one of the Goddard climate models revealed that aerosols blocking sunlight or trapping some of the sun's heat high in the atmosphere were the major driver in 20th-century global dimming. "Much of the dimming trend over the Northern Hemisphere stems from these direct aerosol effects," Romanou said. "Aerosols have other effects that contribute to dimming, such as making clouds more reflective and longer-lasting. These effects were found to be almost as important as the direct effects."

The combined effect of global dimming and warming may account for why one of the major impacts of a warmer climate—the spinning up of the water cycle of evaporation, more cloud formation and more rainfall—has not yet been observed. "Less sunlight reaching the surface counteracts the effect of warmer air temperatures, so evaporation does not change very much," said Gavin Schmidt of GISS, a co-author of the paper. "Increased aerosols probably slowed the expected change in the hydrological cycle."

Whether the recent decline in global aerosols will continue is an open question. A major complicating factor is that aerosols are not uniformly distributed across the world and come from many different sources, some natural and some produced by humans. While global estimates of total aerosols are improving and being extended with new observations by NASA's latest generation of Earth-observing satellites, finding out whether the recent rise and fall of aerosols is due to human activity or natural changes will have to await the planned launch of NASA's Glory Mission in 2008.

"One of Glory's two instruments, the Aerosol Polarimetry Sensor, will have the unique ability to measure globally the properties of natural and human-made aerosols to unprecedented levels of accuracy," said Mishchenko, who is project scientist on the mission.

Stephen E. Cole

COLE, STEPHEN E. "GLOBAL 'SUNSCREEN' HAS LIKELY THINNED, REPORT NASA SCIENTISTS." *NASA*. MARCH 15, 2007. <HTTP://WWW.NASA.GOV/CENTERS/GODDARD/NEWS/TOPSTORY/2007/AEROSOL_DIMMING.HTML> (ACCESSED DECEMBER 2, 2007).

SEE ALSO *Aerosols; Albedo; Clouds and Reflectance; Soot.*

BIBLIOGRAPHY

Books

Weart, Spencer. *The Discovery of Global Warming.* Cambridge, MA: Harvard University Press, 2004.

Periodicals

Argonne National Laboratory. "Global Dimming: A Hot Climate Topic." *Southern Great Plains Newsletter* (July 2004).

Kvalevag, Maria M., et al. "Human Impact on Direct and Diffuse Solar Radiation During the Industrial Era." *Journal of Climate* 20 (2006): 4874-4883.

Schiermeier, Quirin. "Cleaner Skies Leave Global Warming Forecasts Uncertain." *Nature* 435 (2005): 135.

Wild, Martin, et al. "From Dimming to Brightening: Decadal Changes in Solar Radiation at Earth's Surface." *Science* 308 (2005): 847-850.

Web Sites

"Dimming the Sun." *NOVA, PBS*, September 2007. <http://www.pbs.org/wgbh/nova/sun/> (accessed October 26, 2007).

Larry Gilman

Global Warming

Introduction

Global warming is a long-term increase in Earth's average surface temperature. Because global warming does not cause uniform warming in all locations and because many other changes in climate are occurring, scientists often prefer to speak of "global climate change" rather than of global warming when referring to the whole complex of changes being caused by human activities. In such contexts, the phrase "global warming" is reserved for the warming trend as such, apart from other effects. However, in general usage it still refers to the sum of all changes occurring in Earth's climate as a result of human activities. A major effect of global warming is redistribution of climatic zones defined by temperature, precipitation, and associated ecosystems.

Global climate changes, including episodes of global cooling and warming, have occurred many times throughout Earth's history as a result of natural variations in solar radiation, atmospheric chemistry, oceanic and atmospheric circulations, volcanic eruptions, and other causes. Global warming in the last few decades, however, is primarily caused by human activities that started during the Industrial Revolution, when human burning of fossil fuels began to increase drastically, releasing large amounts of carbon dioxide (CO_2). Atmospheric concentrations of carbon CO_2, methane (CH_4), nitrous oxide (N_2O), and artificial chemicals called halocarbons have long been increasing as a result of emissions from fossil-fuel burning and other activities. Increased atmospheric concentrations of these gases are causing Earth to warm.

As of 2005, atmospheric levels of CO_2 and CH_4 were far higher than at any time in at least 650,000 years. Global average surface temperature has increased by about 0.23°F (0.13°C) per decade for the last 50 years, and has increased more rapidly in recent years: the two warmest years since instrumental records began in the 1800s were 1998 and 2005.

Although a few scientists continue to dispute that recent climate warming is caused by human activities, this is no longer doubted by the great majority of climatologists, meteorologists, and geophysicists. The 2007 Assessment Report of the United Nations' Intergovernmental Panel on Climate Change (IPCC) pronounced that global warming is now "unequivocal" and that it is more than 90% likely that most of the global warming observed since the mid-twentieth century is caused by anthropogenic (human-caused) releases of greenhouse gases.

Historical Background and Scientific Foundations

Solar radiation is the major source of energy arriving at Earth's surface (a small fraction, about 1/7500, comes from Earth's interior). Much of this incoming energy is at short wavelengths (ultraviolet and visible light). These forms of light are absorbed by Earth's surface, where their energy drives atmospheric and oceanic circulation and fuels biological processes like photosynthesis. The surface then re-radiates most of this energy as longer-wavelength light, variously termed heat radiation, infrared light, or infrared radiation. If Earth's atmosphere were completely transparent to infrared radiation, this re-radiated energy would simply shine out into space and the planet would have an average surface temperature of about 0°F (–18°C).

However, certain gases in the atmosphere, mostly notably CO_2, CH_4, and water vapor (H_2O), absorb some of the infrared energy and are warmed, warming the air and, by conduction, the land and sea. These gases are termed greenhouse gases because, like the glass windows of a greenhouse, they allow solar energy to enter Earth's atmosphere but impede the loss of heat. Earth's atmosphere naturally contains certain levels of these greenhouse gases. The resulting natural greenhouse effect maintains the planet's average temperature at a livable 59°F (15°C). This natural greenhouse effect has been crucial to the evolution and survival of life on Earth. A surface temperature of 59°F is sufficient to maintain most of Earth's water in liquid form, whereas 0°F (−18°C) is too cold for most organisms to live or for ecological processes to function.

The cover of *Time* magazine on October 19, 1987 was devoted to Earth's warming climate, including the greenhouse effect and the hole in the ozone layer. *Time & Life Pictures/Getty Images.*

Although atmospheric levels of CO_2 and CH_4 have varied greatly over deep geologic time (hundreds of millions of years), by the time of the beginning of the Industrial Revolution around 1750, their concentrations had been fairly stable for hundreds of thousands of years. The pre-industrial concentration of CO_2 was about 280 ppm (parts per million by volume), of CH_4 about 0.7 ppm, and of NO_2 about 0.285 ppm. (The special case of water vapor is discussed below.) As of 2005, the atmospheric concentration of CO_2 had increased to about 379 ppm, a 36% increase. The concentration of CH_4 has gone up to 1.8 ppm and that of N_2O to 0.319 ppm. Concentrations of chlorofluorocarbons (CFCs) and other completely human-made greenhouse gases have increased from zero in pre-industrial times to about 0.7 ppb (parts per billion by volume).

Atmospheric concentrations of greenhouse gases have increased more quickly since the middle of the twentieth century, coinciding with rapid human population growth and global industrialization. The largest

WORDS TO KNOW

ANTHROPOGENIC: Made by people or resulting from human activities. Usually used in the context of emissions that are produced as a result of human activities.

CARBON SINK: Any process or collection of processes that is removing more carbon from the atmosphere than it is emitting. A forest, for example, is a carbon sink if more carbon is accumulating in its soil, wood, and other biomass than is being released by fire, forestry, and decay. The opposite of a carbon sink is a carbon source.

INDUSTRIAL REVOLUTION: The period, beginning about the middle of the eighteenth century, during which humans began to use steam engines as a major source of power.

INFRARED RADIATION: Electromagnetic radiation of a wavelength shorter than radio waves but longer than visible light that takes the form of heat.

KYOTO PROTOCOL: Extension in 1997 of the 1992 United Nations Framework Convention on Climate Change (UNFCCC), an international treaty signed by almost all countries with the goal of mitigating climate change. The United States, as of early 2008, was the only industrialized country to have not ratified the Kyoto Protocol, which is due to be replaced by an improved and updated agreement starting in 2012.

PHOTOSYNTHESIS: The process by which green plants use light to synthesize organic compounds from carbon dioxide and water. In the process, oxygen and water are released. Increased levels of carbon dioxide can increase net photosynthesis in some plants. Plants create a very important reservoir for carbon dioxide.

RADIATIVE FORCING: A change in the balance between incoming solar radiation and outgoing infrared radiation. Without any radiative forcing, solar radiation coming to Earth would continue to be approximately equal to the infrared radiation emitted from Earth. The addition of greenhouse gases traps an increased fraction of the infrared radiation, reradiating it back toward the surface and creating a warming influence (i.e., positive radiative forcing because incoming solar radiation will exceed outgoing infrared radiation).

RESIDENCE TIME: For a greenhouse gas, the average amount of time a given amount of the gas stays in the atmosphere before being absorbed or chemically altered. The residence time of a greenhouse gas is relevant to policy decisions about emitting the gas: emissions of a gas with a long residence time create global warming over decades, centuries, or millennia, until natural processes remove the emitted quantities of gas.

single cause of greenhouse gas releases has been the burning of fossil fuels, including oil, natural gas, and coal. Fossil fuels contain carbon which, in burning, combines with oxygen to create CO_2.

Deforestation is the second-largest emitter of greenhouse gases. As they grow, trees, like all plants, take in CO_2, incorporate the carbon in their structure, and release O_2 into the atmosphere. Deforestation therefore destroys a carbon sink that tends to lower the atmospheric concentration of CO_2. Deforestation also releases CO_2 and CH_4 through the burning and decay of trees and soil carbon in deforested areas.

Agriculture is a primary cause of deforestation. It also releases large quantities of CH_4 through decomposition of organic materials in livestock digestive systems, livestock waste-treatment facilities, and flooded rice fields. Agricultural fertilizers and combustion of fossil fuels and solid wastes are the primary sources of increased N_2O emissions.

Industrial processes emit a variety of powerful synthetic greenhouse gases, including chlorofluorocarbons (CFCs), hydrofluorocarbons (HFCs), perfluorocarbons (PFCs), and sulfur hexafluoride (SF_6). Many of these chemicals also attack the ozone layer, and their manufacture and use has been limited by the Montreal Protocol treaty of 1987.

Greenhouse gases vary greatly in their ability to absorb infrared radiation. On a per-molecule basis, methane is about 21–40 times more absorptive than carbon dioxide, nitrous oxide is 200–270 times more absorptive, and CFCs are 3–15 thousand times more absorptive. CO_2, however, is present in the atmosphere at much higher concentrations than the more-powerful greenhouse gases and has experienced the greatest increases due to human activity. CO_2 is responsible for about 60% of the human contribution to increased atmospheric heat retention, termed radiative forcing.

Water vapor is responsible for more of the greenhouse effect than any other single gas, but is not a primary driver of global warming. This apparent contradiction is explained by the fact that the residence time of water vapor in the atmosphere is short. Increased water vapor concentrations, left to themselves, would disappear in a matter of weeks as rain or snow fall from clouds. The other greenhouse gases, however, linger for decades: these gases force the atmosphere to warm, increasing its ability to hold water vapor. This increased level of water vapor acts to warm the world further. Therefore, water vapor acts as a feedback factor in global warming.

■ Impacts and Issues

Most scientists today are convinced that an increase in greenhouse gases is resulting in an intensification of Earth's greenhouse effect, with resulting global warming and climate change. The exact climatic response to increased concentrations of greenhouse gases and its potential effects on humans are difficult to predict, but can be foretold in outline. Moreover, computer models of climate are constantly improving, allowing more realistic predictions to be made about future climate change. If global climate change proceeds as recent scientific studies forecast, it will likely have substantial negative ecological consequences.

Earth's surface temperature is variable from place to place and over time. Furthermore, the systems that interact to maintain the planet's temperature and climate are complex; cause-and-effect relationships between the oceans, atmosphere, land, living things, and nonliving chemistry are numerous. In spite of these scientific challenges, there is overwhelming evidence that Earth has warmed significantly during the past 150 years or so. Climate records show that from 1906 to 2005 there was a 1.3°F (0.74°C) increase in Earth's average surface temperature, with twice as much warming as this global average occurring in the Arctic and over the West Antarctic Peninsula. The oceans have also warmed, though not as much.

Recent observations show dramatic melting of the Arctic sea ice, accelerated melting of the Greenland ice cap, and accelerated seaward flow of glaciers on the western peninsula of Antarctica. Global sea-level has risen since the mid 1800s; for 1961–2003, at about 0.07 in (1.8 mm) per year. Alterations have been seen in large-scale weather phenomena like the southeast Indian monsoon season, Atlantic hurricane season, El Niño/Southern Oscillation, and North African drought cycle. Wildlife and plants are shifting their ranges to higher latitudes (closer to the poles, to track cool weather as climate warms) and to higher altitudes in mountainous regions. Many other types of observations, some of which are reviewed elsewhere in this book, independently demonstrate the reality of global warming and the complex changes it has produced in regional climates.

The aforementioned observations agree with predictions by computerized mathematical models of global climate processes. These three-dimensional general circulation models (GCMs) simulate the complex movements of energy and mass involved in the global circulation of the atmosphere, oceans, and living things. Scientists use GCMs to predict the effects of changes in specific variables, like the atmospheric concentration of CO_2, on the rest of the global climate system. Because of the complexity of the computational problem, GCMs that attempt to predict global climate change have somewhat variable results. However, modelers now agree that increased concentrations of atmospheric greenhouse gases have resulted, and will continue to result, in global warming. For example, one GCM that doubles the present CO_2 concentration to about 700 ppm predicts a 2–6°F (1.1–3.3°C) rise in global temperature and suggests that the warming would be 2–3 times more intense at high latitudes—nearer the poles—than in the tropics. These predictions are consistent with changes seen so far, or, if anything, tend to underestimate the pace of warming.

IN CONTEXT: GREENHOUSE-GAS EMISSION TRENDS

According to the Intergovernmental Panel on Climate Change (IPCC): "Global greenhouse gas (GHG) emissions have grown since pre-industrial times, with an increase of 70% between 1970 and 2004 *(high agreement, much evidence)*."

- "Since pre-industrial times, increasing emissions of GHGs due to human activities have led to a marked increase in atmospheric GHG concentrations. . . ."
- "Between 1970 and 2004, global emissions of CO_2, CH_4, N_2O, HFCs, PFCs, and SF_6, weighted by their global warming potential (GWP), have increased by 70% (24% between 1990 and 2004). . . . The emissions of these gases have increased at different rates. CO_2 emissions have grown between 1970 and 2004 by about 80% (28% between 1990 and 2004) and represented 77% of total anthropogenic GHG emissions in 2004. . . ."

SOURCE: *Metz, B., et al., eds.* Climate Change 2007: Mitigation of Climate Change. Contribution of Working Group III to the Fourth Assessment Report of the Intergovernmental Panel on Climate Change. *New York: Cambridge University Press, 2007.*

Other predicted consequences of warming include large-scale shifts in atmospheric and oceanographic circulation patterns, partial melting of the polar ice caps, global sea-level rise, reorganization of Earth's climatic zones, and establishment of new large-scale weather patterns. Changes in the distribution of heat, precipitation, and weather phenomena like storms and floods will affect the productivity and distribution of natural and managed vegetation. Animals and microorganisms will experience dramatic changes in their habitats and face higher rates of species extinction. According to the Intergovernmental Panel on Climate Change, a rise of about 4°F (2.2°C) above the pre-industrial average will likely commit 15–37% of all plant and animal species to extinction. Most biologists agree that global warming is a serious threat to biodiversity and to the health of ecosystems worldwide.

The predicted climatic and biological changes associated with anthropogenic global warming are likely to have negative outcomes for Earth's human population, although it is likely that some regions will be more severely affected than others. As of 2000, about 400 million people lived within 65.6 ft (20 m) of sea level and within 12.4 mi (20 km) of a coast. Even small increases in global sea level, accompanied by increasingly intense coastal storms and floods, can therefore threaten the lives and property of a large number of people. Changes in regional temperature, precipitation, and extreme weather events would affect the managed agriculture, fishing, and forestry that provide food and shelter for Earth's human population.

A cyclist rides past a climate change mosaic outside the United Nations Climate Change Conference in Montreal, Quebec, Canada, in 2005. *AP Images.*

Most scientists and many international policymakers now consider global warming to be a credible threat to Earth's natural environment and human population. However, attempts to prevent anthropogenic global warming, especially measures that require socioeconomic sacrifice or unfamiliar forms of behavior, have been controversial. The 1997 Kyoto Protocol—an extension of the 1992 United Nations Framework Convention on Climate Change (UNFCCC), an international treaty signed by most nations—acknowledges that human activities can alter global climate and commits signatory nations to reducing their greenhouse-gas emissions. As of June 2007, 172 nations had ratified the Kyoto Protocol. As of early 2008, the United States, the world's second-largest producer of greenhouse gases (edged out by China in 2007) and one of the biggest emitters on a person-by-person basis, had not ratified the treaty. Neither, up to that time, had Australia, the world's largest per capita producer of greenhouse gases as of 2005. These governments argued that the science of global warming remains inconclusive and that the economic consequences of action would be too great. Negotiations for a follow-up treaty to the Kyoto Protocol began in November 2007. However, in December of that year, Australia (with a new prime minister after a general election) announced that it was going to ratify the protocol.

Although there are ongoing research efforts to study methods of carbon capture and storage to reduce greenhouse gas emissions, most proposals to inject large quantities of CO_2 into underground reservoirs or oceans have, so far, met with economic, technological, and environmental barriers.

Increased extreme weather is a predicted consequence of global climate change. In August 2007, scientists at the World Meteorological Organization, an agency of the United Nations, announced that during the first half of 2007, Earth showed significant increases above long-term global averages in both high temperatures and frequency of extreme weather events, including heavy rainfalls, cyclones, and wind storms. Global average land temperatures for January and April of 2007 were the warmest recorded keeping for those two months since recordkeeping began in the 1800s.

■ Primary Source Connection

This article details comments about global climate change made to schoolchildren by Ban Ki-moon, Secretary-General of the United Nations. Ban states that global climate change will have major impacts on agriculture, economic activity, and migration patterns. Poor people living in developing countries will be among those who suffer the greatest impact from global climate change. Ban asserts that international cooperation will be necessary to win the "war" on global climate change.

William M. Reilly is the U.N. Correspondent for the United Press International (UPI).

THE U.N.'S WAR ON GLOBAL WARMING

UNITED NATIONS, March 5 (UPI) —U.N. Secretary-General Ban Ki-moon reached back to his past to tell the students he was addressing how as a child he first became aware of the world organization he now heads and how his experience shapes the way he equates the fight against global warming with war.

He spoke last Thursday at the U.N. International School in New York debating climate change.

Having taken over as secretary-general only 59 days earlier, from Kofi Annan, Ban said the speech was the first at the GA podium.

"A child of the Korean war, I grew up viewing the United Nations as a savior; an organization which helped my country, the Republic of Korea, recover and rebuild from a devastating conflict," the secretary-general said, referring to the 1950–1953 war. "Because of decisions taken in this building, my country was able to grow and prosper in peace," he said.

The prosperity helped Ban, from a farming village, rise up through his country's diplomatic ranks and become secretary-general.

But Ban said the big difference between the era in which he grew up and the world his audience would inherit was "the relative dangers we face."

"Yet there is one crucial difference," he said. "For my generation, coming of age at the height of the cold war, fear of a nuclear winter seemed the leading existential threat on the horizon."

"Today, war continues to threaten countless men, women and children across the globe," the secretary-general said. "It is the source of untold suffering and loss and the majority of the U.N.'s work still focuses on preventing and ending conflict. But the danger posed by war to all of humanity—and to our planet—is at least matched by the climate crisis and global warming."

Said Ban, "I believe that the world has reached a critical stage in its efforts to exercise responsible environmental stewardship. Despite our best intentions and some admirable efforts to date, degradation of the global environment continues unabated, and the world's natural resource base is being used in an unsustainable manner."

"Moreover, the effects of climate change are being felt around the world," he said. "The latest assessment by the Intergovernmental Panel on Climate Change has established a strong link between human activity and climate change. The panel's projections suggest that all countries will feel the adverse impact."

Not for the first time, Ban warned, "It is the poor—in Africa, small-island developing states and elsewhere—who will suffer most, even though they are the least responsible for global warming."

He has mentioned such consequences previously, when listing climate change among his top priorities as new secretary-general.

"I am encouraged to know that, in the industrialized countries from which leadership is most needed, awareness is growing," Ban said. "In increasing numbers, decision makers are recognizing that the cost of inaction or delayed action will far exceed the short-term investments needed to address this challenge."

One of the issues he hoped the students would consider is "that there is an inextricable, mutually dependent relationship between environmental sustainability and economic development" around the world.

"Global warming has profound implications for jobs, growth and poverty. It affects agricultural output, the spread of disease and migration patterns," Ban said. "It determines the ferocity and frequency of natural disasters. It can prompt water shortages, degrade land and lead to the loss of biodiversity."

The secretary-general said in coming decades, "changes in our environment and the resulting upheavals—from droughts to inundated coastal areas to loss of arable lands—are likely to become a major driver of war and conflict."

"These issues transcend borders. That is why protecting the world's environment is largely beyond the capacity of individual countries," he said, arguing the need for concerted and coordinated international action will mean "the natural arena for such action is the United Nations."

Said the secretary-general, "We are all complicit in the process of global warming. Unsustainable practices are deeply entrenched in our everyday lives. But in the absence of decisive measures, the true cost of our actions will be borne by succeeding generations, starting with yours."

"That would be an unconscionable legacy; one which we must all join hands to avert," he said. "As it stands, the damage already inflicted on our ecosystem will take decades, perhaps centuries, to reverse, if we act now."

"Unfortunately, my generation has been somewhat careless in looking after our one and only planet," Ban said. "But, I am hopeful that is finally changing and I am also hopeful that your generation will prove far better stewards of our environment."

William M. Reilly

REILLY, WILLIAM M. "THE U.N.'S WAR ON GLOBAL WARMING." UPI. MARCH 5, 2007. <HTTP://WWW.UPI.COM/INTERNATIONAL_INTELLIGENCE/ANALYSIS/2007/03/05/ANALYSIS_THE_UNS_WAR_ON_GLOBAL_WARMING/6584/> (ACCESSED NOVEMBER 29, 2007).

SEE ALSO *Feedback Factors; Greenhouse Effect; Greenhouse Gases.*

BIBLIOGRAPHY

Books

Gore, Al. *An Inconvenient Truth: The Planetary Emergency of Global Warming and What We Can Do About It.* Emmaus, PA: Rodale Press, 2006.

Lal, Rattan, et al. *Climate Change and Global Food Security.* New York: CRC, 2005.

McCaffrey, Paul. *Global Climate Change.* Minneapolis, MN: H. W. Wilson, 2006.

Metz, B., et al, eds. *Climate Change 2007: Mitigation of Climate Change: Contribution of Working Group III to the Fourth Assessment Report of the Intergovernmental Panel on Climate Change.* New York: Cambridge University Press, 2007.

Parry, M. L., et al, eds. *Climate Change 2007: Impacts, Adaptation and Vulnerability: Contribution of Working Group II to the Fourth Assessment Report of the Intergovernmental Panel on Climate Change.* New York: Cambridge University Press, 2007.

Solomon, S., et al, eds. *Climate Change 2007: The Physical Science Basis: Contribution of Working Group I to the Fourth Assessment Report of the Intergovernmental Panel on Climate Change.* New York: Cambridge University Press, 2007.

Weart, Spencer. *The Discovery of Global Warming.* Cambridge, MA: Harvard University Press, 2004.

Web Sites

"Climate Change." *U.S. Environmental Protection Agency.* <http://epa.gov/climatechange/index.html> (accessed November 20, 2007).

Intergovernmental Panel on Climate Change. <http://www.ipcc.ch> (accessed November 19, 2007).

United Nations Framework Convention on Climate Change. <http://unfccc.int/2860.php> (accessed November 1, 2007).

Larry Gilman

Great Barrier Reef

Introduction

The Great Barrier Reef, which lies off the northeastern coast of Australia, has been described as "the most complex and perhaps the most productive biological system in the world." The Great Barrier Reef is the largest structure ever made by living organisms, including human beings. It consists of the skeletons of coral polyps and hydrocorals bounded together by the soft remains of coralline algae and microorganisms.

Historical Background and Scientific Foundations

The Great Barrier Reef is over 1,250 mi (2,000 km) long and occupies 80,000 sq mi (207,000 sq km) of surface area, which is larger than the island of Great Britain. It lies off Australia, roughly parallel to the coast of the state of Queensland, at distances ranging from 10–100 mi (16–160 km) offshore. The reef is located on the continental shelf that forms the perimeter of the Australian landmass,

In this image released by the Australian Institute of Marine Science, dying coral on the Great Barrier Reef is shown in August 2007. Decline of the reef is driven by climate change, disease, and coastal development. *AP Images.*

WORDS TO KNOW

ABORIGINES: Native peoples or indigenous peoples: the cultural or racial group that are the oldest inhabitants of a region.

ALGAE: Single–celled or multicellular plants or plantlike organisms that contain chlorophyll, thus making their own food by photosynthesis. Algae grow mainly in water.

BIODIVERSITY: Literally, "life diversity": the number of different kinds of living things. The wide range of organisms—plants and animals—that exist within any given geographical region.

CAY: A low-lying reef of sand or coral.

CONTINENTAL SHELF: A gently sloping, submerged ledge of a continent.

EL NIÑO: Regularly-recurring warming of the surface waters of the eastern Pacific that has affects on global climate; part of the El Niño–Southern Oscillation Cycle (ENSO). In some contexts, "El Niño" refers to the entire ENSO cycle.

PHYTOPLANKTON: Microscopic marine organisms (mostly algae and diatoms) that are responsible for most of the photosynthetic activity in the oceans.

SYMBIOSIS: A pattern in which two or more organisms live in close connection with each other, often to the benefit of both or all organisms.

ZOOPLANKTON: Animal plankton. Small herbivores that float or drift near the surface of aquatic systems and that feed on plant plankton (phytoplankton and nanoplankton).

where the ocean water is warm and clear. At the edge of the continental shelf and the reef, the shelf becomes a range of steep cliffs that plunge to great depths and colder water. The coral polyps require a temperature of at least 70°F (21°C), and the water temperature around the reef often reaches 100°F (38°C).

The Great Barrier Reef is, in reality, a string of 2,900 reefs, cays, inlets, 900 islands, lagoons, and shoals, some with beaches of sand made of pulverized coral. Coral polyps began building the reef in the Miocene Epoch, which began 23.7 million years ago and ended 5.3 million years ago. The continental shelf has subsided almost continually since the Miocene Epoch, so the reef has grown upward with the living additions to the reef in the shallow, warm water near the surface. Coral cannot live below a depth of about 25 fathoms (150 ft or 46 m) and is also sensitive to the salt content in seawater. As the hydrocorals and polyps died and became cemented together by algae, the spaces between the skeletons were filled in by wave action that forced in other debris, called infill, to create a nearly solid mass at depth. The upper reaches of the reef are open and are riddled with grottoes, canyons, caves, holes bored by mollusks, and other cavities that provide natural homes and breeding grounds for thousands of other species of sea life.

Discovery and Exploration

The Aborigines (the native people of Australia) undoubtedly were the first discoverers of the Great Barrier Reef. The Chinese probably explored it about 2,000 years ago while searching for marine creatures like the sea cucumber, which are believed to have medicinal properties. During his voyage across the Pacific Ocean in 1520, Ferdinand Magellan (1480–1521) missed Australia and its reef. Captain James Cook (1728–1779), the British explorer credited with discovering Australia, discovered the Great Barrier Reef by accident when his ship, the Endeavour, ran aground on the reef on June 11, 1770. Cook's crew unloaded ballast (including cannon now imprisoned in the coral growth) and waited for a high tide that dislodged the ship from the reef. After extensive repairs, it took Cook and his crew three months to navigate through the mazelike construction of the Great Barrier Reef. These obstacles did not discourage Cook from exploring and charting the extent of the reef and its cays, passages, and other intricacies on this first of three expeditions of discovery he undertook to the reef.

In 1835, Charles Darwin's voyage of scientific discovery on the British ship the *Beagle* included extensive study of the reef. Mapping continued throughout the nineteenth century, and, in 1928, the Great Barrier Reef Expedition began as a scientific study of coral lifestyles, reef construction, and reef ecology. The Expedition's work concluded in 1929, but a permanent marine laboratory on Heron Island within the reef was founded for scientific explorations and environmental monitoring. The reef is also the final resting place of a number of ships that sank during World War II (1939–1945).

Biology

The reef is the product of over 350 species of coral and red and green algae. The number of coral species in the northern section of the reef exceeds the number (65) of coral species found in the entire Atlantic Ocean. Polyps are the live organisms inside the coral, and most are less than 0.3 in (8 mm) in diameter. They feed at night by extending frond-like fingers to wave zooplankton toward their mouths. In 1981, marine biologists discovered that the coral polyps spawn at the same time on one or two nights in November. Their eggs and sperm form an orange and pink cloud that coats hundreds of square miles of the ocean surface. As the polyps attach to the

IN CONTEXT: REEFS IN PERIL

Evidence is growing that coral destruction is a sensitive indicator of global warming. Although there are vigorous scientific and political arguments about assigning proportions of blame to global warming, pollution, or other more direct forms of danger and destruction, recent data indicate that at the present rate of destruction, Australia's Great Barrier Reef could be doomed by mid-century. According to researchers at the Queensland University Center for Marine Studies, the reef may be in greater and more imminent peril than previously assumed. The study, titled "Implications of Climate Change for Australia's Great Barrier Reef," was commissioned by the Worldwide Fund for Nature. If the reef reaches a critical stage of destruction, it could take hundreds of years to recover if and when environmental conditions improve.

The stress experienced by corals comprising the Great Barrier Reef reflects a global pattern of damage and destruction.

Studies on more than 300 reefs in the Atlantic, Caribbean, and Indo-Pacific regions reveal similar evidence of damage to corals and/or associated indicators of ecological stress such as the absence of lobsters, eels, sea urchins, and declining populations of native fish species.

Estimates peg the worldwide loss of coral reefs at approximately 10% over the last five to seven years. Some interpretations of data, however, assert that the global loss was sudden and due to widespread bleaching of corals in 1997 and 1998, and that corals in some areas are on the rebound. Severely stressed corals lose their symbiotic algae (zooxanthellae) and appear white or "bleached."

U.S. National Oceanic and Atmospheric Administration (NOAA) officials acknowledge that corals have experienced a recent worldwide decline, but that making accurate assessments is difficult because of a lack of baseline data that can tell scientists how far the observed changes deviate from normal cyclic patterns (i.e., how much of the damage is due to stress from human (anthropogenic) activity.

reef, they secrete calcium carbonate around themselves to build secure turrets or cups that protect the living organisms. The daisy- or feather-like polyps leave limestone skeletons when they die. The creation of a 1 in (2.5 cm) thick layer of coral takes five years.

Animal life flourishes on and along the reef, but plants are rare. The Great Barrier Reef has a distinctive purple fringe that is made of the coralline or encrusting algae *Lithothamnion* (also called stony seaweed) and the green algae *Halimeda discodea* that has a creeping form and excretes lime. The algae are microscopic and give the coral its many colors; this is a symbiotic relationship in which both partners, the coral and the algae, benefit.

Other animal life includes worms, crabs, prawns, crayfish, lobsters, anemones, sea cucumbers, starfish, gastropods, sharks, 22 species of whales, dolphins, eels, sea snakes, octopus, squid, dugongs (sea cows), 1,500 species of fish including the largest black marlin in the world, and birds such as the shearwater, which migrates from Siberia to lay its eggs in the hot coral sand. The starfish *Acanthaster planci*, nicknamed the crown-of-thorns, is destructive to the reef because it eats live coral. The starfish ravages the coral during periodic infestations then all but vanishes for nearly twenty years at a time. The crown-of-thorns has lived on the Great Barrier Reef for ages (again according to the history shown in the drilling cores), but scientists are concerned that human activities may be making the plague-like infestations worse. Giant clams that grow to more than 4 ft (1.2 m) across and 500 lb (187 kg) in weight are the largest molluscs in the world. Of the seven species of sea turtles in the world, six nest on Raine Island within the reef and lay over 11,000 eggs in a single reproductive night.

This biodiversity makes the reef a unique ecosystem. Fish shelter in the reef's intricacies, find their food there, and spawn there. Other marine life experience the same benefits. The coastline is protected from waves and the battering of storms, so life on the shore also thrives.

■ Impacts and Issues

Scientists have found that variations in water temperature stress the coral causing them to evict the resident algae. The loss of color is called coral bleaching, and it may indicate global warming or other effects such as El Niño.

The coral is a laboratory of the living and once-living. Scientists have found that coral grows in bands that can be read much like the rings in trees or the icecaps in polar regions. By drilling cores 25 ft (7.6 m) down into the coral, 1,000 years of coral history can be interpreted from the density, skeleton size, band thickness, and chemical makeup of the formation. The drilling program also proved that the reef has died and revived at least a dozen times during its 25-million-year history (although it must be understood that this resiliency predated human activities). The reef as we know it is about 8,000 years old and rests on its ancestors. In the early 1990s, study of the coral cores yielded data about temperature ranges, rainfall, and other climate changes in sufficient detail to use long-term rainfall data for design of a dam.

Coral also shows considerable promise in the field of medicine. Corals produce chemicals that block ultraviolet rays from the sun, and the Australian Institute has applied for a patent to copy these chemicals as potential cancer inhibitors. Chemicals in the coral may also yield analgesics (pain relievers) and anti-AIDS medications.

More than 1.5 million visitors per year visit the area, and development along the Australian coast to accommodate the tourists was largely uncontrolled until 1990. In the 1980s, the island resort of Hamilton was built following the dredging of harbors, leveling of hills, construction of hotels and an airport, and the creation of artificial beaches. About 25 resorts like this dot the reef. Fishing has also decimated local fish populations; fish that are prized include not only those for food, but also tropical fish for home aquariums. Fishing nets, boat anchors, and waste from fishing and pleasure boats all do damage. Prospectors have mined the coral because it can be reduced to lime for the manufacture of cement and for soil improvement in the sugar cane fields.

Environmental hazards such as oil spills have seriously threatened the reef. The maze of reefs includes a narrow, shallow shipping channel that is used by oil and chemical tankers and that has a high accident rate. Lagoons have collected waste runoff from towns, agriculture, and tourist development; and the waste has allowed algae (beyond the natural population) to flourish and strangle the live coral. Pesticides and fertilizers also change the balance between the coral and algae and zoo- and phyto-plankton, and the coral serves as an indicator of chemical damage by accumulating PCBs, metals, and other contaminants. Sediment also washes off the land from agricultural activities and development; it clouds the water and limits photosynthesis. Periodic burning off of the sugar cane fields fills the air with smoke that settles on reef waters, overfishing of particular species of fish shifts the balance of power in the undersea world, and shells and coral are harvested (both within and beyond legal limits) and sold to tourists.

The government of Australia has declared the Great Barrier Reef a national park, and activities like explorations for gold and oil and spearfishing were permanently banned with the reef's new status. The United Nations Educational, Scientific, and Cultural Organization (UNESCO) has named it a world heritage site in attempts to encourage awareness and protect the area.

See Also *El Niño and La Niña; Environmental Pollution; Ocean Circulation and Currents; Oceans and Seas.*

BIBLIOGRAPHY

Books

Aronson, R. B., ed. *Geological Approaches to Coral Reef Ecology.* Berlin: Springer, 2006.

Cote, I. M., and J. D. Reynold, eds. *Coral Reef Conservation.* Cambridge: Cambridge University Press, 2006.

Great Conveyor Belt

Introduction

The great conveyor belt is the circulation of waters throughout the world's oceans. It is also called the ocean conveyor belt, the great ocean conveyor, and similar phrases, but it is known technically as the thermohaline circulation. This circulation is driven by differences of temperature (which affect water density) and halinity (saltiness), hence the term thermohaline. Warm surface currents, which are less dense, move along the ocean's surface, conveying heat from the tropics to the poles, where it is radiated away to space. Near the poles, cooler and therefore denser water sinks and moves back toward the tropics in currents flowing at the bottom of the sea.

The great conveyor belt is not a single loop, but a complex network of loops whose courses are determined by the positions of the continents, the temperature differences between the poles and tropics, and by underwater land features such as the Greenland-Scotland ridge. Among its many effects on world climate, the conveyor moderates Europe's climate, which would otherwise be colder. Increased melting of Greenland and Antarctic ice, as well as atmospheric warming of the polar regions, may slow the conveyor down. Scientific understanding of the conveyor is still not good enough to say whether such slowing is already happening or whether more severe changes may be expected in the next century.

Historical Background and Scientific Foundations

The existence of surface ocean currents, such as the Gulf Stream, has been observed and exploited by sailors for centuries. In the nineteenth century, scientists debated what elements might cause these currents, such as the wind or differences in temperature. In 1908, Swedish oceanographer Johan Sandström performed a series of classic experiments on ocean-water samples, heating and cooling them and blowing air over them to disentangle the effects of what were then called wind-driven and thermal circulation. By the 1920s, the term "thermal" had been changed to the word "thermohaline," since it was learned that the currents were driven by differences not only in temperature but also in halinity (that is, saltiness or salinity when speaking of ocean water). Cooler water is denser than warmer water of the same halinity; saltier water is denser than less-salty water at the same temperature.

The thermohaline circulation is powered by a complex global machinery of polar cooling, freshwater input from land sources, tidal mixing, and other factors. Cooling near the poles, for example, tends to cause water to become denser and to sink; this deep water then flows along the ocean floor to various upwelling points. In general, water is warmed in the tropics and moved toward the poles by the conveyor belt, where it gives up its heat, becomes denser, and sinks again. Some water can take many centuries to complete its journey through the belt, while other water moves more quickly.

Impacts and Issues

The Atlantic part of the conveyor belt has received much scientific and popular-media attention because its warm surface component, the Gulf Stream, makes the climate in northeastern North America and in Europe warmer than it would be otherwise. Computer studies show that if the Atlantic conveyor were shut down, average temperatures in Europe might decrease by several degrees or more, with possibly extreme cooling over the Scandinavian seas.

In 2005, researchers believed that they had detected a 30% slowdown in the North Atlantic conveyor since the 1950s. However, since 2004, a line of buoys containing scientific instruments has been moored across the Atlantic, roughly in a line from Florida to Africa. The buoys float at different depths and measure temperature, salinity, and pressure, allowing water flow to be calculated. The measurements showed that the conveyor circulation is more variable over short time-scales than was thought, so what

WORDS TO KNOW

GREENHOUSE GASES: Gases that cause Earth to retain more thermal energy by absorbing infrared light emitted by Earth's surface. The most important greenhouse gases are water vapor, carbon dioxide, methane, nitrous oxide, and various artificial chemicals such as chlorofluorocarbons. All but the latter are naturally occurring, but human activity over the last several centuries has significantly increased the amounts of carbon dioxide, methane, and nitrous oxide in Earth's atmosphere, causing global warming and global climate change.

GREENLAND-SCOTLAND RIDGE: Underwater ridge connecting Greenland to Scotland in the North Atlantic that separates the Nordic seas, where North Atlantic deep water formation occurs. Below a depth of 2755 ft (840 m), the ridge forms a continuous barrier between the two basins; in some areas it rises to shallower depths or islands. Formation of the North Atlantic deep water is an essential part of the great conveyor belt or thermohaline circulation of the oceans, a key component of the global climate mechanism.

HALINITY: Salt content of seawater: synonym for salinity or saltiness. Not all seawater is equally salty (haline). Halinity can be increased by evaporation and decreased by the addition of freshwater from rivers or melting glaciers. More haline water is denser and tends to sink.

NONLINEAR: Something that cannot be represented by a straight line: jagged, erratic.

THERMOHALINE CIRCULATION: Large-scale circulation of the world ocean that exchanges warm, low-density surface waters with cooler, higher-density deep waters. Driven by differences in temperature and saltiness (halinity) as well as, to a lesser degree, winds and tides. Also termed meridional overturning circulation.

TIDAL MIXING: Mixing of different layers of ocean water due to the rising and falling of the tides. Tidal mixing occurs along shorelines but also in some parts of the deep ocean, where it is crucial to the mixing of deep waters with warmer surface waters in the tropics; this allows water to flow to the surface as part of the global thermohaline circulation of the ocean, one of the defining systems of Earth's climate mechanism.

UPWELLING: The vertical motion of water in the ocean by which subsurface water of lower temperature and greater density moves toward the surface of the ocean. Upwelling occurs most commonly among the western coastlines of continents, but may occur anywhere in the ocean. Upwelling results when winds blowing nearly parallel to a continental coastline transport the light surface water away from the coast. Subsurface water of greater density and lower temperature replaces the surface water and exerts a considerable influence on the weather of coastal regions. Carbon dioxide is transferred to the atmosphere in regions of upwelling.

appeared to have been long-term changes may have only been short-term changes.

Scientific debate continues over how vulnerable the conveyor is to climate change and whether the conveyor might suddenly stop working, with drastic consequences for Earth's climate. Computer models of climate change predict that quadrupling of the carbon dioxide in Earth's atmosphere over the next 140 years, such as might occur if no effective measures are taken to reduce greenhouse-gas emissions, could slow the conveyor by 10-50%.

Other researchers, such as Michael Schlesinger at the University of Illinois at Urbana-Champaign, assert that these estimates are too conservative, and that if greenhouse-gas emissions continue unabated, accelerated melting of Greenland's ice cap and other factors will create a 70% chance of shutting down the North Atlantic conveyor during the next 200 years and a 45% chance of shutting it down during the twenty-first century. Even with strict control of greenhouse-gas emissions, Schlesinger and colleagues state a 25% chance of conveyor shutdown. As of 2007, however, most climate scientists did not set the risks so high.

Scientists agree that the thermohaline circulation is nonlinear—that is, can turn on and off suddenly, rather than ramping smoothly up and down. However, they disagree about whether it is vulnerable to actual shutdown by climate change or merely to being slowed. There is also agreement that climate change has been adding more freshwater to Arctic seas, which may have the effect of slowing the thermohaline circulation. In 2006, it was announced by other scientists that the rate of melting of Greenland's ice had more than doubled from 2002 to 2004 to 2004 to 2006.

The conveyor belt is being more closely observed than ever before, and computer models of climate are improving. Uncertainties about the present and future behavior of the great conveyor belt will likely decrease over the next few years, as this is a rapidly changing area of climate knowledge.

■ Primary Source Connection

This article gives a glimpse of the scientific debate about whether global warming might have drastic effects on the conveyor belt circulation of the Atlantic Ocean. Although some scientists assert that there is scant chance of a shutdown of the circulation by global warming, the scientist interviewed in this online NASA journal thinks that global climate change could disrupt the ocean conveyer belts and drastically alter some climates.

GLOBAL WARMING COULD HALT OCEAN CIRCULATION, WITH HARMFUL RESULTS

Absent any climate policy, scientists have found a 70 percent chance of shutting down the thermohaline circulation in the North Atlantic Ocean over the next 200 years, with a 45 percent probability of this occurring in this century. The likelihood decreases with mitigation, but even the most rigorous immediate climate policy would still leave a 25 percent chance of a thermohaline collapse.

"This is a dangerous, human-induced climate change," said Michael Schlesinger, a professor of atmospheric sciences at the University of Illinois at Urbana-Champaign. "The shutdown of the thermohaline circulation has been characterized as a high-consequence, low-probability event. Our analysis, including the uncertainties in the problem, indicates it is a high-consequence, high-probability event."

Schlesinger will present a talk "Assessing the Risk of a Collapse of the Atlantic Thermohaline Circulation" on Dec. 8 at the United Nations Climate Control Conference in Montreal. He will discuss recent work he and his colleagues performed on simulating and understanding the thermohaline circulation in the North Atlantic Ocean.

The thermohaline circulation is driven by differences in seawater density, caused by temperature and salinity. Like a great conveyor belt, the circulation pattern moves warm surface water from the southern hemisphere toward the North Pole. The water cools between Greenland and Norway, sinks into the deep ocean, and begins flowing back to the south.

"This movement carries a tremendous amount of heat northward, and plays a vital role in maintaining the current climate," Schlesinger said. "If the thermohaline circulation shuts down, the southern hemisphere would become warmer and the northern hemisphere would become colder. The heavily populated regions of eastern North America and western Europe would experience a significant shift in climate."

Higher temperatures caused by global warming could add fresh water to the northern North Atlantic by increasing the precipitation and by melting nearby sea ice, mountain glaciers and the Greenland ice sheet. This influx of fresh water could reduce the surface salinity and density, leading to a shutdown of the thermohaline circulation.

"We already have evidence dating back to 1965 that shows a drop in salinity around the North Atlantic," Schlesinger said. "The change is small, compared to what our model needs to shut down the thermohaline, but we could be standing at the brink of an abrupt and irreversible climate change."

To analyze the problem, Schlesinger and his colleagues first used an uncoupled ocean general circulation model and a coupled atmosphere-ocean general circulation model to simulate the present-day thermohaline circulation and explore how it would behave in response to the addition of fresh water.

They then used an extended, but simplified, model to represent the wide range of behavior of the thermohaline circulation. By combining the simple model with an economic model, they could estimate the likelihood of a shutdown between now and 2205, both with and without the policy intervention of a carbon tax on fossil fuels. The carbon tax started out at $10 per ton of carbon (about five cents per gallon of gasoline) and gradually increased.

"We found that there is a 70 percent likelihood of a thermohaline collapse, absent any climate policy," Schlesinger said. "Although this likelihood can be reduced by the policy intervention, it still exceeds 25 percent even with maximal policy intervention."

Because the risk of a thermohaline collapse is unacceptably large, Schlesinger said, "measures over and above the policy intervention of a carbon tax—such as carbon capture and sequestration—should be given serious consideration."

James E. Kloeppel

KLOEPPEL, JAMES E. "GLOBAL WARMING COULD HALT OCEAN CIRCULATION, WITH HARMFUL RESULTS." *NASA EARTH OBSERVATORY.* DECEMBER 7, 2005. <HTTP://EARTHOBSERVATORY.NASA.GOV/NEWSROOM/MEDIAALERTS/2005/2005120721155.HTML> (ACCESSED DECEMBER 3, 2007).

SEE ALSO *Antarctica: Melting; Arctic Melting: Greenland Ice Cap; Arctic Melting: Polar Ice Cap; Greenland: Global Implications of Accelerated Melting.*

BIBLIOGRAPHY

Books

Solomon, S., et al, eds. *Climate Change 2007: The Physical Science Basis: Contribution of Working Group I to the Fourth Assessment Report of the Intergovernmental Panel on Climate Change.* New York: Cambridge University Press, 2007.

Periodicals

Church, John A. "Oceans: A Change in Circulation?" *Science* 317 (2007): 908-909.

Clark, Peter U. "The Role of the Thermohaline Circulation in Abrupt Climate Change." *Nature* 425 (2002): 863-869.

Quadfasel, Detlief. "The Atlantic Heat Conveyor Slows." *Nature* 438 (2005): 555-556.

Web Sites

Dickson, Bob, and Steven Dye. "Interrogating the 'Great Ocean Conveyor.'" *Oceanus* (Woods Hole Oceanographic Institute), September 6, 2007. <http://www.whoi.edu/oceanus/viewArticle.do?id=20727> (accessed October 8, 2007).

Kloeppel, James E. "Global Warming Could Halt Ocean Circulation, With Harmful Results." *NASA Earth Observatory* (U.S. National Aeronautics and Space Administration), December 7, 2005. <http://earthobservatory.nasa.gov/Newsroom/MediaAlerts/2005/2005120721155.html> (accessed October 8, 2007).

Larry Gilman

Great Lakes

Introduction

The five Great Lakes—Erie, Huron, Michigan, Ontario, and Superior—are environmentally unique and invaluable to the United States and Canada. Together, the lakes extend over 750 mi (1,200 km) from east to west, comprising 18% of Earth's freshwater (its largest single source) and 95% of North America's freshwater, dominating the eight states and two Canadian provinces that border them. The lakes' influence on transportation, industry, and agriculture in this area is pervasive, and over a quarter of the population of Canada and one tenth of the United States population live in the Great Lakes watershed.

As the issue of global warming gains widespread acceptance, its effects on the Great Lakes and the surrounding region have become better documented and of more immediate concern to scientists, policymakers, and concerned members of the public. All need access to information on how further regional changes may affect specific aspects of life in the Great Lakes area—including potentially drastic shifts in economy, ecology, and public health—in order to make decisions that will help alleviate the changes or to better adapt to them.

Michigan Technical University chemist Dr. Noel Urban examines a water sample aboard a research vessel off the coast of Marquette in June 2007. The chemist is helping to develop models that will enable scientists to predict how climate change will affect Lake Superior's ecosystems. *AP Images.*

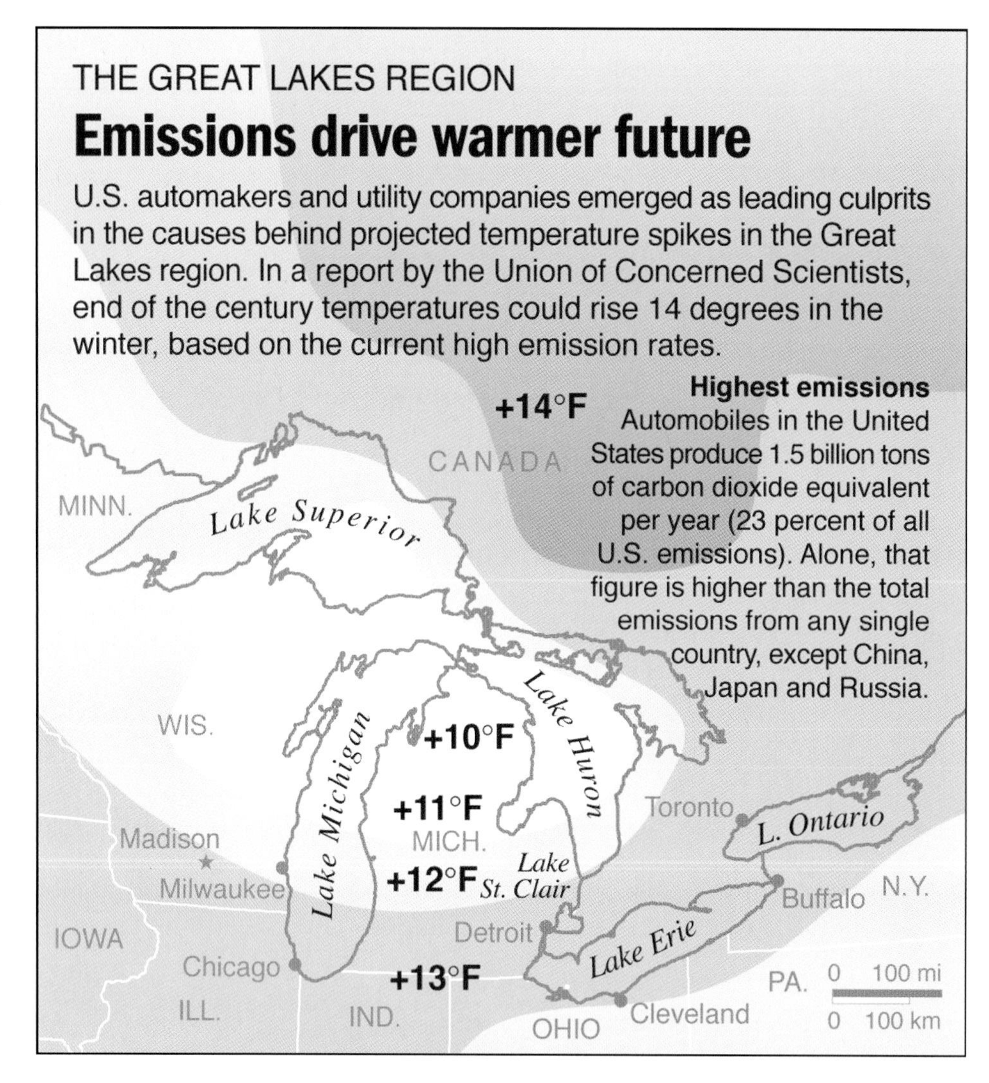

In this graphic from the Associated Press, the effect of emissions on the Great Lakes is explored. [Sources: Union of Concerned Scientists, ESRI, Associated Press.] *AP Images.*

Historical Background and Scientific Foundations

The United States Global Change Research Program (USGCRP) initiated its national assessment of the consequences of climate change in 1997, and the Great Lakes Regional Assessment Group issued its report, *Preparing for a Changing Climate: The Potential Consequences of Climate Variability and Change in the Great Lakes Region* in 2000, with contributions from over thirty specialists in different fields. Researchers have continued to build on this assessment, determining the probable impacts of climate change on Great Lakes communities and ecosystems.

The Great Lakes and surrounding region already suffer from serious air and water pollution, burgeoning populations of invasive species, and continued habitat loss. It is likely that ongoing climate change will intensify current problems as well as create new ones.

Impacts and Issues

In general, temperatures are expected to warm in the Great Lakes area, ranging from 5-12°F (3-6°C) warmer in winter, and anywhere from 5-20°F (3-11°C) warmer in summer. This would greatly reduce the length of winter and the time that lakes (including numerous smaller inland lakes) are covered with ice. Great Lakes water levels may decline anywhere from 2 to 8 ft (0.6 to 2.4 m) despite modest increases in rainfall, mainly because of the increase in evapotranspiration associated with the loss of winter ice cover and warmer temperatures. Lower lake levels will have major impacts on shipping, hydroelectric power, and water use. Water shortages and conflicts over supply are expected to increase.

Hotter summer temperatures will increase the duration of summer stratification (seasonal changes in temperature and density) in all lakes, and according to Peter

Sousounis and Patty Glick (2000), "will reduce the seasonal mixing that replenishes critical oxygen to biologically productive lake zones, possibly shrinking lake biomass productivity by around 20%. This will include losses of zooplankton and phytoplankton that form the very base of aquatic food chains, and are critical to the survival of the many species of fish that live in the Great Lakes." An associated increase in the number of dead zones (areas without living organisms), especially in the deeper portions of Lakes Superior, Michigan, and Huron, is likely.

Instances of extreme heat and rain, producing increased river flooding, agricultural runoff, and sewage overflows, will likely become more common in the summer months. Lake-effect snow may become heavier or more frequent in areas already prone to it. There is evidence that high lake-effect snowfall rates in the twentieth century already reflect warmer lake temperatures.

Associated effects on Great Lakes ecosystems may be profound. Distributions of many species are expected to shift north (or out of the lakes altogether) as temperatures warm, wetlands dry up, and the shoreline breeding grounds that are important for fish, migratory birds, and amphibians suffer. Forest composition is also predicted to change, and many of the invasive plant, insect, and fish species currently found in regions to the south of the Great Lakes may further stress native species. Hotter and drier summer conditions will also promote more frequent wildfires, especially in northern forested areas.

Although earlier studies predicted that agriculture in the area would not be seriously disrupted, new studies show that the increased variability in rainfall now predicted (more rain, but less evenly spread out) may be very detrimental, especially to corn and soybeans.

The anticipated effects of climate change on human health are significant, if not as grave as in some other regions. Asthma and other respiratory diseases may be exacerbated by high ground-level ozone levels and the increased air pollution that accompanies heat waves. A higher risk of waterborne infectious diseases and diseases carried by ticks and mosquitoes is also predicted.

Finally, tourism and local recreation—which are significant to local economies and the sense of place that defines the region—are likely to change as the lakes themselves and the landscapes around them alter. Although this aspect of change is harder to predict than fish resources or forest composition, it is still of great concern to the 60 million inhabitants who live in the Great Lakes region.

SEE ALSO *Agriculture: Vulnerability to Climate Change; Biodiversity; Blizzards; Economics of Climate Change; Fisheries; Hydrologic Cycle; Lake Effect Snows; North America: Climate Change Impacts; Water Shortages.*

WORDS TO KNOW

BIOMASS: The sum total of living and once-living matter contained within a given geographic area. Plant and animal materials that are used as fuel sources.

EVAPOTRANSPIRATION: Transfer of water to the atmosphere from an area of land, combining water transfer from foliage (transpiration) and evaporation from non-living surfaces such as soils and bodies of water.

LAKE-EFFECT SNOW: Snow that falls downwind of a large lake. Cold air moving over relatively warm lake water is warmed and moistened; after leaving the lake, the air cools again and some of its moisture precipitates out as snow.

PHYTOPLANKTON: Microscopic marine organisms (mostly algae and diatoms) that are responsible for most of the photosynthetic activity in the oceans.

STRATIFICATION AND TURNOVER: Two processes in lakes that have to do with the mixing of waters at different depths. Stratification is layering, which occurs when upper layers are warmer and float stably on the deeper water. Stratification opposes mixing or turnover, which is the exchange of water (and thus of dissolved chemicals and suspended particles) between deep and surface waters.

ZOOPLANKTON: Animal plankton. Small herbivores that float or drift near the surface of aquatic systems and that feed on plant plankton (phytoplankton and nanoplankton).

BIBLIOGRAPHY

Books

Dempsey, Dave. *On the Brink: The Great Lakes in the 21st Century.* East Lansing, MI: Michigan State University Press, 2004.

Great Lakes Regional Assessment Group for the U.S. Global Change Research Program. *Preparing for a Changing Climate: The Potential Consequences of Climate Variability and Change in the Great Lakes Region.* Ann Arbor, MI: University of Michigan, Department of Atmospheric, Oceanic and Space Sciences, 2000.

Kling, George W., Katharine Hayhoe, L. B. Johnson, et al. *Confronting Climate Change in the Great Lakes Region: Impacts on Our Communities and Ecosystems.* Cambridge, MA: Union of Concerned Scientists and Ecological Society of America, 2003.

Periodicals

Burnett, Adam W., Matthew E. Kirby, et al. "Increasing Great Lake-Effect Snowfall During the Twentieth Century: A Regional Response to Global Warming?" *Journal of Climate* 16, 21 (2003): 3535-3542.

Oakes, Larry. "Lake Superior Called 'Early Victim of Climate Change.'" *Minneapolis Star Tribune* (October 30, 2007).

Santos, Fernanda. "Inch by Inch, Great Lakes Shrink, and Cargo Carriers Face Losses." *The New York Times* (October 22, 2007).

Scheraga, J. D., and J. Furlow. "Preface to the Potential Impacts of Climate Change in the Great Lakes Region." *Journal of Great Lakes Research* 28 (2002): 493-495.

Web Sites

"Climate Change Impacts on the United States: The Potential Consequences of Climate Variability and Change, Overview: Midwest." *U.S. Global Change Research Program*, October 12, 2003. <http://www.usgcrp.gov/usgcrp/Library/nationalassessment/overviewmidwest.htm> (accessed November 14, 2007).

"Global Warning: Great Lakes Communities and Ecosystems at Risk." *Union of Concerned Scientists*, August 27, 2007. <http://www.ucsusa.org/greatlakes> (accessed November 14, 2007).

"The Great Lakes: An Environmental Atlas and Resource Book." *Government of Canada and U.S. Environmental Protection Agency, Great Lakes National Program Office*, March 9, 2006. <http://www.epa.gov/glnpo/atlas/index.html> (accessed November 14, 2007).

"Great Lakes Regional Assessment." *Michigan State University*. <http://www.geo.msu.edu/glra/index.html> (accessed November 14, 2007).

Sousounis, Peter, and Patty Glick. "The Potential Impacts of Global Warming on the Great Lakes Region: Critical Findings for the Great Lakes Region from the First National Assessment of the Potential Consequences of Climate Variability and Change," 2000. <http://www.climatehotmap. org/impacts/greatlakes.html> (accessed November 14, 2007).

Sandra L. Dunavan

Green Movement

Introduction

The green movement is a diverse scientific, social, conservation, and political movement that broadly addresses the concerns of environmentalism. It encompasses an array of political parties, organizations, and individual advocates operating on international, national, and local levels. Unified only by a desire to protect the environment, but otherwise diverse in philosophy and strategy, the various factions of the green movement have succeeded in heightening public awareness of environmental issues.

Historical Background and Scientific Foundations

The modern green movement emerged in the 1960s and 1970s. Its growth reflected popular and scientific concerns about local and global degradation of the physical environment. However, history is dotted with incidents of environmental protest. Moreover, conservational groups have long campaigned to preserve natural environments and wild species. In Europe, organizations such as Friends of Nature and the National Trust date back to the nineteenth century, while the World Wide Fund for Nature (WWF) was founded in 1963.

The relative newness of a broad green movement can be seen by examining citations under "environment" listed in *The New York Times* index. In 1955, the word is not indexed; the newspaper did not discuss environmental issues. In 1960, there is a single citation; by 1965 there are two. By 1970, the year of the first Earth Day and National Environmental Policy Act (NEPA), there are 86 citations for the environmental movement. In 1990, there were 172; by 2001, the year President George W. Bush withdrew final U.S. support for the Kyoto Protocol, there were more than 3,000. The increased media coverage of environmental issues indicates a growing public concern for the environment and a blossoming green movement.

Although numerous conservation societies predated the 1960s, it was the ensuing decade-and-a-half that saw the birth of hundreds of grassroots green organizations and ultimately large-scale national and international groups such as Greenpeace and Friends of the Earth. Localized concerns—nuclear power, toxic waste, acid rain, road building—found an expression in the era's increased political activism. Works such as Rachel Carson's *Silent Spring* (1962) and Barry Commoner's *Science and Survival* (1965), which found popular audiences, heightened the educational efforts of conservationists and raised awareness of human-made degradation of the natural environment.

In the late 1960s and early 1970s, mass environmental advocacy organizations, such as Greenpeace and Friends of the Earth, started to emerge. Greenpeace was initially an extension of the hippy movement, first emerging in Vancouver, British Columbia, Canada, in 1969 as the Don't Make a Wave Committee of activists opposed to U.S. underground nuclear testing in Amchitka, Alaska. The committee's motivations were a mix of environmental concerns about the fate of the island and its wildlife and a philosophical opposition to nuclear weapons. A mission to disrupt a nuclear test failed but excited huge interest, eventually leading the United States to end nuclear testing on the island. In 1971, the committee was renamed Greenpeace. Daring stunts to prevent French nuclear testing in the Pacific Ocean, including mooring a vessel within the testing exclusion area, brought Greenpeace global fame when they resulted in French abandonment of atmospheric testing.

David McTaggart, a Canadian businessman who led the Pacific seaborne protests, utilized this publicity and his business acumen to transform the Vancouver outfit into a global organization. McTaggart formed Greenpeace in Europe, finding like-minded souls to set up national organizations. This array of national associations was unified in 1979 as Greenpeace International.

In a city widely recognized for its mass transit system, Portland officials took going green one step farther in 2007 with the launch of an aerial tram to move people from its Oregon Heath and Science University to its new River Blocks neighborhood on the waterfront. According to estimates, some 2 million vehicle miles will be saved, resulting in more than 93,000 gallons of gas annually. In addition, OHSU believes that over 1,000 tons of greenhouse gases will not enter the atmosphere as a result. *AP Images.*

Greenpeace's second great cause, from the mid-1970s, was Save the Whale, a campaign that captured the global imagination. In a series of spectacular protests, Greenpeace activists would chase whaling fleets and interpose themselves between the harpoon of a catcher ship and a fleeing whale. As with the protests against French nuclear testing, film and photographic footage was widely circulated, prompting huge numbers to join its membership rolls. Green causes began to permeate mainstream politics in the 1980s, when Greenpeace and other groups brought issues such as species extinction, the destruction of the Amazonian rainforests, and climate change to the global agenda. Worldwide support for Greenpeace (defined as people who donated money) peaked in 1991 at 4.8 million, but has since fallen by up to half.

In the United Kingdom, Friends of the Earth was formed in 1971, initially to protest against a lemonade manufacturer's use of non-recyclable bottles. Like Greenpeace, its protests captured popular imagination, enabling it to expand its interests to such issues as road building, whaling, and climate change. Friends of the Earth now claims to be the largest international network of environmental groups in the world, covering more than 54 countries worldwide. Many other international environmental groups—such as the Environmental Investigation Agency—have also emerged from local origins to assume international reputations.

Others, such as the World Watch Institute, originate from environmentally conscious philanthropists; or, like, Conservation International, as non-governmental organizations that obtain funding from a mixture of private and public sources.

Operating beneath these large organizations are tens of thousands of protest and campaign groups operating on a regional, national, or localized basis. Over the past two decades, support for green causes has greatly shifted to these grassroots groups who espouse a range of local environmental causes from opposition to highway building to water pollution. Moreover, operation on a smaller scale has allowed nascent environmental groups to prosper in places like China, where larger organizations would usually fall foul of state controls.

In the early twenty-first century, the green movement has evolved in different ways. The decline in support experienced by larger environmentalist organizations has been accompanied by the rise of so-called envirocelebrities, such as former U.S. Vice President Al Gore and rock stars such as Bono and Sting. These individuals have used their public prominence to raise awareness of climate change and related environmental issues. Gore, who jointly won the 2007 Nobel Peace Prize for his work in promoting the understanding of climate change, produced an Oscar-winning film about climate change—*An Inconvenient Truth*—and helped organize the Live Earth concerts in 2007.

In the 1970s and early 1980s, the growth in mainstream green organizations precipitated the rise in eco-conscious political parties. Britain had the Ecology Party from the mid–1970s. Former West Germany was among the first countries to have its Green Party—Die Grünen—around 1980. Some 60 national Green Parties have been formed since then, most in developed countries, but also in nations such as Saudi Arabia (where it has been forced underground) and Somalia. Though all national Green Parties have unique policies and agendas, they are unified by a commitment to environmentalism, grassroots advocacy, nonviolence, and social justice.

■ Impacts and Issues

Previous favorite causes, such as whaling and nuclear testing, seem insignificant in comparison, yet groups like

Greenpeace and national Green Parties have failed to influence governments to reach agreement on the radical action necessary to counteract global warming. As a consequence, membership has ebbed, while being further eroded by mainstream political parties, which have increasingly adopted green agendas as part of their electoral strategies. Although membership in large environmental protest groups has declined, green issues have become more visible in the media. More people describe themselves as an "environmentalist" today than ever before, but most engage environmental issues through personal action like recycling at home, carpooling, or eschewing the use of plastic grocery bags. More U.S. voters than ever before rank global climate change and environmental regulation as key issues.

It is now normal for mainstream political parties to adopt green issues into their political platforms. For example, Britain's Conservative Party, traditionally considered anti-regulation, now extols environmentalism as one of the central tenets of its political platform. However, governments and environmental protest groups have sometimes been at extreme odds. In 1985, as Greenpeace prepared another flotilla to try and avert French nuclear testing at Moruroa atoll, French special forces, acting under the direct orders of President François Mitterand, attached two bombs to the hull of the Greenpeace ship, Rainbow Warrior. When the two bombs detonated they sank the ship, killing a Portuguese photographer. The bombing of the Rainbow Warrior caused international outrage, and the French government went on to pay the New Zealand government considerable compensation, as well as offering a formal apology.

Elsewhere, environmental groups have suffered direct state repression. In Saudi Arabia, the local chapter of Greenpeace has been forced underground. In China, where such groups are generally tolerated, state crackdowns on green protests and demonstrations are not uncommon, particularly in its more outlying provinces. Nevertheless, a minority of political analysts have expressed optimism in the growth of the green movement in spite of repressive regimes.

Sections of the green movement are also considered extremist by many governments and law enforcement agencies. Eco-terrorist organizations have splintered from the peaceable, mainstream Green Movement and turned to direct action campaigns that employ vandalism, property destruction, violence, and life-endangering sabotage. Some of these extremist branches of the environmental movement formed after becoming frustrated by a lack of radicalization within the Green Movement. Examples of these include Earth First!, which was founded by disaffected mainstream environmentalists in 1979 and boasted at least 10,000 members worldwide by the late 1980s. Although Earth First! has always publicly disavowed violence, its members have been linked to sabotage at ski resorts and nuclear power plants, including a plot to simultaneously attack power transmission lines in Arizona, California, and Colorado. In general, mainstream members of the Green Movement support several of the same environmental causes as the movement's extreme fringe, but condemn the use of violence and vandalism as means of protest.

More radical still is the Earth Liberation Front (ELF). Formed in Great Britain in 1992, and also active in Canada and the United States, it is modeled after its sister organization, the Animal Liberation Front (ALF) and is a loose-knit amalgam of cells, with no centralized leadership. It is bound together only by a common set of core guidelines, as published on the ELF Web site:

- To inflict economic damage on those profiting from the destruction and exploitation of the natural environment.
- To reveal and educate the public on the atrocities committed against the earth and all species that populate it.
- To take necessary precautions against harming any animal, human and non-human.

WORDS TO KNOW

EARTH LIBERATION FRONT (ELF): An underground group in North America that describes itself as "an international underground organization that uses direct action in the form of economic sabotage to stop the destruction of the natural environment." The group claims to have destroyed $100 million worth of property. The U.S. Federal Bureau of Investigation declared it the number one domestic terror threat as of March 2001.

FACTION: A dissenting group within a larger group such as a political party. A faction is usually outnumbered by other members of the larger group.

GREEN PARTY: Any of a number of political parties in the United States, Europe, and elsewhere, whose policies are centered on environmentalism, participatory democracy, and social justice. About 70 countries have Green Parties: in Europe, Australia, and New Zealand, a number of Greens have been elected to Parliament. The German Green Party is particularly powerful.

GREENPEACE: Nonprofit environmental group formed in 1971, originally to protest nuclear testing and whaling. The group remains active, now addressing a wide range of issues, including climate change.

NATIONAL ENVIRONMENTAL POLICY ACT (NEPA): U.S. federal law signed by President Richard Nixon on January 1, 1970. The act requires federal agencies to produce Environmental Impact Statements describing the impact on the environment of projects they propose to carry out.

ELF engages in sabotage against companies involved in logging, energy production, and construction and is described by the FBI as "one of the most active extremist elements in the United States." ELF members are accused of over $100 million in property damage since 1997. An overwhelming majority of environmentalists reject the methods employed by fringe groups like ELF. Most use the media to distribute information and raise awareness, to work within the confines of the political process, or to employ tactics of peaceful protest and civil disobedience.

Although environmental issues have entered into mainstream politics in many areas, the electoral and legislative success of dedicated Green Parties has been mixed. In some countries, they are important elements in the political landscape, partaking in coalition governments and exacting a huge influence upon their country's politics. In Germany, which has one of the oldest and the most influential Green Party in the world, members have participated in several coalition governments and been responsible for huge policy shifts, such as the phasing out of nuclear power stations. In other large countries, such as Britain and France, Green Parties are on the peripheries of mainstream politics, rarely gaining more than a couple percent of votes in national elections. Britain has never had a Green Party member of parliament (MP), although its public has elected a number of local councilors since the party's formation in 1985. The addition of a Green Party candidate to the ballot made a significant impact on the outcome of the 2000 U.S. presidential election.

SEE ALSO *Environmental Policy; Environmental Protests; Sustainability.*

BIBLIOGRAPHY

Books

Burchell, Jon. *The Evolution of Green Politics: Development and Change within European Green Parties.* London: Earthscan, 2002.

Dryzek, John S., et al. *Green States and Social Movements: Environmentalism in the U.S., U.K., Germany and Norway.* Oxford: Oxford University Press, 2003

Liddick, Donald R. *Eco Terrorism: Radical Environmental and Animal Liberation Movements.* Westport, CT: Praeger, 2006.

Rootes, Christopher, ed. *Environmental Protest in Western Europe.* Oxford: Oxford University Press, 2003

Rubin, Charles T. *The Green Movement: Rethinking the Roots of Environmentalism.* New York: Free Press, 1994.

James Corbett

Greenhouse Ameliorators

Introduction

Greenhouse ameliorators refers to approaches or compounds that are intended to reduce the warming of the atmosphere that is caused by the accumulation of greenhouse gases—gases that increase the retention of heat entering the atmosphere in the form of sunlight.

Greenhouse ameliorators can be direct, involving the alteration of the atmosphere. For example, one scenario that is still conceptual and not yet real involves the introduction of tiny particles into equatorial regions of the atmosphere. The particles, which would increase the reflection of sunlight out of that area of the atmosphere, would lessen the heating of the tropical atmosphere.

Other greenhouse ameliorators are indirect. Reducing the use of fossil fuels and increasing the use energy sources such as solar and wind power that do not emit greenhouse gases is one example.

The use of greenhouse ameliorators has been spurred by the recognition that human activities are an important influence on atmospheric warming.

Historical Background and Scientific Foundations

The term greenhouse ameliorators is derived from the greenhouse effect. The latter is the documented warming of the portions of the atmosphere called the troposphere and the stratosphere, as well as the immediate surface of Earth. The warming is being caused by interactions between sunlight entering the atmosphere, particles suspended in

During a protest at the gate of the Hong Kong government's headquarters, members of Greenpeace placed Doomsday Clocks and a banner reading "Time's Up! Save the Climate." The protesters urged the government to create a greenhouse-gases reduction policy and to set targets. *AP Images.*

WORDS TO KNOW

AGROFORESTRY: The practice of sustainably combining forestry with agriculture by combining trees with shrubs, crops, or livestock. Two forms of agroforestry are alley cropping (strips of crops alternating with strips of woodland) and silvopasture (pasturing grazing livestock under widely spaced trees). Benefits include greater soil retention, biodiversity encouragement, higher monetary returns per acre, and more diverse product marketing.

FOSSIL FUELS: Fuels formed by biological processes and transformed into solid or fluid minerals over geological time. Fossil fuels include coal, petroleum, and natural gas. Fossil fuels are non-renewable on the timescale of human civilization, because their natural replenishment would take many millions of years.

KYOTO PROTOCOL: Extension in 1997 of the 1992 United Nations Framework Convention on Climate Change (UNFCCC), an international treaty signed by almost all countries with the goal of mitigating climate change. The United States, as of early 2008, was the only industrialized country to have not ratified the Kyoto Protocol, which is due to be replaced by an improved and updated agreement starting in 2012.

PLANKTON: Floating animal and plant life.

STRATOSPHERE: The region of Earth's atmosphere ranging between about 9 and 30 mi (15 and 50 km) above Earth's surface.

TROPOSPHERE: The lowest layer of Earth's atmosphere, ranging to an altitude of about 9 mi (15 km) above Earth's surface.

the atmosphere, and gases that are dissolved in the atmosphere. The result of these interactions is that heat is hindered from leaving the atmosphere. The name of the effect refers to the warming of the air in a greenhouse when sunlight is restricted from reflecting back out through the glass enclosure.

The overwhelming scientific consensus is that, beginning in the mid-eighteenth century when industrialization increased, gases (including carbon dioxide, methane, and nitrous oxide) began to accumulate in the atmosphere. These so-called greenhouse gases trap the heat of the sun, restricting its exit from the atmosphere into space.

Greenhouse gases can persist in the atmosphere for decades to centuries. Thus, even if all greenhouse gas production were to cease, the atmosphere could continue to warm for an undetermined period of time. However, eventually the warming would slow and perhaps even reverse.

There are two greenhouse amelioration strategies. The first tries to reduce the amount of greenhouse gases escaping to the atmosphere by reducing energy use and using sources of energy that are less likely to generate the gases. Examples of this strategy include the use of hybrid vehicles that are electrically operated some of the time, using more recycled materials rather than original materials in daily life, and using energy sources that do not involve the burning of fossil fuels.

The second amelioration approach adds compounds to the atmosphere or on Earth's surface that act to reduce the retention of the sun's heat. One example is known as agroforestry—establishing forests to soak up carbon dioxide from the air (in the language of climate change, forests are described as being carbon sinks). As of 2008, research is also being done in Japan and the United States to investigate the feasibility of trapping carbon dioxide released from the chimneys of coal-burning facilities, converting the gas to a liquid, and pumping the liquid deep underground or to the bottom of the ocean.

Other examples of the alteration of the atmosphere that are still conceptual include spreading material on the ocean surface that increases the reflection of sunlight, seeding the atmosphere to increase the formation of light reflecting clouds, adding food sources to the ocean to promote the growth of plankton that would absorb carbon and transport it to the ocean floor when they die, and introducing tiny light-reflecting particles into the atmosphere.

■ Impacts and Issues

The need for greenhouse amelioration is real. According to the Fourth Assessment Report released in 2007 by the United Nation's Intergovernmental Panel on Climate Change (IPCC), the evidence for the warming of Earth's climate is "unequivocal," with human activities responsible for the marked increase in greenhouse gases since the year 1750.

Attempts to reduce the volume of greenhouse gases reaching the atmosphere go to the individual level. Driving more fuel-efficient cars and using energy sources such as wind and solar power can be productive and economical. Seeking alternate energy sources also goes to the national level, with governments of developed countries undertaking projects to establish wind and solar farms—facilities containing many wind turbines or solar panels to generate considerable quantities of electricity.

Such efforts are supported by most people, although opposition to the noise and claimed adverse health affects of the low frequency sound emitted by wind turbines has been raised by those living nearby. Other alternative energy sources such as nuclear power are more controversial.

Despite the present efforts, more needs to be done if the greenhouse effect is to be halted or reversed, according to the IPCC. Hindering this is the economic disparity between developed and developing nations, which makes amelioration efforts by poorer countries harder, and the reluctance of some developed nations to implement international amelioration plans such as the Kyoto Protocol. For example, Canada, which signed the Kyoto Protocol and supported the protocol in past federal governments, began to distance itself from the agreement as of 2007, arguing that the amelioration targets are unattainable and would result in an economic disadvantage to the country. The United States and Australia opted not to ratify the protocol initially, but Australia announced it would in December 2007.

SEE ALSO *Abrupt Climate Change; Atmospheric Structure; Climate Change; Infectious Disease and Climate Change.*

BIBLIOGRAPHY

Books

Hillman, Mayer, Tina Fawcett, and Sudhir Chella Rajan. *The Suicidal Planet: How to Prevent Global Climate Catastrophe.* New York: Thomas Dunne Books, 2007.

Lovelock, James. *The Revenge of Gaia: Earth's Climate Crisis and the Fate of Humanity.* New York: Perseus Books, 2007.

Periodicals

Flemming, James R. "The Climate Engineers." *The Wilson Quarterly* (Spring 2007).

Web Sites

"Climate Engineering Is Doable, as Long as We Never Stop." *Wired*, July 25, 2007. <http://www.wired.com/science/planetearth/news/2007/07/geoengineering> (accessed November 10, 2007).

Greenhouse Effect

Introduction

The greenhouse effect occurs when Earth, or some other planet, retains more heat because of the blanketing effect of its atmosphere. When the sun warms any planet, solar energy arrives at the planet's surface as light (electromagnetic waves) and is retained as heat (random molecular motion). A greenhouse effect causes more heat to be retained than lost. This occurs because certain gases, including carbon dioxide (CO_2) and methane (CH_4) are transparent to most of the wavelengths of light from the sun but are relatively opaque to infrared or heat radiation, which is radiated by Earth's surface and atmosphere. Light arriving from the sun passes through Earth's atmosphere, is converted to heat by absorption in the atmosphere at the surface, and is re-radiated into space more slowly because of the greenhouse effect.

The same process is used to heat a solar greenhouse, only using glass, rather than greenhouse gas, as the heat-trapping material. (In an actual greenhouse, the glass traps heat by preventing warm interior air from rising and mixing with the environment, rather than by blocking re-radiation of infrared light; the effect, however, is similar.) The greenhouse effect maintains Earth's surface temperature within a range appropriate for living things; without it, all of Earth's surface would be below the freezing point of water.

The greenhouse effect is a natural phenomenon. However, its intensity has increased because human activities since the beginning of the Industrial Revolution around 1750—especially the burning of fossil fuels and the clearing of forests (which remove CO_2 from the atmosphere, store the carbon in cellulose, and release the oxygen back to the atmosphere)—have raised atmospheric concentrations of CO_2 and other greenhouse gases. An observed consequence of Earth's artificially intensified greenhouse effect is a significant warming of the atmosphere.

All but a few scientists agree that this anthropogenic (human-caused) atmospheric warming is already occurring and is accelerating. This, in turn, is causing warmer seas, rising sea levels, and changes in patterns of precipitation and weather such as less rain in some places, more in others; more droughts and floods; more violent storms; melting of glaciers; shifts in ranges of plants and animals; extinctions; and more.

Historical Background and Scientific Foundations

How the Greenhouse Effect Works

The habitability of Earth depends both on its distance from the sun—neither too close nor too far—and on the balance between energy arriving from the sun and being radiated away into space. In the absence of the greenhouse effect, Earth's surface temperature would average about 0°F (−18°C), which is well below the freezing point of water and colder than life could tolerate over the long term, except for organisms deriving their energy from Earth's interior heat via hot deep-sea vents. The greenhouse effect maintains Earth's surface at an average of about 59°F (15°C).

On average, 35% of incoming solar radiation is reflected back to space by Earth's atmosphere (7%), clouds (24%), and surface (4%). The albedo (reflectivity) of any given area of Earth's surface is dependent on cloud cover, the density of tiny particulates in the atmosphere (aerosols), and the brightness of the surface, which is high for snow and ice, moderate for water, and relatively low for soil and vegetation.

Another 18% of incoming solar radiation is absorbed by gases (16%) and clouds (2%) in Earth's atmosphere. Forty-seven percent of incoming solar radiation reaches Earth's surface and is absorbed. Upon absorption, its energy is transformed into thermal kinetic energy (i.e., heat, the energy of random molecular motion). The warmed surface and atmosphere reradiate most of this energy in all directions as longer-wavelength (7–14 æm) infrared radiation (a small fraction drives winds, currents, and chemical processes). Much of this re-radiated energy escapes to

A herdsman tends to his sheep grazing on the Tibetan plateau, which is heating up because of the global greenhouse effect. Climate change could have potentially severe effects on the environment as temperatures rise. *AFP/Getty Images.*

space, and some is re-absorbed by the atmosphere—where the greenhouse effect occurs.

The amount of energy arriving as short-wavelength sunlight is approximately the amount being reflected as short-wavelength light or radiated as long-wavelength infrared radiation. While Earth is warming, as is happening now, it is radiating slightly less energy than it is absorbing.

Infrared radiation emitted by the land and sea are absorbed by a number of gases in Earth's atmosphere. The two main constituents of the atmosphere, molecular oxygen (O_2, 21% of the atmosphere by volume) and nitrogen (N_2, 78%), do not absorb infrared light: they are transparent to it, as they are to visible light. Water vapor (H_2O) and CO_2, though present in relatively small fractions—0.25% and 0.038%, respectively—are the most important absorbers of infrared light in Earth's atmosphere. Methane (CH_4), nitrous oxide (N_2O), ozone (O_3), and chlorofluorocarbons (CFCs) play lesser but still significant roles. Some of these gases are much more effective absorbers of infrared than either water or CO_2, molecule for molecule, but they contribute less to the greenhouse effect because they are present in much lower concentrations.

Absorbing infrared light warms the greenhouse gases. All warm matter radiates infrared light, so the warmed gases radiate infrared light. Some of this infrared is radiated into space while shining down on Earth's surface, warming it.

This process has been called the "greenhouse effect" because its mechanism is analogous to that by which a glass-enclosed space is heated by sunlight. A greenhouse's glass and humid atmosphere are transparent to incoming solar radiation, but absorb much of the long-wave infrared light that is radiated from plants and other objects inside the greenhouse. This slows down the rate of cooling of the structure, which reaches equilibrium (gets into balance) with its environment at a higher temperature than it would otherwise.

Other than water vapor, the atmospheric concentrations of all of the greenhouse gases have increased in the past century because of human activities. Prior to 1850, the concentration of CO_2 in the atmosphere was about 280 parts per million (ppm); by 2005 it had gone up to 379 ppm, a 36% increase. During the same period, CH_4 increased from 0.7 ppm to 1.8 ppm, N_2O from 0.285 ppm to 0.319 ppm, and CFCs from nothing to 0.7 parts per billion. These increased concentrations of greenhouses gases have caused a significant increase in the greenhouse effect. Overall, CO_2 is estimated to account for about 60% of this enhancement of the greenhouse effect, CH_4 for 15%, N_2O for 5%, O_3 for 8%, and CFCs for 12%.

Water vapor is a special case because although most of the greenhouse effect is caused by water vapor, it enters and leaves the atmosphere very rapidly compared to the other gases. The concentration of water vapor is thus controlled by the amount of other greenhouse gases in

WORDS TO KNOW

AEROSOL: Particles of liquid or solid dispersed as a suspension in gas.

ALBEDO: A numerical expression describing the ability of an object or planet to reflect light.

BIOMASS: The sum total of living and once-living matter contained within a given geographic area. Plant and animal materials that are used as fuel sources.

CHLOROFLUOROCARBONS: Members of the larger group of compounds termed halocarbons. All halocarbons contain carbon and halons (chlorine, fluorine, or bromine). When released into the atmosphere, CFCs and other halocarbons deplete the ozone layer and have high global warming potential.

CORAL BLEACHING: Decoloration or whitening of coral from the loss, temporary or permanent, of symbiotic algae (zooxanthellae) living in the coral. The algae give corals their living color and, through photosynthesis, supply most of their food needs. High sea surface temperatures can cause coral bleaching.

EL NIÑO/SOUTHERN OSCILLATION: Global climate cycle that arises from interaction of ocean and atmospheric circulations. Every 2 to 7 years, westward-blowing winds over the Pacific subside, allowing warm water to migrate across the Pacific from west to east. This suppresses normal upwelling of cold, nutrient-rich waters in the eastern Pacific, shrinking fish populations and changing weather patterns around the world.

GREENHOUSE GASES: Gases that cause Earth to retain more thermal energy by absorbing infrared light emitted by Earth's surface. The most important greenhouse gases are water vapor, carbon dioxide, methane, nitrous oxide, and various artificial chemicals such as chlorofluorocarbons. All but the latter are naturally occurring, but human activity over the last several centuries has significantly increased the amounts of carbon dioxide, methane, and nitrous oxide in Earth's atmosphere, causing global warming and global climate change.

HABITABILITY: The degree to which a given environment can be lived in by human beings. Highly habitable environments can support higher population densities, that is, more people per square mile or kilometer.

INFRARED RADIATION: Electromagnetic radiation of a wavelength shorter than radio waves but longer than visible light that takes the form of heat.

OZONE: An almost colorless, gaseous form of oxygen with an odor similar to weak chlorine. A relatively unstable compound of three atoms of oxygen, ozone constitutes, on average, less than one part per million (ppm) of the gases in the atmosphere. (Peak ozone concentration in the stratosphere can get as high as 10 ppm.) Yet ozone in the stratosphere absorbs nearly all of the biologically damaging solar ultraviolet radiation before it reaches Earth's surface, where it can cause skin cancer, cataracts, and immune deficiencies, and can harm crops and aquatic ecosystems.

PLEISTOCENE EPOCH: Geologic period characterized by ice ages in the Northern Hemisphere, from 1.8 million to 10,000 years ago.

WATER VAPOR: The most abundant greenhouse gas, it is the water present in the atmosphere in gaseous form. Water vapor is an important part of the natural greenhouse effect. Although humans are not significantly increasing its concentration, it contributes to the enhanced greenhouse effect because the warming influence of greenhouse gases leads to a positive water vapor feedback. In addition to its role as a natural greenhouse gas, water vapor plays an important role in regulating the temperature of the planet because clouds form when excess water vapor in the atmosphere condenses to form ice and water droplets and precipitation.

the atmosphere. The other greenhouse gases are said to contribute forcing to global warming, while water vapor contributes feedback.

Absorption of solar high-frequency radiation heats air, sea, and land in Earth's blanketing atmosphere. By slowing Earth's loss of this energy to outer space through the increased density of the greenhouse gases, the heat increases.

The extra greenhouse effect caused by anthropogenic (human-caused) greenhouse gases is often called global warming. Scientists usually prefer the term "global climate change," however, because not all parts of Earth's surface are actually warming (only the great majority of them) and because warming is causing many other changes in climate, some of them superficially contradicted by warming. For example, melting of glaciers around the southern coasts of Greenland have increased, but so has snowfall in the center of the island: both effects are predicted by scientists studying global climate change.

The Greenhouse Effect and Climate Change

The mechanism of the greenhouse effect is simple, but the details of global warming are complicated by atmospheric and oceanic circulations and by various feedback mechanisms. Although less than 1% of the solar radiation absorbed by Earth drives mass-transport processes in the oceans and lower atmosphere (currents and winds), these spread some of Earth's unevenly distributed thermal energy and are a crucial part of Earth's weather and climate system. Feedback occurs whenever some outcome of a process affects an earlier part of that process. There are many feedbacks involved with the greenhouse effect and global warming. For example, some of the heat at Earth's surface causes water to evaporate from plants and open water. Water vapor is a greenhouse gas, so this increased evaporation tends to increase global warming, which tends to increase evaporation further, and so on.

Scientists have long understood that the greenhouse effect keeps Earth's temperature within a livable range for most life. They have also known that concentrations of CO_2 and other greenhouse relevant gases have increased in Earth's atmosphere and will continue to increase for many decades to come. However, until the 1990s it was difficult to show that the observed warming of Earth's surface and lower atmosphere were being caused mostly by anthropogenic enhancement of the greenhouse effect rather than by some still-unknown process of natural climate change, such as the increased brightness of the sun. Earth's climate has changed thousands of times over hundreds of millions of years, and it was not scientifically out of the question that such natural processes might be at work today.

Since about 1990, however, many types of independent evidence have shown not only that the climate is changing rapidly but that human activities, especially emissions of greenhouse gases and deforestation, are the cause of most of this change. Since the beginning of instrumental recordings of surface temperatures around 1850, almost all of the warmest years on record have occurred since the late 1980s. Typically, these warm years have averaged about 1.4–1.8°F (0.8–1.0°C) warmer than occurred during the decade of the 1880s. Overall, as of 2007, the temperature of Earth's atmosphere near the surface had risen about 1.33°F (0.74°C) since 1906. More recent years have seen more rapid increases.

The temperature data are not simple to interpret. First, air temperature is variable in time and space, making it necessary to collect data over wide areas and long time periods to detect significant long-term trends. Second, older data are generally less accurate than modern records. Third, many weather stations are in urban areas, which warm the air around them independently of global climate: this effect must be accounted for when analyzing the data. Finally, climate can change for reasons other than a greenhouse response to human-caused increases in greenhouse gases, including volcanic emissions of sulfur dioxide, sulfate, and fine particulates into the upper atmosphere and naturally shifting ocean and atmospheric circulations.

However, there are observations linking prehistoric variations of atmospheric CO_2 (the most important greenhouse gas) and climate warming. Important evidence comes, for example, from cores of ancient ice from Antarctica and Greenland—glacial ice that contained layered ice from annual snowfalls dating back hundreds of thousands of years. Concentrations of CO_2 in the ice are determined by analysis of air bubbles in ice layers of known age (determined by counting snowfall layers back from the present, much like tree-rings). Changes in air temperature are inferred from ratios of oxygen isotopes in the ancient ice, so the temperature record can be compared to the CO_2 record over long periods of time. Ice cores show that CO_2 levels are higher today than at any time in at least 650,000 years.

The fact that changes in CO_2 and surface temperature are positively correlated—that is, increased CO_2 almost always accompanies warming of climate—suggests a greenhouse mechanism. However, correlation alone, the fact that two things happen together, does not show which thing caused which, or whether both were caused by some other phenomenon. In this case, scientists have argued that increased CO_2 might have resulted in warming through an intensified greenhouse effect; others have argued that warming (caused by some unknown process) could have accelerated CO_2 release from ecosystems by increasing the rate of decomposition of biomass, especially in cold regions. It is also possible that both effects may occur, either separately or together, one triggering or accelerating the other.

Such questions are resolved using computer models based on the laws of physics that predict climatic changes caused by increases in greenhouse gases or other factors. The most sophisticated simulations are three-dimensional general circulation models (GCMs), which are run on supercomputers. GCMs simulate global atmospheric and ocean circulations and the interaction of these with other variables that contribute to climate. To perform a simulation experiment with a GCM, numbers representing concentrations of CO_2 and other greenhouse relevant gases, along with other factors (e.g., solar brightness, vegetation cover, ice melting, and many more), are fed into equations that represent the laws of physics governing circulation, heat, absorption, life processes, atmospheric chemistry, and the like. The goal is to produce the most realistic possible mathematical description of Earth's climate machine and its inputs, and to use that description or model to understand the behavior of Earth's past and future climate.

Many experiments have been performed using a variety of GCM models. Their results have varied according to the specifics of the experiment, but a central tendency of experiments using a common CO_2 scenario (i.e., a doubling of CO_2 from its recent concentration of 379 ppm) is an increase in average surface temperature of 1.8–7.2°F (1–4°C). This warming is predicted to be especially great in polar regions, where temperature increases could be two or three times greater than in the tropics. Such accelerated warming of the Arctic (north polar region) and the West Antarctic Peninsula has already been observed; central Antarctica is not predicted to warm in this way. Increased CO_2 can, therefore, cause global warming, whether or not the reverse process (warming causing increased CO_2) may also occur.

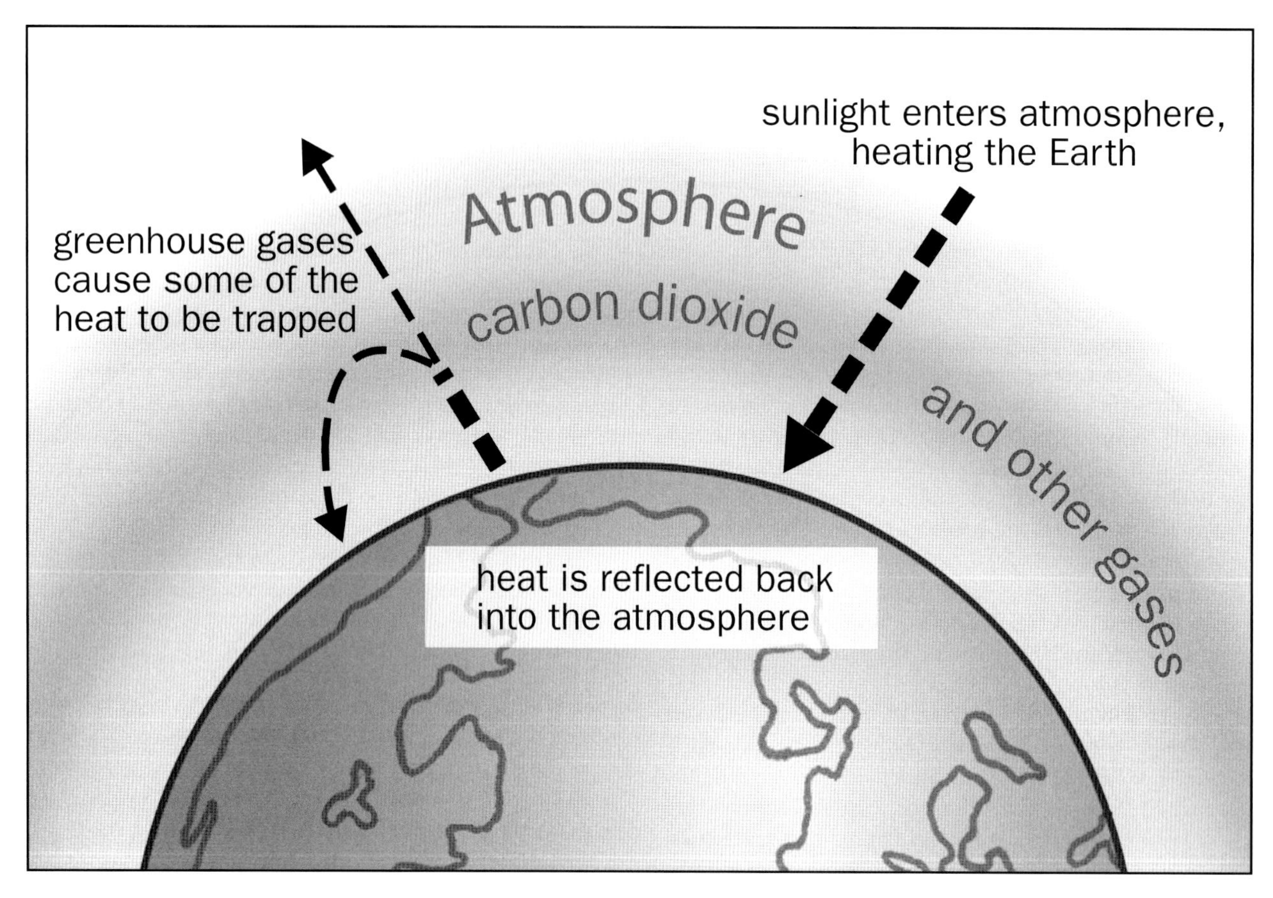

Without the natural greenhouse effect, Earth's surface would be too cold to allow liquid water. *The Gale Group.*

■ Impacts and Issues

Increased warming of global climate is causing significant changes in the quantities, distribution, and timing of precipitation. This, in turn, has a large effect on vegetation, on which all animal life ultimately depends. There is, however, more uncertainty about potential changes in rainfall patterns than about changes in global average temperature. Regional effects of changed precipitation on soil moisture and vegetation are also difficult to predict precisely. Studies of changes in vegetation during the warming climate that followed the most recent (Pleistocene) glaciation suggest that plant species respond to climate change according to the differing tolerances of various species to ranges of temperature, rainfall, and seasonal timing, and their different abilities to colonize newly available habitat.

In a region where the climate becomes drier because of increased evaporation, decreased precipitation, or both, one result can be decreased forest and expanded savanna or prairie. A landscape change of this character is believed to have occurred in the New World tropics during the Pleistocene glaciations. Because of the relatively dry climate at that time, presently continuous rainforest may have been constricted into relatively small, isolated patches. These forest remnants would probably have existed within a landscape matrix of savanna and grassland. Such an enormous restructuring of the character of the tropical landscape must have had a tremendous effect on the multitude of rare species that live in that region. Likewise, climate change potentially associated with an intensification of the greenhouse effect would have a profound effect on Earth's natural ecosystems and the species that they sustain.

There will also be important changes in the ability of some areas to support crop plants. This would be particularly true of regions that are marginal in terms of rainfall or temperature, and are vulnerable to drought and desertification. For example, important crops such as wheat are grown in regions of the western interior of North America that formerly supported natural shortgrass prairie. Ecologists estimate that about 40% of this semiarid region, measuring 988 million acres (400 million hectares), has already been desertified by agricultural activities, and crop-limiting droughts occur there sporadically. This climatic handicap can be partially offset by irrigation. However, there is a shortage of water for irrigation (which may be enhanced by climate change). The practice of irrigation can cause its own environmental problems, such as salinization of (salt buildup in) soils. Clearly, in many areas substantial changes in climate would place the present agricultural systems at great risk. In a few areas, agriculture will probably be benefited by climate change, but the global balance is likely to be negative.

Patterns of wildfire are also influenced by changes in precipitation patterns. For example, based on the predictions of climate models, there could be a 50% increase in the area of forest annually burned in Canada, presently about 2.5–4.9 million acres (1–2 million hectares) in typical years. These wildfires would be due, primarily, to the dryness of the forests, with the change in precipitation levels.

Some shallow marine ecosystems are being affected by increases in seawater temperature and by increased ocean acidity caused by higher CO_2 concentrations. Corals are especially vulnerable to increases in water temperature, which can deprive them of their symbiotic algae and may kill them. Widespread coral bleaching has been observed, partly due to natural weather cycles such as the El Niño/Southern Oscillation and partly due to anthropogenic climate change.

Another observed effect of warming is rising sea level. This is caused by a combination of a thermal expansion of warmed seawater—water, like most materials, expands slightly when warmed—and the melting of alpine and polar glaciers. GCMs consulted by the United Nations' Intergovernmental Panel on Climate Change (IPCC) predicted in 2005 that sea level in 2100 could be 7–23 in (0.18–59 m) higher than today, with greater rises to follow over centuries or millennia. Since 2005, many scientists have argued that the upper possible range of sea-level rise by 2100 should be much higher than the IPCC's estimate, perhaps as high as 6.5 ft (2 m), because data have since shown surprisingly fast melting of parts of Greenland's ice cap. Depending on the rate and magnitude of change in sea level, it could be problematic or even disastrous for low-lying islands and coastal populations.

Most GCMs predict that high latitudes—regions closer to the poles—will experience the greatest intensity of climatic warming, and such areas have, in fact, been observed already to be warming at about twice the average rate of the rest of the world. The warming of northern ecosystems adds positive feedbacks to climate change. These are caused partly by a change of great expanses of arctic tundra from sinks for atmospheric CO_2 into sources of CO_2 and of methane. The climate warming caused by the rise in greenhouse gases is already increasing the depth of annual thawing of frozen soils, exposing large quantities of carbon-rich organic materials in the permafrost to microbial decomposition, and thereby increasing the emission of CO_2 and CH_4 to the atmosphere. Melting of Arctic snow and sea ice also decreases albedo, which speeds warming. Arctic sea-ice melting in the summer of 2007 was the greatest ever observed.

It is likely that a further intensification of Earth's greenhouse effect will occur and have large climatic and ecological consequences. Strategies for managing the causes and consequences of the anthropogenic greenhouse effect include reducing emissions of CO_2 and other greenhouse gases. It will also be necessary, at the very least, to adapt human society to whatever amount of climate change can no longer be avoided because of past emissions.

Any strategy to reduce greenhouse emissions will require adjustments by societies and economies. Because large quantities of CO_2 are emitted through the burning of fossil fuels, to mitigate (reduce the severity of) global climate change would require societies to use different, possibly new, technologies to generate energy, to increase the efficiency of energy use, and possibly to decrease total energy use. Such a strategy of mitigation will be difficult, especially in poorer countries, because of the changes required in economic systems, resource use, technologies, and lifestyles.

Various international negotiations have been undertaken to try to get nations to agree to decisive actions to reduce their emissions of greenhouse gases. The most recent major agreement came out of a large meeting held in Kyoto, Japan, in 1997. There, most of the world's industrial countries agreed to reduce their CO_2 emissions to 5.2% below 1990 levels by the year 2012. The United States, which has about 5% of the world's population but produces about 24% of its greenhouse emissions, signed the Kyoto Protocol in 1998 (that is to say, its ambassador to the United Nations signed the plan) but never ratified it as a binding treaty, which would have needed approval from both the president and Congress. International negotiations for a new, more effective treaty to replace the Kyoto Protocol began in late 2007.

As of January 2007, the warmest year in recorded weather history in terms of global average temperature was 2005, which was 0.94°F (0.52°C) above the average for 1961–1990. The warmest year on record so far for the United States, prior to 2007, was actually 1934. So far, the United States has experienced less warming than many other parts of the world.

Increased extreme weather is another predicted consequence of global climate change. In August 2007, scientists at the World Meteorological Organization, an agency of the United Nations, announced that during the first half of 2007, Earth showed significant increases above long-term global averages in both high temperatures (e.g., heat waves) and frequency of extreme weather events, including heavy rainfalls, cyclones, and wind storms. Global average land temperatures for January and April of 2007 were the warmest recorded for those two months since recordkeeping began in the 1880s.

SEE ALSO *Global Warming; Greenhouse Gases.*

BIBLIOGRAPHY

Books

Gore, Al. *An Inconvenient Truth: The Planetary Emergency of Global Warming and What We Can Do About It.* Emmaus, PA: Rodale Press, 2006.

Hocking, Colin. *Global Warming & the Greenhouse Effect.* Berkeley, CA: GEMS, 2002.

Houghton, John. *Global Warming: The Complete Briefing* Cambridge: Cambridge University Press, 2004.

Morganstein, Stanley. *The Greenhouse Effect.* Cheshire, UK: Trafford, 2003.

Solomon, S., et al, eds. Climate Change 2007: *The Physical Science Basis: Contribution of Working Group I to the Fourth Assessment Report of the Intergovernmental Panel on Climate Change.* New York: Cambridge University Press, 2007.

Periodicals

Karl, Thomas R., and Kevin R. Trenberth. "Modern Global Climate Change." *Science* 302 (2003): 1,719–1,723.

Web Sites

"Climate Change." *U.S. Environmental Protection Agency (EPA)*, November 19, 2007. <www.epa. gov/ climatechange/> (accessed November 25, 2007).

Bill Freedman

Greenhouse Gases

Introduction

Certain gases in Earth's atmosphere—particularly carbon dioxide (CO_2), methane (CH_4), and water vapor—trap energy from solar radiation and thus, keep Earth warmer than it would be otherwise. These gases are termed greenhouse gases and the warming they create is termed the greenhouse effect or greenhouse warming. About 65% of greenhouse warming is caused by CO_2, the most abundant greenhouse gas.

A certain amount of greenhouse warming occurs naturally; without it, Earth would be too cold to sustain most life. However, the intensity of the greenhouse effect is being increased by human activities that add greenhouse gases to the atmosphere. For example, burning fossil fuels (coal, petroleum, and natural gas) was adding over 6 billion metric tons of carbon dioxide to the atmosphere every year as of 2006, with world usage of such fuels increasing rapidly. About half of the CO_2 released each year remains in the atmosphere and intensifies Earth's natural greenhouse effect; the rest is absorbed by plants and the oceans. The atmospheric concentration of CO_2 is also increased by deforestation, because living trees absorb CO_2 while burning and decaying wood emits it.

Because of anthropogenic (human-caused) intensification of Earth's greenhouse effect, the global climate is warming. This is causing secondary changes such as a rise in sea level, changed patterns of precipitation, shifted ranges for plant and animal populations, and more frequent extreme weather. Because the effects of global warming are various, scientists prefer the phrase global climate change, reserving the phrase global warming to refer to the average global temperature increase as such, apart from other changes.

Without the greenhouse effect, Earth's surface temperature would average about 0°F (−18°C), well below the freezing point of water. By slowing the rate at which Earth loses energy, the greenhouse effect helps maintain Earth's surface at an average temperature of about 59°F (15°C), allowing the oceans to be liquid and life to flourish.

Historical Background and Scientific Foundations

The Greenhouse Effect

Greenhouses gases alter Earth's temperature by altering its energy budget. The energy budget of any system, such as Earth, is the sum of the energy entering the system, the energy leaving the system, and any difference between the two, which corresponds to energy that is internally transformed or stored. Almost all of the energy entering Earth's climate system comes from the sun, which, like all objects, emits electromagnetic energy at a rate and of a spectral character determined by its surface temperature, about 10,800°F (6,000°C). Because of this high temperature, about one-half of the sun's emitted energy is radiated as visible light, that is, light rays having wavelengths between 400 and 700 nanometers (nm). Most of the other half of the sun's energy output is radiated in the near-infrared wavelength range between about 700 and 2,000 nm.

There is a nearly even balance between the amount of electromagnetic energy coming to Earth and the amount eventually radiated back to space. The myriad ways in which the incoming energy is absorbed, dispersed, transformed, stored, and re-radiated make up the details of Earth's energy budget. At present, because Earth is warming, it is radiating slightly less energy than it is absorbing: the difference is going to heat the atmosphere, land, and (mostly) oceans.

Incoming solar radiation is termed insolation. On average, 35% of insolation is reflected to space by Earth's atmosphere (7%), surface (4%), and clouds (24%). About 18% of insolation is absorbed by atmospheric gases (16%) and clouds (2%); 47% of insolation is absorbed by Earth's surface. Upon absorption by matter, light's energy is usually transformed into heat, that is, random molecular

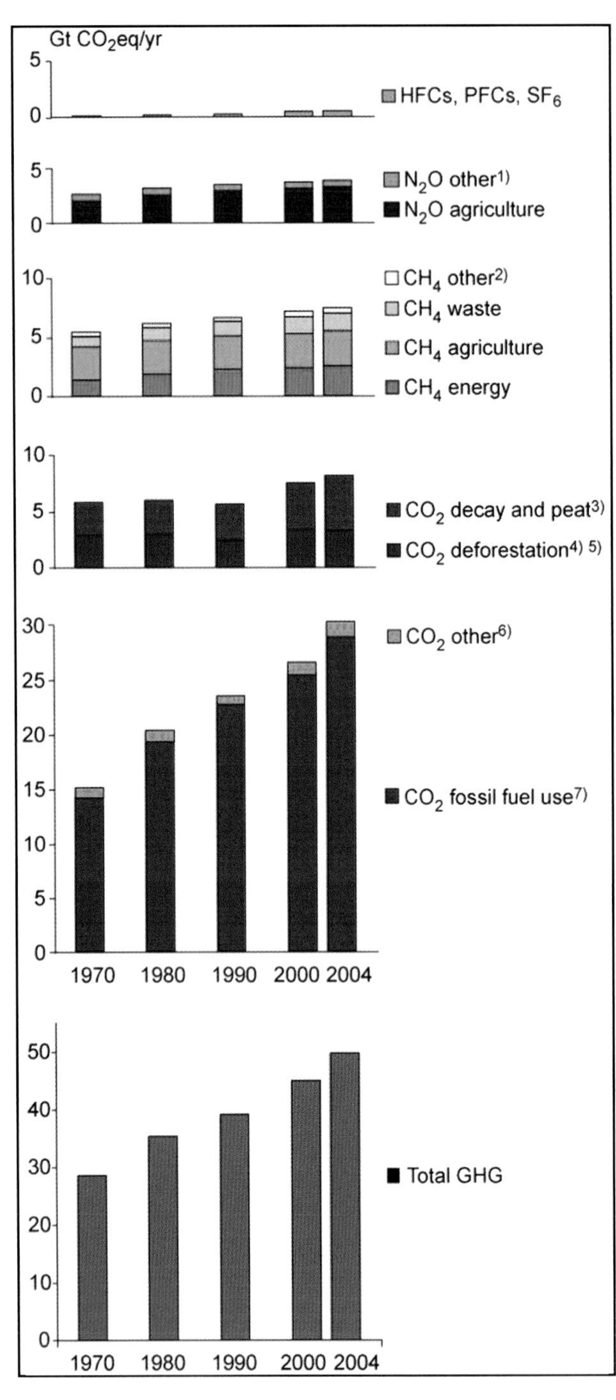

The image shows the Global Warming Potential (GWP) weighted global greenhouse-gas emissions from 1970–2004. One hundred-year GWPs (from the IPCC's 1996 Second Assessment Report) were used to convert emissions to CO_2-eq. CO_2, CH_4, N_2O, HFCs, PFCs and SF_6 from all sources are included. The two CO_2 emission categories reflect CO_2 emissions from energy production and use (second from bottom) and from land use changes (third from the bottom). *Climate Change 2007: Mitigation of Climate Change, Summary for Policymakers, Intergovernmental Panel on Climate Change.*

motion. Land, sea, and atmosphere re-radiate most of this energy as longer-wavelength infrared radiation; a small fraction drives winds, currents, and chemical processes. Some re-radiated infrared energy escapes to space, but some is re-absorbed by the planet's surface or atmosphere. Re-absorption in the atmosphere is accomplished by greenhouse gases and is the basis of the greenhouse effect.

The term "greenhouse effect" is used because this mechanism is analogous to a glass-enclosed greenhouse heated by sunlight. A greenhouse's glass panels are transparent to insolation, but they prevent warm interior air from rising and mixing with the outside air, slowing heat loss and causing the structure to achieve equilibrium or balance with its environment at a higher temperature than it would otherwise. Similarly, when greenhouse gas concentrations in Earth's atmosphere are increased, Earth warms until it achieves a new, warmer equilibrium with its environment. As greenhouse gases are being continually added to the atmosphere, however, Earth cannot achieve thermal (heat) equilibrium, but continues to warm as long as the gases are added. If at some point greenhouse-gas concentrations were to level off, Earth would eventually, at some point, cease to get warmer.

Major Greenhouse Gases

Water vapor and CO_2 are the most important greenhouse gases in Earth's atmosphere; CO_2 also has important greenhouse effects on Mars and Venus. Methane (CH_4), nitrous oxide (N_2O), ozone (O_3), and chlorofluorocarbons and related compounds (used mostly as refrigerants) play a significant role. On a per-molecule basis, most of these gases cause more warming than CO_2, but they cause less warming overall because their atmospheric concentrations are much lower. Compared with carbon dioxide, a molecule of methane is 21–40 times more effective at absorbing infrared wavelengths, nitrous oxide is 200–270 times more effective, ozone 2,000 times, and CFCs and related compounds 3,000–15,000 times more effective.

Other than water vapor, the atmospheric concentrations of all of these gases have increased in the past century because of emissions associated with human activities. Prior to 1850, the concentration of CO_2 in the atmosphere was about 280 parts per million (ppm), while in 1994 it was 355 ppm, and by 2006 it had risen to 379 ppm. During the same period, CH_4 increased from 0.7 ppm to 1.7 ppm, N_2O from 0.285 ppm to over .3 ppm; and CFCs from zero to 0.7 ppb. These increased concentrations of greenhouse gases have been the main cause of the observed warming of Earth's climate in recent years. CO_2 causes about 60% of the anthropogenic greenhouse effect, CH_4 15%, N_2O 5%, O_3 8%, and CFCs 12%. Water vapor is a special case because it does not remain in the atmosphere very long; its atmospheric concentration is controlled by the warming effect of longer-lived greenhouse gases, and so it is termed a feedback (as opposed to a forcing) factor in global warming.

GREENHOUSE GASES AND GREENHOUSE GAS INDEXES

Gas	*Lifetime in Atmosphere*	*Direct Effect on sunlight, 20 years*
Carbon Dioxide	Variable	1
Methane	12 ± 3 years	56*
Nitrous Oxide	120	280
HFCs, PFCs, and Sulfur Hexafluoride		
HFC -23	264	9,200
HFC -125	33	4,800
HFC -134a	15	3,300
HFC -152a	2	460
HFC -227ea	37	4,300
Perfluoromethane	50,000	4,400
Perfluoroethane	10,000	6,200
Sulfur Hexafluoride	3,200	16,300

*Methane is 56 times more reflective than CO_2

Data Source: Energy Information Administration Publication, "Emissions of Greenhouse Gases in the United States, 1999."

Table from the Energy Information Administration showing greenhouse gases and greenhouse gas indexes. *The Gale Group.*

Other Processes Powered by Insolation

Not all energy from absorbed light, whether visible or infrared, is re-radiated at once as infrared light. Energy can also be transformed or stored by various physical processes. A few that are important to climate are as follows:

- Some thermal energy causes water to evaporate from plant and water surfaces or melts ice and snow. Evaporation causes local climatic cooling; condensation of water vapor into snow or rain releases heat. Both of these effects are important to regional climate and weather.
- A small amount (less than 1%) of the absorbed solar radiation generates mass-transport processes in the oceans and lower atmosphere, which disperse some of Earth's unevenly distributed thermal energy. The most important of these processes are winds and storms, ocean currents, and waves on the surface of the oceans and lakes.
- A small but ecologically critical quantity of solar energy (averaging less than 1% of the total) is absorbed by plant pigments, especially chlorophyll. This energy is used to drive photosynthesis, resulting in temporary storage of energy in the chemical bonds of biological compounds. Notably, plants use energy from sunlight to separate the carbon and oxygen in CO_2, keeping the carbon to build their tissues and releasing the oxygen.

Scientific Disputes and Progress

Globally, all of the warmest years since instrumental measurements of surface temperature began to be recorded (around 1880) have occurred since the late 1980s. Typically, these warm years have averaged about 1.5–2.0°F (0.8–1.0°C) warmer than those during the decade of the 1880s. As of 2007, the average temperature of Earth's atmosphere near the surface had risen about 1.33°F (0.74°C) since 1906. This warming closely matches increases in greenhouse gases emitted by human activities.

When human activity was first cited as the cause of increasing greenhouse gas concentrations in the late 1950s and 1960s and global warming was attributed to increasing concentrations, some scientists questioned these conclusions. They based their skepticism on several apparent deficiencies in the argument. These included: 1) air temperature is variable in time and space, making measurements from many locations over long time periods necessary to detect long-term trends; 2) older data are less accurate than modern instrumental records; 3) many

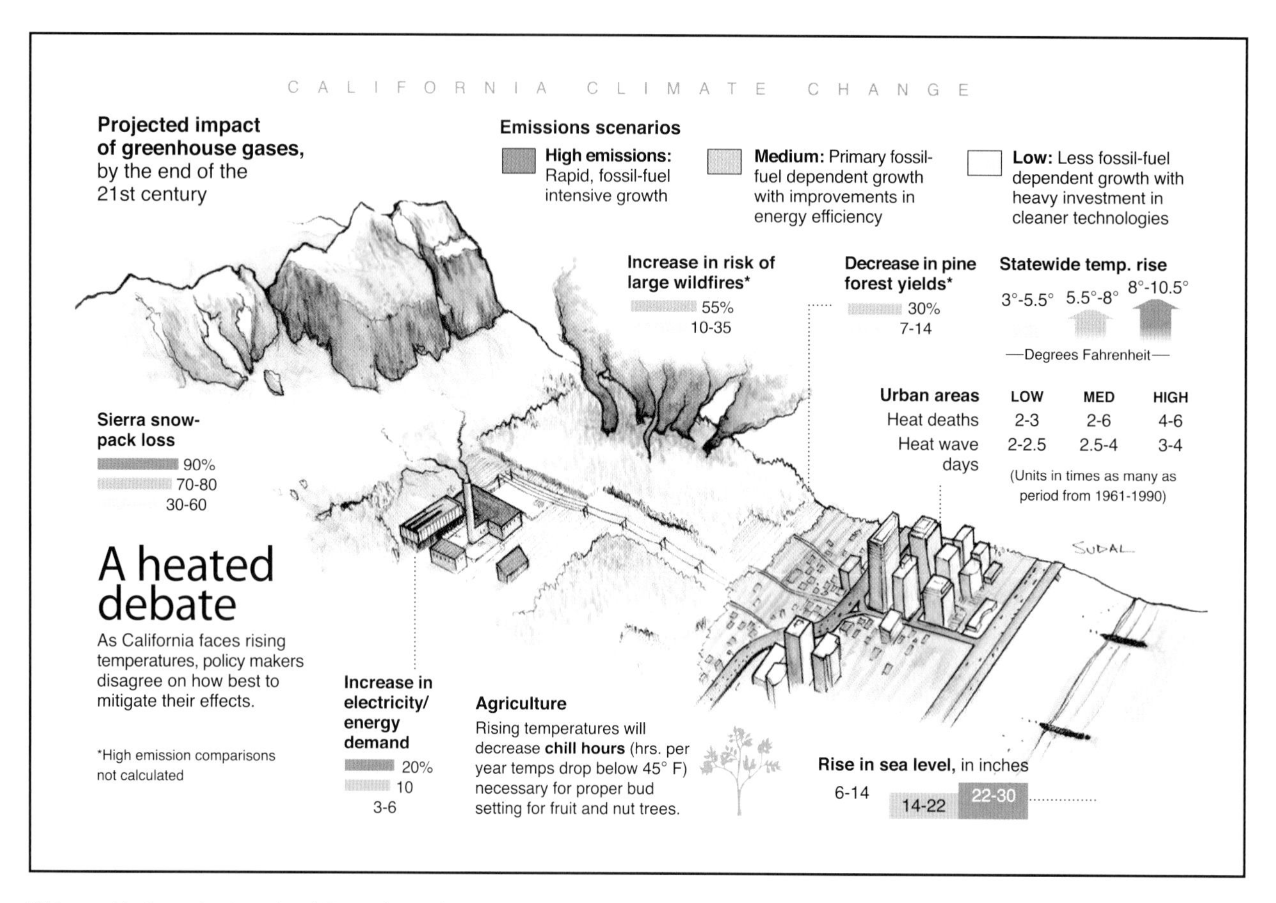

This graphic from the Associated Press shows the projected impacts of greenhouse gases on California by the end of the 21st century. [Source: Union of Concerned Scientists.] *AP Images.*

weather stations are in urban areas that are warmed by increasing quantities of buildings and pavement; and 4) climate can change for reasons other than increased greenhouse gases, including cyclical changes in solar radiation and volcanic emissions. When satellite data became available in the late 1970s, there were apparent inconsistencies between ground-level instrumental data and satellite observations.

However, by the early 2000s, all these causes of skepticism had been resolved to the satisfaction of the great majority of scientists. For example, atmospheric CO_2 and surface temperature had been linked by many independent bodies of proxy data, that is, natural biological or geological records such as tree rings, snow layers, and ocean-floor sediment layers that preserve information about ancient climate. An important form of proxy evidence for ancient temperature changes—and of direct evidence for ancient CO_2 concentrations—are ice cores from Antarctica and Greenland. These cylinders of ice, retrieved by drills from deep inside Earth's two major ice caps, preserve bubbles of ancient air which reveal ancient CO_2 concentrations and, by their ratios of oxygen isotopes, air temperatures at the time the snow that compacted to form the ice fell. Observed changes in CO_2 and surface temperature are matched, each increasing when the other increases, suggesting that either increased CO_2 causes warming, or warming causes increased CO_2, or both.

A 2005 ice-core study by National Aeronautics and Space Administration (NASA) scientists found that atmospheric carbon dioxide was already 27% higher than the highest CO_2 level found during the last 650,000 years. All data confirmed a step-wise correlation between atmospheric carbon dioxide concentrations and the warming of the global climate. Geological data show that atmospheric CO_2 was much higher at much more remote times, on the order of hundreds of millions of years ago, when Earth's climate was radically different than today's.

The other main scientific grounds for doubting the reality of anthropogenic climate change have also dissolved. The urban heat island effect has been taken into careful account in analyzing temperature records; more careful analysis of astronomical influences on climate shows that increased energy output from the sun is not causing today's warming; analytic errors that had created apparent disagreement between satellite and surface data have been discovered and fixed. Satellite data, improved

WORDS TO KNOW

1750 VALUE: Concentrations of greenhouse gases in the atmosphere are often compared to their values in 1750. That year is just before the beginning of the Industrial Revolution, when burning of fossil fuels began to greatly increase, changing atmospheric gas concentrations. The number 1750 is arbitrary: any earlier year in the last several thousand years would do just as well, since greenhouse gas concentrations were stable during that period.

CHLOROFLUOROCARBONS: Members of the larger group of compounds termed halocarbons. All halocarbons contain carbon and halons (chlorine, fluorine, or bromine). When released into the atmosphere, CFCs and other halocarbons deplete the ozone layer and have high global warming potential.

DEFORESTATION: Those practices or processes that result in the change of forested lands to non-forest uses. This is often cited as one of the major causes of the enhanced greenhouse effect for two reasons: 1) the burning or decomposition of the wood releases carbon dioxide; and 2) trees that once removed carbon dioxide from the atmosphere in the process of photosynthesis are no longer present and contributing to carbon storage.

ELECTROMAGNETIC ENERGY: Energy conveyed by electromagnetic waves, which are paired electric and magnetic fields propagating together through space. X rays, visible light, and radio waves are all electromagnetic waves.

FOSSIL FUELS: Fuels formed by biological processes and transformed into solid or fluid minerals over geological time. Fossil fuels include coal, petroleum, and natural gas. Fossil fuels are non-renewable on the timescale of human civilization, because their natural replenishment would take many millions of years.

INSOLATION: Solar radiation incident upon a unit horizontal surface on or above Earth's surface.

METHANE: A compound of one hydrogen atom combined with four hydrogen atoms, formula CH_4. It is the simplest hydrocarbon compound. Methane is a burnable gas that is found as a fossil fuel (in natural gas) and is given off by rotting excrement.

PHOTOSYNTHESIS: The process by which green plants use light to synthesize organic compounds from carbon dioxide and water. In the process, oxygen and water are released. Increased levels of carbon dioxide can increase net photosynthesis in some plants. Plants create a very important reservoir for carbon dioxide.

SPECTRAL: Relating to a spectrum, which is an ordered range of possible vibrational frequencies for a type of wave. The spectrum of visible light, for example, orders colors from red (slowest vibrations visible) to violet (fastest vibrations visible) and is itself a small segment of the much larger electromagnetic spectrum.

URBAN HEAT ISLAND EFFECT: Warming of atmosphere in and immediately around a built-up area. Occurs because pavement and buildings absorb solar energy while being little cooled by evaporation compared to vegetation-covered ground. Skeptics of global climate change at one time argued that the expansion of urban heat islands near and around weather stations has caused an illusion of global warming by biasing temperature measurements. Although urban heat islands do exist, the argument that they produce an illusion of global warming has been discredited.

WATER VAPOR: The most abundant greenhouse gas, it is the water present in the atmosphere in gaseous form. Water vapor is an important part of the natural greenhouse effect. Although humans are not significantly increasing its concentration, it contributes to the enhanced greenhouse effect because the warming influence of greenhouse gases leads to a positive water vapor feedback. In addition to its role as a natural greenhouse gas, water vapor plays an important role in regulating the temperature of the planet because clouds form when excess water vapor in the atmosphere condenses to form ice and water droplets and precipitation.

surface data, proxy data, and computer models have all converged to support the conclusion the world is indeed warming as greenhouse gas concentrations increase. Since we know that human culture, not warming climate, has caused the expansion of fossil-fuel burning since 1750, on this occasion it is definitely increased greenhouse-gas concentrations that have caused global warming, not the other way around. Modern climate warming may, however, cause accelerated releases of some greenhouse gases, such as methane, enhancing warming still further.

Computer models are used to analyze the many complex factors that affect climate change. The most sophisticated simulations are three-dimensional general circulation models (GCMs), which are run on supercomputers. GCM models simulate the complex interactions of clouds, ocean currents, atmospheric circulations, seasons, biological processes, greenhouse gas emissions, and atmospheric chemistry to create a mathematical picture or model of how Earth's climate works.

In 2007, the United Nations' Intergovernmental Panel on Climate Change (IPCC) issued its Fourth Assessment Report on anthropogenic emissions of greenhouse gases and climate change. The IPCC found that, despite ratification by most of the world's nations of the Kyoto Protocol, which calls for reductions in greenhouse emissions, most nations' emissions were increasing. In 2005, CH_4 abundance was up to 1,774 parts per billion, more than twice its 1750 value; N_2O abundance was at 319 parts per billion, 18% higher than its 1750 value.

IN CONTEXT: POST-INDUSTRIAL RISE IN GREENHOUSE GASES

The IPCC asserts that "The post-industrial rise in greenhouse gases is 'unprecedented' and does not stem from natural mechanisms."

"Current concentrations of atmospheric CO_2 and CH_4 far exceed pre-industrial values found in polar ice core records of atmospheric composition dating back 650,000 years. Multiple lines of evidence confirm that the post-industrial rise in these gases does not stem from natural mechanisms. . . ."

"The total radiative forcing of the Earth's climate due to increases in the concentrations of the LLGHGs [Long-lived greenhouse gases] CO_2, CH_4 and N_2O, and *very likely* the rate of increase in the total forcing due to these gases over the period since 1750, are unprecedented in more than 10,000 years."

IPCC scientists further assert that it is "*very likely*" that the rate of increase "over the past four decades is at least six times faster than at any time during the two millennia before the Industrial Era. . . ."

Note: Emphasis shown above exists in the original published text.

SOURCE: *Solomon, S., et al., eds. In* Climate Change 2007: The Physical Science Basis. Contribution of Working Group I to the Fourth Assessment Report of the Intergovernmental Panel on Climate Change. *New York: Cambridge University Press. 2007.*

The IPCC's 2007 report on climate change also presented revised predictions for how much the global climate might warm if emissions of greenhouse gases continue. Depending on whether human societies reduce their emissions or continue with business as usual, by 2100 world climate is likely to warm by several degrees, with a minimum of 1.98°F (1.1°C) and a maximum, barring unforeseen feedbacks or tipping points, of 11.52°F (6.4°C).

Under the auspices of the United Nations, various international negotiations have been undertaken to try to get nations to agree to decisive actions to reduce their emissions of greenhouse gases, beginning with the United Nations Framework Convention on Climate Change (1992). A major modification to that treaty came out of a meeting held in Kyoto, Japan, in 1997. Most industrialized countries (with the notable exceptions of the United States and Australia) agreed to reduce their CO_2 by 5–7% or more below their 1990 levels by the year 2012. Australia announced plans to ratify in late 2007. Kyoto was a first step, and its shortcomings have been much debated; discussions to design a follow-up treaty to take effect after Kyoto's expiration began in late 2007.

Keeping greenhouse gases out of the atmosphere is the most reliable, least hazardous way to mitigate global warming. Researchers are also testing ways to sequester, or store, carbon dioxide emitted by large centralized sources such as power plants, preventing it from entering the atmosphere. Some scientists are considering geo-engineering efforts to deflect sunlight from Earth or otherwise alter the planet's energy budget. Replanting tropical forests and decreasing their destruction, adjusting agriculture practices, generating energy from affordable, rapidly deployable low-carbon sources, ceasing to manufacture CFCs and related compounds, and other measures can reduce greenhouse-gas emissions and eventually stabilize Earth's climate in a less-altered state than will occur if emissions continue to grow.

In 2007, UN experts announced that the most cost-effective measure of all—the most greenhouse-gas abatement achieved per dollar spent—was increased energy efficiency. Much, probably most of the energy consumed by modern industrial society is wasted, and most energy is generated by methods that add greenhouse gases to the atmosphere; decreasing usage of energy and materials reduces greenhouse-gas emissions and mitigates global climate change.

Greenhouse gases already in the atmosphere commit the world to a certain amount of future climate change regardless of how successful society may be at reducing emissions. Even given sudden stabilization of greenhouse-gas concentrations, it would still take centuries for the Earth's atmosphere to get into thermal equilibrium with the waters of the deep oceans, and during this period, climate would continue to change, albeit more slowly than today. Reducing greenhouse-gas emissions cannot halt global climate change, but would slow it and reduce its future extent.

SEE ALSO *Global Warming; Greenhouse Effect.*

BIBLIOGRAPHY

Books

Gore, Al. *An Inconvenient Truth: The Planetary Emergency of Global Warming and What We Can Do About It.* Emmaus, PA: Rodale Press, 2006.

Morganstein, Stanley. *The Greenhouse Effect.* Cheshire, UK: Trafford, 2003.

Solomon, S., et al, eds. *Climate Change 2007: The Physical Science Basis: Contribution of Working Group I to the Fourth Assessment Report of the Intergovernmental Panel on Climate Change.* New York: Cambridge University Press, 2007.

Wuebbles, Donald J., and Jae Edmonds. *Primer on Greenhouse Gases.* Boca Raton, FL: CRC, 1991.

Periodicals

Karl, Thomas R., and Kevin E. Trenberth. "Modern Global Climate Change." *Science* 302 (2003): 1,719–1,723.

Web Sites

"Climate Change." *U.S. Environmental Protection Agency (EPA)*, November 19, 2007. <www.epa.gov/climatechange/> (accessed November 25, 2007).

"Greenhouse Gases: Frequently Asked Questions." *National Oceanic and Atmospheric Administration (NOAA)*. <http://www.ncdc.noaa.gov/oa/climate/gases.html> (accessed November 21, 2007).

Larry Gilman

Greenland: Global Implications of Accelerated Melting

Introduction

Greenland, the world's largest island, bears the world's second-largest mass of ice, the Greenland ice cap or sheet. (Antarctica's is about 10 times larger.) In 2006, satellite observations confirmed that Greenland's ice is melting faster than ever before. Scientists disagree about whether the melting will continue to accelerate, but some specialists think further rapid melting is very likely.

Meltwater from ice that sits on land, such as mountain glaciers and ice sheets, raises sea levels. The addition of large amounts of freshwater to the North Atlantic also may slow the oceans' thermohaline (literally, "heat-and-salt") overturning circulation, which moves water throughout the world's oceans and influences world climate. Melting in Greenland, therefore, has two major effects, rising sea level and slowed thermohaline circulation.

Historical Background and Scientific Foundations

About 80% of Greenland is covered with an ice sheet about 1 mi (1.6 km) deep at its deepest point. This sheet is the accumulation of hundreds of thousands of years of winter snows. Each year's snow has added a thin layer, which packs down to ice and is covered by the next year's snow. The oldest ice in the cap is about 120,000 years old, but the cap has been there for about a million years;

Melting ice in Greenland has people worried about the impact that climate change will have on the area. *AP Images.*

An American doctoral student installs a GPS receiver on Kangerdlugssuaq Glacier in Greenland in July 2005. The student and her professor spent five weeks in Greenland studying climate change by tracking the movement of five glaciers. Kangerdlugssuaq was discovered to be moving at the rate of nearly 9 mi (14.5 km) a year. In the late 1990s, its rate was about 3.5 mi (5.5 km) annually. *AP Images.*

the older, deeper ice has long since flowed away to the sea. If all the ice in the Greenland sheet were to melt, it would raise the oceans by about 21 ft (6.5 m) or possibly more. Although most scientists do not presently think that complete melting of the ice cap is likely in the next century or so, greatly accelerated melting due to global warming has recently been measured by satellites and may force scientists to reconsider the possibility of more extreme melting in Greenland in the next century.

■ Impacts and Issues

Although it is not easy to predict how much of Greenland's ice will melt and how fast—past forecasts appear to have underestimated both—the consequences for rising sea levels of any given amount of melting are fairly easy to predict: given so many cubic miles of meltwater, the sea must rise by so many inches. Data from the last few years have shown that Greenland's contribution to global sea-level rise has risen from 0.006 in (0.15 mm) per year in the six years leading up to 2003 to 0.02 inches (0.5 mm) per year from 2004–2006 and probably since then as well. Two-thirds of the increased loss of ice mass is due to faster glacier flow, and one-third is due to increased surface melting and runoff.

Greenland's meltwater is also an important part of the global thermohaline circulation that transports water from the south polar regions to the north polar regions and back again, moving heat from the tropics to the polar zones. This circulation warms Europe and has other effects on global climate.

In the North Atlantic Ocean, the thermohaline circulation delivers a large surface current of warm water from the American tropics—the Gulf Stream—across the North Atlantic. This warm water swoops northward along the eastern coast of Europe. Westerly winds deliver heat from it to Europe, moderating its climate. As the water moves toward the North Pole, it cools and therefore becomes denser, until it finally sinks. This denser, colder water forms a returning current that follows the western coast of Greenland southward, eventually tracing the westward edge of North and South America all the way down to the South Atlantic Ocean.

Because freshwater is less dense than salty water, the addition of large amounts of freshwater to the sea from Greenland's melting ice tends to slow the sinking of cool Arctic waters. With slowed sinking, the return current must slow, and so must the northward surface current of warm water. (All the circulations must balance out because the ocean can't build up in one place.) The accelerated melting of Greenland's ice sheet, therefore, tends to slow down the entire Atlantic circulation system.

WORDS TO KNOW

GREENLAND-SCOTLAND RIDGE: Underwater ridge connecting Greenland to Scotland in the North Atlantic that separates the Nordic seas, where North Atlantic deep water formation occurs. Below a depth of 2,755 ft (840 m), the ridge forms a continuous barrier between the two basins; in some areas it rises to shallower depths or islands. Formation of the North Atlantic deep water is an essential part of the great conveyor belt or thermohaline circulation of the oceans, a key component of the global climate mechanism.

ICE SHEET: Glacial ice that covers at least 19,500 square mi (50,000 square km) of land and that flows in all directions, covering and obscuring the landscape below it.

THERMOHALINE CIRCULATION: Large-scale circulation of the world ocean that exchanges warm, low-density surface waters with cooler, higher-density deep waters. Driven by differences in temperature and saltiness (halinity) as well as, to a lesser degree, winds and tides. Also termed meridional overturning circulation.

One possible consequence is a slow or sudden cooling of European climate. Data on prehistoric climate show that air temperatures in the North can drop by as much as 18°F (10°C) in only a few decades, and that such changes are linked to sudden changes in the ocean circulation. Scientists caution that the ocean circulation system is highly nonlinear, meaning that it does not always react in a smooth, steady way. Rather, when a certain (as-yet unknown) threshold is reached, it is possible that the circulation will switch suddenly to a new state that transports little or no heat to the North. Also, even when the flow of freshwater weakens again, as when the Arctic re-cooled or when most of the Greenland ice finished melting, the system might not quickly shift back to its earlier pattern.

In 2005, scientists reported that the conveyor has slowed by about 30% over the past 50 years. However, in 2007 the same group of scientists, after making more complete measurements of ocean flow over the Greenland-Scotland subsea ridge (which the conveyor circulation must cross to get to the Arctic), reported that they had found that the magnitude of the conveyor current fluctuates so much on a seasonal timescale that they cannot be sure that the conveyor has, in fact, slowed since the 1950s. If there has been a long-term slowing trend, recently accelerated Greenland warming cannot have been its cause but may accelerate it. A complete breakdown of the North Atlantic circulation due to human-caused climate change cannot be ruled out by current scientific knowledge, although such a breakdown could probably not occur for at least a century, if ever.

The consequences of further, possibly sudden, slowing of the ocean circulation are hard to predict. The result might or might not be a net cooling of Europe. The tropics would probably warm, since the circulation removing heat to the polar regions would be less effective.

SEE ALSO *Antarctica: Melting; Arctic Melting: Greenland Ice Cap; Arctic Melting: Polar Ice Cap; Great Conveyor Belt.*

BIBLIOGRAPHY

Periodicals

Alley, Richard B., et al. "Ice-Sheet and Sea-Level Changes." *Science* 310 (October 21, 2005): 456–460.

Dowdeswell, Julian A. "The Greenland Ice Sheet and Global Sea-Level Rise." *Science* 311 (February 17, 2006): 963–964.

Luthcke, S. B., et al. "Recent Greenland Ice Mass Loss by Drainage System from Satellite Gravity Observations." *Science* 314 (November 24, 2006): 1286–1289.

Murray, Tavi. "Greenland's Ice on the Scales." *Nature* 443 (September 21, 2006): 277–278.

Overpeck, Jonathan T., et al. "Paleoclimatic Evidence for Future Ice-Sheet Instability and Rapid Sea-Level Rise." *Science* 311 (March 24, 2006): 1747–1750.

Pollitz, Fred F. "A New Class of Earthquake Observations." *Science* 313 (August 4, 2006): 619–620.

Quadfasel, Detlief. "The Atlantic Heat Conveyor Slows." *Nature* 438 (December 1, 2005): 555–556.

Larry Gilman

Ground Data and Satellite Data Discrepancy

Introduction

Measurements taken over time can be very useful in revealing changes. This is certainly true for the atmosphere that surrounds Earth. Temperature measurements of the atmosphere at Earth's surface have been taken for more than 150 years. Whereas surface temperatures will vary from place to place due to variations in the local geography and land use, data from such a long time period have been useful in revealing air temperature changes.

Since the late 1950s, atmospheric measurements of the lower layers of the atmosphere—the regions in which weather occurs and the influences of surface activities will be felt—have been obtained using instruments sent aloft in weather balloons. Beginning in 1978, satellite measurements of the atmosphere have been possible.

A NASA ER-2 pilot makes his way to his aircraft at a naval air station in Florida in July 2002. The ER-2 flies at such high altitudes that the pilot must wear a space suit. He is participating in a program that seeks to use aircraft to calibrate cloud measurements made from satellites so that cloud characteristics can be observed more accurately from orbiting spacecraft. The new system was created so there would be less uncertainty in climate change predictions. *Bill Ingalls/NASA/Getty Images.*

WORDS TO KNOW

ANTHROPOGENIC: Made by people or resulting from human activities. Usually used in the context of emissions that are produced as a result of human activities.

CLIMATE MODELS: Mathematical representations of climate processes. Climate models are computer programs that describe the structure of Earth's land, ocean, atmospheric, and biological systems and the laws of nature that govern the behavior of those systems. Detail and accuracy of models are limited by scientific understanding of the climate system and by computer power. Climate models are essential to understanding paleoclimate, present-day climate, and future climate.

KYOTO PROTOCOL: Extension in 1997 of the 1992 United Nations Framework Convention on Climate Change (UNFCCC), an international treaty signed by almost all countries with the goal of mitigating climate change. The United States, as of early 2008, was the only industrialized country to have not ratified the Kyoto Protocol, which is due to be replaced by an improved and updated agreement starting in 2012.

TROPOSPHERE: The lowest layer of Earth's atmosphere, ranging to an altitude of about 9 mi (15 km) above Earth's surface.

WAVELENGTHS: Distances between the crests or troughs of repeating waves. Sound waves, water waves, and electromagnetic waves (e.g., light) all have characteristic wavelengths. The greenhouse effect depends on the fact that some gases, such as carbon dioxide, are transparent to some wavelengths of light (those radiated by the sun) but not to others (those radiated by warm substances such as land, sea, and air). Energy from the sun arrives at wavelengths that are allowed to enter Earth's climate system, but once at the surface or near the surface is transformed into heat and re-radiated as energy that cannot so easily leave.

Comparison of the data obtained nearer the ground with the satellite data have not been consistent. Although the ground data indicated the gradual warming of the atmosphere, as would be expected in global warming, satellite data have instead revealed that little if any temperature change has occurred in the troposphere.

Is the atmosphere warming or not? Is global warming real? Scientists have been working hard to find explanations for the discrepancy between the ground and satellite data, which are crucial to making climate models that can accurately predict atmospheric changes and the role of human activities.

Historical Background and Scientific Foundations

People have been measuring surface temperatures using thermometers for more than 150 years. Although measurements from any single point over time can provide a meaningful assessment of climate change (at least the climate very close to the ground), trying to piece together the countless measurements taken all over the world to assess the state of Earth's surface since the 1850s has been very complicated.

However, the task is not impossible. For scientists who have toiled to piece together the millions of ground data measurements, the global patterns obtained from various studies have been similar and have revealed that since the 1960s the nighttime ground temperatures have been increasing at approximately twice the rate as daytime temperatures. Because the sunlit daytime air tends to mix more than nighttime air, the greater warming measured at night supports the idea that the ground temperature is warming more quickly than the rest of the atmosphere.

A big criticism of ground data is that the temperature readings tend to be collected in urban areas, which are warmer than rural areas. Thus, the increasing ground temperatures could just reflect the growth of cities during the last half of the twentieth century. However, data collected over the ocean and in rural areas have also revealed a temperature increase, although not as great as that obtained in cities. Overall, the ground data indicate a warming of approximately 2.7°F (1.5°C) per decade. This is consistent with the predictions of atmospheric warming from climate models of global warming.

The discrepancy in this ground data and data obtained from weather balloons began around 1979. The balloon measurements have indicated that the temperature of the troposphere (a layer of the atmosphere in which most weather occurs) has been fluctuating a bit, but has not been increasing decade by decade.

The difference has become even more significant with the coming of the satellite era. Atmospheric measurements that have been obtained with orbiting satellites beginning in 1979 agree with the measurements obtained from weather balloons.

For climate scientists, the discrepancy between ground data and satellite/balloon data is serious, because it goes to the heart of the global warming debate. Climate models of global warming predict that the troposphere should be increasing in temperature more than the atmosphere very near the ground. If it is not, then either the models of global warming are incorrect, or global warming itself is a myth rather than a reality.

However, a number of studies that have been published since 2005 indicate that the discrepancy may not be real, that it is simply a result of how the balloon and satellite measurements were made.

Upper atmosphere temperature measurements are not direct recordings of temperature. Rather, they measure differences in the release of energy at different wavelengths, with the data being mathematically treated to provide an indication of temperature. The detection of different wavelengths uses different channels of the instrumentation, essentially windows that let in certain wavelengths while excluding others. The discrepancy with ground data is mostly due to the data obtained using one of the windows. Re-analysis of the balloon/satellite data has demonstrated a good match between these data and the ground data. As of 2007, the discrepancy appears to have been resolved.

■ Impacts and Issues

The debate that has arisen over the discrepancy between ground data and satellite/balloon data has been contentious. Some scientists have argued that only ground data from developed countries such as the United States can be trusted. Other scientists point out that this excludes large regions of the globe and could indicate that global warming is not occurring when it really is. Arguing that climate science was not complete, the U.S. government under President George W. Bush opted out of the Kyoto Protocol.

The re-analysis of the balloon satellite data now supports the idea that the atmosphere is indeed warming at a rate predicted by models of global warming. Although a majority of climate scientists have accepted the re-analysis, a small minority of scientists continue to argue that the atmosphere is not warming due to anthropogenic (human-caused) activities. Scientists continue to collect data from satellites in an effort to learn more.

SEE ALSO *Satellite Measurements.*

BIBLIOGRAPHY

Books

DiMento, Joseph F. C., and Pamela M. Doughman. *Climate Change: What It Means for Us, Our Children, and Our Grandchildren.* Boston: MIT Press, 2007.

Gore, Al. *An Inconvenient Truth: The Planetary Emergency of Global Warming and What We Can Do About It.* New York: Rodale Books, 2006.

Seinfeld, John H., and Spyros N. Pandis. *Atmospheric Chemistry and Physics: From Air Pollution to Climate Change.* New York: Wiley Interscience, 2006.

Periodicals

Fu, Qiang, Celeste M. Johanson, Stephen G. Warren, and Dian J. Seidel. "Contribution of Stratospheric Cooling to Satellite-Inferred Tropospheric Temperature Trends." *Nature* 429, no. 6987 (May 6, 2004): 55–58.

Gulf of Mexico

Introduction

The Gulf of Mexico is a geographic area and a body of water that forms the so-called third coast of the contiguous United States. The Gulf of Mexico is surrounded on the United States side by coastlines of western and northern Florida, Alabama, Mississippi, Louisiana, and Texas. In Mexico, the Gulf is bordered by the states of Tamaulipas, Veracruz, Tabasco, Campeche, and Yucatán. The western end of the island of Cuba forms a partial barrier to the eastern Gulf of Mexico, where it joins the Caribbean Sea. The Gulf of Mexico is roughly oval-shaped with a long dimension of about 950 mi (1,500 km). The area of the Gulf of Mexico is about 615,000 square mi (1.6 million square km). For more than 500 years, the Gulf of Mexico has played a key role in the economic and political development of the United States, Cuba, and Mexico.

Bleached fire coral (center) contrasts against healthy coral in the Gulf of Mexico. Bleaching—caused by high temperatures, pollution, or disease—can kill the coral. Scientists anticipate more bleaching due to climate change and other factors. *AP Images.*

WORDS TO KNOW

RIVER DELTA: Flat area of fine-grained sediments that forms where a river meets a larger, stiller body of water such as the ocean. Rivers carry particles in their turbulent waters that settle out (sink) when the water mixes with quieter water and slows down; these particles build the delta. Deltas are named after the Greek letter delta, which looks like a triangle. Very large deltas are termed megadeltas and are often thickly settled by human beings. Rising sea levels threaten settlements on megadeltas.

SEDIMENT: Solid unconsolidated rock and mineral fragments that come from the weathering of rocks and are transported by water, air, or ice and form layers on Earth's surface. Sediments can also result from chemical precipitation or secretion by organisms.

SHORELINE: The band or belt of land surrounding a large body of surface water, such as a lake or ocean.

TECTONIC PLATE: Rigid unit of Earth's crust that moves about over geological time, merging with and separating from other tectonic plates as the continents rearrange but retaining its identity through these encounters. There are seven major tectonic plates on Earth and a number of smaller ones.

UPWELLING: The vertical motion of water in the ocean by which subsurface water of lower temperature and greater density moves toward the surface of the ocean. Upwelling occurs most commonly among the western coastlines of continents, but may occur anywhere in the ocean. Upwelling results when winds blowing nearly parallel to a continental coastline transport the light surface water away from the coast. Subsurface water of greater density and lower temperature replaces the surface water and exerts a considerable influence on the weather of coastal regions. Carbon dioxide is transferred to the atmosphere in regions of upwelling.

Historical Background and Scientific Foundations

The Gulf of Mexico is thought to have originated about 250 million years ago with the rifting or breaking apart of the tectonic plates of North America, South America, and Africa. When these components separated, an area of ocean floor developed between North and South America, which became the basin for the Gulf of Mexico. Not long after this basin was formed, sea water access to the basin was restricted and much of the sea water evaporated. We know this because today a large part of the deeper Gulf of Mexico basin is underlain by a thick layer of salt from this evaporation event. After an early dynamic history of crustal plate movement, the Gulf of Mexico basin became a stable area of Earth's crust and has remained so ever since.

Features of the Gulf of Mexico

The Gulf of Mexico is generally characterized by wide continental shelves around most of its periphery. These shelves, where water depths are at most a few hundred feet, have been the sites of intensive oil exploration in the past and to the present. The shelves give way to continental slopes, which lead down to the deeper plain of the Gulf of Mexico floor, which is known as the Sigsbee Deep. On the northern shelf of the Gulf of Mexico, an enormous pile of sediment from the Mississippi River delta has built a feature called the Mississippi sedimentary cone. This cone extends across the continental shelf and down the continental slope in front of the mouth of the Mississippi River.

Plastic and pliable salt from deep within the Gulf of Mexico basin has been squeezed up over time due to the weight of continental shelf and slope sediments. The upward movement of this buoyant salt has formed a submarine ridge on the northern continental shelf's outer margin, which is called the Sigsbee Ridge. In addition, numerous salt domes (conical salt intrusions) rise through the continental shelf sediments and in some places on land as a result of this pressure-related salt mobilization from below.

Geological History of the Gulf of Mexico

After salt deposition early in the Gulf's history, sediment began to gradually fill in this stable basinal area between continents. Sand, clays, and muds from the adjacent land areas of the northern Gulf of Mexico rim were continually washed into the Gulf basin by the Mississippi and other rivers. Much of this sediment came from sources in the Appalachian, Ouachita, and other adjacent uplifted mountains. In the Florida area and in the Yucatán area of Mexico, sediments of a chemical nature (calcium-carbonate precipitates and organic remains) filled in the basin.

Starting about 160 million years ago, organic rich sediments were deposited in the Gulf of Mexico, which eventually became a key source layer for the Gulf's rich petroleum industry. Much of this petroleum became entrapped in a formation found through much of the northern Gulf rim called the Smackover Formation. Around 120 million years ago, a large reef system rimmed the western and northern Gulf of Mexico. These reefs, which were composed of now-extinct clams and associated shell fish, eventually formed some of the highly productive oil fields of eastern Mexico. Over the past 100 million years, the Gulf has remained a stable area, which is gradually being filled, mainly from the north and west, by sediment from sand- and clay-laden rivers.

Gulf of Mexico Economics

In addition to the oil production mentioned earlier, there is associated gas production from wells drilled into sediments of the Gulf of Mexico. Further, the Gulf has a highly valuable fishing production, both shell fish (for example, oysters) and swimming fish. The fishing industries of the Gulf coastal United States, Cuba, and Gulf coastal Mexico are supported by abundant living resources of the Gulf area. The continental margins of Florida and the Yucatán (Mexico) are situated in areas where deeper, cold, nutrient-rich waters in the Gulf rise through a process called upwelling. This provides for abundant growth of marine plankton, which in turn supports fish, shrimp, and squid harvesting.

The Gulf of Mexico has historically been an important avenue for shipping and there are many key ports on the Gulf, including New Orleans, Louisiana; Houston, Texas; and others. Gulf shores are well known as resort areas in Florida, Mississippi, Alabama, and parts of the Texas coast.

Gulf of Mexico Waters and Islands

Warm waters of the Gulf of Mexico are both a blessing and a curse. They give rise to the waters of the Gulf Stream, which flows north out of the Gulf and brings warmer waters to northern areas of the Atlantic. Such waters are a key factor in the success of the tourism industry mentioned earlier. Warm waters of the Gulf help fuel the intensity of tropical storms and hurricanes, which commonly enter the Gulf from sites in the western tropical Atlantic Ocean. Gulf of Mexico hurricanes, especially some in recent years such as Katrina, are famous for their potential for heavy damage and loss of life.

The Gulf of Mexico shoreline is notable for its barrier islands, which form end-to-end chains from Florida to eastern Louisiana and eastern Texas to eastern Mexico. These barrier islands are separated from the mainland by a narrow body of water such as a lagoon, bay, or estuary. Barrier islands are low-lying narrow strips of land that represent a delicate balance between sand availability, sea level, and coastal wave energy. The only parts of the Gulf shoreline that are not part of this barrier island trend are the marshy coasts of Louisiana and the Mexican coast (for example, along the Yucatán coast), where sand is not readily available.

■ Impacts and Issues

Like all bodies of water on Earth, the Gulf of Mexico responds to climatic change. For example, during times of warming climates, as today, higher sea surface temperatures cause intensification of cyclonic storms in the Gulf of Mexico. Communities and ecosystems along the Gulf of Mexico still remain especially vulnerable to disruption from storms after the record hurricane season of 2005 that included Hurricanes Katrina and Rita. The 2005 Atlantic hurricane season was the most active in recorded history. Also, sea level is rising in the Gulf of Mexico as it is globally today. Alternatively, during past times of much cooler global climates, the Gulf of Mexico was a much smaller body of water (due to lower sea level) and probably had far fewer cyclonic storms than today.

SEE ALSO *Beach and Shoreline; Hurricanes; Sea Level Rise.*

BIBLIOGRAPHY

Books

Gore, R. H. *The Gulf of Mexico: A Treasury of Resources in the American Mediterranean.* Sarasota, FL: Pineapple Press, 1992.

Web Sites

"Gulf of Mexico Integrated Science Data Information Management System." *U.S. Geological Survey*, July 25, 2007. <http://gulfsci.usgs.gov> (accessed December 3, 2007).

GulfBase: Resource Database for Gulf of Mexico Research, 2007. <http://www.gulfbase.org> (accessed December 3, 2007).

David T. King Jr.

Gulf Stream

Introduction

The Gulf Stream is a strong, fast-running, and clockwise-rotating system of warm ocean currents that begins in the Gulf of Mexico; proceeds through the Straits of Florida; follows the eastern coast of the United States and the Grand Banks of Newfoundland, Canada, in a general northeasterly direction; and then crosses easterly the North Atlantic Ocean. During its travels along the North American coast, the Gulf Stream appears much bluer in color when compared to other waters and has a high salt content. On its way east across the Atlantic Ocean, it splits into the northern stream (the North Atlantic Drift [NAD]), which flows to the coast of Western Europe and the islands of the Arctic Ocean, and the southern stream (the Canary Current), which veers off toward northwestern Africa.

Based on past and current meteorological studies, global climate change may affect the Gulf Stream. Whether this is the case or not is very important to the world's climate, especially with regard to Western Europe and other land masses adjacent to the Gulf Stream.

Historical Background and Scientific Foundations

The two equatorial sources of the Gulf Stream are the North Equatorial Current (NEC), which flows generally westward along the Tropic of Cancer, and the South Equatorial Current (SEC), which flows westward from southwestern Africa to South America and then northward to

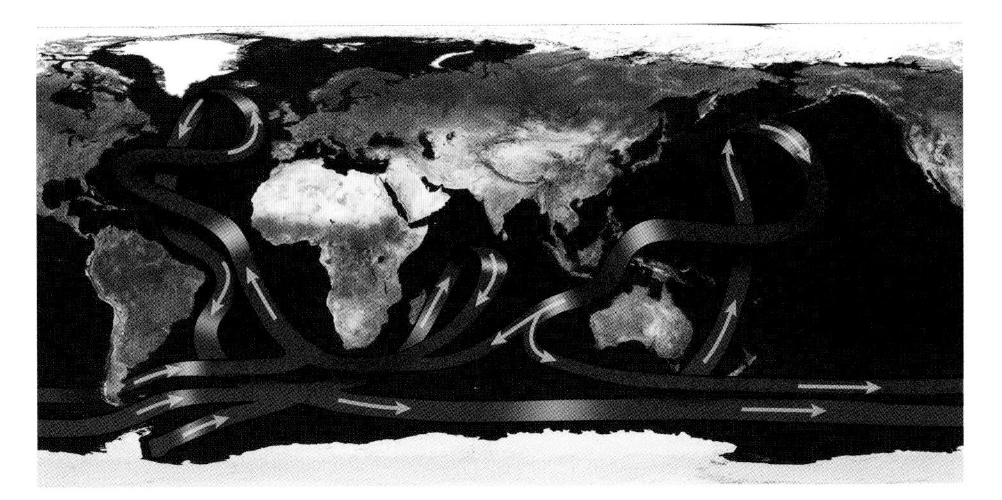

This satellite image of Earth from space shows global ocean circulation. Ocean currents flow around the world due to differences in temperature and salinity. A current of warm water (red) from the Pacific Ocean travels westward on the ocean surface. As it flows, it evaporates and becomes saltier. The Gulf Stream carries the warm, salty water up along the East Coast of the United States, then toward Europe. At colder northern latitudes, the water becomes so dense that it sinks to the sea floor and travels south (blue). Disruption of the Gulf Stream from global warming could result in ice-age conditions in Europe. *SPL/Photo Researchers, Inc.*

WORDS TO KNOW

LATITUDE: The angular distance north or south of Earth's equator measured in degrees.

THERMOHALINE CIRCULATION: Large-scale circulation of the world ocean that exchanges warm, low-density surface waters with cooler, higher-density deep waters. Driven by differences in temperature and saltiness (halinity) as well as, to a lesser degree, winds and tides. Also termed meridional overturning circulation.

TROPIC OF CANCER: In geography, a tropic is one of the two lines of latitude (lines of equal distance from the equator on Earth's surface) at 23° 26′ north and 23° 26′ south. The northern tropic is the Tropic of Cancer and the southern tropic is the Tropic of Capricorn. The belt of Earth's surface between these lines is the tropic.

the Caribbean Sea. Together, these two warm currents, along with waters from the Gulf of Mexico, form the Gulf Stream.

At the Florida Straits, the Gulf Stream transports its waters at a rate of about 39 million cubic yd (30 million cubic m) per second at a depth of about 1 mi (1.6 km) and a width of less than 50 mi (80 km). At this location, its temperature averages about 80°F (27°C). Further north, off of North Carolina, the flow rate of water increases to about 105 million cubic yd (80 million cubic m) per second. It averages a depth of up to 0.75 mi (1.2 km) and a width of about 50 to 93 mi (80 to 150 km). The Gulf Stream surface waters have a maximum speed of around 6.5 ft (2 m) per second at its northern boundary but overall average about 4 ft (1.2 m) per second. At Cape Hatteras, North Carolina, the cold NEC meets the warm SEC, resulting in some of the world's largest and most severe storms.

■ Impacts and Issues

The steady increase in the average temperature of Earth, what is commonly called global warming, could affect the Gulf Stream. Scientists studying this association speculate that global warming could decrease the circulation of the North Atlantic Drift if the pattern remains in effect over the next several decades. If such a decrease in circulation happens, the North Atlantic Ocean could become cooler, which could likely cause colder climates along coastal areas of Europe. However, diminished circulation of the Gulf Stream in recent years could just be a natural cycle that will eventually reverse itself.

The Gulf Stream is extremely important to the global climate because it provides moderating temperatures on neighboring land areas of the east coast of North America, the coasts of Western Europe and northwestern Africa, and other coastal areas along its path. In the United States, for instance, it allows the southeastern coast of Florida to maintain warmer temperatures during the winter months than neighboring off-coastal states. Further north, the waters off the Grand Banks contain some of the best commercial fishing areas in the world due to the Gulf Stream. The adjacent Gulf Stream along the coastal areas of Western Europe, especially the southwestern United Kingdom, provides for much milder winters when compared to other areas on similar northern latitudes.

The Gulf Stream is also a primary ingredient in moderating temperature differences between equatorial regions and areas about the north and south poles. It releases its heat into the atmosphere of the colder northern latitudes. Then, the cooler Gulf Stream waters begin to sink under colder water because of its heavier density (caused by its high salt content) when compared to other waters. In a process called thermohaline circulation, this downward movement helps to bring in warmer tropical waters to these colder waters.

However, geological and meteorological studies of ice-sheet cores and deep-sea sediments have described past eras where these moderating effects associated with thermohaline circulation have been reduced in intensity. This activity caused much cooler conditions throughout the northern climates. If today's thermohaline circulation were slowed or halted, the global climate patterns could likewise be changed. Computer models have shown that if this condition becomes reality, then northern latitude winters, such as those in the northeastern United States, would be much colder than normal as the ocean waters of the North Atlantic cool down.

According to the Woods Hole Oceanographic Institution, the cooling of the North Atlantic Ocean could be in a range of 5-9°F (3-5°C) if thermohaline circulation were completely stopped. With global trends showing warming temperatures around the planet, more regional trends within the areas of the Gulf Stream could be just the opposite, growing colder over the next several decades or longer.

See Also *Africa: Climate Change Impacts; Global Warming; Meteorology; Ocean Circulation and Currents; Oceans and Seas; Sea Temperatures and Storm Intensity.*

BIBLIOGRAPHY

Books

Permetta, John. *Guide to the Oceans.* Richmond Hill, Ontario, Canada: Firefly Books, 2003.

Voituriez, Bruno. *The Changing Ocean: Its Effects on Climate and Living Resources.* Paris, France: UNESCO, 2003.

Voituriez, Bruno. *The Gulf Stream.* Paris, France: UNESCO, 2006.

Wang, Chunzia, Shang-Ping Xie, and James A. Carton, eds. *Earth's Climate: The Ocean-Atmosphere Interaction.* Washington, DC: American Geophysical Union, 2004.

Web Sites

"Abrupt Climate Change: Should We Be Worried?" *Woods Hole Oceanographic Institution*, January 27, 2003. <http://www.whoi.edu/page.do?pid=12455&tid=282&cid=9986> (accessed November 4, 2007).